Barron's Regents Exams and Answers

Earth Science

The Physical Setting

EDWARD J. DENECKE, JR., M.A.
Formerly, Science Department
William H. Carr J.H.S. 194Q
Whitestone, N.Y.

Barron's Educational Series, Inc.

© Copyright 2017, 2016, 2015, 2014, 2013, 2012, 2011, 2010, 2009,
2008, 2007, 2006, 2005, 2004, 2003, 2002, 2001, 2000, 1999, 1997
by Barron's Educational Series, Inc.

All rights reserved.
No part of this book may be reproduced or distributed
in any form or by any means without the written permission
of the copyright owner.

All inquiries should be addressed to:
Barron's Educational Series, Inc.
250 Wireless Boulevard
Hauppauge, New York 11788
www.barronseduc.com

ISBN: 978-0-8120-3165-2
ISSN: 1069-2959

PRINTED IN THE UNITED STATES OF AMERICA
9 8 7 6 5 4 3 2 1

Contents

Preface   v

How to Use This Book   1
Format of the Physical Setting/Earth Science
    Regents Examination   1
Contents of This Book   4

Test-Taking Tips   6
General Helpful Tips   6
Specific Helpful Tips for the Multiple-Choice
    Questions   10
Helpful Tips for Questions on Parts B and C   20

Topic Outline and Question Index   28
How to Use the Topic Outline for the Physical Setting/
    Earth Science Examination   28
Topic Outline and Question Index—Physical Setting/
    Earth Science Examination   29

**Earth Science Reference Tables
and Charts**   41
2011 Edition   43

Glossary of Earth Science Terms   60

Regents Examinations, Answers, and Self-Analysis Charts   73

PHYSICAL SETTING/EARTH SCIENCE EXAM
August 2014 .... .75
June 2015 .... .164
August 2015 .... .259
June 2016 .... .355
August 2016 .... .450

Preface

This book is designed to strengthen your understanding and mastery of the material in the New York State Physical Setting/Earth Science Core Curriculum, a comprehensive course of study in earth science on the secondary level. This book has been specifically written to assist you in preparing for the Regents examination covering this course. It contains the following special features that will help you in studying for this exam:

(1) *Complete sets of questions from previous Physical Setting/Earth Science Regents examinations are provided.* Attempting to answer these questions will make you familiar with the topics tested on the examination and the degree of difficulty to which you are expected to master each topic. It will also increase your confidence as you become familiar with the nature and language of the various question types that will appear on the examination.

(2) *Correct answers to all Regents questions are fully explained.* Explanations of correct answers are detailed, yet simply and clearly stated. Each explanation has been written to help you grasp the reasoning used to arrive at the correct answer rather than just stating the correct choice. Careful study of the step-by-step explanations of how the correct answer is arrived at for each question will improve your mastery of the subject. Even if you have answered the question correctly, you should read the explanation to check your reasoning and gain insight into the topic. This insight may be valuable when answering a more difficult question on the same topic.

(3) *Wrong choices are explained.* If you've ever answered a test question incorrectly and not understood *why* your answer was wrong, you will find this feature helpful. Understanding *why* a choice is incorrect will help you to clear up misconceptions you may have and to see errors in your reasoning.

(4) *Reference Tables for Physical Setting/Earth Science.* These tables are provided to all students for use while taking the Physical Setting/Earth Science Regents examination. Some questions *require* the use of these tables, and that is usually indicated in the question. When answering other questions, you may find information in the tables *useful*. These tables are included in this book so that you may become *thoroughly familiar* with all of the information contained in them.

(5) *Topic Outline and Self-Analysis Charts.* The New York State Physical Setting/Earth Science Core Curriculum has been cross-referenced to questions from past Physical Setting/Earth Science Regents examinations. This will help you to identify your areas of weakness and to locate questions to review particular earth science concepts. Refer to "How to Use The Topic Outline" and Self-Analysis Charts for more detailed instructions.

(6) *Test-Taking Tips.* These helpful tips and strategies will help you to increase your grade on the Physical Setting/Earth Science Regents examination.

The author would like to acknowledge his editor, Linda Turner, for her invaluable assistance in writing and producing this book; his wife, Gerry, for her infinite patience; and his children, Merry, Abby, and Ben, for being a constant source of joy.

How to Use This Book

FORMAT OF THE PHYSICAL SETTING/EARTH SCIENCE REGENTS EXAMINATION

The examination consists of a performance portion containing three tasks to measure laboratory skills and a written portion containing multiple-choice, constructed-response, and extended constructed-response questions. You will be required to answer ALL of the questions on the Physical Setting/Earth Science Regents Examination.

THE LABORATORY PERFORMANCE TEST

Since laboratory experiences are an essential part of a science course, a portion of the Physical Setting/Earth Science Regents Examination is devoted to assessing laboratory skills. Tasks have been identified from laboratory experiments that you will have performed during the school year. These tasks, which represent skills that you are expected to have mastered, change only slightly, if at all, from year to year.

The performance portion of the examination is administered separately from the written portion, normally two weeks earlier. Arrangements for administering the performance exam are made at each school in accordance with guidelines set by the New York State Education Department.

The scoring for each task is based upon accuracy. Values within a certain range are granted the full point value allotted to each task. It is possible to accumulate a maximum of 16 points on the performance portion of the examination.

Additional information regarding the performance test, including an indication of the three tasks to be completed, will be provided by your teacher when this portion of the examination is given. The following is an outline of these tasks. The time allowed for completing the tasks at each station is 9 minutes.

Note: The following description represents the information that the State Education Department has stated may be shared with students before taking the performance part of the examination. You should be familiar with the skills being assessed because you have used them in laboratory activities throughout the year. However, you will not be allowed to practice the entire test or any of the individual stations before this performance component is administered.

Station 1 . . *Mineral and Rock Identification*
The student determines the properties of a mineral and identifies that mineral using a flowchart. Then the student classifies two different rock samples and states a reason for each classification based on observed characteristics.

Station 2 . . *Locating an Epicenter*
The student determines the location of an earthquake epicenter using various types of data that were recorded at three seismic stations.

Station 3 . . *Constructing and Analyzing an Asteroid's Elliptical Orbit*
The student constructs a model of an asteroid's elliptical orbit and compares the eccentricity of the orbit with that of a given planet.

THE WRITTEN TEST

The written portion of The Physical Setting/Earth Science Regents Examination represents 85 points of the total score and has three parts: A, B, and C. You should be prepared to answer questions in multiple-choice, constructed-response, and extended constructed-response formats. Questions will be content- and skills-based and may require you to graph data, complete a data table, label or draw diagrams, design experiments, make calculations, or write short or extended responses. In addition, you may be required to hypothesize, to interpret, analyze, or evaluate data, or to apply your scientific knowledge and skills to real-world situations. Some of the questions will require use of the 2011 Edition of the *Reference Tables for Physical Setting/Earth Science*.

Format of the Physical Setting/Earth Science Regents Exam 3

> You will be required to answer ALL of the questions in the Physical Setting/Earth Science Regents examination!

The following is a brief description of the types of questions comprising each part of the examination:

Part A—Multiple-Choice A multiple-choice question offers several answers from which you choose the one that best answers the question or completes the statement. Part A of the exam focuses on earth science content from Standard 4 (see Topic Outline) and represents 30–40 percent of the examination. Many practice questions of this type from previous Regents examinations are included at the end of each chapter.

Part B—Multiple-Choice and Constructed-Response In a constructed-response question there is no list of choices from which to choose an answer; rather you are required to provide the answer. Constructed-response questions test skills ranging from constructing graphs or topographic maps to formulating hypotheses, evaluating experimental designs, and drawing conclusions based upon data. In Part B, which represents 25–35 percent of the examination, you are asked to demonstrate skills identified in Standards 1, 2, 6, and 7 in the context of earth science. Practice questions of this type are also included at the end of each chapter.

Part C—Extended Constructed-Response These are constructed-response questions that require more time (15–20 minutes per item) and effort on your part to answer. Questions in Part C require you to apply your earth science knowledge and skills to real-world problems and applications. You may be asked to write short essays, design controlled experiments, predict outcomes, or analyze the risks and benefits of various solutions to a problem. Part C represents 15–25 percent of the examination. Again, practice questions are provided at the end of most chapters.

You should review constructed-response and extended constructed-response questions from previous examinations to familiarize yourself with these types of questions. Parts B and C are recent additions, and

each examination may contain questions unlike those in previous exams. Keep these points in mind:
- Write complete sentences whenever you are asked to answer in your own words.
- Show all work, and use correct units in mathematical problems.
- Be able to interpret map scales and legends, to construct isolines, and to draw profiles.

Although the questions on each written test are unique, they are similar to questions on past Regents examinations. By studying questions that have been asked in the past, therefore, you can strengthen your skills and knowledge in preparation for the examination you will take.

CONTENTS OF THIS BOOK

THE TOPIC OUTLINE AND THE REFERENCE TABLES FOR PHYSICAL SETTING/EARTH SCIENCE

The first part of this book consists of two special features, the Topic Outline and the Earth Science Reference Tables. For detailed instructions on how to use the *Topic Outline*, see page 28. The *Reference Tables for Physical Setting/Earth Science* in this book are the same ones that are provided to students for use while taking the Regents examination. You should be familiar with these tables at the time of the exam since you will have used them during the school year. Some questions require the use of these tables, and this fact is usually indicated in the question. For other questions you may find information in the tables helpful. It is therefore a good idea to become generally familiar with what is contained in the tables. For the Physical Setting/Earth Science exam, use the tables beginning on page 43.

REGENTS EXAMINATIONS AND ANSWERS

In the second, much longer part of the book there are five Physical Setting/Earth Science Regents examinations with Answer Keys. Every correct answer is fully explained. Also, to help clear up misconceptions that may lead students astray, in many cases explanations for wrong answers are provided as well.

THE SELF-ANALYSIS CHART

At the end of the *Answers* section for each examination, a *Self-Analysis Chart* is provided to aid you in pinpointing your areas of weakness. The upper portion of the chart, labeled "Standards 1, 2, 6, and 7: Skills and Applications" lists the numbers of questions requiring the use of specific skills, or the ability to apply knowledge. These skills and abilities are described more completely in the *Major Understandings* of Standards 1, 2, 6, and 7 listed in the Topic Outline. The lower portion of the chart, labeled "Standard 4: The Physical Setting/Earth Science" lists the numbers of questions requiring knowledge of specific earth science concepts and is subdivided into sections labeled *"Astronomy," "Geology,"* and *"Meteorology."* These concepts are described more completely in the *Major Understandings* of Standard 4 listed in the Topic Outline.

Many question numbers may appear in both portions of the chart because many questions involve using a skill, such as deductive reasoning, together with your knowledge of specific earth science concepts to answer the question. Therefore, you should consider these two sections of the chart independently, using the upper section of the Self-Analysis Chart to pinpoint areas of weakness in your skills, and the lower section of the chart to pinpoint areas of weakness in your knowledge of earth science concepts.

Test-Taking Tips

The following pages contain several tips to help you achieve a good grade on the Physical Setting/Earth Science Regents exam. They are divided into GENERAL HELPFUL TIPS and SPECIFIC HELPFUL TIPS.

GENERAL HELPFUL TIPS

TIP 1
Be Confident and Prepared

SUGGESTIONS
- Review previous tests.
- Use a clock or watch, and take previous exams at home under examination conditions, (i.e., don't have the radio or television on.)
- Get a review book. (The preferred book is Barron's *Let's Review: Earth Science*.)
- Talk over the answers to questions on these tests with someone else, such as another student in your class or someone at home.
- Finish all your homework assignments.
- Look over classroom exams that your teacher gave during the term.
- Take class notes carefully.
- Practice good study habits.
- Know that there are answers for every question.

General Helpful Tips

- Be aware that the people who made up the Regents exam want you to pass.
- Remember that thousands of students over the last few years have taken and passed an Earth Science Regents. You can pass too!
- On the night prior to the exam day: lay out all the things you will need, such as clothing, pens, and admission cards.
- Go to bed early; eat wisely.
- Bring at least two pens to the exam room.
- Bring your favorite good luck charm/jewelry to the exam.
- Once you are in the exam room, arrange things, get comfortable, be relaxed, attend to personal needs (the bathroom).
- Keep your eyes on your own paper; do not let them wander over to anyone else's paper.
- Be polite in making any reasonable requests of the exam room proctor, such as changing your seat or having window shades raised or lowered.

TIP 2

Read Test Instructions and Questions Carefully

SUGGESTIONS

- Be familiar with the test directions ahead of time.
- Decide upon the task(s) that you have to complete.
- Know how the test will be graded.
- Know which question or questions are worth the most points.
- Give only the information that is requested.
- Where a choice of questions exists, read all of them and answer *only* the number requested.
- Underline important words and phrases.
- Ask for assistance from the exam room proctor if you do not understand the directions.

TIP 3

Budget Your Test Time in a Balanced Manner

SUGGESTIONS
- Bring a watch or clock to the test.
- Know how much time is allowed.
- Arrive on time; leave your home earlier than usual.
- Prepare a time schedule and try to stick to it. Remember that Regents exams are longer than classroom tests, so you will need to pace yourself accordingly.
- Answer the easier questions first.
- Devote more time to the harder questions and to those worth more credit.
- Don't get "hung up" on a question that is proving to be very difficult; go on to another question and return later to the difficult one.
- Ask the exam room proctor for permission to go to the lavatory, if necessary, or if only to "take a break" from sitting in the room.
- Plan to stay in the room for the entire three hours. If you finish early, read over your work—there may be some things that you omitted or that you may wish to add. You also may wish to refine your grammar, spelling, and penmanship.

TIP 4

Be "Kind" to the Exam Grader/Evaluator

SUGGESTIONS
- Assume that you are the teacher grading/evaluating your test paper.
- Answer questions in an orderly sequence.
- Write legibly.
- Answer Part C questions with complete sentences.
- Proofread your answers prior to submitting your exam paper. Have you answered all the Part A and (if appropriate) Part C questions and the required number of Part B questions?

TIP 5
Use Your Reasoning Skills

SUGGESTIONS
- Answer *all* questions.
- Relate (connect) the question to anything that you studied, wrote in your notebook, or heard your teacher say in class.
- Relate (connect) the question to any film you saw in class, any project you did, or to anything you may have learned from newspapers, magazines, or television.
- Decide whether your answers would be approved by your teacher.
- Look over the entire test to see whether one part of it can help you answer another part.
- Be cautious when changing an answer. Try to remember why you selected the first answer to be sure that the new answer is better.

TIP 6
Don't Be Afraid to Guess

SUGGESTIONS
- In general, go with your first answer choice.
- Eliminate obvious incorrect choices.
- If you are still unsure of an answer, make an educated guess.
- There is no penalty for guessing; therefore, answer ALL questions. An omitted answer gets no credit.

Let's now review the six GENERAL HELPFUL TIPS for short-answer questions:

SUMMARY OF TIPS
1. Be confident and prepared.
2. Read test instructions and questions carefully.
3. Budget your test time in a balanced manner.
4. Be "kind" to the exam grader/evaluator.
5. Use your reasoning skills.
6. Don't be afraid to guess.

SPECIFIC HELPFUL TIPS FOR THE MULTIPLE-CHOICE QUESTIONS

TIP 1

Read the Question and Try to Answer It Before Looking at the Possible Choices

Thinking out the answer to a question before looking at the possible choices stimulates your memory. Key facts, concepts, and relationships come to mind. If the answer you anticipated is among the possible choices, it is likely to be the correct one. But **keep reading** through all the choices! By thinking of the answer first, you are less likely to be fooled by a wrong answer. But if you don't read all of the choices, you might miss a more correct or more complete choice farther down the list.

EXAMPLE

Earth's hydrosphere is best described as the

(1) solid outer layer of Earth
(2) liquid outer layer of Earth
(3) magma layer located below Earth's stiffer mantle
(4) gaseous layer extending several hundred kilometers from Earth into space

As you read the question, you will probably think of the terms *lithosphere*, *hydrosphere*, and *atmosphere*. You may recall that "hydro" refers to water and remember that the hydrosphere is all the liquid water resting on the lithosphere. Therefore, you anticipate an answer referring to Earth's layer of liquid water. With this in mind, you recognize that the correct answer is choice 2, because choice 1 refers to a solid layer, choice 4 refers to a gaseous layer, and choice 3 refers to a layer of magma, which is liquid rock—not liquid water.

Specific Helpful Tips for the Multiple-Choice Questions 11

TIP 2

Always Read *All* of the Choices Before Choosing an Answer

Read the question and all of the choices carefully. Make sure you are reading exactly what has been written, and not just skimming for words that you are anticipating. Continue reading even if you find your anticipated answer because there may be a more complete or more correct answer farther down the list.

EXAMPLE

The length of an Earth year is based on Earth's

(1) rotation of 15°/hr
(2) revolution of 15°/hr
(3) rotation of approximately 1°/day
(4) revolution of approximately 1°/day

When reading this question you might anticipate that Earth's year is based on its revolution around the Sun. If you rushed through the answers and chose the first one to mention revolution—choice 2—you would miss this question. If you carefully read *all* of the choices, you realize that both choice 2 and choice 4 refer to revolution, and that choice 4 also correctly states the rate at which Earth revolves.

TIP 3

Answer the Easy Questions First; Do Not Spend Too Much Time on Any One Question

In this part of the test, all of the questions have the same value. You get as much credit for an easy question as a hard one. Don't waste a lot of time on questions you can't answer quickly. Answer all of the easy questions first, then go back and figure out the hard ones in the time left over. Easy questions usually contain short sentences and answers with few words. The question can usually be answered quickly from the information presented.

EXAMPLE

An air mass classified as mT usually forms over which type of Earth surface?

(1) cool land (3) warm land
(2) cool water (4) warm water

The correct answer, choice 4, is obvious if you know that mT stands for maritime tropical. Since tropical means warm and maritime refers to the sea, such an air mass usually forms over warm water. But if you don't know what mT means, and don't realize you can find its meaning in the *Reference Tables for Physical Setting/Earth Science*, don't waste a lot of time on the question.

TIP 4

Look for and Underline Key Words or Phrases in the Question That Tell You What the Questioner Wants You to Do

Read every word of the question. Make sure you understand what is being asked before you move to the possible answers. Sometimes there are key words or phrases that will help you pick the correct answer. Some examples of key words are *all*, *none*, *some*, *most*, *never*, *always*, *least*, *greatest*, *less*, *more*, *maximum*, *minimum*, *highest*, *lowest*, *average*, *most nearly*, and *best*. Pay careful attention to these words. If you overlook a key word, you may miss a question to which you really know the answer.

EXAMPLE

A student read in a newspaper that the maximum length of the daylight period for the year in Syracuse, New York, had just been reached. What was the date of the newspaper?

(1) March 22 (3) September 22
(2) June 22 (4) December 22

The key phrase in the question is "the maximum length of the daylight period for the year" and the question asks for the "date" on which

Specific Helpful Tips for the Multiple-Choice Questions 13

this was reported in a newspaper. The questioner wants you to choose that *date* on which the *daylight period* is at a *maximum*. Since the longest period of daylight in New York State occurs at the summer solstice, June 21, the newspaper probably reported it the next day, June 22—choice 2.

TIP 5

Use Partial Knowledge to Eliminate Wrong Choices

When you have been through all the questions once, go back and find questions about which you have some knowledge. Read all of the possible choices and use your partial knowledge to eliminate one or two of the answers. Cross off incorrect choices to narrow possible answers. Then choose between the remaining answers. If you can eliminate two wrong answers, your chance of guessing the right answer is greater.

EXAMPLE

In which list are celestial features correctly shown in order of increasing size?

(1) galaxy → solar system → universe → planet
(2) solar system → galaxy → planet → universe
(3) planet → solar system → galaxy → universe
(4) universe → galaxy → solar system → planet

Let's suppose you *know* that the universe is the biggest feature because it contains all the others. You *think* a planet might be the smallest of the features, but you are not absolutely sure. You have no idea of the sizes of the other features. You can use this partial knowledge to greatly increase your chances of guessing correctly. Because the question asks which list shows the features in *order of increasing size*, the list should end with the largest feature—the universe. Therefore, you can eliminate choices 1 and 4. Of the remaining two choices, you would then guess the answer to be choice 3, since you think a planet might be the smallest feature. And choice 3 is the correct answer.

TIP 6

Look for Clues Among the Choices in the Question, and Use Information from One Part of the Test to Help You with Others

By reading questions and choices carefully, you may often find words or phrases that provide clues to an answer. This tip is important because, if you are alert, you may find links and connections between various questions on the exam. One question may give you a clue (or even an answer) to another question.

EXAMPLE

The diagram below represents part of Earth's latitude-longitude system.

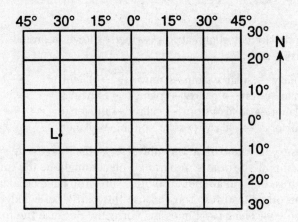

What is the latitude and longitude of point *L*?

(1) 5° E 30° N (3) 5° N 30° E
(2) 5° W 30° S (4) 5° S 30° W

The first clue is that all of the possible answers give 5° and 30° as the coordinates of point *L*. If you look at the axes, you see that 5° is the coordinate of point *L* on the vertical axis to the right. If you are

Specific Helpful Tips for the Multiple-Choice Questions 15

alert, you will notice that the very next question shows part of North America with the vertical axis labeled 20° N near the bottom and 40° N near the top. If you make the connection between the two questions, you will realize that the vertical axis is a North-South axis, and you can eliminate choices 1 and 2 because they give the 5° coordinates as East or West. If the vertical axis is a North-South axis, coordinates above the 0° mark (the Equator) are N; those below it are S. Since point L is 5° below the 0° mark, it is an S coordinate as shown in only one of the remaining choices—choice 4.

TIP 7

Know What Is in the *Reference Tables for Physical Setting/Earth Science* and Carefully Examine Any Diagrams, Graphs, or Tables Given with a Question

Many questions expect you to apply what these graphs, tables, and diagrams contain. If you can't answer a question with the information given, there is a good chance that you will find the information you need in the *Reference Tables for Physical Setting/Earth Science*.

EXAMPLE 1: You need information given in the *Reference Tables for Physical Setting/Earth Science* to answer the question.

> The air outside a classroom has a dry-bulb temperature of 10°C and a wet-bulb temperature of 4°C. What is the relative humidity of the air?
>
> (1) 1%   (3) 33%
> (2) 14%   (4) 54%

In this question, you must know that there is a chart in the *Reference Tables for Physical Setting/Earth Science* that can be used to determine relative humidity given wet- and dry-bulb temperatures. If you know how to use this chart, you look for the intersection of the dry-bulb temperature and the difference between the wet- and dry-bulb temperatures, that is, the intersection of the row for 10°C and the column for 6°C, where you find the value for the % Relative Humidity—33—choice 3.

EXAMPLE 2: You need to apply information in a chart, graph, or diagram given with the question.

The table below shows the noontime data for air pressure and air temperature at a location over a period of 1 week.

Date	Nov. 9	Nov. 10	Nov. 11	Nov. 12	Nov. 13	Nov. 14	Nov. 15
Air Temperature (°C)	1	6	0	−2	−4	5	10
Air pressure (mb)	1,024	998	1,015	1,021	1,030	1,013	?

Based on the data provided, which air pressure would most likely occur at noon on November 15?

(1) 987 mb (3) 1,017 mb
(2) 1,015 mb (4) 1,022 mb

In this question, you need to apply the information in the chart. By carefully examining the data in the table, you see that as air temperature increases, air pressure decreases. Note that as temperature goes from 1°C to 5°C to 6°C, the air pressure goes from 1,024 mb to 1,013 mb to 998 mb. Therefore, if the temperature increases to 10°C, the air pressure should drop even lower than 998 mb. Now read the choices and note that only one is lower than 998 mb, choice 1—987 mb.

TIP 8

Do Not Choose an Answer That Contains an Element of Truth but That Is Incorrect as It Relates to the Question

---

It is very important to remember that just because a statement is true, or contains an element of truth, does not mean that it is the correct answer to the question. Statements that are true, or partially true,

Specific Helpful Tips for the Multiple-Choice Questions 17

make attractive distractors (wrong answers). Several answers may contain an element of truth, but you must decide which one is related most directly to the question.

EXAMPLE

Major ocean and air currents appear to curve to the right in the Northern Hemisphere because

(1) Earth has seasons
(2) Earth's axis is tilted
(3) Earth rotates on its axis
(4) Earth revolves around the Sun

Each of the answer choices is a true statement. However, the question focuses on the curving of major ocean and air currents, that is, the Coriolis Effect. If you understand the Coriolis Effect, you know that its primary cause is Earth's rotation on its axis—choice 3.

TIP 9

Beware of Broad, Sweeping Generalizations in the Answer Choices—They Are Generally Incorrect

Pay careful attention to the use of words such as *always, never, only, must*, and so on. Such words usually make general statements improbable. This tip is particularly helpful when two or more choices may be correct. Often, the broadest, most inclusive choice is correct because it may include the other choice(s).

EXAMPLE

According to the Rock Cycle in Earth's Crust diagram in the *Reference Tables for Physical Setting/Earth Science*, which type(s) of rock can be the source of deposited sediments?

(1) igneous and metamorphic rocks, only
(2) metamorphic and sedimentary rocks, only
(3) sedimentary rocks, only
(4) igneous, metamorphic, and sedimentary rocks

In this question, the best answer is choice 4—it is the least restrictive, includes all the other choices, and gives the broadest possibilities.

TIP 10

Don't Be Afraid to Guess If You're Not Sure of the Correct Answer, but Do Not Keep Changing Your Answer

---

If you are not sure of an answer, don't be afraid to guess. You have nothing to lose if you guess wrong. A wrong answer and no answer are both worth zero points. However, the correct answer is right there on the page, and if you guess correctly, you can increase your point score. If you can eliminate one or more wrong answers before guessing, your chance of choosing the correct answer jumps from one in four to one in three, or one in two. But don't think that you can pass a Regents exam by guessing; the odds are that if you guessed on every question you would get one in four correct—and score a 25 on the exam! There is no substitute for careful, diligent preparation for the exam.

If you do have to guess, make an educated guess. An educated guess is made by considering the options and focusing on what you *do* know. Use what you *do* know to eliminate improbable answers. Then read the question and answer as a complete sentence and ask yourself if it is true, or sounds true. Often, when reading the choices to yourself, one of the answers will "sound right."

Once you have made your choice, be slow to change your answer without a good reason, because research shows that a person's first choices are usually correct. Most people who change their answers change from a correct one to a wrong one. Choose the answer that seems best to you and move on to the next question. Change your answer only if you are absolutely sure you made a mistake (for example, if another question on the test reminds you of the correct answer).

Specific Helpful Tips for the Multiple-Choice Questions

EXAMPLE

In Earth's geologic past there were long periods that were much warmer than the present climate. What is the primary evidence that these long warm periods existed?

(1) United States National Weather Service records
(2) polar magnetic directions preserved in the rock record
(3) radioactive decay rates
(4) plant and animal fossils

Suppose you had no idea how to answer this question. You could use what you *do* know to make an educated guess. You do know that Earth's geologic past covers billions of years. You do know that the United States didn't exist until the late 1700s. So you can eliminate choice 1 because it is unlikely that USNWS records go back far enough to tell you about long warm periods in Earth's geologic past. You also know that compasses work in both winter and summer, so you have the idea that Earth's magnetic field doesn't really change with temperature. So now you eliminate choice 2. You remember reading something about radioactive decay and heat, and you know that the plants and animals that live in places with warm climates are very different from those that live in cold climates. At least, though, you've narrowed it down to two choices. Because you are more sure of the connection between climate, plants, and animals, you select choice 4 as your answer—and get the question right!

SUMMARY OF TIPS

1. Read the question and try to answer it before looking at the possible choices.
2. Always read *all* of the choices before choosing an answer.
3. Answer the easy questions first; do not spend too much time on any one question.
4. Look for and underline key words or phrases in the question that tell you what the questioner wants you to do.
5. Use partial knowledge to eliminate wrong choices.

Continued

6. Look for clues among the choices in the question, and use information from one part of the test to help you with others.
7. Know what is in the *Reference Tables for Physical Setting/Earth Science* and carefully examine any diagrams, graphs, or tables given with a question.
8. Do not choose an answer that contains an element of truth but that is incorrect as it relates to the question.
9. Beware of broad, sweeping generalizations in the answer choices—they are generally incorrect.
10. Don't be afraid to guess if you're not sure of the correct answer, but do not keep changing your answer.

HELPFUL TIPS FOR QUESTIONS ON PARTS B AND C

Part B includes both multiple-choice and constructed-response questions and represents 25–35 percent of the examination. Part C consists of extended constructed-response questions and represents 15–25 percent of the examination. The precise number of questions in each part may vary from year to year. Some questions may require the use of the *Reference Tables for Physical Setting/Earth Science*.

TIP 1

Know What a Constructed-Response Question Is

"Constructed-response" means that you must provide, in writing, your own response to a question, rather than choosing your answer from a list of possible choices. There are two types of constructed-response questions: open-response and free-response. For an open-response question there is only one correct answer, while a free-response question has many alternative answers.

Example of an open-response question:

Helpful Tips for Questions on Parts B and C

Base your answers to questions 109 and 110 on the diagram below and on your knowledge of earth science. The diagram represents the apparent path of the Sun on the dates indicated for an observer in New York State. The diagram also shows the angle of Polaris above the horizon.

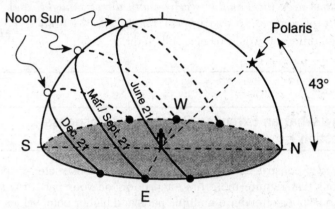

109 State the latitude of the location represented by the diagram to the *nearest degree*. Include the latitude direction in your answer. [2]

This is an open-response question; it has only one possible answer: 43°N.

Example of a free-response question.

110 The climate of most locations near the Equator is warm and moist.

(a) Explain why the climate is warm. [1]

___

___

(b) Explain why the climate is moist. [1]

___

___

Although an acceptable response must be scientifically correct, there are many alternative ways of expressing your answer to this question. Therefore, it is a free-response question.

Acceptable answers to (a) might include statements such as these: *"Locations near the Equator receive more direct sunlight"* and *"The intensity of sunlight is greater near the Equator."* Both responses, although different, would receive full credit.

TIP 2

Know What an Extended Constructed-Response Question Is

Extended constructed-response questions are constructed-response questions that require more time and effort on your part to answer. These questions often have multiple parts and higher point values, and may require you to graph data, complete a data table, label or draw diagrams, design experiments, make calculations, or write short or extended responses. In addition, you may be required to hypothesize, interpret, analyze, or evaluate data, or apply your scientific knowledge and skills to real-world situations.

TIP 3

Understand Why These Kinds of Questions Are on the Examination

Constructed-response and extended constructed-response questions are designed to:
- Assess your *understanding* of earth science concepts rather than straight recall of factual information.
- Require you to demonstrate that you can "pull all the pieces together," that is, integrate data, to solve a problem.
- Provide insight into how you think by requiring you to defend your answer in writing or to show all of your work.
- Assess knowledge and skills that cannot easily be assessed using a multiple-choice format.

Helpful Tips for Questions on Parts B and C 23

TIP 4

Know What Will Be Expected of You for These Parts of the Examination

Parts B and C of the examination will require you to demonstrate in a variety of ways that you have certain knowledge, skills, and abilities. You may be asked to construct a graph or draw a diagram. You may be asked to solve a mathematical problem, showing all of your work. You may be asked to provide explanations for physical phenomena or to write an essay on a topic in earth science.

Some of the skills and abilities that may be assessed in Parts B and C of the examination are:
- Apply the concepts of eccentricity, rate, gradient, standard error of measurement, and density in context.
- Determine the relationships among velocity, slope, sediment size, channel shape, and volume of a stream.
- Understand the relationships among the planets' distances from the Sun, gravitational forces, periods of revolution, and speeds of revolution.
- In a field, use isolines to determine a source of pollution.
- Show how observations of celestial motions supports the early idea that stars moved around a stationary Earth (the geocentric model), but further investigation led scientists to understand that most of these motions are a result of Earth's revolution around the Sun (the heliocentric model).
- Design an experiment to test sediment properties and the rate of deposition of particles.
- Graph the changing length of a shadow based on the motion of the Sun.
- Analyze patterns of topography and drainage, and design solutions to deal effectively with runoff.
- Analyze weather maps to predict future weather events.
- Critique printed or electronic materials that exemplify miscommunication and/or misconceptions of current commonly accepted scientific knowledge.
- Discuss how early warning systems can protect society and the environment from natural disasters such as hurricanes, tornadoes, earthquakes, tsunamis, floods, and volcanoes.

- Analyze a depositional-erosional system of a stream.
- Draw a simple contour map of a model landform.
- Construct and interpret a profile based on an isoline map.
- Use flowcharts to identify rocks and minerals.
- Develop a scale model to represent planet size and/or distance from the Sun.
- Develop a scale model of units of geologic time.
- Use topographical maps to determine distances and elevations.
- Analyze the interrelationship between gravity and inertia and its effects on the orbits of planets or satellites.
- Graph and interpret the nature of cyclic changes such as sunspots, tides, and atmospheric carbon dioxide concentrations.
- Determine, based on current data of plate movements, past and future positions of land masses.
- Using given weather data, identify the interface between air masses, such as cold fronts, warm fronts, and stationary fronts.
- Compare and contrast the effects of human activities as they relate to quality of life on Earth systems (global warming, land use, preservation of natural resources, pollution).
- Analyze the issues related to local energy needs, and develop a viable energy-generation plan for a community.
- Evaluate the political, economic, and environmental impacts of global distribution and use of mineral resources and fossil fuels.
- Consider the environmental and social implications of various solutions to a specific environmental Earth resources problem.
- Using a topographic map, determine the safest and most efficient route for rescue purposes.

A complete list of skills and abilities is described in greater detail in Standards 1, 2, 6, and 7 of the Topic Outline and Index.

TIP 5

Work to Improve Your Score on These Types of Questions

1. **Determine what the question is asking you to do. Underline key words in the question.**

    Most questions of this type have an action word that identifies what you must do to earn points.

Helpful Tips for Questions on Parts B and C

Below, the action words in a question you looked at earlier are bold and underlined. Other key words are underlined.

State the <u>latitude</u> of the location represented by the diagram to the <u>nearest degree</u>. **Include** the <u>latitude direction</u> in your answer. [2]

2. **Check to see what points are being awarded for.**
    The points are listed in brackets after each question. If more than one point is being awarded, you can usually make a good guess about what each point will be given for if you read the question again carefully. In the example above, you have probably guessed that one point is given for stating the latitude to the nearest degree, and one point is given for including the correct direction. The answer is 43° North. If you wrote just 43°, you would get only one point. For full credit you must state both 43° and North (or N).

3. **Always include units in your answers, and check to make sure that the form of your answer matches the form asked for in the question.**
    Consider this example:

Air Temperature	21°C
Barometric Pressure	993.1 mb
Wind Direction	From the east
Windspeed	25 knots

State the <u>barometric pressure</u> in its proper form, <u>as used on a station model</u>.
    What you are asked to do is state the barometric pressure. However, if you write 993.1 mb, your answer will be marked incorrect because the form asked for is "as used on a station model." On a station model, the decimal point is dropped and the last three digits only are listed. Your answer should be given as 931 because that is what would appear on a station model.

4. **When asked to respond in the form of a statement or essay, use complete sentences and organize your answer carefully.**
    A complete sentence begins with a capital letter, has a subject and a verb, and ends with a period. Very often, a point is awarded solely for answering in a complete sentence—even if the answer is wrong!! Therefore, read over your statements or essays carefully.

5. Examine the answer sheet for Part B-2 or C for clues to how the question is to be answered.

The answer sheet for Part B-2 or C will often give you a format for answering the question. For example, the answer sheet for the question below provides several blank lines, so you are probably being asked to write a short essay.

115 Using one or more complete sentences, state one reason that ultraviolet rays are dangerous. [2]

_____

_____

_____

If the answer sheet has a blank graph, you are probably being asked to construct a graph. Be sure to include all elements asked for in the question. For example, on the graph below you must not only plot points, but also plot them with specific shapes, label axes, and choose an appropriate scale for each axis.

101 Plot the data for rock sample *A* for the 20 minutes of the investigation. Surround each point with a small circle and connect the points. [1]

   Example:

102 Plot the data for rock sample *B* for the 20 minutes of the investigation. Surround each point with a small triangle and connect the points. [1]

   Example:

103 Plot the data for rock sample *C* for the 20 minutes of the investigation. Surround each point with a small square and connect the points. [1]

   Example:

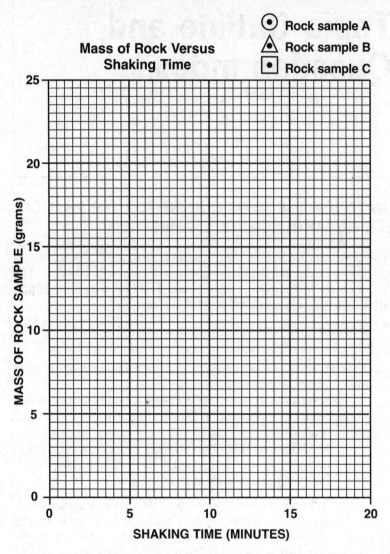

6. **Review questions from Parts B-2 and C of the sample exam and from Part III of previous exams.**

   Use the Answers Explained section of this book to review how questions of these types are answered and how point values are awarded. Focus on the kinds of questions being asked, rather than on specific questions.

Topic Outline and Question Index

HOW TO USE THE TOPIC OUTLINE FOR THE PHYSICAL SETTING/EARTH SCIENCE EXAMINATION

The topic outline follows the organization of the Physical Setting/Earth Science Core Curriculum, which was written to accompany The New York State Mathematics, Science, and Technology Learning Standards, which are organized in the following way:

STANDARDS – statements of what you are expected to know and be able to do

1. **KEY IDEAS** – broad, unifying, general statements of the most important ideas relating to the Standard
    2. **PERFORMANCE INDICATORS** – statements describing how you should be able to demonstrate that you understand the Key Ideas
        3. **MAJOR UNDERSTANDINGS** – specific statements of underlying concepts needed to do the things described in the Performance Indicators

The topic outline is divided into the five Standards (Standards 1, 2, 4, 6, and 7) that will be addressed on the Physical Setting/Earth Science Regents Examination. Each Standard is subdivided into Key Ideas. Each Key Idea is further subdivided into Performance Indicators and their associated Major Understandings so that you may locate questions for individual concepts. Suppose, for example, you wish to check your mastery of concepts related to tectonic plate boundaries. If you look in the topic outline under Standard 4, Key Idea 2, Performance Indicator 2.1, Major Understanding 2.1ℓ, you will find *"The lithosphere consists* of separate plates that ride on the more fluid asthenosphere and move slowly in relationship to one another, creating convergent, divergent, and transform plate boundaries." By following a parallel line across the page, you can locate appropriate questions from past examinations. On the June 2012 exam, for example, question 85 tests concepts related to plate boundaries. In a similar manner you can locate questions on this topic from other past examinations.

When you have identified questions from the topic outline, turn to the appropriate examination in the book for the first keyed question. Try to answer it and then check your response, reviewing explanations where desirable. Continue in this manner for other keyed questions.

You will notice that for some headings in the topic outline there is more than one test item on a particular examination. This usually occurs because more than one concept is involved. To find a question on a particular concept, you may find it necessary to check the keyed questions on several examinations.

TOPIC OUTLINE AND QUESTION INDEX—
PHYSICAL SETTING/EARTH SCIENCE EXAMINATION

STANDARD 1—Analysis, Inquiry, and Design: Students will use mathematical analysis, scientific inquiry, and engineering design, as appropriate, to pose questions, seek answers, and develop solutions.

	EXAMINATION QUESTIONS				
	August 2014	June 2015	August 2015	June 2016	August 2016
MATHEMATICAL ANALYSIS **Key Idea 1:** Abstraction and symbolic representation	62, 70, 75	6, 68, 71, 77	44, 55, 70, 76, 78, 81	1, 40, 68, 70, 76, 77	60, 67, 82, 83
Key Idea 2: Deductive and inductive reasoning	2, 49, 59, 66	11, 14, 21, 30, 41, 48, 50, 62, 64, 75, 78–80, 84	5, 15, 25, 27, 32, 34, 37, 38, 40, 42, 43, 49, 59, 75, 84	3, 30, 66, 67, 72, 79	18, 26, 66, 78, 81, 84
Key Idea 3: Critical thinking skills	72	5, 28, 55, 57, 67, 70, 72, 81	71	58, 62, 75	55
SCIENTIFIC INQUIRY **Key Idea 1:** The central purpose of scientific inquiry is to develop explanations of natural phenomena in a continuing, creative process.	15, 18, 37, 41, 51, 53, 54, 56, 58, 67, 69, 71, 76, 77, 81	12, 15, 17, 18–20, 24, 33, 34, 37, 42, 43, 45, 52, 53, 56, 59, 61, 63, 66, 85	6, 27, 36, 37, 41, 45, 47, 51, 55, 66, 68, 72, 74, 79, 83, 85	2, 4, 5, 12, 16, 24, 29, 31, 32, 36–39, 46, 59, 64	57, 58, 68, 69, 85
Key Idea 2: Beyond the use of reasoning and consensus, scientific inquiry involves the testing of proposed explanations involving the use of conventional techniques and procedures and usually requiring considerable ingenuity.		1, 21, 58			79

Topic Outline and Question Index

	EXAMINATION QUESTIONS				
	August 2014	June 2015	August 2015	June 2016	August 2016
Key Idea 3: The observations made while testing proposed explanations, when analyzed using conventional and invented methods, provide new insights into phenomena.	3, 8, 9, 11–14, 16, 17, 19–23, 31, 35, 36, 38, 43–46, 48, 53, 55, 57, 58, 62, 70, 72–75, 78–80, 83, 85	1, 5–11, 13, 14, 18, 22–25, 31–35, 37, 41, 44, 45, 49, 50, 52, 53, 57, 62, 63, 65, 71, 74, 75, 80	1, 2, 7, 8, 12–15, 17–21, 23, 32, 33–35, 38, 39, 50, 53–58, 61–65, 67, 69, 75, 76, 82, 85	3, 11, 13, 17, 19–21, 25, 33, 35, 38, 42, 45, 47, 50, 52, 61, 63, 65, 68, 73, 74, 77, 80–83	1, 4, 7, 10, 13–16, 18, 20, 22, 25, 30, 35–43, 49, 51–53, 59, 64, 69–71, 75, 77, 80, 82
ENGINEERING DESIGN **Key Idea 1:** Engineering design is an iterative process involving modeling and optimization (finding the best solution within given constraints); this process is used to develop technological solutions to problems within given constraints.					
STANDARD 2—Students will access, generate, process, and transfer information, using appropriate technologies.					
INFORMATION SYSTEMS **Key Idea 1:** Information technology is used to retrieve, process, and communicate information as a tool to enhance learning.		36		60	
Key Idea 2: Knowledge of the impacts and limitations of information systems is essential to its effective and ethical use.					
Key Idea 3: Information technology can have positive and negative impacts on society, depending upon how it is used.		36			62
STANDARD 6—Interconnectedness: Common Themes: Students will understand the relationships and common themes that connect mathematics, science, and technology and apply the themes to these and other areas of learning.					
INTERCONNECTEDNESS COMMON THEMES **Key Idea 1: Systems Thinking** Through systems thinking, people can recognize the commonalities that exist among all systems and how parts of a system interrelate and combine to perform specific functions.	4, 7, 25, 27, 29, 55, 58, 63, 78, 80, 81	25, 27, 29, 31, 32, 39, 42, 43, 46, 51, 52, 54, 58, 60, 61, 69	8, 39, 40	16, 23, 26, 28, 29, 34, 39, 43, 49, 53, 54, 60, 64, 80, 85	54
Key Idea 2: Models Models are simplified representations of objects, structures, or systems used in analysis, explanation, interpretation, or design.	6, 14, 19, 22, 24–35, 38–40, 42–48, 50, 52–54, 60, 61, 63–65, 77–80, 83–85	3, 6, 8, 9, 13, 16, 18–21, 23–29, 31, 32, 35–40, 43–50, 53–56, 58, 60, 63, 65–70, 72, 74, 82–85	7, 11, 12, 14, 15, 17, 18, 22, 25–31, 33, 35–50, 52, 53, 57–62, 66, 68, 73, 77, 81	6, 8, 9, 15, 17, 26, 28, 31, 32, 34, 35, 37, 40, 42–46, 48, 49, 51, 54, 55, 58, 60, 62, 70, 71, 73, 74, 76, 78, 79, 84, 85	16, 19–21, 23–35, 39–57, 59, 61, 63, 65–67, 72–74, 76–79, 81, 83, 84
Key Idea 3: Magnitude and Scale The grouping of magnitudes of size, time, frequency, and pressures or other units of measurement into a series of relative order provides a useful way to deal with the immense range and the changes in scale that affect the behavior and design of systems.	36, 84	10, 18, 41, 67, 73	56, 57	50, 70, 75	60

Physical Setting/Earth Science Examination

	EXAMINATION QUESTIONS				
	August 2014	June 2015	August 2015	June 2016	August 2016
Key Idea 4: Equilibrium and Stability Equilibrium is a state of stability due either to a lack of change (static equilibrium) or a balance between opposing forces (dynamic equilibrium).				57	
Key Idea 5: Patterns of Change Identifying patterns of change is necessary for making predictions about future behavior and conditions.	4, 7, 15, 19, 27, 29, 40, 51, 54, 56, 68	3, 7, 16, 22, 26, 36, 40, 51, 54, 59, 61, 63, 82–84	3, 46, 49, 53, 65, 75, 83, 84	34, 41, 47, 55, 56, 67–69, 72, 78, 84	26, 65, 72–74
Key Idea 6: Optimization In order to arrive at the best solution that meets criteria within constraints, it is often necessary to make trade-offs.					12

STANDARD 7—Interdisciplinary Problem Solving: Students will apply the knowledge and thinking skills of mathematics, science, and technology to address real-life problems and make informed decisions.

INTERDISCIPLINARY PROBLEM SOLVING **Key Idea 1: Connections** The knowledge and skills of mathematics, science, and technology are used together to make informed decisions and solve problems, especially those relating to issues of science/technology/society, consumer decision making, design, and inquiry into phenomena.		79			
Key Idea 2: Strategies Solving interdisciplinary problems involves a variety of skills and strategies, including effective work habits; gathering and processing information; generating and analyzing ideas; realizing ideas; making connections among the common themes of mathematics, science, and technology; and presenting results.	82	76	80		62

STANDARD 4—Students will understand and apply scientific concepts, principles, and theories pertaining to the physical setting and living environment and recognize the historical development of ideas in science.

Physical Setting Key Idea 1: EARTH AND CELESTIAL PHENOMENA CAN BE DESCRIBED BY PRINCIPLES OF RELATIVE MOTION AND PERSPECTIVE.					
Intro 1: People have observed the stars for thousands of years, using them to find direction, note the passage of time, and to express their values and traditions. As our technology has progressed, so has understanding of celestial objects and events.					
Intro 2: Theories of the universe have developed over many centuries. Although to a casual observer celestial bodies appeared to orbit a stationary Earth, scientific discoveries led us to the understanding that Earth is one planet that orbits the Sun, a typical star in a vast and ancient universe. We now infer an origin and an age and evolution of the universe, as we speculate about its future.					
Intro 3: As we look at Earth, we find clues to its origin and how it has changed through nearly five billion years, as well as the evolution of life on Earth.					

Topic Outline and Question Index

	EXAMINATION QUESTIONS				
	August 2014	June 2015	August 2015	June 2016	August 2016
Physical Setting Performance Indicator 1.1: Explain complex phenomena, such as tides, variations in day length, solar insolation, apparent motion of the planets, and annual traverse of the constellations.					
MAJOR UNDERSTANDINGS	4, 27, 28, 29, 67	83–85	2, 27, 44–47	30, 54–57	45, 46, 72, 73
1.1a Most objects in the solar system are in regular and predictable motion. • These motions explain such phenomena as the day, the year, seasons, phases of the moon, eclipses, and tides. • Gravity influences the motions of celestial objects. The force of gravity between two objects in the universe depends on their masses and the distance between them. *See Standard 6, Key Idea 4*					
1.1b Nine planets move around the Sun in nearly circular orbits. • The orbit of each planet is an ellipse with the Sun located at one of the foci. • Earth is orbited by one Moon and many artificial satellites. *See Standard 1: Mathematical Analysis, Key Idea 2, and Standard 6, Key Ideas 3 and 4*			55	32	
1.1c Earth's coordinate system of latitude and longitude, with the Equator and Prime Meridian as reference lines, is based upon Earth's rotation and our observation of the Sun and stars.			5, 60		
1.1d Earth rotates on an imaginary axis at a rate of 15 degrees per hour. To people on Earth, this turning of the planet makes it seem as though the Sun, the Moon, and the stars are moving around Earth once a day. Rotation provides a basis for our system of local time; meridians of longitude are the basis for time zones. *See Standard 1: Scientific Inquiry, Key Idea 3*	30	54, 55		40	48, 55, 58
1.1e The Foucault pendulum and the Coriolis effect provide evidence of Earth's rotation.	5	3, 19	70–73	2, 6	3, 29
1.1f Earth's changing position with regard to the Sun and the Moon has noticeable effects. • Earth revolves around the Sun with its rotational axis tilted at 23.5 degrees to a line perpendicular to the plane of its orbit, with the North Pole aligned with Polaris. • During Earth's one-year period of revolution, the tilt of its axis results in changes in the angle of incidence of the Sun's rays at a given latitude; these changes cause variation in the heating of the surface. This produces seasonal variation in weather.	63–65	16, 56, 57, 77, 78	3, 7, 25	1, 8, 36, 42, 67–69	24, 47, 49, 56
1.1g Seasonal changes in the apparent positions of constellations provide evidence of Earth's revolution.		20			
1.1h The Sun's apparent path through the sky varies with latitude and season.				41	50, 57

Physical Setting/Earth Science Examination

		EXAMINATION QUESTIONS				
		August 2014	June 2015	August 2015	June 2016	August 2016
1.1i	Approximately 70 percent of Earth's surface is covered by a relatively thin layer of water, which responds to the gravitational attraction of the Moon and the Sun with a daily cycle of high and low tides.					26, 74
Physical Setting Performance Indicator 1.2: Describe current theories about the origin of the universe and solar system.						
1.2a	The universe is vast and estimated to be over ten billion years old. The current theory is that the universe was created from an explosion called the big bang. Evidence for this theory includes: • cosmic background radiation; • a redshift (the Doppler effect) in the light from very distant galaxies.	1, 3	18, 37	6	51, 52	2, 66, 67
1.2b	Stars form when gravity causes clouds of molecules to contract until nuclear fusion of light elements into heavier ones occurs. Fusion releases great amounts of energy over millions of years. • The stars differ from each other in size, temperature, and age. • Our Sun is a medium-sized star within a spiral galaxy of stars known as the Milky Way. Our galaxy contains billions of stars, and the universe contains billions of such galaxies.	2, 76	2, 39–42	1, 4	3, 53	1, 68
1.2c	Our solar system formed about five billion years ago from a giant cloud of gas and debris. Gravity caused Earth and the other planets to become layered according to density differences in their materials. • The characteristics of the planets of the solar system are affected by each planet's location in relationship to the Sun. • The terrestrial planets are small, rocky, and dense. The Jovian planets are large and gaseous, and have low density.	43, 73–75	1	56, 57	4	28, 75
1.2d	Asteroids, comets, and meteors are components of our solar system. • Impact events have been correlated with mass extinction and global climatic change. • Impact craters can be identified in Earth's crust.	36, 37, 72		58, 65	18	36, 38
1.2e	Earth's early atmosphere formed as a result of the out-gassing of water vapor, carbon dioxide, nitrogen, and lesser amounts of other gases from its interior.				7	
1.2f	Earth's oceans formed as a result of precipitation over millions of years. The presence of an early ocean is indicated by sedimentary rocks of marine origin, dating back about four billion years.					

Topic Outline and Question Index

		EXAMINATION QUESTIONS				
		August 2014	June 2015	August 2015	June 2016	August 2016
1.2g	Earth has continuously been recycling water since the outgassing of water early in its history. This constant recirculation of water at and near Earth's surface is described by the hydrologic (water) cycle. • Water is returned from the atmosphere to Earth's surface by precipitation. Water returns to the atmosphere by evaporation or transpiration from plants. A portion of the precipitation becomes runoff over the land or infiltrates into the ground to become stored in the soil or groundwater below the water table. Soil capillarity influences these processes. • The amount of precipitation that seeps into the ground or runs off is influenced by climate, slope of the land, soil, rock type, vegetation, land use, and degree of saturation. • Porosity, permeability, and water retention affect runoff and infiltration.	7, 8, 49, 50	4, 21, 22	9, 36, 37	5, 43, 44	5
1.2h	The evolution of life caused dramatic changes in the composition of Earth's atmosphere. Free oxygen did not form in the atmosphere until oxygen-producing organisms evolved.			19		
1.2i	The pattern of evolution of life-forms on Earth is at least partially preserved in the rock record. • Fossil evidence indicates that a wide variety of lifeforms has existed in the past and that most of these forms have become extinct. • Human existence has been very brief compared to the expanse of geologic time.	18	79–82			14
1.2j	Geologic history can be reconstructed by observing sequences of rock types and fossils to correlate bedrock at various locations. • The characteristics of rocks indicate the processes by which they formed and the environments in which 69 these processes took place. • Fossils preserved in rocks provide information about past environmental conditions. • Geologists have divided Earth history into time units based upon the fossil record. • Age relationships among bodies of rocks can be determined using principles of original horizontality, superposition, inclusions, cross-cutting relationships, contact metamorphism, and unconformities. The presence of volcanic ash layers, index fossils, and meteoritic debris can provide additional information. • The regular rate of nuclear decay (half-life time period) of radioactive isotopes allows geologists to determine the absolute age of materials found in some rocks.	17, 19, 44, 46, 51, 52, 54	5, 23, 25, 26	18, 20, 22, 24, 33, 62, 64, 66, 68, 69	16, 19, 27, 45, 48, 49, 62–65	13, 27, 31, 51, 54, 69–71
Physical Setting Key Idea 2: MANY OF THE PHENOMENA THAT WE OBSERVE ON EARTH INVOLVE INTERACTIONS AMONG COMPONENTS OF AIR, WATER, AND LAND.						

Physical Setting/Earth Science Examination

	EXAMINATION QUESTIONS					
	August 2014	June 2015	August 2015	June 2016	August 2016
Intro 1: Earth may be considered a huge machine driven by two engines, one internal and one external. These heat engines convert heat energy into mechanical energy.					
Intro 2: Earth's external heat engine is powered primarily by solar energy and influenced by gravity. Nearly all the energy for circulating the atmosphere and oceans is supplied by the Sun. As insolation strikes the atmosphere, a small percentage is directly absorbed, especially by gases such as ozone, carbon dioxide, and water vapor. Clouds and Earth's surface reflect some energy back to space, and Earth's surface absorbs some energy. Energy is transferred between Earth's surface and the atmosphere by radiation, conduction, evaporation, and convection. Temperature variations within the atmosphere cause differences in density that cause atmospheric circulation, which is affected by Earth's rotation. The interaction of these processes results in the complex atmospheric occurrence known as weather.					
Intro 3: Average temperatures on Earth are the result of the total amount of insolation absorbed by Earth's surface and its atmosphere and the amount of long-wave energy radiated back into space. However, throughout geologic time, ice ages occurred in the middle latitudes. In addition, average temperatures may have been significantly warmer at times in the geologic past. This suggests that Earth had climate changes that were most likely associated with long periods of imbalances of its heat budget.					
Intro 4: Earth's internal heat engine is powered by heat from the decay of radioactive materials and residual heat from Earth's formation. Differences in density resulting from heat flow within Earth's interior caused the changes explained by the theory of plate tectonics: movement of the lithospheric plates, earthquakes, volcanoes, and the deformation and metamorphism of rocks during the formation of young mountains.					
Intro 5: Precipitation resulting from the external heat engine's weather systems supplies moisture to Earth's surface that contributes to the weathering of rocks. Running water erodes mountains that were originally uplifted by Earth's internal heat engine and transports sediments to other locations, where they are deposited and may undergo the processes that transform them into sedimentary rocks.					
Intro 6: Global climate is determined by the interaction of solar energy with Earth's surface and atmosphere. This energy transfer is influenced by dynamic processes such as cloud cover and Earth rotation, and the positions of mountain ranges and oceans.					
Physical Setting Performance Indicator 2.1: Use the concepts of density and heat energy to explain observations of weather patterns, seasonal changes, and the movements of Earth's plates.					

Topic Outline and Question Index

	EXAMINATION QUESTIONS				
	August 2014	June 2015	August 2015	June 2016	August 2016
MAJOR UNDERSTANDINGS					
2.1a Earth systems have internal and external sources of energy, both of which create heat.					
2.1b The transfer of heat energy within the atmosphere, the the hydrosphere, and Earth's interior results in the formation of regions of different densities. These density differences result in motion.	57	8	82		37
2.1c Weather patterns become evident when weather variables are observed, measured, and recorded. These variables include air temperature, air pressure, moisture (relative humidity and dewpoint), precipitation (rain, snow, hail, sleet, etc.), wind speed and direction, and cloud cover.	9, 12			11	64
2.1d Weather variables are measured using instruments such as thermometers, barometers, psychrometers, precipitation gauges, anemometers, and wind vanes.		6, 38		12, 59, 61	63
2.1e Weather variables are interrelated. For example: • Temperature and humidity affect air pressure and probability of precipitation. • Air pressure gradient controls wind velocity.		28			
2.1f Air temperature, dewpoint, cloud formation, and precipitation are affected by the expansion and contraction of air due to vertical atmospheric movement.	10			39	8, 65
2.1g Weather variables can be represented in a variety of formats including radar and satellite images, weather maps (including station models, isobars, and fronts), atmospheric cross sections, and computer models. *See Standard 2, Information Systems, Key Idea 1*	11, 31, 32, 38, 70	9, 36	12	9, 34, 58	30, 76–78
2.1h Atmospheric moisture, temperature, and pressure distributions; jet streams; wind; air masses and frontal boundaries; and the movement of cyclonic systems and associated tornadoes, thunderstorms, and hurricanes occur in observable patterns. Loss of property, personal injury, and loss of life can be reduced by effective emergency preparedness. *Also see Standard 6, Key Idea 5, and Standard 7, Key Idea 2*	33, 39, 40, 82	7, 52, 53	10, 28, 77, 78, 80	10, 37, 60	
2.1i Seasonal changes can be explained using concepts of density and heat energy. These changes include the shifting of global temperature zones, the shifting of planetary wind and ocean current patterns, the occurrence of monsoons, hurricanes, flooding, and severe weather.	81	82	29		6, 7
2.1j Properties of Earth's internal structure (crust, mantle, inner core, and outer core) can be inferred from the analysis of the behavior of seismic waves, including velocity and refraction. • Analysis of seismic waves allows the determination of the location of earthquake epicenters, and the measurement of earthquake magnitude; this analysis leads to the inference that Earth's interior is composed of layers that differ in composition and states of matter.	20, 41, 42	11, 44, 72, 73, 75	15, 38–40, 59	20–22	18, 61

Physical Setting/Earth Science Examination

		EXAMINATION QUESTIONS			
		August 2014	June 2015	August 2015	June 2016	August 2016

		August 2014	June 2015	August 2015	June 2016	August 2016
2.1k	The outward transfer of Earth's internal heat drives convective circulation in the mantle that moves the lithospheric plates comprising Earth's surface.		43		73	15
2.1l	The lithosphere consists of separate plates that ride on the more fluid asthenosphere and move slowly in relationship to one another, creating convergent, divergent, and transform plate boundaries. These motions indicate Earth is a dynamic geologic system. • These plate boundaries are the sites of most earthquakes, volcanoes, and young mountain ranges. • Compared to continental crust, ocean crust is thinner and denser. New ocean crust continues to form at mid-ocean ridges. • Earthquakes and volcanoes present geologic hazards to humans. Loss of property, personal injury, and loss of life can be reduced by effective emergency preparedness. *See Standard 7, Key Idea 2*	16, 21, 83–85	74, 76	17, 61	35	59, 60, 62
2.1m	Many processes of the rock cycle are consequences of plate dynamics. These include the production of magma (and subsequent igneous rock formation and contact metamorphism) at both subduction and rifting regions, regional metamorphism within subduction zones, and the creation of major depositional basins through downwarping of the crust.					
2.1n	Many of Earth's surface features, such as mid-ocean ridges/rifts, trenches/subduction zones/island arcs, mountain ranges (folded, faulted, and volcanic), hot spots, and the magnetic and age patterns in surface bedrock, are a consequence of forces associated with plate motion and interaction.	35	32	16, 85	70–72, 74	17
2.1o	Plate motions have resulted in global changes in geography, climate, and the patterns of organic evolution. *See Standard 6, Key Idea 5*		49		17, 33	16
2.1p	Landforms are the result of the interaction of tectonic forces and the processes of weathering, erosion, and deposition.				46	
2.1q	Topographic maps represent landforms through the use of contour lines that are isolines connecting points of equal elevation. Gradients and profiles can be determined from changes in elevation over a given distance. *See Standard 6, Key Ideas 2 and 3, and Standard 7, Key Idea 2*	59–62	67–71	81	75–78	21, 32
2.1r	Climate variations, structure, and characteristics of bedrock influence the development of landscape features including mountains, plateaus, plains, valleys, ridges, escarpments, and stream drainage patterns.	22, 24, 34	12, 13	13, 14	23, 26	35, 41

Topic Outline and Question Index

		EXAMINATION QUESTIONS			
	August 2014	June 2015	August 2015	June 2016	August 2016
2.1s Weathering is the physical and chemical breakdown of rocks at or near Earth's surface. Soils are the result of weathering and biological activity over long periods of time.		24			
2.1t Natural agents of erosion, generally driven by gravity, remove, transport, and deposit weathered rock particles. Each agent of erosion produces distinctive changes in the material that it transports and creates characteristic surface features and landscapes. In certain erosional situations, loss of property, personal injury, and loss of life can be reduced by effective emergency preparedness. *See Standard 7, Key Idea 2*	56			79	19
2.1u The natural agents of erosion include:					
• **Streams (running water):** Gradient, discharge, and channel shape influence a stream's velocity and the erosion and deposition of sediments. Sediments transported by streams tend to become rounded as a result of abrasion. Stream features include V-shaped valleys, deltas, flood plains, and meanders. A watershed is the area drained by a stream and its tributaries. *See Standard 1: Engineering Design, Key Idea 1; Standard 1: Mathematical Analysis, Key Idea 2; and Standard 6, Key Idea 1*	25	14, 58–61	11, 74–76	28, 47	33, 34, 39, 40, 80, 81
• **Glaciers (moving ice):** Glacial erosional processes include the formation of U-shaped valleys, parallel scratches, and grooves in bedrock. Glacial features include moraines, drumlins, kettle lakes, finger lakes, and outwash plains.	26		30	24	23
• **Wave Action:** Erosion and deposition cause changes in shoreline features, including beaches, sandbars, and barrier islands. Wave action rounds sediments as a result of abrasion. Waves approaching a shoreline move sand parallel to the shore within the zone of breaking waves.			48, 49		
Other • **Wind:** Erosion of sediments by wind is most common in arid climates and along shorelines. Wind-generated features include dunes and sand-blasted bedrock. • **Mass Movement:** Earth materials move downslope under the influence of gravity.		27			
2.1v Patterns of deposition result from a loss of energy within the transporting system and are influenced by the size, shape, and density of the transported particles. Sediment deposits may be sorted or unsorted. *See Standard 1: Scientific Inquiry, Key Idea 2*	48, 58		31, 84	29	

Physical Setting/Earth Science Examination

	August 2014	June 2015	EXAMINATION QUESTIONS August 2015	June 2016	August 2016
2.1w Sediments of inorganic and organic origin often accumulate in depositional environments. Sedimentary rocks form when sediments are compacted and/or cemented after burial or as the result of chemical precipitation from seawater.	47	31			
Physical Setting Performance Indicator 2.2: **Explain how incoming solar radiation, ocean currents, and land masses affect weather and climate.**					
2.2a Insolation (solar radiation) heats Earth's surface and atmosphere unequally due to variations in: • the intensity caused by differences in atmospheric transparency and angle of incidence, which vary with time of day, latitude, and season; • characteristics of the materials absorbing the energy, such as color, texture, transparency, state of matter, and specific heat; • duration, which varies with seasons and latitude.	6, 66, 68, 69	15, 30	8, 41–43, 54, 83	13, 15, 38, 66	4, 11, 83–85
2.2b The transfer of heat energy within the atmosphere, the hydrosphere, and Earth's surface occurs as the result of radiation, convection, and conduction. • Heating of Earth's surface and atmosphere by the Sun drives convection within the atmosphere and oceans, producing winds and ocean currents.		17		31	10
2.2c A location's climate is influenced by latitude, proximity to large bodies of water, ocean currents, prevailing winds, vegetative cover, elevation, and mountain ranges.	13, 14	29, 62–66	50–53	84, 85	9
2.2d Temperature and precipitation patterns are altered by: • natural events, such as El Niño and volcanic eruptions; • human influences, including deforestation, urbanization, and the production of greenhouse gases such as carbon dioxide and methane.	15, 55, 71		26	14	12
Physical Setting Key Idea 3: **MATTER IS MADE UP OF PARTICLES WHOSE PROPERTIES DETERMINE THE OBSERVABLE CHARACTERISTICS OF MATTER AND ITS REACTIVITY.**					
Intro 1: Observation and classification have helped us understand the great variety and complexity of Earth materials. Minerals are the naturally occurring inorganic solid elements, compounds, and mixtures from which rocks are made. We classify minerals on the basis of their chemical composition and observable properties. Rocks are generally classified by their origin (igneous, metamorphic, and sedimentary), texture, and mineral content.					

Topic Outline and Question Index

	EXAMINATION QUESTIONS				
	August 2014	June 2015	August 2015	June 2016	August 2016
Intro 2: Rocks and minerals help us understand Earth's historical development and its dynamics. They are important to us because of their availability and properties. The use and distribution of mineral resources and fossil fuels have important economic and environmental impacts. As limited resources, they must be used wisely.					
Physical Setting Performance Indicator 3.1: Explain the properties of materials in terms of the arrangement and properties of the atoms that compose them.					
3.1a Minerals have physical properties determined by their chemical composition and crystal structure. • Minerals can be identified by well-defined physical and chemical properties, such as cleavage, fracture, color, density, hardness, streak, luster, crystal shape, and reaction with acid. • Chemical composition and physical properties determine how minerals are used by humans. *See Standard 6, Key Idea 2*	23, 78	10, 34, 35			42–44, 53, 79, 82
3.1b Minerals are formed inorganically by the process of crystallization as a result of specific environmental conditions. These include: • cooling and solidification of magma; • precipitation from water caused by such processes as evaporation, chemical reactions, and temperature changes; • rearrangement of atoms in existing minerals subjected to conditions of high temperature and pressure.		33		25	
3.1c Rocks are usually composed of one or more minerals. • Rocks are classified by their origin, mineral content, and texture. • Conditions that existed when a rock formed can be inferred from the rock's mineral content and texture. • The properties of rocks determine how they are used and also influence land usage by humans.	45, 53, 77, 79, 80	45–48, 50	21, 23, 32, 34, 35, 63, 67	50, 80–83	20, 22, 25, 52
Reference Tables for Physical Settings/Earth Science 2011 Edition (Revised)	2, 3, 8, 9, 11–14, 16, 17, 19–22, 36, 38, 43–46, 48, 53, 57, 58, 62, 72–75, 77–80, 83	1, 5–11, 13, 14, 18, 23–25, 31–35, 37, 41, 44, 45, 49, 50, 52, 53, 57, 62, 63, 65, 71, 74, 75, 80		3, 9–11, 13, 17, 19–21, 25, 27, 33, 35–38, 43, 45, 50, 52, 61–63, 65, 73, 74, 77, 79–83	1, 4, 7, 10, 13–16, 18, 20, 22, 25, 30, 35–43, 51–53, 59, 64, 69–71, 75, 77, 80, 82

Earth Science Reference Tables and Charts

The University of the State of New York • THE STATE EDUCATION DEPARTMENT • Albany, New York 12234 • www.nysed.gov

Reference Tables for Physical Setting/EARTH SCIENCE

Radioactive Decay Data

RADIOACTIVE ISOTOPE	DISINTEGRATION	HALF-LIFE (years)
Carbon-14	$^{14}C \rightarrow {}^{14}N$	5.7×10^3
Potassium-40	$^{40}K \rightarrow {}^{40}Ar$ / ^{40}Ca	1.3×10^9
Uranium-238	$^{238}U \rightarrow {}^{206}Pb$	4.5×10^9
Rubidium-87	$^{87}Rb \rightarrow {}^{87}Sr$	4.9×10^{10}

Equations

$$\text{Eccentricity} = \frac{\text{distance between foci}}{\text{length of major axis}}$$

$$\text{Gradient} = \frac{\text{change in field value}}{\text{distance}}$$

$$\text{Rate of change} = \frac{\text{change in value}}{\text{time}}$$

$$\text{Density} = \frac{\text{mass}}{\text{volume}}$$

Specific Heats of Common Materials

MATERIAL	SPECIFIC HEAT (Joules/gram • °C)
Liquid water	4.18
Solid water (ice)	2.11
Water vapor	2.00
Dry air	1.01
Basalt	0.84
Granite	0.79
Iron	0.45
Copper	0.38
Lead	0.13

Properties of Water

Heat energy gained during melting	334 J/g
Heat energy released during freezing	334 J/g
Heat energy gained during vaporization	2260 J/g
Heat energy released during condensation	2260 J/g
Density at 3.98°C	1.0 g/mL

Average Chemical Composition of Earth's Crust, Hydrosphere, and Troposphere

ELEMENT (symbol)	CRUST		HYDROSPHERE	TROPOSPHERE
	Percent by mass	Percent by volume	Percent by volume	Percent by volume
Oxygen (O)	46.10	94.04	33.0	21.0
Silicon (Si)	28.20	0.88		
Aluminum (Al)	8.23	0.48		
Iron (Fe)	5.63	0.49		
Calcium (Ca)	4.15	1.18		
Sodium (Na)	2.36	1.11		
Magnesium (Mg)	2.33	0.33		
Potassium (K)	2.09	1.42		
Nitrogen (N)				78.0
Hydrogen (H)			66.0	
Other	0.91	0.07	1.0	1.0

2011 EDITION

This edition of the Earth Science Reference Tables should be used in the classroom beginning in the 2011–12 school year. The first examination for which these tables will be used is the January 2012 Regents Examination in Physical Setting/Earth Science.

Eurypterus remipes

New York State Fossil

Generalized Bedrock Geology of New York State

modified from
GEOLOGICAL SURVEY
NEW YORK STATE MUSEUM
1989

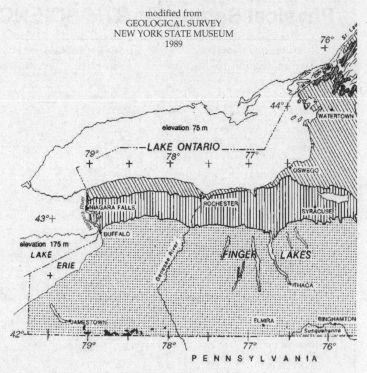

GEOLOGIC PERIODS AND ERAS IN NEW YORK

- CRETACEOUS and PLEISTOCENE (Epoch) weakly consolidated to unconsolidated gravels, sands, and clays
- LATE TRIASSIC and EARLY JURASSIC conglomerates, red sandstones, red shales, basalt, and diabase (Palisades sill)
- PENNSYLVANIAN and MISSISSIPPIAN conglomerates, sandstones, and shales
- DEVONIAN } limestones, shales, sandstones, and conglomerates
- SILURIAN   SILURIAN *also contains salt, gypsum, and hematite.*
- ORDOVICIAN } limestones, shales, sandstones, and dolostones
- CAMBRIAN
- CAMBRIAN and EARLY ORDOVICIAN sandstones and dolostones
  moderately to intensely metamorphosed east of the Hudson River
- CAMBRIAN and ORDOVICIAN (undifferentiated) quartzites, dolostones, marbles, and schists
  intensely metamorphosed; includes portions of the Taconic Sequence and Cortlandt Complex
- TACONIC SEQUENCE sandstones, shales, and slates
  slightly to intensely metamorphosed rocks of CAMBRIAN through MIDDLE ORDOVICIAN ages
- MIDDLE PROTEROZOIC gneisses, quartzites, and marbles
  Lines are generalized structure trends.
- MIDDLE PROTEROZOIC anorthositic rocks

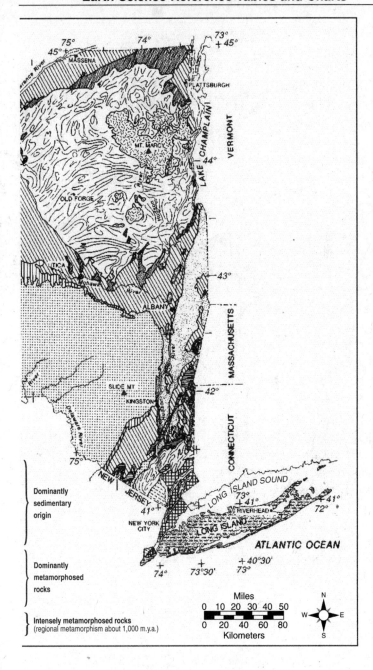

Earth Science Reference Tables and Charts

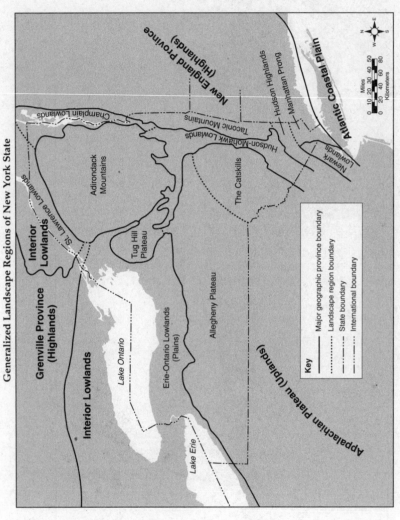

Generalized Landscape Regions of New York State

Earth Science Reference Tables and Charts 47

Surface Ocean Currents

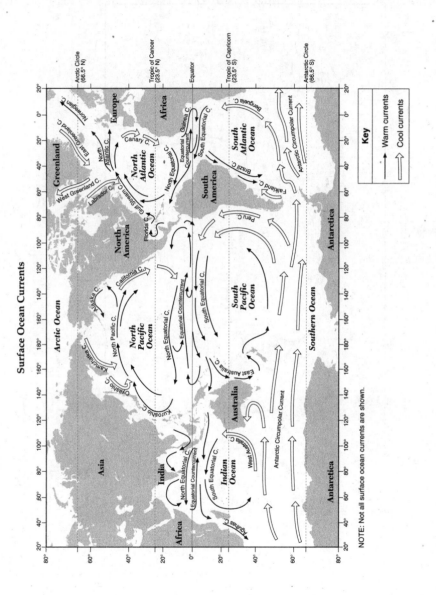

NOTE: Not all surface ocean currents are shown.

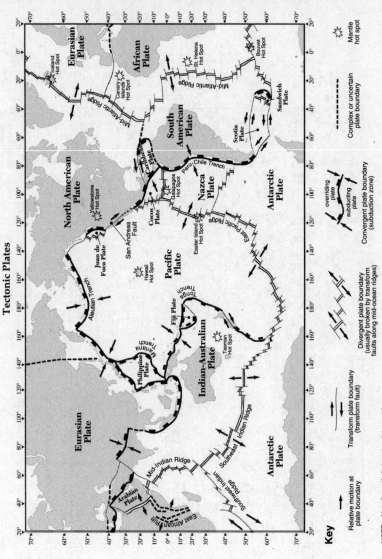

Earth Science Reference Tables and Charts

Rock Cycle in Earth's Crust

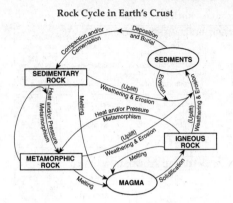

Relationship of Transported Particle Size to Water Velocity

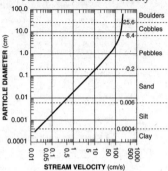

This generalized graph shows the water velocity needed to maintain, but not start, movement. Variations occur due to differences in particle density and shape.

Scheme for Igneous Rock Identification

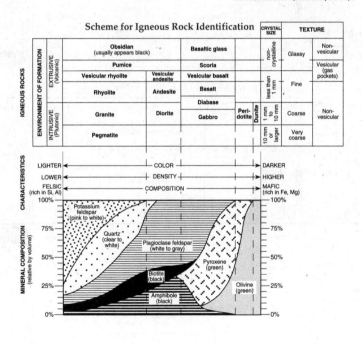

Scheme for Sedimentary Rock Identification

INORGANIC LAND-DERIVED SEDIMENTARY ROCKS

TEXTURE	GRAIN SIZE	COMPOSITION	COMMENTS	ROCK NAME	MAP SYMBOL
Clastic (fragmental)	Pebbles, cobbles, and/or boulders embedded in sand, silt, and/or clay	Mostly quartz, feldspar, and clay minerals; may contain fragments of other rocks and minerals	Rounded fragments	Conglomerate	
			Angular fragments	Breccia	
	Sand (0.006 to 0.2 cm)		Fine to coarse	Sandstone	
	Silt (0.0004 to 0.006 cm)		Very fine grain	Siltstone	
	Clay (less than 0.0004 cm)		Compact; may split easily	Shale	

CHEMICALLY AND/OR ORGANICALLY FORMED SEDIMENTARY ROCKS

TEXTURE	GRAIN SIZE	COMPOSITION	COMMENTS	ROCK NAME	MAP SYMBOL
Crystalline	Fine to coarse crystals	Halite	Crystals from chemical precipitates and evaporites	Rock salt	
		Gypsum		Rock gypsum	
		Dolomite		Dolostone	
Crystalline or bioclastic	Microscopic to very coarse	Calcite	Precipitates of biologic origin or cemented shell fragments	Limestone	
Bioclastic		Carbon	Compacted plant remains	Bituminous coal	

Scheme for Metamorphic Rock Identification

TEXTURE	GRAIN SIZE	COMPOSITION	TYPE OF METAMORPHISM	COMMENTS	ROCK NAME	MAP SYMBOL
FOLIATED — MINERAL ALIGNMENT	Fine	MICA, QUARTZ, FELDSPAR, AMPHIBOLE, GARNET, PYROXENE	Regional (Heat and pressure increases)	Low-grade metamorphism of shale	Slate	
	Fine to medium			Foliation surfaces shiny from microscopic mica crystals	Phyllite	
				Platy mica crystals visible from metamorphism of clay or feldspars	Schist	
FOLIATED — BANDING	Medium to coarse			High-grade metamorphism; mineral types segregated into bands	Gneiss	
NONFOLIATED	Fine	Carbon	Regional	Metamorphism of bituminous coal	Anthracite coal	
	Fine	Various minerals	Contact (heat)	Various rocks changed by heat from nearby magma/lava	Hornfels	
	Fine to coarse	Quartz	Regional or contact	Metamorphism of quartz sandstone	Quartzite	
		Calcite and/or dolomite		Metamorphism of limestone or dolostone	Marble	
	Coarse	Various minerals		Pebbles may be distorted or stretched	Metaconglomerate	

Earth Science Reference Tables and Charts

Inferred Properties of Earth's Interior

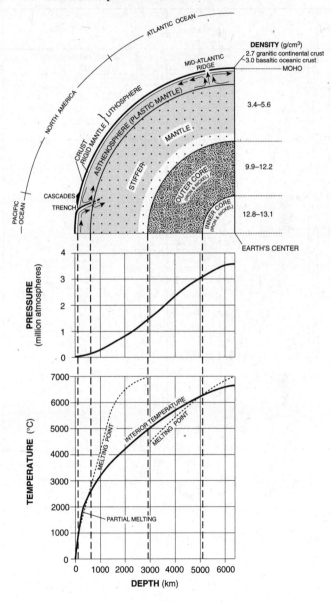

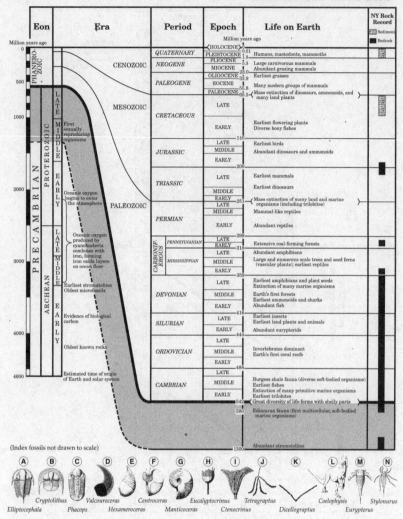

Earth Science Reference Tables and Charts 53

OF NEW YORK STATE

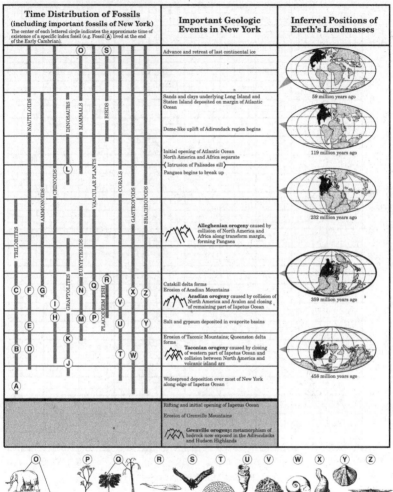

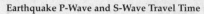

Earthquake P-Wave and S-Wave Travel Time

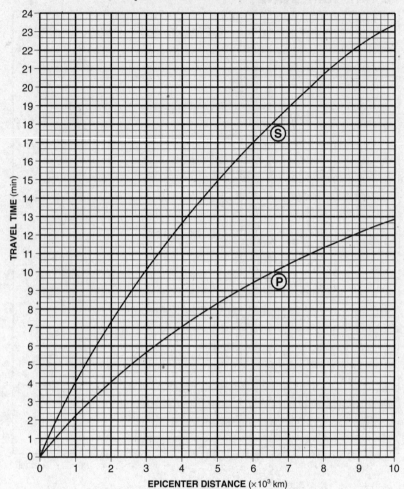

Earth Science Reference Tables and Charts

Dewpoint (°C)

Dry-Bulb Temperature (°C)	Difference Between Wet-Bulb and Dry-Bulb Temperatures (C°)															
	0	1	2	3	4	5	6	7	8	9	10	11	12	13	14	15
−20	−20	−33														
−18	−18	−28														
−16	−16	−24														
−14	−14	−21	−36													
−12	−12	−18	−28													
−10	−10	−14	−22													
−8	−8	−12	−18	−29												
−6	−6	−10	−14	−22												
−4	−4	−7	−12	−17	−29											
−2	−2	−5	−8	−13	−20											
0	0	−3	−6	−9	−15	−24										
2	2	−1	−3	−6	−11	−17										
4	4	1	−1	−4	−7	−11	−19									
6	6	4	1	−1	−4	−7	−13	−21								
8	8	6	3	1	−2	−5	−9	−14								
10	10	8	6	4	1	−2	−5	−9	−14	−28						
12	12	10	8	6	4	1	−2	−5	−9	−16						
14	14	12	11	9	6	4	1	−2	−5	−10	−17					
16	16	14	13	11	9	7	4	1	−1	−6	−10	−17				
18	18	16	15	13	11	9	7	4	2	−2	−5	−10	−19			
20	20	19	17	15	14	12	10	7	4	2	−2	−5	−10	−19		
22	22	21	19	17	16	14	12	10	8	5	3	−1	−5	−10	−19	
24	24	23	21	20	18	16	14	12	10	8	6	2	−1	−5	−10	−18
26	26	25	23	22	20	18	17	15	13	11	9	6	3	0	−4	−9
28	28	27	25	24	22	21	19	17	16	14	11	9	7	4	1	−3
30	30	29	27	26	24	23	21	19	18	16	14	12	10	8	5	1

Relative Humidity (%)

Dry-Bulb Temperature (°C)	Difference Between Wet-Bulb and Dry-Bulb Temperatures (C°)															
	0	1	2	3	4	5	6	7	8	9	10	11	12	13	14	15
−20	100	28														
−18	100	40														
−16	100	48														
−14	100	55	11													
−12	100	61	23													
−10	100	66	33													
−8	100	71	41	13												
−6	100	73	48	20												
−4	100	77	54	32	11											
−2	100	79	58	37	20	1										
0	100	81	63	45	28	11										
2	100	83	67	51	36	20	6									
4	100	85	70	56	42	27	14									
6	100	86	72	59	46	35	22	10								
8	100	87	74	62	51	39	28	17	6							
10	100	88	76	65	54	43	33	24	13	4						
12	100	88	78	67	57	48	38	28	19	10	2					
14	100	89	79	69	60	50	41	33	25	16	8	1				
16	100	90	80	71	62	54	45	37	29	21	14	7	1			
18	100	91	81	72	64	56	48	40	33	26	19	12	6			
20	100	91	82	74	66	58	51	44	36	30	23	17	11	5		
22	100	92	83	75	68	60	53	46	40	33	27	21	15	10	4	
24	100	92	84	76	69	62	55	49	42	36	30	25	20	14	9	4
26	100	92	85	77	70	64	57	51	45	39	34	28	23	18	13	9
28	100	93	86	78	71	65	59	53	47	42	36	31	26	21	17	12
30	100	93	86	79	72	66	61	55	49	44	39	34	29	25	20	16

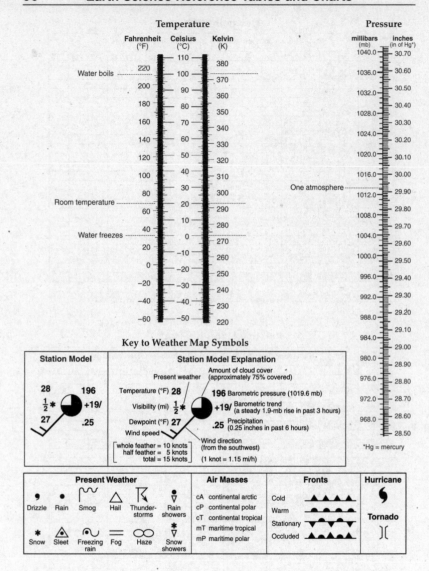

Earth Science Reference Tables and Charts

Selected Properties of Earth's Atmosphere

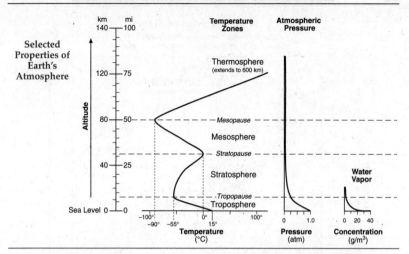

Planetary Wind and Moisture Belts in the Troposphere

The drawing on the right shows the locations of the belts near the time of an equinox. The locations shift somewhat with the changing latitude of the Sun's vertical ray. In the Northern Hemisphere, the belts shift northward in the summer and southward in the winter.

(Not drawn to scale)

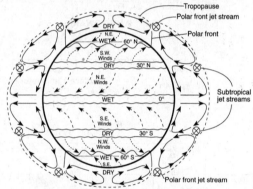

Electromagnetic Spectrum

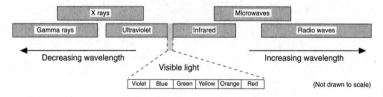

(Not drawn to scale)

Earth Science Reference Tables and Charts

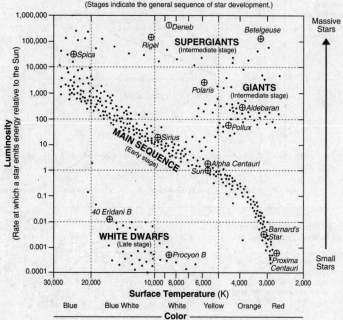

Characteristics of Stars
(Name in italics refers to star represented by a ⊕.)
(Stages indicate the general sequence of star development.)

Solar System Data

Celestial Object	Mean Distance from Sun (million km)	Period of Revolution (d=days) (y=years)	Period of Rotation at Equator	Eccentricity of Orbit	Equatorial Diameter (km)	Mass (Earth = 1)	Density (g/cm³)
SUN	—	—	27 d	—	1,392,000	333,000.00	1.4
MERCURY	57.9	88 d	59 d	0.206	4,879	0.06	5.4
VENUS	108.2	224.7 d	243 d	0.007	12,104	0.82	5.2
EARTH	149.6	365.26 d	23 h 56 min 4 s	0.017	12,756	1.00	5.5
MARS	227.9	687 d	24 h 37 min 23 s	0.093	6,794	0.11	3.9
JUPITER	778.4	11.9 y	9 h 50 min 30 s	0.048	142,984	317.83	1.3
SATURN	1,426.7	29.5 y	10 h 14 min	0.054	120,536	95.16	0.7
URANUS	2,871.0	84.0 y	17 h 14 min	0.047	51,118	14.54	1.3
NEPTUNE	4,498.3	164.8 y	16 h	0.009	49,528	17.15	1.8
EARTH'S MOON	149.6 (0.386 from Earth)	27.3 d	27.3 d	0.055	3,476	0.01	3.3

Properties of Common Minerals

LUSTER	HARD-NESS	CLEAVAGE	FRACTURE	COMMON COLORS	DISTINGUISHING CHARACTERISTICS	USE(S)	COMPOSITION*	MINERAL NAME
Metallic luster	1–2	✓		silver to gray	black streak, greasy feel	pencil lead, lubricants	C	Graphite
Metallic luster	2.5	✓		metallic silver	gray-black streak, cubic cleavage, density = 7.6 g/cm³	ore of lead, batteries	PbS	Galena
Metallic luster	5.5–6.5		✓	black to silver	black streak, magnetic	ore of iron, steel	Fe_3O_4	Magnetite
Metallic luster	6.5		✓	brassy yellow	green-black streak, (fool's gold)	ore of sulfur	FeS_2	Pyrite
Either	5.5–6.5 or 1		✓	metallic silver or earthy red	red-brown streak	ore of iron, jewelry	Fe_2O_3	Hematite
Nonmetallic luster	1	✓		white to green	greasy feel	ceramics, paper	$Mg_3Si_4O_{10}(OH)_2$	Talc
Nonmetallic luster	2		✓	yellow to amber	white-yellow streak	sulfuric acid	S	Sulfur
Nonmetallic luster	2	✓		white to pink or gray	easily scratched by fingernail	plaster of paris, drywall	$CaSO_4 \cdot 2H_2O$	Selenite gypsum
Nonmetallic luster	2–2.5	✓		colorless to yellow	flexible in thin sheets	paint, roofing	$KAl_3Si_3O_{10}(OH)_2$	Muscovite mica
Nonmetallic luster	2.5	✓		colorless to white	cubic cleavage, salty taste	food additive, melts ice	NaCl	Halite
Nonmetallic luster	2.5–3	✓		black to dark brown	flexible in thin sheets	construction materials	$K(Mg,Fe)_3$ $AlSi_3O_{10}(OH)_2$	Biotite mica
Nonmetallic luster	3	✓		colorless or variable	bubbles with acid, rhombohedral cleavage	cement, lime	$CaCO_3$	Calcite
Nonmetallic luster	3.5	✓		colorless or variable	bubbles with acid when powdered	building stones	$CaMg(CO_3)_2$	Dolomite
Nonmetallic luster	4	✓		colorless or variable	cleaves in 4 directions	hydrofluoric acid	CaF_2	Fluorite
Nonmetallic luster	5–6	✓		black to dark green	cleaves in 2 directions at 90°	mineral collections, jewelry	$(Ca,Na)(Mg,Fe,Al)(Si,Al)_2O_6$	Pyroxene (commonly augite)
Nonmetallic luster	5.5	✓		black to dark green	cleaves at 56° and 124°	mineral collections, jewelry	$CaNa(Mg,Fe)_4(Al,Fe,Ti)_3$ $Si_6O_{22}(O,OH)_2$	Amphibole (commonly hornblende)
Nonmetallic luster	6	✓		white to pink	cleaves in 2 directions at 90°	ceramics, glass	$KAlSi_3O_8$	Potassium feldspar (commonly orthoclase)
Nonmetallic luster	6	✓		white to gray	cleaves in 2 directions, striations visible	ceramics, glass	$(Na,Ca)AlSi_3O_8$	Plagioclase feldspar
Nonmetallic luster	6.5		✓	green to gray or brown	commonly light green and granular	furnace bricks, jewelry	$(Fe,Mg)_2SiO_4$	Olivine
Nonmetallic luster	7		✓	colorless or variable	glassy luster, may form hexagonal crystals	glass, jewelry, electronics	SiO_2	Quartz
Nonmetallic luster	6.5–7.5		✓	dark red to green	often seen as red glassy grains in NYS metamorphic rocks	jewelry (NYS gem), abrasives	$Fe_3Al_2Si_3O_{12}$	Garnet

*Chemical symbols:
Al = aluminum, C = carbon, Ca = calcium, Cl = chlorine, F = fluorine, Fe = iron, H = hydrogen, K = potassium, Mg = magnesium, Na = sodium, O = oxygen, Pb = lead, S = sulfur, Si = silicon, Ti = titanium

✓ = dominant form of breakage

Glossary of Earth Science Terms

abrasion the wearing away of rock material that results when one rock particle strikes another. In stream abrasion, particles carried by the stream may strike other particles, causing pieces to break off.

adiabatic change the change in temperature of a gas caused by expansion or compression. When a gas expands, for example, it cools and its temperature drops.

air mass a large area in the atmosphere in which the temperature and moisture conditions are similar.

altitude the angle between the line of sight to a star and the horizon; also, the elevation above sea level.

angle of insolation the angle at which the Sun's rays strike the Earth's surface.

anticline a series of folded rock layers that bends upward near the center. *See also* **syncline**.

apparent diameter the size that an object appears to an observer. When objects are close by, they appear larger than when they are farther away.

asthenosphere a region of the upper mantle between 100 and 350 kilometers in depth that behaves like a fluid.

atmosphere the envelope of air that encircles Earth. It has been divided into zones based largely on temperature differences.

axis an imaginary line around which an object rotates. Earth's axis extends from the North Pole to the South Pole.

barometer an instrument for measuring atmospheric pressure.

barometric pressure the amount of force exerted by the air per square inch or centimeter at a particular location.

bedrock the rock layer nearest Earth's surface, lying directly below any soil layers.

big bang theory the idea that the universe started out with all of its matter in a small volume and then expanded outward in all directions.

bioclastic a sedimentary rock consisting of fragmental or broken remains of organisms, such as limestone composed of shell fragments.

boiling point the temperature at which a liquid changes to a vapor.

capillarity (capillary action) the rising of water against gravity, as when water rises above the water table in soil because of the attraction between the water and the soil.

cementation the process by which sediments are bonded together by material dissolved in water when the water evaporates.

chemical weathering weathering that occurs because of the chemical reaction between material dissolved in water and local rock material. *See also* **physical weathering**.

cleavage the splitting of a mineral in distinctive directions caused by the arrangement of the atoms in a mineral. For example, the mineral mica cleaves in layers because the atoms in the crystal structure of mica are arranged in layers.

climate the average weather conditions, in terms of temperature and moisture, of an area over a long period of time.

clouds masses of water droplets suspended in the air. They form when air is cooled and moisture condenses.

cold front the boundary between two air masses of different temperatures, the point where the colder air mass moves under and pushes up the warmer air mass. *See also* **warm front**.

compaction the process that results when buried sediments are subjected to pressure, which packs them together. Together with cementation, this process causes sediments to be converted to rock.

condensation the process whereby, when moisture in the atmosphere is cooled, it changes from a vapor to a liquid.

conduction the method of heat transfer in solids in which faster moving molecules strike other molecules, causing them to speed up.

contact metamorphism changes in rock that result from the extreme heat produced by contact with magma or lava.

continental air mass an air mass that forms over land and therefore is relatively dry.

continental drift a hypothetical slow movement of the continents, which forms the basis of the theory that the present continents were once part of one large landmass that broke up. Since that time, the continents have been drifting apart.

continental glacier a large sheet of ice that covered much of a continent during a period of Earth's past geologic history.

contour interval the differences in elevation between contour lines on a contour or topographic map.

contour line a line on a topographic map that connects points having the same elevation.

contour map *See* **topographic map**.

convection the method of energy transfer in fluids in which the fluid expands when it is heated and rises. When the fluid cools, it contracts and sinks.

convection cell the circular pattern of movement in a fluid caused by the rising and sinking of the fluid due to differences in density caused by differences in temperature.

convergent boundary places where edges of adjacent plates are colliding.

core the innermost zone of Earth's interior. A solid inner core is surrounded by a molten outer core.

coriolis effect an apparent force, due to the rotation of the Earth, that deflects winds toward the right in the Northern Hemisphere and toward the left in the Southern Hemisphere.

correlation the matching of rock layers at different locations, based on composition, thickness, and in some cases fossil content, for the purpose of establishing that they represent the same rock layer.

cosmic background radiation a remnant of radiation left over from the original big bang that fills the universe.

crust the outermost solid layer of Earth, extending across the continents and under the oceans. It is thicker under the continents than under the oceans.

density the mass per unit volume of a substance.

deposition the process by which material carried by running water, glaciers, or wind settles out when the velocity of the carrier slows down.

dew moisture that condenses at Earth's surface when moist air touches a cool area.

dewpoint temperature the temperature at which the air becomes saturated and excess moisture begins to condense.

Glossary of Earth Science Terms

dike a rock layer that forms when molten rock material flows through breaks in rock layers and then cools and hardens.

discharge the total volume of water flowing in a river or stream per unit of time.

divergent boundary places where edges of adjacent plates are spreading apart.

Doppler effect the shift in wavelength of a spectrum line from its normal position due to relative motion between the source and the observer.

duration of insolation the number of hours that the Sun's rays strike Earth's surface over a 24-hour period; the number of hours of daylight.

earthquake the large-scale and rapid motion of rock layers that occurs when pressure is released.

electromagnetic spectrum the range of wavelengths of energy released by the Sun. A small range of wavelengths represents visible light. Other bands within the spectrum include ultraviolet radiation, cosmic rays, infrared radiation, and radio waves.

elevation the height above sea level.

ellipse a curve that has two centers, or foci. The sum of the distance between either focus and any point on the ellipse and the distance between the other focus and that point is a constant. The shape of Earth's orbit around the Sun is an ellipse with the Sun at one focus.

epicenter the location on Earth's surface directly above the point of origin or focus of an earthquake.

equator an imaginary circle around Earth lying halfway between the poles and dividing the Earth's surface into the Northern and Southern hemispheres.

equinox the two times a year when the Sun is directly overhead at the equator at noon. At these times, about March 21 and September 23, the number of hours of day and of night is the same.

erosion the process by which weathered rock material is carried away by agents such as running water, ice, wind, and gravity.

evaporation the process by which water is converted from a liquid to a vapor.

evapotranspiration the combined processes of evaporation and transpiration, representing the method by which water vapor returns to the atmosphere.

Glossary of Earth Science Terms

extinct referring to a plant or animal species that lived in the past but is no longer found alive on Earth.

fault movement within rock layers where pressure has caused the layers to break. The layers on one side of the break move up, while the opposite layers move down.

felsic referring to igneous rocks composed of minerals with a high content of aluminum. These rocks tend to be lighter in color than others.

focus the point of origin within Earth's crust or mantle of an earthquake.

foliated a metamorphic rock texture characterized by thin, leaflike layers caused by the flattening of mineral grains under heat and pressure.

formation (rock) a sequence of rock layers that cover a large area and were formed over a period of time.

fossil the preserved remains or traces of an animal or plant that lived in the past.

fossil record groups of fossils found in a rock layer that are used to interpret how and when the rock layer was formed and what the environment was like at that time.

Foucault pendulum a freely swinging pendulum whose path appears to change direction relative to Earth's surface in a predictable manner due to Earth's rotation.

fracture as term used to describe the irregular way in which some minerals break.

freezing point the temperature at which a liquid changes to a solid.

front the boundary between two different air masses. *See also* **cold front; warm front**.

frost moisture that condenses directly from a vapor to a solid when moist air touches a cold surface. When the air temperature is below freezing, frost forms instead of dew.

galaxy a system consisting of hundreds of billions of stars.

gradient the slope of the land or of a river or stream.

greenhouse effect the process by which longer wavelength radiation emitted from Earth's surface is absorbed by carbon dioxide in the atmosphere, causing a rise in air temperature.

Glossary of Earth Science Terms

half-life the amount of time it takes for half the mass of a radioactive substance to decay. For example, the half-life of carbon-14 is 5.6×10^3 years.

high-pressure center a location on a weather map where winds blow outward, in a clockwise direction, away from the center. This occurs because the air pressure is lower at the center than over the surrounding area. *See also* **low-pressure center**.

humidity the amount of moisture in the air.

hydrosphere the outer zone of Earth, which consists of the oceans and seas.

hypothesis the attempt to explain a scientific phenomenon on the basis of observations and other relevant information.

igneous referring to rocks that form when molten rock material cools and solidifies.

impermeable referring to a layer of material through which water cannot pass. Most soil layers are permeable, while most rock layers are not.

index fossil a fossil that is found over widespread areas and that formed from an organism that existed for a relatively short geologic period of time.

inference an interpretation or explanation of a natural phenomenon based on observations.

infiltration the process by which water at Earth's surface penetrates and filters down through porous soil and rock layers.

insolation the radiation reaching Earth's surface from the Sun. This term is a contraction of "incoming solar radiation."

intrusion the forcible entry of solidified molten rock material between rock layers or into breaks within a rock layer.

isobar a line on a weather map that connects points having the same air or barometric pressure.

land breeze a wind blowing from over the land to over a lake or ocean. It occurs when the water temperature is warmer than the land temperature. *See also* **sea breeze**.

landscape region a grouping of landscapes with similar relief, stream patterns and soil associations.

latent heat of fusion the amount of energy required to convert one gram of ice at 0°C to water at 0°C.

latent heat of vaporization the amount of energy required to convert one gram of water at 100°C to vapor at 100°C.

latitude imaginary circles around Earth parallel to the equator and between the North and South poles.

lava molten rock material that reaches Earth's surface through volcanoes.

light-year the distance light travels in one year; roughly 9.5 trillion kilometers, or 6 trillion miles.

lithosphere the solid outer portion of Earth.

longitude imaginary circles around Earth that pass through the North and South poles.

low-pressure center a location on a weather map where winds blow inward, in a counterclockwise direction, into a center. This occurs because the air pressure is lower at the center than in the surrounding area. *See also* **high-pressure center**.

luminosity the total amount of energy radiated by a star in 1 second.

lunar eclipse the phenomenon that occurs when Earth passes between the Moon and the Sun, causing light from the Sun to be blocked from reaching the surface of the Moon. *See also* **solar eclipse**.

luster the type of shine exhibited by a mineral, based on the way its surface reflects light. Examples are metallic, glassy, and dull lusters.

mafic referring to igneous rocks composed of minerals with a high content of magnesium and iron. They tend to be darker in color than other rocks.

magma molten rock material beneath the Earth's surface.

magnetic north (pole) the point near the geographic North Pole toward which the needle on a compass points.

mantle the zone within Earth that lies between the crust and the outer core.

maritime air mass an air mass that forms over water and therefore has a relatively high moisture content.

mass the measure of the amount of matter contained in a sample.

meander a curve in a river or stream.

meridian a line of longitude measured in degrees. The prime meridian (0°) passes through Greenwich, England.

Glossary of Earth Science Terms

metamorphic referring to rocks that form when existing rocks are subjected to enough heat and pressure to cause partial melting of the minerals present.

meter the standard unit of length in the metric system; 1 meter = 39.37 inches.

mid-ocean ridge a chain of undersea mountains running through the center of an ocean.

millibar a unit of air pressure in the atmosphere, commonly used on weather maps.

mineral a naturally occurring substance that is always made of the same elements in a fixed proportion.

modified Mercalli scale a system that measures the strength of an earthquake based upon perception of motion by human observers and damage to structures built by humans.

Moho the boundary between the dense rock of the mantle and the less dense rock of the crust. It was named for the seismologist Andrija Mohorovicic, who recognized its existence after analyzing seismic wave behavior inside Earth.

moraine a large deposit formed when a glacier melts and leaves behind the material it is carrying.

mountain any part of Earth's crust that projects at least 300 m (1,000 ft) above the surrounding land, has a limited summit area (as opposed to a plateau), steep sides, and considerable bare-rock surface.

observation a description of what is perceived by the senses. It can often be made more accurate by using instruments.

occluded front a weather front that forms when a cold front moves in behind a warm front, lifting the warm front off the ground.

orbit the path followed by one object as it revolves around another; for example, the orbit of Earth around the Sun.

outgassing the release of gases and water vapor from molten rocks, leading to the formation of Earth's atmosphere and oceans.

permeability the ability of water to penetrate through a material. The permeability of a soil depends upon the size of the soil grains and the amount of space between the grains.

physical weathering the breakdown of rocks due to physical changes, such as changes in temperature or the freezing and thawing of water, that fills cracks in the rock. *See also* **chemical weathering**.

plain a flat area at low elevation.

plateau a large, flat region elevated more than 150–300 m (500–1,000 ft.) above the surrounding land, or above sea level.

polar air mass an air mass that forms over the polar regions and therefore contains relatively cold air.

pollution a condition of the air, water, or land in which there is a surplus of materials present that may be harmful to living things.

porosity the percentage of open space in a soil sample.

precipitation all forms of moisture that reach the Earth's surface from the atmosphere, including rain, snow, hail, and sleet.

prevailing westerlies the wind belt stretching across the United States in which the general direction of wind movement is from southwest to northeast.

P-wave (primary wave) the compression type of wave emitted by an earthquake that travels through both liquids and solids. *See also* **S-wave**.

radiation the method of energy transfer by which energy from the Sun reaches the Earth.

radioactive dating the use of the radioactive isotope to determine the age of a fossil or rock layer. This technique is possible because each radioactive isotope decays at a unique and fixed rate.

radioactive decay the process by which a radioactive element breaks down to emit particles and radiation that form a new element.

radioactive isotope an unstable form of an element that breaks down by radioactive decay. For example, carbon-14 is a radioactive isotope of carbon that breaks down by emitting electrons to form nitrogen-14.

rain gauge an instrument that collects atmospheric precipitation so that the amount of rainfall can be measured.

regional metamorphism changes in rock over an extensive area that occur due to the pressure and high temperatures associated with either deep burial or movements of Earth's crust.

relative humidity the percentage of moisture present in the air as compared to the maximum amount of moisture the air can hold at the prevailing temperature.

residual soil soil that is formed by the weathering of local rock material.

Richter scale a scale with a range of 1 to 10 that indicates the magnitude of an earthquake. For each successive number the magnitude increases by a factor of 10.

Glossary of Earth Science Terms

rock naturally occurring materials that are composed of one or more minerals.

rock cycle the process by which each of the three rock types (igneous, metamorphic, and sedimentary) can be converted into each other type.

runoff excess precipitation reaching Earth's surface that flows into rivers and streams because the surface soil is saturated.

salinity a measure of the amount of salt dissolved in water.

saturation temperature *See* **dewpoint temperature**.

sea breeze a wind blowing from over a lake or ocean to over land. It occurs when the water temperature is cooler than the land temperature. *See also* **land breeze**.

sea floor spreading the concept that portions of the ocean floor are moving away from a central ridge because new material moving upward at the ridge is pushing the old material outward.

sedimentary referring to rocks that form by the compaction and cementation of sediments.

seismograph an instrument used to measure the disturbances caused by an earthquake.

sill a rock layer that forms when molten rock material forces its way between existing rock layers and then cools and solidifies.

sling psychrometer an instrument used to determine relative humidity.

solar eclipse the phenomenon that occurs when the Moon passes between Earth and the Sun, causing light from the Sun to be blocked from reaching the surface of Earth. *See also* **lunar eclipse**.

solar system the Sun and the various objects that orbit it, including the planets and their moons, comets, and asteroids.

solstice one of the two times a year, about June 22 and December 22, when the Sun is directly overhead at $23\frac{1}{2}°$ north and south latitude. The solstices represent the longest and shortest days of the year.

specific heat the relative amount of energy needed to raise the temperature of one gram of a material by one degree Celsius.

spectroscope an instrument used to break down the visible portion of the electromagnetic spectrum into individual wavelengths or colors.

spectrum the band of colors formed when a beam of white light is passed through a prism; each wavelength of light in the beam is refracted at a slightly different angle, causing the light to be arrayed in order of its constituent wavelengths.

Glossary of Earth Science Terms

stationary front a front that forms when the opposing warm and cold air masses are of equal energy.

station model a pattern of symbols on a weather map that is used to describe local weather conditions such as temperature, pressure, humidity, wind speed and direction, and cloud cover.

stratosphere the layer of the atmosphere directly above the troposphere.

streak a more accurate method of identifying the color of a mineral by rubbing it against a plate, causing powder to form.

stream load the total amount of material carried by a stream, including material that is carried in solution or suspension or is pushed along the bottom.

subduction the sliding of a denser ocean plate beneath a less dense continental plate resulting in the melting of the ocean plate as it plunges into the hot mantle.

subsoil the layer of soil just below the topsoil. It does not contain the organic matter found in the topsoil and has not been as extensively weathered.

S-wave (shear wave) the transverse type of wave emitted by an earthquake that travels through solids but is absorbed by liquids. *See also* **P-wave**.

syncline a series of folded rock layers that dips downward near the center. *See also* **anticline**.

terminal moraine the material deposited by the meltwater of a glacier at the point of its farthest advance.

texture the grain size of a rock. A course texture represents large grains; a fine texture, small grains.

theory an explanation for scientific observations. A theory is formed from a hypothesis when there is substantial evidence to support it.

till the material found in a glacial moraine and usually representing a wide range of particle sizes.

time zone a region in which the same time is used throughout, instead of the local time at each place within it.

topographic (contour) map a map of an area with contour lines to show the elevations at all locations, as well as other geologic features such as rivers, streams, and mountain peaks.

topsoil the uppermost layer of soil. It contains the most highly weathered rock fragments, as well as organic remains.

Glossary of Earth Science Terms

transform boundary places where edges of plates are sliding laterally past each other.

transpiration the process by which plants release moisture to the atmosphere.

transported soils soils formed from weathered rock material that have been carried from other locations and deposited.

trench a large valley on the ocean floor.

tropical air mass an air mass that has formed over the tropics and therefore contains relatively warm air.

Tropic of Cancer the line of latitude at 23½° north, which represents the farthest north that the noon Sun can be overhead.

Tropic of Capricorn the line of latitude at 23½° south, which represents the farthest south that the noon Sun can be overhead.

troposphere the layer of the atmosphere closest to Earth's surface. All weather phenomena occur within this zone.

unconformity a gap in a sequence of rock layers, resulting either from the removal of layers by erosion or failure of layers to form for long periods.

uniformitarianism the principle that the geologic processes acting today are the same as those that occurred in the past.

valley glacier a glacier that forms in a valley between mountain peaks.

vesicular a rock texture characterized by cavities formed by gas bubbles escaping from a lava as it cools and solidifies.

volcano a mountain formed when lava erupts from beneath the surface and then cools and solidifies.

warm front the boundary between two air masses of different temperatures; the point where the warmer air mass moves over the colder air mass and pushes it backward. *See also* **cold front**.

water cycle the cyclic process during which water leaves Earth's surface by evaporation and transpiration to enter the atmosphere. It returns to the surface as precipitation.

watershed the area drained by a stream and its tributaries.

water table the upper boundary of saturated rock or soil beneath Earth's surface.

weathering the processes in nature by which rock materials are broken down into smaller pieces to form sand, soil, and so on. *See also* **chemical weathering; physical weathering**.

wind belts zones around the Earth in which the winds blow in the same general direction.

Regents Examinations, Answers, and Self-Analysis Charts

Examination August 2014
Physical Setting/Earth Science

PART A
Answer all questions in this part.

Directions (1–35): For *each* statement or question, choose the word or expression that, of those given, best completes the statement or answers the question. Some questions may require the use of the *2011 Edition Reference Tables for Physical Setting/Earth Science*. Record your answers in the space provided.

1 Which evidence best supports the theory that the universe was created by an explosion called the Big Bang?
  (1) impact craters found on Earth
  (2) cosmic background radiation
  (3) the different compositions of terrestrial and Jovian planets
  (4) the blue shift of light from distant galaxies   1 _____

2 Which star is more massive than our Sun, but has a lower surface temperature?
  (1) 40 Eridani B   (3) Aldebaran
  (2) Sirius   (4) Barnard's Star   2 _____

3 Which color of visible light has the *shortest* wavelength?
   (1) violet
   (2) green
   (3) yellow
   (4) red

4 The table below shows the times of ocean high tides and low tides on a certain date at a New York State location.

Ocean Tides

Type of Tide	Time
high	4:45 a.m.
low	10:58 a.m.
high	5:15 p.m.
low	11:22 p.m.

At approximately what time on the following day did the next high tide occur at this location?

   (1) 4:40 a.m.
   (2) 5:40 a.m.
   (3) 4:40 p.m.
   (4) 5:40 p.m.

5 The best evidence of Earth's rotation is provided by the
   (1) shape of Earth's orbit
   (2) shape of the Milky Way galaxy
   (3) changes in the total yearly duration of insolation at a location on Earth
   (4) apparent changes in the direction of swing of a Foucault pendulum

6 The model below shows the apparent path of the Sun as seen by an observer in New York State on the first day of one of the four seasons.

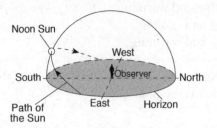

This apparent path of the Sun was observed on the first day of

(1) spring (3) fall
(2) summer (4) winter 6____

7 Which processes are most likely to cause a rise in the water table?

(1) runoff and erosion
(2) precipitation and infiltration
(3) deposition and burial
(4) solidification and condensation 7____

8 During which phase change does water release the most heat energy?

(1) freezing (3) condensation
(2) melting (4) vaporization 8____

9 What is the average air pressure exerted by Earth's atmosphere at sea level, expressed in millibars and inches of mercury?

(1) 1013.25 mb and 29.92 in of Hg
(2) 29.92 mb and 1013.25 in of Hg
(3) 1012.65 mb and 29.91 in of Hg
(4) 29.91 mb and 1012.65 in of Hg 9____

10 Which two processes lead to cloud formation in rising air?

(1) compressing and cooling
(2) compressing and warming
(3) expanding and cooling
(4) expanding and warming

11 The weather station model below shows some of the weather data for a certain location.

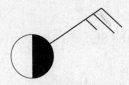

What is the wind speed shown on the station model and which instrument is used to measure the wind speed?

(1) 15 knots, measured by a wind vane
(2) 15 knots, measured by an anemometer
(3) 25 knots, measured by a wind vane
(4) 25 knots, measured by an anemometer

12 If air has a dry-bulb temperature of 2°C and a wet-bulb temperature of –2°C, what is the relative humidity?

(1) 11%   (3) 36%
(2) 20%   (4) 67%

13 Which current has a cooling effect on the climate of the west coast of South America?

(1) Falkland Current   (3) Benguela Current
(2) Peru Current   (4) Brazil Current

14 Near which two latitudes are most of Earth's dry climate regions found?

(1) 0° and 60° N
(2) 0° and 30° S
(3) 30° N and 60° N
(4) 30° N and 30° S   14 ____

15 Which event followed a massive volcanic eruption and led to the cooling of global temperatures?

(1) thunderstorms that developed near the eruption
(2) the release of carbon dioxide and methane gases
(3) the outflow of magma over Earth's surface
(4) the addition of ash particles into the atmosphere   15 ____

16 Rifting of tectonic plates in eastern North America during the Jurassic Period was responsible for the

(1) formation of the Catskill delta
(2) first uplift of the Adirondack Mountains
(3) Alleghenian orogeny
(4) opening of the Atlantic Ocean   16 ____

17 The surface bedrock of Mt. Marcy, New York, is composed primarily of which rock?

(1) anorthosite
(2) marble
(3) quartzite
(4) hornfels   17 ____

18 Much of the evidence for the evolution of life-forms on Earth has been obtained by

(1) studying the life spans of present-day animals
(2) radioactive dating of metamorphic rock
(3) correlating widespread igneous ash deposits
(4) examining fossils preserved in the rock record   18 ____

19 The table below shows the radioactive decay of carbon-14. Part of the table has been left blank.

Half-Life	Original Carbon-14 Remaining (%)	Number of Years
0	100	0
1	50	5,700
2	25	11,400
3		17,100
4		
5		

After 22,800 years, approximately what percentage of the original carbon-14 remains?

(1) 15%   (3) 6.25%
(2) 12.5%   (4) 3.125%   19 _____

20 The diagram below represents a model of Earth's surface and internal structure. Letters A, B, C, and D represent four different layers. Some depths below Earth's surface are shown.

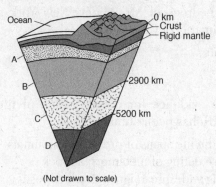

(Not drawn to scale)

Which Earth layer is inferred to be composed of solid nickel and iron?

(1) A   (3) C
(2) B   (4) D   20 _____

21 Oceanic crust is sliding beneath the Aleutian Islands in the North Pacific Ocean, forming the Aleutian Trench at a
(1) convergent plate boundary between the Pacific Plate and the North American Plate
(2) convergent plate boundary between the Pacific Plate and the Juan de Fuca Plate
(3) divergent plate boundary between the Pacific Plate and the North American Plate
(4) divergent plate boundary between the Pacific Plate and the Juan de Fuca Plate 21 _____

22 Which New York State landscape region is composed of mostly horizontal sedimentary bedrock and has a high elevation?
(1) Hudson Highlands (3) the Catskills
(2) Manhattan Prong   (4) Taconic Mountains 22 _____

23 Which mineral is commonly mined as a source of the element lead (Pb)?
(1) galena   (3) magnetite
(2) quartz   (4) gypsum 23 _____

24 The block diagram below represents a landscape where caverns and sinkholes have gradually developed over a long period of time.

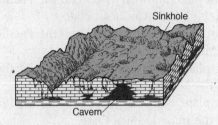

Why did these caverns and sinkholes form?
(1) The bedrock chemically reacted with acidic groundwater.
(2) This type of bedrock contained large amounts of oxygen and silicon.
(3) Glacial deposits altered the shape of the bedrock.
(4) Crustal uplift formed gaps in the bedrock.

24 ____

25 The block diagram below represents a stream flowing from a mountain region.

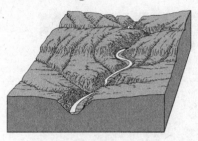

A brief, heavy rainstorm occurs in the mountains. How will the volume of water and the rate of erosion in the stream change shortly after the rainstorm?

(1) The volume of water will decrease and the rate of erosion will increase.
(2) The volume of water will increase and the rate of erosion will decrease.
(3) Both the volume of water and the rate of erosion will decrease.
(4) Both the volume of water and the rate of erosion will increase.

26 The photograph below shows scratched and polished bedrock produced by weathering and erosion.

Which agent of erosion most likely carried sediment that scratched and polished this bedrock surface?

(1) a moving glacier
(2) running water
(3) wave action
(4) wind

Base your answers to questions 27 and 28 on the diagram below and on your knowledge of Earth science. The diagram represents the Moon at different positions, labeled A, B, C, and D, in its orbit around Earth.

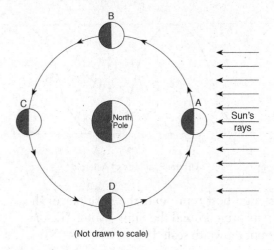

(Not drawn to scale)

27 At which two Moon positions would an observer on Earth most likely experience the highest high tides and the lowest low tides?

(1) A and B
(2) B and C
(3) C and A
(4) D and B

27 _____

28 During which Moon phase could an observer on Earth see a lunar eclipse occur?

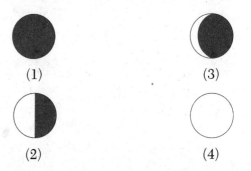

28 _____

29 The graph below shows the varying amount of gravitational attraction between the Sun and an asteroid in our solar system. Letters *A*, *B*, *C*, and *D* indicate four positions in the asteroid's orbit.

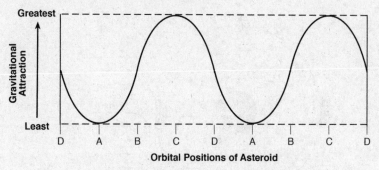

Which diagram best represents the positions of the asteroid in its orbit around the Sun? [Note: The diagrams are not drawn to scale.]

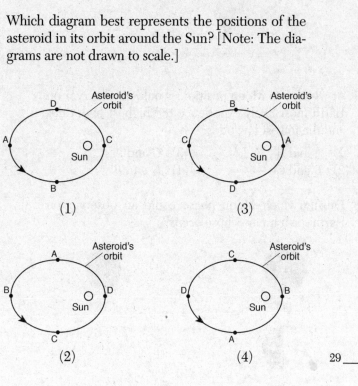

29 _____

30 The map below shows four major time zones of the United States. The dashed lines represent meridians of longitude. The locations of New York City and Denver are shown.

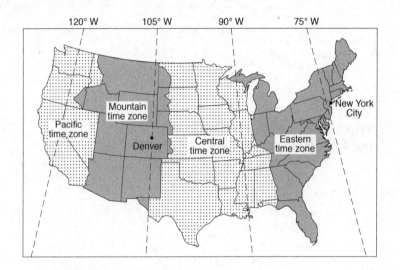

What is the time in New York City when it is noon in Denver?

(1) 10 a.m. (3) 3 p.m.
(2) 2 p.m. (4) noon   30 ____

31 Which station model shows an air temperature of 75°F and a barometric pressure of 996.3 mb?

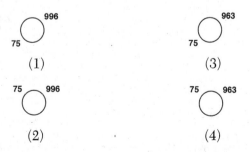

31 ____

32 The map below shows air pressures recorded in millibars (mb).

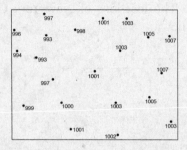

Which map shows the correct location of the 996-mb, 1000-mb, and 1004-mb isobars?

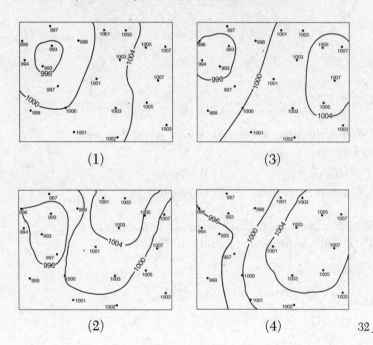

32 _____

33 The map below shows the amount of snowfall, in inches, produced by a lake-effect snowstorm in central New York State.

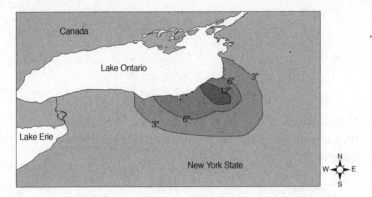

The wind that produced this snowfall pattern most likely came from the

(1) northeast
(2) northwest
(3) southeast
(4) southwest

34 The block diagram below represents an igneous dome that uplifted overlying rock layers, which were then weathered and eroded.

Which stream drainage pattern is most likely found on the surface of the area represented by the block diagram?

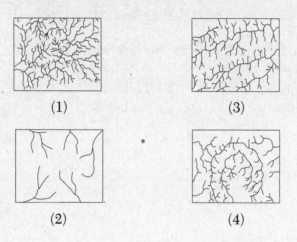

34 _____

35 On the map below, points A through D represent locations on Earth's surface.

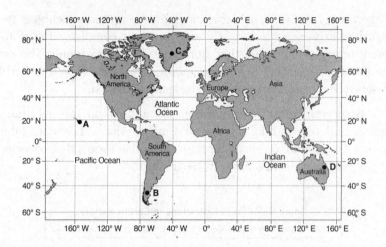

Which location is positioned over a mantle hot spot?

(1) A
(2) B
(3) C
(4) D

35 ____

PART B-1
Answer all questions in this part.

Directions (36–50): For *each* statement or question, choose the word or expression that, of those given, best completes the statement or answers the question. Some questions may require the use of the *2011 Edition Reference Tables for Physical Setting/Earth Science*. Record your answers in the space provided.

Base your answers to questions 36 and 37 on the data table below and on your knowledge of Earth science. The data table shows information on six major mass extinction events that occurred many million years ago (mya) in Earth's history.

Some Major Mass Extinctions in Earth's History

Approximate Time (mya)	Certain Life-Forms That Became Extinct
65.5	all dinosaurs and all ammonoids
200	many species of nautiloids, ammonoids, mammal-like reptiles, and early dinosaurs
251	all trilobites and 90% of other marine species and 70% of land species
376	many species of corals, brachiopods, and trilobites
444	more than half of brachiopod species, many trilobite species, and some coral species
520	small shelly fossil species and some early trilobite species

36 More than half of brachiopod species became extinct at the end of the

(1) Devonian Period   (3) Ordovician Period
(2) Silurian Period   (4) Cambrian Period   36 _____

37 Which event is generally accepted as the cause of the mass extinction that occurred 65.5 million years ago?

(1) volcanic eruption (3) asteroid impact
(2) continental collision (4) sea-level change   37 ____

Base your answers to questions 38 through 40 on the weather map below and on your knowledge of Earth science. The map of a portion of eastern North America shows a high-pressure center (**H**) and a low-pressure center (**L**), frontal boundaries, and present weather conditions.

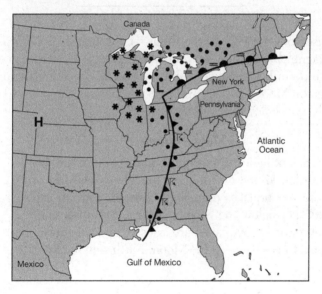

38 Which weather condition is shown along the cold front?

(1) fog   (3) haze
(2) snow   (4) thunderstorms   38 ____

39 What was the most likely source region for the air mass over Pennsylvania?

(1) New York State
(2) Pacific Ocean
(3) Gulf of Mexico
(4) Canada

39 _____

40 The general surface wind circulation associated with the high-pressure center (**H**) is most likely

(1) clockwise and outward
(2) clockwise and inward
(3) counterclockwise and outward
(4) counterclockwise and inward

40 _____

Base your answers to questions 41 through 43 on the passage and diagram below and on your knowledge of Earth science. The passage describes geologic studies of the Moon. The diagram represents the Moon's surface and interior, showing the inferred depth of each layer below the Moon's surface.

Moon Studies

Scientific instruments left on the Moon's surface recorded 12,558 moonquakes in eight years. Most of these moonquakes originated between 700 km and 1200 km below the Moon's surface. Scientists infer that most moonquakes are caused by the gravitational forces between the Moon, Earth, and the Sun.

Layers of the Moon

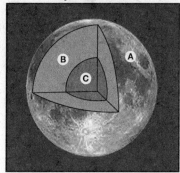

Key
Inferred Depth Below the Surface:
A Crust: 0 km to 60 km
B Mantle: 60 km to 1100 km
C Core: 1100 km to 1738 km

(Not drawn to scale)

41 The same type of evidence was used to find the inferred depths of both the Moon's interior layers and Earth's interior layers. What evidence was used to determine the inferred depth of the boundary between the Moon's mantle and core?

(1) seismic data recorded on the Moon's surface
(2) magnetic data measured on the Moon's surface
(3) convection currents mapped in the Moon's mantle and core
(4) temperatures measured in the Moon's mantle and core   41 _____

42 What is the inferred thickness of the Moon's mantle?

(1) 60 km   (3) 1040 km
(2) 638 km   (4) 1738 km   42 _____

43 Which planet has an average density most similar to the average density of the Moon?

(1) Mercury   (3) Jupiter
(2) Mars   (4) Neptune   43 _____

Base your answers to questions 44 through 47 on the geologic cross section below and on your knowledge of Earth science. The cross section represents rock and sediment layers, labeled *A* through *F*. Each layer contains fossil remains, which formed in different depositional environments. Some layers contain index fossils. The layers have *not* been overturned.

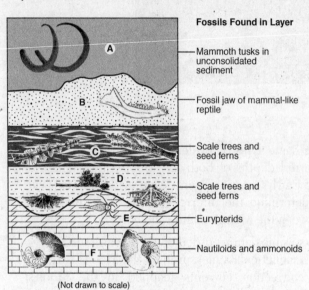

44 Which pair of organisms existed when the unconsolidated sediment in layer *A* was deposited?
 (1) birds and trilobites
 (2) dinosaurs and mastodonts
 (3) ammonoids and grasses
 (4) humans and vascular plants   44 _____

45 Which rock layer formed mainly from the compaction of plant remains?
 (1) *E*   (3) *C*
 (2) *B*   (4) *F*   45 _____

46 During which geologic epoch was layer *F* deposited?
  (1) Late Devonian   (3) Early Devonian
  (2) Middle Devonian   (4) Late Silurian   46 _____

47 The depositional environment during the time these layers and fossils were deposited
  (1) was consistently marine
  (2) was consistently terrestrial (land)
  (3) changed from marine to terrestrial (land)
  (4) changed from terrestrial (land) to marine   47 _____

Base your answers to questions 48 through 50 on the diagram below and on your knowledge of Earth science. The diagram represents setups of laboratory equipment, labeled *A*, *B*, *C*, and *D*. This equipment was used to test the infiltration rate and water retention of four different particle sizes. Each column was filled to the same level with uniform-sized dry, spherical particles. Water was poured into each column until the water level rose to the top of the particles. Then, the clamp was opened to allow the water to drain into the beaker beneath each column.

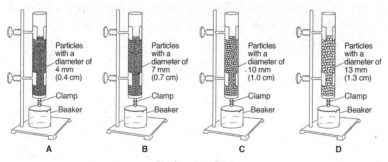

(Not drawn to scale)

48 All of the particles in these four columns are classified as

(1) clay (3) sand
(2) silt (4) pebbles   48 ____

49 Which graph best shows the rate of infiltration of water through the particles in these four columns?

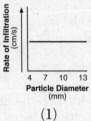
(1)

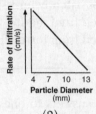
(3)

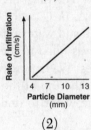
(2)

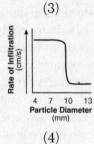
(4)   49 ____

50 Which column of particles retained the most water after the clamps were opened and the water was drained into the beakers?

(1) A (3) C
(2) B (4) D   50 ____

PART B–2
Answer all questions in this part.

Directions (51–65): Record your answers in the spaces provided. Some questions may require the use of the *2011 Edition Reference Tables for Physical Setting/Earth Science*.

Base your answers to questions 51 through 54 on the cross section of part of Earth's crust below and on your knowledge of Earth science. On the cross section, some rock units are labeled with letters *A* through *I*. The rock units have *not* been overturned. Line *XY* represents a fault. Line *UV* represents an unconformity.

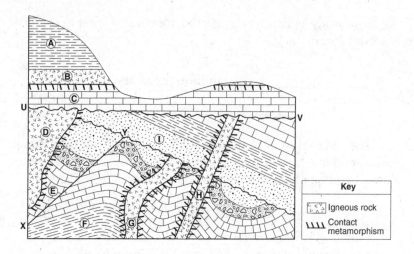

51 On the cross section *above*, draw *two* arrows, one on *each* side of line *XY*, to show the direction of relative movement that has occurred along the fault. [1]

52 Write the letter of the oldest rock unit in the cross section. [1]

_____

53 Identify the contact metamorphic rock that formed between rock units *B* and *C*. [1]

_____

54 The table below shows the ages of the igneous rock units, determined by radioactive dating. [1]

Rock Unit	D	G	H	B
Age (million years)	420	454	420	140

How many million years ago did rock unit *I* most likely form? [1]

_____ **million years ago**

_____

Base your answers to questions 55 through 58 on the passage and map below and on your knowledge of Earth science. The map shows a portion of the Dust Bowl in the southern Great Plains.

The Dust Bowl

In the 1930s, several years of drought affected over 100 million acres in the Great Plains from North Dakota to Texas. For several decades before this drought, farmers had plowed the prairie and loosened the soil. When the soil became extremely dry from lack of rain, strong prairie winds easily removed huge amounts of soil from the farms, forming dust storms. This region was called the Dust Bowl.

In the spring of 1934, a windstorm lasting a day and a half created a dust cloud nearly 2000 kilometers long and caused "muddy rains" in New York

State and "black snow" in Vermont. Months later, a Colorado storm carried dust approximately 3 kilometers up into the atmosphere and transported it 3000 kilometers, creating twilight conditions at midday in New York State.

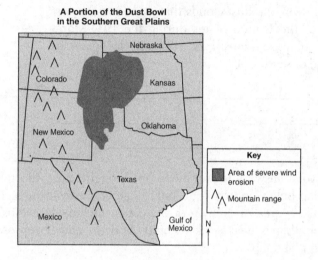

A Portion of the Dust Bowl in the Southern Great Plains

55 Identify *one* human activity that was a major cause of the huge dust storms that formed in the Great Plains during the 1930s. [1]

_____

56 Describe *one* change in the appearance of the sand particles that were abraded when transported by winds within the Dust Bowl region. [1]

_____
_____

57 Identify the name of the layer of the atmosphere in which the dust particles were transported by the Colorado storm to New York State. [1]

_____

58 Explain why the dust clouds that moved to the east coast of the United States during the 1934 storm were composed mostly of silt and clay particles instead of sand. [1]

_____

_____

Base your answers to questions 59 through 62 on the topographic map below and on your knowledge of Earth science. The map shows an area of New York State that includes a campsite, trail, and buildings near a lake. Points A, B, C, and D represent locations on the map.

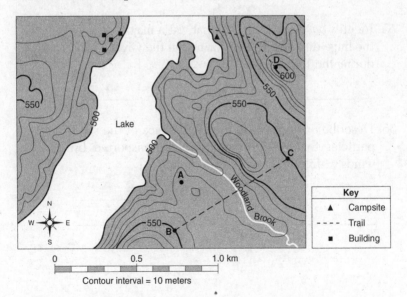

59 Point *A* on the topographic map *on the previous page* indicates a certain elevation on the east side of the lake. Place an **X** at the same elevation on the west side of the lake. [1]

60 On the grid *below*, construct a topographic profile along line *BC*. Plot the elevation of *each* contour line that crosses line *BC*. Connect *all seven* plots with a line to complete the profile. [1]

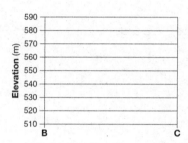

61 Circle the phrase *below* that indicates the direction of flow of Woodland Brook. Describe the contour-line evidence that supports your answer. [1]

Circle one: into the lake out of the lake

Contour-line evidence: _____

_____

62 Campers hiked along the trail from the shoreline of the lake to point *D* to view the landscape. Determine the average gradient, in meters per kilometer, of the route they took on their hike. [1]

_____ **m/km**

Base your answers to questions 63 through 65 on the diagram below and on your knowledge of Earth science. The diagram represents a model of Earth's orbit around the Sun. Arrows represent two motions of Earth. Distances from the center of the Sun to the center of Earth are indicated in kilometers. Earth is represented when it is closest to the Sun and when it is farthest from the Sun.

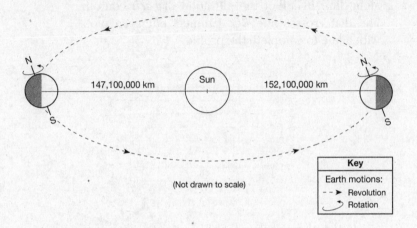

(Not drawn to scale)

Key
Earth motions:
- - ▶ Revolution
↷ Rotation

63 On the diagram *above*, place an **X** on Earth's orbit at *one* location where Earth's Northern Hemisphere is in winter. [1]

64 How many degrees is Earth's axis tilted to a line perpendicular to the plane of Earth's orbit? [1]

_____°

65 The diagram *below* represents Earth at one position in its orbit around the Sun. Starting at the North Pole, draw a straight arrow that points to the location of *Polaris*. [1]

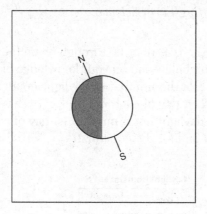

PART C
Answer all questions in this part.

Directions (66–85): Record your answers in the spaces provided. Some questions may require the use of the *2011 Edition Reference Tables for Physical Setting/Earth Science*.

Base your answers to questions 66 through 68 on the data table below and on the graph below and on your knowledge of Earth science. The data table lists the number of daylight hours for a location at 50° N on the 21st day of each month for 1 year. The graph shows the number of daylight hours on the 21st day of each month for a location at 70° N and for the equator, 0°.

Daylight Hours at 50° N

Date	Daylight (h)
January 21	8.4
February 21	10.0
March 21	12.0
April 21	13.8
May 21	15.5
June 21	16.2
July 21	15.5
August 21	14.0
September 21	12.0
October 21	10.2
November 21	8.4
December 21	7.5

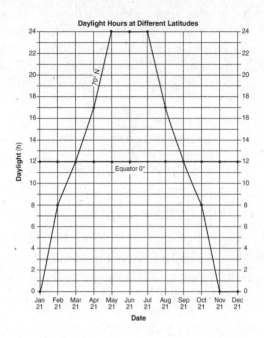

66 On the graph *above*, plot the number of daylight hours for the 21st day of *each* month listed on the data table. Connect *all* of your plotted data with a line. [1]

67 Explain why the number of daylight hours for all three latitudes was 12 hours on March 21 and September 21. [1]

_____

_____

68 Predict the number of daylight hours that occur at 70° S on June 21. [1]

_____ h

Base your answers to questions 69 through 71 on the data table below and on your knowledge of Earth science. The table shows air temperatures recorded under identical conditions at 2-hour intervals on a sunny day. Data were recorded 1 meter above ground level both inside and outside of a glass greenhouse.

Data Table

Time	Inside Air Temperature (°C)	Outside Air Temperature (°C)
8 a.m.	15	15
10 a.m.	18	16
12 noon	21	17
2 p.m.	24	18
4 p.m.	24	17

69 Describe the color and texture of the surfaces inside the greenhouse that would most likely absorb the greatest amount of visible light. [1]

Color: _____

Texture: _____

70 Calculate the rate of change in the outside air temperature from 8 a.m. to 2 p.m. in Celsius degrees per hour. [1]

_____ C°/h

71 Most atmospheric scientists infer that global warming is occurring due to an increase in greenhouse gases. [1]

(1) _____

(2) _____

Base your answers to questions 72 through 76 on the side-view model of the solar system below and on your knowledge of Earth science. The planets are shown in their relative order of distance from the Sun. Letter A indicates one of the planets.

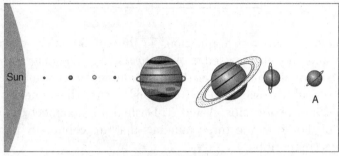

(Not drawn to scale)

72 The center of the asteroid belt is approximately 503 million kilometers from the Sun. On the model *above*, draw an **X** between two planets to indicate the center of the asteroid belt. [1]

73 State the period of rotation at the equator of planet A. Label your answer with the correct units. [1]

_____

74 How many million years ago did Earth and the solar system form? [1]

_____ **million years ago**

75 Calculate how many times larger the equatorial diameter of the Sun is than the equatorial diameter of Venus. [1]

_____ **times larger**

76 Identify the process that occurs within the Sun that converts mass into large amounts of energy. [1]

_____

_____

Base your answers to questions 77 through 80 on the diagram below and on your knowledge of Earth science. The diagram represents several common rock-forming minerals and some of the igneous rocks in which they commonly occur. The minerals are divided into two groups, A and B. Dashed lines connect the diagram of diorite to the three minerals that are commonly part of diorite's composition.

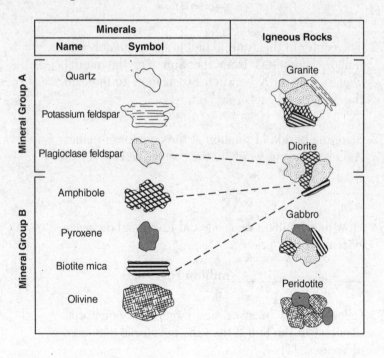

77 On the diagram *above*, draw *five* lines to connect the diagram of granite to the symbols of the minerals that are commonly part of granite's composition. [1]

78 Describe *one* characteristic of the minerals in group A that makes them different from the minerals in group B. [1]

Group A: _____

79 Based on the *Earth Science Reference Tables*, identify *one* other mineral found in some samples of diorite that is not shown in the diorite sample in the diagram. [1]

_____

80 A sedimentary rock sample has the same basic mineral composition as granite. Describe *one* observable characteristic of the sedimentary rock that is different from granite. [1]

_____

Base your answers to questions 81 and 82 on the Atlantic hurricane map below and on your knowledge of Earth science. The arrows on the map show the tracks of various hurricanes that occurred during late summer and early fall.

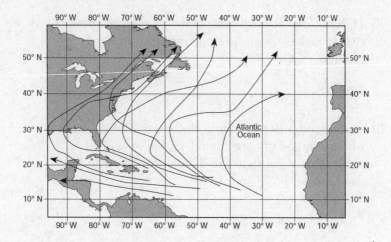

81 Describe *one* ocean surface condition or atmospheric condition that makes the area over the Atlantic Ocean between 10° N latitude and 20° N latitude ideal for these hurricanes to form. [1]

_____

_____

82 Several of these hurricanes have affected land areas. Describe *two* actions that people who live in hurricane-prone areas should take in order to prepare for future hurricanes. [1]

Action 1: _____

_____

Action 2: _____

_____

Base your answers to questions 83 through 85 on the maps and table below and on your knowledge of Earth science. The maps show earthquake intensities (IV to IX), according to the table of the Modified Mercalli Intensity Scale, for the 1906 and 1989 earthquakes at several locations in California. The asterisk (*) on each map is the location of each epicenter. The dashed line represents the location of a major fault.

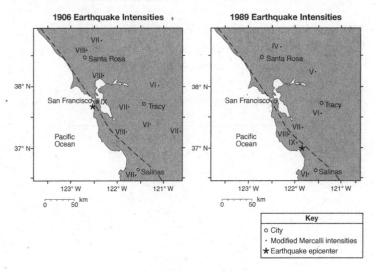

83 Name the major fault along which both of these earthquakes occurred and identify the type of plate tectonic boundary that is located along this fault. [1]

_____ **Fault**

Type of plate boundary: _____

84 Based on the Modified Mercalli Intensity Scale, identify the perceived shaking and the observed damage that occurred in the San Francisco area during the 1906 earthquake. [1]

Perceived shaking: _____

Observed damage: _____

85 Explain why Santa Rosa experienced a lower Modified Mercalli intensity shaking than Salinas experienced during the 1989 earthquake. [1]

_____

_____

Answers August 2014
Physical Setting/Earth Science

Answer Key

PART A

1. 2	8. 3	15. 4	22. 3	29. 1
2. 3	9. 1	16. 4	23. 1	30. 2
3. 1	10. 3	17. 1	24. 1	31. 4
4. 2	11. 4	18. 4	25. 4	32. 3
5. 4	12. 3	19. 3	26. 1	33. 2
6. 4	13. 2	20. 4	27. 3	34. 4
7. 2	14. 4	21. 1	28. 4	35. 1

PART B–1

36. 3	39. 3	42. 3	45. 3	48. 4
37. 3	40. 1	43. 2	46. 2	49. 2
38. 4	41. 1	44. 4	47. 3	50. 1

PART B–2 and **PART C.** *See* **Answers Explained.**

Answers Explained

PART A

1. **2** Observations indicate that the universe is expanding. If the motion of galaxies is traced back in time, there is a point about 13–14 billion years ago at which the galaxies were very close to each other. Thus, we have a model in which the universe started out with all its matter and energy in a very small volume and then expanded outward in all directions. This motion is similar to an explosion—hence the name the Big Bang Theory. Initially, the universe was smaller, denser, much hotter, and filled with a uniform glowing fog of hydrogen plasma. As the universe exploded outward in all directions from its origin point, it formed a sphere. Both the plasma and the radiation filling it spread out and grew cooler. Since an observer on Earth would be looking at the radiation from within the sphere of the expanding universe, that radiation should be coming at us from all directions, even from what may appear to be "empty space." Such radiation, observed coming from all directions in the universe, was detected in the mid-1960s by Arno Penzias and Robert Wilson at Bell Laboratories in New Jersey. They named this "cosmic background radiation." Thus, the existence of cosmic background radiation in space best supports the theory that the universe began with an explosion called the Big Bang.

WRONG CHOICES EXPLAINED:

(1) Impact craters found on Earth are evidence that chunks of matter in our solar system have collided with Earth. The most recent estimates are that the Big Bang occurred more than 13 billion years ago. Find the Geologic History of New York State chart in the *Reference Tables for Physical Setting/Earth Science*. Locate the time scale along the left edge of the column labeled "Eon." Locate 4600 million years ago (4.6 billion) on the time scale, trace right to the column labeled "Era," and note the statement "Estimated time of origin of Earth and solar system." Therefore, impact craters on Earth are unlikely to provide evidence of the Big Bang because Earth did not come into existence until long after the Big Bang occurred.

(3) During the formation of the solar system, gravity caused Earth and the other planets to become layered according to the density of the materials of which they are composed. The distance of each planet from the Sun was a key factor in determining a planet's characteristics. Powerful emissions from the Sun drove off most of the nearby gases, leaving behind the small, dense, rocky terrestrial planets. The more distant Jovian planets had sufficient gravity to retain much of the gas that surrounded them. Thus, they evolved into large, gaseous low-density planets. As a result, the different compositions of terrestrial and Jovian planets are evidence of the formation of the solar sys-

tem, not the theory that the universe was created by an explosion called the Big Bang. Additionally, as explained above, the solar system formed long after the Big Bang created the universe.

(4) A blue shift of light from distant galaxies indicates that the galaxies are moving toward Earth. If the universe is expanding (which is a key part of the Big Bang theory), distant galaxies would be moving away from Earth, not toward it. Thus, a blue shift in light from distant galaxies would contradict the theory that the universe was created by an explosion called the Big Bang, not support it.

2. **3** Find the Characteristics of Stars chart in the *Reference Tables for Physical Setting/Earth Science*. Locate the star labeled "*Sun*." Note the arrow along the right side of the chart leading from "Small Stars" to "Massive Stars." This arrow indicates that the higher up on the chart a star is located, the more massive the star is. Thus, a star that is more massive than the Sun will appear higher up on the chart than the Sun. Of the choices given, only *Sirius* and *Aldebaran* are located higher up on the chart than our Sun. Thus, those stars are more massive than our Sun. Note that the horizontal scale at the bottom of the chart labeled "Surface Temperature (K)" increases in temperature from right to left. So stars on the chart that are located to the right of our Sun have a lower surface temperature than our Sun; stars to the left of our Sun have a higher surface temperature than our Sun. Note that of the stars that are more massive than the Sun, *Sirius* is located to the left of our Sun on the chart and *Aldebaran* is located to the right of our Sun. Therefore, the star that is more massive than our Sun, but has a lower surface temperature, is *Aldebaran*.

3. **1** Find the Electromagnetic Spectrum chart in the *Reference Tables for Physical Setting/Earth Science*. Note the arrows indicating that on the chart, wavelength decreases to the left and increases to the right. Locate the section of the chart labeled "Visible light." Note that of the colors of visible light, violet light is farthest to the left, which means it has the shortest wavelength. Thus, the color of visible light that has the shortest wavelength is violet.

Ocean Tides

Type of Tide	Time
high	4:45 a.m.
low	10:58 a.m.
high	5:15 p.m.
low	11:22 p.m.

4. **2** According to the Ocean Tides table, on the date in the question, high tides occurred at 4:45 a.m. and 5:15 p.m. Thus, the time from one high tide to the next was 12 hours 30 minutes. It is reasonable to infer that the next high tide will occur approximately 12 hours 30 minutes after the 5:15 p.m.

high tide, or at about 5:45 a.m. the next day. Of the choices given, the closest to this time is 5:40 a.m.

5. **4** In 1851, French physicist Jean Foucault devised a way of conclusively proving Earth's rotation. He used a pendulum. He suspended a heavy iron ball on a long steel wire from the top of the dome of the Pantheon in Paris. As the pendulum swung back and forth, it appeared to change direction, slowly passing over different lines on the floor until eventually coming full circle to its original position after 24 hours. Since he knew that the path of a freely swinging pendulum would not change on its own, Foucault concluded that the apparent shift in the direction of swing of the pendulum was due to the floor (Earth's surface) rotating beneath the pendulum. Thus, the best evidence of Earth's rotation is provided by the apparent changes in the direction of swing of a Foucault pendulum.

WRONG CHOICES EXPLAINED:
(1) The direction of swing of a Foucault pendulum is measured relative to Earth's surface, not to its position relative to the Sun. Therefore, the shape of Earth's orbit would not cause the direction of swing of a Foucault pendulum to change relative to Earth's surface and, thus, would not provide evidence of Earth's rotation.

(2) The direction of swing of a Foucault pendulum is measured relative to Earth's surface, not to its position relative to the Milky Way galaxy. Therefore, the shape of the Milky Way galaxy would not cause the direction of swing of a Foucault pendulum to change relative to Earth's surface and, thus, would not provide evidence of Earth's rotation.

(3) The direction of swing of a Foucault pendulum is measured relative to Earth's surface, not to changes in the total yearly duration of insolation. Therefore, such changes in yearly duration of insolation would not cause the direction of swing of a Foucault pendulum to change relative to Earth's surface and, thus, would not provide evidence of Earth's rotation.

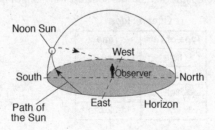

6. **4** It is given that the day shown is the first day of one of the four seasons. According to the diagram, on the first day of this season, the Sun rises south of east and sets south of west. In New York State, the Sun rises due east and sets due west on the equinoxes in March and September at the beginning

of the spring and fall seasons. On the summer solstice in June, the Sun rises farthest north of east and sets farthest north of west. On the winter solstice in December, the Sun rises farthest south of east and sets farthest south of east. On the day shown in the diagram, the Sun rises south of east and sets south of west. Therefore, the day of the year on which this path was observed was December 21, the first day of winter.

7. **2** The water table is the upper surface of underground pore spaces that are filled with groundwater. Precipitation delivers water to the surface, where some of that water will infiltrate before running off. Water that infiltrates will continue to seep downward through the soil. When it reaches the water table, where the pore spaces in the soil are already filled with water, the water that infiltrated will begin to fill the pore spaces in the soil above the water table, causing the level of the water table to rise. Thus, the processes most likely to cause a rise in the water table are precipitation and infiltration.

WRONG CHOICES EXPLAINED:
(1) Runoff is precipitation that does not evaporate or sink into the ground but, instead, runs downhill along Earth's surface. Since runoff does not sink into the ground, it does not reach the water table and does not cause the water table to rise. Erosion is any process that transports sediments from one location to another. Erosion may involve the movement of water. However, erosion is the movement of sediment along Earth's surface due to the force exerted by moving water, not infiltration that would cause the water to reach the water table and cause the water table to rise.

(3) Deposition is the process by which transported sediment is dropped in a new place. Burial is the process by which something is covered by sediment that is deposited on top of it. Neither is a process by which water infiltrates and reaches the water table, causing the water table to rise.

(4) Solidification is the process by which a liquid changes into a solid. Condensation is the process by which a gas changes into a liquid. Neither is a process that would cause water to infiltrate and reach the water table, causing the water table to rise.

8. **3** Find the Properties of Water table in the *Reference Tables for Physical Setting/Earth Science*. Note that heat energy is released during freezing and condensation. During condensation, 2260 J/g of heat energy are released. During freezing, 334 J/g of heat energy are released. Thus, the phase change during which water releases the most heat energy is condensation.

9. **1** The average air pressure exerted by Earth's atmosphere at sea level is defined as one atmosphere. Find the Pressure scale in the *Reference Tables for Physical Setting/Earth Science*. Locate "One atmosphere" along the left side of the scale. Trace right to the portion of the scale labeled at the top "millibars (mb)." Note that one atmosphere corresponds to 1013.25 mb. Now

trace right to the portion of the scale labeled at the top "inches (in of Hg*)." Note at the bottom of the scale that *Hg = mercury. So one atmosphere corresponds to 29.92 in of Hg, or 29.92 inches of mercury. Thus, the average air pressure exerted by Earth's atmosphere at sea level, expressed in millibars and inches of mercury, is 1013.25 mb and 29.92 in of Hg.

10. **3** In the troposphere, air pressure decreases with elevation. Therefore when air rises to higher elevations, the pressure confining the air decreases, causing the air to expand. When air expands, it cools adiabatically. When the air is cooled to its dew point, water vapor in the air condenses into water droplets, forming clouds. Therefore, the two processes that lead to cloud formation in rising air are expanding and cooling.

WRONG CHOICES EXPLAINED:
(1) When air sinks to lower elevations, the pressure confining the air increases and the air is compressed. Thus, compression occurs in sinking air, not in rising air. Compression causes adiabatic warming, not cooling.
(2) Compression occurs in sinking air, not in rising air. Warming would increase the likelihood that liquid water would evaporate to form water vapor, not the likelihood that water vapor would condense to form the liquid droplets that make up clouds.
(4) Expanding causes adiabatic cooling, not warming. Warming would increase the likelihood that liquid water would evaporate to form water vapor, not that water vapor would condense to form the liquid droplets that make up clouds.

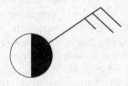

11. **4** Wind speed is measured using an anemometer. Wind direction is measured by a wind vane. Find the Key to Weather Map Symbols in the *Reference Tables for Physical Setting/Earth Science*. Locate the section labeled "Station Model Explanation." Note that wind speed is indicated by the number of whole and half feathers on the wind direction line (staff). Note, too, that each whole feather corresponds to 10 knots and that each half feather corresponds to 5 knots. The weather station model shown in the question shows two whole feathers and one half feather. Thus, the wind speed is 25 knots, measured by an anemometer.

12. **3** It is given that the dry-bulb temperature is 2°C and the wet-bulb temperature is –2°C. Thus, the difference between the wet-bulb and dry-bulb temperatures is 4°C. Find the Relative Humidity (%) chart in the

Reference Tables for Physical Setting/Earth Science. Locate the column headed "4" in the "Difference Between Wet-Bulb and Dry-Bulb Temperatures (°C)" scale along the top of the chart. Find the "Dry-Bulb Temperature (°C)" scale along the left side of the chart. Locate the row labeled "2," trace to the right until it intersects the column headed "4," and note the value of 36. Thus, when the air has a dry-bulb temperature of 2°C and a wet-Bulb temperature of –2°C, the relative humidity is 36%.

13. **2** Find the Surface Ocean Currents map in the *Reference Tables for Physical Setting/Earth Science.* Locate the western coast of South America. Note that the surface ocean current flowing along the western coast of South America is indicated by a white arrow labeled "Peru C." According to the key at the bottom of the map, a white arrow corresponds to a cool current. Thus, the current that has a cooling effect on the climate of the west coast of South America is the cool Peru Current.

14. **4** Find the Planetary Wind and Moisture Belts in the Troposphere diagram in the *Reference Tables for Physical Setting/Earth Science.* Note the belts labeled "Dry" at 30° N and 30° S latitude and at 90° N and 90° S latitude. Thus of the choices given, the two latitudes near which most of Earth's dry climate regions are found are 30° N and 30° S.

15. **4** During a massive volcanic eruption, volcanic ash is ejected high into the atmosphere. While in the atmosphere, the ash is carried by global winds and jet streams, quickly spreading throughout the atmosphere. Volcanic ash is opaque and blocks or reflects solar radiation, thereby decreasing the amount of solar radiation reaching Earth's surface. The less solar radiation that reaches Earth's surface, the less energy Earth's surface absorbs and the less the surface is warmed. Therefore, the addition of ash particles into the atmosphere would lead to the cooling of global temperatures.

WRONG CHOICES EXPLAINED:
(1) Thunderstorms that developed near the eruption would affect only that immediate area, not the entire globe. Thus, it is unlikely that thunderstorms near the eruption would affect global temperatures.

(2) Carbon dioxide and methane gas are greenhouse gases. Greenhouse gases allow short-wavelength radiation, such as visible light, ultraviolet light, and gamma rays, to pass through freely but absorb long-wavelength infrared radiation re-emitted from Earth (terrestrial radiation). As a result, the atmosphere warms. Thus, the release of carbon dioxide and methane would have a warming, not a cooling, effect on global temperatures.

(3) Magma is molten rock and typically has a temperature of more than 1,000° F. An outflow of magma over Earth's surface would heat the atmosphere that came in contact with the magma, not cool the atmosphere. Furthermore, that heating would be localized near the magma flow and

would not affect global temperatures. Thus, an outflow of magma on Earth's surface would not lead to a cooling of global temperatures.

16. **4** Rifting refers to the splitting apart of tectonic plates as they separate at a divergent plate boundary. Rifting is the major process that results in seafloor spreading. Magma wells up in the rift. The magma hardens to form new ocean floor. Then new magma wells up, splitting the just-formed rock and pushing it aside. The new magma hardens. As this process is repeated, new ocean floor is constantly formed and the plates on either side of the ridge are pushed farther apart. Find the Geologic History of New York State chart in the *Reference Tables for Physical Setting/Earth Science*. In the column labeled "Period," locate "Jurassic." Trace right to the column labeled "Important Geologic Events in New York." Note the events that occurred during the Jurassic: "Initial opening of the Atlantic Ocean" and "North America and Africa separate." Thus, the rifting of tectonic plates in eastern North America during the Jurassic Period was responsible for the opening of the Atlantic Ocean.

17. **1** Find the Generalized Bedrock Geology of New York State map in the *Reference Tables for Physical Setting/Earth Science*. Locate Mt. Marcy. Note the symbol for the surface bedrock at Mt. Marcy. Find the section of the map labeled "Geologic Periods and Eras in New York." Locate the symbol corresponding to the surface bedrock at Mt. Marcy. Note that this symbol corresponds to "Middle Proterozoic anorthositic rocks." Thus, the surface bedrock of Mt. Marcy, New York, is composed primarily of anorthosite.

18. **4** Evolution is the concept that existing life-forms developed from earlier, different ones. In order to trace the development of life, it is necessary to examine the remains of earlier life-forms, or fossils, found in rocks. The fossil record indicates that many new life-forms have appeared and most old life-forms have disappeared. It also suggests that life-forms have changed gradually over time, or evolved, so that current life-forms differ significantly from the earliest ones. Thus, much of the evidence for the evolution of life-forms on Earth has been obtained by examining fossils preserved in the rock record.

WRONG CHOICES EXPLAINED:
(1) In order for evidence to support the idea that life-forms have evolved over time, that evidence would have to include information about both past and present-day life-forms, not just present-day life-forms.
(2) The great heat and pressure under which metamorphic rocks form would destroy any evidence of past life-forms they may have contained. Therefore, radioactive dating of metamorphic rocks does not produce evidence of past life-forms and would be of little use as evidence for the evolution of life-forms on Earth.
(3) Correlating widespread igneous ash deposits allows the relative ages of rock layers to be inferred. However, simply correlating the ash deposits pro-

vides no information about earlier life-forms or about how they may have changed over time, or evolved.

19. **3** Find the Radioactive Decay Data chart in the *Reference Tables for Physical Setting/Earth Science*. Locate carbon-14 in the column labeled "Radioactive Isotope." Trace right to the column labeled "Half-life (years)," and note that carbon-14 has a half-life of 5.7×10^3 (5,700) years. Half-life is the time required for one-half of the unstable radioactive isotope to change into a stable decay product. The table in the question can be completed by decreasing the value in the column labeled "Original Carbon-14 Remaining (%)" by half at the end of each half-life and increasing the number of years in the column labeled "Number of Years" by 5,700 at the end of each half-life. A completed table is shown below:

Half-Life	Original Carbon-14 Remaining (%)	Number of Years
0	100	0
1	50	5,700
2	25	11,400
3	12.5	17,100
4	6.25	22,800
5	3.125	28,500

The completed table shows that after 22,800 years, 6.25% of the original carbon-14 remains.

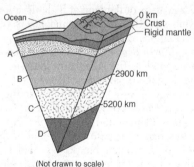

(Not drawn to scale)

20. **4** Find the Inferred Properties of Earth's Interior chart in the *Reference Tables for Physical Setting/Earth Science*. In the cross section, note the regions labeled "Inner Core (iron & nickel)" and "Outer Core (iron & nickel)." Thus, both the inner core and the outer core are composed of nickel

and iron. Now trace downward from the region labeled "Outer Core" to the graph labeled "Temperature (°C)." Note that in this region, the bold line labeled "Interior Temperature" is higher than the dotted line labeled "Melting Point." This indicates that the interior temperature of the rocks is greater than their melting point and the rocks are therefore inferred to be liquid in the outer core. Finally, trace downward from the region labeled "Inner Core" to the graph labeled "Temperature (°C)." Note that in this region, the bold line labeled "Interior Temperature" is lower than the dotted line labeled "Melting Point." This indicates that the interior temperature of the rocks in the inner core is less than their melting point and the rocks are therefore inferred to be solid. Therefore, the Earth layer inferred to be composed of solid nickel and iron is the inner core.

21. **1** Find the Tectonic Plates map in the *Reference Tables for Physical Setting/Earth Science*. Locate the Aleutian Trench. Note that the Aleutian Trench marks the boundary between the Pacific Plate and the North American Plate. (The North American Plate continues past the Aleutian Islands to the dashed line representing the boundary between the North American and Eurasian Plates.) Note the symbol at the Aleutian Trench and the arrows pointing toward one another on either side of the trench. According to the key, the symbol at the Aleutian Trench corresponds to a "Convergent plate boundary (subduction zone)." Note that in the symbol, the black rectangles are located on the side of the boundary labeled "overriding plate" and the opposite side of the boundary is labeled "subducting plate." Thus, at the Aleutian Trench, the Pacific Plate is the subducting plate and the North American Plate is the overriding plate. Therefore, oceanic crust is sliding beneath the Aleutian Islands in the North Pacific Ocean, forming the Aleutian Trench at a convergent boundary between the Pacific Plate and the North American Plate.

22. **3** Plateaus are large areas of flat land at high elevations. Plateaus generally have an underlying structure of horizontal layers of rock. Find the Generalized Landscape Regions of New York State map in the *Reference Tables for Physical Setting/Earth Science*. Note that the major plateau regions of New York State correspond to the Alleghany Plateau and the Catskills, which are part of the larger Appalachian Plateau (Uplands). Find the Generalized Bedrock Geology of New York State map in the *Reference Tables for Physical Setting/Earth Science*. Locate the region in New York State corresponding to the Alleghany Plateau and the Catskills. Note the symbol for the surface bedrock in this region. Locate that symbol in the key. Note that the symbol corresponds to "Devonian limestones, shales, sandstones, and conglomerates." Find the Scheme for Sedimentary Rock Identification in the *Reference Tables for Physical Setting/Earth Science*. In the column labeled "Rock Name," note that limestone, shale, sandstone, and conglomerate are all sedimentary rocks. Thus, of the choices given, only the Catskills are a New York Landscape region composed of mostly horizontal sedimentary bedrock and has a high elevation.

23. **1** Find the Properties of Common Minerals chart in the *Reference Tables for Physical Setting/Earth Science*. In the column labeled "Uses," locate the statement "ore of lead." Trace right to the column labeled "Mineral Name." Note that the mineral that is an ore of lead is galena. An ore is a naturally occurring solid material from which a metal or valuable mineral can be profitably extracted. Thus, the mineral commonly mined as a source of the element lead (Pb) is galena.

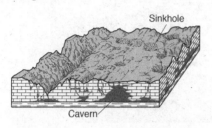

24. **1** Find the Scheme for Sedimentary Rock Identification in the *Reference Tables for Physical Setting/Earth Science*. In the column labeled "Map Symbol," locate the symbol corresponding to the rock layer in which the caves are located in the block diagram. From this symbol, trace left to the column labeled "Rock Name." Note that the rock layer in which the caves are located is limestone. From limestone, trace left to the column labeled "Composition." Note that limestone is composed of calcite. Find the Properties of Common Minerals chart in the *Reference Tables for Physical Setting/Earth Science*. In the column labeled "Mineral Name," locate calcite. From calcite, trace left to the column labeled "Distinguishing Characteristics," and note the statement "bubbles with acid." This indicates that calcite chemically reacts with acid. Acid can completely dissolve the mineral calcite. Carbon dioxide in the atmosphere reacts with water to form carbonic acid. Carbonic acid completely dissolves the mineral calcite. Therefore, carbonic acid in rainwater and groundwater that seeps through limestone bedrock may almost completely dissolve away the bedrock, leaving large holes or caverns. If the roof of the cavern collapses, a depression called a sinkhole is formed in the ground above it. Thus, the caverns and sinkholes in the block diagram formed because the bedrock chemically reacted with acidic groundwater.

WRONG CHOICES EXPLAINED:

(2) As explained above, the bedrock in the block diagram is limestone and limestone is composed of calcite. Find the Properties of Common Minerals chart in the *Reference Tables for Physical Setting/Earth Science*. In the column labeled "Mineral Name," locate calcite. From calcite, trace left to the column labeled "Composition," and note that calcite is composed of the compound $CaCO_3$. Find the key labeled "Chemical symbols" at the bottom of the table. Note that Ca is the symbol for calcium, C is the symbol for carbon, and O is the symbol for oxygen. Note, too, that the symbol for silicon is Si. Thus,

calcite does not contain silicon. Therefore, this type of bedrock does not contain large amounts of silicon.

(3) When a glacier melts, sediments carried in and on the ice fall directly to the ground and are deposited on top of the bedrock. Therefore, glacial deposits would not alter the shape of the underlying bedrock.

(4) Crustal uplift would cause the entire bedrock unit to be uplifted. It would not cause gaps to form within the bedrock.

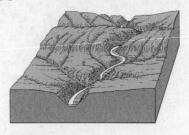

25. **4** Note that in the block diagram, the stream is located at the base of the surrounding slopes. If a brief, heavy rainstorm occurs in the mountains, the rate at which water builds up on the surface of the land would most likely exceed the rate at which the water can seep into the ground. Water that accumulates at the surface but does not seep into the ground moves downhill as runoff. Since the stream is located at the base of the surrounding slopes, that runoff would enter the stream. This process would increase the volume of water flowing in the stream. In general, increasing the volume of water flowing in a stream increases the stream velocity. Water in motion can exert a force on sediment particles, causing them to move, or be transported. The higher the stream velocity, the greater the force exerted by the water and the greater the stream's ability to transport sediment particles. As the stream's ability to transport sediment particles increases, the rate of erosion increases. Thus, shortly after the rainstorm, both the volume of water and the rate of erosion will increase.

26. **1** Note that the scratches in the polished surface of the bedrock are roughly parallel to one another. As a glacier moves along, loose rock may

freeze into the ice at the bottom of the glacier. As the glacier moves over bedrock, these rocks scrape parallel grooves or scratches, called striations, in the bedrock beneath the glacier. Small sediments frozen into the ice can act as sandpaper, smoothing and polishing the surface of the bedrock. Thus, the agent of erosion that most likely carried sediment that scratched and polished this bedrock surface was a moving glacier.

WRONG CHOICES EXPLAINED:
(2) Sediment carried by running water can smooth the surface of bedrock over which it flows. However, large particles transported by a stream typically bounce and roll along the streambed and do not exert enough downward pressure on the bedrock to create the deep scratches shown in the photograph. Additionally, the turbulence of a stream flowing fast enough to transport particles large enough to make the scratches shown in the photograph would make it unlikely that any scratches left behind would be straight and parallel to one another.

(3) Particles transported by wave action can smooth the surface of exposed bedrock. However, to make the deep, parallel scratches shown in the photograph, a large sediment particle would have to be pushed down against the bedrock with sufficient force to create a deep scratch and held in place as it moved along the bedrock surface long enough to create a long, straight scratch. This would then have to be repeated to create other parallel scratches. Wave action is turbulent. It frequently changes direction due to storms or changes in wind direction. Thus, it is unlikely that any scratches left by larger particles would leave the long, deep, parallel scratches shown in the photograph.

(4) Wind is able to transport only fine particles. It would not be able to transport particles large enough to create the large scratches shown in the photograph.

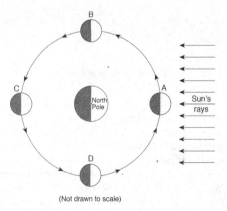

(Not drawn to scale)

27. **3** Both the Sun and Moon exert a force of gravity on Earth that causes tidal bulges in the oceans. The highest high tides occur when the grav-

ity of the Sun and Moon pull in the same direction so that the tidal bulges of the Sun align and combine with the tidal bulges of the Moon. The combined tidal bulges form very high tides. The additional water drawn into these very high tides decreases the water levels in between them to form very low tides. This occurs when the centers of the Sun, Moon, and Earth align during the new moon and full moon phases. The arrows labeled "Sun's rays" in the diagram indicate that the Sun is located to the right of Earth. Thus, the two Moon positions in which the Sun and Moon and Earth are aligned and an observer on Earth would most likely experience the highest high tides and the lowest low tides are positions C and A.

28. **4** A celestial body is said to be eclipsed when it passes into the shadow of another celestial body. Therefore, an event in which the Moon passes into Earth's shadow is known as a lunar eclipse. A lunar eclipse can be viewed when Earth is directly between the Moon and the Sun, and the Moon moves through Earth's shadow. According to the diagram, Earth is located directly between the Sun and Moon when the Moon is in position C. Thus, a lunar eclipse could occur when the Moon is at position C, which corresponds to the full moon phase. During the full moon phase, the side of the Moon facing Earth is entirely illuminated. Thus, the moon phase during which an observer on Earth could see a lunar eclipse is best represented by the diagram shown in choice (4).

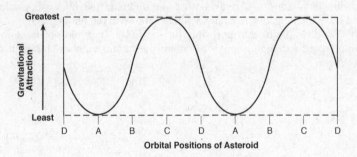

29. **1** Gravitational attraction between any two bodies is greatest when the distance between them is smallest. Conversely, gravitational attraction between any two bodies is least when the distance between them is greatest. According to the graph, gravitational attraction between the Sun and the asteroid is greatest at orbital position C and gravitational attraction is least at orbital position A. Thus, the diagram that best represents the positions of the asteroid in its orbit around the Sun will show the asteroid closest to the Sun at orbital position C and farthest from the Sun at orbital position A, as shown in choice (1).

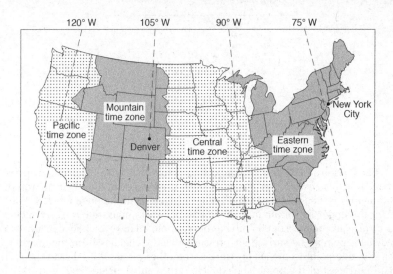

30. **2** Earth rotates 15° per hour, so time zones were set up around a series of meridians spaced at 15° intervals starting at the Prime Meridian, which passes through Greenwich, England. Earth rotates from west to east. So for every time zone you travel west of the Prime Meridian, the clocks are set one hour earlier. Note that Denver is located within the shaded region labeled "Mountain time zone" and New York City is located within the shaded region labeled "Eastern time zone." The Mountain Time Zone is two time zones west of the Eastern Time Zone. Therefore, clocks in the Mountain Time Zone are set 2 hours earlier than those in the Eastern Time Zone. Thus, when it is noon (12 p.m.) in Denver, it is 2 p.m. in New York City.

31. **4** Find the Key to Weather Map Symbols in the *Reference Tables for Physical Setting/Earth Science*. Locate the section labeled "Station Model Explanation." Note that on a station model, the value representing temperature (°F) is shown to the upper left of the circle. The value representing barometric pressure is shown to the upper right of the circle. Note that the barometric pressure value lists only the last three digits of the barometric pressure. Thus, the station model that shows an air temperature of 75°F and a barometric pressure of 996.3 mb will have 75 to the upper left of the circle and 963 to the upper right of the circle as shown in choice (4).

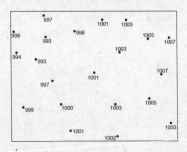

32. **3** Isobars are isolines connecting points that have the same air pressure. For this answer explanation, we will use the convention that the top of the map is north. To draw the 996-mb isobar, begin at the point labeled "996" near the upper left of the map. Note that not every point that is experiencing a surface air pressure of 996 mb has been labeled. However, between any two adjacent points, the surface air pressure steadily changes from the air pressure at one point to the air pressure at the other point. Note that to the east and northeast of the point labeled "996" are points labeled "993" and "997." Thus, it can be inferred that somewhere between these two points will be a point with an air pressure of 996 millibars. Therefore, from the point labeled "996," extend the isobar to the right between these two points. Now extend the isobar south between the points labeled "993" and "998." Next extend the isobar farther south between the points labeled "993" and "997." Continue to curve the isobar to the west between the points labeled "994" and "999." Complete the isobar by extending it to the edges of the map at both ends. Similarly, construct the 1000-mb line by beginning at the point labeled "1000" and extending the isobar southwest between the points labeled "999" and "1001" to the edge of the map. Then extend the 1000-mb isobar north-northeast between the points labeled "997" and "1001" and the points farther north labeled "998" and "1001" to the edge of the map. Finally, note that no point on the map is labeled "1004." However, it is reasonable to infer that between each of the points labeled "1003" and the adjacent points labeled "1005" are points with an air pressure of 1004 mb. Draw the 1004-mb isobar by extending a curving line between these 1003-mb and 1005-mb points to the edges of the map. Thus, the map that best shows the correct location of the 996-mb, 1000-mb, and 1004-mb isobars is map (3).

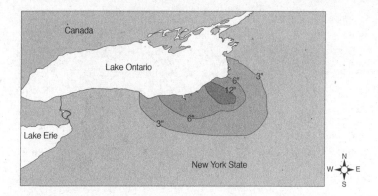

33. **2** Winter lake-effect snowstorms occur when cold winds move across large stretches of warm lake water. Evaporation of the warm lake water adds water vapor to the air. The water vapor is picked up by the cold wind, freezes, and is deposited as snow on the cold land of the leeward shores of the lake. Note that the shaded regions representing different snowfall amounts are on the southeast shore of the lake and are elongated to the southeast. The leeward shore of a lake is downwind. Therefore, the wind that produced this snowfall pattern on the southeast shore of the lake most likely came from the northwest.

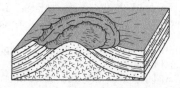

34. **4** Note that the block diagram shows a series of concentric circular ridges and valleys. Water falling on these circular ridges would flow downhill on both sides of the ridges and end up forming streams in the circular valleys between the ridges. Thus, the stream drainage pattern most likely found on the surface of the area represented by the block diagram would consist of a series of concentric circular streams as shown in pattern (4).

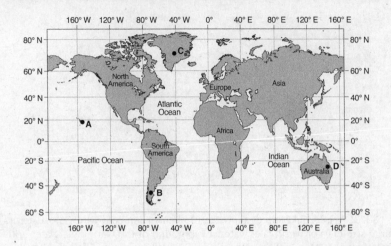

35. **1** Find the Tectonic Plates map in the *Reference Tables for Physical Setting/Earth Science.* In the key, locate the symbol labeled "Mantle hot spot." Locate on the Tectonic Plates map the locations corresponding to points A through D on the map in the question. Note that only location A is positioned over a mantle hot spot.

WRONG CHOICES EXPLAINED:

(2) Location B is on land in the center of the southern tip of South America. The Tectonic Plates map does not show a mantle hot spot at this location.

(3) Location C is in the center of Greenland. The nearest mantle hot spot is located on Iceland, not Greenland.

(4) Location D is on land near the east coast of Australia. The nearest mantle hot spot is the Tasman Hot Spot, but it is located in the ocean east of Australia.

PART B–1

Some Major Mass Extinctions in Earth's History

Approximate Time (mya)	Certain Life-Forms That Became Extinct
65.5	all dinosaurs and all ammonoids
200	many species of nautiloids, ammonoids, mammal-like reptiles, and early dinosaurs
251	all trilobites and 90% of other marine species and 70% of land species
376	many species of corals, brachiopods, and trilobites
444	more than half of brachiopod species, many trilobite species, and some coral species
520	small shelly fossil species and some early trilobite species

36. **3** Find the column labeled "Certain Life-Forms That Became Extinct," and locate the statement "more than half of brachiopod species." Trace left to the column labeled "Approximate Time (mya)," and note that more than half of brachiopod species became extinct 444 million years ago. Find the Geologic History of New York State chart in the *Reference Tables for Physical Setting/Earth Science*. Locate the time scale labeled "Million years ago" to the right of the column labeled "Epoch." On the time scale, locate "444." Trace left to the column labeled "Period," and note that 444 million years ago corresponds to the end of the Ordovician Period. Thus, more than half of brachiopod species became extinct at the end of the Ordovician Period.

37. **3** Evidence suggests that an impact event was responsible for the extinction of many life-forms at the end of the Cretaceous Period 65.5 million years ago. The dividing line between the Cretaceous and the Tertiary is a thin layer of clay that has been identified in sediments worldwide. This boundary of clay contains numerous indications of a massive impact event: levels of iridium higher than those found in Earth's crust but similar to those found in meteorites, shocked quartz grains, melted spherules, soot from the widespread forest fires ignited by the meteorite, and evidence of large waves such as would be caused by an ocean impact. The impact of asteroids or large meteorites would generate a massive shock wave and spew large amounts of dust high into the atmosphere. The resulting decrease in sunlight would cause global temperatures to decrease markedly and have devastating effects on photosynthetic life and all of the life-forms that depend on them for food. Thus, some scientists have inferred that the mass extinction that occurred 65.5 million years ago was caused by an asteroid impact.

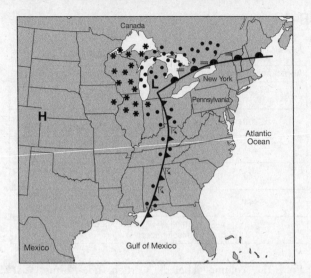

38. **4** Find the Key to Weather Map Symbols in the *Reference Tables for Physical Setting/Earth Science*. In the section labeled "Fronts," note that the symbol for a cold front is a bold line with triangles. Locate the symbol corresponding to a cold front on the map. Note the symbols along the cold front. In the Key to Weather Map Symbols, find the section labeled "Present Weather," locate the symbols corresponding to those along the cold front. Note that these symbols correspond to rain and thunderstorms. Thus, of the choices given, the weather condition shown along the cold front is thunderstorms.

39. **3** Find the Key to Weather Map Symbols in the *Reference Tables for Physical Setting/Earth Science*. In the section labeled "Fronts," locate the symbols corresponding to the fronts on the map. Note that a cold front extends southward from the low-pressure center (**L**) and that a warm front extends eastward from the low-pressure center. A cold front marks the boundary along which a cold air mass is advancing against a warm air mass. The side of the front on which the symbols are drawn indicates the direction in which the cold air is advancing. Thus, the air mass west of the cold front is a cold air mass and the air mass east of the cold front is a warm air mass. Similarly, a warm front marks the boundary along which a warm air mass is advancing against a cold air mass. Thus, the air mass south of the warm front is a warm air mass and the air mass north of the warm front is a cold air mass. The presence of rain and thunderstorms along the advancing cold front indicates that the warm air contains a lot of moisture. Therefore, the air mass over Pennsylvania is a warm, moist air mass. The prevailing southwest winds in the United States carry weather systems from southwest to northeast. So the most likely source region for a warm, moist air mass over the

eastern United States is a body of warm water near the equator to the southwest of Pennsylvania, such as the Gulf of Mexico.

40. **1** Winds blow from regions of high pressure toward regions of low pressure. Thus, winds blow outward from a high-pressure center. In the Northern Hemisphere, the Coriolis effect causes winds to curve toward the right. So the winds blowing outward from a high-pressure center curve in a clockwise direction. Thus, the general surface wind circulation associated with the high-pressure center (**H**) is clockwise and outward.

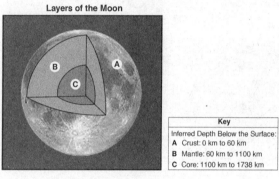

(Not drawn to scale)

41. **1** Increasing pressure and temperature with depth has foiled all attempts to drill or dig deep enough to reach Earth's mantle. Therefore, no one has directly observed Earth's interior. However, whenever an earthquake occurs, seismic waves travel through Earth and are recorded by seismographs around the world. Seismic waves are reflected from the surfaces of materials and refracted as they cross boundaries between substances of different densities. Analysis of the behavior of seismic waves as they travel through Earth reveals much about the structure of Earth's interior. Thus, most inferences about the structure of Earth's interior are based on the analysis of seismic data. It is given that the same type of evidence was used to find the inferred depths of both the Moon's interior layers and Earth's interior layers. Therefore, the evidence used to determine the inferred depth of the boundary between the Moon's mantle and core is seismic data recorded on the Moon's surface.

42. **3** According to the key, the Moon's mantle extends from 60 km to 1100 km below the surface. Thus, the inferred thickness of the Moon's mantle is 1040 km (1100 km − 60 km = 1040 km).

43. **2** Find the Solar System Data table in the *Reference Tables for Physical Setting/Earth Science*. In the column labeled "Celestial Object," locate "Earth's Moon." Trace right to the column labeled "Density (g/cm^3)," and note that Earth's Moon has a density of 3.3. Examine the other values in

the column labeled "Density (g/cm³)." Note that the value closest to Moon's 3.3 is 3.9. From 3.9, trace left to the column labeled "Celestial Object." Note that the planet Mars has a density of 3.9. Thus, the planet that has an average density most similar to the average density of the Moon is Mars.

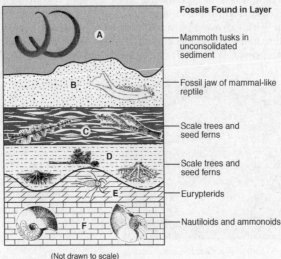

(Not drawn to scale)

44. **4** According to the cross section, mammoth tusks are found in layer A. Thus, mammoths existed when layer A was deposited. Find the Geologic History of New York State chart in the *Reference Tables for Physical Setting/Earth Science*. In the column labeled "Life on Earth," locate the statement "Humans, mastodonts, mammoths." Thus, humans and mastodonts existed along with mammoths when layer A was deposited. Trace left from this statement to the column labeled "Period," and note that this corresponds to the Quaternary Period. Trace right from the Quaternary to the column labeled "Time Distribution of Fossils." Note the bold, gray lines labeled with the names of different types of organisms. Note that the line labeled "Vascular Plants" extends throughout the Quaternary Period. Thus, vascular plants also existed when layer A was deposited. Therefore, the pair of organisms that existed when the unconsolidated sediment in layer A was deposited is humans and vascular plants.

45. **3** Find the Scheme for Sedimentary Rock Identification in the *Reference Tables for Physical Setting/Earth Science*. In the column labeled "Comments," find the statement "compacted plant remains." From this statement, trace right to the column labeled "Rock Name," and note that the rock composed of compacted plant remains is bituminous coal. From bituminous coal, trace right to the column labeled "Map Symbol," and note the symbol for bituminous coal. Now note that the layer in the geologic cross section

whose map symbol corresponds to bituminous coal is layer C. Thus, the rock layer that formed mainly from the compaction of plant remains was layer C.

46. **2** According to the cross section, layer F contains fossils of nautiloids and ammonoids. Find the Geologic History of New York State chart in the *Reference Tables for Physical Setting/Earth Science*. Among the diagrams of index fossils at the bottom of the chart, locate the diagrams corresponding to those shown in layer F. Note that the index fossil diagrams in layer F correspond to circled letters Ⓕ *Centroceras* and Ⓖ *Manticoceras*. Now locate the column headed "Time Distribution of Fossils." Note the bold gray lines labeled with names and circled letters. The names are of types of fossil organisms. The circled letters are keyed to the illustrations of index fossils printed along the bottom of the chart and are placed on the lines to indicate the approximate time of existence of each specific important fossil. Locate circled letter Ⓕ on the line labeled "Nautiloids" and circled letter Ⓖ on the line labeled "Ammonoids." From each letter, trace left to column labeled "Period." Note that index fossils Ⓕ *Centroceras* and Ⓖ *Manticoceras* both existed during the Middle Devonian Period. Thus, the geologic epoch during which layer F was deposited was the Middle Devonian.

47. **3** Whether the depositional environment was marine or terrestrial when each of the layers was formed can be inferred from the fossils found in the layers. Eurypterids, nautiloids, and ammonoids are marine organisms. Thus, it can be inferred that layers D, E, and F formed in a marine depositional environment. Scale trees, seed ferns, mammal-like reptiles, and mammoths are terrestrial organisms. Therefore, it can be inferred that layers A, B, and C formed in a terrestrial depositional environment. According to the Law of Superposition, the top layer of a series of sedimentary layers is the youngest unless it has been overturned or older rock has been thrust over it. Thus, the youngest rock layer in the cross section is rock layer A and the oldest rock layer is rock layer F. Therefore, the depositional environment during the time these layers and fossils were deposited changed from marine to terrestrial (land).

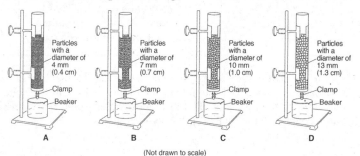

(Not drawn to scale)

48. **4** According to the diagram, the particles in these four columns range from 0.4 cm to 1.3 cm in diameter. Find the Relationship of Transported

Particle Size to Stream Velocity graph in the *Reference Tables for Physical Setting/Earth Science*. Note that particles in the range of 0.2 cm to 6.4 cm in diameter are classified as pebbles. All of the particles in these four columns fall within that range. Therefore, they are all classified as pebbles.

49. **2** The infiltration rate of a substance depends on two factors: the size of its pores and the degree to which the pores are interconnected. Small pores constrict the flow of water, slowing the rate of infiltration. If water cannot get from one pore to another, it cannot flow through a substance. Since all four sediment samples have uniform-sized, dry spherical particles, all four will have interconnecting pores. Thus, the size of the pores determines the infiltration rate. In general, the larger the sediment particles, the larger the pore spaces between them. Therefore, as particle diameter increases, the rate of infiltration will increase. This relationship is best shown by graph (2).

50. **1** The amount of water that adheres to the surface of the particles or is retained in the columns after the columns are drained depends on the total surface area of the particles. The total surface area of the particles increases as particle size decreases as shown below.

Figure A   Figure B

The cube in figure A has 6 sides, each with a surface area of 4 cm^2. So the total surface area is 24 cm^2.	Each of the eight small cubes in figure B has 6 sides, each with a surface area of 1 cm^2. So the surface area of each cube is 6 cm^2. Thus, the total surface area of the eight small cubes is 48 cm^2.

Therefore, as particle size in the columns decreases, the total surface area of the particles increases and the amount of water retained in the columns after they are drained increases. Thus, the most water would be retained in the column containing the smallest particles—column A.

PART B–2

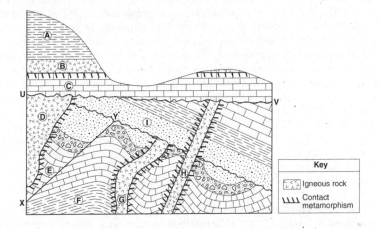

51. Locate the fault represented by line *XY* in the cross section. Note that the upper surfaces of the layers to the left of the fault are lower than the upper surfaces of the corresponding layers to the right of the fault. Thus, the rock layers to the left of the fault have moved downward relative to the rock layers to the right of the fault. Therefore, the arrow on the left side of the fault should be drawn pointing downward and the arrow on the right side of the fault should be pointing upward.

One credit is allowed for **one arrow pointing downward on the left side of line XY and one arrow pointing upward on the right side of line XY.**

Example of a 1-credit response:

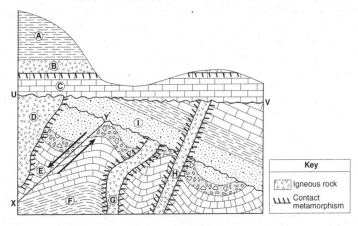

52. According to the Law of Superposition, the bottom layer of a series of sedimentary layers is the oldest unless it has been overturned or older rock has been thrust over it. The introductory paragraph states that the rock units have not been overturned. The rock units that touch the bottom of the cross section are E, F, G, and H. According to the key, rock units G and H are igneous rock and have caused contact metamorphism in the rocks adjacent to them. In order to undergo contact metamorphism, a rock must already exist when the igneous rock intrudes. Thus, all of the rock units in contact with rock units G and H that have undergone contact metamorphism are older than rock units G and H. Although rock unit E is the bottom layer on the right side of the cross section, rock unit F is the bottom layer on the left side of the cross section. Note that rock layer F lies beneath rock layer E. Therefore, rock layer F is older than rock layer E. Thus, the oldest rock unit in the cross section is F.

One credit is allowed for **F**.

53. Contact metamorphism describes changes in rock that result from the extreme heat produced by contact with magma or lava. When magma and lava cool, they form igneous rock. Thus, a zone of contact metamorphism would most likely be found where one of two adjoining bedrock types is igneous. The rock that will undergo contact metamorphism is the one that came in contact with the igneous rock when it was still molten magma or lava. According to the cross section and key, rock unit B is igneous rock and rock unit C has undergone contact metamorphism. Find the Scheme for Sedimentary Rock Identification in the *Reference Tables for Physical Setting/Earth Science*. In the column labeled "Map Symbol," locate the symbol corresponding to rock unit C. Note that rock unit C corresponds to limestone. Now find the Scheme for Metamorphic Rock Identification in the *Reference Tables for Physical Setting/Earth Science*. In the column labeled "Comments," note that metamorphism of limestone or dolostone forms marble, and metamorphism of various rocks by contact with magma or lava forms hornfels. Therefore, the rock formed by contact metamorphism between rock units B and C is most likely marble or hornfels.

One credit is allowed for **marble *or* hornfels**.

Rock Unit	D	G	H	B
Age (million years)	420	454	420	140

54. Note that contact metamorphism has occurred at the boundary between rock unit I and rock unit D and at the boundary between rock unit I and rock unit H. Thus, rock unit I already existed when rock units D and H formed. Therefore, rock unit I is older than rock units D and H. According to

the data table, rock units *D* and *H* are both 420 million years old. Therefore, the age of rock unit *I* is more than 420 million years. Note that rock unit *G* does not cut across rock unit *I* and that there is no contact metamorphism along the boundary between rock unit *G* and rock unit *I*. Note, too, the wavy line along the bottom of rock unit *I*. This represents a buried erosional surface, indicating that all of the rock units directly beneath the wavy line were eroded before the rock layers directly above the wavy line were deposited. Thus, rock unit *I* did not exist when rock unit *G* formed. This means rock unit *G* is older than rock unit *I*. According to the data table, rock unit *G* is 454 million years old. Therefore, the age of rock unit *I* is greater than 420 million years but less than 454 million years.

One credit is allowed for **any value greater than 420 million years ago but less than 454 million years ago**.

55. According to the reading passage, "For several decades before this drought, farmers had plowed the prairie and loosened the soil. When the soil became extremely dry from lack of rain, strong prairie winds easily removed huge amounts of soil from the farms, forming dust storms." Plowing the prairie removed the native grasses and other vegetation that had held the soil in place. These farmers exercised poor farming practices by not replacing this vegetation with drought-resistant plants when crops were not being grown. Loosening the soil made it easier for winds to pick up and transport soil particles. Thus, human activities that were a major cause of the huge dust storms that formed in the Great Plains during the 1930s included poor farming practices such as plowing the prairie and loosening the soil.

One credit is allowed for an acceptable response. Acceptable responses include but are not limited to:

- **plowing large areas of the plains**
- **poor farming practices**
- **farmers loosened the soil**
- **farming**

56. Abrasion is the breaking down of rocks by the rocks rubbing against each other. Abrasion occurs mainly when rock fragments are being transported by agents of erosion. A typical agent of erosion is wind. As the fragments are carried along by the wind, they bounce off one another and rub against one another. This breaks smaller pieces off the surfaces of fragments, particularly at corners that protrude. As a result, the fragments change shape, becoming smaller, smoother, and more rounded. As particles rub against one another, their surfaces become scratched and pitted by collisions with other particles. This causes the particles to take on a "frosted" appearance due to the many tiny scratches on their surfaces. Thus, changes in the appearance of sand particles that were abraded when transported by winds within the Dust

Bowl region include getting smaller, rounder, and smoother with surfaces that are more scratched, pitted, and frosted.

One credit is allowed for an acceptable response. Acceptable responses include but are not limited to:

- **They became more rounded.**
- **They became smaller in size/thinner/finer.**
- **The outside surface became scratched/frosted/pitted.**
- **Sand grains become smoother.**

57. According to the reading passage, "[A] Colorado storm carried dust approximately 3 kilometers up into the atmosphere." Find the Selected Properties of Earth's Atmosphere chart in the *Reference Tables for Physical Setting/Earth Science*. Locate the point corresponding to 3 kilometers on the left (km) side of the scale labeled "Altitude." Trace right, and note that 3 kilometers is within the layer of the atmosphere labeled "troposphere." Thus, the layer of the atmosphere in which the dust particles were transported by the Colorado storm to New York State is the troposphere.

One credit is allowed for **troposphere**.

58. Air flows in wind in much the same way that water flows in a stream, and wind carries particles in much the same way that a stream of water does. However in wind, the medium that carries the particles is air. Air is not very dense and exerts less force on a particle than does a denser medium, such as water, moving at the same speed. Imagine the difference in being hit by a balloon filled with air and being hit by a balloon filled with water! Therefore, winds are usually able to carry only small particles such as sand, silt, clay, and dust. Wind carries very small particles, such as clay and dust, in suspension. The finest particles may be carried high into the atmosphere, where winds carry them for hundreds or even thousands of kilometers before they finally settle to the ground. Wind carries larger particles, such as sand, closer to the ground. They slide, roll, and bounce while skipping along the surface. As the wind dies down, larger, denser, heavier particles settle to the ground first and are not carried very far. The smallest, least dense, lightest particles are carried the farthest before settling to the ground. Thus, the dust clouds that moved to the east coast of the United States during the 1934 storm were composed mostly of silt and clay particles instead of sand because the velocity of the wind was not great enough to carry larger, heavier particles such as sand. The wind could carry only smaller, lighter particles such as silt and clay.

One credit is allowed for an acceptable response. Acceptable responses include but are not limited to:

- The velocity of the wind could carry only small/less dense/flatter particles.
- Sand is heavier and not likely to be carried that far.
- The velocity of the wind was not great enough to carry sand particles.
- Smaller particles are eroded more easily.
- Silt and clay are smaller-sized particles.

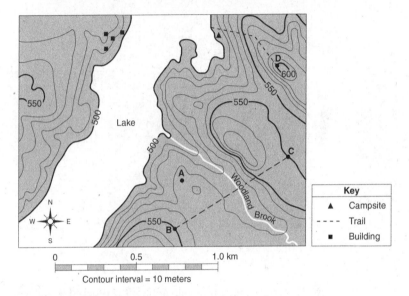

59. Note the bold black contour line labeled "500" that marks the shoreline of the lake. Note the bold black contour lines labeled "550" to the northeast and southwest of point A. Thus the elevation is increasing as you travel eastward from the lake shore to the bold 550 m contour lines. Follow the contour line nearest point A to either side between the bold 500 m lakeshore line and either of the bold 550 m contour lines. Note that point A is located between the third and fourth regular contour line east of the lakeshore. It is given that the contour interval is 10 meters. Therefore, the elevation of point A is greater than 530 meters and less than 540 meters. Note that there is also a bold 550 m contour line on the opposite side of the lake. Thus, the elevation is increasing as you travel westward on the opposite side of the lake. So the same elevation will be found on the opposite side of the lake between the third and fourth contour lines to the west of the lake. Place an **X** on the opposite side of the lake anywhere between the third and fourth regular contour line west of lake shore.

One credit is allowed if **the center of the X is located in the white area between the 530 m and 540 m contour lines on the west side of the lake as shown below.**

Note: Credit is allowed even if a symbol other than an **X** is used. Credit is *not* allowed if the center of the **X** touches either the 530 or the 540 contour lines.

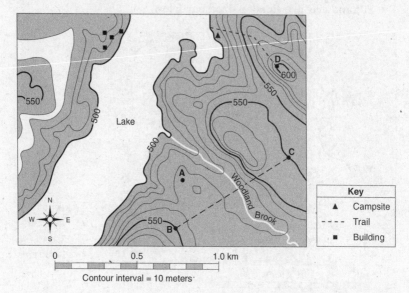

60. To construct a topographic profile along line BC, proceed as follows. Place the straight edge of a piece of scrap paper along the line connecting point B to point C on the map in your answer booklet. Mark the edge of the paper at points B and C and wherever the paper intersects a contour line. Wherever the paper intersects a contour line, label the mark with the elevation of the contour line as shown below.

Then place this paper along the lower edge of the grid provided in your answer booklet so that points B and C on the scrap paper align with points B and C on the grid. Next, at each point where a contour line crosses the edge of the paper, draw a plot on the grid at the appropriate elevation. Finally, connect all of the plots in a smooth curve to form the finished profile as shown below. Note that the section of line BC within the circular bold 550 contour line nearest point C does not touch the 560 m contour line. Therefore, it is higher than 550 m but less than 560 m.

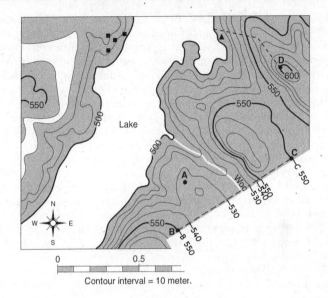

One credit is allowed if **the centers of all seven student plots are within or touching the rectangles shown above and are correctly connected with a line passing within or touching the rectangles. The line must show the lowest elevation between 520 m and 530 m and the highest elevation between 550 m and 560 m.**

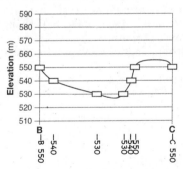

Note: Credit is allowed if the line does not pass through the student plots but is still within or touching the rectangles.

61. Water flows from higher elevations to lower elevations. Recall that the contour lines to the east of the lake increase to the east. Thus, the portion of Woodland Brook in the southeast corner of the map is at a higher elevation

than the point where Woodland Brook enters the lake. Thus, Woodland Brook flows from southeast to northwest into the lake. Additionally, the bed of a stream slopes downhill and is lower than its banks. A person standing on a streambed is at a lower elevation than a person standing on a bank of the stream. A person standing on the streambed would have to walk uphill along the streambed—upstream—to reach the same elevation as a person on the bank. On a topographic map, contour lines connect points of equal elevation. Therefore, a contour line would have to bend upstream of the stream bank to intercept the same elevation on the streambed and then bend downstream on the opposite side to reach the same elevation on the opposite bank. As a result, contour lines bend upstream when they cross a streambed, forming a distinctive V-shaped curve with the apex pointing upstream and the sides opening toward the downstream direction. So contour lines that cross a stream bend in the direction opposite to that of stream flow. Note that where Woodland Brook crosses a contour line, the contour line forms a V-shaped curve with the apex pointing toward the southeast and the sides opening toward the northwest. Thus, the general direction in which Woodland Brook flows is northwest into the lake.

One credit is allowed if both "into the lake" is circled and the contour-line evidence provided is correct. Acceptable evidence includes but is not limited to:

- The contour lines bend away from the lake where they cross the stream.
- The lines do not go straight across but curve to the southeast when they cross Woodland Brook.
- The contour lines that cross Woodland Brook show the lowest elevation where the brook enters the lake.
- Law of the Vs/contour lines make a V shape that points uphill where they cross a stream.
- A river flows from a higher elevation to a lower elevation.

62. Find the Equations section of the *Reference Tables for Physical Setting/Earth Science*. Note the equation for gradient:

$$\text{Gradient} = \frac{\text{change in field value}}{\text{change in distance}}$$

The map shown is a topographic map. The field value on a topographic map is elevation. According to the key, a trail is represented by a dashed line. On the map, locate the trail leading from the shoreline of the lake to point D. Note that the elevation of the shoreline of the lake is 500 meters and the elevation of point D is 600 meters. Thus, the change in elevation when hiking along the trail is 100 meters (600 m − 500 m = 100 m). On a piece of scrap paper, mark off the distance along the dashed line between the shoreline of the lake and point *D*. Compare the distance marked off on the scrap paper to the scale printed beneath the map to determine the distance along the trail between the shoreline of the lake and point *D*. This distance is about 0.5 kilometers. Substitute these values into the equation for gradient and solve.

$$\text{Gradient} = \frac{\text{change in field value}}{\text{change in distance}} = \frac{100 \text{ m}}{0.5 \text{ km}} = 200 \text{ m/km}$$

One credit is allowed for **any value from 185 m/km to 215 m/km**.

63. Earth is closest to the Sun at perihelion (147,100,000 km), which occurs during the first week of January. Earth is farthest from the Sun at aphelion (152,100,000 km), which occurs during the first week of July. The winter season in Earth's Northern Hemisphere begins on the winter solstice and ends on the spring equinox. The winter solstice occurs on or about December 21, shortly before Earth reaches perihelion. The summer solstice occurs on or about June 21, shortly before Earth reaches aphelion. The spring equinox occurs about halfway between the winter solstice, shortly before Earth reaches perihelion, and the summer solstice, shortly before Earth reaches aphelion. Note the arrows indicating the direction in which Earth revolves around the Sun in its orbit. Thus, the winter season extends from shortly before Earth reaches perihelion in its orbit until just before Earth reaches half the distance toward aphelion. Therefore, the **X** should be placed on Earth's orbit somewhere in this region.

One credit is allowed if **the center of the X is within or touching the clear banded region below.**

Note: Credit is allowed if a symbol other than **X** is used.

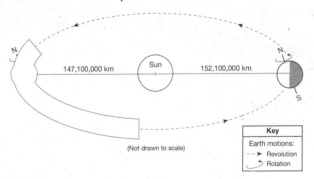

64. Earth's axis is tilted 23.5° from a line perpendicular to the plane of Earth's orbit.

One credit is allowed for **any value from 23.4° to 23.5°**.

Note: Credit is allowed if the value is shown as a fraction, such as 23½.

65. The star *Polaris* is located almost directly in line with the north pole of Earth's axis of rotation. Therefore, a straight arrow should be drawn extending from Earth's North Pole outward from Earth's surface along the line representing Earth's axis of rotation.

One credit is allowed for **an arrow that is aligned with Earth's axis and is within the cone-shaped area shown below**.

Note: One credit is allowed even if the arrow does not start exactly at the North Pole.

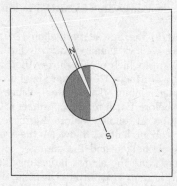

PART C

Daylight Hours at 50° N

Date	Daylight (h)
January 21	8.4
February 21	10.0
March 21	12.0
April 21	13.8
May 21	15.5
June 21	16.2
July 21	15.5
August 21	14.0
September 21	12.0
October 21	10.2
November 21	8.4
December 21	7.5

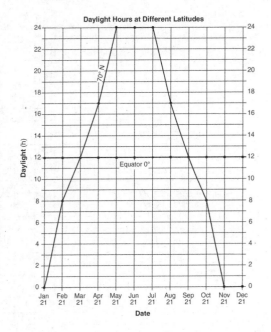

66. To plot the number of daylight hours for the 21st day of each month listed on the data table, do the following. In the data table, note that the number of daylight hours on January 21 corresponds to 8.4. On the graph provided in your answer booklet, locate "Jan 21" on the horizontal axis labeled "Date" and trace vertically upward to the point on the vertical axis labeled "Daylight (h)" corresponding to the value 8.4. Make a mark at this point. Next locate the point corresponding to February 21 on the horizontal axis labeled "Date" and the point corresponding to 10.0 on the vertical axis labeled "Daylight (h)." Trace vertically upward from the Feb 21 date and horizontally to the right from 10.0 until the two lines intersect. Make another mark at this point. Repeat this process for each of the remaining dates given in the data table. When all of the points have been plotted, connect them with a line that passes through every point in order of depth. A graph with all points correctly plotted and connected with a line is shown below.

One credit is allowed if **the centers of *all 12* student plots are within or touching the circles shown below and a correctly drawn line passes within or touches each circle.**

Note: One credit is allowed if the student's line does not pass through the student's plots but is still within or touching the circles.

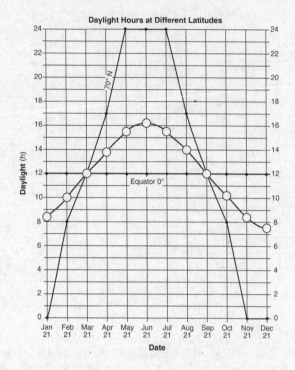

67. March 21 and September 21 represent the first days of spring and fall when Earth is midway in its orbit between the summer and winter solstices. At the summer solstice on June 21, Earth's axis of rotation is tilted farthest toward the Sun. At the winter solstice on December 21, Earth's axis of rotation is tilted farthest away from the Sun. However on both March 21 and September 21, Earth's axis of rotation is still tilted but to the side rather than toward or away from the Sun. The Sun's direct rays are at the equator at solar noon. The boundary between the half of Earth that is illuminated and the half that is in shadow cuts precisely through the poles. As Earth rotates, the Sun rises due east and sets due west at all points on Earth's surface. Additionally, all points spend exactly one-half rotation in daylight and one-half rotation in darkness. Earth completes one rotation in 24 hours. Therefore, each one-half rotation is completed in 12 hours. Thus, on March 21 and September 21, every point on Earth's surface experiences 12 daylight hours and 12 nighttime hours.

71. Many chemical compounds found in the Earth's atmosphere act as "greenhouse gases." These gases allow sunlight to enter the atmosphere and strike Earth's surface. When sunlight strikes Earth's surface, some of it is reflected back toward space as infrared radiation (heat). Some of that sunlight, though, is absorbed and then reradiated as infrared radiation (heat). Greenhouse gases absorb this infrared radiation and trap the heat in the atmosphere. Over time, the amount of energy received from the Sun should be about the same as the amount of energy radiated back into space, leaving the temperature of Earth's surface roughly constant. However, many gases exhibit these greenhouse properties, trapping heat. Thus, these gases are inferred to be responsible for global warming. Some of these greenhouse gases occur in nature (water vapor, carbon dioxide, methane, ozone, and nitrous oxide), while others are exclusively human-made (like the chlorofluorocarbons used in aerosol sprays).

One credit is allowed if *both* responses are acceptable. Acceptable responses include but are not limited to:

- **carbon dioxide (CO_2)**
- **methane (CH_4)**
- **water vapor (H_2O)**
- **chlorofluorocarbons (CFCs)**
- **nitrous oxide (N_2O)**
- **ozone (O_3)**

72. It is given that the asteroid belt is 503 million kilometers from the Sun. Find the Solar System Data table in the *Reference Tables for Physical Setting/Earth Science*. In the column labeled "Mean Distance from Sun (million km)," note that 503 million kilometers falls between the values 227.9 and 778.4. From these two values, trace left to the column labeled "Celestial Object." Note that the planet Mars is located 227.9 million kilometers from the Sun and the planet Jupiter is located 778.4 million kilometers from the Sun. Thus, at 503 million kilometers, the asteroid belt is located about midway between the orbits of Mars and Jupiter. According to the Solar System Data table, Mars is the fourth planet from the Sun and Jupiter is the fifth planet from the Sun. Thus, in the model in your answer booklet, the **X** should be drawn about midway between the fourth (Mars) and fifth (Jupiter) planets from the Sun.

One credit is allowed if **the center of the X is drawn in or touches the box shown below**.

Note: Credit is allowed if a symbol other than **X** is used.

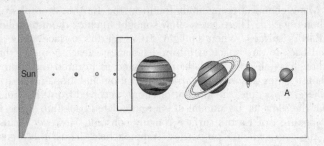

73. According to the model in your answer booklet, planet A is the eighth and most distant planet from the Sun. Find the Solar System Data table in the *Reference Tables for Physical Setting/Earth Science*. Refer to the columns labeled "Celestial Object" and "Mean Distance from Sun (million km)." Note that the eighth planet from the Sun is Neptune. Thus, planet A corresponds to Neptune. From Neptune, trace right to the column labeled "Period of Rotation at Equator," and note that Neptune has a period of rotation of 16 h. Thus, the period of rotation at the equator of planet A is 16 h.

One credit is allowed for **16 with the correct units**. Acceptable units include but are not limited to:

- **h**
- **hrs**
- **hours**

74. Find the Geologic History of New York State chart in the *Reference Tables for Physical Setting/Earth Science*. Find the column labeled "Era," locate the statement "Estimated time of origin of Earth and our solar system." Trace left to the time scale labeled "Million years ago." Note that Earth and the solar system formed approximately 4600 million, or 4.6 billion, years ago.

One credit is allowed for **a value equivalent to 4600 million years ago**.

Note: Credit is allowed if the "million years ago" in the answer booklet is crossed out and an equivalent value is expressed in other units (e.g., 4.6 billion years ago).

75. Find the Solar System Data table in the *Reference Tables for Physical Setting/Earth Science*. In the column labeled "Celestial Object," locate "Sun" and "Venus." From each of these two celestial objects, trace horizontally right to the column labeled "Equatorial Diameter (km)." Note that the equatorial diameter of the Sun is 1,392,000 and the equatorial diameter of Venus is 12,104. To calculate how many times larger the equatorial diameter of the

Sun is than the equatorial diameter of Venus, divide the equatorial diameter of the Sun by the equatorial diameter of Venus and solve:

$$\frac{1{,}392{,}000 \text{ km}}{12{,}104 \text{ km}} = 115.003305$$

Thus, the Sun's equatorial diameter is 115 times larger than the equatorial diameter of Venus.

One credit is allowed for **any value from 115 times larger to 115.003305 times larger**.

76. Within the Sun and other stars, the force of gravity is strong enough to overcome the force of repulsion between atomic nuclei, allowing the nuclei to combine in a process called nuclear fusion. By definition, nuclear fusion involves the combining of several atoms of a lighter element to form a single atom of a heavier element. The single heavier atom typically has less mass than the lighter atoms from which it formed. The mass "missing" from the heavier atom is not lost. Instead, it is converted into energy according to Einstein's formula $E = mc^2$. This formula states that if mass is converted to energy, the amount of energy released, E, is equal to the mass, m, times the speed of light, c, squared. The speed of light squared is a very large number. Therefore, the conversion of even a small amount of mass to energy during nuclear fusion results in the release of a very large amount of energy. Thus, great amounts of energy are released in the core of a star as lighter elements combine and form heavier elements during the process of fusion. The process that occurs within the Sun that converts mass into large amounts of energy is fusion.

One credit is allowed for an acceptable response. Acceptable responses include but are not limited to:

- **fusion**
- **nuclear fusion**
- **conversion of hydrogen to helium/of H to He**

77. Find the Scheme for Igneous Rock Identification in the *Reference Tables for Physical Setting/Earth Science*. Locate granite in the upper portion of the chart. From granite, trace downward in the column to the graph labeled "Mineral Composition (relative by volume)." Note that granite is composed primarily of potassium feldspar (pink to white,) quartz (clear to white), and plagioclase feldspar (white to gray) with smaller amounts of biotite (black) and amphibole (black.) Thus, on the diagram in your answer booklet, lines should be drawn from the diagram of granite to the symbols of quartz, potassium feldspar, plagioclase feldspar, amphibole, and biotite mica as shown below.

One credit is allowed if *all five* lines are drawn from granite to the minerals quartz, potassium feldspar, plagioclase feldspar, amphibole, and biotite mica.

Note: If extra lines are drawn between the minerals and rocks, all lines must be correct in order to receive credit.

Example of a 1-credit response:

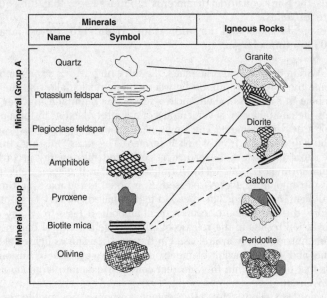

78. Note that group A consists of the minerals quartz, potassium feldspar, and plagioclase feldspar. Note that group B consists of the minerals amphibole, pyroxene, biotite mica, and olivine. Find the Scheme for Igneous Rock Identification in the *Reference Tables for Physical Setting/Earth Science*. Locate the section labeled "Mineral Composition (relative by volume)." Note the colors listed under the name of each mineral on the chart. Note that the minerals in group A are lighter in color (clear, pink, white, and gray) than the minerals in group B (black and green.) Locate the section labeled "Characteristics." Note the arrows indicating the ranges in color, density, and composition. Note that rocks composed primarily of quartz, potassium feldspar, and plagioclase feldspar (all group A minerals) are to the left in the mineral composition section. Thus, group A minerals are on the light end of the color range, are on the lower end of the density range, and are felsic (rich in Si and Al) in composition. Note that rocks composed primarily of amphibole, pyroxene, biotite mica, and olivine (all group B minerals) are to the right in the mineral composition section. Thus, group B minerals are on the dark end of the color range, are on the higher end of the density range, and

are mafic (rich in Fe and Mg) in composition. Thus, group A minerals are lighter in color, lower in density, and more felsic in composition (richer in Si and Al, lower in Fe and Mg) than group B minerals.

One credit is allowed for an acceptable response. Acceptable responses include but are not limited to:

- **Group A:**
  - **lighter colored**
  - **more felsic**
  - **lower in density**
  - **lacking in magnesium/Mg/iron/Fe**
  - **rich in silicon/Si/aluminum/Al**

79. Find the Scheme for Igneous Rock Identification in the *Reference Tables for Physical Setting/Earth Science*. Locate the column containing diorite in the upper section of the chart. Trace the column for diorite downward to the mineral composition graph. Note that it intersects portions of the regions representing quartz, plagioclase feldspar, biotite, pyroxene, and amphibole. Note that only small portions of the regions representing quartz and pyroxene lie within the diorite column. Note, too, that neither of these two portions extends across the entire column. Therefore, quartz and pyroxene are found in only some samples of diorite. Note that on the diagram in your answer booklet, lines are drawn from diorite only to three minerals: plagioclase feldspar, amphibole, and biotite mica. Thus, in addition to these three minerals, some samples of diorite may also contain quartz or pyroxene.

One credit is allowed for **quartz *or* pyroxene**.

80. Find the Scheme for Igneous Rock Identification in the *Reference Tables for Physical Setting/Earth Science*. Locate "granite" in the upper section of the chart. Trace the column for granite downward to the mineral composition graph. Note that granite is composed primarily of quartz and feldspars. From "granite," trace right to the columns labeled "Crystal Size" and "Texture." Note that granite is an igneous rock with a coarse, crystalline texture. Now find the Scheme for Sedimentary Rock Identification in the *Reference Tables for Physical Setting/Earth Science*. Find the column labeled "Composition." Note the statement "Mostly quartz, feldspar, and clay minerals; may contain fragments of other rocks and minerals" in the upper part of the chart. Trace left to the column labeled "Texture." Note that the upper part of the chart corresponds to sedimentary rocks that have a clastic texture. A clastic texture means rock consists of grains of rock fragments that have been cemented or compacted together. Thus, clastic sedimentary rocks have a similar composition to granite. This is not surprising because the rock fragments found in clastic rocks were typically formed by the weathering and erosion of the granite that makes up the continents. However, even though

granite and clastic sedimentary rocks have similar compositions, they formed in very different ways and have different observable characteristics. Igneous rocks form by the cooling and solidification of molten rock. As the molten rock cools, it forms a solid consisting of a random pattern of intergrown crystals. Fossils are not found in igneous rock because the great heat associated with forming igneous rocks destroys fossils. On the other hand, the formation of clastic sedimentary rocks involves the processes of deposition, burial, compaction, and cementation. The grains of a clastic sedimentary rock are fragments of rock that have been weathered and eroded from other rocks and then have been compacted or cemented together. These sediments may be angular or rounded. During deposition, sediments can become sorted into layers, so many sedimentary rocks contain layers of particles. Remains of once-living things that are quickly buried during the process of deposition can be preserved. Therefore, sedimentary rocks can contain fossils.

One credit is allowed for an acceptable response. Acceptable responses include but are not limited to:

- **The particles are layered.**
- **The sedimentary rock may have fossils.**
- **There are no intergrown crystals.**
- **The sedimentary rock may have rounded or angular fragments.**
- **The grains are cemented together.**
- **The rock contains different sediments.**
- **Sedimentary rock contains fragments.**

81. When air over warm oceans is heated by solar radiation, the warming of the air causes a decrease in pressure and the evaporation of water from the ocean. This adds humidity to the air, which decreases the air pressure even further. The high temperature and moisture levels result in air with a very low density. This air then rises rapidly. This process results in numerous convection cells and thunderstorms. With very little wind shear, widespread thunderstorm activity can merge into a large updraft that is part of a huge convection cell. The rising air further decreases the pressure. Latent heat released when the moisture in the air condenses also decreases the pressure. These processes create a region of very low pressure. Winds begin to blow toward the low-pressure center as air moves in from surrounding areas of higher pressure. The Coriolis effect deflects the winds, and a cyclone forms. Heat that is released when fresh moisture brought in by winds condenses, feeding the convection cell with new energy. As a result, the convection cell gets larger and stronger. When the winds in the cyclone exceed 119 km/hr, it is called a hurricane.

Hurricanes tend to form during summer months over oceans near the equator when the ocean surface is the warmest. Once formed, heat and moisture from the ocean provide the energy that maintains the hurricane.

The area over the Atlantic Ocean between 10° N and 20° N latitude is located in the tropics near the equator. Thus, by late summer and early fall, the ocean surface temperature has peaked and evaporation rates are high. So the air in this region is very warm and moist—ideal conditions for the formation of hurricanes.

One credit is allowed. Acceptable responses include but are not limited to:

- **The warm waters that give the hurricane its energy are located in this tropical region of the ocean.**
- **Warm ocean waters between 10° N and 20° N fuel hurricanes.**
- **Warm and/or humid atmospheric conditions exist between 10° N and 20° N.**
- **A maritime tropical air mass**
- **Low air pressure**
- **Rising air currents**
- **Low wind shear**

82. According to the National Weather Service, the major hazards associated with hurricanes are storm surge and storm tide, heavy rains, inland flooding, high winds, rip currents, and tornadoes. Actions that people who live in hurricane-prone areas should take in order to prepare for future hurricanes include: find out if you live in an evacuation area and learn your evacuation routes/shelter locations; assess your risks and update your insurance; know your home's vulnerabilities to wind, storm surge, or flooding and have materials needed to secure your home on hand; find out what types of emergencies might occur, learn how you should respond, and obtain emergency equipment; and obtain water and nonperishable food.

One credit is allowed. Acceptable responses include but are not limited to:

- **Learn about hurricane risks for your area.**
- **Learn safe emergency evacuation routes/shelter locations.**
- **Obtain/check emergency equipment (radio, flashlight, first-aid kit).**
- **Have enough water and nonperishable food.**
- **Make sure to have materials to secure your home (plywood, shatter-resistant glass, hurricane shutters/straps, sandbags).**
- **Update your insurance.**

Note: Credit is not allowed for any action that implies a hurricane is imminent.

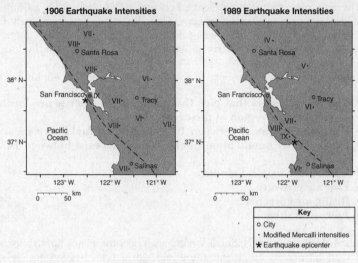

83. Find the key and note the symbol for an earthquake epicenter. Find the epicenter on the 1906 Earthquake Intensities map. From the epicenter, trace horizontally to the left to the vertical latitude scale. Note the latitude: 37°40′ N. From the epicenter, trace vertically downward to the horizontal longitude scale at the bottom of the map. Note the longitude: 122°30′ W. Thus, the 1906 earthquake occurred at coordinates 37°40′ N latitude, 122°30′ W longitude. Repeat these steps for the epicenter of the 1989 earthquake. Note that the 1989 earthquake occurred at coordinates 37° N, 121°50′ W. Find the Tectonic Plates map in the *Reference Tables for Physical Setting/Earth Science*. Locate the coordinates for the epicenters of these two earthquakes on the map. Note that their locations correspond to the San Andreas Fault. Note the bold black arrows pointing in opposite directions on either side of the fault showing that the plates are sliding horizontally past each other. Locate this symbol in the key at the bottom of the map. Note that it corresponds to a transform plate boundary (transform fault). Thus, both of these earthquakes occurred along the San Andreas Fault, which is a transform plate boundary.

One credit is allowed for **San Andreas Fault** *and* an acceptable plate tectonic boundary. Acceptable boundaries include but are not limited to:

- **Transform boundary/transforming**
- **Plates slide horizontally past each other.**

Modified Mercalli Intensity Scale

Level of Intensity	IV	V	VI	VII	VIII	IX
Perceived shaking	light	moderate	strong	very strong	severe	violent
Observed damage	none	very light	light	moderate	moderate to heavy	heavy

84. On the 1906 Earthquake Intensities map, locate San Francisco. Note that an earthquake intensity of IX was recorded in San Francisco. Locate IX on the Modified Mercalli Intensity Scale. Trace downward. Note that when the intensity level is IX, the perceived shaking is "violent" and the observed damage is "heavy."

One credit is allowed if *both* responses are correct:

- **Perceived shaking: violent**
- **Observed damage: heavy**

85. Find the cities labeled "Santa Rosa" and "Salinas" on the 1989 Earthquake Intensities map. Refer to the key, and note the symbol for an earthquake epicenter. Then locate the epicenter of the 1989 earthquake on the map. Note that during the 1989 earthquake, Salinas was located closer to the epicenter than Santa Rosa. When an earthquake occurs, seismic waves travel outward in all directions from the epicenter. As seismic waves travel through Earth, they lose energy. Thus, when seismic waves reach a city close to the epicenter, they have more energy than when they reach a city farther from the epicenter. That is why on the maps, you will note that as distance from the epicenter increases, the intensity decreases. Therefore, Santa Rosa experienced a lower Modified Mercalli intensity shaking than Salinas experienced during the 1989 earthquake because Santa Rosa was farther from the epicenter.

One credit is allowed for an acceptable response. Acceptable responses include but are not limited to:

- **Santa Rosa was farther from the 1989 earthquake epicenter.**
- **Earthquake waves lose energy as they travel outward from the epicenter.**
- **As distance from the epicenter increases, intensity decreases.**
- **Salinas was closer to the origin of the earthquake.**

Note: All responses must correctly refer to the earthquake epicenter or earthquake origin in order to receive credit.

SELF-ANALYSIS CHART August 2014

Topic	Question Numbers (Total)	Wrong Answers (x)	Grade
Standards 1, 2, 6, and 7: Skills and Application			
Skills			
Standard 1 Analysis, Inquiry, and Design	2, 3, 8, 9, 11–23, 31, 35–38, 41, 43–46, 48, 49, 51, 53–59, 62, 66, 67, 69, 70–81, 83, 85		$\frac{100(55-x)}{55} = \%$
Standard 2 Information Systems			$\frac{100(0-x)}{0} = \%$
Standard 6 Interconnectedness, Common Themes	4, 6, 7, 14, 15, 19, 22, 24–36, 38–40, 42–48, 50–56, 58, 60, 61, 63–65, 68, 77–81, 83–85		$\frac{100(52-x)}{52} = \%$
Standard 7 Interdisciplinary Problem Solving	82		$\frac{100(1-x)}{1} = \%$
Standard 4: The Physical Setting/Earth Science			
Astronomy			
The Solar System (MU 1.1a, b; 1.2d)	4, 27–29, 36, 37, 67, 72		$\frac{100(8-x)}{8} = \%$
Earth Motions and Their Effects (MU 1.1c, d, e, f, g, h, i)	5, 30, 63–65		$\frac{100(5-x)}{5} = \%$
Stellar Astronomy (MU 1.2b)	2, 76		$\frac{100(2-x)}{2} = \%$
Origin of Earth's Atmosphere, Hydrosphere, and Lithosphere (MU 1.2e, f, h)			$\frac{100(0-x)}{0} = \%$
Theories of the Origin of the Universe and Solar System (MU 1.2a, c)	1, 3, 43, 73–75		$\frac{100(6-x)}{6} = \%$

SELF-ANALYSIS CHART August 2014

Topic	Question Numbers (Total)	Wrong Answers (x)	Grade
Meteorology			
Energy Sources for Earth Systems (MU 2.1a, b)	57		$\dfrac{100(1-x)}{1} = \%$
Weather (MU 2.1c, d, e, f, g, h)	9–12, 31–33, 38–40, 70, 82		$\dfrac{100(12-x)}{12} = \%$
Insolation and Seasonal Changes (MU 2.1i; 2.2a, b)	6, 66, 68, 69, 81		$\dfrac{100(5-x)}{5} = \%$
The Water Cycle and Climates (MU 1.2g; 2.2c, d)	7, 8, 13–15, 49, 50, 55, 71		$\dfrac{100(9-x)}{9} = \%$
Geology			
Minerals and Rocks (MU 3.1a, b, c)	23, 45, 53, 77–80		$\dfrac{100(7-x)}{7} = \%$
Weathering, Erosion, and Deposition (MU 2.1s, t, u, v, w)	25, 26, 47, 48, 56, 58		$\dfrac{100(6-x)}{6} = \%$
Plate Tectonics and Earth's Interior (MU 2.1j, k, l, m, n, o)	16, 20, 21, 35, 41, 42, 83–85		$\dfrac{100(9-x)}{9} = \%$
Geologic History (MU 1.2i, j)	17–19, 44, 46, 51, 52, 54		$\dfrac{100(8-x)}{8} = \%$
Topographic Maps and Landscapes (MU 2.1p, q, r)	22, 24, 34, 59–62		$\dfrac{100(7-x)}{7} = \%$
ESRT			
Earth Science Reference Tables 2011 Edition	2, 3, 8, 9, 11–14, 16, 17, 19–22, 36, 38, 43–46, 48, 53, 57, 58, 62, 72–75, 77–80, 83		$\dfrac{100(34-x)}{34} = \%$

To further pinpoint your weak areas, use the Topic Outline in the front of the book.
MU = Major Understanding (see Topic Outline)

Examination June 2015
Physical Setting/Earth Science

PART A
Answer all questions in this part.

Directions (1–35): For *each* statement or question, choose the word or expression that, of those given, best completes the statement or answers the question. Some questions may require the use of the *2011 Edition Reference Tables for Physical Setting/Earth Science*. Record your answers in the space provided.

1 Compared to the terrestrial planets, the Jovian planets are
   (1) larger and less dense
   (2) smaller and more dense
   (3) closer to the Sun and less rocky
   (4) farther from the Sun and more rocky   1 _____

2 Earth, the Sun, and billions of stars are contained within
   (1) a single constellation
   (2) the Milky Way galaxy
   (3) the solar system
   (4) a giant cloud of gas   2 _____

3 The diagram below represents a globe that is spinning to represent Earth rotating. The globe is spinning in the direction indicated by the arrow. Points A, B, C, D, X, and Y are locations on the globe.

A student attempted to draw a straight line from point X to point Y on the spinning globe. Due to the Coriolis effect, the student's drawn line most likely passed through point

(1) A (3) C
(2) B (4) D 3 ___

4 Flash flooding is most likely to occur when heavy rain falls on

(1) deforested landscapes with clay soils
(2) deforested landscapes with sandy soils
(3) forested landscapes with clay soils
(4) forested landscapes with sandy soils   4 ___

5 Radioactive decay of ^{40}K atoms in an igneous rock has resulted in a ratio of 25 percent ^{40}K atoms to 75 percent ^{40}Ar and ^{40}Ca atoms. How many years old is this rock?

(1) 0.3×10^9 y
(3) 2.6×10^9 y
(2) 1.3×10^9 y
(4) 3.9×10^9 y

6 A student using a sling psychrometer measured a wet-bulb temperature of 10°C and a dry-bulb temperature of 16°C. What was the dewpoint?

(1) –10°C
(3) 6°C
(2) 45°C
(4) 4°C

7 Most of the hurricanes that affect the east coast of the United States originally form over the

(1) warm waters of the Atlantic Ocean in summer
(2) warm land of the southeastern United States in summer
(3) cool waters of the Atlantic Ocean in spring
(4) cool land of the southeastern United States in spring

8 The ozone layer protects life on Earth by absorbing harmful ultraviolet radiation. The ozone layer is located between 17 kilometers and 35 kilometers above Earth's surface in which atmospheric temperature zone?

(1) troposphere
(3) mesosphere
(2) stratosphere
(4) thermosphere

9 Which weather map symbol is associated with extremely low air pressure?

10 Which two elements make up the greatest percentages by mass in Earth's crust?

(1) oxygen and potassium
(2) oxygen and silicon
(3) aluminum and potassium
(4) aluminum and silicon   10 _____

11 What is the approximate P-wave travel time from an earthquake if the P-wave arrives at the seismic station 8 minutes before the S-wave?

(1) 4 minutes 20 seconds
(2) 6 minutes 30 seconds
(3) 10 minutes 0 seconds
(4) 11 minutes 20 seconds   11 _____

12 Which two factors have the most influence on the development of landscape features?

(1) bedrock age and weathering rates
(2) bedrock structure and climate variations
(3) rate of deposition and thickness of the bedrock
(4) rate of erosion and fossils in the bedrock   12 _____

13 Which two New York State landscape regions have surface bedrock that formed about 1000 million years ago?

(1) Hudson Highlands and Adirondack Mountains
(2) Erie-Ontario Lowlands and Atlantic Coastal Plain
(3) Tug Hill Plateau and Allegheny Plateau
(4) Newark Lowlands and Manhattan Prong   13 _____

14 What is the minimum water velocity necessary to maintain movement of 0.1-centimeter-diameter particles in a stream?

(1) 0.02 cm/s (3) 5.0 cm/s
(2) 0.5 cm/s   (4) 20.0 cm/s

15 Compared to a light-colored rock with a smooth surface, a dark-colored rock with a rough surface will

(1) both absorb and reflect less insolation
(2) both absorb and reflect more insolation
(3) absorb less insolation and reflect more insolation
(4) absorb more insolation and reflect less insolation

16 The diagram below indicates regions of daylight and darkness on Earth on the first day of summer in the Northern Hemisphere. Four latitudes are labeled A, B, C, and D.

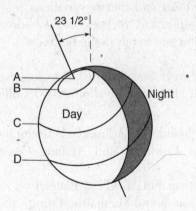

At which latitude is the Sun above the horizon for the *least* number of hours on the day shown?

(1) A   (3) C
(2) B   (4) D

17 Which process is responsible for the greatest loss of energy from Earth's surface into space on a clear night?

(1) condensation (3) radiation
(2) conduction   (4) convection   17 _____

18 The timeline below represents time from the present to 20 billion years ago. Letters A, B, C, and D represent specific times.

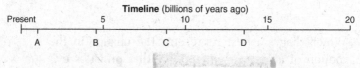

Which letter on the timeline best represents the time when scientists estimate that the Big Bang occurred?

(1) A   (3) C
(2) B   (4) D   18 _____

19 The diagram below represents a Foucault pendulum that is swinging back and forth.

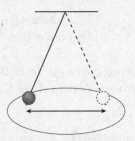

Which diagram best represents the change in the motion of a Foucault pendulum that provides evidence of Earth's rotation?

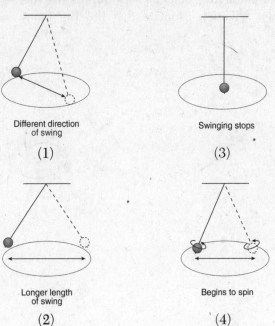

Different direction of swing
(1)

Swinging stops
(3)

Longer length of swing
(2)

Begins to spin
(4)

19 ___

20 The diagram below represents some constellations and one position of Earth in its orbit around the Sun. These constellations are visible to an observer on Earth at different times of the year.

(Not drawn to scale)

When Earth is located in the orbital position shown, two constellations that are both visible to an observer on Earth at midnight are

(1) Libra and Virgo
(2) Gemini and Taurus
(3) Aquarius and Capricorn
(4) Cancer and Sagittarius

21 The cross sections below represent three beakers that were used to test porosity. Beakers A, B, and C each contain a different size of bead. Each beaker holds an equal volume of beads. The amount of water needed to fill the total pore space between the beads in each beaker was measured.

Which statement best describes the porosity that was found for these three samples?

(1) A had a greater porosity than B and C.
(2) B had a greater porosity than A and C.
(3) C had a greater porosity than A and B.
(4) All three samples had the same porosity.   21 _____

22 Which graph best indicates the general relationship between soil particle size and the amount of water retention by a permeable soil?

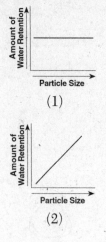
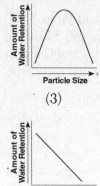

22 _____

23 The cross section below represents surface bedrock where faulting has occurred along line *AB*.

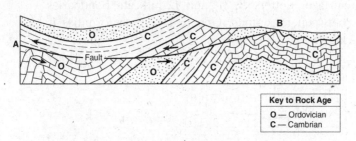

When could this faulting have occurred?

(1) before the Ordovician rocks were deposited
(2) during the Ordovician period
(3) before the Cambrian rocks were deposited
(4) during the Cambrian period   23 _____

24 The two block diagrams below represent the formation of caves.

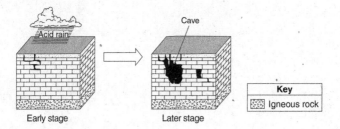

Which types of weathering and erosion are primarily responsible for the formation of caves?

(1) chemical weathering and groundwater flow
(2) chemical weathering and runoff
(3) physical weathering and groundwater flow
(4) physical weathering and runoff   24 _____

25 The cross section below represents several rock units within Earth's crust. Letter *A* represents Earth's surface. Letters *B*, *C*, and *D* indicate boundaries between rock units. One of the unconformities is labeled.

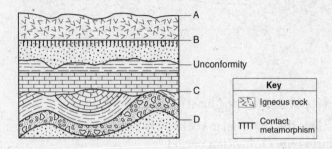

Which lettered boundary is most likely another unconformity?

(1) *A* (3) *C*
(2) *B* (4) *D*   25 ___

26 The map below shows the distribution of ash across the United States as a result of the May 18, 1980 volcanic eruption of Mount St. Helens.

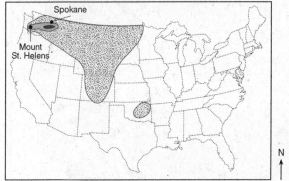

Source: United States Geological Survey, 1990 (adapted)

Volcanic ash deposits such as these are usually excellent geologic time markers because they

(1) occur at regular time intervals
(2) spread over a large area in a short amount of time
(3) represent a time gap in the rock record
(4) contain index fossils from different time periods 26 ____

27 Which diagram represents a side view of a sand dune most commonly formed as a result of the prevailing wind direction shown?

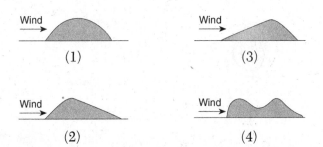

27 ____

28 The map below indicates an air-pressure field over North America. Isobar values are recorded in millibars.

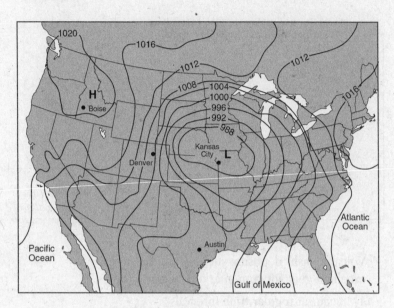

At which city was the greatest wind speed occurring?

(1) Boise
(2) Denver
(3) Kansas City
(4) Austin

28 _____

29 The striped areas on the map below show regions along the Great Lakes that often receive large amounts of snowfall due to lake-effect storms.

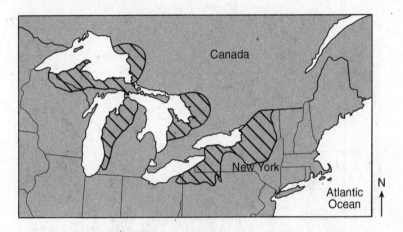

These storms generally develop when

(1) cold air moves to the east over warmer lake water
(2) cold air moves to the west over warmer land regions
(3) warm air moves to the east over colder lake water
(4) warm air moves to the west over colder land regions

29 _____

30 The graph below shows the heating effect that different land uses have on surface air temperatures on a summer afternoon.

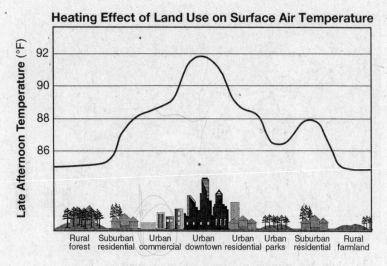

Source: US Global Change Research Program (adapted)

Which land use results in the least heating effect in urban areas?

(1) commercial
(2) downtown
(3) residential
(4) parks

Base your answers to questions 31 and 32 on the flowchart below and on your knowledge of Earth science. The boxes labeled *A* through *G* represent rocks and rock materials. Arrows represent the processes of the rock cycle.

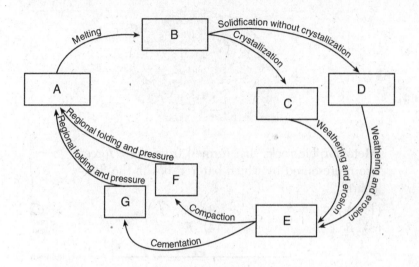

31 Which lettered box could represent the rock conglomerate?

(1) *E* (3) *C*
(2) *G* (4) *D*

31 _____

32 The arrows in the block diagram below represent forces forming mountains in a region of Earth's lithosphere.

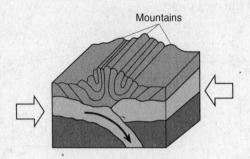

Metamorphic rocks that formed from these forces are represented by which lettered box in the flowchart?

(1) A (3) E
(2) B (4) F   32 ____

---

33 Which rock is composed of the mineral halite that formed when seawater evaporated?

(1) limestone   (3) rock gypsum
(2) dolostone   (4) rock salt   33 ____

34 Which mineral is mined for its iron content?

(1) hematite   (3) galena
(2) fluorite   (4) talc   34 ____

35 The data table below gives characteristics of the gemstone peridot.

Characteristics of Peridot

Luster	nonmetallic
Hardness	6.5
Color	green
Composition	$(Fe,Mg)_2SiO_4$

Peridot is a form of the mineral
(1) pyrite (3) olivine
(2) pyroxene (4) garnet   35 _____

PART B–1

Answer all questions in this part.

Directions (36–50): For *each* statement or question, choose the word or expression that, of those given, best completes the statement or answers the question. Some questions may require the use of the *2011 Edition Reference Tables for Physical Setting/Earth Science*. Record your answers in the space provided.

Base your answers to questions 36 through 38 on the diagrams below and on your knowledge of Earth science. The diagrams represent electromagnetic waves being transmitted (*T*) by a Doppler radar weather instrument and waves being reflected (*R*) by rain showers. This instrument produces computer images that show the movement of rainstorms.

A Stationary Rain Shower

The reflected wavelengths (*R*) from a stationary rain shower are equal to the transmitted wavelengths (*T*).

A Rain Shower Moving Toward the Instrument

The reflected wavelengths (*R*) from a rain shower moving toward the instrument are shorter than the transmitted wavelengths (*T*).

A Rain Shower Moving Away from the Instrument

The reflected wavelengths (*R*) from a rain shower moving away from the instrument are longer than the transmitted wavelengths (*T*).

36 The computer image below shows a rainstorm over Texas. Letters A and B represent locations on Earth's surface.

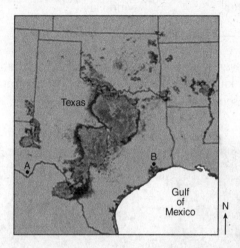

If Doppler radar is used at locations A and B, as this rainstorm moves eastward, reflected wavelengths from this storm will be

(1) shorter at both locations A and B
(2) longer at both locations A and B
(3) shorter at location A and longer at location B
(4) longer at location A and shorter at location B   36 _____

37 This Doppler radar instrument transmits electromagnetic energy in the form of microwaves. Some microwave wavelengths are between the wavelengths of

(1) gamma rays and x rays
(2) infrared and radio waves
(3) ultraviolet and infrared
(4) x rays and ultraviolet   37 _____

38 Which weather instrument was used to measure the amount of rainfall from this storm?

(1) barometer (3) precipitation gauge
(2) anemometer (4) wind vane   38 ____

Base your answers to questions 39 through 42 on the diagram below and on your knowledge of Earth science. The diagram represents two possible sequences in the evolution of stars.

Stages of Star Evolution

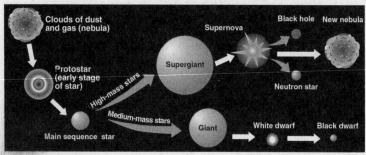

(Not drawn to scale)

39 What causes clouds of dust and gas to form a protostar?

(1) magnetism
(2) gravitational attraction
(3) expansion of matter
(4) cosmic background radiation   39 ____

40 Which property primarily determines whether a giant star or a supergiant star will form?

(1) mass   (3) shape
(2) color   (4) composition   40 ____

41 Which table includes data that are characteristic of the surface temperature and luminosity of some white dwarf stars?

Surface Temperature	5000 K
Luminosity	100

(1)

Surface Temperature	10,000 K
Luminosity	100

(3)

Surface Temperature	5000 K
Luminosity	0.001

(2)

Surface Temperature	10,000 K
Luminosity	0.001

(4)

41 _____

42 Which process generates the energy that is released by stars?
  (1) nuclear fusion
  (2) thermal conduction
  (3) convection currents
  (4) radioactive decay

42 _____

Base your answers to questions 43 through 45 on the passage and cross section below and on your knowledge of Earth science. The cross section represents one theory of the movement of rock materials in Earth's dynamic interior. Some mantle plumes that are slowly rising from the boundary between Earth's outer core and stiffer mantle are indicated.

Hot Spots and Mantle Plumes

Research of mantle hot spots indicates that mantle plumes form in a variety of sizes and shapes. These mantle plumes range in diameter from several hundred kilometers to 1000 kilometers. Some plumes rise as blobs rather than in a continuous streak; however, most plumes are long, slender columns of hot rock slowly rising in Earth's stiffer mantle. One theory is that most plumes form at the boundary between the outer core and the stiffer mantle. They may reach Earth's surface in the center of plates or at plate boundaries, producing volcanoes or large domes.

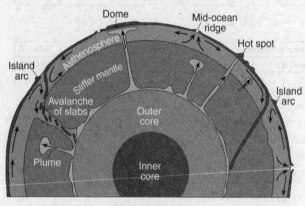

(Not drawn to scale)

43 Compared to the surrounding material, mantle plumes rise toward Earth's surface from the core-mantle boundary because they are

(1) cooler and less dense
(2) cooler and more dense
(3) hotter and less dense
(4) hotter and more dense   43 _____

44 At which depth below Earth's surface is the boundary between Earth's outer core and stiffer mantle located?

(1) 700 km   (3) 2900 km
(2) 2000 km   (4) 5100 km   44 _____

45 The basaltic rock that forms volcanic mountains where mantle plumes reach Earth's surface is usually composed of
(1) fine-grained, dark-colored felsic minerals
(2) fine-grained, dark-colored mafic minerals
(3) coarse-grained, light-colored felsic minerals
(4) coarse-grained, light-colored mafic minerals   45 _____

Base your answers to questions 46 through 50 on the geologic cross section and graph below, and on your knowledge of Earth science. The cross section represents the intrusive igneous rock of the Palisades sill and surrounding bedrock located on the west side of the Hudson River across from New York City. The graph indicates changes in the percentages of the major minerals found in the sill.

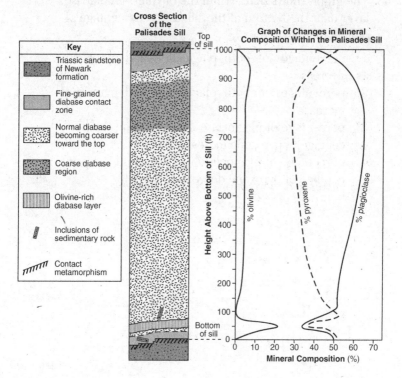

46 The inclusions shown near the bottom of the Palisades sill are pieces of the Triassic sandstone that

(1) formed from deposits of minerals within the sill
(2) crystallized within the sill and were cemented together
(3) were part of the olivine-rich layer that broke apart
(4) broke off from the surrounding bedrock during the intrusion   46 _____

47 Approximately how far above the bottom of the Palisades sill is the coarse diabase region found?

(1) 50 ft   (3) 800 ft
(2) 400 ft   (4) 950 ft   47 _____

48 The graph shows that, within the olivine-rich diabase layer near the bottom of the sill, as the percentage of olivine increases, the

(1) percentages of both plagioclase and pyroxene decrease
(2) percentages of both plagioclase and pyroxene increase
(3) percentage of plagioclase decreases and the percentage of pyroxene increases
(4) percentage of plagioclase increases and the percentage of pyroxene decreases   48 _____

49 The Palisades sill intruded as North America began the process of separating from Africa and Europe as Pangaea was breaking apart. Approximately when did these events occur?

(1) 65 million years ago
(2) 200 million years ago
(3) 299 million years ago
(4) 400 million years ago   49 _____

50 Which two minerals, *not* shown on the Graph of Changes in Mineral Composition Within the Palisades Sill, are also likely to be found in some other samples of diabase?

(1) amphibole and potassium feldspar
(2) potassium feldspar and quartz
(3) quartz and biotite
(4) biotite and amphibole   50 _____

PART B–2
Answer all questions in this part.

Directions (51–65): Record your answers in the spaces provided. Some questions may require the use of the *2011 Edition Reference Tables for Physical Setting/Earth Science*.

51 Describe the effect that global warming most likely will have on *both* present-day glaciers and sea level. [1]

Glaciers: _____

Sea level: _____

Base your answers to questions 52 and 53 on the weather map below and on your knowledge of Earth science. The map indicates the location of a low-pressure system over New York State during late summer. Isobar values are recorded in millibars. Shading indicates regions receiving precipitation. The air masses are labeled mT and cP. The locations of some New York State cities are shown. Points *A* and *B* represent other locations on Earth's surface.

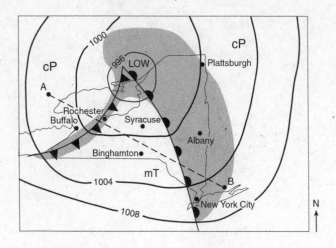

52 An air mass acquires the characteristics of the surface over which it forms. Circle the type of Earth surface (land or ocean) and describe the relative temperature of the surface over which the mT air mass most likely formed. [1]

Circle one: land   ocean

Relative temperature of Earth's surface: _____

53 The cross section below represents the atmosphere along the dashed line from A to B on the map. The warm frontal boundary is already shown on the cross section. Draw a curved line to represent the shape and location of the cold frontal boundary. [1]

Atmospheric Cross Section

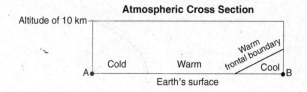

Base your answers to questions 54 through 57 on the diagram below and on your knowledge of Earth science. The diagram represents a time-exposure photograph taken by aiming a camera at *Polaris* in the night sky and leaving the shutter open for a period of time to record star trails. The angular arcs (star trails) show the apparent motions of some stars.

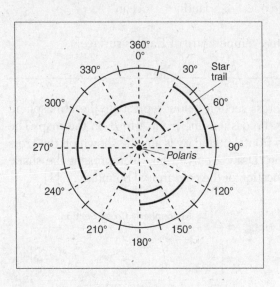

54 Identify the motion of Earth that causes these stars to appear to move in a circular path. [1]

_____

55 Determine the number of hours it took to record the star trails labeled on the diagram. [1]

_____ h

56 The diagram below represents Earth as viewed from space. The dashed line indicates Earth's axis. Some latitudes are labeled. On the diagram below, draw an arrow that points from the North Pole toward *Polaris*. [1]

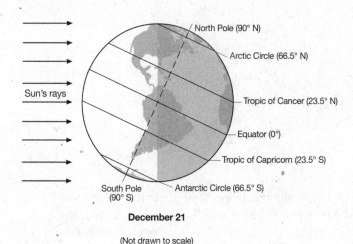

December 21

(Not drawn to scale)

57 Record, to the *nearest whole degree*, the altitude of *Polaris* when it is viewed from the top of New York State's Mt. Marcy. [1]

_____ °

Base your answers to questions 58 through 61 on the block diagram below and on your knowledge of Earth science. The diagram represents a meandering stream flowing into the ocean. Points A and B represent locations along the streambanks. Letter C indicates a triangular-shaped depositional feature where the stream enters the ocean.

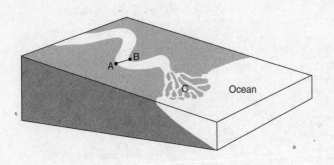

58 The top of the box below represents the stream surface between points A and B. In the box, draw a line from point A to point B to represent a cross-sectional view of the shape of the bottom of the stream channel. [1]

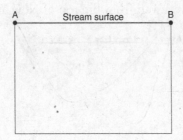

59 Explain how sediments eroded by the water in this stream become smoother and rounder in shape. [1]

_____

_____

60 Identify the triangular-shaped depositional feature indicated by letter *C*. [1]

_____

61 Identify *two* factors that determine the rate of stream erosion. [1]

_____ and _____

_____

Base your answers to questions 62 through 65 on the graph and map below and on your knowledge of Earth science. The average monthly temperatures for Eureka, California, and Omaha, Nebraska, are plotted on the graph. The map indicates the locations of these two cities.

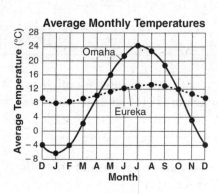

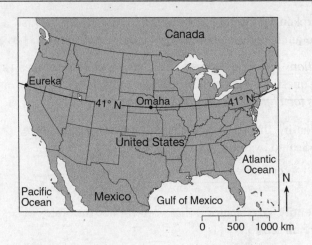

62 Calculate the rate of change in the average monthly temperature for Omaha during the two-month period between October and December, as shown on the graph. [1]

_____ C°/month

63 Explain why Omaha, which is farther inland, has a greater variation in temperatures throughout the year than Eureka, which is closer to the ocean. [1]

64 Identify the month with the greatest difference in the average temperature between these two cities. [1]

65 Identify the surface ocean current that affects the climate of Eureka. [1]

_____ Current

PART C
Answer all questions in this part.

Directions (66–85): Record your answers in the spaces provided. Some questions may require the use of the *2011 Edition Reference Tables for Physical Setting/Earth Science*.

66 The cross section below represents the windward and leeward sides of a mountain range. Arrows show the movement of air over a mountain. Points X and Y represent locations on Earth's surface.

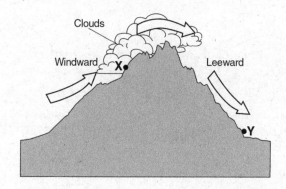

Describe how the air's temperature and water vapor content at point X is different from the air's temperature and water vapor content at point Y. [1]

Air temperature at X: _____

Water vapor content at X: _____

Base your answers to questions 67 through 71 on the topographic map below and on your knowledge of Earth science. Point A represents a location on Earth's surface. Lines BC and XY are reference lines on the map. Points D, E, F, and G represent locations along Coe Creek. Elevations are shown in feet.

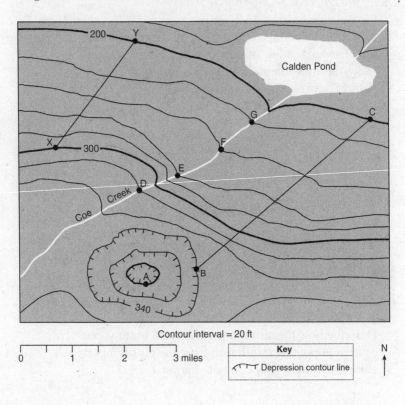

67 What is the elevation of location A? [1]

_____ ft

68 On the grid below, construct a topographic profile of the land surface along the line from point *B* to point *C*. Plot the elevation of *each* contour line that crosses line *BC*. Connect *all nine* plots with a line to complete the profile. [1]

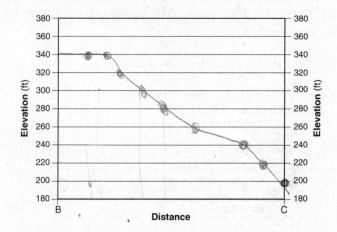

69 Describe the evidence shown on the map that indicates Coe Creek flows toward the northeast. [1]

_____

70 Describe how the contour lines indicate that Coe Creek flows faster between locations *D* and *E* than between locations *F* and *G*. [1]

_____

_____

71 Calculate the gradient along line *XY*. Label your answer with the correct units. [1]

_____

Base your answers to questions 72 through 76 on the modified Mercalli scale of earthquake intensity below, on the map of Japan below the chart, and on your knowledge of Earth science. The modified Mercalli scale classifies earthquake intensity based on observations made during an earthquake. The map indicates the modified Mercalli scale intensity values recorded at several locations in Japan during the March 11, 2011 earthquake, which triggered destructive tsunamis in the Pacific Ocean.

Modified Mercalli Scale of Earthquake Intensity

Intensity Value	Description of Effects
I	Not felt except by a very few under especially favorable conditions.
II	Felt only by a few persons at rest, especially on upper floors of buildings.
III	Felt quite noticeably by persons indoors, especially on upper floors of buildings. Many people do not recognize it as an earthquake. Parked cars may rock slightly. Vibrations similar to the passing of a truck.
IV	Felt indoors by many, outdoors by few during the day. At night, some awakened. Dishes, windows, doors disturbed; walls make cracking sound. Sensation like heavy truck striking building. Parked cars rocked noticeably.
V	Felt by nearly everyone; many awakened. Some dishes, windows broken. Unstable objects overturned. Pendulum clocks may stop.
VI	Felt by all, many frightened. Some heavy furniture moved; a few instances of fallen plaster. Damage slight.
VII	Damage minimal in buildings of good design and construction; slight to moderate in well-built ordinary structures; considerable damage in poorly built or badly designed structures; some chimneys broken.
VIII	Damage slight in specially designed structures; considerable damage with partial collapse in ordinary substantial buildings. Damage great in poorly built structures. Fall of chimneys, factory stacks, columns, monuments, walls. Heavy furniture overturned.
IX	Damage considerable in specially designed structures; well-designed frame structures tilted. Damage great in substantial buildings, with partial collapse. Buildings shifted off foundations.
X	Most masonry and frame structures and foundations are destroyed. Train rails bent.
XI	Few, if any, structures remain standing. Bridges destroyed. Train rails bent greatly.
XII	Damage total. Objects thrown into the air.

March 11, 2011 Earthquake in Japan

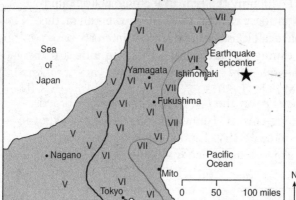

72 On the map above, a line has been drawn to separate regions with Mercalli values of V from regions with Mercalli values of VI. Draw *another* line to separate regions with Mercalli values of VI from regions with Mercalli values of VII. [1]

73 Some observations that might be made during an earthquake according to the modified Mercalli scale are listed below. Place a check mark (✓) in the box if that observation most likely was recorded at Yamagata during the March 11, 2011, earthquake. More than one box may be checked. [1]

☐	parked cars rock
☐	dishes and windows broken
☐	felt by all persons
☐	some heavy furniture moved
☐	chimneys and monuments fall
☐	buildings shifted off foundations

74 The epicenter of this earthquake was located at 38° N 142° E. Identify the type of tectonic plate boundary that is located nearest to the epicenter of this earthquake. [1]

_____

75 Describe *one* way the *P*-waves and *S*-waves recorded on seismograms at Ishinomaki and Nagano were used to indicate that Ishinomaki was closer to the earthquake epicenter than was Nagano. [1]

_____

_____

76 A 25-foot high tsunami hit the Japanese city of Ishinomaki. Describe a precaution the city could take now to protect citizens from tsunamis in future years. [1]

_____

_____

Base your answers to questions 77 and 78 on the diagram below, which represents an exaggerated model of the shape of Earth's orbit, and on your knowledge of Earth science. The positions of Earth in its orbit on December 21 and June 21 are indicated. The positions of perihelion (when Earth is closest to the Sun) and aphelion (when Earth is farthest from the Sun) are also indicated. Both perihelion and aphelion occur approximately two weeks after the dates shown.

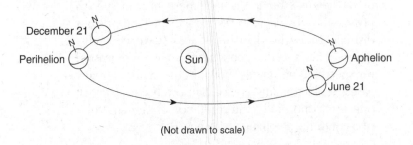

(Not drawn to scale)

77 How many months after Earth's perihelion position does Earth's aphelion position occur? [1]

_____ **months**

78 Explain why warm summer temperatures occur in New York State when Earth is at aphelion. [1]

Base your answers to questions 79 through 82 on the passage and chart below, and on your knowledge of Earth science. The chart identifies some human species and the times when they are believed to have existed.

Human Species

Modern humans, *Homo sapiens*, appear to have evolved through several species of earlier members of the genus *Homo*. Each of these human species possessed specific features that made that species distinct. Many lived in (or at least have been discovered in) specific geographic areas, and existed for specific time ranges shown in the chart. In many cases, fossil remains are partial, often consisting of only teeth and skulls. Interpretation of human evolution continues to change with new discoveries.

Human Species Distributed Through Time

Human Species	Time of Existence from Fossil Evidence (million years ago)
Homo sapiens	0.25 to the present
Homo neanderthalensis	0.35 to 0.03
Homo rhodesiensis	0.6 to 0.1
Homo heidelbergensis	0.6 to 0.3
Homo mauritanicus	1.2 to 0.6
Homo erectus	1.5 to 0.2
Homo ergaster	1.8 to 1.25
Homo habilis	2.25 to 1.4

79 Complete the graph below by drawing a bar to represent the time span that *each* human species existed. The bars for the first four species listed have already been drawn. [1]

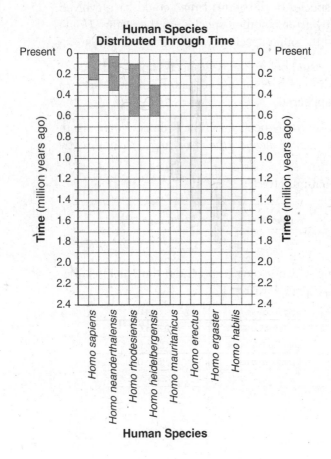

80 Which human species shown in the chart was the first to exist? [1]

_____

81 One species of the genus *Homo* could have evolved directly from another species of the genus *Homo* only if the other species:

- existed before the new species appeared
- did not become extinct before the new species appeared

Identify *two* species of the genus *Homo* from which *Homo neanderthalensis* may have directly evolved. [1]

Homo _____

and *Homo* _____

82 During which geologic epoch did the *Homo mauritanicus* species exist? [1]

_____ **Epoch**

Base your answers to questions 83 through 85 on the diagrams and tables below and on your knowledge of Earth science. Each diagram represents the Moon's orbital position and each table lists times of high and low tides and tide heights, in meters, at New York City for the date shown.

Moon's Orbital Position and Tide Data on May 13

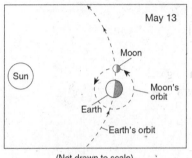

Tide	Time	Height (m)
high	12:59 a.m.	1.92
low	7:15 a.m.	0.37
high	1:32 p.m.	2.07
low	7:59 p.m.	0.27

Moon's Orbital Position and Tide Data on May 20

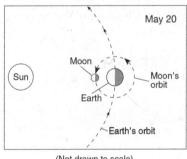

Tide	Time	Height (m)
low	1:22 a.m.	0.06
high	7:50 a.m.	2.47
low	2:10 p.m.	0.09
high	8:10 p.m.	2.21

83 Determine the length of time between the two high tides shown for May 13. [1]

_____ h _____ min

84 On the diagram below, shade the portion of the Moon that is in darkness to observers in New York City on May 13. [1]

85 On the diagram below, place an **X** on the Moon's orbit to represent the location of the Moon on May 28. [1]

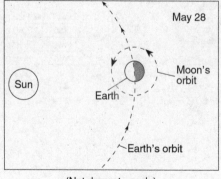

(Not drawn to scale)

Answers June 2015
Physical Setting/Earth Science

Answer Key

PART A

1. 1	8. 2	15. 4	22. 4	29. 1
2. 2	9. 4	16. 4	23. 2	30. 4
3. 2	10. 2	17. 3	24. 1	31. 2
4. 1	11. 3	18. 4	25. 3	32. 1
5. 3	12. 2	19. 1	26. 2	33. 4
6. 4	13. 1	20. 2	27. 3	34. 1
7. 1	14. 3	21. 4	28. 2	35. 3

PART B–1

36. 4	39. 2	42. 1	45. 2	48. 1
37. 2	40. 1	43. 3	46. 4	49. 2
38. 3	41. 4	44. 3	47. 3	50. 4

PART B–2 and **PART C.** *See* **Answers Explained**.

Answers Explained

PART A

1. **1** The planets can be divided by size, density, and composition into the terrestrial (Earthlike) and Jovian (Jupiter-like) planets. The terrestrial planets are small, dense, and rocky. They are composed mostly of metal silicates and iron. The terrestrial planets include the four innermost planets: Mercury, Venus, Earth, and Mars. The Jovian planets are large, low in density, and gaseous. They are composed mainly of hydrogen and helium. The Jovian planets include the outer planets: Jupiter, Saturn, Neptune, and Uranus.

Find the Solar System Data chart in the *Reference Tables for Physical Setting/Earth Science*. Locate the column labeled "Density (g/cm^3)." Note that the density of the terrestrial planets ranges from 3.9 g/cm^3 to 5.5 g/cm^3 while the density of the Jovian planets ranges from 0.7 g/cm^3 to 1.8 g/cm^3. Thus, the Jovian planets have lower densities than the terrestrial planets.

Now locate the column labeled "Equatorial Diameter (km)." Note that the diameter of the terrestrial planets ranges from 4,879 km to 12,756 km while the diameter of the Jovian planets ranges from 49,528 km to 142,984 km. Thus, the Jovian planets are larger than the terrestrial planets. Therefore, compared to the terrestrial planets, the Jovian planets are larger and less dense.

2. **2** Our solar system consists of eight planets (one of which is Earth), fifty or so moons, and numerous smaller objects that orbit the Sun. Thus, Earth and the Sun are contained within our solar system. By definition, a system containing billions of stars is a galaxy. Our Sun is one of the billions of stars contained within the Milky Way galaxy. Thus, Earth, the Sun and billions of stars are contained within the Milky Way galaxy.

3. **2** The Coriolis effect is the apparent deflection of a moving object from a straight-line path due to Earth's rotation. Since Earth rotates from west to east, the deflection is to the right in the Northern Hemisphere and

to the left in the Southern Hemisphere. Note that point X lies in the globe's Northern Hemisphere. Thus, the student's drawn line will be deflected to the right of the straight-line direction from X to Y. In other words, if the student was standing at point X facing toward point Y, the line would be deflected to the student's right. So the student's drawn line will most likely pass through point B.

4. **1** Flash flooding is the rapid flooding of low-lying areas due to heavy rainfall that runs off rather than evaporating or infiltrating. The roots and stems of surface vegetation in forested landscapes slow the downhill flow of water. This allows more time for the water to infiltrate, thereby decreasing the amount of runoff and making flash flooding less likely. Therefore, flash flooding is more likely to occur when heavy rain falls on a deforested landscape.

The amount of, and rate at which, water sinks into the ground depends on the porosity and permeability of the surface materials. These factors, in turn, depend on the soil type and composition. When comparing soil porosity, the more porous soil has more space into which water can infiltrate. As a result on more porous soil, more water will sink into the ground rather than running off. When comparing soil permeability, the more permeable soil allows the water to sink in faster, resulting in less runoff. All other factors being equal, particle size does not affect porosity. The smaller size of the pore spaces between small particles is offset by the greater number of pore spaces. However, particle size does affect permeability—the rate at which water infiltrates (seeps into) soil. As soil particles become smaller, the empty spaces between the soil particles also become smaller. Small spaces restrict the flow of water through soil. Large soil particles have larger spaces between them, which allows water to seep into and flow through the soil more rapidly. Thus, water will infiltrate soils made of small particles more slowly, thereby increasing the amount of runoff and making flash flooding more likely. Find the Relationship of Transported Particle Size to Water Velocity graph in the *Reference Tables for Physical Setting/Earth Science*. Note the sediments named along the right side of the chart. Trace left from each sediment to the vertical scale labeled "Particle Diameter." Note that the sediments increase in size from the bottom to the top of the chart and that clay is smaller in diameter than sand. Therefore, heavy rain will infiltrate clay soils more slowly than sandy soils, thereby increasing runoff and making flash flooding more likely in landscapes with clay soils. Thus, flash flooding is most likely to occur when heavy rain falls on deforested landscapes with clay soils.

5. **3** Find the Radioactive Decay Data chart in the *Reference Tables for Physical Setting/Earth Science*. Note that potassium-40 has a half-life of 1.3×10^9 years (1.3 billion years). In 1.3 billion years, half of the original potassium-40, or 50%, would remain. After another 1.3 billion years, half of that 50%, or 25% of the original potassium-40, would remain. Thus, an igneous rock that contains 25% of its original potassium-40 and 75% of the decay

products argon-40 and calcium-40 would have undergone 2 half-lives and would be 2.6 billion years (2.6×10^9 y) old.

6. **4** Find the Dewpoint Temperatures chart in the *Reference Tables for Physical Setting/Earth Science*. The difference between the dry-bulb and wet-bulb temperatures is 16°C – 10°C, or 6°C. At the top of the chart, find the column that shows a 6°C difference between wet-bulb and dry-bulb temperatures. Along the left side of the chart, find the row that shows a 16°C dry-bulb temperature. Where the row and column meet, read the dewpoint temperature: 4°C.

7. **1** Most hurricanes develop in late summer when ocean waters reach temperatures of 27°C or higher. The high temperature and moisture levels result in air with a very low density. This wet, warm air rises rapidly, reducing pressure near the surface and encouraging a rapid inflow of air. This process results in numerous convection cells and thunderstorms. With very little wind shear, widespread thunderstorm activity can merge into a large updraft that is part of a huge convection cell. The rising air further decreases the pressure. Latent heat released when the moisture in the air condenses also decreases the pressure. These processes create a region of very low pressure. Winds begin to blow toward the very low-pressure center as air moves in from surrounding areas of higher pressure. The Coriolis effect deflects the winds, and a cyclone forms. Heat is released when fresh moisture brought in by winds condenses, feeding the convection cell with new energy. As a result, the convection cell gets larger and stronger. When the winds in the cyclone exceed 119 km/h, it is called a hurricane.

Hurricanes tend to form during summer months over oceans near the equator when the ocean surface is the warmest. Once formed, heat and moisture from the ocean provide the energy that maintains the hurricane. The area over the Atlantic Ocean between 10° N and 20° N latitude is located in the tropics near the equator. Thus, by late summer and early fall, the ocean surface temperature has peaked and evaporation rates are high. So the air in this region is very warm and moist—ideal conditions for the formation of hurricanes. Therefore, most of the hurricanes that affect the east coast of the United States originally form over the warm waters of the Atlantic Ocean in summer.

8. **2** Find the Selected Properties of Earth's Atmosphere chart in the *Reference Tables for Physical Setting/Earth Science*. Locate 17–35 kilometers on the altitude scale (km are listed on the left-hand side of the scale). Trace right to the section labeled "Temperature Zones," and note that this altitude range corresponds to the layer of the atmospheric temperature zone labeled "stratosphere." Thus, the ozone layer is located in the stratosphere.

9. **4** Find the Key to Weather Map Symbols in the *Reference Tables for Physical Setting/Earth Science*. In the section labeled "Present Weather" and in the section to its far right, note that the symbols corresponding to the

choices are (1) fog, (2) haze, (3) smog, and (4) hurricane. Hurricanes are large-scale storms surrounding an area of extremely low pressure. Therefore, the weather map symbol associated with extremely low air pressure is choice (4)—hurricane.

WRONG CHOICES EXPLAINED:
(1) Fog is a cloud whose base is at or near ground level. Fog forms when air at or near ground level is cooled to its dewpoint and moisture in the air condenses to form tiny liquid water droplets suspended in the air. The formation of fog depends on the air temperature relative to the dewpoint and can occur at a range of air pressures. Thus, fog is not necessarily associated with extremely low air pressure.

(2) Haze is a phenomenon in which the clarity of the sky is obscured by dust, smoke, and other dry particles. The presence of these particles in the atmosphere is independent of air pressure. Thus haze is not associated with extremely low air pressure.

(3) Originally smog meant a mixture of smoke and fog. It now means that air has restricted visibility due to the presence of air pollutants. The presence of these pollutants is independent of air pressure. Thus, smog is not associated with extremely low air pressure.

10. **2** Find the Average Chemical Composition of Earth's Crust, Hydrosphere, and Troposphere chart in the *Reference Tables for Physical Setting/Earth Science*. In the section labeled "Crust," locate the column headed "Percent by mass." According to the chart, the two elements that make up the largest percentage by mass of Earth's crust are oxygen (46.10%) and silicon (28.20%).

11. **3** Find the Earthquake P-Wave and S-Wave Travel Time graph in the *Reference Tables for Physical Setting/Earth Science*. It is given that the difference in P-wave and S-wave arrival times is 8 minutes. Hold the straight edge of a piece of scrap paper along the vertical axis labeled "Travel Time (min)," and mark off a distance corresponding to 8 minutes. Keeping this edge of the paper vertical, find the place where this distance just fits between the P-wave and S-wave lines. From this place, trace horizontally left from the P-wave line until you intersect the vertical "Travel Time (min)" axis. Read the P-wave travel time: 10 minutes 0 seconds.

12. **2** A landscape is a region in which the landforms are related by their structure, the processes that formed the landforms, and the way in which the landforms developed. Landscapes are the product of the interaction of geological processes, climate, and human activities acting on the land over a long period of time. Thus, of the choices given, the two factors having the most influence on the development of landscape features are bedrock structure and climate variations.

WRONG CHOICES EXPLAINED:

(1) Bedrock age alone does not determine how a landscape feature was formed, its structure, and the processes that shape it. For example, bedrock of many different ages may have been formed by volcanic activity or by deposition in water. Similarly, weathering rates influence only the exposed surface of a landscape, not how that landscape was formed or its structure.

(3) Although the rate of deposition and thickness of the bedrock are related to how the bedrock was formed, these alone do not determine the overall structure of the landscape nor are they processes that shape the landscape. For example, landscape features that differ greatly, such as mountains and plateaus, may both contain thick layers of bedrock that were deposited at similar rates.

(4) The rate of erosion may affect the rate at which a landscape develops over time. However, the fossils in the bedrock do not influence the landscape's structure, the processes that formed the landscape, or the way in which the landscape developed.

13. **1** Find the Geologic History of New York State chart in the *Reference Tables for Physical Setting/Earth Science*. Locate the time scale along the left edge of the column labeled "Eon." Locate 1000 million years (1 billion) on the scale. Note that it corresponds to the middle and late Proterozoic era. Now find the Generalized Bedrock Geology of New York State map in the *Reference Tables for Physical Setting/Earth Science*. Locate the "Geologic Periods and Eras in New York" key at the bottom of the map. Find the symbol corresponding to "Middle Proterozoic—gneisses, quartzites, and marbles." Note the location of the regions on the map in which this symbol appears. Then find the Generalized Landscape Regions of New York State map in the *Reference Tables for Physical Setting/Earth Science*. Note that Middle Protoerozoic—gneisses, quartzites, and marbles are found in locations that correspond to the Hudson Highlands and the Adirondack Mountains.

14. **3** Find the Relationship of Transported Particle Size to Water Velocity graph in the *Reference Tables for Physical Setting/Earth Science*. Locate 0.1 cm on the vertical axis labeled "Particle Diameter (cm)." Trace horizontally to the right until you intersect the bold curve. At this point, trace down to the horizontal axis labeled "Stream Velocity (cm/s)." Read the minimum stream velocity that can maintain the transportation of a 0.1-centimeter-diameter particle, which is 5.0 cm/s.

15. **4** When insolation strikes an opaque surface, such as a rock, it may be absorbed or reflected. All other factors being equal, dark-colored substances absorb more insolation than light-colored ones. The more insolation that is absorbed, the less that is reflected. Thus, dark-colored surfaces reflect less insolation than light-colored surfaces.

Rough surfaces have a greater surface area through which insolation can be absorbed than smooth surfaces; therefore, rough surfaces absorb more insolation than smooth surfaces. So rough surfaces reflect less insolation than smooth surfaces. Therefore, compared to a light-colored rock with a smooth surface, a dark-colored rock with a rough surface will absorb more insolation and reflect less insolation.

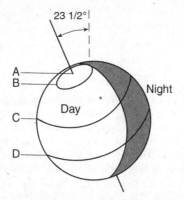

16. **4** The Sun is above the horizon during the daylight hours. Thus, the Sun will be above the horizon the least number of hours at the latitude with the greatest number of hours of darkness. As Earth rotates, observers at each of the four latitudes would travel along a path corresponding to their latitude line and arrive back at their starting position 24 hours later. The number of those 24 hours during which the observers would experience daylight or darkness at any given latitude would correspond to the fraction of the latitude line that extends through daylight and the fraction that extends through darkness. Note that at latitude *D*, the fraction of the latitude line that extends through darkness is the greatest of the four choices. Therefore, at latitude *D* on the day shown, an observer would experience the greatest number of hours of darkness, that is, the least number of hours of daylight. Thus, the Sun would be above the horizon for the least number of hours at latitude *D*.

17. **3** Energy can be transferred from one place to another in three ways: conduction, radiation, and convection. In conduction, fast molecules collide with slow molecules. Some of the kinetic energy of the fast molecules passes to the slow molecules. In convection, a volume of hot fluid (gas or liquid) moves from one region to another, carrying heat energy along with it. In radiation, energy is carried by electromagnetic waves. Energy can be lost from Earth to the atmosphere by conduction and convection because the atmosphere is composed of matter. However, space is nearly a vacuum and does not consist of matter. Thus, the only method by which energy can be transferred from Earth to space must be one that does not involve matter as

a medium. Electromagnetic waves do not require a medium and can travel through space. In radiation, energy is carried by electromagnetic waves that do not require the presence of a medium (solid, liquid, or gas) in order for the energy transfer to occur. So radiation is the primary method by which energy from Earth is lost into space. Thus, the process responsible for the greatest loss of energy from Earth's surface into space on a clear night is radiation.

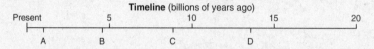

18. **4** The explosion associated with the Big Bang theory and the formation of the universe occurred *before* the formation of Earth and the solar system. In other words, the universe is older than either Earth or the solar system. Find the Geologic History of New York State chart in the *Reference Tables for Physical Setting/Earth Science*. Locate the column labeled "Eon." Note that the entry "Estimated time of origin of Earth and solar system" corresponds to 4600 millions of years ago (4.6 billion years ago) on the time scale. Thus, the explosion associated with the Big Bang theory took place more than 4.6 billion years ago. The most recent estimates are that the Big Bang occurred more than 13 billion years ago. Thus, the letter on the timeline that best represents the time when scientists estimate the Big Bang occurred is *D*.

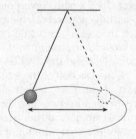

19. **1** In 1851, French physicist Jean Foucault devised a way of conclusively proving Earth's rotation. He used a pendulum. He suspended a heavy iron ball on a long steel wire from the top of the dome of the Pantheon in Paris. As the pendulum swung back and forth, it appeared to change direction, slowly passing over different lines on the floor until eventually coming full circle to its original position after 24 hours. Since he knew that the path of a freely swinging pendulum would not change on its own, Foucault concluded that the apparent shift in the direction of swing of the pendulum was due to the floor (Earth's surface) rotating beneath the pendulum. Thus, the

diagram that best represents the change in the motion of a Foucault pendulum that provides evidence of Earth's rotation is the different direction of swing shown in diagram (1).

20. **2** As Earth revolves around the Sun, the side of Earth facing the Sun experiences day and the side facing away experiences night. Midnight occurs when a location on Earth is directly opposite the Sun. Stars are visible only at night. Therefore, the only constellations visible at midnight to an observer in New York State will be those that are facing the side of Earth directly opposite the Sun on that date. Note that only choice (2), Gemini and Taurus, pairs two constellations that are facing the side of Earth opposite the Sun.

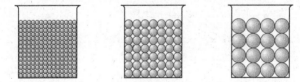

21. **4** Porosity is affected by the shape of the particles, how well sorted they are, and the degree to which they are packed together. However, all other factors being equal, particle size does not affect porosity. The smaller size of the pore spaces between small particles is offset by the greater number of pore spaces. Examine the diagram carefully. Note that the beads in beakers *A*, *B*, and *C* are round, well sorted, and packed together the same way. Therefore, the only difference between the three beakers is the size of the beads. Thus, since particle size does not affect porosity, the statement

that best describes the porosity found in these three samples is that all three samples had the same porosity.

22. **4** Water retention is the amount of water that adheres to the surface of the soil particles, or is retained in the soil, after water has drained out of the pores between the particles. Thus, the amount of water retained by a permeable soil depends on the total surface area of the soil particles. The total surface area of the soil particles increases as particle size decreases as shown below:

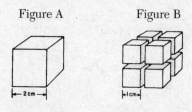

Each of the six sides of the cube in figure A has a surface area of 4 cm² for a total surface area of 24 cm².	Each of the eight small cubes in figure B has a surface area of 6 cm² for a total surface area of 48 cm².

Therefore, as particle size decreases, the amount of water retained by a permeable soil increases. This relationship is best shown by graph (4).

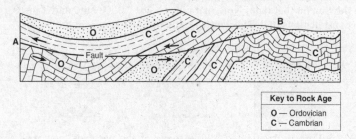

23. **2** Note that rocks of both Cambrian and Ordovician age were displaced along the fault. However, a layer of Ordovician rock at the surface was not displaced along the fault. Therefore, the faulting occurred after both the Cambrian and Ordovician rocks that were displaced were formed but before the Ordovician rock that was not displaced formed. Thus, this faulting could have occurred during the Ordovician period.

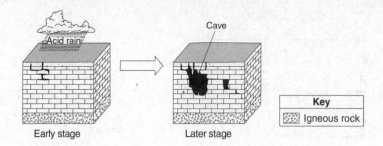

Early stage   Later stage

24. **1** Find the Scheme for Sedimentary Rock Identification in the *Reference Tables for Physical Setting/Earth Science*. In the column labeled "Map Symbol," locate the symbol corresponding to the rock layer in which the caves are located in the block diagram. From this symbol, trace left to the column labeled "Rock Name." Note that the rock layer in which the caves are located is limestone. From limestone, trace left to the column labeled "Composition." Note that limestone is composed of calcite.

The first cross section shows cracks in the limestone through which the acid rain can seep into the limestone. Now, find the Properties of Common Minerals chart in the *Reference Tables for Physical Setting/Earth Science*. In the column labeled "Mineral Name," locate calcite. Trace left to the column labeled "Distinguishing Characteristics." Note that calcite bubbles with acid. The bubbling of calcite with acid indicates the formation of a new substance, the gas that causes the bubbles to form. A change that results in the formation of a new substance is called a chemical change. Therefore, a chemical change occurs when limestone is exposed to acidic groundwater. Chemical weathering processes break down rock by changing its chemical composition. Thus, the reaction of limestone with acidic groundwater is a chemical weathering process.

The limestone bedrock in the block diagrams is composed mainly of calcite. Calcite reacts with acid to produce substances that dissolve in groundwater. Therefore, acid in groundwater that seeps through limestone bedrock may almost completely dissolve the limestone, leaving large holes or caves in the bedrock. Thus, the types of weathering and erosion primarily responsible for the formation of caves are chemical weathering and groundwater flow.

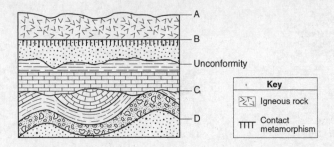

25. **3** Layers of rock are generally deposited in an unbroken sequence. However if forces within Earth uplift rocks, deposition ceases. Erosion may wear away layers of rock before the land surface is low enough for another layer to be deposited. The uplift may cause rock layers to tilt or fold. The result is an unconformity, a break or gap in the sequence of a series of rock layers. Thus, the rocks above an unconformity are quite a bit younger than those below it. The erosional surface marking the gap between the older and younger rocks is generally drawn as a wavy line.

There are several types of unconformities. *Disconformities* are irregular erosional surfaces between parallel layers of rock. The labeled unconformity is a disconformity. Disconformities occur when deposition stops and layers are eroded but no tilting or folding occurs. These surfaces are not easy to discern and are often found when fossils of very different ages are discovered in adjacent layers. *Nonconformities* are places where sedimentary layers lie on top of igneous or metamorphic rocks and are not metamorphosed in any way. *Angular unconformities* form when rock layers are tilted or folded before being eroded. When new layers are deposited, these new layers form horizontally. The layers below the unconformity are at an angle to those above it. Note that lettered boundary *C* is a wavy line and that the layers above the wavy line are horizontal while those below it are folded. Therefore, lettered boundary *C* is most likely the other unconformity.

WRONG CHOICES EXPLAINED:

(1) Although lettered boundary *A* is wavy, indicating that erosion is taking place, no rock layers have been deposited on top of it. Therefore, no age gap between rock layers yet exists. So there is no unconformity.

(2) Note that at lettered boundary *B*, igneous rock has been deposited on top of sedimentary rock and there has been contact metamorphism. Note, too, that the boundary between the sedimentary and igneous rock is flat, not wavy. Therefore, the layers of sedimentary rock below boundary *B* were not eroded before the igneous rock was deposited. Thus, no age gap exists between the sedimentary and igneous rock. So there is no unconformity.

(4) Although lettered boundary *D* has curves, the layers above and below the boundary are sedimentary and curve parallel to boundary *D* and

to one another. This indicates that they all underwent folding together as a unit, not that there is a gap in the sequence of rock layers. So there is no unconformity.

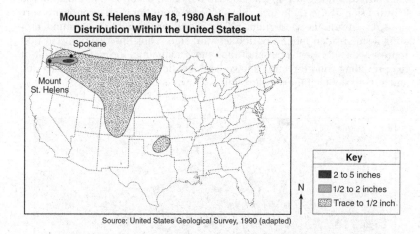

Source: United States Geological Survey, 1990 (adapted)

26. **2** Volcanic ash from a volcano is ejected high into the atmosphere during a volcanic eruption. The ash is carried by global winds, quickly spreading out and then settling to the ground to form a volcanic ash layer over a wide area. Once the volcanic eruption is over, the source of volcanic ash is cut off and deposition quickly tapers off. Thus, a volcanic ash layer is considered an excellent geologic time marker because it is spread over a large area in a short amount of time.

WRONG CHOICES EXPLAINED:
(1) If volcanic ash deposits occurred at regular time intervals, they would be found in rock of many different ages. This would make them poor choices for marking a specific geologic time.

(3) Volcanic ash deposits do not remove rock and do not create a gap in the time record. Instead, volcanic ash deposits add to the rock and time record.

(4) The ash from a volcanic eruption forms directly from molten rock. Therefore, the ash does not contain any fossils. Furthermore, since the ash was deposited over a short period of time, it would not contain index fossils from different time periods.

27. **3** In the cross sections, note that the prevailing wind is from the west (left). Dunes are mounds of sand deposited by wind. A dune often forms when windblown sand meets an obstacle such as a rock or a bush. A dune

develops as follows. On the windward side of the dune (the side facing the wind), sand that strikes an obstacle falls to the ground. In time, the sand forms a mound that *slopes gently* up toward the tip of the obstacle. Wind then pushes sand grains up this slope to the crest, or top of the dune. On the leeward side of the dune (the side sheltered from the wind), winds are very slow. As a result, the sand grains rolling over the crest fall to the ground, creating a steep-sided mound on the leeward side. Thus, the diagram that best represents a side view of a sand dune most commonly formed as a result of the prevailing wind direction shown has a gentle slope on the windward side and a steep slope on the leeward side as shown in choice (3).

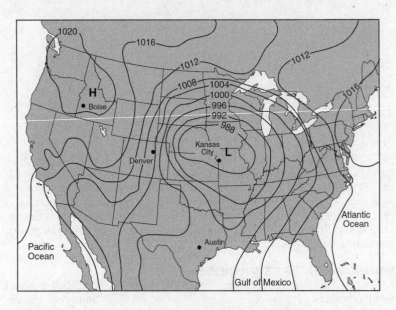

28. **2** Winds blow from regions of high air pressure toward regions of low air pressure. The greater the pressure gradient between the two regions, the higher the wind speed is. The greater the pressure gradient, the more closely spaced the isobars are. Note that the isobars are most closely spaced near Denver. This indicates that the pressure gradient is greatest near Denver. Therefore, the wind speed was greatest in Denver.

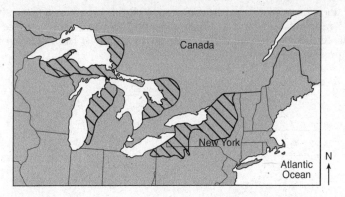

29. **1** Winter lake-effect snowstorms occur when cold winds move across large stretches of warm lake water. Evaporation of the warm lake water adds water vapor to the air. The water vapor is picked up by the cold wind, freezes, and is deposited as snow on the cold land of the leeward shores of the lake. Note that the striped areas indicating regions that receive large amounts of snowfall due to lake-effect storms are located on the eastern shores of the Great Lakes. Thus, the eastern shores of the lake are the leeward shores and the cold winds are moving from west to east over the lakes. Therefore, these lake-effect storms generally develop when cold air moves to the east over warmer lake water.

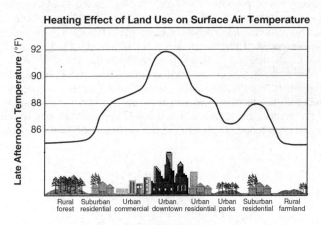

30. **4** From each of the Urban land uses listed along the horizontal axis (urban commercial, urban downtown, urban residential, and urban parks) trace vertically upward to the bold temperature line. From each intersection, trace horizontally to the axis labeled "Late Afternoon Temperature (°F)."

Note the temperatures: urban commercial 88–89, urban downtown 90–92, urban residential 88–89, and urban parks 86.5–87. Thus, the land use that results in the least heating effect in urban areas (lowest late afternoon temperatures) is parks.

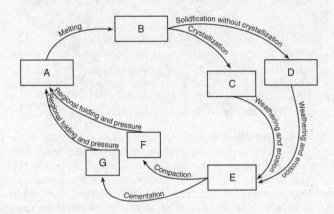

31. **2** Find the Scheme for Sedimentary Rock Identification in the *Reference Tables for Physical Setting/Earth Science*. In the "Rock Name" column, locate "conglomerate." Trace left to the "Grain Size" column. Note that conglomerate is composed of "pebbles, cobbles, and/or boulders embedded in sand, silt, and/or clay." Continue tracing left to the "Texture" column, and note that conglomerate has a "clastic (fragmental)" texture. Clastic sedimentary rocks are composed of fragments of rock or sediments. Find the Rock Cycle in Earth's Crust diagram in the *Reference Tables for Physical Setting/Earth Science*. Note along the arrow leading from "Sediments" to "Sedimentary Rock" the labels "compaction and/or cementation." Thus, two processes that lead directly to the formation of conglomerate are compaction and cementation. Of the choices given, only lettered box *G* is formed by one of these two processes—cementation. Thus, lettered box *G* could represent conglomerate.

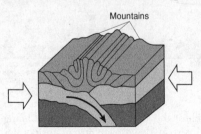

32. **1** Note that the arrows representing the forces forming the mountains are facing toward one another. Forces acting toward one another create

pressure. Note, too, that the pressure has caused the rock layers in the mountains to undergo folding. In the flowchart, locate the arrows labeled "Regional folding and pressure." Both of these arrows lead to lettered box A. Thus, metamorphic rocks that formed from these forces underwent regional folding and pressure and are represented by lettered box A.

33. **4** Find the Scheme for Sedimentary Rock Identification in the *Reference Tables for Physical Setting/Earth Science*. In the "Composition" column, locate "halite." Trace right to the "Comments" column. Note the statement "crystals from chemical precipitates or evaporites." Evaporites are sediments chemically precipitated due to evaporation of an aqueous solution (such as seawater). Trace right to the "Rock Name" column. Note that the rock composed of halite that formed when seawater evaporated is rock salt.

34. **1** Find the Properties of Common Minerals chart in the *Reference Tables for Physical Setting/Earth Science*. In the column labeled "Use(s)," locate the statement "ore of iron." Trace right to the column labeled "Mineral Name." Note that the mineral that is an ore of iron is hematite. An ore is a naturally occurring solid material from which a metal or valuable mineral can be profitably extracted or mined. Thus, the mineral commonly mined for its iron content is hematite.

Characteristics of Peridot

Luster	nonmetallic
Hardness	6.5
Color	green
Composition	$(Fe,Mg)_2SiO_4$

35. **3** Note that the composition of peridot is given as $(Fe,Mg)_2SiO_4$. Find the Properties of Common Minerals chart in the *Reference Tables for Physical Setting/Earth Science*. In the column labeled "Composition," locate "$(Fe,Mg)_2SiO_4$." Trace right to the "Mineral Name" column. Note that the mineral whose composition is $(Fe,Mg)_2SiO_4$ is olivine. From olivine, trace left to the columns labeled "Common Colors," "Hardness," and "Luster." Note that olivine has a green to gray or brown color, a hardness of 6.5, and a nonmetallic luster. According to the data table, these characteristics match those of peridot. Thus, peridot is a form of the mineral olivine.

PART B-1

A Stationary Rain Shower

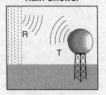

The reflected wavelengths (R) from a stationary rain shower are equal to the transmitted wavelengths (T).

A Rain Shower Moving Toward the Instrument

The reflected wavelengths (R) from a rain shower moving toward the instrument are shorter than the transmitted wavelengths (T).

A Rain Shower Moving Away from the Instrument

The reflected wavelengths (R) from a rain shower moving away from the instrument are longer than the transmitted wavelengths (T).

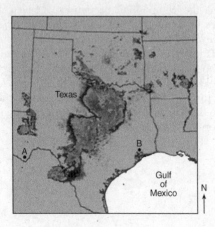

36. **4** If the rainstorm is moving east, it is moving away from location *A* and toward location *B*. According to the diagrams, the reflected wavelengths from a rain shower moving away from the Doppler radar weather instrument are longer than the transmitted wavelengths. The rainstorm is moving away from location *A*. Therefore, the reflected wavelengths will be longer at location *A*. The diagrams also show that the reflected wavelengths from a rain shower moving toward the instrument are shorter than the transmitted wavelengths. The rainstorm is moving toward location *B*; therefore, the reflected wavelengths will be shorter at location *B*. Thus, if Doppler radar is used at locations *A* and *B* as the rainstorm moves eastward, the reflected wavelengths will be longer at location *A* and shorter at location *B*.

37. **2** Find the chart labeled Electromagnetic Spectrum in the *Reference Tables for Physical Setting/Earth Science*. Locate the shaded region labeled "Microwaves." Note that microwaves extend past the shaded regions labeled "Infrared" and "Radio waves" but that there is a portion of the microwave region that corresponds to the gap between the infrared and radio wave regions. Thus, some microwave wavelengths are between the wavelengths of infrared and radio waves.

38. **3** Rainfall is a type of precipitation. The weather instrument used to measure precipitation is a precipitation gauge.

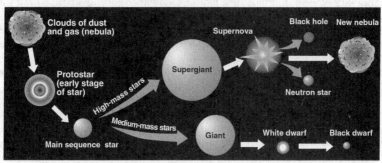

(Not drawn to scale)

39. **2** All matter exerts a mutual gravitational attraction on all other matter. Thus, matter has a tendency to draw together, or contract, over time. When particles of matter, such as dust and gas molecules, draw close enough to one another, their mutual gravitational attraction may cause them to hold together, forming a larger clump of matter. This larger clump of matter exerts a greater gravitational attraction than nearby individual gas molecules because the clump has more mass. Therefore, the clump of matter more strongly attracts nearby gas molecules and holds them to itself, causing the

clump to become even larger. Over time, the result is further contraction and the formation of larger and larger clumps of molecules. Thus, clouds of dust and gas form a protostar because those molecules draw together due to gravitational attraction.

40. **1** In the diagram, the arrows leading from a main sequence star to a giant star or to a supergiant star are labeled "Medium-mass stars" and "High-mass stars," respectively. Thus, the property that primarily determines whether a giant star or a supergiant star will form from a main sequence star is mass.

41. **4** Find the Characteristics of Stars chart in the *Reference Tables for Physical Setting/Earth Science*. On the chart, locate the region labeled "White Dwarfs." From the white dwarf region, trace left to the axis labeled "Luminosity (rate at which a star emits energy relative to the Sun)." Note that white dwarfs range in luminosity from 0.0001 to about 1. From the white dwarf region, trace down to the axis labeled "Surface Temperature (K)." Note that white dwarfs range in temperature from about 6,500 to 20,000 K. Thus, the table that includes data characteristic of the surface temperature and luminosity of some white dwarf stars is table (4): surface temperature 10,000 K, luminosity 0.001.

42. **1** Within stars, the force of gravity is strong enough to overcome the force of repulsion between atomic nuclei, allowing the nuclei to combine in a process called nuclear fusion. By definition, nuclear fusion involves the combining of several atoms of a lighter element to form a single atom of a heavier element. The single heavier atom typically has less mass than the several lighter atoms from which it formed. The mass "missing" from the heavier atom is not lost. Instead, it is converted into energy according to Einstein's formula $E = mc^2$. This formula means that if mass is converted to energy, the amount of energy released, E, is equal to the mass, m, times the speed of light, c, squared. The speed of light squared is a very large number. Therefore, the conversion of even a small amount of mass to energy during nuclear fusion results in the release of a very large amount of energy. Thus, great amounts of energy are released in the core of a star as lighter elements combine and form heavier elements during the process of fusion. Therefore, the process that generates the energy that is released by stars is fusion.

WRONG CHOICES EXPLAINED:

(2) Thermal conduction is a method by which energy is transferred, not a process that generates energy.

(3) Convection involves the transfer of heat due to density differences within a fluid. Thus, it is a method by which energy is transferred, not a process that generates energy.

(4) Although radioactive decay does occur in stars, the energy released by radioactive decay in a star is tiny compared with that released by nuclear fusion. Some reasons for this include: only a small percentage of atoms of any

given element are radioisotopes and radioisotopes decay at a constant rate over time. For example, stars are composed primarily of hydrogen. Only a trillionth of hydrogen atoms are tritium (the most common radioisotope of hydrogen), and the half-life of tritium is 12.32 years. Thus, over 12.32 years, only a half-trillionth of the hydrogen atoms in the Sun would undergo radioactive decay. In reality, there are no such constraints on the number of hydrogen atoms that can undergo fusion nor the rate at which fusion can occur in the Sun. Compared with the estimated 9.29×10^{37} fusion reactions *per second* of normal hydrogen atoms in the Sun, the energy released by the radioactive decay of a half-trillionth of the hydrogen atoms spread over 12.32 years is miniscule. Thus, radioactive decay could not account for the vast amounts of energy released by stars.

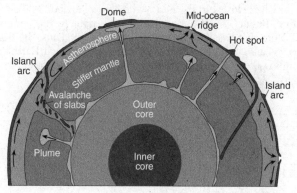

43. **3** According to the reading passage, "One theory is that most plumes form at the boundary between the outer core and the stiffer mantle." Find the Inferred Properties of Earth's Interior in the *Reference Tables for Physical Setting/Earth Science*. In the lower graph labeled "Temperature (°C)," note that temperatures are hotter at the bottom of the stiffer mantle than at the top. Thus, the mantle plumes form in a region that is hotter than the region through which they rise.

Heating causes the molecules of a material to move more rapidly and collide more energetically. This causes the molecules to spread farther apart, thereby increasing the volume of the material and causing the density of the material to decrease. Thus, heating of the mantle material at the outer core/stiffer mantle boundary causes the material to become less dense. So, compared to the surrounding material, mantle plumes rise toward Earth's surface from the core-mantle boundary because they are hotter and less dense.

44. **3** Find the Inferred Properties of Earth's Interior in the *Reference Tables for Physical Setting/Earth Science*. In the cross section, locate the

boundary between the outer core and the stiffer mantle. Now follow the dashed line at this boundary down to the horizontal scale labeled "Depth (km)" at the bottom of the temperature graph. Read the inferred depth of this region: 2900 km below Earth's surface. Thus, the boundary between Earth's outer core and stiffer mantle is located at a depth below Earth's surface of 2900 km.

45. **2** It is given that the rock that forms is basaltic. Find the Scheme for Igneous Rock Identification in the *Reference Tables for Physical Setting/ Earth Science*. Locate "Basalt" in the upper section of the chart. Trace this column down to where it intersects the arrow labeled "Color." Note that basalt is closest to the end of the arrow labeled "Darker." Continue tracing down to where the column intersects the arrow labeled "Composition." Note that basalt is closest to the end of the arrow labeled "Mafic (rich in Fe, Mg)." From "Basalt," trace horizontally to the right to the column labeled "Texture." Note that basalt has a fine-grained texture. Thus, the basaltic rock that forms volcanic mountains where mantle plumes reach Earth's surface is usually composed of fine-grained, dark-colored mafic minerals.

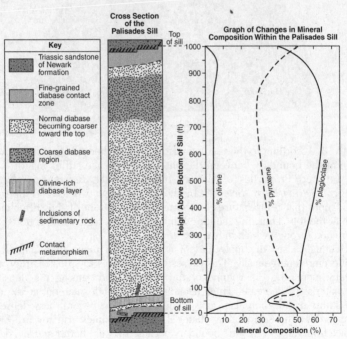

46. **4** It is given that the inclusions are pieces of Triassic sandstone and that they are located near the bottom of the Palisades sill. Note that the rock

layers directly above and below the Palisades sill are shaded with the same rock symbol. Locate this symbol in the key, and note that it corresponds to "Triassic sandstone of the Newark formation." Thus, the Palisades sill is surrounded by Triassic sandstone bedrock. Therefore, it is reasonable to infer that the inclusions shown near the bottom of the Palisades sill are pieces of the Triassic sandstone that broke off from the surrounding bedrock during the intrusion.

WRONG CHOICES EXPLAINED:
(1) and (3) It is given that the inclusions are pieces of Triassic sandstone. Find the Scheme for Sedimentary Rock Identification in the *Reference Tables for Physical Setting/Earth Science*. In the column labeled "Rock Name," locate sandstone. From sandstone, trace left to the column labeled "Composition." Note that sandstone is composed of mostly quartz, feldspar, and clay minerals. According to the key, the Palisades sill is composed of diabase. Find the Scheme for Igneous Rock Identification in the *Reference Tables for Physical Setting/Earth Science*. In the upper section of the chart, locate diabase. Trace down from the diabase column to the graph labeled "Mineral Composition (relative by volume)." Note that diabase is composed of plagioclase feldspar, pyroxene, biotite, amphibole, and olivine. Therefore, it is unlikely that the Triassic sandstone inclusions formed from either deposits of minerals within the sill or the olivine-rich layer in the sill because the sill is composed of diabase, which does not contain the quartz or clay minerals found in the sandstone.

(2) Find the Scheme for Sedimentary Rock Identification in the *Reference Tables for Physical Setting/Earth Science*. In the column labeled "Rock Name," locate sandstone. From sandstone, trace left to the column labeled "Texture." Note that sandstone has a clastic (fragmental) texture. Thus, sandstone forms from sediments that are fragments of rock, not by crystallization like the sedimentary rocks in the lower section of the scheme.

47. **3** In the key, find the symbol labeled "Coarse diabase region." In the cross section of the Palisades sill, locate the region corresponding to this symbol. From the top and bottom of the coarse diabase region, trace right to the vertical scale labeled "Height Above Bottom of Sill (ft)." Note that the coarse diabase region extends from just over 700 feet to just under 900 feet above the bottom of the sill. Of the choices, only 800 feet lies within this range. Thus, the coarse diabase region is found approximately 800 feet above the bottom of the Palisades sill.

48. **1** In the key, find the symbol labeled "Olivine-rich diabase layer." In the cross section of the Palisades sill, locate the region corresponding to this symbol. (It is about 50 feet above the bottom of the sill.) From this region, trace right to the graph labeled "Graph of Changes in Mineral Composition Within the Palisades Sill." Note that in this region, the line

labeled "% olivine" curves sharply to the right. Note, too, that the lines labeled "% pyroxene" and "% plagioclase" curve sharply to the left. At the bottom of the graph, locate the horizontal axis labeled "Mineral Composition (%)." Note that the percentages increase to the right and decrease to the left. Thus, the olivine line curving sharply to the right in this region indicates that the percentage of olivine is increasing. The pyroxene and plagioclase lines curving sharply to the left in this region indicate that the percentages of both pyroxene and plagioclase are decreasing. Therefore, within the olivine-rich diabase layer near the bottom of the sill as the percentage of olivine increases, the percentages of both plagioclase and pyroxene decrease.

49. **2** Find the Geologic History of New York State chart in the *Reference Tables for Physical Setting/Earth Science*. In the column labeled "Important Geologic Events in New York," find the line labeled "Intrusion of Palisades sill." Trace this line left to the vertical scale labeled "Million years ago." Note that the intrusion of the Palisades sill occurred approximately 200 million years ago.

50. **4** Find the Scheme for Igneous Rock Identification in the *Reference Tables for Physical Setting/Earth Science*. In the upper section of the chart, locate diabase. Trace the diabase column downward to the graph labeled "Mineral Composition (relative by volume)." Note that diabase is composed of plagioclase feldspar, pyroxene, biotite, amphibole, and olivine. The Graph of Changes in Mineral Composition Within the Palisades Sill shows the minerals plagioclase, pyroxene, and olivine. Thus, the two minerals found in some other samples of diabase not shown on this graph are biotite and amphibole.

PART B-2

51. Global warming refers to an increase in average global temperature. An increase in average global temperature will most likely cause present-day glaciers to melt. This will cause glaciers to shrink in size and retreat. As the glacial ice melts, it will run off and eventually reach the oceans. The glacial meltwater added to the oceans will cause sea level to rise, most likely resulting in coastal flooding.

One credit is allowed if *both* responses are correct. Acceptable responses include but are not limited to:

Glaciers:
- **will melt**
- **will retreat**
- **decrease in size**
- **become smaller**
- **shrink**
- **the rate of melting will increase**

Sea level:
- **will rise**
- **increase**
- **higher**
- **coastal flooding**

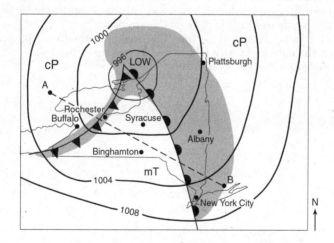

52. Find the Key to Weather Map Symbols in the *Reference Tables for Physical Setting/Earth Science*. Find the section labeled "Air Masses." Note that mT is the symbol for a maritime tropical air mass. In general, air masses

that form over land are dry and are called continental; those that form over water are moist and are called maritime. Air masses that form near the poles are cold and are called polar; those that form near the equator are warm and are called tropical. Thus, an mT air mass formed over the ocean, and the relative temperature of the surface over which it formed was warm.

One credit is allowed for circling **ocean** *and* for correctly describing the relative temperature of Earth's surface. Acceptable descriptions include but are not limited to:

- **warmer**
- **hot**
- **a tropical temperature**

Note: Credit is *not* allowed for a numerical answer because there are no temperatures indicated for comparison.

53. Cold fronts form along the leading edge of a cold air mass advancing against a warmer air mass. The cold air mass is denser than the warmer air ahead of it. So the cold air mass bulges into the warm air, pushing against and under it like a wedge. The curved slope of the front is rather steep, especially if it is moving fast. Thus, a cold front slopes downward from the cold air mass toward the warm air mass in a steep-sided, curved bulge. The frontal boundary is located between the cold and warm regions on the cross section and curves sharply downward toward the warm region. To represent the shape and location of the cold frontal boundary, start at a point on line *AB* between the regions labeled "Cold" and "Warm." From this point, draw a line that curves upward and to the left as shown below.

One credit is allowed for **a line that starts from line *AB*, passes between the cold and warm labels,** *and* **curves up to the left.**

Example of a 1-credit response:

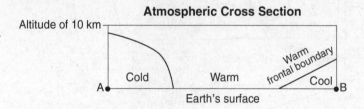

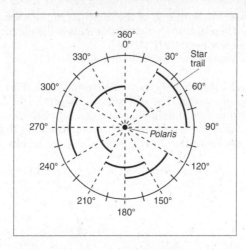

54. When you observe the sky, you find that every celestial object changes position over time, or is in motion. If you keep track of this motion, you find that with very few exceptions, all celestial objects appear to move across the sky from east to west along a path that is an arc, or part of a circle. If you measure the rate at which celestial objects are moving, they appear to follow a circular path at a constant rate of 15° per hour, or one complete circle every day (24 h/day × 15°/h = 360°/day). Thus, this motion is called apparent daily motion. In the Northern Hemisphere, all the circles formed by completing the arcs along which celestial objects move are centered very near the star *Polaris*. *Polaris* is located directly above the North Pole and is roughly aligned with Earth's axis of rotation. Therefore, as Earth spins on its axis, *Polaris* remains in a fixed position in the sky relative to an observer in the Northern Hemisphere while the celestial objects around it appear to move in circular paths around *Polaris*. So, it is reasonable to conclude that the motion of Earth that causes the stars to appear to move in a circular path is rotation.

One credit is allowed for a correct response. Acceptable responses include but are not limited to:

- **rotation**
- **spinning/turning on its axis**

Note: Credit is *not* allowed for "Earth's rotation around the Sun" because this confuses rotation with revolution.

55. Earth completes one 360° rotation in 24 hours. Therefore, the rate at which celestial objects appear to move across the sky is constant at 15° per hour (360°/24 h = 15°/h). Long photographic exposures of stars show that the stars form arcs called star trails. A 1-hour time exposure photo of a star will show an arc that is 15° in angular length because celestial objects appear to

move 15° per hour. Note that all of the star trails in the photograph are 60° of arc in length. Therefore, to create the star trails in the photograph, the lens was kept open for 4 hours (60°/15°/h = 4 h).

One credit is allowed for a response that indicates **4 h**.

56. *Polaris* is located directly above the North Pole and is roughly aligned with Earth's axis of rotation. Therefore, the arrow should be drawn pointing away from the North Pole and should be aligned with Earth's axis of rotation.

One credit is allowed for **an arrow within or touching the zone shown that points away from the North Pole and is generally aligned with Earth's axis**.

Note: A line without an arrowhead is *not* accepted because it does not show direction.

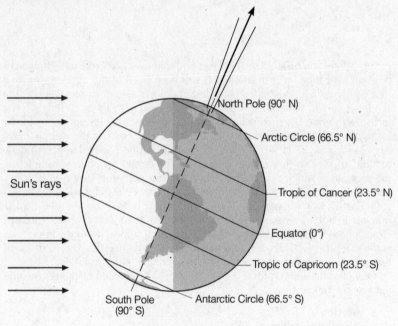

57. Simple geometry shows that to an observer in the Northern Hemisphere, the altitude of *Polaris* is the same as the observer's latitude. (Refer to Figure 4.5 in *Let's Review: Earth Science, The Physical Setting* for a fuller explanation.) Thus, the altitude of *Polaris* when viewed from Mt.

Marcy will be the same as the latitude of Mt. Marcy in New York State. Find the Generalized Bedrock Geology of New York State map in the *Reference Tables for Physical Setting/Earth Science*. Locate Mt. Marcy (to the SSW of Plattsburgh). From Mt. Marcy, trace right to the vertical latitude scale along the side of the map. Note that Mt. Marcy is located at latitude 44.1°N. Therefore, to the nearest whole degree, the altitude of Polaris when it is viewed from the top of New York State's Mt. Marcy is 44°.

One credit is allowed for **44°**.

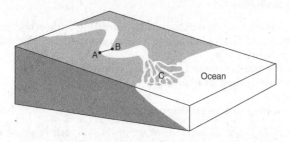

58. When a stream curves, or meanders, water velocity is greatest along the outside of the curve because the water is forced to cover a greater distance along the outside of the curve. Stream erosion is greatest where the velocity of the water is greatest. Therefore, the outside of the curve of a meandering channel experiences more erosion than the inside of the curve. Thus, the bottom of the stream channel will be more deeply eroded at the outside of the curve near point A than it is at the inside of the curve near point B. In other words, the stream channel will be deeper near point A and shallower near point B. Therefore, the line representing a cross-sectional view of the shape of the bottom of the stream channel should be drawn from point A to point B and should show that the stream is deeper near point A and shallower near point B as shown below.

One credit is allowed if **the student's line is drawn from point A to point B and shows that the stream channel is deeper near side A**.

Example of a 1-credit response:

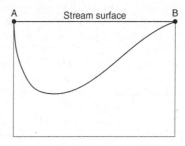

59. As sediments are carried along by the moving water in a stream, they bounce off of and rub against one another. These collisions cause smaller pieces to break off of sharp corners and edges on the surface of the sediments, particularly at corners that protrude—a process called abrasion. The sediments are also abraded as they bounce, roll, and scrape against the bottom of the streambed. As a result, the sediments eroded by the water in this stream change in size and shape, becoming smaller, smoother, and more rounded.

One credit is allowed for a correct response. Acceptable responses include but are not limited to:

- **The particles were weathered by abrading with other particles.**
- **Rolling and bouncing along the streambed breaks off corners and polishes rock.**
- **The sediments scrape against the streambed.**
- **They rub against one another.**
- **Abrasion**
- **Weathering due to collision of particles**

Note: Credit is not allowed for water or erosion, acting alone, to smooth rocks because this is restating the question, and water alone, without sediments, does not abrade rock.

60. It is given that letter C indicates a triangular-shaped depositional feature where the stream enters the ocean. When a stream flows into a standing body of water, such as the ocean, it stops moving and most of the stream's sediment is deposited. The resulting landform is a large, flat, fan-shaped (triangular) pile of sediment at the mouth of the stream called a delta. (It is called a delta because its triangular shape is similar to that of the Greek letter delta, Δ.) Thus, a delta is a depositional feature, is located where a stream enters an ocean, and is triangular in shape. Therefore, the triangular-shaped depositional feature indicated by letter C is a delta.

One credit is allowed for **delta** *or* **any specific type of delta**.

61. Pure water flowing over rock barely wears away the rock. However, water that carries sediments acts like a cutting tool. The process of crushing, grinding, and wearing away rock by the impact of sediments is called abrasion. It is the main process by which streams erode. Thus, one important factor that determines the rate of stream erosion is the nature of the stream's load, i.e., the amount, size, shape, density, and hardness of the sediment particles being transported by the stream. No matter the stream's load, as a stream's velocity increases, the stream's cutting power increases. As a stream's cutting power increases, the rate of stream erosion also increases. Thus, any factor that affects stream velocity will affect the rate of stream erosion. Such factors include stream gradient, the shape of and location within the stream channel, and the depth and volume of water flowing in the stream (discharge). Furthermore, the rate of erosion is also affected by the nature of the material being eroded, specifically, its resistance to erosion.

Factors that affect the material's resistance to erosion include the type of bedrock over which the stream flows and the nature of the materials found in or along the streambed (i.e., vegetation, trees, sediments).

One credit is allowed for *two* correct responses. Acceptable responses include but are not limited to:

- **stream velocity/speed**
- **gradient/slope of the stream**
- **location within a meander/stream channel**
- **volume/amount of stream discharge**
- **shape of stream channel (straight vs. meandering)**
- **water depth**
- **material found in the stream or along the streambed (vegetation, trees, sediments)**
- **type of bedrock**
- **particle size/shape/density**

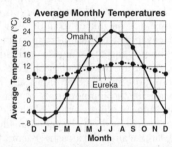

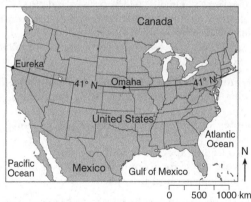

62. Find the Equations table in the *Reference Tables for Physical Setting/Earth Science*. Note the equation for rate of change.

$$\text{Rate of change} = \frac{\text{change in value}}{\text{time}}$$

In this case, the value is average monthly temperature. It is given that the time is 2 months (between October and December). Locate October (O) on the horizontal axis labeled "Month," and trace upward until you intersect the line labeled "Omaha." From this intersection, trace left to the vertical axis labeled "Average Temperature (°C)." Read the value: 12. Now locate December (D), trace upward to the "Omaha" line, and then trace left to the "Average Temperature (°C)" axis. Read the value: –4. Thus, the change in value is 16°C (12– (–4) =16). Substitute these values into the equation and solve.

$$\text{Rate of change} = \frac{16 \text{ degrees Celsius}}{2 \text{ months}} = 8 \text{ degrees Celsius/month}$$

One credit is allowed for **8°C/mo** *or* **–8°C/mo**.

63. According to the map, Eureka, California, and Omaha, Nebraska, are located at the same latitude: 41° N. Therefore, both cities receive the same intensity of insolation. However, Eureka is located on the coast next to the Pacific Ocean while Omaha is an inland location. Due to the high specific heat of water, land surfaces increase in temperature more than water surfaces when insolation strikes them. Land surfaces also cool more rapidly than water surfaces. Furthermore, moist air has a higher specific heat than dry air, so dry air will heat and cool more rapidly than moist air. Therefore, large bodies of water tend to moderate the temperatures of coastal regions of nearby landmasses by warming them in winter and cooling them in summer. Thus, Eureka, which is a coastal city, will experience a smaller range of temperatures throughout the year because the nearby Pacific Ocean moderates its temperatures. Conversely, Omaha, which is an inland city, will experience a greater range of temperatures throughout the year because it is surrounded by land and relatively drier air, both of which have a low specific heat.

One credit is allowed for an acceptable response. Acceptable responses include but are not limited to:

- **Omaha is surrounded by land, which has a low specific heat.**
- **The Pacific Ocean moderates the temperature/climate of Eureka.**
- **Large bodies of water change temperature more slowly than land does.**
- **Water has a higher specific heat than land.**
- **The relatively drier air around Omaha has a lower specific heat than the moist air around Eureka.**

Note: Credit is not allowed for "Eureka is closer to water so temperatures remain constant" because this just restates the question without explaining the role water plays in causing constant temperatures.

64. The difference in average temperature between these two cities is indicated by the vertical distance between the line labeled "Eureka" and the line labeled "Omaha" on the graph. Locate the point at which there is the greatest vertical distance between the two lines. At this point, trace down to the axis labeled "Month." The greatest difference in the average temperature occurs in January.

One credit is allowed for **January** or **Jan**.

65. According to the map, Eureka is located along the western coast of North America at latitude 41° N. Find the Surface Ocean Currents map in the *Reference Tables for Physical Setting/Earth Science*. Locate the western coast of North America at 41° N. Note that the surface ocean current flowing along the western coast of North America in this region is indicated by a white arrow labeled "California C." Thus, the surface ocean current that affects the climate of Eureka is the California Current.

One credit is allowed for **California Current**.

PART C

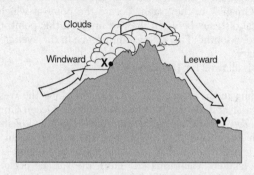

66. According to the diagram, winds are forcing air to rise on the windward side and descend on the leeward side of the mountain range. As the air on the windward side rises to higher elevations, the air pressure decreases, causing the air to expand and cool. When the air temperature reaches the dew point, moisture begins to condense and form clouds. On the leeward side, the cool air descends to lower elevations and the air pressure increases. The increased air pressure causes the air to be compressed. The compression causes warming, resulting in warm, dry air. Note that point **X** is on the windward side of the mountain at a high elevation where temperatures have dropped to the dew point and clouds have formed. Note that point **Y** is at a lower elevation on the leeward side, where the air is clear because the air has been warmed by compression to above its dew point. Thus, the air temperature at point **X** will be cooler than the air temperature at point **Y**. The water vapor content of the air at point **X** will be higher than the water vapor content of the air at point **Y**.

One credit is allowed for an acceptable response. Acceptable responses include but are not limited to:

Air temperature at **X**:

- **cooler**
- **lower/less**
- **decreased**
- **colder than Y**

Water vapor content at **X**:

- **higher/more**
- **100% relative humidity**
- **wetter**
- **saturated**
- **more humid than Y**

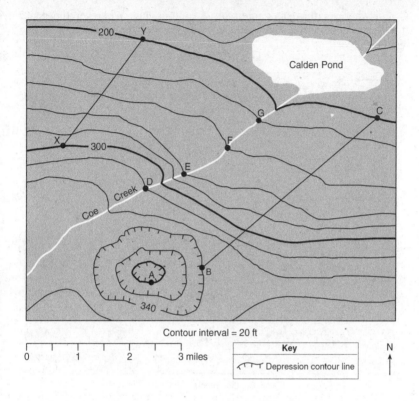

67. Note that location A lies directly on a contour line with hachure marks. According to the key, this is the symbol for a depression contour line. Contour lines around an enclosed depression are printed with hachure marks pointing into the depression in order to distinguish them from small hills. The contour interval for depression contour lines is the same as for regular contour lines. However instead of indicating increasing elevation, they indicate decreasing elevation. Note that the contour interval of the map is 20 feet. Note that the depression contour line on which B is located is labeled "340." Also note that two contour intervals separate the depression contour line on which A is located and the depression contour line on which B is located. Thus the elevation at A is two contour intervals below 340 feet, or 300 feet.

One credit is allowed for **300 ft**.

68. To construct a topographic profile along line BC, proceed as follows. Place the straight edge of a piece of scrap paper along the line connecting point B to point C on the map. Mark the edge of the paper at points B and

C and wherever the paper intersects a contour line. Wherever the paper intersects a contour line, label the mark with the elevation of the contour line as shown below.

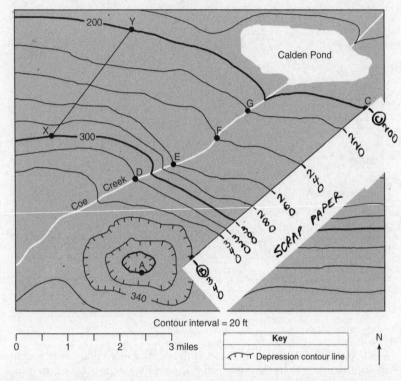

Then place this paper along the lower edge of the grid provided for the answer to this quesiton in the exam section of this book so that points *B* and *C* on the scrap paper align with points *B* and *C* on the grid. At each point where a contour line crosses the edge of the paper, draw a plot on the grid at the appropriate elevation. Finally, connect all of the plots in a smooth curve to form the finished profile as shown below. Note that the depression contour line at *B* and the adjacent regular contour line both have an elevation of 340 feet. The elevation between these two points on the profile has an elevation greater than 340 feet but less than 360 feet. Therefore, the line must show a hill higher than 340 feet but less than 360 feet between these two points.

One credit is allowed if **the centers of *all nine* plots are within or touch the rectangles shown and are correctly connected with a line that passes within or touches the rectangles. The line must show a hill higher than 340 feet but lower than 360 feet.**

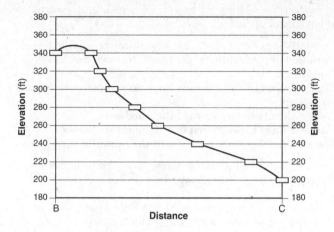

69. One way to determine the direction of flow of Coe Creek is to note the elevation of the creek when it crosses labeled contour lines. Water flows downhill from higher elevations to lower elevations. Therefore, the direction in which the elevation of the contour lines decreases is the direction in which the water will flow. Note the bold contour line labeled "300" between points D and E on Coe Creek. Then note the bold contour line labeled "200" between point G on Coe Creek and Calden Pond. Thus Coe Creek flows from the higher elevation near point D toward the lower elevation near Calden Pond. Therefore, Coe Creek flows toward the northeast.

Another way to determine the direction of flow of Coe Creek is to note the shape of contour lines as they cross Coe Creek. The bed of a stream slopes downhill and is lower than its banks. A person standing on a streambed would be at a lower elevation than a person standing on the stream's bank. A person standing on the streambed would have to walk uphill along the streambed, that is, upstream, to reach the same elevation as a person on the bank. On a topographic map, contour lines connect points of equal elevation. Therefore, the contour line would have to bend upstream of the stream's bank to intercept the same elevation at the center of the streambed and then downstream on the opposite side to reach the same elevation on the opposite bank. Therefore, contour lines bend upstream when they cross a streambed, forming a distinctive V-shaped curve with the apex pointing upstream and the sides opening toward the downstream direction. Note that where the contour lines cross Coe Creek, they bend with the apex of each bend pointing southwest and the sides opening toward the northeast. Therefore, Coe Creek is flowing toward the northeast.

One credit is allowed for an acceptable response. Acceptable responses include but are not limited to:

- **Contour lines bend upstream when they cross Coe Creek.**
- **Contour line elevations decrease toward the northeast along Coe Creek.**
- **The V shapes of the contour lines point upstream toward higher elevations.**
- **Lower elevations are toward the northeast.**
- **Contour lines make V shapes that point southwest.**
- **The contour lines are bending in the opposite direction.**

Note: Credit is *not* allowed for "water flows downhill" because this does not indicate how contour lines show stream direction.

70. The map is a topographic map, which shows elevation using contour lines. The spacing of contour lines indicates the nature of the slope. The closer the contour lines are spaced, the greater the elevation is changing in a given distance and the steeper the slope. The steeper the slope, or gradient of a stream's channel, the faster the water flows. The wider the spacing of the contour lines, the gentler the slope is and the slower the water flows. Note that near points D and E, the contour lines are closely spaced. Near points F and G, the contour lines are widely spaced. Thus the slope is steeper and Coe Creek flows faster between points D and E than between points F and G.

One credit is allowed for an acceptable response. Acceptable responses include but are not limited to:

- **Contour lines between D and E are closer together.**
- **Contour lines between F and G are farther apart, indicating a slower stream velocity.**
- **Contour lines that are closer together indicate a steeper slope/gradient.**
- **There is a greater elevation change between D and E.**

Note: Credit is *not* allowed for "D and E are closer" because F and G are the same distance apart. Credit is *not* allowed for "the slope or gradient is steeper" alone because this does not indicate how contour lines show a steeper slope.

71. Find the Equations table in the *Reference Tables for Physical Setting/Earth Science*. Locate the equation for gradient:

$$\text{Gradient} = \frac{\text{change in field value}}{\text{distance}}$$

The map shown is a topographic map. The field value on a topographic map is elevation. It is given that the contour interval is 20 feet. Note that point X lies directly on the bold 300-foot contour line. Note that point Y lies directly on the bold 200-foot contour line. On a piece of scrap paper, mark off the distance between points X and Y. Compare the distance marked off on the scrap paper to the scale printed beneath the map to determine the distance between points X and Y. The distance is about 2.5 miles. Substitute these values into the equation as shown below:

$$\text{Gradient} = \frac{300 \text{ ft} - 200 \text{ ft}}{2.5 \text{ mi}}$$

Solve the equation as shown below:

$$\text{Gradient} = \frac{100 \text{ ft}}{2.5 \text{ mi}} = 40 \frac{\text{ft}}{\text{mi}}$$

One credit is allowed for **any value from 38 to 42 with acceptable units**. Acceptable units include but are not limited to:

- **ft/mi**
- **feet/mi**
- **feet/mile**

72. To draw a line separating regions with Mercalli values of VI from regions with Mercalli values of VII, do the following. Start at the VII near the upper right-hand edge of the shaded area indicating the landmass of Japan. Draw a line southward between the VII and the VI to its lower left. Continue the line south between the VII at Ishinomaki and the VI to its west. Next continue the line southward between the VI and VII north of Fukushima and the two VIs and VII south-southeast of Fukushima. Next find the VII that is about 50 miles northwest of Mito. The line can loop around this VII and then curve back out to the shore south of Mito. Alternatively, this line can curve to the shore north of the VII and then pick up along the shoreline south of the VII before looping inland around the VII. A completed line separating all values of VI from VII is shown below.

One credit is allowed for **any acceptable line separating all values of VI from VII**.

Note: Credit is allowed even if the line passes through water. Credit is *not* allowed if the line touches or passes through any Mercalli value.

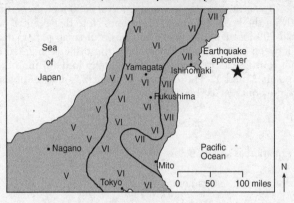

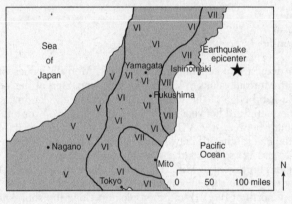

73. Locate Yamagata on the map. Note that it is located in a region that experienced a Mercalli value of VI. On the Modified Mercalli Scale of Earthquake Intensity, locate the intensity value VI. Note the description of effects, "Felt by all, many frightened. Some heavy furniture moved; a few instances of fallen plaster. Damage slight." Note that when an earthquake of Mercalli intensity VI occurs, all of the effects of lower intensities may also occur but none of the effects of higher intensities will occur. For example, if an intensity VI earthquake can cause heavy furniture to move, it can certainly cause parked cars to rock slightly as in IV or cause dishes to break as in a V. However, an intensity VI earthquake would not cause chimneys to fall as in VIII or shift buildings off their foundations as in IX. Thus on the chart provided for the answer to this question in the June 2015 exam section of this book, the first four boxes should be checked.

One credit is allowed if **only the first four boxes are checked as shown below**.

√	parked cars rock
√	dishes and windows broken
√	felt by all persons
√	some heavy furniture moved
	chimneys and monuments fall
	buildings shifted off foundations

Note: Credit is allowed if a symbol other than a check mark is used.

74. Find the Tectonic Plates map in the *Reference Tables for Physical Setting/Earth Science*. Locate 38° N on the latitude scales along the left and right edges of the map. Draw a horizontal line across the map representing latitude 38° N. Locate 142° E on the longitude scales along the top and bottom of the map. Draw a vertical line across the map representing longitude 142° E. The point where the 38° N latitude line and 142° E longitude line intersect marks the location of the epicenter. Note the symbol at this location. According to the key along the bottom of the Tectonic Plates map, this symbol represents a "Convergent plate boundary (subduction zone)." Thus, the type of plate boundary that is located nearest to the epicenter of this earthquake is a convergent boundary or subduction zone.

One credit is allowed for an acceptable response. Acceptable responses include but are not limited to:

- **convergent boundary**
- **subduction zone**

75. *P*-wave and *S*-wave recordings from a single event can be used to find the distance to the epicenter of an earthquake. One way the *P*-waves and *S*-waves recorded at Ishinomaki and Nagano could be used to indicate that Ishinomaki was closer to the earthquake epicenter than Nagano would be to compare the arrival time of the *P*-waves in the two cities. The closer the city to the epicenter, the sooner the *P*-waves would arrive. Therefore, if the *P*-waves arrived earlier in Ishinomaki than in Nagano, this would indicate that Ishinomaki was closer to the epicenter than Nagano.

Another way the *P*-waves and *S*-waves recorded at Ishinomaki and Nagano could be used to indicate that Ishinomaki was closer to the earthquake epicenter than Nagano would be to compare the difference in the arrival times of the *P*-waves and *S*-waves in the two cities. When an earthquake occurs, both types of seismic waves start moving outward from the focus at the same time. However since they travel at different speeds, these two different types of seismic waves do not arrive at a seismic station at the same time. The faster *P*-waves will arrive first, followed by the slower *S*-waves some time later. The farther a seismic station is located from the epicenter, the greater the difference between the arrival times of the *P*-waves and the *S*-waves recorded on a seismogram. Thus, a greater difference in the arrival times of the *P*-waves and the *S*-waves recorded at Nagano compared to those recorded at Ishinomaki would indicate that Nagano was farther from the epicenter than Ishinomaki.

Yet another way the *P*-waves and *S*-waves recorded at Ishinomaki and Nagano could be used to indicate that Ishinomaki was closer to the earthquake epicenter than Nagano would be to compare the amplitude of the seismic waves recorded in the two cities. The amplitude of the seismic waves is related to the amount of energy the earthquake released, or its magnitude. The greater the amplitude of the seismic waves, the greater the amount of energy released. However, as seismic waves travel through Earth, they lose energy. The farther from the epicenter seismic waves travel, the more energy they lose and the smaller the amplitude of the waves. Thus, a greater seismic wave amplitude at Ishinomaki than at Nagano would indicate that Ishinomaki is closer to the earthquake epicenter.

One credit is allowed for an acceptable response. Acceptable responses include but are not limited to:

- *P*-waves arrived earlier at Ishinomaki than at Nagano.
- The difference in arrival times was less at Ishinomaki.
- The *P*-wave and *S*-wave arrival time interval was greater at Nagano.
- The amplitude/magnitude of seismic waves was greater/bigger/stronger at Ishinomaki.
- There is less time difference between the *P*-waves and *S*-waves at the closer location.
- *P*-waves and *S*-waves were closer together.

76. According to the National Weather Service, a tsunami can strike any ocean coast at any time. There is no season for tsunamis. Meteorologists cannot predict where, when, or how destructive the next tsunami will be. Although tsunamis cannot be prevented, some things can be done before a tsunami hits that could save lives. A city could install a tsunami monitoring and warning system. Tsunami warnings could then be broadcast through local radio and television, via wireless emergency alerts, or by the use of outdoor sirens. Seawalls, barricades, or barriers could be built to block or slow a tsunami. Buildings in low-lying areas could be relocated to higher ground. The city could require that structures built in high-risk locations be taller and built on strong foundations so the structure would survive the tsunami and occupants could escape to higher floors. The city could also develop emer-

gency plans, including identifying evacuation routes. The city could also prepare emergency kits and supplies for use in the event of a tsunami. Most important, the city could educate its citizens about how to prepare for a tsunami and how to practice all-hazard preparedness.

One credit is allowed for an acceptable response. Acceptable responses include but are not limited to:

- **Install a tsunami monitoring and warning system.**
- **Build a seawall/barricade/barrier.**
- **Build tall structures on strong foundations.**
- **Designate or plan evacuation routes.**
- **Prepare emergency kits/supplies.**
- **Relocate buildings to higher ground.**

Note: Credit is *not* allowed for an action indicating an imminent tsunami (e.g., evacuate to higher ground).

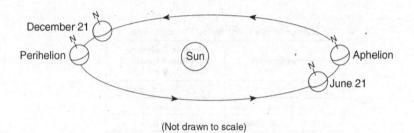

77. It is given that both perihelion and aphelion occur approximately two weeks after the dates shown. According to the diagram, perihelion occurs two weeks after December 21 on January 4 and aphelion occurs two weeks after June 21 on July 4. January 4 and July 4 are six months apart. Therefore, aphelion and perihelion are six months apart.

One credit is allowed for **6** *or* **six months**.

78. According to the diagram, at aphelion the north end of Earth's axis of rotation is tilted toward the Sun. Thus, Earth's Northern Hemisphere is tilted toward the Sun. New York State is located in Earth's Northern Hemisphere. When Earth's Northern Hemisphere is tilted toward the Sun, both the angle of insolation and the duration of insolation (number of daylight hours) in New York State are greater than when the Northern Hemisphere is tilted away from the Sun. High-angle insolation is more intense than low-angle insolation and has a greater warming effect. Therefore, when Earth's Northern Hemisphere is tilted toward the Sun, New York State receives more intense insolation for more daylight hours than when the Northern

Hemisphere is tilted away from the Sun. Thus, warm summer temperatures occur in New York State when Earth is at aphelion because at aphelion, Earth's Northern Hemisphere is tilted toward the Sun and New York State has more hours of daylight and receives high-angle insolation.

One credit is allowed for an acceptable response. Acceptable responses include but are not limited to:

- **New York State is receiving higher angles of insolation.**
- **The Northern Hemisphere is experiencing longer duration of insolation.**
- **The North Pole axis or Northern Hemisphere is tilted toward the Sun.**
- **The Sun appears higher in the sky.**

Note: Credit is *not* allowed for "Earth is tilted toward the Sun" because only the Northern Hemisphere is tilted toward the Sun. Credit is *not* allowed for "the Northern Hemisphere faces the Sun" because part of the Southern Hemisphere also faces the Sun during the daylight hours.

Human Species Distributed Through Time

Human Species	Time of Existence from Fossil Evidence (million years ago)
Homo sapiens	0.25 to the present
Homo neanderthalensis	0.35 to 0.03
Homo rhodesiensis	0.6 to 0.1
Homo heidelbergensis	0.6 to 0.3
Homo mauritanicus	1.2 to 0.6
Homo erectus	1.5 to 0.2
Homo ergaster	1.8 to 1.25
Homo habilis	2.25 to 1.4

79. According to the Human Species Distributed Through Time table, *Homo mauritanicus* existed from 1.2 to 0.6 million years ago. On the graph provided for the answer to this question in the June 2015 exam section of this book, locate "1.2" on the right vertical axis labeled "Time (million years ago)." Trace left to the column labeled "*Homo mauritanicus*" on the horizontal axis labeled "Human Species," and draw a horizontal line. Now locate 0.6 on the right vertical axis labeled "Time (million years ago)." Trace left to the column labeled "*Homo mauritanicus*" on the horizontal axis labeled "Human Species," and draw a horizontal line. Shade the region between these two lines. Repeat this procedure for each of the remaining human species. A correctly completed graph is shown below.

One credit is allowed if **all four bars are drawn in the correct columns and the ends of the bars are within or touch the rectangular areas shown at the end of each bar.**

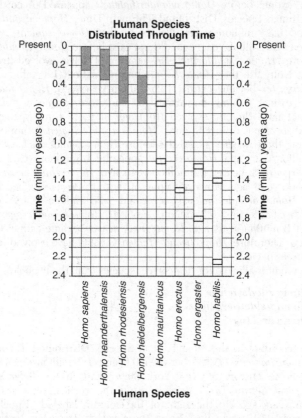

80. Of the human species shown on the chart Human Species Distributed Through Time, the first to exist was the species that existed the greatest number of years ago. According to the chart, *Homo habilis* existed 2.25 million years ago, the greatest number of years ago of any of the human species shown. Thus, the human species shown in the chart that was the first to exist was *Homo habilis*.

One credit is allowed for ***Homo habilis* or *habilis* or *H. habilis***.

81. According to the information provided in the question, *Homo neanderthalensis* could have evolved from another species that existed before *Homo neanderthalensis* appeared as long as that species did not become extinct before *Homo neanderthalensis* appeared. According to the chart Human Species Distributed Through Time, *Homo neanderthalensis* appeared 0.35 million years ago. Note that *Homo sapiens*, which first appeared 0.25 million years ago, did not exist before *Homo neanderthalensis*. Therefore, *Homo neanderthalensis* could not have evolved from *Homo sapiens*. Note also that *Homo habilis* became extinct 1.4 million years ago. *Homo ergaster* became extinct 1.25 million years ago. *Homo mauritanicus* became extinct 0.6 million years ago. Therefore, *Homo neanderthalensis* could not have directly evolved from any of these three species because they all became extinct before *Homo neanderthalensis* appeared. The remaining three species all existed before *Homo neanderthalensis* appeared 0.35 million years ago. *Homo erectus* appeared 1.5 million years ago. *Homo heidelbergensis* appeared 0.6 million years ago. *Homo rhodesiensis* appeared 0.6 million years ago. However, none of these three species became extinct before *Homo neanderthalensis* appeared 0.35 million years ago. *Homo erectus* became extinct 0.2 million years ago. *Homo heidelbergensis* became extinct 0.3 million years ago. *Homo rhodesiensis* became extinct 0.1 million years ago. Therefore, *Homo neanderthalensis* could have evolved from any of these three species.

One credit is allowed for *two* correct species from the list below:

- **Homo rhodesiensis**
- **Homo heidelbergensis**
- **Homo erectus**

82. According to the table Human Species Distributed Through Time, *Homo mauritanicus* existed between 0.6 and 1.2 million years ago. Find the Geologic History of New York State chart in the *Reference Tables for Physical Setting/Earth Science*. Find the vertical time scale labeled "Million years ago" to the right of the column labeled "Epoch." Locate the values "0.01" and "1.8." From these values, trace left to the column labeled "Epoch." Note that the time range from 0.01–1.8 million years ago corresponds to the Pleistocene epoch. Note that the time range of 0.6–1.2 million years ago lies within this range. Thus, *Homo mauritanicus* existed during the Pleistocene epoch.

One credit is allowed for **Pleistocene epoch**.

Moon's Orbital Position and Tide Data on May 13

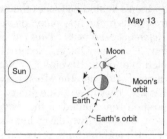

Tide	Time	Height (m)
high	12:59 a.m.	1.92
low	7:15 a.m.	0.37
high	1:32 p.m.	2.07
low	7:59 p.m.	0.27

(Not drawn to scale)

Moon's Orbital Position and Tide Data on May 20

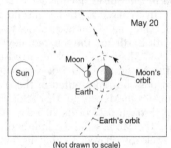

Tide	Time	Height (m)
low	1:22 a.m.	0.06
high	7:50 a.m.	2.47
low	2:10 p.m.	0.09
high	8:10 p.m.	2.21

(Not drawn to scale)

83. According to the table, the first high tide on May 13 occurred at 12:59 A.M. and the next high tide on May 13 occurred at 1:32 P.M. One way to determine the length of time from 12:59 A.M. to 1:32 P.M. is as follows. From 12:59 A.M. to 1:00 A.M. is 1 min. From 1:00 A.M. to 1:00 P.M. is 12 h, and from 1:00 P.M. to 1:32 P.M. is 32 min. Add these together. The total time from 12:59 A.M. to 1:32 P.M. is 12 h 33 min (1 min + 12 h + 32 min).

One credit is allowed for **12 h 33 min**.

84. An observer on Earth can see only that portion of the Moon facing Earth that is illuminated by the Sun's rays and only from the side of Earth that is in darkness. In the diagram for May 13, only half of the side of the Moon facing Earth is illuminated and the dark portion of the Moon is to the right. Therefore, on the diagram provided for the answer to this question in the June 2015 exam section of this book, the right half of the Moon should be shaded.

One credit is allowed if **the student shades the right half of the Moon to show a last-quarter moon as shown below**.

Examples of 1-credit responses:

85. According to the diagram, on May 20 the Moon is located directly between Earth and the Sun so that the Moon's lighted half faces the Sun and its unlighted half faces Earth. This corresponds to the new moon phase. Find the Solar System Data chart in the *Reference Tables for Physical Setting/Earth Science*. In the column labeled "Celestial Object," locate Earth's Moon. Trace right to the column labeled "Period of Revolution." Note that the Moon takes 27.3 days to orbit Earth once. Although the Moon makes one revolution in 27.32 days, the Moon takes 29.5 days to go through a complete cycle of phases from the new moon position represented in the diagram to the new moon the following month. Why the extra two days? Let's start at new moon, when the Moon is between Earth and the Sun. At the same time that the Moon revolves around Earth, Earth is revolving around the Sun at a rate of about 1° per day. Thus, by the time the Moon has completed one revolution, Earth's position in relation to the Sun is not the same as it was when the Moon started its revolution. In 27 days, Earth has moved 27° in its orbit. Since it moves at about 13° per day, the Moon takes about two days to catch up to Earth and align again with Earth and the Sun in a new moon phase. The word *month* has its origin in "Moon-th," which referred to this 29.5-day cycle of phases. Therefore, the Moon takes 29.5 days to complete a cycle of phases from the new moon position represented in the diagram to the new moon the following month. From May 20 to May 28 is 8 days. In 8 days, the Moon will have moved through 0.27 of its 29.5-day cycle of phases (8 days/29.5 days = 0.27). Thus, the Moon will be a little more than one-quarter of the way through its cycle of phases. In the diagram, the arrows showing the Moon's orbit show the Moon moving a counterclockwise direction. Thus, on the diagram for May 28 provided for the answer to this question in the June 2015 exam section of this book, the **X** should be drawn a little more than one-quarter of an orbit in a counterclockwise direction past the new moon position. In other words, the **X** should be a little past the point where the Moon's orbit first crosses Earth's orbit moving counterclockwise from the new moon phase.

One credit is allowed if **the center of the X falls within or touches the band in the Moon's orbit shown below**.

Note: One credit is allowed if a symbol other than an **X** is used.

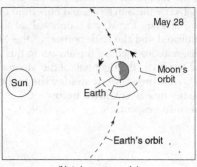

(Not drawn to scale)

SELF-ANALYSIS CHART June 2015

Topic	Question Numbers (Total)	Wrong Answers (x)	Grade
Standards 1, 2, 6, and 7: Skills and Application			
Skills			
Standard 1 Analysis, Inquiry, and Design	1, 5–15, 17–25, 28, 30–35, 37, 41–45, 48–50, 52, 53, 55–59, 61–68, 70–72, 74, 75, 77–81, 84, 85		$\dfrac{100(64-x)}{64} = \%$
Standard 2 Information Systems	36		$\dfrac{100(1-x)}{1} = \%$
Standard 6 Interconnectedness, Common Themes	3, 6–10, 13, 16, 18–29, 31, 32, 35–61, 63, 65–70, 72–74, 82–85		$\dfrac{100(63-x)}{63} = \%$
Standard 7 Interdisciplinary Problem Solving	76, 79		$\dfrac{100(2-x)}{2} = \%$
Standard 4: The Physical Setting/Earth Science			
Astronomy			
The Solar System (MU 1.1a, b; 1.2d)	83–85		$\dfrac{100(3-x)}{3} = \%$
Earth Motions and Their Effects (MU 1.1c, d, e, f, g, h, i)	3, 16, 19, 20, 54–57, 77, 78		$\dfrac{100(10-x)}{10} = \%$
Stellar Astronomy (MU 1.2b)	2, 39–42		$\dfrac{100(5-x)}{5} = \%$
Origin of Earth's Atmosphere, Hydrosphere, and Lithosphere (MU 1.2e, f, h)			
Theories of the Origin of the Universe and Solar System (MU 1.2a, c)	1, 18, 37		$\dfrac{100(3-x)}{3} = \%$

SELF-ANALYSIS CHART June 2015

Topic	Question Numbers (Total)	Wrong Answers (x)	Grade
Meteorology			
Energy Sources for Earth Systems (MU 2.1a, b)	8		$\dfrac{100(1-x)}{1} = $ %
Weather (MU 2.1c, d, e, f, g, h)	6, 7, 9, 28, 36, 38, 52, 53		$\dfrac{100(8-x)}{8} = $ %
Insolation and Seasonal Changes (MU 2.1i; 2.2a, b)	15, 17, 30, 51		$\dfrac{100(4-x)}{4} = $ %
The Water Cycle and Climates (MU 1.2g; 2.2c, d)	4, 21, 22, 29, 62–66		$\dfrac{100(9-x)}{9} = $ %
Geology			
Minerals and Rocks (MU 3.1a, b, c)	10, 33–35, 45–48, 50		$\dfrac{100(9-x)}{9} = $ %
Weathering, Erosion, and Deposition (MU 2.1s, t, u, v, w)	14, 24, 27, 31, 58–61		$\dfrac{100(8-x)}{8} = $ %
Plate Tectonics and Earth's Interior (MU 2.1j, k, l, m, n, o)	11, 32, 43, 44, 49, 72–76		$\dfrac{100(10-x)}{10} = $ %
Geologic History (MU 1.2i, j)	5, 23, 25, 26, 79–82		$\dfrac{100(8-x)}{8} = $ %
Topographic Maps and Landscapes (MU 2.1p, q, r)	12, 13, 67–71		$\dfrac{100(7-x)}{7} = $ %
ESRT			
Earth Science Reference Tables 2011 Edition	1, 5–11, 13, 14, 18, 23–25, 31–35, 37, 41, 44, 45, 49, 50, 52, 53, 57, 62, 63, 65, 71, 74, 75, 80		$\dfrac{100(35-x)}{35} = $ %

To further pinpoint your weak areas, use the Topic Outline in the front of the book.
MU = Major Understanding (see Topic Outline)

Examination August 2015
Physical Setting/Earth Science

PART A
Answer all questions in this part.

Directions (1–35): For *each* statement or question, choose the word or expression that, of those given, best completes the statement or answers the question. Some questions may require the use of the *2011 Edition Reference Tables for Physical Setting/Earth Science*. Record your answers in the space provided.

1 Which characteristics best describe the star *Betelgeuse*?
  (1) reddish orange with low luminosity and high surface temperature
  (2) reddish orange with high luminosity and low surface temperature
  (3) blue white with low luminosity and low surface temperature
  (4) blue white with high luminosity and high surface temperature   1 _____

2 Which motion occurs at a rate of approximately one degree per day?
  (1) the Moon revolving around Earth
  (2) the Moon rotating on its axis
  (3) Earth revolving around the Sun
  (4) Earth rotating on its axis   2 _____

3 If the tilt of Earth's axis were increased from 23.5° to 30°, summers in New York State would become

(1) cooler, and winters would become cooler
(2) cooler, and winters would become warmer
(3) warmer, and winters would become cooler
(4) warmer, and winters would become warmer   3 _____

4 Which object in space emits light because it releases energy produced by nuclear fusion?

(1) Earth's Moon   (3) Venus
(2) Halley's comet   (4) *Polaris*   4 _____

5 Since Denver's longitude is 105° W and Utica's longitude is 75° W, sunrise in Denver occurs

(1) 2 hours earlier   (3) 3 hours earlier
(2) 2 hours later   (4) 3 hours later   5 _____

6 A major piece of evidence supporting the Big Bang theory is the observation that wavelengths of light from stars in distant galaxies show a

(1) redshift, appearing to be shorter
(2) redshift, appearing to be longer
(3) blueshift, appearing to be shorter
(4) blueshift, appearing to be longer   6 _____

7 During the month of January, at which location in New York State is the Sun lowest in the sky at solar noon?

   (1) Massena   (3) Utica
   (2) Niagara Falls (4) New York City

8 Which process releases 2260 joules of heat energy per gram of water into the environment?

   (1) melting   (3) condensation
   (2) freezing   (4) evaporation

9 When snow cover on the land melts, the water will most likely become surface runoff if the land surface is

   (1) frozen
   (2) porous
   (3) grass covered
   (4) unconsolidated gravel

10 Which area is the most common source region for cold, dry air masses that move over New York State?

   (1) North Atlantic Ocean
   (2) Gulf of Mexico
   (3) central Canada
   (4) central Mexico

11 The map below shows a portion of the Hudson River and three tributaries: Catskill Creek, Fishkill Creek, and Wallkill River.

The greatest discharge of the Hudson River is generally observed near

(1) Albany
(2) Kingston
(3) Poughkeepsie
(4) Ossining

11 ____

12 Which station model represents a location that has the greatest chance of precipitation?

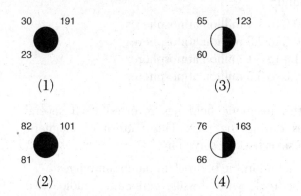

12 ____

13 The Adirondacks are classified as mountains because of the high elevation and bedrock that consists mainly of

(1) deformed and intensely metamorphosed rocks
(2) glacial deposits of unconsolidated gravels, sands, and clays
(3) Cambrian and Ordovician quartzites and marbles
(4) horizontal sedimentary rocks of marine origin

13 ____

14 In which landscape region are New York State's Finger Lakes primarily located?

(1) Adirondack Mountains
(2) Allegheny Plateau
(3) Atlantic Coastal Plain
(4) Erie-Ontario Lowlands

14 ____

15 What is the range of pressure in Earth's interior where rock with a density range of 9.9 to 12.2 g/cm^3 is found?

(1) 0.2 to 1.4 million atmospheres
(2) 0.8 to 2.3 million atmospheres
(3) 1.4 to 3.1 million atmospheres
(4) 2.3 to 3.5 million atmospheres   15 _____

16 Earth's magnetic field has reversed itself several times during the past. This pattern of magnetic reversal is best preserved in

(1) metamorphic bedrock in mountain ranges
(2) bedrock with fossils containing radioactive carbon-14
(3) layers of sedimentary bedrock of the Grand Canyon
(4) igneous bedrock of the oceanic crust   16 _____

17 Which two features are commonly found at divergent plate boundaries?

(1) mid-ocean ridges and rift valleys
(2) wide valleys and deltas
(3) ocean trenches and subduction zones
(4) hot spots and island arcs   17 _____

18 New York State bedrock of which age contains salt, gypsum, and hematite?

(1) Cambrian   (3) Mississippian
(2) Devonian   (4) Silurian   18 _____

19 Scientists infer that oxygen in Earth's atmosphere did *not* exist in large quantities until after
  (1) the first multicellular, soft-bodied marine organisms appeared on Earth
  (2) the initial opening of the Atlantic Ocean
  (3) the first sexually reproducing organisms appeared on Earth
  (4) photosynthetic cyanobacteria evolved in Earth's oceans   19 ____

20 Which organisms were alive when New York State was last covered by a continental ice sheet?
  (1) *Eurypterus* and *Cooksonia*
  (2) *Aneurophyton* and Naples Tree
  (3) mastodont and Beluga whale
  (4) *Coelophysis* and *Elliptocephala*   20 ____

21 One difference between a breccia rock and a conglomerate rock is that the particles in a breccia rock are
  (1) more aligned
  (2) more angular
  (3) harder
  (4) land derived   21 ____

22 The photograph below shows rock layers separated by unconformity XY.

Which sequence of events most likely produced this unconformity?

(1) uplift and erosion of bedrock, followed by subsidence and more deposition
(2) intrusion of magma into preexisting rock, causing contact metamorphism
(3) eruption of a volcano, spreading lava over horizontal sedimentary rock layers
(4) separation of one rock layer, by movement along a plate boundary   22 _____

23 The igneous rock gabbro most likely formed from molten material that cooled

(1) rapidly at Earth's surface
(2) slowly at Earth's surface
(3) rapidly, deep underground
(4) slowly, deep underground   23 _____

24 Which statement best supports the inference that most of Earth's present-day land surfaces have, at one time, been covered by water?

(1) Volcanic eruptions contain large amounts of water vapor.
(2) Coral reefs formed, in the past, along the edges of many continents.
(3) Seafloor spreading has pulled landmasses apart and pushed them together.
(4) Sedimentary bedrock of marine origin covers large areas of Earth's continents.

24 _____

25 Which diagram best represents the correct orientation of the North Pole [NP] as Earth revolves around the Sun? [Diagrams are not drawn to scale.]

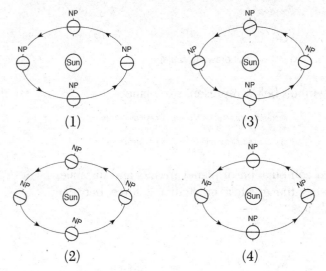

25 _____

26 Which diagram best represents how greenhouse gases in our atmosphere trap heat energy?

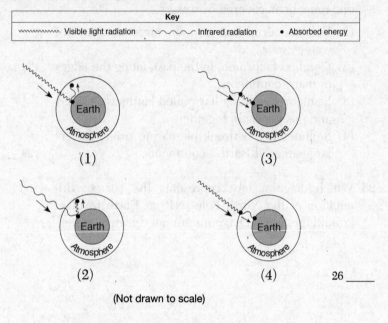

(Not drawn to scale)

27 The symbols below represent two planets.

⑤ represents a planet with a mass 5 times Earth's mass.

⑨ represents a planet with a mass 9 times Earth's mass.

Which combination of planet masses and distances produces the greatest gravitational force between the planets?

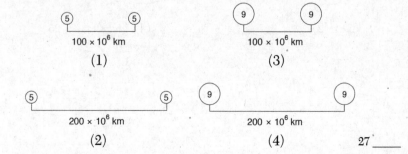

28 The diagram below represents the circulation of air above Earth's surface at a coastal location during the day and at night.

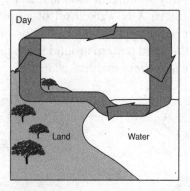

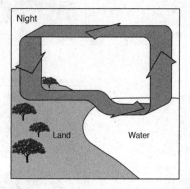

This local air movement is best described as an example of

(1) conduction between Earth's surface and the atmosphere above it
(2) condensation of water vapor during the day, and evaporation of water during the night
(3) convection resulting from temperature and pressure differences above land and water
(4) greater radiation from the warmer ocean during the day and from the warmer land at night

28 _____

29 A change in the type and location of large high-pressure systems (**H**) and large low-pressure systems (**L**) over Asia creates shifts in prevailing winds that cause a rainy summer season and a dry winter season in southern Asia. Which set of maps below best represents the type and location of pressure systems and the wind pattern around these pressure systems that cause these seasonal changes?

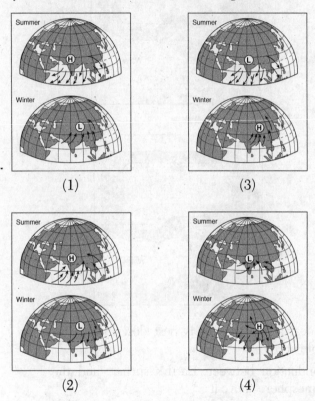

29 ____

30 The photograph below shows both erosional and depositional features formed by an agent of erosion.

Which agent of erosion produced the features shown in the photograph?

(1) running water (3) ocean waves
(2) glacial ice   (4) prevailing wind   30 ____

31 Which cross section best represents the pattern of sediments deposited on the bottom of a lake as the velocity of the stream entering the lake steadily decreased?

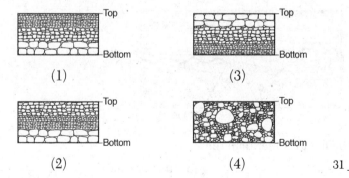

31 ____

32 Which graph best shows the relationship between the compositions of different igneous rocks and their densities?

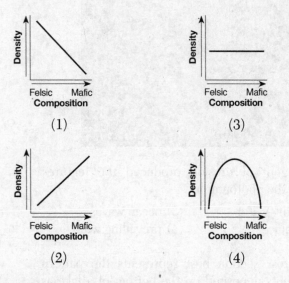

33 The geologic cross section below shows rock layers that have not been overturned.

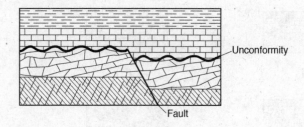

The fault is older than the

(1) slate
(2) marble
(3) unconformity
(4) shale

34 The pie graph below represents the composition, in percent by mass, of the chemical elements found in an Earth layer.

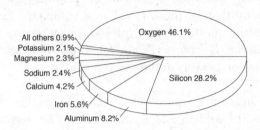

The composition of which Earth layer is represented by the pie graph?

(1) crust
(2) outer core
(3) troposphere
(4) hydrosphere

34 ____

35 The diagram below indicates physical changes that accompany the conversion of shale to gneiss.

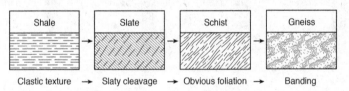

Which geologic process is occurring to cause this conversion?

(1) sedimentary layering
(2) intrusion of magma
(3) metamorphism
(4) weathering

35 ____

PART B–1
Answer all questions in this part.

Directions (36–50): For *each* statement or question, choose the word or expression that, of those given, best completes the statement or answers the question. Some questions may require the use of the *2011 Edition Reference Tables for Physical Setting/Earth Science*. Record your answers in the space provided.

Base your answers to questions 36 and 37 on the diagram below and on your knowledge of Earth science. The diagram represents four tubes, labeled *A*, *B*, *C*, and *D*, each containing 150 mL of sediments. Tubes *A*, *B*, and *C* contain well-sorted, closely packed sediments of uniform shape and size. Tube *D* contains uniformly shaped, closely packed sediments of mixed sizes. The particle size of the sediment in each tube is labeled.

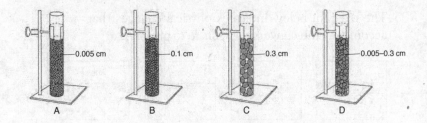

36 Water was added to each tube to just cover the sediments and the volumes of water added were recorded. These data can best be used to determine the

(1) particle size of the sediments
(2) particle shape of the sediments
(3) water retention of the sediments
(4) porosity of the sediments   36 _____

37 If tubes A, B, and C were set up to test for capillarity, the data would show that capillarity is
(1) greatest in tube A
(2) greatest in tube B
(3) greatest in tube C
(4) the same for tubes A, B, and C

Base your answers to questions 38 through 40 on the diagram below and on your knowledge of Earth science. The diagram represents a cut-away view of Earth's interior and the paths of some of the seismic waves produced by an earthquake that originated below Earth's surface. Points A, B, and C represent seismic stations on Earth's surface. Point D represents a location at the boundary between the core and the mantle.

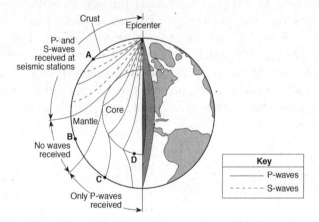

38 Seismic station A is 5000 kilometers from the epicenter. What is the difference between the arrival time of the first P-wave and the arrival time of the first S-wave recorded at this station?
(1) 2 minutes 20 seconds
(2) 6 minutes 40 seconds
(3) 8 minutes 20 seconds
(4) 15 minutes 00 second

39 Which process prevented *P*-waves from arriving at seismic station *B*?

(1) refraction
(2) reflection
(3) convection
(4) conduction

39 ____

40 Only *P*-waves were recorded at seismic station *C* because *P*-waves travel

(1) only through Earth's interior, and *S*-waves travel only on Earth's surface
(2) fast enough to penetrate the core, and *S*-waves travel too slowly
(3) through iron and nickel, while *S*-waves cannot
(4) through liquids, while *S*-waves cannot

40 ____

Base your answers to questions 41 through 43 on the diagrams below and on your knowledge of Earth science. The diagrams, labeled *A*, *B*, and *C*, represent equal-sized portions of the Sun's rays striking Earth's surface at 23.5° N latitude at noon at three different times of the year. The angle at which the Sun's rays hit Earth's surface and the relative areas of Earth's surface receiving the rays at the three different angles of insolation are shown.

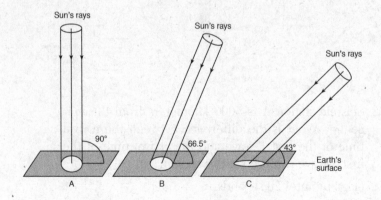

41 As viewed in sequence from A to B to C, these diagrams represent which months and which change in the intensity of insolation?

(1) December → March → June; and decreasing intensity
(2) December → March → June; and increasing intensity
(3) June → September → December; and decreasing intensity
(4) June → September → December; and increasing intensity

41 _____

42 As the angle of the Sun's rays striking Earth's surface at noon changes from 90° to 43°, the length of a shadow cast by an object will

(1) decrease
(2) increase
(3) decrease, then increase
(4) increase, then decrease

42 _____

43 Which graph best shows the duration of insolation at this location as the angle of insolation changes?

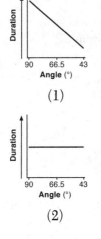
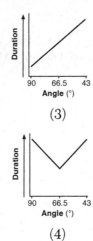

43 _____

Base your answers to questions 44 through 47 on the diagram below and on your knowledge of Earth science. The diagram represents the Moon at four positions, labeled A, B, C, and D, in its orbit around Earth. The position of the full-Moon phase is labeled.

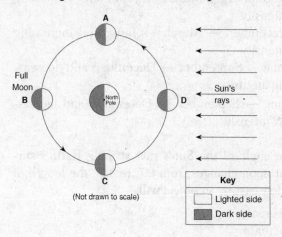

44 Approximately how many days (d) does it take for the Moon to move from the phase shown at position A to the full-Moon phase?

(1) 7.4 d (3) 27.3 d
(2) 14.7 d (4) 29.5 d

44 _____

45 Which phase of the Moon could be observed from New York State when the Moon is at position C?

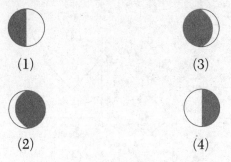

45 _____

41 As viewed in sequence from *A* to *B* to *C*, these diagrams represent which months and which change in the intensity of insolation?

(1) December → March → June; and decreasing intensity
(2) December → March → June; and increasing intensity
(3) June → September → December; and decreasing intensity
(4) June → September → December; and increasing intensity   41 ____

42 As the angle of the Sun's rays striking Earth's surface at noon changes from 90° to 43°, the length of a shadow cast by an object will

(1) decrease
(2) increase
(3) decrease, then increase
(4) increase, then decrease   42 ____

43 Which graph best shows the duration of insolation at this location as the angle of insolation changes?

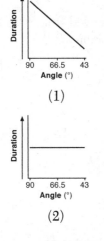

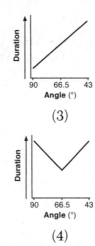

43 ____

Base your answers to questions 44 through 47 on the diagram below and on your knowledge of Earth science. The diagram represents the Moon at four positions, labeled A, B, C, and D, in its orbit around Earth. The position of the full-Moon phase is labeled.

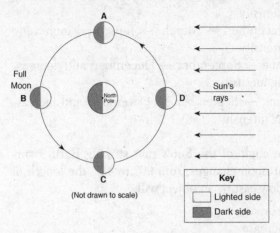

44 Approximately how many days (d) does it take for the Moon to move from the phase shown at position A to the full-Moon phase?

(1) 7.4 d (3) 27.3 d
(2) 14.7 d (4) 29.5 d

44 _____

45 Which phase of the Moon could be observed from New York State when the Moon is at position C?

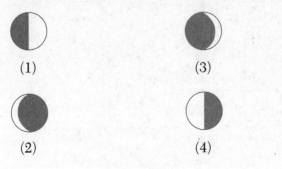

45 _____

46 The same side of the Moon always faces Earth because the Moon's period of revolution
   (1) is longer than the Moon's period of rotation
   (2) equals the Moon's period of rotation
   (3) is longer than Earth's period of rotation
   (4) equals Earth's period of rotation   46 _____

47 Solar and lunar eclipses rarely happen during a cycle of phases because the
   (1) Moon's orbit is circular and Earth's orbit is elliptical
   (2) Moon's orbit is elliptical and Earth's orbit is elliptical
   (3) plane of the Moon's orbit is different from the plane of Earth's orbit
   (4) plane of the Moon's orbit is the same as the plane of Earth's orbit   47 _____

Base your answers to questions 48 through 50 on the maps and data table below and on your knowledge of Earth science. Map I shows the Outer Banks and part of North Carolina along the southeastern coast of the United States. Maps II and III show enlargements of the Avon-Buxton section of the Outer Banks indicated by box **X** on map I. Map II shows the land and shoreline in 1852. Map III shows the land and shoreline in 1998. The dotted line on map III shows the location of the 1852 shoreline. The data table shows the average width, in meters, at various years, of the Avon-Buxton section.

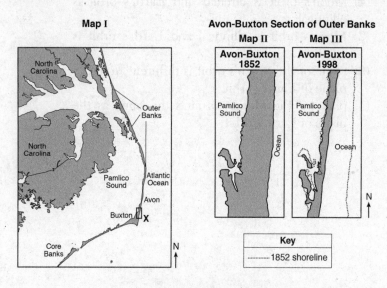

Avon-Buxton Section Width from 1852 to 1998

Year	Average Width (m)
1852	813
1917	547
1940	426
1962	284
1974	284
1998	219

48 The Outer Banks were formed primarily from sediments eroded and deposited by ocean waves. Which type of landform are the Outer Banks?

(1) outwash plains
(2) moraine deposits
(3) river deltas
(4) barrier islands

49 Which bar graph best shows the average width of the Avon-Buxton section of the Outer Banks from 1852 to 1998?

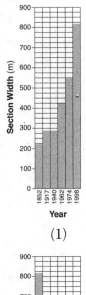
(1)

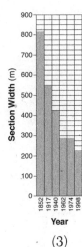
(3)

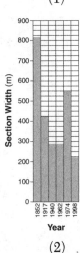
(2)

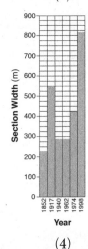
(4)

50 Which ocean current has the greatest warming influence on the climate of the Outer Banks of North Carolina?

(1) Gulf Stream Current
(2) North Atlantic Current
(3) Labrador Current
(4) Canary Current

PART B–2
Answer all questions in this part.

Directions (51–65): Record your answers in the spaces provided. Some questions may require the use of the *2011 Edition Reference Tables for Physical Setting/Earth Science.*

Base your answers to questions 51 through 54 on the passage and the graph below and on your knowledge of Earth science.

Great Lake Effects

The Great Lakes influence the weather and climate of nearby land regions at all times of the year. Much of this lake effect is determined by the relative temperatures of surface lake water compared to the surface air temperatures over those land areas. The graph below shows the average monthly temperature of the surface water of Lake Erie and the surface air temperature at Buffalo, New York.

In an average year, four lake-effect seasons are experienced. When surface lake temperatures are colder than surface air temperatures, a stable season occurs. The cooler lake waters suppress cloud development and reduce the strength of rainstorms. As a result, late spring and early summer in the Buffalo region tends to be very sunny.

A season of lake-effect rains follows. August is usually a time of heavy nighttime rains, and much of the rainy season is marked by heavy, localized rainstorms downwind from the lake. Gradually, during late October, lake-effect rains are replaced by snows. Generally, the longer the time the wind travels over the lake, the heavier the lake effect becomes in Buffalo.

Finally, conditions stabilize again, as the relatively shallow Lake Erie freezes over, usually near the end of January. Very few lake-effect storms occur during this time period.

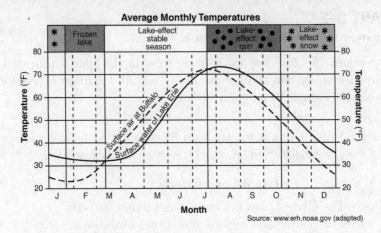

51 The passage states, "The cooler lake waters suppress cloud development..." because the water cools the air above its surface. Explain why this cool air above the lake surface reduces the amount of cloud development. [1]

_____

_____

52 Identify *one* weather variable that determines whether Buffalo receives rain or snow from a lake-effect storm in October. [1]

_____

53 On the map *below*, draw *one* straight arrow in Lake Erie to show the winter wind direction most likely to bring the heaviest lake-effect snows to Buffalo. [1]

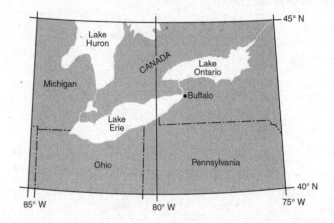

54 Explain why the Buffalo surface air temperatures increase faster and earlier in the year than do the surface water temperatures of Lake Erie. [1]

_____

_____

Base your answers to questions 55 through 58 on the graph below and on your knowledge of Earth science. The graph shows planet equatorial diameters and planet mean distances from the Sun. Neptune is *not* shown.

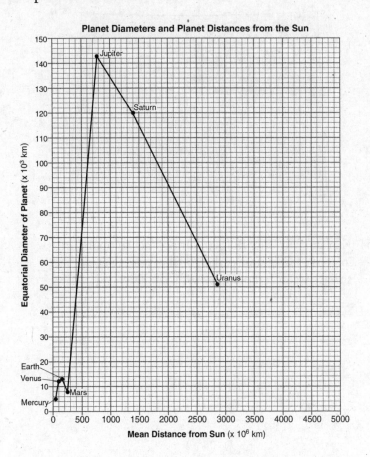

55 Place an **X** on the graph to indicate where Neptune would be plotted, based on its mean distance from the Sun and its equatorial diameter. [1]

56 The diagram below represents Earth drawn to a scale of 1 cm = 2000 km. Centimeter markings along the equatorial diameter of Earth are also shown on the diagram. On this diagram, shade in the space between the centimeter markings to represent the equatorial diameter of Earth's Moon at this same scale. [1]

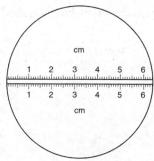

Scale: 1 cm = 2000 km

57 Compared to the periods of revolution and periods of rotation of the terrestrial planets, how are the periods of revolution and periods of rotation for the Jovian planets different? [1]

Jovian periods of revolution:

_____

Jovian periods of rotation:

_____

58 The center of the asteroid belt is approximately 404 million kilometers from the Sun. State the name of the planet that is closest to the center of the asteroid belt. [1]

_____

Base your answers to questions 59 through 61 on the map of Haiti's location and portion of the Modified Mercalli Intensity Scale below, on the Haiti Earthquake Intensity Map below, and on your knowledge of Earth science. The map shows the location of Haiti in the Atlantic Ocean. The Modified Mercalli Intensity Scale describes the amount and type of damage caused by an earthquake on a scale from I to XII. A portion of this scale is shown below. Modified Mercalli intensity values for the January 12, 2010, earthquake in Haiti are represented on the Haiti Earthquake Intensity Map below.

Map of Haiti's Location

A Portion of the Modified Mercalli Intensity Scale

Intensity	Description of Effects
IV	Generally felt by people in motion, loose objects disturbed
V	Felt by nearly everyone; some dishes, windows broken
VI	Felt by all; slight damage to ordinary structures
VII	Damage rare in buildings of good design and construction; slight to moderate in ordinary structures; considerable damage in poorly built structures; some chimneys broken
VIII	Damage slight in specially designed structures; considerable damage in ordinary substantial buildings with partial collapse; damage great in poorly built structures; falling chimneys, columns, monuments, walls
IX	Damage considerable in specially designed structures; damage great in substantial buildings, with partial collapse; buildings shifted off foundations
X	Some well-built wooden structures destroyed; most concrete and frame structures destroyed along with foundations

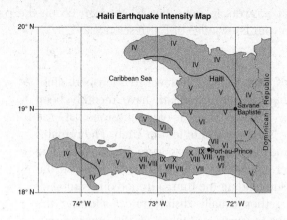

59 On the earthquake intensity map, boundary lines have been drawn between the Modified Mercalli intensity values of IV and V. On this map, draw boundary lines between the Modified Mercalli intensity values of V and VI. [1]

60 State the latitude and longitude of Savane Baptiste. Include the units and compass directions in your answer. [1]

Latitude: _____

Longitude: _____

61 Haiti is located at a transform boundary between which two tectonic plates? [1]

_____ **Plate** and _____ **Plate**

Base your answers to questions 62 through 65 on the passage below and on your knowledge of Earth science.

Dinosaur Fossils

Bones of juvenile long-necked sauropod dinosaurs, *Abydosaurus mcintoshi*, have recently been found in 105-million-year-old sandstone at the Dinosaur National Monument in Utah. The remains of four individual dinosaurs were found, including two intact skulls. This find is unusual because the softer tissue holding the thin sauropod dinosaur skull bones together usually disintegrates, allowing the skull bones to separate. Only 8 of 120 types of sauropods discovered have complete skull specimens. These dinosaurs were herbivores, with large numbers of sharp teeth that were probably replaced five to six times each year. These teeth allowed only for the harvesting of plant material, but not for chewing it afterward. The plant-harvesting teeth and long neck identify *Abydosaurus mcintoshi* as a descendant of the brachiosaurs.

62 On the timeline below, place an **X** on line *AB* to indicate the time when *Abydosaurus mcintoshi* lived. [1]

```
              B
              ┌Oligocene
   Paleogene  │Eocene
              └Paleocene
              ┌Late
   Cretaceous │
              └Early
              ┌Late
   Jurassic   │Middle
              └Early
              ┌Late
   Triassic   │Middle
              └Early
              ┌Late
              │Middle
   Permian    └Early
              A
```

63 Indicate the range of grain sizes in the type of bedrock in which *Abydosaurus mcintoshi* bones were found. [1]

_____ **cm** to _____ **cm**

64 Identify *one* group of organisms that was a likely food source for *Abydosaurus mcintoshi*. [1]

_____

65 State a natural event that is inferred by most scientists to be the cause of extinction of the last of the dinosaurs. [1]

_____

PART C
Answer all questions in this part.

Directions (66–85): Record your answers in the spaces provided. Some questions may require the use of the *2011 Edition Reference Tables for Physical Setting/Earth Science*.

Base your answers to questions 66 through 69 on the block diagram below and on your knowledge of Earth science. The diagram represents an igneous intrusion that solidified between some layers of sedimentary rock. Letter X represents an index fossil in a sedimentary rock layer. The rock layers have *not* been overturned.

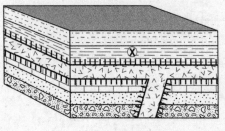 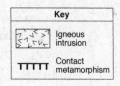

(Not drawn to scale)

66 Describe the evidence represented in the diagram that indicates that the shale layer and the limestone layer are older than the igneous intrusion. [1]

_____

_____

67 The limestone layer is composed mostly of what mineral? [1]

_____

68 Describe *one* characteristic of fossil X that makes it a good index fossil. [1]

_____

69 The igneous intrusion contains the radioactive isotope potassium-40, which is used in radioactive dating to determine the age of rocks. State *one* property of potassium-40 that allows it to be useful in the radioactive dating of rocks. [1]

_____

_____

Base your answers to questions 70 through 73 on the data table below and on your knowledge of Earth science. The data show the rate of change in the apparent direction of the swing of a Foucault pendulum at various latitudes on Earth, in degrees per hour.

A Foucault Pendulum's Swing

Latitude (°)	Rate of Change in Apparent Direction of Swing (°/h)
0	0.0
10	2.6
20	5.1
30	7.5
40	9.6
50	11.5
60	13.0
70	14.1
80	14.8
90	15.0

70 On the grid *below*, plot the hourly change in a Foucault pendulum's apparent direction of swing at the latitudes shown on the data table. Connect the plots with a line. [1]

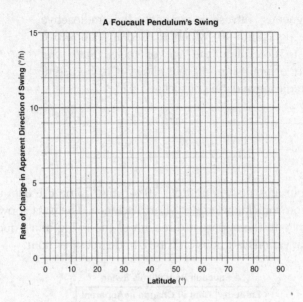

71 Calculate how many hours are needed for a Foucault pendulum located at the North Pole to complete a 360° change in its apparent direction of swing. [1]

_____ h

72 If a Foucault pendulum were set up on Mars, it would most likely show similar changes in the pendulum's apparent direction of swing. Identify the motion of the planet Mars that would cause this change. [1]

_____

73 The Coriolis force results from the same motion that causes the Foucault pendulum to change its apparent direction of swing. The diagram below represents the relative strength of the Coriolis force acting on air moving over Earth's surface.

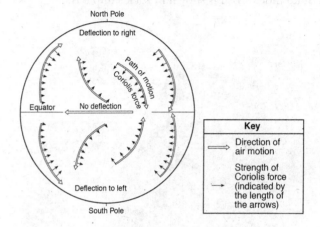

Describe how the strength of the Coriolis force changes with latitude. [1]

_____

_____

Base your answers to questions 74 through 76 on the block diagram below, which represents a landscape drained by a stream system, and on your knowledge of Earth science. The actual sizes and shapes of three rock samples, labeled *A*, *B*, and *C*, and the locations where they were found in the stream are indicated in the diagram. A New York State index fossil is shown in rock sample *A*.

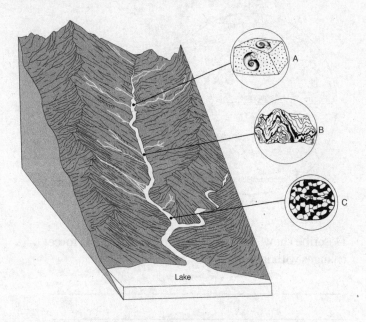

74 Explain how the appearance of rock sample *A* indicates that the sample has spent very little time being transported by the stream. [1]

_____

_____

75 Rock sample C has a diameter of 2 centimeters. Determine the minimum stream velocity needed to transport rock sample C to its present location. [1]

_____ cm/s

76 The stream profile below shows the locations of rock samples A, B, and C in the streambed.

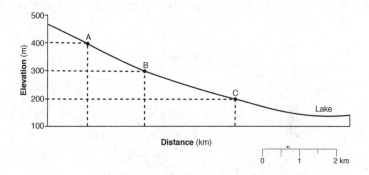

Calculate the stream gradient between the locations of rock sample A and rock sample C. [1]

_____ m/km

Base your answers to questions 77 through 80 on the map below and on your knowledge of Earth science. The map shows the path of a tornado that moved through a portion of Nebraska on May 22, 2004 between 7:30 p.m. and 9:10 p.m. The path of the tornado along the ground is indicated by the shaded region. The width of the shading indicates the width of destruction on the ground. Numbers on the tornado's path indicate the Fujita intensity at those locations. The Fujita Intensity Scale (F-Scale), in the left corner of the map, provides information about wind speed and damage at various F-Scale intensities.

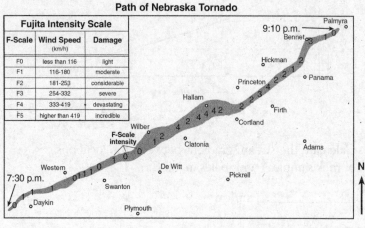

77 On the map above, place an **X** at a location where the tornado damage was greatest. [1]

78 State a possible wind speed of the tornado, in kilometers per hour (km/h), when it was moving through the town of Bennet. [1]

_____ **km/h**

79 Identify the weather instrument usually used to measure wind speed. [1]

_____

80 Describe *one* safety precaution that should be taken if a tornado has been sighted approaching your home. [1]

_____

Base your answers to questions 81 through 85 on the reading passage and map below and on your knowledge of Earth science. The passage provides information regarding the eruption of a volcano in Iceland. The map shows the thickness of ash deposits, in centimeters (cm), during the first three days of the eruption. Point A, representing the volcano's location, and point B, representing a location on Earth's surface, are connected with a reference line.

Iceland Volcano Eruption Spreads Ash Cloud over Europe

On April 14, 2010, Eyjafjallajökull volcano, located in southern Iceland, explosively erupted, sending large volumes of volcanic ash high into the atmosphere. Much of the ash fell quickly to Earth, as seen in the map, but large quantities remained airborne and spread over Europe. Most of the ash was transported within the atmosphere below 10 kilometers. Air traffic across the Atlantic and throughout Europe was severely disrupted, as airlines were forced to keep jet aircraft on the ground.

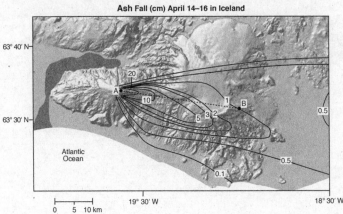

Source: Ash generation and distribution from the April-May 2010 eruption of Eyjafjallajökull, Iceland, Gudmundsson et al., *Scientific Reports*, August 14, 2012 (adapted)

81 On the grid *below*, construct a profile of the thickness of the volcanic ash deposits by plotting the ash fall along line *AB*. Plot *each* point where an isoline showing thickness is crossed by line *AB*. Ash thickness at location *A* has been plotted. Complete the profile by connecting *all seven* plots with a line. [1]

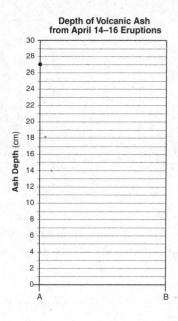

82 Identify the atmospheric layer within which most of the volcanic ash was transported. [1]

_____

83 Describe *one* way the volcanic ash cloud may have contributed to cooler weather conditions in Europe. [1]

_____

_____

84 The graphs below indicate the percent by mass of different diameters of ash particles deposited at 2 kilometers and 60 kilometers from the volcanic eruption.

Volcanic Ash Deposited from the April 14-16 Eruption

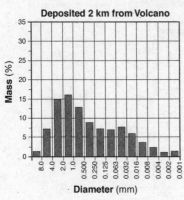
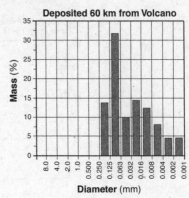

Describe how the size of the deposited ash particles changed with increased distance from the volcano. [1]

_____

85 Explain why the lithosphere in the vicinity of Eyjafjallajökull is more volcanically active than most other regions of Earth's surface. [1]

_____

_____

Answers August 2015
Physical Setting/Earth Science

Answer Key

PART A

1. 2	8. 3	15. 3	22. 1	29. 4
2. 3	9. 1	16. 4	23. 4	30. 2
3. 3	10. 3	17. 1	24. 4	31. 1
4. 4	11. 4	18. 4	25. 2	32. 2
5. 2	12. 2	19. 4	26. 1	33. 4
6. 2	13. 1	20. 3	27. 3	34. 1
7. 1	14. 2	21. 2	28. 3	35. 3

PART B–1

36. 4	39. 1	42. 2	45. 4	48. 4
37. 1	40. 4	43. 1	46. 2	49. 3
38. 2	41. 3	44. 1	47. 3	50. 1

PART B–2 and **PART C**. *See* **Answers Explained**.

Answers Explained

PART A

1. **2** Find the Characteristics of Stars chart in the *Reference Tables for Physical Setting/Earth Science* and locate *Betelgeuse*. From *Betelgeuse*, trace downward to the horizontal axis labeled "Surface Temperature (K)." Note that the temperature of *Betelgeuse* is slightly more than 3,000K and that this is near the low end of the surface temperature scale. Continue tracing downward to the horizontal scale labeled "Color." Note that *Betelgeuse* is orange but on the side of the orange range nearest red. From *Betelgeuse*, trace left horizontally to the vertical axis labeled "Luminosity (Rate at which a star emits energy relative to the Sun)." Note that *Betelgeuse* is slightly more than 100,000 times as luminous as the Sun, which is near the high end of the luminosity scale. Thus, the characteristics that best describe the star *Betelgeuse* are reddish orange with high luminosity and low surface temperature.

2. **3** Note that the choices refer only to the rotation or revolution of Earth and the rotation or revolution of the Moon (Earth's Moon). One complete rotation or revolution consists of motion through 360 degrees. At a rate of motion of approximately one degree per day, approximately 360 days would be required to complete one rotation or one revolution. Find the Solar System Data chart in the *Reference Tables for Physical Setting/ Earth Science*. In the column labeled "Celestial Object," locate "Earth" and "Earth's Moon." From each, trace right to the columns labeled "Period of Revolution (d = days) (y = years)" and "Period of Rotation at Equator." Note that the period of rotation or revolution nearest 360 days is Earth's period of revolution—365.26 days. Thus, the motion that occurs at a rate of approximately one degree per day is Earth revolving around the Sun.

3. **3** If the tilt of Earth's axis were to increase from 23½° to 30°, Earth would tilt farther toward the Sun in the summer and would tilt farther away from the Sun in the winter. Thus, New York State would receive more direct sunlight during the summer and less direct sunlight during the winter. More direct sunlight has greater intensity and warms the surface more. Conversely, less direct sunlight has less intensity and warms the surface less. Therefore, if the tilt of Earth's axis were to increase, summers in New York State would become warmer and winters would become cooler.

4. **4** For nuclear fusion to occur, atomic nuclei must come together. However, the protons in each nucleus repel each other because they have the same charge (positive). High temperatures (about 100 million Kelvin) are needed to give the nuclei enough energy to overcome these forces

of repulsion. High pressures are also needed to squeeze the nuclei close together enough to fuse. In the interior of stars, gravity is strong enough to create temperatures and pressures high enough to overcome the force of repulsion between atomic nuclei so that the nuclei can combine by the process of nuclear fusion. Thus, the object in space that emits light because it releases energy produced by nuclear fusion is the star *Polaris*.

WRONG CHOICES EXPLAINED:

(1), (2), (3) Comets, planets, and moons do not contain enough mass for gravitational attraction to result in core pressures and temperatures high enough to trigger nuclear fusion. Therefore, Earth's Moon, Halley's comet, and Venus do not emit light produced by nuclear fusion. They simply reflect sunlight from their surfaces.

5. **2** Earth rotates through 360° of longitude in 24 hours, or 15° of longitude per hour (360 ÷ 24 = 15). Thus, every 15° of longitude that separates two locations corresponds to a difference of 1 hour in solar time. Since Denver's longitude is 105°W and Utica's longitude is 75°W, they are separated by 30° of longitude and therefore experience a 2-hour difference in solar time. Thus, Denver and Utica experience sunrise 2 hours apart. As Earth rotates from west to east, Utica, NY, in the eastern part of the United States is carried from darkness into daylight (experience sunrise) before Denver, CO, in the western part of the United States. Therefore, sunrise in Denver occurs 2 hours later than sunrise in Utica.

6. **2** According to the Big Bang theory, the universe started out with all of its matter in a small volume and then expanded outward in all directions, a motion very much like an explosion. Observations that indicate that the universe is expanding directly support this model. Find the Electromagnetic Spectrum chart in the *Reference Tables for Physical Setting/Earth Science*. Locate the section of the chart labeled "Visible light." Note that light at the red end of the spectrum has a longer wavelength than light at the blue end of the spectrum. If a source of electromagnetic waves is moving *away* from an observer at the same time as it is emitting light of a particular wavelength, fewer wave crests will reach the eye of the observer each second. The eye will interpret this as meaning that the light has a longer wavelength than it actually has. In other words, the light will appear shifted toward the red end of the spectrum, or redshifted. When the light from distant galaxies is observed, all are seen to be redshifted, indicating that they are all moving away from us. Furthermore, the farther away the galaxy, the more it is redshifted. This indicates that distant galaxies are moving away from us faster than nearer galaxies. The only explanation for such an observation is that the universe is expanding in all directions. Thus, a major piece of evidence supporting the Big Bang theory is the observation that wavelengths of light from stars in distant galaxies show a redshift, appearing to be longer.

7. **1** In January, shortly after the winter solstice, Earth's northern hemisphere is tilted away from the Sun and the Sun's direct rays strike Earth's surface south of the equator at solar noon. Since Earth is a sphere, for every degree of latitude north of the point at which the Sun's direct rays strike Earth's surface at solar noon, the altitude of the Sun at solar noon decreases by one degree. Thus, within New York State, the Sun would be lowest in the sky at solar noon at the location with the northernmost latitude. Find the Generalized Bedrock Geology of New York State map in the *Reference Tables for Physical Setting/Earth Science*. Note that the north latitude increases toward the top of the map. Therefore, the northernmost location is located nearest the top of the map. Of the choices given, Massena is located nearest the top of the map and is therefore the northernmost location. Thus, during the month of January, the Sun is lowest in the sky at solar noon at Massena.

8. **3** Find the Properties of Water table in the *Reference Tables for Physical Setting/Earth Science*. Note that the heat energy released during condensation is 2260 J/g. Therefore, the process that releases 2260 joules of heat energy per gram of water into the environment is condensation.

9. **1** Surface runoff is precipitation that does not evaporate or sink into the ground but, instead, runs downhill along Earth's surface. If the land surface is frozen, water has solidified in the pore spaces between soil particles at the surface, blocking any openings through which liquid water could sink into the ground. Thus, when snow cover on the land melts, the water will most likely become surface runoff if the land surface is frozen.

WRONG CHOICES EXPLAINED:
(2) If a land surface is porous, it has open spaces through which water can sink into the ground. This increases the likelihood that water will sink into the ground rather than running off, thereby decreasing the amount of surface runoff.

(3) On a grass-covered surface, the roots and stems of the vegetation slow the downhill flow of water. This allows more time for the water to sink into the ground, thereby decreasing the amount of surface runoff.

(4) Unconsolidated sediments are loose materials that are not cemented together into a solid rock. In unconsolidated sediments, water can seep into and flow through spaces between the grains. Gravel consists of particles of rock intermediate in size between sand and boulders. Therefore, if the land surface is unconsolidated gravel, water is able to seep into the ground, thereby decreasing the amount that runs off.

10. **3** The characteristics of an air mass are the result of the geographical region over which it formed, or its *source region*. Air resting on or moving very slowly over a region tends to take on the characteristics of that region. In general, air masses that form near the poles are cold and are called polar air

masses. Air masses that form near the equator are warm and are called tropical air masses. Air masses that form over water are moist and are called maritime air masses. Air masses that form over land are dry and are called continental air masses. Thus, cold, dry air masses that move over New York State are continental polar air masses. The source region for continental polar air masses is land near the poles. The closest body of land near a pole, from which an air mass could move into New York State, without first moving over a large body of water and becoming a maritime air mass, is central Canada.

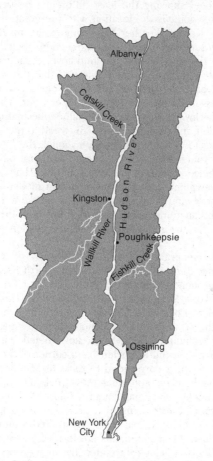

11. **4** Stream discharge is the volume of water that flows through a stream in a given amount of time, measured in units like cubic meters per second. Every tributary flowing into a river increases its discharge. According to the map, the greatest number of tributaries has flowed into the river and discharge is greatest at Ossining.

12. **2** Several weather factors serve as indicators of the chance of precipitation. First, precipitation is associated with clouds. Therefore, as cloud cover increases, the chance of precipitation increases. Second, moist air is less dense than dry air and thus exerts less air pressure. Therefore, as air pressure decreases, the chance of precipitation increases. Third, as the difference between the air temperature and the dewpoint decreases, the relative humidity increases. As the relative humidity increases, the likelihood of condensation that forms precipitation increases. Therefore, the location with the station model showing the least difference between the air temperature and the dewpoint is the location with the greatest chance of precipitation. Find the Key to Weather Map Symbols in the *Reference Tables for Physical Setting/Earth Science*. Locate the section labeled "Station Model Explanation." Note that the amount of cloud cover is indicated by the percentage of the central circle that is filled in and that the station models in choices (1) and (2) have the greatest amount of cloud cover. Note too that the barometric pressure is indicated to the upper right. The station model in choice (2) has the lowest air pressure. Finally, note that dewpoint (°F) is listed to the lower left and temperature (°F) is listed to the upper left of the station model. Note, too, that on the station model, the °F is dropped and only the numerical values of the dewpoint and temperature are shown. The station model in choice (2) shows the least difference between the air temperature and the dewpoint. Thus, since the station model in choice (2) has the greatest amount of cloud cover, the lowest air pressure, and the least difference between the air temperature and the dewpoint, the location represented in choice (2) has the greatest chance of precipitation.

13. **1** Find the Generalized Landscape Regions of New York State map in the *Reference Tables for Physical Setting/Earth Science*. Locate the region labeled "Adirondack Mountains." Now find the Generalized Bedrock Geology of New York State map in the *Reference Tables for Physical Setting/Earth Science*. Locate the region corresponding to the Adirondack Mountains. Note the symbols used to denote the two types of rocks found in the Adirondack Mountains. Refer to the key labeled "Geological Periods and Eras in New York" in the lower left-hand corner of the map. Note that the symbols for these rocks correspond to rocks of Middle Proterozoic age. To the right of the bracket, note that these rocks are labeled "Intensely Metamorphosed Rocks (regional metamorphism about 1,000 mya)." Mountains are parts of Earth's crust that project at least 300 meters above the surrounding land and have great relief, a restricted summit, and considerable bare-rock surface. Most mountains are formed by crustal motions that fold or fault rock. Thus, it is logical to infer that the bedrock beneath mountains is deformed and intensely metamorphosed by the forces that cause rock to fold or fault. Therefore, the Adirondacks are classified as mountains because of the high elevation and bedrock that consists mainly of deformed and intensely metamorphosed rocks.

14. **2** Find the Generalized Bedrock Geology of New York State map in the *Reference Tables for Physical Setting/Earth Science*. Locate the Finger Lakes. Now find the Generalized Landscape Regions of New York State map in the *Reference Tables for Physical Setting/Earth Science*. Locate the region corresponding to the Finger Lakes, and note that it lies primarily within the region labeled "Allegheny Plateau." Thus, the Finger Lakes are located primarily in the Allegheny Plateau landscape region.

15. **3** Find the Inferred Properties of Earth's Interior in the *Reference Tables for Physical Setting/Earth Science*. Locate the column labeled "Density (g/cm^3)" along the right side of the cross section, and note that a density of 9.9–12.2 g/cm^3 corresponds to the outer core. Next trace the dashed lines marking the upper and lower boundaries of the outer core downward to the bold black curve on the pressure graph in the center of the diagram. Where each of the dashed lines intersects the pressure curve, trace left to the vertical axis labeled "Pressure (million atmospheres)." Note that in the section corresponding to the outer core, pressures range from about 1.4 to 3.1 million atmospheres. Thus, the range of pressure in Earth's interior where rock with a density of 9.9 to 12.2 g/cm^3 is found is 1.4 to 3.1 million atmospheres.

16. **4** Along divergent boundaries, adjacent plates move apart and open rifts through which molten rock can rise to the surface and solidify to form new ocean floor. When molten rock is extruded in the mid-ocean rifts, crystals of minerals in the molten rock that are affected by magnetism align with Earth's magnetic field. When the molten rock cools and solidifies, the crystals are locked in place and the rock contains a record of the direction of Earth's magnetic field in its crystals. Studies of ancient rocks show that Earth's magnetic field has reversed many times in the past. As the plates continue to move apart, this process is repeated, constantly creating new ocean floor and pushing the older floor on either side of the mid-ocean ridge away from the ridge. The result is a striking pattern of parallel bands of normal and reversed magnetic polarity that are identical but mirror images of one another in the igneous bedrock of the oceanic crust on either side of the mid-ocean ridges. Thus, the pattern of magnetic reversal of Earth's magnetic field is best preserved in the igneous bedrock of the oceanic crust.

WRONG CHOICES EXPLAINED:

(1) Metamorphic bedrock forms by metamorphism. By definition, metamorphism refers to the changes that occur in solid rock due to great temperature, great pressure, and chemical activity. Metamorphism occurs while rocks are still in the solid state and does not involve the melting of rock. Therefore, crystals of minerals that are affected by magnetism are still in the solid state and are not free to move into alignment with Earth's magnetic field during

metamorphism. Thus, the pattern of magnetic reversal is unlikely to be preserved in metamorphic bedrock in mountain ranges.

(2) Radioactive carbon-14 is not significantly affected by magnetism. Even if it was, carbon-14 is present in fossils in very small quantities. Thus, there is little chance that magnetic forces would be strong enough to cause a dead organism to become aligned with Earth's magnetic field as it was deposited. Thus, the pattern of magnetic reversal is unlikely to be preserved in bedrock with fossils containing radioactive carbon-14.

(3) Layers of sedimentary bedrock are formed by the process of deposition. Magnetic grains in sediments may align with Earth's magnetic field during or shortly after deposition. However, anything that disturbs the sedimentary rock after deposition, such as tilting, faulting, or folding, will also disturb the alignment of the magnetic grains. Thus, the pattern of magnetic reversal in sedimentary rock is not as reliable as the pattern in igneous bedrock.

17. **1** Find the Tectonic Plates map in the *Reference Tables for Physical Setting/Earth Science*. In the key, note the symbol for a divergent plate boundary and the statement "(usually broken by transform faults along mid-ocean ridges.)" Note, too, that most divergent boundaries on the map are labeled as mid-ocean ridges. Thus, mid-ocean ridges are commonly found at divergent plate boundaries.

On the Tectonic Plates map, note that the arrows indicating plate motion along mid-ocean ridges show that divergent plate boundaries are places where plates are moving apart. This creates tension stresses that pull rock apart and cause the rock to fracture, causing rifts. Fractures result in earthquakes and faulting. When diverging plates pull apart, a block of crust drops down into the opening between the two plates, forming a rift valley. Fractures also create breaks in the crust through which molten rock can rise to the surface. Mid-ocean ridges are volcanic mountains on either side of a rift valley; mountains that formed from molten rock emerging from the rifts. Thus, two features commonly found at divergent plate boundaries are mid-ocean ridges and rift valleys.

WRONG CHOICES EXPLAINED:

(2) Find the Tectonic Plates map in the *Reference Tables for Physical Setting/Earth Science*. In the key, note the symbol for a divergent plate boundary and the statement "(usually broken by transform faults along mid-ocean ridges.)" Note, too, that most divergent boundaries on the map are labeled as mid-ocean ridges. Thus, most divergent boundaries occur beneath oceans. Wide valleys are landforms created by erosion of land surfaces, not beneath oceans. Deltas are coastal features formed by deposition of sediments where streams enter larger bodies of water such as oceans. Thus, deltas form at the margins of oceans, not mid-ocean. Furthermore, erosion and deposition may occur at locations far from plate boundaries. Therefore,

wide valleys and deltas are not features commonly found at divergent plate boundaries.

(3) Find the Tectonic Plates map in the *Reference Tables for Physical Setting/Earth Science*. In the key, note the symbol for a convergent plate boundary (subduction zone). Note, too, that trenches occur along many convergent boundaries. Trenches are deep crevices in the ocean floor where one crustal plate bends downward sharply and plunges beneath another (subducts) as the two plates converge. Thus, trenches and subduction zones are commonly found along convergent boundaries, not divergent boundaries.

(4) Find the Tectonic Plates map in the *Reference Tables for Physical Setting/Earth Science*. In the key, note the symbol for a hot spot. Note that hot spots occur near the centers of plates, not near plate boundaries. In subduction zones, two plates collide and one plate (the subducted plate) is pushed beneath the other. The subducted plate bends downward sharply, creating a deep trench. Friction between the subducted plate and the adjacent plate melts rock along the interface and produces volcanic activity directly above and parallel to the trench. This is why most trenches are bordered by volcanic island arcs or volcanic mountain chains. Island arcs are associated with convergent boundaries where subduction is taking place, not divergent plate boundaries. Thus, hot spots and island arcs are not commonly found near divergent plate boundaries.

18. **4** Find the Generalized Bedrock Geology of New York State chart in the *Reference Tables for Physical Setting/Earth Science*. In the key labeled "Geologic Periods and Eras in New York," locate the statement "Silurian *also contains salt, gypsum and hematite*." Thus, New York State bedrock of Silurian age contains salt, gypsum, and hematite.

19. **4** Earth's early atmosphere most likely formed by outgassing, the release of water and trace gases trapped in rocks in the interior of the hot, young Earth. Another likely source of gases was comets that collided with Earth. Comets are rich in ices of water, carbon dioxide, carbon monoxide, ammonia, and organic compounds. Thus, Earth's early atmosphere contained virtually no free oxygen. However, over time, life-forms evolved that were able to carry out photosynthesis. Photosynthetic life-forms use energy from sunlight to convert carbon dioxide and water into carbohydrates (food) and oxygen. Therefore, once photosynthetic life evolved, these organisms released oxygen into the environment. Find the Geologic History of New York State chart in the *Reference Tables for Physical Setting/Earth Science*. In the column labeled "Era," locate the statements "Oceanic oxygen produced by cyanobacteria combines with iron, forming iron oxide layers on ocean floor," and "Oceanic oxygen begins to enter the atmosphere." Cyanobacteria, or blue-green algae, are single-celled organisms that produce gaseous oxygen as a by-product of photosynthesis. Oxygen is highly reactive when combined with substances dissolved in seawater. Some, such as carbon-

ates, are soluble and remained in solution in the oceans. Others, such as iron oxides, are insoluble and settled out of the oceans to form layers of iron oxide on the ocean floor. Eventually, oxygen levels grew high enough in the oceans that excess oxygen began to escape into the atmosphere. As photosynthetic organisms multiplied over time, photosynthesis steadily increased. Over millions of years, the oxygen released by these organisms steadily increased the oxygen in Earth's atmosphere to its present level. Thus, scientists infer that oxygen in Earth's atmosphere did not exist in large quantities until after photosynthetic cyanobacteria evolved in Earth's oceans.

WRONG CHOICES EXPLAINED:

(1) Find the Geologic History of New York State chart in the *Reference Tables for Physical Setting/Earth Science*. In the column labeled "Era," locate the statement "Oceanic oxygen begins to enter the atmosphere." Trace left from this statement to the vertical time scale labeled "Million years ago," and note that this event occurred about 2,200 million (2.2 billion) years ago. Now find the column labeled "Life on Earth," and find the statement "Ediacaran fauna (first multicellular, soft-bodied marine organisms)." Trace left from this statement to the vertical time scale labeled "Million years ago," and note that this event occurred about 580 million years ago. Thus, oxygen in Earth's atmosphere existed in large quantities before the first multicellular, soft-bodied marine organisms appeared on Earth.

(2) Find the Geologic History of New York State chart in the *Reference Tables for Physical Setting/Earth Science*. In the column labeled "Era," locate the statement "Oceanic oxygen begins to enter the atmosphere." Trace left from this statement to the vertical time scale labeled "Million years ago," and note that this event occurred about 2,200 million (2.2 billion) years ago. Now find the column labeled "Important Geologic Events in New York," and locate the statement "Initial opening of Atlantic Ocean." Trace left from this statement to the vertical time scale labeled "Million years ago," and note that this event occurred about 180 million years ago. Thus, oxygen in Earth's atmosphere existed in large quantities before the initial opening of the Atlantic Ocean.

(3) Find the Geologic History of New York State chart in the *Reference Tables for Physical Setting/Earth Science*. In the column labeled "Era," locate the statement "Oceanic oxygen begins to enter the atmosphere." Trace left from this statement to the vertical time scale labeled "Million years ago," and note that this event occurred about 2,200 million (2.2 billion) years ago. Now find the column labeled "Era," and locate the statement "First sexually reproducing organisms." Trace left from this statement to the vertical time scale labeled "Million years ago," and note that this event occurred about 1,100 million years ago. Thus, oxygen in Earth's atmosphere existed in large quantities before the first sexually reproducing organisms appeared.

20. **3** Find the Geologic History of New York State chart in the *Reference Tables for Physical Setting/Earth Science*. In the column labeled "Important Geologic Events in New York," locate the statement "Advance and retreat of last continental ice." Trace left from this statement to the column labeled "Time Distribution of Fossils." Note that at the time of the last continental ice sheet, the index fossils represented by the circled letters "O" and "S" existed in New York State. The circled letters are keyed to the illustrations of specific important fossils printed along the bottom of the chart. Locate the circled letters "O" and "S" along the bottom of the chart. Note that circled letter "O" corresponds to "Mastodont and Beluga Whale." Note that the circled letter "S" corresponds to "Condor." Thus, of the choices given, the organisms that were alive when New York State was last covered by a continental ice sheet were the mastodon and Beluga whale.

WRONG CHOICES EXPLAINED:

(1) Find the Geologic History of New York State chart in the *Reference Tables for Physical Setting/Earth Science*. Among the drawings of index fossils along the bottom of the chart, locate *Eurypterus* and *Cooksonia*. Note that *Eurypterus* corresponds to the circled letter "M" and *Cooksonia* corresponds to the circled letter "P." In the column labeled "Time Distribution of Fossils," locate the circled letters "M" and "P." From each of these letters, trace left to the vertical time scale labeled "Million years ago." Note that "M" *Eurypterus* and "P" *Cooksonia* existed between 416 and 444 million years ago. In the column labeled "Important Geologic Events in New York," locate the statement "Advance and retreat of last continental ice." Trace left from this statement to the vertical time scale labeled "Million years ago." Note that New York State was last covered by a continental ice sheet 0.01 million years ago. Thus, *Eurypterus* and *Cooksonia* existed and became extinct between 416 and 444 million years ago—long before New York State was last covered by a continental ice sheet 0.01 million years ago.

(2) Find the Geologic History of New York State chart in the *Reference Tables for Physical Setting/Earth Science*. Among the drawings of index fossils along the bottom of the chart, locate *Aneurophyton* and Naples tree. Note that these index fossils correspond to the circled letter "Q." In the column labeled "Time Distribution of Fossils," locate the circled letter "Q" and trace left to the vertical time scale labeled "Million years ago." Note that "Q" *Aneurophyton* and Naples tree existed between 369 and 416 million years ago—long before New York State was last covered by a continental ice sheet 0.01 million years ago.

(4) Find the Geologic History of New York State chart in the *Reference Tables for Physical Setting/Earth Science*. Among the drawings of index fossils along the bottom of the chart, locate *Coelophysis* and *Elliptocephala*. Note that *Coelophysis* corresponds to circled letter "L" and *Elliptocephala* corresponds to circled letter "A." In the column labeled "Time Distribution of Fossils," locate the circled letters "A" and "L." From each of these let-

ters, trace left to the vertical time scale labeled "Million years ago." Note that "A" *Elliptocephala* existed between 488 and 542 million years ago and "L" *Coelophysis* existed between 200 and 251 million years ago. Thus, *Coelophysis* and *Elliptocephala* existed long before New York State was last covered by a continental ice sheet 0.01 million years ago.

21. **2** Find the Scheme for Sedimentary Rock Identification in the *Reference Tables for Physical Setting/Earth Science*. In the column labeled "Rock Name," locate conglomerate and breccia. From each of these rocks, trace left to the column labeled "Comments." Note that conglomerate consists of rounded particles and breccia consists of angular particles. Thus, one difference between a breccia rock and a conglomerate rock is that the particles in breccia are more angular.

22. **1** Layers of rock are generally deposited in an unbroken sequence. However if forces within Earth uplift rocks, deposition ceases. The uplift may cause rock layers to tilt or fold. Erosion may wear away layers of rock before the land surface is low enough for another layer to be deposited, or the land may subside until it is low enough for another layer to be deposited. The result is an unconformity, a break or gap in the sequence of a series of rock layers. Thus, the rocks above an unconformity are quite a bit younger than those below it. The erosional surface marking the gap between the older and younger rocks is generally drawn as a wavy line.

There are several types of unconformities. *Disconformities* are irregular erosional surfaces between parallel layers of rock. The labeled unconformity is a disconformity. Disconformities occur when deposition stops and layers are eroded but no tilting or folding occurs. These surfaces are not easy to discern and are often found when fossils of very different ages are discovered in adjacent layers. *Nonconformities* are places where sedimentary layers lie on top of igneous or metamorphic rocks and are not metamorphosed in any way. *Angular unconformities* form when rock layers are tilted or folded before being eroded. When new layers are deposited, these new layers form horizontally. The layers below the unconformity are at an angle to those above it. Note that the rock layers below the wavy line representing unconformity **X** are tilted and folded while the layers above the wavy line are horizontal. Thus, unconformity **X** is an angular unconformity. The sequence of events that most likely produced this unconformity was uplift and erosion of bedrock, followed by subsidence and more deposition.

WRONG CHOICES EXPLAINED:

(2) There is no indication in the photograph that any of the layers are igneous rock or that any of the layers were changed by contact metamorphism. Furthermore, an intrusion and contact metamorphism could not account for the folding and tilting of layers below the unconformity and horizontal layers above the unconformity. Thus, there is no basis for inferring that magma intruded and caused contact metamorphism.

(3) Lava is a fluid. If it spreads out over horizontal sedimentary rock layers, it will tend to form horizontal layers. The photograph shows horizontal layers of rock over folded and tilted layers of rock, not horizontal layers of lava over horizontal layers of sedimentary rock.

(4) If one rock layer separated to form the unconformity, part of that layer would be above the unconformity and part would be below the unconformity. Multiple layers below the unconformity are in contact with the layer directly above the unconformity. Therefore, it is unlikely that separation of one rock layer produced this unconformity.

23. **4** Find the Scheme for Igneous Rock Identification in the *Reference Tables for Physical Setting/Earth Science*. Locate gabbro in the upper section of the scheme and trace left to the column labeled "Environment of Formation." Note that gabbro is an intrusive igneous rock. From gabbro, trace right to the column labeled "Texture," and note that gabbro has a coarse texture. Intrusive igneous rocks form by the cooling and solidification of molten magma beneath Earth's surface. Insulated by surrounding rock beneath Earth's surface, intrusive rocks cool slowly. The slow cooling time allows large crystals to form, resulting in the coarse texture of gabbro. Thus, the igneous rock gabbro most likely formed from molten material that cooled slowly, deep underground.

24. **4** "Marine origin" means that the sedimentary bedrock formed from layers of sediment that were deposited beneath an ocean. Thus, if sedimentary bedrock of marine origin is now exposed on land surfaces, it is logical to infer that those land surfaces were underwater at the time when the sediments that formed the bedrock were deposited. Thus, the inference that most of Earth's present-day land surfaces have, at one time, been covered with water is best supported by the statement "Sedimentary bedrock of marine origin covers large areas of Earth's continents."

WRONG CHOICES EXPLAINED:

(1) Water vapor released by volcanic eruptions is gaseous and enters the atmosphere. Even if that water vapor condensed and fell as precipitation, it would run off into the oceans. It would not cover with liquid water an area of land surface equal to the area of most of Earth's present-day land surfaces. Thus, it does not support the inference that most of Earth's present-day land surfaces have, at one time, been covered by water.

(2) The edges of many continents in the past make up only a fraction of the area of "most of Earth's present-day land surfaces." Thus, the statement that in the past, coral reefs formed along the edges of many continents does not support an inference that most of Earth's present-day land surfaces have, at one time, been covered by water.

(3) Landmasses that are pulled apart and pushed together are still landmasses. Seafloor spreading does not cause these landmasses to be submerged beneath an ocean. Thus, seafloor spreading that has pulled landmasses apart and pushed them together does not support the inference that most of Earth's present-day land surfaces have, at one time, been covered by water.

25. **2** Earth's axis is tilted 23½° from a line perpendicular to the plane of its orbit. As Earth orbits the Sun, Earth's axis always points in the same direction in space. In other words, the direction of its axis at any given time is parallel to its direction at any other time. Thus, the diagram that best represents the correct orientation of the North Pole as Earth revolves around the Sun is diagram (2).

26. **1** Many chemical compounds found in Earth's atmosphere act as greenhouse gases. These gases allow short wavelength visible light radiation to enter the atmosphere and strike Earth's surface. When the visible light radiation strikes Earth's surface, some of it is absorbed and then reradiated as longer wavelength infrared radiation (heat). Greenhouse gases in the atmosphere absorb this infrared radiation and trap the heat in the atmosphere. Note the key showing the symbols for visible light radiation, infrared radiation, and absorbed energy. Thus, the best representation of how greenhouse gases in our atmosphere trap heat energy shows visible light radiation coming in through Earth's atmosphere, absorbed energy at Earth's surface, outgoing infrared radiation, and absorbed energy within the atmosphere as shown in diagram (1).

27. **3** The force of gravitational attraction between any two objects is directly proportional to their masses and indirectly proportional to the square of the distance between their centers. Newton expressed this relationship in a simple mathematical expression:

$$F \propto \frac{m_1 \cdot m_2}{d^2}$$

Gravitational force is directly proportional to the product of the masses of the two planets. Thus, the force is greatest where the masses of the both planets are greatest as shown in choices (3) and (4). Since the gravitational force is inversely proportional to distance between the two planets squared, the gravitational force between the two planets decreases as the distance between them increases. Thus, the force is greatest where the distance between the two planets is smallest, as shown in choices (1) and (3). The combination of planet masses and distances that produces the greatest gravitational force between the planets is the greatest mass for each of the two planets and the smallest distance between the two planets as shown in choice (3).

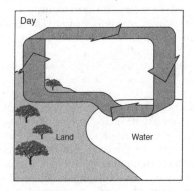
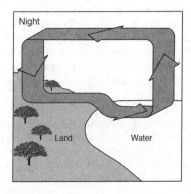

28. **3** Note that the air movements in the diagrams show a circular pattern of motion. Such a circular pattern of motion in a fluid (such as air) is called a convection current. Surface winds blow from regions of high pressure toward regions of low pressure. According to the diagram, during the day, the surface winds blow from the water toward the land. Thus, during the day, the air pressure is higher over the water than over the land. According to the diagram, at night, the surface winds blow from the land toward the water. Thus, during the night, the air pressure is higher over the land than over the water. Find the Specific Heats of Common Materials chart in the *Reference Tables for Physical Setting/Earth Science*. Note that liquid water has a specific heat of 4.18 joules/gram • C° and that basalt and granite (common materials from which land is derived) have specific heats of 0.84 and 0.79 joules/gram • C°, respectively. This means that if 4.18 joules of heat are added to 1 gram of water, the water's temperature will increase by 1°C. However if 4.18 joules of heat are added to 1 gram of basalt, the basalt's temperature will increase by almost 5°C (4.18 joules/g ÷ 0.84 cal/g • °C = 4.98°C). Thus, when exposed to the same amount of sunlight, landmasses will become warmer than adjacent bodies of water. Additionally, the air over the water will become cooler than the air over the land. Since the cooler air over the water exerts more pressure than the warmer air over the land, there is a

movement of air from the water toward the land. The lower specific heat of land means it also cools off faster than water. Thus, at night, the land surface will become cooler than adjacent bodies of water. Since the cooler air over the land exerts more pressure than the warmer air over the water, there is a movement of air at night from the land toward the water. Thus, the local air movement shown in the diagrams is best described as an example of convection resulting from temperature and pressure differences above land and water.

WRONG CHOICES EXPLAINED:

(1) Conduction is the transfer of heat energy from one molecule to another through collision, not a circular pattern of local air movements.

(2) Condensation and evaporation are changes in state of matter. The local air movements shown in the diagrams are examples of a pattern of motion of the air, not changes from one state of matter to another.

(4) Radiation is the transfer of heat energy by electromagnetic waves, not a circular pattern of local air movements.

29. **4** The rainy summer season and dry winter season in southern Asia is caused by a monsoon. A monsoon is a seasonally reversing system of surface winds caused by temperature differences between continents and neighboring oceans. A monsoon is a large-scale version of the type of daily changes in circulation we see in land and sea breezes. In a monsoon, though, we see seasonal changes in wind direction between a large landmass and surrounding oceans. For example, during the summer, the large landmass of Asia warms up and a low-pressure center develops in the southern part of the continent. This low-pressure center results in winds that blow inland from the surrounding oceans. These moist summer monsoon winds bring clouds and heavy rains. During the winter, intense cooling produces a mass of very cold, very dry air in the northern part of the continent. The result is a high-pressure center in the northern part of the continent that causes the winds to reverse direction and blow from the continental interior outward to the coast. Although these winter monsoon winds may warm up as they descend toward the coast, they pick up little moisture as they blow over the land and bring clear, dry weather. Thus, the set of maps that best represents the type and location of pressure systems and wind patterns that cause these seasonal changes shows the following: low pressure over the continent during the summer with winds blowing inland from the surrounding oceans and high pressure over the continent in the winter with winds blowing from the continental interior toward the coast. This is best shown by the maps in choice (4).

30. **2** Note that the scratches in the polished surface of the bedrock are roughly parallel to one another. As a glacier moves along, loose rock may freeze into the ice at the bottom of the glacier. As the glacier moves over bedrock, these loose rocks frozen into the bottom of the glacier scrape parallel grooves or scratches, called striations, into the bedrock beneath the glacier. Small sediments frozen in the ice can act as sandpaper, smoothing and polishing the surface of the bedrock. Note, too, the many parallel wedge-shaped marks on the surface of the rock. These are called chatter marks and form when the bedrock surface is chipped by rock carried in the base of a glacier. Thus, the agent of erosion that most likely produced the features seen in the photograph is glacial ice.

WRONG CHOICES EXPLAINED:
(1) Sediment carried by running water can smooth the surface of bedrock over which it flows. However, large particles transported by a stream typically bounce and roll along the streambed and do not exert enough downward pressure on the bedrock for a long enough period of time to create the deep, long scratches shown in the photograph. Particles carried in a stream also do not move perpendicular to the flow of water to create the wedge-shaped marks perpendicular to the striations. Additionally, the turbulence of a stream flowing fast enough to transport particles large enough to make the scratches shown in the photograph make it unlikely that any scratches left behind would be straight and parallel to one another.

(3) Particles transported by wave action can smooth the surface of exposed bedrock. However, to make the deep, parallel scratches shown in the photograph, a large sediment particle would have to be pushed down against the bedrock with sufficient force to create a deep scratch and held in place as it moved along the bedrock surface long enough to create a long, straight scratch. This would then have to be repeated to create other parallel scratches. Wave action is turbulent. It frequently changes direction due to storms or changes in wind direction. Thus, it is unlikely that any scratches left by larger particles would leave the long, deep, parallel scratches or the

wedge-shaped marks perpendicular to those scratches as shown in the photograph.

(4) Wind is able to transport only fine particles. It cannot transport particles large enough to create the large scratches shown in the photograph.

31. **1** When a stream flows into a standing body of water, such as a lake, the velocity of the water steadily decreases and any particles carried by the stream settle to the bottom of the lake. When sediments are deposited in a fluid medium, such as the water in a lake, sediments of different sizes settle through the water at different rates. For example, as water slows down, large particles settle first, followed by smaller and smaller particles. The smallest particles settle last. The result is a separation of different particles into layers ranging from largest at the bottom to smallest at the top as shown in cross section (1).

32. **2** Find the Scheme for Igneous Rock Identification in the *Reference Tables for Physical Setting/Earth Science*. Locate the "Characteristics" section in the middle of the scheme, and note the arrows labeled "Density" and "Composition." Note that lower density corresponds with a felsic composition and higher density corresponds with a mafic composition. Thus, as density increases from lower to higher, the composition of the igneous rocks changes from felsic to mafic. This relationship is best shown by graph (2).

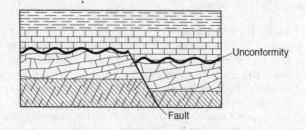

33. **4** Find the Scheme for Sedimentary Rock Identification in the *Reference Tables for Physical Setting/Earth Science*. In the column labeled "Map Symbol," locate the symbols corresponding to the top two layers shown in the cross section. Trace left from each of these two symbols to the column labeled "Rock Name." Note that the top layer in the cross section is shale and the layer beneath it is limestone. Now find the Scheme for Metamorphic Rock Identification in the *Reference Tables for Physical Setting/Earth Science*. In the column labeled "Map Symbol," locate the symbols corresponding to the bottom two layers shown in the cross section. Trace left from each of these two symbols to the column labeled "Rock Name." Note that the bottom layer is slate and the layer above the slate is marble. Thus, from top to bottom, the rock layers in the cross section are shale, limestone, marble, and slate. Note that the slate layer, the marble layer, and the unconformity have all been displaced

along the fault. In order to have been displaced along the fault, these must already have existed when the faulting occurred. Thus, the fault is younger than the slate layer, the marble layer, and the unconformity. However, the limestone and shale layers have not been displaced along the fault. This means that these layers did not yet exist when the faulting occurred and are younger than the fault. Thus, the fault is older than the shale.

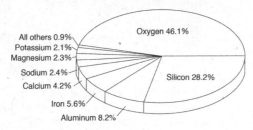

34. **1** According to the pie graph, the percent by mass of the two most abundant elements in this Earth layer are oxygen (46.1%) and silicon (28.2%). Find the Average Chemical Composition of Earth's Crust, Hydrosphere, and Troposphere chart in the *Reference Tables for Physical Setting/Earth Science*. Note that in the column labeled "Crust," the percent by mass of oxygen is 46.1% and of silicon is 28.2%. Thus, the layer of Earth represented by the composition in the pie graph is the crust.

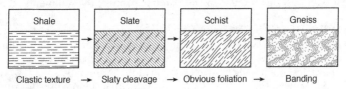

35. **3** According to the diagram, as shale is converted first to slate, then to schist, and finally to gneiss, it undergoes changes in its texture. Metamorphism is the change in the mineral composition or texture of a rock due to heat, pressure, or chemical activity without the rock undergoing melting into liquid magma. Find the Scheme for Metamorphic Rock Identification in the *Reference Tables for Physical Setting/Earth Science*. In the column labeled "Rock Name," locate slate. Trace left to the column labeled "Comments," and note the statement "Low-grade metamorphism of shale." Thus, shale is converted to slate by low-grade metamorphism. Now locate "Schist" and "Gneiss" in the column labeled "Rock Name." For each, trace left to the column labeled "Comments." Note that schist contains "Platy mica crystals visible from metamorphism of clay or feldspars" and gneiss has undergone "High-grade metamorphism" and has "mineral types segregated into bands." Thus, the geologic process that is occurring in the conversion of shale to gneiss is metamorphism.

PART B-1

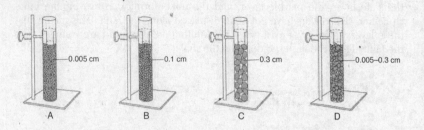

36. **4** Porosity is a measure of the amount of empty space in a rock or sediment sample. It is often expressed as the percentage of the total volume of a rock or sediment sample that is empty space. One way to measure porosity is to measure the volume of water needed to fill the empty spaces in a rock or sediment sample. In the diagram, water is added to sediments in a tube with a closed bottom. The water seeps down to the bottom of the container and then begins to fill the open spaces between sediment particles. When the water has filled all of the open spaces in the sample, the water will be level with the top of the sample. The volume of water added is equivalent to the volume of empty space in the sample. Thus, when water is added to the tubes to just cover the sediments and the volumes of water are recorded, the data can best be used to determine the porosity of the sediments.

WRONG CHOICES EXPLAINED:
(1) and (2) Knowing only the volume of water added to the sediments provides information about only the volume of empty space between the sediment particles, not about the size or shape of the sediments.

(3) Water retention is the amount of water that adheres to the surfaces of the sediment particles after the water filling the empty spaces between them has drained out. To measure water retention, the volume of water that drained out of the empty spaces in each tube would have to be measured. This volume could then be compared with the volume that filled the spaces to determine how much remained behind adhering to the surfaces of the sediment particles. These data were not collected.

37. **1** Capillarity is the tendency for water to rise in narrow openings in earth materials. The water rises because its attraction to the surrounding earth materials pulls it upward and because surface tension drags the water surface between the materials upward along with it. The narrower the opening, the greater the force of attraction is between the water and the surrounding earth materials compared to the weight of the water column and the higher the water rises. Smaller particles fit closer together than larger ones. Therefore, the smaller the sediment particles, the narrower the open-

ings are between particles and the higher the water rises due to capillarity. According to the diagram, the particles in tube A have the smallest diameter. Therefore, the data would show that capillarity is greatest in tube A.

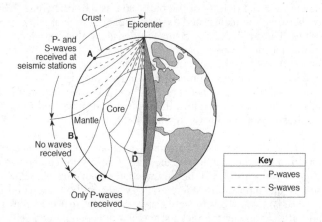

38. **2** Find the Earthquake *P*-Wave and *S*-Wave Travel Time graph in the *Reference Tables for Physical Setting/Earth Science*. Locate 5000 kilometers (5×0^3 km) along the horizontal "Epicenter Distance ($\times 10^3$)" axis at the bottom of the graph. Trace vertically until you intersect the bold line labeled "P." Then trace horizontally to the left to the "Travel Time (min)" axis. Read the *P*-wave travel time—about 8 minutes 20 seconds. Now trace vertically from 5000 kilometers until you intersect the bold line labeled "S." Then trace horizontally to the left to the "Travel Time (min)" axis. Read the *S*-wave travel time—about 15 minutes 00 seconds. Use subtraction to determine that the *S*-waves arrive about 6 minutes 40 seconds after the *P*-waves. Thus, the difference between the arrival time of the first *P*-wave and the arrival time of the first *S*-wave recorded at this station was 6 minutes 40 seconds.

39. **1** *P*-waves are compression waves. Compression waves travel faster in more dense substances than they do in less dense substances. When waves are transmitted from a substance with one density to a substance with another density, the change in speed causes the transmitted wave to travel in a different direction. The amount of change in direction depends on the ratio of the wave velocities of the two different substances. This change in the direction of wave motion is called refraction. In general, pressure steadily increases with depth beneath Earth's surface, causing a steady increase in the density of rock. The steady increase in density causes seismic waves to bend steadily, or refract. This is why the paths of the seismic waves in the diagram steadily curve as they move away from the epicenter. However, note

that in order to arrive at seismic station B, P-waves have to cross the boundary between the mantle and the core. Find the Inferred Properties of Earth's Interior chart in the *Reference Tables for Physical Setting/Earth Science*. In the column labeled "Density (g/cm^3)," note that the density of the mantle is 3.4–5.6 g/cm^3 and the density of the outer core is 9.9–12.2 g/cm^3. Thus, the boundary between the mantle and the core separates rock of markedly different densities. The sudden, large change in density at this boundary produces strong refraction of P-waves that cross it. Up to 103° from the epicenter, waves arrive directly from the epicenter. The strong refraction as soon as the boundary is encountered immediately angles the waves away to 143° so no P-waves reach the region between 103° and 143°. Seismic station B most likely falls within this region. Thus, the process that prevented P-waves from arriving at seismic station B is refraction.

40. **4** In order to reach seismic station C, seismic waves from the earthquake must travel through the mantle and the core. P-waves are compression waves that compress and expand rock in the direction of wave travel. Solids, liquids, and gases are all compressible; therefore, P-waves can be transmitted through solids, liquids, and gases. S-waves are transverse waves that twist rock material back and forth, deforming the rock's shape in a direction perpendicular to that of wave travel. S-waves can be transmitted only through solids. S-waves cannot travel through liquids and gases because when liquids and gases are deformed, they do not return to their original shape. Find the Inferred Properties of Earth's Interior in the *Reference Tables for Physical Setting/Earth Science*. Locate the outer core in the cross section at the top of the diagram. From the outer core, trace downward between the dashed lines to the "Temperature (°C)" graph. Note that the actual temperature of the outer core is higher than its melting temperature. Thus, the outer core is a liquid. Note, too, that this is the only place on the "Temperature (°C)" graph where this is the case, so the outer core is the only layer of Earth's interior that is a liquid. Therefore, S-waves from this earthquake cannot pass through the liquid outer core and will not reach seismic station C. However, P-waves from this earthquake will be able to pass through the liquid outer core as well as the solid mantle and inner core to reach seismic station C. Thus, only P-waves were recorded at seismic station C because P-waves travel through liquids, while S-waves cannot.

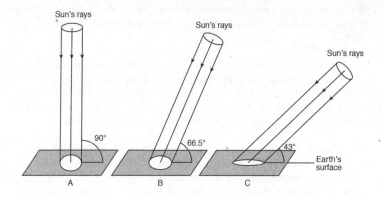

41. **3** It is given that the diagrams represent locations at 23.5° N latitude. On the summer solstice in June, the northern pole of Earth's axis of rotation is inclined farthest toward the Sun and the Sun's noontime rays strike perpendicular to Earth's surface at 23½° N latitude. Therefore, diagram A represents the summer solstice in June.

On the winter solstice in December, the northern pole of Earth's axis of rotation is inclined farthest away from the Sun and the Sun's noontime rays strike perpendicular to Earth's surface at 23½° S latitude. For every degree of latitude away from the latitude where the Sun is directly overhead, the angle at which the Sun's noontime rays strike Earth decreases by one degree. On the winter solstice in December, latitude 23.5° N is 47° of latitude away from where the Sun is directly overhead at latitude 23.5° S. Therefore, on the winter solstice in December, the angle at which the Sun's rays strike Earth's surface at 23.5 N latitude is 43° (90°− 47° = 43°). Thus, diagram C represents the winter solstice in December.

On the equinoxes in March and September, Earth's axis of rotation is not inclined toward or away from the Sun and the Sun's noontime rays strike perpendicular to Earth's surface at the equator (0° latitude.) Therefore, on the equinoxes, latitude 23.5° N is 23.5° of latitude away from where the Sun is directly overhead at latitude 0°. Therefore, on the March and September equinoxes, the angle at which the Sun's rays strike Earth's surface at 23.5° N latitude is 66.5° (90° − 23.5° = 66.5°). Since diagram B represents an equinox that occurs in sequence between diagram A in June and diagram C in December, diagram B represents the equinox in September.

The greatest intensity of insolation occurs where the Sun's rays strike perpendicular to Earth's surface. As the angle at which the Sun's rays strikes Earth's surface decreases, the intensity of the insolation also decreases. As viewed in sequence from A to B to C, the angle at which the Sun's rays strikes Earth's surface decreases from 90° to 66.5° to 43°. Thus, the insolation becomes less intense.

Therefore, as viewed in sequence from A to B to C, the diagrams represent the months of June → September → December; and decreasing intensity.

42. **2** When the Sun's rays strike at 90° to Earth's surface, an object will cast no shadow. However, as the angle at which the Sun's rays strike Earth's surface decreases, the length of the shadow cast by an object will increase. Thus, as the angle of the Sun's rays striking Earth's surface at noon changes from 90° to 43°, the length of a shadow cast by an object will increase.

43. **1** As explained in the answer to question 41, diagram A represents the summer solstice in June, diagram B represents the equinox in September, and diagram C represents the winter solstice in December. The maximum duration of insolation (number of daylight hours) occurs at the summer solstice in June and then decreases steadily through its midpoint at the equinox in September until it reaches a minimum at the winter solstice in December. Note that the angles corresponding to these months are June 90°, September 66.5°, and December 43°. Thus, the graph that best shows the duration of insolation shows maximum duration at 90° and decreases steadily to a minimum at 43° as shown in graph (1).

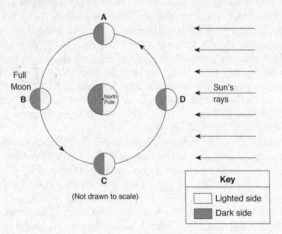

44. **1** The time it takes for the Moon to complete one cycle of phases is 29.5 days. In moving from the phase shown at position A to the full-Moon phase at position B, the Moon moves through one-quarter of a complete cycle of phases. Thus, it will require one-quarter of 29.5 days, or about 7.4 days, for the Moon to move from the phase shown at position A to the full-Moon phase.

45. **4** An observer on Earth looking at the Moon at position C would see one-half of the Moon illuminated and that illuminated half would be to the observer's left. This phase is best shown in diagram (4).

46. **2** Find the Solar System Data table in the *Reference Tables for Physical Setting/Earth Science*. In the column labeled "Celestial Object," locate "Earth's Moon." Trace right to the columns labeled "Period of Revolution (d=days) (y=years)" and "Period of Rotation at Equator." Note that Earth's Moon has a period of revolution of 27.3 days and a period of rotation of 27.3 days. Thus, the Moon's period of rotation and period of revolution are equal. That is, whenever the Moon revolves one-quarter of the way around Earth, it also rotates one-quarter of the way on its axis. Thus, the same side of the Moon always faces Earth. See the diagram below.

47. **3** The shadows cast by Earth and the Moon are extremely long and narrow. Additionally, the plane of the Moon's orbit is tilted at an angle of about 5° to the plane of Earth's orbit. Solar and lunar eclipses can occur when the Moon intersects the plane of Earth's orbit in the new-Moon or full-Moon positions. However, if the Moon is above or below the plane of Earth's orbit during these phases, no eclipse occurs. Thus, eclipses rarely happen during a cycle of phases because the plane of the Moon's orbit is different from the plane of Earth's orbit.

Map I

Avon-Buxton Section of Outer Banks

Map II — Avon-Buxton 1852

Map III — Avon-Buxton 1998

Key
.... 1852 shoreline

Avon-Buxton Section Width from 1852 to 1998

Year	Average Width (m)
1852	813
1917	547
1940	426
1962	284
1974	284
1998	219

48. **4** According to map I, the Outer Banks are long, narrow, offshore islands. They are separated from the mainland by a sound. It is given that they were formed primarily from sediments eroded and deposited by ocean waves. By definition, barrier islands are long, narrow, offshore deposits of sand or sediment that run parallel to the coastline. They are separated from the mainland by a sound or a bay. The islands themselves are separated by narrow tidal inlets. Thus, the Outer Banks are a type of landform called a barrier island.

49. **3** According to the data table, the Avon-Burton section width decreased from 813 feet in 1852 to 219 feet in 1998 except for two years (1962, 1974) when the width stayed the same. At no time did the section width increase during the time period from 1852 to 1998. Thus, the bar graph that best shows the average width of the Avon-Buxton section of the

Outer Banks from 1852 to 1998 shows the highest bar to the left and the bars decreasing in height to the right as shown in graph (3).

50. **1** It is given that the Outer Banks are located along the southeastern coast of the United States. Find the Surface Ocean Currents map in the *Reference Tables for Physical Setting/Earth Science*. Locate the southeastern coast of the United States. Note that the nearest ocean currents are labeled "Florida C." and "Gulf Stream C." Note that both of these currents are represented by a black arrow. According to the key, a black arrow represents a warm current. Thus, of the choices given, the ocean current that has the greatest warming influence on the climate of the Outer Banks of North Carolina is the Gulf Stream Current.

PART B-2

51. In the troposphere, air temperature and air pressure drop steadily with altitude. So as warm air rises, it expands and cools. When it has been cooled to its dewpoint, the water vapor in the air condenses. This condensation results in clouds made of a great many tiny water droplets or ice crystals. Cool air is denser than warm air. Since the air above the lake is cooler, it is denser than the surrounding air. Therefore, the cool air over Lake Erie is less likely to rise and result in condensation that forms clouds.

Additionally, the less water vapor the air contains, the less water that is available to condense to form clouds. The molecules in cool air move slower than the molecules in warm air. If the air above the lake water is cool, its slower-moving air molecules impart less energy to the liquid lake water. Fewer water molecules are able to break free of the liquid and enter the gas phase, or evaporate. If less water evaporates, the air contains less water vapor and there is less cloud development. Thus, cool air above the lake surface reduces the amount of cloud development because less evaporation occurs when the air is colder.

One credit is allowed for an acceptable response. Acceptable responses include but are not limited to:

- **Cooler air near the lake remains close to the surface because it is more dense than the surrounding air.**
- **Cold air over the lake is more dense.**
- **Cooler air over Lake Erie is less likely to rise.**
- **Convection is reduced.**
- **Less evaporation occurs when the air is colder.**
- **Lack of moisture**
- **Warm air rises to form clouds.**

52. When water vapor condenses, it forms water droplets if the air temperature is above freezing or forms ice crystals (snow) if the air temperature is below freezing. Thus, one weather variable that determines whether Buffalo receives rain or snow is air temperature.

One credit is allowed for an acceptable response. Acceptable responses include but are not limited to:

- **Temperature/air temperature**
- **The average temperature of the air is colder when Buffalo receives snow.**

53. According to the passage, "Generally, the longer the time the wind travels over the lake, the heavier the lake effect becomes in Buffalo." Refer to the map on page 457. Note that Buffalo is located on the northeastern tip of Lake Erie. In order for wind to travel the longest over Lake Erie, it has to travel across the long axis of Lake Erie from its southwestern tip to its northeastern tip at Buffalo. Thus, the arrow should be drawn on the map from the southwest to the northeast over Lake Erie toward Buffalo.

One credit is allowed for **any arrow drawn from a southwest to northeast orientation on Lake Erie pointing toward Buffalo.**

Note: One credit is allowed even if the arrow extends before Lake Erie or beyond Buffalo. If additional arrows are drawn, they need not be over Lake Erie but must have a general SW to NE direction.

Example of a 1-credit response:

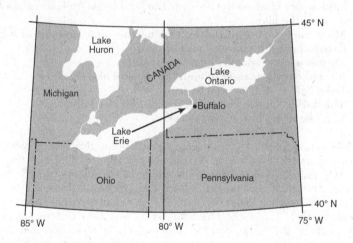

54. Find the Specific Heats of Common Materials chart in the *Reference Tables for Physical Setting/Earth Science*. Note that liquid water has a specific heat of 4.18 joules/gram • C°. Note, too, that basalt and granite (common materials from which land is derived) have specific heats of 0.84 and 0.79 joules/gram • C°, respectively. This means that if 1 gram of water absorbs 4.18 Joules of heat, the temperature of the water will increase by 1 °C. However, if 1 gram of basalt absorbs 4.18 Joules of heat, the temperature of the basalt will increase by about 5°C (4.18 j/g ÷ 0.84 joule/g • °C = 4.98°C). Thus, when both a land surface and a water surface absorb the same amount of heat, the land will become warmer than the water. Therefore, Buffalo surface air temperatures increase faster and earlier in the year than do the surface water temperatures of Lake Erie because water has a higher specific heat than land. So the air over land heats up faster than the air over the lake.

Some additional factors also contribute to Buffalo surface air temperatures increasing faster and earlier in the year than the surface water temperatures of Lake Erie. Early in the spring, even after snow has melted off of the land, Lake Erie is still covered by ice. The darker land surface absorbs more energy than the ice, so the land heats up faster. When ice melts, it absorbs a lot of energy without changing temperature. Thus, early in the spring, insolation absorbed by the land raises the land's temperature. In contrast, insolation absorbed by Lake Erie melts the ice without increasing the temperature of the ice.

One credit is allowed for an acceptable response. Acceptable responses include but are not limited to:

- **The specific heat of water is greater than the specific heat of land or dry land, so the air over the land heats up faster than the air over the lake.**
- **More energy is required to heat up the same amount of water than to heat the same amount of land.**
- Air has a lower specific heat than water.
- A lot of energy is used to melt the ice on Lake Erie.
- **Lake Erie is still covered by ice.**
- **The darker land surface absorbs greater insolation.**
- Land heats up faster than water.

55. Find the Solar System Data table in the *Reference Tables for Physical Setting/Earth Science*. In the column labeled "Celestial Object," locate "Neptune." From Neptune, trace right to the column labeled "Mean Distance from Sun (million km)." Note that Neptune's distance from the Sun is 4,498.3 million km (4498.3 × 10^6 km). Continue tracing right to the column labeled "Equatorial Diameter (km)," and note that the diameter of Neptune is 49,528 km. On the graph of Planet Diameters and Planet Distances from the Sun, locate the point corresponding to 4,498 on the horizontal axis labeled "Mean Distance from Sun (× 10^6 km)." At this point, draw a vertical line across the graph. Now locate the point corresponding to 49.498 on the vertical axis labeled "Equatorial Diameter of Planet (× 10^3 km)." At this point, draw a horizontal line across the graph. Place an **X** on the graph where the horizontal and vertical lines you drew intersect.

One credit is allowed if **the center of the X for Neptune is plotted within or touches the grid square that is circled as shown below.**

Note: Credit is allowed if a symbol other than an **X** is used. Neptune need not be labeled.

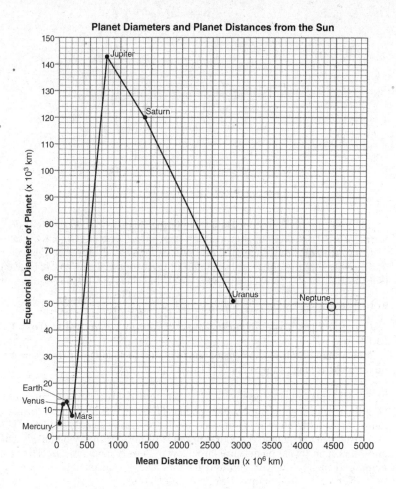

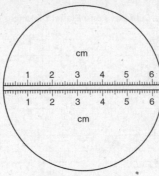

Scale: 1 cm = 2000 km

56. Find the Solar System Data table in the *Reference Tables for Physical Setting/Earth Science*. In the column labeled "Celestial Object," locate "Earth's Moon." Trace right to the column labeled "Equatorial Diameter (km)." Note that the equatorial diameter of the Moon is 3,476 km. It is given that the scale in the drawing is 1 cm = 2000 km. Thus, 3,476 km is equivalent to 1.74 cm ($\frac{1}{2,000} = \frac{x}{3,476}$; $2000x = 3476$; $x = \frac{3,476}{2,000}$; $x = 1.738$).

One credit is allowed for **indicating a diameter of any value from 1.6 to 1.9 cm.**

Note: One credit is allowed for shading anywhere along or on the centimeter scale as long as the shading is 1.6 to 1.9 cm long.

57. The planets can be divided by mass and density into the terrestrial (Earthlike) and Jovian (Jupiter-like) planets. The terrestrial planets are small, dense, and rocky. They are composed mostly of metal silicates and iron. The terrestrial planets include the four innermost planets: Mercury, Venus, Earth, and Mars. The Jovian planets are large, low in density, and gaseous. They are composed mainly of hydrogen and helium. The Jovian planets include the outer planets: Jupiter, Saturn, Neptune, and Uranus.

Find the Solar System Data chart in the *Reference Tables for Physical Setting/Earth Science*. Locate the column labeled "Period of Revolution (d=days) (y=years)." Note that the periods of revolution of the four terrestrial planets are measured in days and range from 88–687 days. Note that the periods of revolution of the four Jovian planets are measured in years and range from 11.9–164.8 years. Thus, the Jovian planets have longer periods of revolution than the terrestrial planets. Now find the column labeled "Period of Rotation at Equator." Note that the periods of rotation for the terrestrial planets range from 23 h 56 min 4 s to 243 days. Note that the periods of rotation for the Jovian planets range from 9 h 50 min 30 s to 17 h 14 min. Thus, the Jovian planets have shorter periods of rotation than the terrestrial planets.

Thus, compared to the periods of revolution and periods of rotation for the terrestrial planets, the Jovian planets have longer periods of revolution and shorter periods of rotation.

One credit is allowed if both responses are correct. Acceptable responses include but are not limited to:

Jovian periods of revolution:

- **longer**
- **greater**
- **more time**

Jovian periods of rotation:

- **shorter**
- **less time**

58. Find the Solar System Data chart in the *Reference Tables for Physical Setting/Earth Science*. Locate the column labeled "Mean Distance from Sun (million km)." Note that the distance from the Sun in million km that is closest to 404 is 227.9. From 227.9, trace left to the column labeled "Celestial Object," and note that 227.9 million km corresponds to Mars. Thus, the name of the planet that is closest to the center of the asteroid belt is Mars.

One credit is allowed for **Mars**.

59. To draw a line separating regions with Mercalli values of V from regions with Mercalli values of VI, do the following. Start at the V near the coast to the west of Savane Baptiste at 19° N latitude and draw a line northeastward midway between the V and the VI to its southeast. Continue the line southeast midway between the V just west of Savane Baptiste at 19° N and the VI to the northwest of Port-au-Prince. Continue the line southeast midway between the V south of Savane Baptiste and VIs to the east and southeast of Port-au-Prince. Next find the V on the peninsula that is about midway between 73° W and 74° W. Draw a line from coast to coast midway between the V and the VIs to the east. Finally, find the V and VI on the island just south of 19° N, 73° W, and draw a line from coast to coast midway between these two values. A completed line separating all values of V from VI is shown below.

One credit is allowed if **boundaries between V and VI are correctly drawn.**

Note: Credit is allowed even if the line passes through water.

Credit is *not* allowed if the line touches or passes through any Mercalli value.

If extra Mercalli lines are drawn, all must be correct to receive credit.

Example of a 1-credit response:

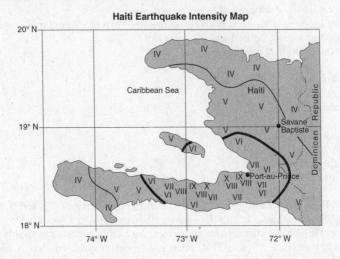

60. Lines of latitude run east-west (horizontally), and lines of longitude run north-south (vertically). Note that on this map, north latitudes are listed along the left side of the map and increase toward the north (top) of the map. West longitudes are listed along the bottom of the map and increase toward the west (left) on the map. On the map, locate Savane Baptiste. Note that Savane Baptiste is located at the intersection of the horizontal latitude line labeled "19° N" and the vertical longitude line labeled "72° W." Thus, Savane Baptiste is located at latitude 19° N, longitude 72° W.

One credit is allowed if *both* responses are correct.

- Latitude: 19° N
- Longitude: 72° W

61. According to the map, Haiti is located roughly between latitudes 18° N to 20° N and between longitudes 72° W to 74° W. Find the Tectonic Plates map in the *Reference Tables for Physical Setting/Earth Science*. Along the right side of the map, locate the position corresponding to latitudes 18° N to 20° N, and draw a horizontal line across the map. Along the top of the map, locate the position corresponding to longitudes 72° W to 74° W, and draw a vertical line across the map. Haiti is located at the intersection of these two lines. Note that Haiti is located at the boundary between the North American Plate and the Caribbean Plate.

One credit is allowed for **North American Plate and Caribbean Plate**.

62. Fossils are inferred to be of the same age as the rock in which they are found. According to the reading passage, "Bones of juvenile long-necked sauropod dinosaurs, *Abydosaurus mcintoshi*, have recently been found in 105-million-year-old sandstone." Thus, *Abydosaurus mcintoshi* are assumed to have lived 105 million years ago. Find the Geologic History of New York State chart in the *Reference Tables for Physical Setting/Earth Science*. On the time scale labeled "Million years ago" to the right of the column labeled "Epoch," locate the point corresponding to 105. Trace left to the columns labeled "Epoch" and "Period." Note that 105 million years ago occurred between the early and late epochs of the Cretaceous period. On the timeline, place an **X** on line *AB* between the Early and Late Cretaceous.

One credit is allowed for **the center of an X within or touching the box shown below.**

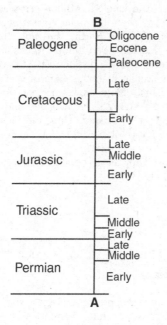

Note: Credit is allowed if a symbol other than an **X** is used.

63. The reading passage states that the bones of *Abydosaurus mcintoshi* were found in sandstone. Find the Scheme for Sedimentary Rock Identification in the *Reference Tables for Physical Setting/Earth Science*. In the column labeled "Rock Name," locate sandstone. From sandstone, trace left to the column labeled "Grain Size." Note that sandstone consists of grains of "Sand (0.006 to 0.2 cm)." Thus, the range of grain sizes in the type of bedrock in which *Abydosaurus mcintoshi* bones were found is 0.006 to 0.2 cm.

One credit is allowed for an acceptable response. Acceptable responses include but are not limited to:

- 0.006 cm to 0.2 cm
- 0.2 cm to 0.006 cm

64. As explained in the answer to question 62, *Abydosaurus mcintoshi* are inferred to have existed 105 million years ago. The reading passage states, "These dinosaurs were herbivores," and their teeth "allowed only for the harvesting of plant material." Thus, the most likely food source for *Abydosaurus mcintoshi* was plants or plant material that existed at the same time these dinosaurs existed 105 million years ago. Find the Geologic History of New York State chart in the *Reference Tables for Physical Setting/Earth Science*. On the time scale labeled "Million years ago" to the right of the column labeled "Epoch," locate the point corresponding to 105. Trace right to the column labeled "Life on Earth," and note the following statements concerning plant life that appeared before 105 million years ago: "Earliest land plants and animals," "Earth's first forests," "Large and numerous scale trees and seed ferns (vascular plants)," "Extensive coal-forming forests," and "Earliest flowering plants." Note, however, that the "Earliest grasses" did not appear until 23.0 to 33.9 million years ago and therefore did not exist when *Abydosaurus mcintoshi* existed. Thus, the most likely food source for *Abydosaurus mcintoshi* was plants and plant materials, such as flowering plants, vascular plants, trees, and ferns.

One credit is allowed for an acceptable response. Acceptable responses include but are not limited to:

- vascular plants
- flowering plants
- trees
- ferns
- plants
- plant materials

Note: Grasses are *not* accepted because the earliest grasses appeared in the Oligocene. The name of any index fossil found in the *Reference Tables for Physical Setting/Earth Science* is not accepted because these organisms did not live during the time that *Abydosaurus mcintoshi* lived.

65. Find the Geologic History of New York State chart in the *Reference Tables for Physical Setting/Earth Science*. In the column labeled "Life on Earth," locate the statement "Mass extinction of dinosaurs, ammonoids, and many land plants." From this statement, trace left to the time scale labeled "Million years ago." Note that the extinction of the last dinosaurs occurred 65.5 million years ago. Trace left to the column labeled "Period." Note that

65.5 million years ago marks the end of the Cretaceous period and the beginning of the Paleogene period.

Evidence suggests that an impact event was responsible for the extinction of many life-forms at the end of the Cretaceous period 65.5 million years ago. The dividing line between the Cretaceous and the Paleogene is a thin layer of clay that has been identified in sediments worldwide. This boundary of clay contains numerous indications of a massive impact event: levels of iridium higher than those found in Earth's crust but similar to those found in meteorites, shocked quartz grains, melted spherules, soot from the widespread forest fires ignited by the meteorite, and evidence of large waves such as would be caused by an ocean impact. The impact of asteroids or large meteorites would generate a massive shock wave and spew large amounts of dust high into the atmosphere. The resulting decrease in sunlight would cause global temperatures to decrease markedly and have devastating effects on photosynthetic life and all of the life-forms that depend on them for food. Thus, some scientists have inferred that the extinction of the last of the dinosaurs that occurred 65.5 million years ago was caused by an asteroid impact and its associated climate changes.

One credit is allowed for an acceptable response. Acceptable responses include but are not limited to:

- **an asteroid impact**
- **an impact event**
- **a meteorite/meteor/meteoroid collision with Earth**
- **climate change**
- **a disruption of food chains/food webs**
- **comet impact**

Note: Credit is *not* allowed for "meteorite," "meteor," "meteoroid," or "comet" alone because they do not describe a natural event.

Credit is *not* allowed for "volcanic eruption" because this is not the most widely inferred cause of dinosaur extinction.

PART C

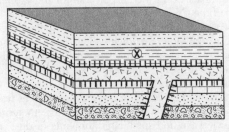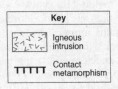

(Not drawn to scale)

66. Find the Scheme for Sedimentary Rock Identification in the *Reference Tables for Physical Setting/Earth Science*. In the column labeled "Rock Name," locate shale and limestone. Trace right to the column labeled "Map Symbol," and note the map symbols for shale and limestone. On the block diagram, locate the layers corresponding to shale and limestone. In the block diagram key, locate the map symbols for an igneous intrusion and contact metamorphism. Note that the layer directly between the limestone and shale layers corresponds to an igneous intrusion. Note that the limestone and shale layers have undergone contact metamorphism along their boundary with the igneous intrusion. In order to undergo contact metamorphism, a rock must already exist when the igneous rock intrudes. Thus, all of the rock layers in contact with the igneous intrusion that have undergone contact metamorphism are older than the igneous intrusion. Therefore, the evidence represented in the diagram that indicates that the shale layer and the limestone layer are older than the igneous intrusion is the presence of contact metamorphism along the boundaries between these layers and the igneous intrusion.

One credit is allowed for an acceptable response. Acceptable responses include but are not limited to:

- **Both layers have contact metamorphism.**
- **The layers were changed by the igneous intrusion.**
- **The intrusion could metamorphose only layers that were already existing.**
- **Contact metamorphism**

Note: Credit is *not* allowed for "igneous intrusions are younger than the rock they cut across" because the intrusion does not cut across the shale layer.

67. Find the Scheme for Sedimentary Rock Identification in the *Reference Tables for Physical Setting/Earth Science*. In the column labeled "Rock Name," locate limestone. Trace left to the column labeled "Composition," and note that limestone is composed of calcite.
One credit is allowed for **calcite**.

68. Index fossils are fossils of an organism that had distinctive body features and was abundant over a wide geographic area but existed for only a short period of geologic time. The organism's distinctive body features make it easy to recognize. Abundance over a wide geographic area allows the index fossils to be used to correlate rocks in widely separated places. Existing for only a short period of time allows one to date the age of the rocks in which they are found roughly. Thus, these are the characteristics that fossil X must have in order to be considered a good index fossil.
One credit is allowed for an acceptable response. Acceptable responses include but are not limited to:

- **widespread distribution**
- **lived for a short time**
- **easily identified**

69. The most common isotopes of the elements that comprise rocks are stable, meaning their atoms do not change. However, some isotopes are unstable. These isotopes are called radioactive isotopes. Radioactive isotopes break apart, or decay, at a steady, constant rate that is not affected by outside factors such as changes in temperature, pressure, or chemical state. Therefore, the decay process may be used as a clock to determine the actual age of these rocks. Radioactive isotopes start to decay when the rock forms. The decay of a radioactive isotope occurs at a statistically predictable rate known as its half-life. Half-life is the time required for one-half of the unstable radioisotope to change into a stable decay product. By comparing the relative amounts of radioactive isotope and stable decay product in a rock, the number of half-lives that have elapsed may be determined and used to determine the actual ages of these rocks. Potassium-40 is a radioactive isotope. Therefore, the property of potassium-40 that allows it to be useful in the radioactive dating of rocks is its constant rate of decay, or half-life. Find the Radioactive Decay Data chart in the *Reference Tables for Physical Setting/Earth Science*. In the column labeled "Radioactive Isotope," locate potassium-40. Trace right to the column labeled "Half-Life (years)." Note that the half-life of potassium-40 is 1.3×10^9 years. Potassium-40 can be used to date rocks ranging in age from about 100,000 to 4.3 billion years. Thus, potassium-40 can be used to date rocks across a wide range of ages (including some of the oldest rocks on Earth) because it has a very long half-life.

One credit is allowed for an acceptable response. Acceptable responses include but are not limited to:

- Potassium-40 decays at a specific rate.
- K-40 has a known half-life.
- K-40 has a constant rate of decay.
- K-40 decays at a rate independent of external factors.
- Potassium-40 has a long half-life.
- The half-life of K-40 is 1.3×10^9 years.

A Foucault Pendulum's Swing

Latitude (°)	Rate of Change in Apparent Direction of Swing (°/h)
0	0.0
10	2.6
20	5.1
30	7.5
40	9.6
50	11.5
60	13.0
70	14.1
80	14.8
90	15.0

70. To plot the hourly change in a Foucault pendulum's apparent direction of swing at the latitudes shown in the data table, do the following. In the data table labeled "A Foucault Pendulum's Swing," note that at 0° latitude, the rate of change in apparent direction of swing is 0.0°/h. On the provided grid, locate "0" on the horizontal axis labeled "Latitude (°)." Now locate "0" on the vertical axis labeled "Rate of Change in Apparent Direction of Swing (°/h)." Note that these two values correspond to a single point—the origin of the graph. Make a mark at this point on the grid. Next, note that at latitude 10°, the rate of change in apparent direction of swing is 2.6°/h. On the grid, locate "10" on the horizontal axis labeled "Latitude (°)." Now locate "2.6" on the vertical axis labeled "Rate of Change in Apparent Direction of Swing (°/h)." Trace vertically upward from 0° latitude and horizontally to the right from 2.6°/h to the point where these two values intersect. At the intersection, make a mark on the grid. Repeat this process for each of the remaining latitudes given in the data table. When all of the points have been plotted, connect them with a line that passes through every point in order of latitude. A graph with all points correctly plotted and connected with a line is shown on the following page.

One credit is allowed if **the centers of *all ten* plots are within or touch the circles shown below and are correctly connected with a line passing within or touching the circles**.

Note: Credit is allowed if a line misses a plot but is still within or touches the circle.

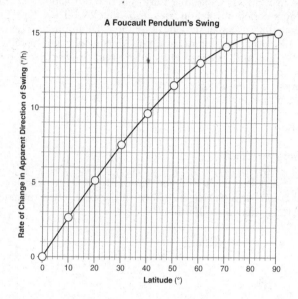

71. The latitude of the North Pole is 90° N. On the data table "A Foucault Pendulum's Swing," locate "90" in the column labeled "Latitude (°)." Trace right to the column labeled "Rate of Change in Apparent Direction of Swing (°/h)" and note the value "15.0." Therefore, at the North Pole, a Foucault pendulum will change direction at a rate of 15°/h. In order to complete a 360° change in its apparent direction of swing, the Foucault pendulum would require 24 hours (360° ÷ 15°/h = 24 h).

One credit is allowed for **24 h**.

72. In 1851, French physicist Jean Foucault devised a way of conclusively proving Earth's rotation. He used a pendulum. He suspended a heavy iron ball on a long steel wire from the top of the dome of the Pantheon in Paris. As the pendulum swung back and forth, it appeared to change direction, slowly passing over different lines on the floor until eventually coming full circle to its original position after 24 hours. Since he knew that the path of a freely swinging pendulum would not change on its own, Foucault concluded that the apparent shift in the direction of swing of the pendulum was due to the floor (Earth's surface) rotating beneath the pendulum. Thus, Earth's

rotation causes the change in the apparent direction of swing of a Foucault pendulum.

Find the Solar System Data chart in the *Reference Tables for Physical Setting/Earth Science*. In the column labeled "Celestial Object," locate Earth and Mars. From these two planets, trace right to the column labeled "Period of Rotation at Equator." Note that Earth's period of rotation is 23 hr 56 min 4 s and Mars's period of rotation is 24 hr 37 min 23 s. Thus, Earth and Mars both rotate and have similar periods of rotation. Therefore, it is reasonable to infer that a Foucault pendulum set up on Mars would show similar changes to one set up on Earth because Mars rotates and also because it rotates at a rate similar to that of Earth.

One credit is allowed for an acceptable response. Acceptable responses include but are not limited to:

- **rotation**
- **spin**
- **turning on its axis**

Note: Credit is not allowed for "faster rotation" or "shorter rotation" because these are scientifically incorrect with respect to the rotation of Mars compared to the rotation of Earth.

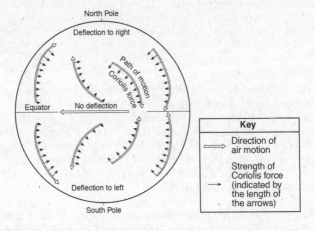

73. According to the key, the strength of the Coriolis force is indicated by the length of the black arrows on the diagram. Note that along all of the white arrows that indicate direction of air motion, the black arrows representing strength of Coriolis force increase in length with distance north or south of the equator. The equator corresponds to latitude 0°. The North and South Poles correspond to 90° latitude. As you go from the equator toward either pole, latitude increases from 0° toward 90°. Thus, latitude increases with

distance north or south of the equator. Therefore, as latitude increases, the strength of the Coriolis force increases—a direct relationship.

One credit is allowed for an acceptable response. Acceptable responses include but are not limited to:

- **As latitude increases, the Coriolis force increases.**
- **The closer to the equator, the weaker the Coriolis force**
- **The Coriolis force is strongest by the poles.**
- **Direct relationship**

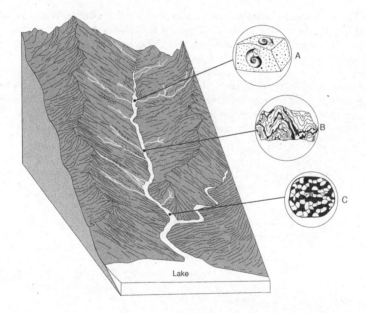

74. Note that rock sample A has an angular shape with sharp edges and corners. As sediments are carried along by the moving water in a stream, they bounce off of and rub against one another. The sediments also bounce, roll, and scrape against the bottom of the streambed. These collisions cause smaller pieces to break off of sharp corners and edges on the surface of the sediments, particularly at corners that protrude—a process called abrasion. As a result, the sediments eroded by the water in this stream become more rounded. The longer a particle is transported in a stream, the more it is abraded and the more rounded the particle becomes. The fact that rock sample A still has an angular shape with sharp corners and edges indicates that it has spent very little time being transported by the stream and undergoing abrasion.

One credit is allowed for an acceptable response. Acceptable responses include but are not limited to:

- **The sample is angular in appearance.**
- **It is not rounded.**
- **The edges are not worn off.**

Note: The responses "not very weathered" or "no abrasion" are not accepted because they do not describe an appearance or observable characteristic of the rock sample.

75. Find the Relationship of Transported Particle Size to Water Velocity graph in the *Reference Tables for Physical Setting/Earth Science*. On the vertical axis labeled "Particle Diameter (cm)," locate the point corresponding to 2.0 cm. From this point, trace right until you intersect the bold curve. From the intersection, trace vertically downward to the horizontal axis labeled "Stream Velocity (cm/s)." Note that in order to transport a 2.0 cm particle, the stream velocity must be 90 cm/s. Thus, the minimum velocity this stream needs to transport a 2.0 cm particle is 90 cm/s.

One credit is allowed for **any value from 70 cm/s to 110 cm/s**.

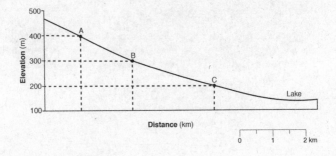

76. Find the Equations table in the *Reference Tables for Physical Setting/Earth Science*. Locate the equation for gradient:

$$\text{Gradient} = \frac{\text{change in field value}}{\text{distance}}$$

The field value on the stream profile is elevation. Find the location corresponding to rock sample A on the stream profile. Trace horizontally to the left to the axis labeled "Elevation (m)," and note that the elevation of rock sample A is 400 m. Find the location corresponding to rock sample C on the stream profile. Trace horizontally to the left to the axis labeled "Elevation (m)," and note that the elevation of rock sample C is 200 m. Note where the dashed vertical lines extending downward from points A and C intersect the horizontal axis labeled "Distance (km)." On a piece of scrap paper, mark off the distance between the dashed lines corresponding to A and C on the horizontal axis labeled "Distance (km)." Compare the distance marked off on the scrap paper to the scale printed beneath the stream profile. Note that the distance between points A and C is 4 km. Substitute these values into the equation as shown below:

$$\text{Gradient} = \frac{400 \text{ m} - 200 \text{ m}}{4 \text{ km}}$$

Solve the equation:

$$\text{Gradient} = \frac{200 \text{ m}}{4 \text{ km}} = 50 \frac{\text{m}}{\text{km}}$$

One credit is allowed for **any value from 48 m/km to 52 m/km**.

77. It is given that the width of the shaded path indicates the width of destruction on the ground and that the numbers on the shaded path indicate the Fujita intensity at those locations. According to the Fujita Intensity Scale chart on the map, as F-scale increases, damage increases. Thus, the tornado damage was greatest at the location along the path where the F-scale number is highest and the width is greatest. Note that the highest F-scale number on the path is 4. Thus, an **X** indicating where the tornado damage was greatest should be placed on the map where the F-scale number is 4. Note that of the locations where the F-scale numbers are 4, the path is widest near Hallam.

One credit is allowed if **the center of the X is within or touches any of the clear areas along the path of the tornado shown below**.

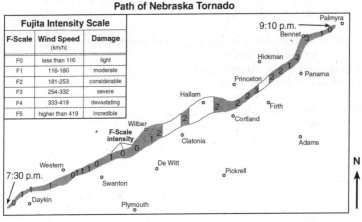

78. On the map labeled "Path of Nebraska Tornado," locate Bennet. Note that the F-scale number nearest Bennet is 3. On the Fujita Intensity Scale chart, locate "3" in the column labeled "F-scale." Trace right to the column labeled "Wind Speed (km/h)." Note that an F3 tornado has wind speeds ranging from 254 km/h to 332 km/h. Thus, when it was moving through Bennet, the wind speed of the tornado was between 254 km/h and 332 km/h.

One credit is allowed for **any value from 254 km/h to 332 km/h**.

79. Wind speed is measured with an instrument consisting of cups or fan-like blades attached to a shaft in such a way that when wind blows against the cups or blades, they move and cause the shaft to spin. The greater the wind speed, the faster the shaft spins. An instrument consisting of cups attached to a shaft that spins is called an anemometer. Other instruments that have blades attached to a shaft that spins are called wind speed meters.

One credit is allowed for **anemometer** *or* **wind speed meter**.

80. When a tornado has been sighted or indicated by weather radar, the National Weather Service issues a tornado warning. According to "A Preparedness Guide Including Safety Information for Schools" issued by the U.S. Department of Commerce National Oceanic and Atmospheric Administration National Weather Service, if a tornado warning is issued or if threatening weather approaches, you should do the following. If you are in a school, go to a predesignated shelter area, such as a basement, or to the structurally strongest area in the school nearest your location. If an underground shelter is not available, move to an interior room or hallway on the lowest floor and get under a sturdy piece of furniture. Stay away from windows.

One credit is allowed for an acceptable response. Acceptable responses include but are not limited to:

- **go into a basement or underground storm shelter**
- **go to an interior room**
- **stay away from windows**
- **get under something sturdy**

Note: Credit is *not* allowed for a response that indicates a safety precaution to prepare for a future tornado.

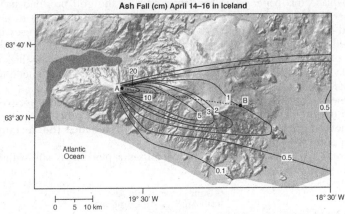

81. In order to construct a profile of the thickness of the volcanic ash deposits along line *AB*, do the following. Place the straight edge of a piece of scrap paper along the dashed line connecting location *A* to location *B* on the map. Mark the edge of the paper at locations *A* and *B* and wherever the paper intersects an ash thickness isoline. Wherever the paper intersects an ash thickness isoline, label the mark with the thickness of the ash as shown below. (Note that the ash thickness at location *A* has been plotted, so you do not need to mark the ash thickness at point *A* on the scrap paper.)

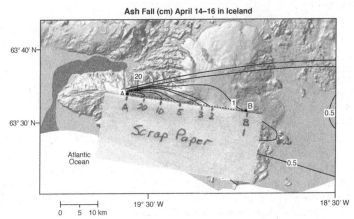

Then place this paper along the lower edge of the grid provided in your answer booklet so that points *A* and *B* on the scrap paper align with points *A* and *B* on the grid. Note that the ash thickness at point *A* has already been plotted. At each of the other points where an ash thickness isoline crossed the edge of the scrap paper, draw a plot on the grid at the appropriate thickness. Finally, connect all of the plots in a smooth curve to form the finished profile as shown below.

One credit is allowed if **the centers of *all six* student plots are within or touch the rectangles shown below and *all seven* plots are correctly connected with a line that passes within or touches the rectangles from point *A* to point *B*.**

Note: Credit is allowed if the line misses a plot but is still within or touches the rectangle.

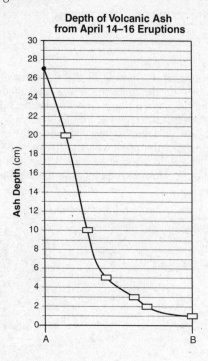

82. The reading passage states "Most of the ash was transported within the atmosphere below 10 kilometers." Find the Selected Properties of Earth's Atmosphere chart in the *Reference Tables for Physical Setting/Earth Science*. Locate 10 kilometers on the "Altitude" scale (km are listed on the left-hand side of the scale). Trace right to the section labeled "Temperature Zones." Note that an altitude of less than 10 km corresponds to the atmo-

spheric temperature zone layer labeled "troposphere." Thus, the atmospheric layer in which most of the volcanic ash was transported is the troposphere.

One credit is allowed for **troposphere**.

83. Volcanic ash is opaque and blocks, reflects, or scatters incoming solar radiation. Thus, a volcanic ash cloud in the atmosphere decreases atmospheric transparency and results in less incoming solar radiation reaching Earth's surface. If less solar radiation reaches Earth's surface, less energy is absorbed by the surface. The less energy that Earth's surface absorbs, the less the surface is warmed. Thus, a volcanic ash cloud in Earth's atmosphere would cause Earth's temperature to decrease. The reading passage states that "Much of the ash fell quickly to Earth, as seen in the map, but large quantities remained airborne and spread over Europe." If large quantities of ash spread over Europe, the ash cloud would block incoming sunlight and result in cooler weather conditions in Europe.

One credit is allowed for an acceptable response. Acceptable responses include but are not limited to:

- **The ash most likely reflected/scattered the incoming solar radiation.**
- **Less sunlight was received at Earth's surface.**
- **The ash most likely blocked some of the sunlight.**
- **Atmospheric transparency was reduced.**
- **The cloud blocked the Sun.**

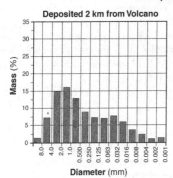
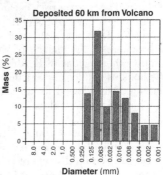

Volcanic Ash Deposited from the April 14-16 Eruption

84. In the graph labeled "Deposited 2 km from Volcano," note that the greatest percentages by mass of ash particles were in the range of 0.500 to 4.0 mm in diameter. In the graph labeled "Deposited 60 km from Volcano," note that the greatest percentages by mass of ash particles were in the range

of 0.063 to 0.125 mm in diameter. Also note that on the graph labeled "Deposited 60 km from Volcano," no particles greater than 0.250 mm in diameter were deposited. Thus, the size of deposited ash particles decreased with distance from the volcano. In other words, larger particles were carried shorter distances while greater percentages by mass of small particles were carried farther.

One credit is allowed for an acceptable response. Acceptable responses include but are not limited to:

- A greater percentage by mass of smaller ash particles was carried farther.
- The larger particles were carried shorter distances.
- They tend to be a smaller size.

85. The reading passage states that Eyjafjallajökull volcano is located in southern Iceland. Find the Tectonic Plates map in the *Reference Tables for Physical Setting/Earth Science*, and locate Iceland. (If you have trouble finding Iceland, note that on the ash fall map, Eyjafjallajökull volcano is located at roughly 63° 35′ N latitude, 19° 35′ W longitude.) Note that Iceland is located on the Mid-Atlantic Ridge that, according to the key, is a divergent plate boundary. Divergent boundaries mark places where plates are moving away from one another. Divergent plate boundaries create tension stresses, causing rock to fracture and open rifts through which magma can rise to the surface. Thus, divergent boundaries are often associated with volcanic activity. Also note that Iceland is labeled "Iceland Hot Spot." A mantle hot spot is a long-lasting zone of rising hot magma beneath moving plates. Large batches of magma rise from these hot spots and work their way upward through the plate above. The magma rises because it is more buoyant than the surrounding rock, wedges apart cracks in the plate, and melts through to erupt, forming a volcano. Thus, hot spots are associated with volcanic activity. Therefore, the lithosphere in the vicinity of Eyjafjallajökull is more volcanically active than most other regions of Earth's surface because it is both located over a mantle hot spot and located at a divergent plate boundary.

One credit is allowed for an acceptable response. Acceptable responses include but are not limited to:

- Iceland is located above a mantle hot spot.
- A tectonic plate boundary passes through Iceland.
- Iceland is on a divergent plate boundary.
- Iceland is located on a mid-ocean ridge.

SELF-ANALYSIS CHART August 2015

Topic	Question Numbers (Total)	Wrong Answers (x)	Grade
Standards 1, 2, 6, and 7: Skills and Application			
Skills			
Standard 1 Analysis, Inquiry, and Design	1, 2, 5–8, 12–15, 17–21, 23, 25, 27, 32–45, 47, 49–59, 61–72, 74–76, 78, 79, 81–85		$\dfrac{100(66-x)}{66} = \%$
Standard 2 Information Systems			
Standard 6 Interconnectedness, Common Themes	3, 7, 8, 11, 12, 14, 15, 17, 18, 22, 25–31, 33, 35–50, 52, 53, 56–62, 65, 66, 68, 73, 75, 77, 81, 83, 84		$\dfrac{100(52-x)}{52} = \%$
Standard 7 Interdisciplinary Problem Solving	80		$\dfrac{100(1-x)}{1} = \%$
Standard 4: The Physical Setting/Earth Science			
Astronomy			
The Solar System (MU 1.1a, b; 1.2d)	2, 27, 44–47, 55, 58, 65		$\dfrac{100(9-x)}{9} = \%$
Earth Motions and Their Effects (MU 1.1c, d, e, f, g, h, i)	3, 5, 7, 25, 60, 70–73		$\dfrac{100(9-x)}{9} = \%$
Stellar Astronomy (MU 1.2b)	1, 4		$\dfrac{100(2-x)}{2} = \%$
Origin of Earth's Atmosphere, Hydrosphere, and Lithosphere (MU 1.2e, f, h)	19		$\dfrac{100(1-x)}{1} = \%$
Theories of the Origin of the Universe and Solar System (MU 1.2a, c)	6, 56, 57		$\dfrac{100(3-x)}{3} = \%$

SELF-ANALYSIS CHART August 2015

Topic	Question Numbers (Total)	Wrong Answers (x)	Grade
Meteorology			
Energy Sources for Earth Systems (MU 2.1a, b)	82		$\dfrac{100(1-x)}{1} = \%$
Weather (MU 2.1c, d, e, f, g, h)	10, 12, 28, 77–80		$\dfrac{100(7-x)}{7} = \%$
Insolation and Seasonal Changes (MU 2.1i; 2.2a, b)	8, 29, 41–43, 54, 83		$\dfrac{100(7-x)}{7} = \%$
The Water Cycle and Climates (MU 1.2g; 2.2c, d)	9, 26, 36, 37, 50–53		$\dfrac{100(8-x)}{8} = \%$
Geology			
Minerals and Rocks (MU 3.1a, b, c)	21, 23, 32, 34, 35, 63, 67		$\dfrac{100(7-x)}{7} = \%$
Weathering, Erosion, and Deposition (MU 2.1s, t, u, v, w)	11, 30, 31, 48, 49, 74–76, 84		$\dfrac{100(9-x)}{9} = \%$
Plate Tectonics and Earth's Interior (MU 2.1j, k, l, m, n, o)	15–17, 38–40, 59, 61, 85		$\dfrac{100(9-x)}{9} = \%$
Geologic History (MU 1.2i, j)	18, 20, 22, 24, 33, 62, 64, 66–69		$\dfrac{100(10-x)}{10} = \%$
Topographic Maps and Landscapes (MU 2.1p, q, r)	13, 14, 81		$\dfrac{100(3-x)}{3} = \%$
ESRT			
Earth Science Reference Tables 2011 Edition	1, 2, 7, 8, 12–15, 17–21, 23, 32–35, 38, 46, 50, 53–58, 61–65, 67, 69, 75, 76, 82, 85		$\dfrac{100(38-x)}{38} = \%$

To further pinpoint your weak areas, use the Topic Outline in the front of the book.
MU = Major Understanding (see Topic Outline)

Examination June 2016
Physical Setting/Earth Science

PART A
Answer all questions in this part.

Directions (1–35): For *each* statement or question, choose the word or expression that, of those given, best completes the statement or answers the question. Some questions may require the use of the *2011 Edition Reference Tables for Physical Setting/Earth Science*. Record your answers in the space provided.

1 Earth's approximate rate of revolution is
 (1) 1° per day
 (2) 15° per day
 (3) 180° per day
 (4) 360° per day

2 Planetary winds in the Northern Hemisphere are deflected to the right due to the
 (1) Doppler effect
 (2) Coriolis effect
 (3) tilt of Earth's axis
 (4) polar front jet stream

3 Which star is hotter, but less luminous, than *Polaris*?
 (1) *Deneb*
 (2) *Aldebaran*
 (3) *Sirius*
 (4) *Pollux*

4 Which statement best explains why Earth and the other planets of our solar system became layered as they were being formed?

(1) Gravity caused less-dense material to move toward the center of each planet.
(2) Gravity caused more-dense material to move toward the center of each planet.
(3) Materials that cooled quickly stayed at the surface of each planet.
(4) Materials that cooled slowly stayed at the surface of each planet.

5 Which conditions on Earth's surface will allow for the greatest amount of water to seep into the ground?

(1) gentle slope and permeable
(2) gentle slope and impermeable
(3) steep slope and permeable
(4) steep slope and impermeable

6 The photograph below shows a Foucault pendulum at a museum. The pendulum knocks over pins in a regular pattern as it swings back and forth.

This pendulum movement, and the pattern of knocked-over pins, is evidence of Earth's

(1) nearly spherical shape
(2) gravitational attraction to the Sun
(3) rotation on its axis
(4) nearly circular orbit around the Sun   6 _____

7 Earth's early atmosphere contained carbon dioxide, sulfur dioxide, hydrogen, nitrogen, water vapor, methane, and ammonia. These gases were present in the atmosphere primarily because

(1) radioactive decay products produced in Earth's core were released from Earth's surface
(2) evolving Earth life-forms produced these gases through their activity
(3) Earth's growing gravitational field attracted these gases from space
(4) volcanic eruptions on Earth's surface released these gases from the interior   7 _____

8 The diagram below represents the apparent positions of the Big Dipper, with respect to *Polaris*, as seen by an observer in New York State at midnight on the first day of summer and on the first day of winter.

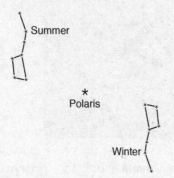

The change in the apparent position of the Big Dipper between the first day of summer and the first day of winter is best explained by Earth

(1) rotating for 12 hours
(2) rotating for 1 day
(3) revolving for 6 months
(4) revolving for 1 year

9 The weather station model shown below indicates that winds are coming from the

(1) southeast at 10 knots
(2) northwest at 10 knots
(3) southeast at 20 knots
(4) northwest at 20 knots

10 Which type of air mass most likely has high humidity and high temperature?

(1) cP (3) mT
(2) cT (4) mP

10 ____

11 What is the relative humidity if the dry-bulb temperature is 16°C and the wet-bulb temperature is 10°C?

(1) 45% (3) 14%
(2) 33% (4) 4%

11 ____

12 The table below shows the air temperature and dewpoint at each of four locations, A, B, C, and D.

Location	A	B	C	D
Air temperature (°F)	80	60	45	35
Dewpoint (°F)	60	43	35	33

Based on these measurements, which location has the greatest chance of precipitation?

(1) A (3) C
(2) B (4) D

12 ____

13 Which type of electromagnetic radiation has the shortest wavelength?

(1) ultraviolet (3) radio waves
(2) gamma rays (4) visible light

13 ____

14 Which gas is considered a major greenhouse gas?

(1) methane (3) oxygen
(2) hydrogen (4) nitrogen

14 ____

15 The diagram below represents Earth and the Sun's incoming rays. Letters A, B, C, and D represent locations on Earth's surface.

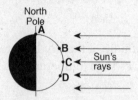

Which two locations are receiving the same intensity of insolation?

(1) A and B   (3) C and D
(2) B and C   (4) D and B   15 ____

16 Most of the sand that makes up the sandstone found in New York State was originally deposited in which type of layers?

(1) tilted   (3) faulted
(2) horizontal   (4) folded   16 ____

17 The map below shows the current location of New York State in North America.

Approximately how many million years ago (mya) was this New York State region located at the equator?

(1) 59 mya   (3) 359 mya
(2) 119 mya   (4) 458 mya   17 ____

18 Many scientists infer that one cause of the mass extinction of dinosaurs and ammonoids that occurred approximately 65.5 million years ago was

(1) tectonic plate subduction of most of the continents
(2) an asteroid impact that resulted in climate change
(3) a disease spreading among many groups of organisms
(4) severe damage produced by worldwide earthquakes   18 ____

19 During which geologic epoch do scientists infer that the earliest grasses first appeared on Earth?

(1) Holocene   (3) Oligocene
(2) Pleistocene   (4) Eocene   19 ____

20 What are the inferred pressure and temperature at the boundary of Earth's stiffer mantle and outer core?

(1) 1.5 million atmospheres pressure and an interior temperature of 4950°C
(2) 1.5 million atmospheres pressure and an interior temperature of 6200°C
(3) 3.1 million atmospheres pressure and an interior temperature of 4950°C
(4) 3.1 million atmospheres pressure and an interior temperature of 6200°C 20 ____

21 A seismic P-wave is recorded at 2:25 p.m. at a seismic station located 7600 kilometers from the epicenter of an earthquake. At what time did the earthquake occur?

(1) 2:05 p.m.   (3) 2:14 p.m.
(2) 2:11 p.m.   (4) 2:36 p.m.   21 ____

22 A seismic station recorded the *P*-waves, but no *S*-waves, from an earthquake because *S*-waves were

(1) absorbed by Earth's outer core
(2) transmitted only through liquids
(3) weak and detected only at nearby locations
(4) not produced by this earthquake   22 _____

23 The Catskills of New York State are best described as a plateau, while the Adirondacks are best described as mountains. Which factor is most responsible for the difference in landscape classification of these two regions?

(1) climate variations   (3) vegetation type
(2) bedrock structure   (4) bedrock age   23 _____

24 An elongated hill that is composed of unsorted sediments deposited by a glacier is called

(1) a delta   (3) a sand dune
(2) a drumlin   (4) an outwash plain   24 _____

25 Which rock was subjected to intense heat and pressure but did *not* solidify from magma?

(1) sandstone   (3) gabbro
(2) schist   (4) rhyolite   25 _____

26 The map below shows a stream drainage pattern where the streams radiate outward from the center.

Which landscape feature would produce this stream drainage pattern?

(1) steep cliff
(2) glacial kettle lake
(3) volcanic mountain
(4) flat plain

26 _____

27 The map below shows the area that, at one time, was covered by ancient Lake Bonneville. Evidence of ancient shorelines indicates that, near the end of the last ice age, Lake Bonneville existed in western Utah and eastern Nevada. The Great Salt Lake in Utah is a remnant of the former Lake Bonneville.

Which material that was formerly on the bottom of Lake Bonneville is most likely exposed on the land surface today?

(1) folded metamorphic bedrock
(2) flat-lying evaporite deposits
(3) coarse-grained coal beds
(4) fine-grained layers of volcanic lava

27 _____

28 The cross section below represents a portion of a meandering stream. Points X and Y represent two positions on opposite sides of the stream.

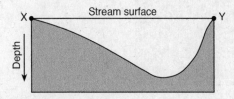

Based on the cross section, which map of a meandering stream best shows the positions of points X and Y?

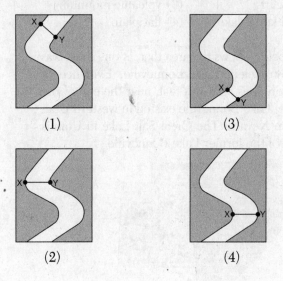

28 _____

29 When wind and running water gradually decrease in velocity, the transported sediments are deposited
  (1) all at once, and are unsorted
  (2) all at once, and are sorted by size and density
  (3) over a period of time, and are unsorted
  (4) over a period of time, and are sorted by size and density   29 ____

30 The graph below shows ocean water levels for a shoreline location on Long Island, New York. The graph also indicates the dates and times of high and low tides.

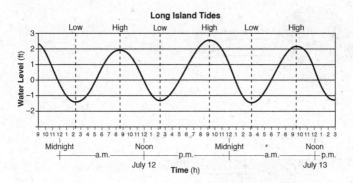

Based on the data, the next high tide occurred at approximately

  (1) 4 p.m. on July 13   (3) 4 p.m. on July 14
  (2) 10 p.m. on July 13   (4) 10 p.m. on July 14   30 ____

31 Which diagram best represents heat transfer mainly by the process of conduction?

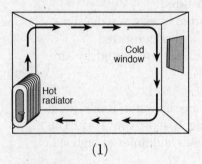

(1)

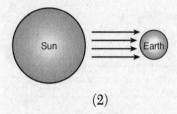

(2)

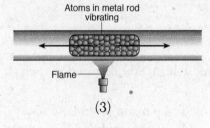

(3)

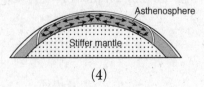

(4)

31 _____

32 The diagram below represents the position of Earth in its orbit and the position of a comet in its orbit around the Sun.

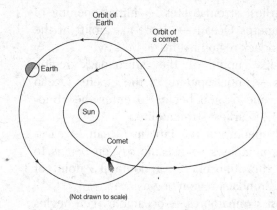

(Not drawn to scale)

Which inference can be made about the comet's orbit, when it is compared to Earth's orbit?

(1) Earth's orbit and the comet's orbit have the same distance between foci.
(2) Earth's orbit has a greater distance between foci than the comet's orbit.
(3) The comet's orbit has one focus, while Earth's orbit has two foci.
(4) The comet's orbit has a greater distance between foci than Earth's orbit.

32 _____

33 Which sequence of geologic events is in the correct order, from oldest to most recent?

(1) oceanic oxygen begins to enter the atmosphere → earliest stromatolites → initial opening of the Iapetus Ocean → dome-like uplift of the Adirondack region begins
(2) dome-like uplift of the Adirondack region begins → initial opening of the Iapetus Ocean → oceanic oxygen begins to enter the atmosphere → earliest stromatolites
(3) initial opening of the Iapetus Ocean → earliest stromatolites → oceanic oxygen begins to enter the atmosphere → dome-like uplift of the Adirondack region begins
(4) earliest stromatolites → oceanic oxygen begins to enter the atmosphere → initial opening of the Iapetus Ocean → dome-like uplift of the Adirondack region begins   33 _____

34 The cross section of the atmosphere below represents the air motion near two frontal boundaries along reference line *XY* on Earth's surface.

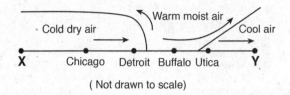

Which weather map correctly identifies these fronts and indicates the direction that these fronts are moving?

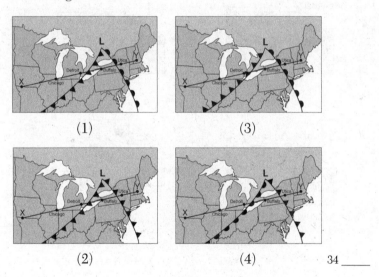

34 _____

35 Which block diagram represents the plate motion that causes the earthquakes that occur along the San Andreas Fault in California?

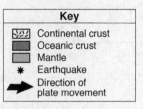

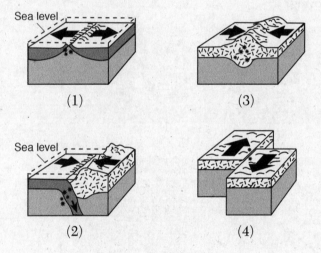

35 ____

PART B–1
Answer all questions in this part.

Directions (36–50): For *each* statement or question, choose the word or expression that, of those given, best completes the statement or answers the question. Some questions may require the use of the *2011 Edition Reference Tables for Physical Setting/Earth Science*. Record your answers in the space provided.

Base your answers to questions 36 through 39 on the map and the passage below and on your knowledge of Earth science. The map shows four different locations in India, labeled *A*, *B*, *C*, and *D*, where vertical sticks were placed in the ground on the same clear day. The locations of two cities in India are also shown.

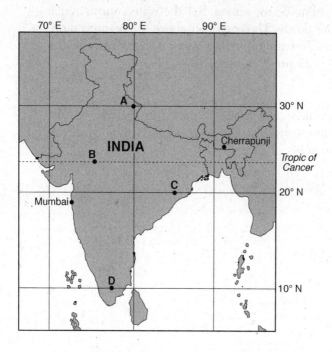

Monsoons in India

A monsoon season is caused by a seasonal shift in the wind direction, which produces excessive rainfall in many parts of the world, most notably India. Cherrapunji, in northeast India, received a record 30.5 feet of rain during July 1861. During the monsoon season from early June into September, Mumbai, India averages 6.8 feet of rain. Mumbai's total average rainfall for the other eight months of the year is only 3.9 inches.

Monsoons are caused by unequal heating rates of land and water. As the land heats throughout the summer, a large low-pressure system forms over India. The heat from the Sun also warms the surrounding ocean waters, but the water warms much more slowly. The cooler air above the ocean is more dense, creating a higher air pressure relative to the lower air pressure over India.

36 At which map location would no shadow be cast by the vertical stick at solar noon on the first day of summer?

(1) A   (3) C
(2) B   (4) D   36 _____

37 Which map shows both the dominant air pressure system that forms over India in the summer and the direction of surface winds around this air pressure system? [High pressure = **H**, Low pressure = **L**]

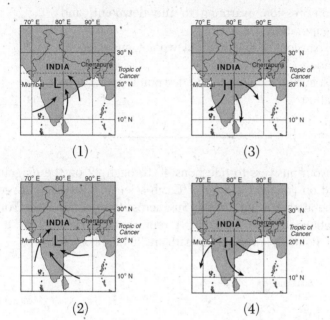

37 _____

38 The unequal heating rates of India's land and water are caused by

(1) land having a higher density than water
(2) water having a higher density than land
(3) land having a higher specific heat than water
(4) water having a higher specific heat than land

38 _____

39 Which processes lead to cloud formation when humid air rises over India?

(1) compression, warming to the dewpoint, and condensation
(2) compression, warming to the dewpoint, and evaporation
(3) expansion, cooling to the dewpoint, and condensation
(4) expansion, cooling to the dewpoint, and evaporation

39 _____

Base your answers to questions 40 through 42 on the diagram below and on your knowledge of Earth science. The diagram represents the apparent path of the Sun across the sky at a New York State location on June 21. Point A represents the position of the noon Sun. Points A and B on the path are 45 degrees apart.

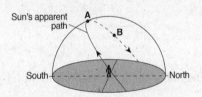

40 How many hours (h) will it take for the apparent position of the Sun to change from point A to point B?

(1) 1 h   (3) 3 h
(2) 2 h   (4) 4 h

40 _____

41 Compared to the Sun's apparent path on June 21, the Sun's apparent path on December 21 at this location will

(1) be shorter, and the noon Sun will be lower in the sky
(2) be longer, and the noon Sun will be higher in the sky
(3) remain the same length, and the noon Sun will be lower in the sky
(4) remain the same length, and the noon Sun will be higher in the sky

41 _____

42 Which diagram represents the correct position of *Polaris* as viewed from this New York State location on a clear night?

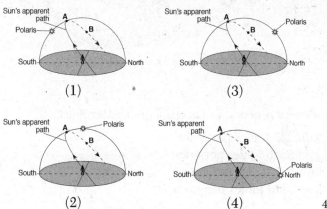

Base your answers to questions 43 and 44 on the diagram below and on your knowledge of Earth science. The diagram represents the water cycle. Letters A through C represent different processes in the water cycle.

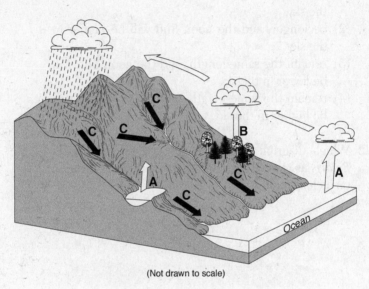

(Not drawn to scale)

43 In order for process A to occur, liquid water must

(1) gain 334 Joules per gram
(2) gain 2260 Joules per gram
(3) lose 334 Joules per gram
(4) lose 2260 Joules per gram   43 _____

44 Which process is represented by letter B?

(1) capillarity
(2) transpiration
(3) infiltration
(4) precipitation   44 _____

Base your answers to questions 45 through 47 on the photograph below and on your knowledge of Earth science. The photograph shows a small waterfall located on the Tug Hill Plateau.

45 During which geologic time period was the surface bedrock at this location formed?

(1) Cretaceous   (3) Devonian
(2) Triassic   (4) Ordovician   45 ____

46 Compared to the bedrock layers above and below the rock ledge shown at the waterfall, the characteristic that is primarily responsible for the existence of the rock ledge is its greater

(1) resistance to weathering
(2) abundance of fossils
(3) thickness
(4) age   46 ____

47 Rock fragments that are tumbled and carried over long distances by this stream are most likely becoming

(1) less dense, harder, and smaller
(2) less rounded, jagged, and larger
(3) more dense, angular, and smaller
(4) more rounded, smoother, and smaller   47 ____

Base your answers to questions 48 through 50 on the rock columns below and on your knowledge of Earth science. The rock columns represent four widely separated locations, W, X, Y, and Z. Numbers 1, 2, 3, and 4 represent fossils. The rock layers have *not* been overturned.

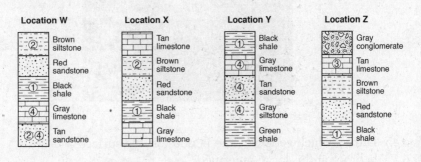

48 Which numbered fossil best represents an index fossil?

(1) 1   (3) 3
(2) 2   (4) 4   48 ____

49 Which rock layer is the oldest?
   (1) tan sandstone   (3) green shale
   (2) gray limestone   (4) black shale   49 ____

50 Which rock layer formed from the deposition of land-derived sediments that had a uniform particle size of about 0.01 cm in diameter?
   (1) brown siltstone   (3) gray conglomerate
   (2) black shale   (4) red sandstone   50 ____

PART B–2
Answer all questions in this part.

Directions (51–65): Record your answers in the spaces provided. Some questions may require the use of the *2011 Edition Reference Tables for Physical Setting/Earth Science*.

Base your answers to questions 51 through 53 on the data table below and on your knowledge of Earth science. The data table lists four constellations in which star clusters are seen from Earth. A star cluster is a group of stars near each other in space. Stars in the same cluster move at the same velocity. The length of the arrows in the table represents the amount of redshift of two wavelengths of visible light emitted by these star clusters.

Data Table

Constellation in which star cluster is seen from Earth	Redshift of two wavelengths of light absorbed by calcium	Distance from Earth (billion light years)	Velocity of star cluster moving away from Earth (km/s)
Ursa Major	Violet [→] Red	1.0	15,000
Corona Borealis	Violet [——→] Red	1.4	22,000
Boötes	Violet [———→] Red	2.5	39,000
Hydra	Violet [—————→] Red	4.0	61,000

Note: One light year is the distance light travels in one year.

51 Describe the evidence shown by the light from these star clusters that indicates that these clusters are moving away from Earth. [1]

_____

_____

52 Write the chemical symbol for the element, shown in the table, that absorbs the two wavelengths of light. [1]

_____

53 Identify the name of the nuclear process that is primarily responsible for producing energy in stars. [1]

_____

Base your answers to questions 54 through 57 on the diagram below and on your knowledge of Earth science. The diagram represents the Moon in eight positions in its orbit around Earth. One position is labeled A.

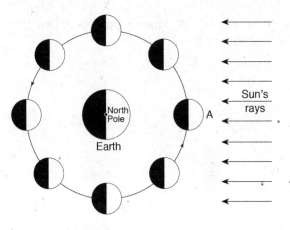

(Not drawn to scale)

54 Circle the type of eclipse that may occur when the Moon is at position A. Explain why this type of eclipse may occur when the Moon is at this position. [1]

Circle one: **lunar eclipse**   **solar eclipse**

Explanation: _____

_____

55 The diagram below represents one phase of the Moon as observed from New York State.

On the diagram *below*, place an **X** on the Moon's orbit to represent the Moon's position when this phase was observed. [1]

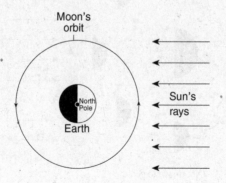

56 State the number of days needed for the Moon to show a complete cycle of phases from one full Moon to the next full Moon when viewed from New York State. [1]

_____ **days**

57 Explain why the Moon's revolution and rotation cause the same side of the Moon to always face Earth. [1]

_____

_____

Base your answers to questions 58 through 61 on the weather map below and on your knowledge of Earth science. The weather map shows atmospheric pressures, recorded in millibars (mb), at locations around a low-pressure center (**L**) in the eastern United States. Isobars indicate air pressures in the western portion of the mapped area. Point *A* represents a location on Earth's surface.

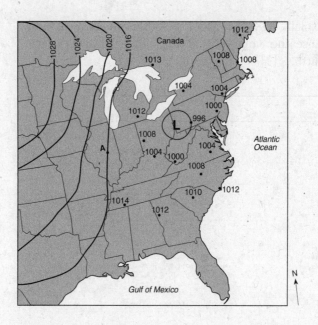

58 On the weather map *above*, draw the 1012 millibar and the 1008 millibar isobars. Extend the isobars to the east coast of the United States. [1]

59 Identify the weather instrument that was used to measure the air pressures recorded on the map. [1]

60 Identify the compass direction toward which the center of the low-pressure system will move if it follows a typical storm track. [1]

_____

61 Convert the air pressure at location A from millibars to inches of mercury. [1]

_____ in of Hg

_____

Base your answers to questions 62 through 65 on the graph below and on your knowledge of Earth science. The graph shows the rate of decay of the radioactive isotope carbon-14 (^{14}C).

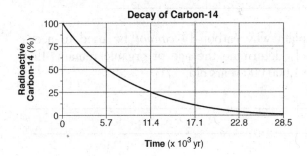

62 Complete the flow chart *below* by filling in the boxes to indicate the percentage of carbon-14 remaining and the time that has passed at the end of each half-life. [1]

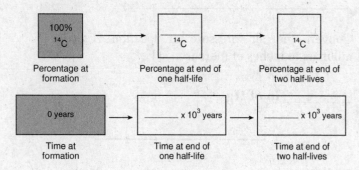

63 Identify the decay product formed by the disintegration of carbon-14. [1]

_____

64 Explain why carbon-14 *cannot* be used to accurately determine the age of organic remains that are 1,000,000 years old. [1]

_____

_____

65 State the name of the radioactive isotope that has a half-life that is approximately the same as the estimated time of the origin of Earth. [1]

_____

PART C

Answer all questions in this part.

Directions (66–85): Record your answers in the spaces provided. Some questions may require the use of the *2011 Edition Reference Tables for Physical Setting/Earth Science*.

Base your answers to questions 66 through 69 on the graph below and on your knowledge of Earth science. The graph shows changes in hours of daylight during the year at the latitudes of 0°, 30° N, 50° N and 60° N.

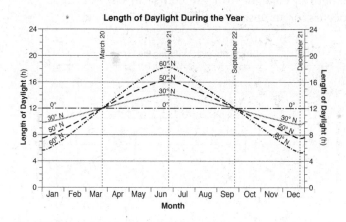

66 Estimate the number of daylight hours that occur on January 1 at 40° N latitude. [1]

_____ h

67 Identify the latitude shown on the graph that has the earliest sunrise on June 21. Include the units and compass direction in your answer. [1]

_____

68 Explain why all four latitudes have the same number of hours of daylight on March 20 and September 22. [1]

_____

_____

69 The graph *below* shows a curve for the changing length of daylight over the course of one year that occurs for an observer at 50° N latitude. On this same graph, draw a line to show the changing length of daylight over the course of one year that occurs for an observer at 50° S latitude. [1]

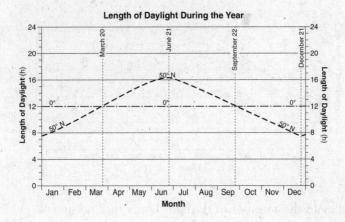

Base your answers to questions 70 through 74 on the passage and data tables below, on the map below, and on your knowledge of Earth science. The data tables show trends (patterns) of two lines of Hawaiian island volcanoes, the Loa trend and the Kea trend. For these trends, ages and distances of the Hawaiian island volcanoes are shown. The map shows the locations of volcanoes, labeled with Xs, that make up each trend line.

Hawaiian Volcano Trends

The Hawaiian volcanic island chain, located on the Pacific Plate, stretches over 600 kilometers. This chain of large volcanoes has grown from the seafloor to heights of over 4000 meters. Geologists have noted that there appear to be two lines, or "trends," of volcanoes—one that includes Mauna Loa and one that includes Mauna Kea. Loihi and Kilauea are the most recent active volcanoes on the two trends shown on the map.

Loa Trend

Loa Trend Volcanoes	Volcano Age (million years)	Distance from Loihi (km)
Kauai	4.6	575
Waianae	3.7	465
Koolau	2.2	375
West Molokai	1.7	350
Lanai	1.2	300
Kahoolawe	1.1	250
Hualalai	0.3	130
Mauna Loa	0.2	70
Loihi	0	0

Kea Trend

Kea Trend Volcanoes	Volcano Age (million years)	Distance from Kilauea (km)
East Molokai	1.7	256
West Maui	1.5	221
Haleakala	0.9	182
Kohala	0.5	100
Mauna Kea	0.4	54
Kilauea	0.1	0

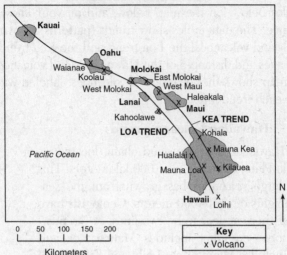

Volcanoes and Islands of Hawaii

70 The average distance between the volcanoes along the Kea trend is 51.2 kilometers. Place an **X** on the map *above* to identify the location on the seafloor where the next volcano will most likely form as a part of the Kea trend. [1]

71 Identify the *two* volcanoes, one from each trend, that have the same age. [1]

_____ and _____

72 State the general relationship between the age of the volcanoes and the distance from Loihi. [1]

_____

_____

73 Identify the tectonic feature beneath the moving Pacific Plate that caused volcanoes to form in *both* the Loa and Kea trends. [1]

_____

74 Identify the compass direction in which the Pacific Plate has moved during the last 4.6 million years. [1]

_____

Base your answers to questions 75 through 79 on the topographic map below and on your knowledge of Earth science. Lines *AB* and *CD* are reference lines on the map. Letter *E* indicates a location in a stream.

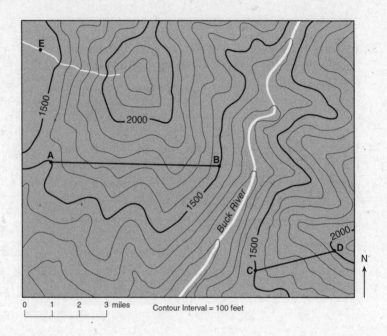

75 On the map *above*, draw an **X** on the location with the highest elevation. [1]

76 Using the grid *below*, construct a topographic profile along line AB by plotting the elevation of each contour line that crosses line AB. Points A and B have already been plotted on the grid. Connect all plots with a line from A to B to complete the profile. [1]

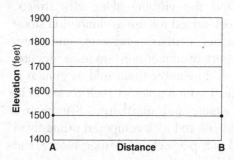

77 Calculate the gradient along line CD. [1]

_____ ft/mi

78 Describe how the contour lines indicate the direction in which Buck River flows. [1]

_____

_____

79 Determine the velocity of the stream at location E where the largest particle being carried at location E has a diameter of 10.0 centimeters. [1]

_____ cm/s

Base your answers to questions 80 through 83 on the passage below and on your knowledge of Earth science.

Dimension Stone: Granite

Dimension stone is any rock mined and cut for specific purposes, such as kitchen countertops, monuments, and the curbing along city streets. Examples of rock mined for use as dimension stone include limestone, marble, sandstone, and slate. The most important dimension stone is granite; however, not all dimension stone sold as granite is actually granite. Two examples of such rock sold as "granite" are syenite and anorthosite. Syenite is a crystalline, light-colored rock composed primarily of potassium feldspar, plagioclase feldspar, biotite, and amphibole, while anorthosite is composed almost entirely of plagioclase feldspar. Like actual granite, both syenite and anorthosite have large, interlocking crystals.

80 Explain why syenite is classified as a plutonic igneous rock. [1]

81 State *one* reason why anorthosite is likely to be white to gray in color. [1]

Base your answers to questions 59 through 62 on the passage and map below and on your knowledge of Earth science. The passage describes the March 13, 2013, earthquake that occurred off the coast of Japan and the tsunami that it generated. The map shows the location of the earthquake epicenter and the tsunami travel times across the Pacific Ocean.

Earthquake and Tsunami Rattle the Pacific

At 2:46 p.m., in Japan, on Friday, March 13, 2013, a magnitude 9.0 earthquake occurred below the ocean floor, at a depth of 18.6 miles under the ocean surface. The epicenter was located approximately 80 miles off Japan's eastern coast at the approximate coordinates of 38° N 142° E. The earthquake shook buildings across Japan and generated a 7-meter-high tsunami that killed thousands of people as it engulfed towns on the northern coast of Japan. The tsunami also occurred along the coasts of other countries and islands in the Pacific Ocean. The tsunami first arrived at the Hawaiian island of Maui seven hours after the earthquake occurred.

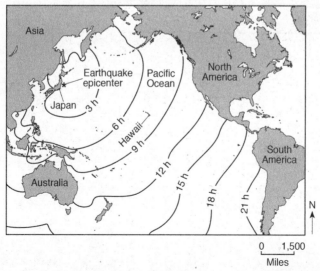

55 State the number of degrees of longitude that separates New York City from Reno, Nevada, and the time difference, in hours, between these two cities. [1]

Longitude difference: _____ °

Time difference: _____ h

56 Identify *two* cities on the map where measurements of the altitude of *Polaris* are within one degree of each other. [1]

_____ and _____

57 Identify the city labeled on the map where sunrise occurs first each day. [1]

_____

58 Identify the Earth motion that provides the basis for our system of local time and time zones. [1]

_____

54 Identify *two* processes that formed the unconformity at WZ. [1]

Process 1: _____

Process 2: _____

Base your answers to questions 55 through 58 on the map below and on your knowledge of Earth science. The map shows the four time zones and some latitude and longitude lines across the continental United States. Some cities are labeled on the map.

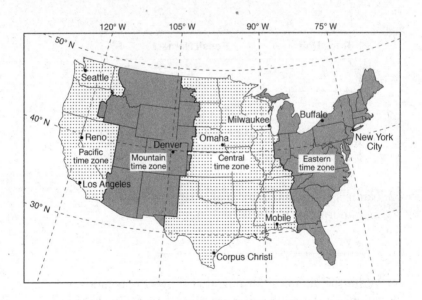

51 The three index fossils below are found within rock units 1, 2, and 3. Since the rock units were deposited during different geologic time periods, each fossil is found in a different rock unit.

Hexameroceras　　　**Centroceras**　　　**Cryptolithus**

Write the name of each of these index fossils next to the rock unit where the fossil is most likely found. [1]

Rock Unit	Fossil Name
1	
2	
3	

52 Name *one* sedimentary rock that was most likely metamorphosed to form rock unit 6. [1]

_____

53 Write the chemical formula that shows the composition for the most common mineral found in rock unit 3. [1]

Chemical formula: _____

PART B–2
Answer all questions in this part.

Directions (51–65): Record your answers in the spaces provided. Some questions may require the use of the *2013 Edition Reference Tables for Physical Setting/Earth Science*.

Base your answers to questions 51 through 54 on the cross section below and on your knowledge of Earth science. On the cross section, numbers 1 through 7 represent rock units in which overturning has *not* occurred. Line XY represents a fault and line WZ represents the location of an unconformity.

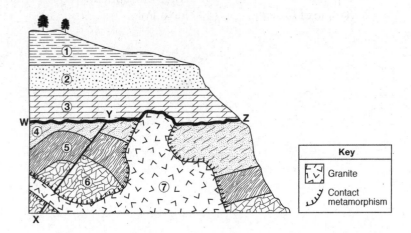

48 What is the approximate time of day represented at each location?
  (1) 6:00 a.m.   (3) 3:00 p.m.
  (2) 9:00 a.m.   (4) 6:00 p.m.   48 ____

49 During the course of the day, which location had the greatest intensity of insolation at solar noon?
  (1) A   (3) C
  (2) B   (4) D   49 ____

50 Based on the Sun's apparent path, where is location D?
  (1) equator   (3) Tropic of Capricorn
  (2) Tropic of Cancer   (4) North Pole   50 ____

46 Which change in seasons occurs in the Northern Hemisphere at position *D*?

(1) Winter is ending and spring is beginning.
(2) Spring is ending and summer is beginning.
(3) Summer is ending and fall is beginning.
(4) Fall is ending and winter is beginning.   46 ____

47 At all four positions, the northern end of Earth's axis points toward

(1) the Sun   (3) *Betelgeuse*
(2) the Moon   (4) *Polaris*   47 ____

Base your answers to questions 48 through 50 on the diagram below and on your knowledge of Earth science. The diagram represents the apparent path of the Sun as observed at four locations, *A* through *D*, on Earth's surface on the same date. The present positions of the Sun represent the same time of day at each location. The zenith (the position directly overhead) is shown for an observer at each location. [Diagrams are not drawn to scale.]

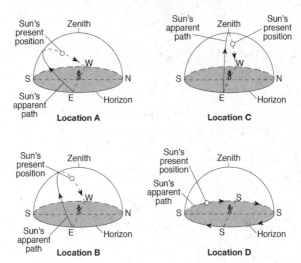

44 Which mineral has a different common color from its color in powdered form?

(1) brucite (3) magnesite
(2) carnallite (4) olivine   44 _____

Base your answers to questions 45 through 47 on the diagram below and on your knowledge of Earth science. The diagram represents Earth in its orbit around the Sun. Locations A through D represent four positions of Earth in its orbit. Earth is closest to the Sun (perihelion) at position A, and farthest from the Sun (aphelion) at position C.

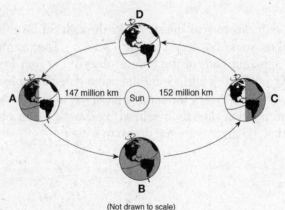

(Not drawn to scale)

45 At which position is the gravitational attraction between the Sun and Earth the greatest?

(1) A (3) C
(2) B (4) D   45 _____

41 Over which two landscape regions do the streams in watershed *D* flow?

(1) Tug Hill Plateau and the Catskills
(2) Tug Hill Plateau and Erie-Ontario Lowlands
(3) Adirondack Mountains and Champlain Lowlands
(4) Adirondack Mountains and St. Lawrence Lowlands   41 _____

Base your answers to questions 42 through 44 on the mineral chart below and on your knowledge of Earth science. The mineral chart lists some properties of five minerals that are the major sources of the same metallic element that is used by many industries.

Mineral Chart

Mineral Name	Composition	Density (g/cm³)	Hardness	Streak	Nonmetallic Luster	Common Colors
brucite	$Mg(OH)_2$	2.4	2.5-3	white	glassy to waxy	white
carnallite	$KMgCl_3 \cdot 6H_2O$	1.6	2.5	white	greasy	white
dolomite	$CaMg(CO_3)_2$	2.8	3.5-4	white	glassy to waxy	shades of pink
magnesite	$MgCO_3$	3.1	3.5-4.5	white	glassy	white
olivine	$(Fe,Mg)_2SiO_4$	3.3	6.5	white	glassy	green

42 Which two minerals have compositions that are most similar to calcite?

(1) brucite and carnallite
(2) carnallite and dolomite
(3) dolomite and magnesite
(4) magnesite and olivine   42 _____

43 Which mineral might scratch the mineral fluorite, but would *not* scratch the mineral amphibole?

(1) brucite   (3) carnallite
(2) magnesite   (4) olivine   43 _____

38 The extinction of which group of animals 65.5 million years ago is thought to have been due to an impact event and global climate change?

(1) ammonoids (3) trilobites
(2) brachiopods (4) placoderm fish   38 ____

Base your answers to questions 39 through 41 on the map below and on your knowledge of Earth science. The map shows the locations of major watersheds in New York State. Letters A through K represent individual watersheds.

39 In which major watershed is the Susquehanna River located?

(1) F   (3) I
(2) H   (4) J   39 ____

40 The Genesee River in watershed A generally flows in which direction?

(1) north   (3) east
(2) south   (4) west   40 ____

PART B–1
Answer all questions in this part.

Directions (36–50): For *each* statement or question, choose the word or expression that, of those given, best completes the statement or answers the question. Some questions may require the use of the 2013 Edition *Reference Tables for Physical Setting/Earth Science*. Record your answers in the space provided.

Base your answers to questions 36 through 38 on the passage below and on your knowledge of Earth science.

Comets and Asteroids

Since comets and asteroids orbit the Sun, both are part of our solar system. Asteroids are rocky objects that vary greatly in size. Most asteroids follow orbits between 300 and 600 million kilometers from the Sun, but several have been pulled from this region by the gravitational attraction of nearby planets. Many of these dislodged asteroids have struck both Earth and the Moon, causing the large impact craters that are visible on the surfaces of both bodies.

Comets have often been described as "dirty snowballs" and occupy highly eccentric orbits, traveling from near the Sun to far beyond the orbits of the outer planets. As they move through space, comets leave a debris trail of mostly dust-sized particles. When Earth passes through this debris, a meteor shower occurs, often filling the night sky with "shooting star" trails as they burn up in the atmosphere 50 to 80 kilometers above Earth's surface.

36 Between which two planets are most asteroids located?

 (1) Earth and Mars (3) Jupiter and Saturn
 (2) Mars and Jupiter (4) Saturn and Uranus   36 _____

37 In which temperature zone of Earth's atmosphere will most meteors burn up?

 (1) troposphere (3) mesosphere
 (2) stratosphere (4) thermosphere   37 _____

35 Which chart best describes the landscape category and the general bedrock structure, type, and composition of New York State's Catskills?

Landscape Category	plateau
Bedrock Structure	horizontal
Bedrock Type	sedimentary
Bedrock Composition	limestone, shale, sandstone

(1)

Landscape Category	mountain
Bedrock Structure	folded
Bedrock Type	sedimentary
Bedrock Composition	sandstone, dolostone, schist

(2)

Landscape Category	mountain
Bedrock Structure	horizontal
Bedrock Type	metamorphic
Bedrock Composition	gneiss, quartzite, marble

(3)

Landscape Category	plateau
Bedrock Structure	folded
Bedrock Type	metamorphic
Bedrock Composition	shale, slate, dunite

(4)

35 _____

Base your answers to questions 33 and 34 on the block diagram below and on your knowledge of Earth science. The block diagram represents a landscape that was produced by a meandering stream. One landscape feature is labeled X. Letters A, B, C, and D represent locations on the stream banks.

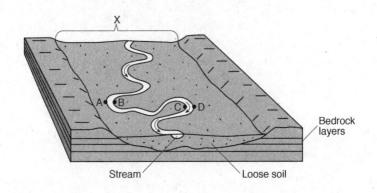

33 The landscape feature labeled X is best described as
  (1) a flood plain (3) a delta
  (2) a sand bar (4) an escarpment 33 ____

34 Erosion is most likely greatest at locations
  (1) A and B (3) C and D
  (2) B and C (4) D and A 34 ____

32 The topographic map below has a contour interval of 20 feet. Points A and B represent locations on Earth's surface.

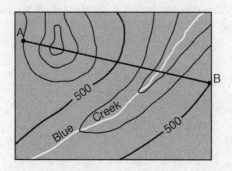

Which profile best represents the topographic cross section along the line from A to B?

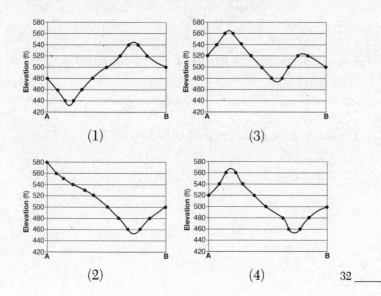

32 _____

31 The cross sections below represent three geologic columns, I, II, and III, exposed at three different locations. The rock layers have *not* been overturned. Letters A through E represent different fossils.

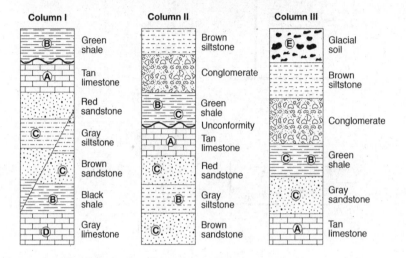

Which rock layer is the youngest?

(1) green shale containing fossil B in column I
(2) glacial soil containing fossil E in column III
(3) brown sandstone containing fossil C in column II
(4) gray limestone containing fossil D in column I

29 Which device provides evidence that Earth rotates on its axis?

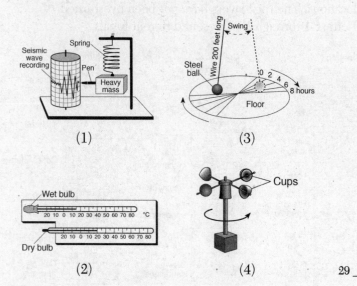

30 Which weather map symbol is used to represent violently rotating winds that have the appearance of a funnel-shaped cloud?

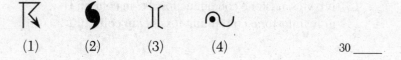

28 The diagram below represents the interiors of three planets in our solar system.

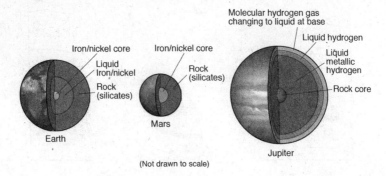

(Not drawn to scale)

Which inference best describes the interiors of the planets in our solar system?

(1) Both terrestrial and Jovian planets have layered interiors, with density decreasing toward the center.
(2) Both terrestrial and Jovian planets have layered interiors, with density increasing toward the center.
(3) Only terrestrial planets have layered interiors, with density decreasing toward the center.
(4) Only Jovian planets have layered interiors, with density increasing toward the center.

28 _____

27 The cross sections below represent three bedrock outcrops found several kilometers apart.

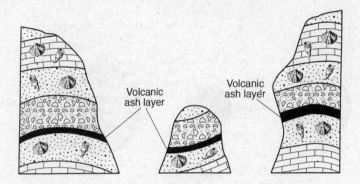

Which statement best explains why the volcanic ash layers are useful for correlating the relative ages of the bedrock in the three outcrops?

(1) The ash was deposited over a large area when a volcano erupted.
(2) There are no fossils found within the volcanic ash.
(3) The volcanic eruptions that produced the ash layer occurred over a long period of geologic time.
(4) The volcanic ash is found between many different layers of bedrock.

26 The graph below shows ocean tide height in feet (ft) over a 44-hour period for a coastal location in the northeastern United States. The dots represent either high or low tides.

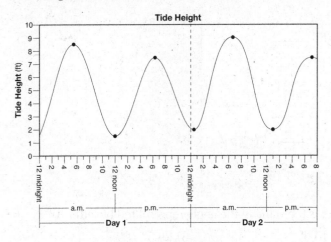

If the pattern shown continued, the next low tide occurred on Day 3 at approximately

(1) 12 midnight
(2) 1:30 a.m.
(3) 1:00 p.m.
(4) 6:00 p.m.

26 _____

24 The diagram below represents Earth as viewed from space. Letter A represents the approximate angle of tilt between Earth's rotational axis and a line (XY) perpendicular to the plane of Earth's orbit.

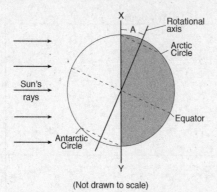

(Not drawn to scale)

What is the value of the angle represented by letter A?

(1) 15.0° (3) 24.5°
(2) 23.5° (4) 30.0°

25 What are the rock name and map symbol used to represent the sedimentary rock that has a grain size of 0.006 to 0.2 centimeters?

23 The topographic map below shows three drumlins located in New York State.

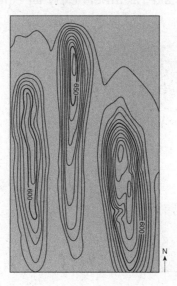

What was the direction of the advancing ice movement that created these drumlins, and what is the most likely arrangement of sediments in the drumlins?

(1) north to south ice movement, and unsorted sediments
(2) north to south ice movement, and sorted sediments
(3) south to north ice movement, and unsorted sediments
(4) south to north ice movement, and sorted sediments

23 _____

21 The contour map below shows a lake and river system. The Birch and Elk Rivers carry an equal volume of water.

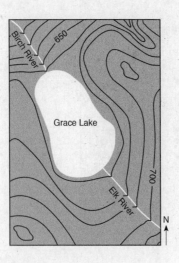

Compared to the Elk River, the Birch River can best be described as flowing

(1) faster, and in the same general compass direction
(2) faster, and in the opposite general compass direction
(3) slower, and in the same general compass direction
(4) slower, and in the opposite general compass direction

22 Which rock would be the best source of the mineral garnet?

(1) basalt
(2) limestone
(3) schist
(4) slate

20 The photograph below shows an igneous rock with mineral crystals ranging in size from 2 to 6 millimeters. The rock is composed of 58% plagioclase feldspar, 26% amphibole, and 16% biotite.

What is the name of this rock?
(1) diorite
(2) gabbro
(3) andesite
(4) pumice

19 The photograph below shows a steep-sided rock formation that is over 100 meters high. This landscape feature is located in an arid region.

What would happen to this landscape feature if the climate became more humid?

(1) less weathering and erosion, producing a more rounded landscape feature
(2) less weathering and erosion, producing a more angular landscape feature
(3) more weathering and erosion, producing a more rounded landscape feature
(4) more weathering and erosion, producing a more angular landscape feature

19 _____

16 The map below shows the inferred shape of the North American landmass in the past. The location of Florida is labeled.

Florida

Which event was occurring on Earth when Florida was located at the equator?

(1) The dome-like uplift of the Adirondack region began.
(2) The earliest dinosaurs appeared on Earth.
(3) Oceanic oxygen began to enter the atmosphere.
(4) Earth's first coral reefs were forming.   16 ____

17 Which geologic feature is composed of the youngest crustal bedrock?

(1) Peru-Chile Trench
(2) Mid-Atlantic Ridge
(3) Adirondack Mountains
(4) San Andreas Fault   17 ____

18 If a seismic station is 3200 km from an earthquake epicenter, what is the time needed for an S-wave to travel from the epicenter to the seismic station?

(1) 4 min 40 sec   (3) 10 min 40 sec
(2) 6 min 0 sec   (4) 13 min 10 sec   18 ____

11 Large volcanic eruptions sometimes send dust and ash into the stratosphere. After these eruptions, global air temperatures are often

(1) cooler than normal because the atmosphere is less transparent
(2) cooler than normal because the atmosphere is more transparent
(3) warmer than normal because the atmosphere is less transparent
(4) warmer than normal because the atmosphere is more transparent

11 ____

12 Which list contains three major greenhouse gases found in Earth's atmosphere?

(1) carbon dioxide, methane, and water vapor
(2) carbon dioxide, oxygen, and nitrogen
(3) hydrogen, oxygen, and methane
(4) hydrogen, water vapor, and nitrogen

12 ____

13 During which geologic epoch does the New York State rock record consist of weakly consolidated to unconsolidated sediments?

(1) Early Permian   (3) Late Cretaceous
(2) Early Jurassic   (4) Pliocene

13 ____

14 The New York State index fossil *Valcouroceras* is classified as a

(1) coral   (3) eurypterid
(2) crinoid   (4) nautiloid

14 ____

15 The convection currents responsible for moving tectonic plates occur in which Earth layer?

(1) crust   (3) stiffer mantle
(2) rigid mantle   (4) asthenosphere

15 ____

7 The seasonal shifts of Earth's planetary wind and moisture belts are due to changes in the

(1) distance between Earth and the Sun
(2) amount of energy given off by the Sun
(3) latitude that receives the Sun's vertical rays
(4) rate of Earth's rotation on its axis   7 ____

8 Which condition will most likely result in the formation of a cloud?

(1) wind speed decreasing
(2) air pressure increasing
(3) cool, moist air sinking
(4) warm, moist air rising   8 ____

9 Which climate condition generally results from both an increase in distance from the equator and an increase in elevation above sea level?

(1) cooler temperatures
(2) warmer prevailing winds
(3) increased precipitation
(4) increased air pressure   9 ____

10 Most of which type of electromagnetic radiation is given off by Earth's surface at night?

(1) gamma rays   (3) visible light
(2) ultraviolet light   (4) infrared rays   10 ____

3 Earth's planetary winds curve to the right in the Northern Hemisphere due to

(1) the Coriolis effect
(2) the Doppler effect
(3) the tilt of Earth's axis
(4) Earth's gravity

4 Which process releases 334 Joules (J) of energy for each gram of water?

(1) melting
(2) freezing
(3) vaporization
(4) condensation

5 After a heavy rainstorm, vegetation on a hillslope was completely removed. How will this removal of vegetation affect the relative amounts of infiltration and runoff that occur during the next heavy rainstorm?

(1) Infiltration and runoff will both be less.
(2) Infiltration and runoff will both be greater.
(3) Infiltration will be less and runoff will be greater.
(4) Infiltration will be greater and runoff will be less.

6 Hurricane season in the North Atlantic Ocean officially begins in June and ends in November. Which ocean surface conditions are responsible for the development of hurricanes?

(1) warm water temperatures and low evaporation rates
(2) warm water temperatures and high evaporation rates
(3) cool water temperatures and low evaporation rates
(4) cool water temperatures and high evaporation rates

Examination August 2016
Physical Setting/Earth Science

PART A
Answer all questions in this part.

Directions (1–35): For *each* statement or question, choose the word or expression that, of those given, best completes the statement or answers the question. Some questions may require the use of the 2013 Edition Reference Tables for Physical Setting/Earth Science. Record your answers in the space provided.

1 Compared to the Sun, the star *Betelgeuse* is

   (1) less luminous and warmer
   (2) less luminous and cooler
   (3) more luminous and warmer
   (4) more luminous and cooler   1 ____

2 Which evidence best supports scientists' inferences about the origin and age of the universe?

   (1) the existence of planets
   (2) cosmic background radiation
   (3) formation of star constellations
   (4) similar composition of Earth and the Moon   2 ____

SELF-ANALYSIS CHART June 2016

Topic	Question Numbers (Total)	Wrong Answers (x)	Grade
Meteorology			
Energy Sources for Earth Systems (MU 2.1a, b)			
Weather (MU 2.1c, d, e, f, g, h)	9–12, 34, 37, 39, 58–61	$\dfrac{100(11-x)}{11} = \%$	
Insolation and Seasonal Changes (MU 2.1i; 2.2a, b)	13, 15, 31, 38, 66	$\dfrac{100(5-x)}{5} = \%$	
The Water Cycle and Climates (MU 1.2g; 2.2c, d)	5, 14, 43, 44, 84, 85	$\dfrac{100(6-x)}{6} = \%$	
Geology			
Minerals and Rocks (MU 3.1a, b, c)	25, 50, 80–83	$\dfrac{100(6-x)}{6} = \%$	
Weathering, Erosion, and Deposition (MU 2.1s, t, u, v, w)	24, 28, 29, 47, 79	$\dfrac{100(5-x)}{5} = \%$	
Plate Tectonics and Earth's Interior (MU 2.1j, k, l, m, n, o)	17, 20–22, 33, 35, 70–74	$\dfrac{100(11-x)}{11} = \%$	
Geologic History (MU 1.2i, j)	16, 19, 27, 45, 48, 49, 62–65	$\dfrac{100(10-x)}{10} = \%$	
Topographic Maps and Landscapes (MU 2.1p, q, r)	23, 26, 46, 75–78	$\dfrac{100(7-x)}{7} = \%$	
ESRT			
Earth Science Reference Tables 2011 Edition	3, 9–11, 13, 17, 19–21, 25, 27, 33, 35–38, 43, 45, 50, 52, 61–63, 65, 73, 74, 77, 79–83	$\dfrac{100(32-x)}{32} = \%$	

To further pinpoint your weak areas, use the Topic Outline in the front of the book.
MU = Major Understanding (see Topic Outline)

SELF-ANALYSIS CHART June 2016

Topic	Question Numbers (Total)	Wrong Answers (x)	Grade
Standards 1, 2, 6, and 7: Skills and Application			
Skills			
Standard 1 Analysis, Inquiry, and Design	1–5, 11–13, 16, 17, 19–21, 24, 25, 29–33, 35–40, 42, 45–47, 50, 52, 58, 59, 61–68, 70, 72–77, 79–83		$\frac{100(54-x)}{54} = \%$
Standard 2 Information Systems	60		$\frac{100(1-x)}{1} = \%$
Standard 6 Interconnectedness, Common Themes	6, 8, 9, 15–17, 23, 26, 28, 29, 31, 32, 34, 35, 37, 39–51, 53–58, 60, 62, 64, 67–76, 78–80, 84, 85		$\frac{100(52-x)}{52} = \%$
Standard 7 Interdisciplinary Problem Solving			
Standard 4: The Physical Setting/Earth Science			
Astronomy			
The Solar System (MU 1.1a, b; 1.2d)	18, 30, 32, 54–57		$\frac{100(7-x)}{7} = \%$
Earth Motions and Their Effects (MU 1.1c, d, e, f, g, h, i)	1, 2, 6, 8, 36, 40–42, 67–69		$\frac{100(11-x)}{11} = \%$
Stellar Astronomy (MU 1.2b)	3, 53		$\frac{100(2-x)}{2} = \%$
Origin of Earth's Atmosphere, Hydrosphere, and Lithosphere (MU 1.2e, f, h)	7		$\frac{100(1-x)}{1} = \%$
Theories of the Origin of the Universe and Solar System (MU 1.2a, c)	4, 51, 52		$\frac{100(3-x)}{3} = \%$

air temperature causes a decrease in relative humidity, resulting in less precipitation at location *C*. Thus, location *D* has a wetter climate than location *C* because location *D* is close to the ocean on the leeward side of the mountain where moist air is rising, expanding, and being cooled to the dew point, resulting in precipitation.

One credit is allowed for an acceptable response. Acceptable responses include but are not limited to:

- **Location *D* has air that is rising, expanding, and cooling to the dew point.**
- **Location *D* is on the windward side of the mountain.**
- **Location *D* is closer to the ocean.**
- **Location *C* is on the leeward side of the mountain.**

84. According to the map, location A is in the mountains and location B is not. Therefore, location A is at a higher elevation than location B. Find the Selected Properties of Earth's Atmosphere chart in the *Reference Tables for Physical Setting/Earth Science*. Locate the graph labeled "Atmospheric Pressure," and note that pressure decreases rapidly with altitude (elevation) in the troposphere. Now locate the graph labeled "Temperature Zones," and note that temperature decreases rapidly with altitude (elevation) in the troposphere. Thus, both air pressure and air temperature decrease with elevation. At higher elevations, the air is less dense and air molecules are more spread out and less likely to collide. If a thermometer is placed into such air, fewer molecules will collide with it. Fewer collisions means less energy transmitted to the thermometer, causing less expansion of the fluid in the thermometer. So the thermometer registers a lower temperature. Thus, a decrease in air pressure causes a decrease in air temperature. At locations in mountainous regions, temperature is lower year-round than in locations nearer sea level. Therefore, the average yearly temperature in mountainous regions is lower than it is nearer sea level. Thus, location A has a cooler average yearly temperature than location B because location A is in the mountains at a higher elevation than location B.

One credit is allowed for an acceptable response. Acceptable responses include but are not limited to:

- **The higher elevation at A has a cooler temperature.**
- **Location A is at a higher elevation.**
- **Location A is in the mountains.**
- **Location B is not as high in elevation.**

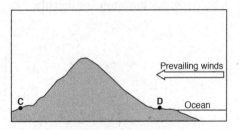

85. According to the cross section, locations C and D are at the same elevation on different sides of the mountain. The arrow labeled "Prevailing winds" indicates that location D is on the windward side of the mountain and location C is on the leeward side of the mountain. As moist air from over the ocean is carried upward and over the windward side of the mountain by the prevailing winds, the moist air expands and cools. When the air temperature reaches the dew point, moisture begins to condense and clouds form, causing precipitation at location D. As the air descends on the leeward side of the mountain, it is compressed and the air temperature increases. The increase in

Note: Credit is *not* allowed if the response lists all of the minerals in gabbro because the question asks how the composition of gabbro is different from that of granite.

83. The reading passage states, "Examples of rock mined for use as dimension stone include limestone, marble, sandstone, and slate. The most important dimension stone is granite; however, not all dimension stone sold as granite is actually granite. Two examples of such rock sold as 'granite' are syenite and anorthosite." Note that the descriptions of the composition of syenite and anorthosite indicate that neither contain calcite. Thus, the dimension stones to consider for this question are limestone, marble, sandstone, slate, and granite. Find the Scheme for Igneous Rock Identification in the *Reference Tables for Physical Setting/Earth Science*. Locate granite. Trace down to the graph labeled "Mineral Composition," and note that calcite is not one of the minerals of which granite is composed. Find the Scheme for Sedimentary Rock Identification in the *Reference Tables for Physical Setting/Earth Science*. In the column labeled "Rock Name," locate limestone and sandstone. From these rocks, trace left to the column labeled "Composition." Note that sandstone is composed of mostly quartz, feldspar, and clay minerals while limestone is composed of calcite. Thus, limestone is one dimension stone composed primarily of calcite. Find the Scheme for Metamorphic Rock Identification in the *Reference Tables for Physical Setting/Earth Science*. In the column labeled "Rock Name," locate marble and slate. From these rocks, trace left to the column labeled "Composition." Note that marble is composed of calcite and/or dolomite, and slate is composed of mica. Thus, marble is another dimension stone composed primarily of calcite.

One credit is allowed for **limestone *or* marble**.

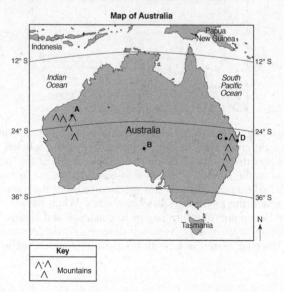

81. The reading passage states, "[A]northosite is composed almost entirely of plagioclase feldspar." Find the Properties of Common Minerals table in the *Reference Tables for Physical Setting/Earth Science*. In the column labeled "Mineral Name," locate plagioclase feldspar. Trace left to the column labeled "Common Colors," and note that plagioclase feldspar is white to gray. Thus, anorthosite is white to gray in color because it is composed almost entirely of plagioclase feldspar, which is white to gray in color.

One credit is allowed for an acceptable response. Acceptable responses include but are not limited to:

- **Anorthosite is made of plagioclase feldspar, which is white to gray in color.**
- **Anorthosite is made of light-colored minerals.**
- **Plagioclase feldspar is white to gray.**
- **Because of anorthosite's mineral composition**

Note: Do *not* allow credit for "anorthosite is felsic rich" because plagioclase feldspar is contained in both felsic-rich and mafic-rich igneous rocks.

82. Find the Scheme for Igneous Rock Identification in the *Reference Tables for Physical Setting/Earth Science*. Locate both granite and gabbro. Trace downward from each of these rocks to the scale labeled "Density" in the section labeled "Characteristics." Note that granite is on the lower side of the scale and basalt is on the higher side of the scale. Thus, gabbro is denser than granite. Trace downward from each of these rocks to the scale labeled "Composition" in the section labeled "Characteristics." Note that granite is on the felsic (rich in Si, Al) side of the scale and gabbro is on the mafic (rich in Fe, Mg) side of the scale. Now, trace the columns for granite and gabbro downward to the graph labeled "Mineral Composition." Note that gabbro contains a high percentage of the green minerals pyroxene and olivine, which are not present in granite. Note that granite contains a high percentage of quartz and potassium feldspar, which are not present in gabbro.

One credit is allowed if **both the density and composition of gabbro are correct.**

Acceptable responses include but are not limited to:

Density of gabbro:

- **higher**
- **greater**

Composition of gabbro:

- **mafic**
- **rich in Fe and Mg**
- **presence of pyroxene and/or olivine**
- **absence of quartz and/or potassium feldspar**

- The V shapes of the contour lines point upstream toward higher elevations.
- The contour lines bend in the opposite direction that the stream flows.

Note: Credit is *not* allowed for simply stating that "water flows downhill" because this does not indicate how contour lines show the direction of stream flow.

79. Find the Relationship of Transported Particle Size to Water Velocity graph in the *Reference Tables for Physical Setting/Earth Science*. On the vertical axis labeled "Particle Diameter (cm)," locate the point corresponding to 10.0 cm. From this point, trace right until you intersect the bold curve. From the intersection, trace vertically downward to the horizontal axis labeled "Stream Velocity (cm/s)." Note that water traveling at 200 cm/s can move particles with diameters of up to 10 cm. Thus, the minimum velocity this stream needs to transport a 2.0 cm particle is 200 cm/s.

One credit is allowed for **any value from 150 cm/s to 250 cm/s.**

80. The reading passage states, "Like actual granite, both syenite and anorthosite have large, interlocking crystals." Find the Scheme for Igneous Rock Identification in the *Reference Tables for Physical Setting/Earth Science*. Locate granite. Trace left to the section labeled "Environment of Formation," and note that granite is an intrusive (plutonic) igneous rock. From granite, trace right to the column labeled "Texture," and note that granite has a coarse texture. Igneous rocks all form from molten rock. The molten rock from which they form was created when other rocks were heated and then melted. Intrusive (plutonic) igneous rocks form by the cooling and solidification of molten magma beneath Earth's surface. Insulated by surrounding rock beneath Earth's surface, intrusive rocks cool slowly. The slow cooling time allows large, interlocking crystals to form, hence the coarse texture of granite. Thus, syenite is also classified as a plutonic igneous rock because its large, interlocking crystals indicate that it formed in an intrusive environment where it cooled slowly underground.

One credit is allowed for an acceptable response. Acceptable responses include but are not limited to:

- **Large crystals form from slow cooling deep underground.**
- **The crystals in syenite formed in an intrusion or an intrusive environment.**
- **The texture is coarse.**
- **Syenite formed by solidification of magma.**
- **Large interlocking crystals**
- **Syenite formed inside of Earth.**

Note: Credit is *not* allowed for writing only "texture," "crystal," or "interlocking crystals" because these terms also describe volcanic igneous rock.

Solve the equation as shown below:

$$\text{gradient} = \frac{500 \text{ ft}}{3 \text{ mi}} = 167 \frac{\text{ft}}{\text{mi}}$$

Thus, the approximate gradient along the straight line between points C and D is 167 feet/mile.

One credit is allowed for **any value from 161 to 173 feet/mile**.

78. Locate the northern end of the Buck River. Note that the elevation decreases from the 2000-foot contour line toward the 1500-foot contour line on both sides of Buck River. Before reaching the edge of the map, Buck River crosses the first regular contour line between the bold 1500-foot contour lines on either side. It is given that the contour interval is 100 feet. Thus, the elevation of this contour line is 1400 feet, and the elevation of Buck River where it crosses this contour line is 1400 feet. Now locate the southern end of the Buck River. Note that here, too, the elevation decreases from the 2000-foot contour line toward the 1500-foot contour line on both sides of Buck River. Before reaching the edge of the map, Buck River crosses the fourth regular contour line between the bold 1500-foot contour lines on either side. Therefore, the elevation of this contour line is 1100 feet, and the elevation of Buck River where it crosses this contour line is 1100 feet. Thus, the elevations of the contour lines decrease from north to south. This indicates that the river is flowing in a southerly direction because water flows from higher to lower elevations.

Alternately, the direction in which Buck River flows can be determined as follows. On a topographic map, contour lines connect points of equal elevation. Therefore, a contour line would have to bend upstream of the stream bank to intercept the same elevation on the streambed and then bend downstream on the opposite side to reach the same elevation on the opposite bank. Therefore, contour lines bend upstream when they cross a streambed, forming a distinctive V-shaped curve with the apex pointing upstream and the sides opening toward the downstream direction. Thus, contour lines that cross a stream bend in the direction opposite to that of stream flow. Note that where Buck River crosses a contour line, the contour line forms a V-shaped curve with the apex pointing toward the north and the sides opening toward the south. Thus, the general direction in which the stream flows is south.

One credit is allowed for an acceptable response. Acceptable responses include but are not limited to:

- **Contour lines that cross a river form "V" shapes that point to the source of the stream.**
- **The elevations of the contour lines decrease from north to south, indicating that the river is flowing in a southerly direction.**
- **The contour lines point upstream.**
- **The contour lines bend upstream when they cross a stream.**

Finally, connect each **X** in a smooth curve to form the finished profile as shown below. Note that line AB crosses the 1800-foot contour line twice but does not cross the 1900-foot contour line. Thus, between these two 1800-foot points, the elevation is greater than 1800 feet but less than 1900 feet.

One credit is allowed if **the centers of *all six* plots are within or touch the rectangles shown below and are connected with a line from A to B that passes within or touches the rectangles. The line must extend above 1800 feet but below 1900 feet.**

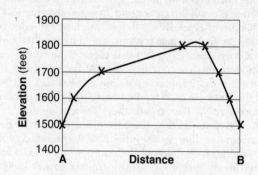

Note: Credit is allowed if the line does not pass through the student plots but is within or touches the boxes shown below.

77. Find the Equations section of the *Reference Tables for Physical Setting/Earth Science*, and note the equation for gradient:

$$\text{gradient} = \frac{\text{change in field value}}{\text{change in distance}}$$

The map shown is a topographic map. The field value on a topographic map is elevation. Note that point C is located directly on the 1500-foot contour line and point D is located directly on the 2000-foot contour line. Thus, the change in field value (elevation) from C to D is 500 feet.

On a piece of scrap paper, mark off the distance along the straight line between points C and D. Compare the distance marked off on the scrap paper to the scale printed beneath the map to determine the distance between points C and D—about 3 miles. Substitute these values in the equation as shown below:

$$\text{gradient} = \frac{2000 \text{ feet} - 1500 \text{ feet}}{3 \text{ miles}}$$

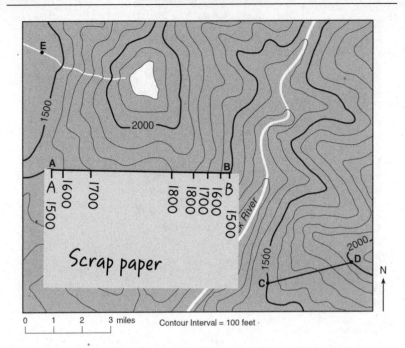

Then place this paper along the lower edge of the grid provided on your answer sheet so that points *A* and *B* on the paper align with points *A* and *B* on the grid. Next, at each point where a contour line crosses the edge of the paper, draw an **X** on the grid at the appropriate elevation.

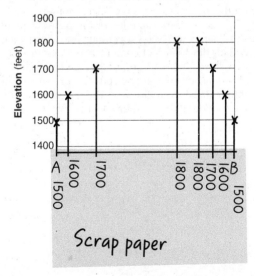

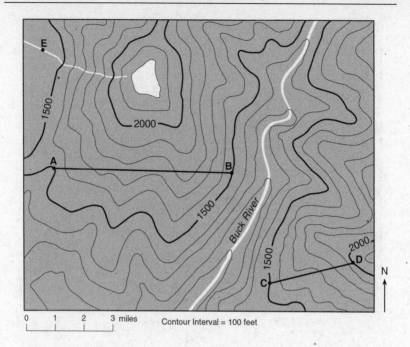

Note: Credit is allowed if a symbol other than an **X** is used. Credit is *not* allowed if the center of the **X** touches the 2200-foot contour line.

76. To construct a topographic profile along line *AB*, proceed as follows. Place the straight edge of a piece of scrap paper along the dashed line connecting point *A* to point *B*. Mark the edge of the paper at point *A*, at point *B*, and wherever the paper intersects a contour line. Note that both points *A* and *B* are located on the 1500-foot contour line. Note that line *AB* crosses the region between the 1500-foot contour line and the 2000-foot contour line. Therefore, elevation is increasing by 100 feet for every contour line past the 1500-foot contour. Wherever the paper intersects a contour line, label the mark with the elevation of the contour line as shown below.

melts through to erupt, forming a volcano. As the plate continues to move, the volcano that formed over the hot spot is carried away from the hot spot in the direction in which the plate is moving. At the same time, the plate motion carries a new section of the plate over the hot spot and a new volcano forms. As this process continues, a series of volcanoes form, with the youngest located nearest the hot spot and the oldest farthest from the hot spot. Thus, the direction of plate movement can be inferred to be in the direction from the youngest to the oldest volcano. According to the Loa Trend data table, the youngest volcano in the Loa Trend is Loihi and the oldest volcano is Kauai. According to the Kea Trend data table, the youngest volcano is Kilauea and the oldest volcano is East Molokai. Locate these volcanoes on the map provided. Note that the general pattern is that volcano age increases as you move northwest along both trend lines. Thus, the compass direction in which the Pacific Plate has moved during the last 4.6 million years is to the northwest, or NW.

One credit is allowed for an acceptable response. Acceptable responses include but are not limited to:

- **to the northwest**
- **NW**
- **from the SE toward the NW**
- **NNW**
- **west northwest**

75. Examine the topographic map in your answer booklet, and note the bold contour lines labeled "1500" and "2000." The elevation increases from the 1500-foot contour line toward the 2000-foot contour line. Note that there is only one regular contour line beyond the 2000-foot contour line in the lower right quadrant of the map in the direction of increasing elevation, but there are two regular contour lines within the 2000-foot contour line in the upper left quadrant of the map in the direction of increasing elevation. It is given that the contour interval is 100 feet. Thus, the elevation of the first contour line past the 2000-foot contour line in the lower right quadrant is 2100 feet and the elevation of the second contour line past the 2000-foot contour in the upper left quadrant is 2200 feet. Therefore, the location with the highest elevation on the map is located within the 2200-foot contour line in the upper right quadrant of the map.

One credit is allowed if **the center of the student's X is within the clear area inside the 2200-foot contour line shown on the map below.**

72. In the Loa Trend data table, find the column labeled "Loa Trend Volcanoes," and locate Loihi. Note that in the data table, Loihi is at the bottom of the chart. Locate the column labeled "Volcano Age (million years)," and note that volcano age increases from the bottom of the chart to the top of the chart. Locate the column labeled "Distance from Loihi (km)," and note that distance from Loihi increases from the bottom of the chart to the top of the chart. Therefore, as volcano age increases, distance from Loihi increases. In other words, there is a direct relationship between the age of the volcanoes and distance from Loihi.

One credit is allowed for an acceptable response. Acceptable responses include but are not limited to:

- **As distance increases, age increases.**
- **Direct relationship**
- **The oldest volcanoes are farthest from Loihi.**
- **The younger the volcano, the closer it is to Loihi.**
- **They both increase.**

73. According to the reading passage, the Hawaiian volcanic island chain is located on the Pacific Plate. Find the Tectonic Plates map in the *Reference Tables for Physical Setting/Earth Science*. Locate the Hawaiian Islands in the middle of the Pacific Plate. Note the symbol labeled "Hawaii Hot Spot" over the island of Hawaii. Locate this symbol in the key along the bottom of the Tectonic Plates map, and note that it corresponds to a mantle hot spot. A hot spot is a long-lasting zone of rising hot magma beneath moving plates. Large plumes of magma rise from the mantle at these hot spots and work their way upward through the plate above. The magma rises because it is more buoyant than the surrounding rock, wedges apart cracks in the plate, and melts through to erupt, forming a volcano. Thus, the tectonic feature beneath the moving Pacific Plate that caused volcanoes to form in *both* the Loa and Kea trends is a mantle hot spot.

One credit is allowed for an acceptable response. Acceptable responses include but are not limited to:

- **Hawaii Hot Spot**
- **mantle plume**
- **hot spot**
- **rising magma**

Note: Do *not* accept "convection" because this is a process, not a tectonic feature.

74. A hot spot is a long-lasting zone of rising hot magma beneath moving plates. Large batches of magma rise from these hot spots and work their way upward through the plate above. The magma rises because it is more buoyant than the surrounding rock, wedges apart cracks in the plate, and

will be located 51.2 kilometers southeast of Kilauea in the direction of the Kea trend line. Extend the Kea trend line to the southeast along the same line as the existing Kea trend line. Using a piece of scrap paper, mark off a distance corresponding to 51.2 kilometers. Use the distance marked on your scrap paper to locate the point 51.2 kilometers to the southeast of Kilauea along the extended Kea trend line. Mark this point with an **X** to represent the location on the seafloor where the next volcano will most likely form as a part of the Kea trend.

One credit is allowed if **the center of the X is within or touches the box shown below.**

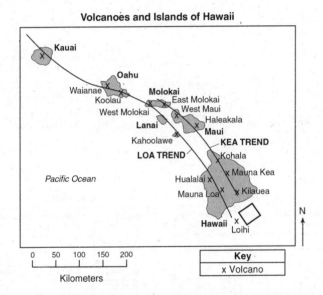

Note: Credit is allowed if a symbol other than an **X** is used.

71. Locate the column labeled "Volcano Age (million years)" in both the Loa Trend and Kea Trend data tables. Note that only one age value appears in both tables—1.7 million years. In the Loa Trend table, trace left from 1.7 million years to the column labeled "Loa Trend Volcanoes," and note that West Molokai is 1.7 million years old. In the Kea Trend table, trace left from 1.7 million years to the column labeled "Kea Trend Volcanoes," and note that East Molokai is 1.7 million years old. Thus, the two volcanoes, one from each trend, that are the same age are West Molokai and East Molokai.

One credit is allowed for ***both* West Molokai and East Molokai**.

One credit is allowed for **any line that extends from the beginning of January to the end of December and is completely within the clear band shown below.**

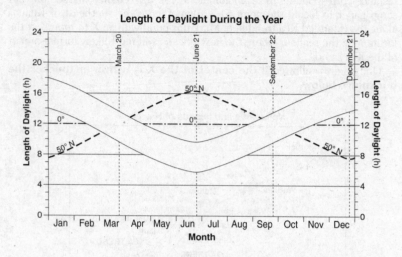

Loa Trend Volcanoes	Volcano Age (million years)	Distance from Loihi (km)
Kauai	4.6	575
Waianae	3.7	465
Koolau	2.2	375
West Molokai	1.7	350
Lanai	1.2	300
Kahoolawe	1.1	250
Hualalai	0.3	130
Mauna Loa	0.2	70
Loihi	0	0

Kea Trend Volcanoes	Volcano Age (million years)	Distance from Kilauea (km)
East Molokai	1.7	256
West Maui	1.5	221
Haleakala	0.9	182
Kohala	0.5	100
Mauna Kea	0.4	54
Kilauea	0.1	0

70. Note in the list of volcanoes in the data table labeled Kea Trend that Molokai is the oldest volcano and Kilauea is the youngest. In the map provided, note that the Kea trend line extends to the southeast from Molokai to Kilauea. Thus, the volcanoes along the Kea trend decrease in age toward the southeast, with Kilauea being the most recent volcano to form. It is given that the average distance between volcanoes along the Kea trend is 51.2 kilometers. Thus, it can be inferred that the next volcano to form in the Kea trend

68. March 20 and September 22 are the spring and fall equinoxes, respectively. When Earth is in an equinox position, its axis of rotation is not either tilted toward or away from the Sun. The boundary between daylight and darkness on the spherical Earth runs directly through both the north and south poles, bisecting Earth along its axis of rotation. So, sunlight strikes Earth perpendicular to the axis of rotation. Sunrise is due east and sunset is due west at all locations on Earth's surface on March 20 and September 22. On those days, the Sun's direct rays strike Earth's surface at noon at the equator. Therefore, as Earth rotates, every point on Earth spends one-half rotation (12 hours) in daylight and one-half rotation (12 hours) in darkness.

One credit is allowed for an acceptable response. Acceptable responses include but are not limited to:

- **Earth's North Pole is not tilted toward or away from the Sun on those dates.**
- **Earth's axis is perpendicular to sunlight on those two dates.**
- **The Sun's direct rays at noon are over the equator.**
- **The Sun rises directly east and sets directly west on those dates.**
- **These dates are equinoxes.**
- **These dates are the first day of spring and the first day of fall.**

Note: The statement "Earth is not tilted" alone is not acceptable because Earth is always tilted on its axis with respect to its orbital plane.

69. As Earth revolves around the Sun, the tilt of its axis of rotation causes the Southern Hemisphere to tilt farthest toward the Sun on December 21 and tilt farthest away from the Sun on June 21. When the Southern Hemisphere is tilted farthest toward the Sun on December 21, the Southern Hemisphere experiences its summer solstice. Thus, on December 21, locations in the Southern Hemisphere experience their greatest length of daylight period. Conversely, when the Southern Hemisphere is tilted farthest away from the Sun on June 21, the Southern Hemisphere experiences its winter solstice. Thus, on December 21, locations in the Southern Hemisphere experience their shortest length of daylight period. Note that the seasons in the Southern Hemisphere differ by 6 months from the corresponding seasons in the Northern Hemisphere. Thus, in order to draw a line showing the changing length of daylight over the course of one year that occurs for an observer at 50° S latitude, do the following. For every month at 50° S latitude, plot the length of daylight value at 50° N latitude 6 months away. For example, for January 1 at 50° S latitude, plot the length of daylight value at 50° N latitude 6 months away on July 1, which is 16 hours. For February 1 at 50° S latitude, plot the length of daylight value at 50° N latitude 6 months away on August 1, which is 14.5 hours. For March 1 at 50° S latitude, plot the length of daylight value at 50° N latitude 6 months away on September 1, which is 13 hours. Continue plotting points representing every month, and then connect the points with a smooth line.

PART C

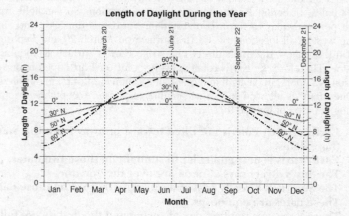

66. The Length of Daylight During the Year graph does not show the length of daylight during the year at latitude 40° N. However, there are lines showing the length of daylight during the year at 30° N and at 50° N. It is reasonable to infer that the length of daylight at 40° N can be interpolated as being midway between the values for 30° N and 50° N. On the graph, locate the point corresponding to January 1 on the horizontal axis labeled "Month" (the left side of the column labeled "Jan"). Trace vertically upward until you intersect the curve labeled "50° N," and then trace left to vertical axis labeled "Length of Daylight (h)." Note the value: 7.5 hours. Now trace vertically upward from January 1 until you intersect the curve labeled 30° N, and then trace left to vertical axis labeled "Length of Daylight (h)." Note the value: 9.9 hours. Thus, the number of daylight hours that occurs on January 1 at 40° N is midway between 7.5 hours and 9.9 hours, which is 8.7 hours.

One credit is allowed for **any value between 8.0 to 9.5 hours**.

67. The daylight period extends from sunrise to sunset, with solar noon marking the midpoint of the daylight period. Therefore, the greater the total length of the daylight period, the greater the number of hours before noon that sunrise will occur, that is, the earlier the sunrise. On the Length of Daylight During the Year graph, locate the vertical dashed line labeled "June 21." Note that on the graph, the length of daylight increases toward the top of the graph. Note that the curve labeled "60° N" crosses the line labeled "June 21" nearest the top of the graph. Thus, on June 21, 60° N latitude will experience the greatest total length of daylight period and the earliest sunrise.

One credit is allowed for a **latitude of 60° N**.

65. Find the Geologic History of New York State chart in the *Reference Tables for Physical Setting/Earth Science*. In the column labeled "Era," locate the statement "Estimated time of origin of Earth and solar system." From this statement, trace left to the time scale labeled "Millions years ago." Note that the estimated time of the origin of Earth is 4600 million years ago, which is equivalent to 4.6 billion, or 4.6×10^9, years ago. Find the Radioactive Decay Data table in the *Reference Tables for Physical Setting/Earth Science*, and locate the column labeled "Half-Life (years)." Note that the half-life closest to the estimated time of Earth's origin (4.6×10^9 years) is 4.5×10^9 years. Trace this row to the left to the column labeled "Radioactive Isotope," and note that a half-life of 4.5×10^9 years corresponds to uranium-238. Thus, the name of the radioactive isotope that has a half-life that is approximately the same as the estimated time of the origin of Earth is uranium-238.

One credit is allowed for an acceptable response. Acceptable responses include but are not limited to:

- **Uranium-238**
- ^{238}U
- **U-238**
- **uranium/U**

umn labeled "Radioactive Isotope," locate carbon-14. Trace this row right to the column labeled "Disintegration," and note that carbon-14 (^{14}C) disintegrates to form ^{14}N, or nitrogen-14. Thus, the decay product formed by the disintegration of carbon-14 is nitrogen-14.

One credit is allowed for an acceptable response. Acceptable responses include but are not limited to:

- ^{14}N
- nitrogen-14
- N-14
- nitrogen/N
- ^{14}C → ^{14}N

64. Compared to all of geologic time, carbon-14 has a very short half-life. So it is used to date only recent organic remains. Carbon-14 is present in the environment and is incorporated into the bodies of living things as they carry out their life processes. When an organism dies, its life processes cease. The carbon-14 in its body that undergoes radioactive decay is no longer replenished from the environment. Therefore, the amount of carbon-14 present in remains decreases over time, and this decrease can be used to measure the time elapsed since the organisms died. However, after 1,000,000 years, it is unlikely that any actual organic materials remain in the rock. Any actual organic materials would have decomposed and/or been replaced by mineral matter. However, if any did remain, they would be too old to be dated using carbon-14. Find the Radioactive Decay Data table in the *Reference Tables for Physical Setting/Earth Science*, and locate carbon-14 in the column labeled "Radioactive Isotope." Trace right to the column labeled "Half-Life (years)," and note that carbon-14 has a half-life of 5.7×10^3 (5700) years. After about 50,000 years (8 to 9 half-lives), the amount of carbon-14 left in a fossil is too small to be measured. During 1,000,000 years, carbon-14 would have gone through more than 175 half-lives (1,000,000/5700 = 175.4)! The amount of carbon-14 remaining would be much too small to be measured. Thus, carbon-14 cannot be accurately used to determine the age of organic remains that are 1,000,000 years old because the remains would no longer contain a measurable amount of carbon-14.

One credit is allowed for an acceptable response. Acceptable responses include but are not limited to:

- **Carbon-14 has a short half-life.**
- **After 1,000,000 years, there would not be enough C-14.**
- **^{14}C decays quickly.**
- **The organic remains are too old to be dated with C-14.**
- **Too little of the original radioactive sample would remain.**

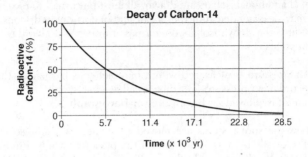

62. Find the table labeled Radioactive Decay Data in the *Reference Tables for Physical Setting/Earth Science*. Note that carbon-14 has a half-life of 5.7×10^3 (5700) years. Enter 5.7 in the box labeled "Time at end of one half-life" on the second row of the flow chart provided. On the Decay of Carbon-14 graph, locate 5.7 on the horizontal axis labeled "Time ($\times 10^3$ yr)." From 5.7, trace vertically upward until you intersect the bold line. From this intersection, trace horizontally to the left to the axis labeled "Radioactive Carbon-14 (%)." Note that the value is "50." Thus, the percentage of carbon-14 remaining at the end of one half-life is 50. Enter 50 in the box labeled "Percentage at end of one half-life" on first row of the flow chart provided. Repeat this procedure for the end of two half-lives. Not that at the end of two half-lives, or 11.4×10^3 years, the percentage of carbon-14 remaining is 25. Enter these values on the flow chart.

One credit is allowed if **all of the percentages and ages are correct, as shown below.**

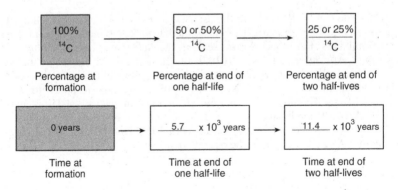

63. When a radioactive isotope decays, its atoms break apart and form more stable atoms of a different element. Find the Radioactive Decay Data table in the *Reference Tables for Physical Setting/Earth Science*. In the col-

Note: If additional isobars are drawn, all must be correct to receive credit. Isobars may be extended into the ocean and/or form closed loops. Do *not* allow credit if the drawn isobars do *not* pass through or touch the 1012 and 1008 data points.

59. Air pressure is measured with a barometer. A barometer that maintains a record of air pressure over time is called a barograph.
One credit is allowed for **barometer *or* barograph**.

60. Low-pressure systems are moved by planetary winds, so the storm track will typically be in the direction of the planetary winds. Find Tectonic Plates map in the *Reference Tables for Physical Setting/Earth Science*. Note that the region of the United States shown on the weather map provided for question 58 in your answer booklet lies roughly between 35° N and 50° N latitude. Now, find the Planetary Wind and Moisture Belts in the Troposphere chart in the *Reference Tables for Physical Setting/Earth Science*. Note that between 35° N and 50° N latitude, the planetary winds blow from the southwest toward the northeast. Toward the northern edge of this zone, the winds tend to curve more toward the east-northeast or toward the east. Therefore, the compass direction toward which the center of the low-pressure system will move if it follows a typical storm track is northeast.

One credit is allowed for an acceptable response. Acceptable responses include but are not limited to:

- **NE**
- **northeast**
- **east**
- **ENE**

61. Point A is located on the isobar labeled "1016." Isobars connect points of equal air pressure. Therefore, the barometric pressure at point A is 1016 millibars. Find the Pressure scale in the *Reference Tables for Physical Setting/Earth Science*. Locate "1016.0" on the left side of the scale labeled "millibars (mb)." Trace right to the scale along the right side of the scale labeled "inches (in of Hg°)." Note that 1016 millibars corresponds to 30.0 inches of Hg. The note at the bottom of the scale indicates that Hg is mercury. Thus, 1016 millibars corresponds to 30 inches of mercury.

One credit is allowed for **any value from 30.00 to 30.01 inches of Hg**.
Note: Also allow credit for 30 or 30.0 inches of Hg.

labeled. However, between any two adjacent points, the surface air pressure steadily changes from the air pressure at one point to the air pressure at the other point. Note that to the southwest and west of the point labeled "1012" are points labeled "1008" and "1013." Thus, it can be inferred that somewhere between these two points will be a point with an air pressure of 1012 millibars. Therefore, from the point labeled "1012," extend the isobar to the left between these two points. Now extend the isobar south to the next point labeled "1012." Next extend the isobar farther south between the point labeled "1008" and the 1016 isobar. Continue to curve the isobar to the east between the points labeled "1014" and "1004" to the next point labeled "1012." Complete the isobar by extending eastward to the south of the point labeled "1010" and to the point labeled "1012" on the coast. Similarly, construct the 1008-millibar line by beginning at the point labeled "1008" on the coast just south of the southern tip of Maine. Extend the isobar west to the next point labeled "1008." From there, extend the isobar to the southwest between the points labeled "1004" and "1012" to the next point labeled "1008." Continue extending the isobar south between the point labeled "1004" and the 1016 isobar. Then curve sharply east between the points labeled "1000" and "1010" to the next point labeled "1008." Finish the 1008 isobar by extending it to the coast between the point labeled "1004" to the north and the point labeled "1012" to the south.

One credit is allowed if ***both*** **isobars are correctly drawn to the east coast of the United States or to the edge of the map.**

Example of a 1-credit response for question 58:

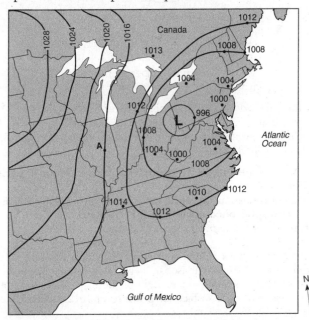

56. The time it takes for the Moon to complete one cycle of phases is 29.5 days.
One credit is allowed for **any value from 29 to 30 days**.

57. Find the Solar System Data table in the *Reference Tables for Physical Setting/Earth Science*. In the column labeled "Celestial Object," locate Earth's Moon. Trace right to the columns labeled "Period of Revolution" and "Period of Rotation at Equator." Note that Earth's Moon has a period of revolution of 27.3 days and a period of rotation of 27.3 days. Thus, the Moon's period of rotation and period of revolution are equal. That is, whenever the Moon revolves one-quarter of the way around Earth, it also rotates one-quarter of the way on its axis. Thus, the same side of the Moon always faces Earth. See the diagram below.

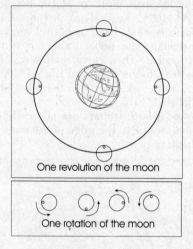

Therefore, the same side of the Moon always faces Earth because the Moon's period of revolution equals the Moon's period of rotation.
One credit is allowed for an acceptable response. Acceptable responses include but are not limited to:

- **The Moon's period of rotation equals the Moon's period of revolution.**
- **The Moon rotates at the same rate that it revolves around Earth.**
- **The Moon spins once during each revolution.**
- **Both motions are completed in 27.3 days.**

58. Isobars are isolines connecting points that have the same air pressure. To draw the 1012-millibar isobar, begin at the point labeled "1012" on the coast of Maine near the upper right of the map. Note that not every point that is experiencing a surface air pressure of 1012 millibars has been

passes directly between the Earth and the Sun and when the Moon casts a shadow on Earth, a solar eclipse occurs. Thus, when the Moon is at position A, a solar eclipse may occur.

One credit is allowed for *both* circling solar eclipse and providing an acceptable explanation. Acceptable responses include but are not limited to:

- **The shadow of the Moon falls on Earth during a solar eclipse.**
- **The Moon blocks some sunlight from reaching Earth.**
- **The Moon is aligned between the Sun and Earth.**
- **Solar eclipses occur only during the New Moon phase.**

Note: Allow credit if neither eclipse is circled but "solar eclipse" is correctly used in the explanation. Do *not* allow "alignment" or "lined up" alone because this occurs in both types of eclipses. The correct sequence of celestial objects in a solar eclipse must be indicated (e.g., "Sun, Moon, Earth" or "Moon in the middle").

55. Note that the diagram shows the Moon's surface almost completely illuminated and the illuminated portion of the Moon is to the observer's left. This corresponds to the old gibbous phase of the Moon, which occurs right after the full moon phase and before the last-quarter phase. The full moon phase occurs when Earth is directly between the Moon and Sun, or directly opposite position A. Note the arrows indicating the direction of motion of the Moon in its orbit. The old gibbous phase occurs when the Moon is at the first position past full moon in its orbit as shown below.

One credit is allowed if **the center of the X is within or touches the clear band shown below.**

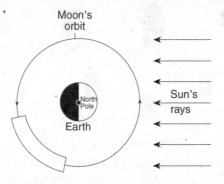

Note: Allow credit if a symbol other than an **X** is used.

Properties of Common Minerals table in the *Reference Tables for Physical Setting/Earth Science*. In the key to chemical symbols at the bottom of the chart, note that the chemical symbol for calcium is Ca. Thus, the chemical symbol for the element, shown in the table, that absorbs the two wavelengths of light is Ca.

One credit is allowed for **Ca**.

53. Within stars, the force of gravity is strong enough to overcome the force of repulsion between atomic nuclei, allowing the nuclei to combine in a process called nuclear fusion. By definition, nuclear fusion involves the combining of several atoms of a lighter element to form a single atom of a heavier element. The single heavier atom typically has less mass than the several lighter atoms from which it formed. The mass "missing" from the heavier atom is not lost. Instead, it is converted into energy according to Einstein's formula $E = mc^2$. This formula means that if mass is converted to energy, the amount of energy released, E, is equal to the mass, m, times the speed of light, c, squared. The speed of light squared is a very large number. Therefore, the conversion of even a small amount of mass to energy during nuclear fusion results in the release of a very large amount of energy. Thus, great amounts of energy are released in the core of a star as lighter elements combine and form heavier elements during the process of fusion. Therefore, the nuclear process that is primarily responsible for producing energy in stars is nuclear fusion, or just fusion.

One credit is allowed for **fusion *or* nuclear fusion**.

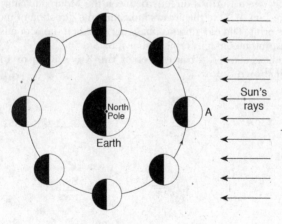

(Not drawn to scale)

54. The arrows in the diagram labeled "Sun's rays" indicate that the Sun is to the right of the diagram. Thus, when the Moon is at position A, or the new moon phase, the Moon is between the Sun and Earth. When the Moon

PART B-2

Data Table

Constellation in which star cluster is seen from Earth	Redshift of two wavelengths of light absorbed by calcium	Distance from Earth (billion light years)	Velocity of star cluster moving away from Earth (km/s)
Ursa Major	Violet → Red	1.0	15,000
Corona Borealis	Violet → Red	1.4	22,000
Boötes	Violet → Red	2.5	39,000
Hydra	Violet → Red	4.0	61,000

Note: One light year is the distance light travels in one year.

51. Find the chart labeled Electromagnetic Spectrum in the *Reference Tables for Physical Setting/Earth Science*. Note that visible light at the red end of the spectrum has a longer wavelength than visible light at the blue end of the spectrum. If a source of electromagnetic waves is moving away from an observer at the same time as it is emitting light of a particular wavelength, fewer wave crests will reach the eye of the observer each second. The eye will interpret this as meaning that the light has a longer wavelength than it actually has. In other words, the light will appear shifted toward the red end of the spectrum. The faster the source is moving away from the observer, the greater the distance between wave crests that reach the eye and the more the light will be shifted toward the red end of the spectrum. The Data Table shows that the light from all of these star clusters has been shifted toward the red end of the spectrum. Therefore, the fact that the light from these star clusters shows a red shift is evidence that these clusters are moving away from Earth.

One credit is allowed for an acceptable response. Acceptable responses include but are not limited to:

- **The wavelengths are shifting toward the red end of the spectrum.**
- **The farther a star cluster is from Earth, the more the redshift.**
- **Redshift of light**
- **The wavelengths of light are getting longer or increasing.**

Note: Credit is *not* allowed for "the more red in color a star is, the more it is moving away" because star color alone does not indicate motion.

52. The second column of the Data Table is headed "Redshift of two wavelengths of light absorbed by calcium." Therefore, the element shown in the table that absorbs the two wavelengths of light is calcium. Find the

50. **4** Find the Relationship of Transported Particle Size to Water Velocity graph in the *Reference Tables for Physical Setting/Earth Science*. Locate 0.01 cm on the vertical axis labeled "Particle Diameter (cm)." Trace horizontally to the right to the vertical axis labeled with the names of particles within different size ranges indicated by the dashed lines. Note that 0.01 cm falls within the size range labeled "Sand" that extends from 0.2 to 0.006 cm. Find the Scheme for Sedimentary Rock Identification in the *Reference Tables for Physical Setting/Earth Science*. In the section labeled "Inorganic, land-derived sedimentary rocks," find the column labeled "Grain Size." Locate "Sand (0.006 to 0.2 cm)." Trace this row right to the column labeled "Rock Name." Note that a sedimentary rock composed of sand is called sandstone. Thus, the rock layer that formed from the deposition of land-derived sediments that had a uniform particle size of about 0.01 cm in diameter was the red sandstone.

graphic range) but present in only one rock layer (short time period) at any location.

49. **3** It is given that the rock layers at these locations have not been overturned. The principle of superposition states that the bottom layer of a sedimentary series is the oldest, unless it was overturned or had older rock thrust over it, because the bottom layer was deposited first. Similarly, in a sequence of rock layers, a rock layer is older than those above it and younger than those below it. Rock layers can sometimes be correlated on the basis of distinct similarities in physical characteristics such as composition, color, thickness, and fossil remains. Thus, it is reasonable to correlate the rock layers at these four locations by their color, composition, and fossil content.

As explained above, fossil 1 best represents an index fossil. The black shale containing fossil 1 is present at all four locations. Therefore, to begin, correlate the black shale across all four locations as shown here:

Location W	Location X	Location Y	Location Z
② Brown siltstone	Tan limestone		Gray conglomerate
	② Brown siltstone		③ Tan limestone
	Red sandstone		Brown siltstone
① Black shale	① Black shale	① Black shale	Red sandstone
④ Gray limestone	Gray limestone	④ Gray limestone	① Black shale
②④ Tan sandstone		④ Tan sandstone	
		④ Gray siltstone	
		Green shale	

Note the gray limestone beneath the black shale at locations W, X, and Y. Note that at locations W and Y, this gray limestone layer contains fossil 4. Thus, it is reasonable to infer that the gray limestone is older than the black shale. Note the tan sandstone containing fossil 4 beneath the gray limestone at locations W and Y. Thus, it can be inferred that the tan sandstone is older than the gray limestone. Finally, note at location Y that beneath the tan sandstone containing fossil 4 there is an older layer of gray siltstone with fossil 4. Beneath that is a layer of green shale. Thus, we can infer that the oldest rock layer is the green shale.

WRONG CHOICES EXPLAINED:

(2) The photograph does not show fossils or their relative abundance in any of the rock layers. Furthermore, a rock's hardness and reactivity determine the rate at which the rock weathers and erodes, not the relative abundance of fossils the rock contains. Therefore, the abundance of fossils is not the characteristic that is primarily responsible for the existence of the rock ledge.

(3) The thickness of a rock layer is not related to its hardness or reactivity. A thick layer may be more resistant or less resistant to weathering than a thin layer, and vice versa. Therefore, thickness is not the characteristic that is primarily responsible for the existence of the rock ledge.

(4) A rock's resistance to weathering and erosion is related to its hardness and reactivity, not its age. More resistant and less resistant rocks may both be of the same age. Therefore, age is not the characteristic that is primarily responsible for the existence of the rock ledge.

47. **4** As the rock fragments are carried along by the stream, they bounce off of and rub against one another. This breaks smaller pieces off the surface, particularly at corners that protrude. As a result, the shape of the fragments changes, becoming smaller and more rounded. As the fragments rub against each other, they also become smoother. Thus, rock fragments that are tumbled and carried over long distances by this stream are most likely becoming more rounded, smoother, and smaller.

Location W	Location X	Location Y	Location Z
② Brown siltstone	Tan limestone	① Black shale	Gray conglomerate
Red sandstone	② Brown siltstone	④ Gray limestone	③ Tan limestone
① Black shale	Red sandstone	④ Tan sandstone	Brown siltstone
④ Gray limestone	① Black shale	④ Gray siltstone	Red sandstone
②④ Tan sandstone	Gray limestone	Green shale	① Black shale

48. **1** Index fossils are remains of organisms that had distinctive body features, were common, were abundant, and had a broad, even worldwide range yet existed for only a short period of time. Their distinctive body and broad distribution makes index fossils easy to find in widely separated rock layers. Their short existence pinpoints the time period during which the rock layer was formed. Thus, organisms that later became good index fossils lived over a wide geographic area and existed for a short time. So, fossil 1 best represents an index fossil because it is present in all of the locations (wide geo-

as water vapor from a plant-covered land surface. Plants remove liquid water from the ground and release it into the atmosphere in the form of water vapor by a process called transpiration. In transpiration, liquid water absorbed by plant roots from the soil increases the water pressure inside the lower parts of the plant. Simultaneously, evaporation of water through plant stems or leaf stomata decreases the water pressure in the upper parts of a plant. This difference in pressure causes the liquid water to move upward from the roots and toward the leaves. When the liquid water reaches the leaves, it evaporates. Then more water is drawn upward from the roots, and the process of transpiration continues. Thus, letter B best represents the process of transpiration because it shows water moving into the atmosphere from a plant-covered land surface.

WRONG CHOICES EXPLAINED:

(1) Capillarity is the tendency for water to rise in narrow openings in earth materials. In the process at B, water is rising from the ground into the atmosphere, not in narrow openings in Earth materials.

(3) Infiltration is the process by which water seeps into the ground. If infiltration was occurring at B, the arrow would be pointed down into the ground, not up into the atmosphere.

(4) Precipitation is condensed moisture in the atmosphere that falls to the ground. If precipitation was occurring at B, the arrow would be pointed from the clouds down toward the ground, not up into the atmosphere.

45. **4** Find the Generalized Landscape Regions of New York State map in the *Reference Tables for Physical Setting/Earth Science*. Locate the region labeled "Tug Hill Plateau." Now find the Generalized Bedrock Geology of New York State map in the *Reference Tables for Physical Setting/Earth Science*. Locate the region corresponding to the Tug Hill Plateau. Note the rock symbol for the surface bedrock in this region. Locate this rock symbol in the key labeled "Geologic Periods and Eras in New York." Note this rock symbol corresponds to limestones, shales, sandstones, and dolostones dating from the Ordovician Period. Thus, the geologic time period during which the surface bedrock at this location was formed was the Ordovician.

46. **1** Note that the rock ledge protrudes outward from the bedrock layers above and below it. Since all of the layers are exposed to the same weathering and erosion processes, the most likely explanation for the difference in the depth is a difference in resistance to weathering. The more resistant a layer, the slower it will weather and the less it will have weathered and eroded back compared to other layers. Thus, compared to the bedrock layers above and below the rock ledge shown at the waterfall, the characteristic that is primarily responsible for the existence of the rock ledge is its greater resistance to weathering.

The Physical Setting for a fuller explanation.) Find the Generalized Bedrock Geology of New York State map in the *Reference Tables for Physical Setting/ Earth Science*. Note that New York State lies between 40° N and 45° N. latitude. Therefore, the altitude of *Polaris* to an observer in New York State will be between 40° and 45° above the horizon. Thus, the diagram that represents the correct position of *Polaris* as viewed from this New York State location on a clear night would show *Polaris* due north of the observer at an altitude of 40° to 45° above the horizon, as shown in diagram (3).

(Not drawn to scale)

43. **2** All processes of the water cycle involve the movement of water. The arrow at letter *A* extends from the ocean surface to the atmosphere. Water in the ocean exists mainly as liquid water. Water in the atmosphere exists mainly as water vapor. Therefore, the process at letter *A* involves liquid water entering the atmosphere as water vapor. The process by which water changes from a liquid to a vapor is called evaporation, or vaporization. Find the Properties of Water chart in the *Reference Tables for Physical Setting/ Earth Science*. Note that during vaporization, water gains 2260 J/g. Thus, in order for process *A* to occur, liquid water must gain 2260 Joules per gram.

44. **2** All processes of the water cycle involve the movement of water. The arrow at letter *B* extends from a land surface covered by trees to the atmosphere. Water on land surfaces and in plants exists mainly as liquid water. Water in the atmosphere exists mainly as water vapor. Therefore, letter *B* represents a process by which liquid water enters the atmosphere

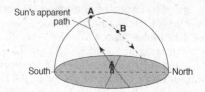

40. **3** Note the arrows on the diagram indicating the direction of apparent motion along the Sun's path. Thus, the Sun moves from point A to point B. It is given that points A and B are 45° apart. As Earth rotates on its axis from west to east at 15° per hour, the Sun appears to move through the sky from east to west at 15° per hour. Therefore, it would take 3 hours (45°/15°/h = 3h) for the apparent position of the Sun to change from point A to point B.

41. **1** On June 21, which is the summer solstice, an observer in New York State will see the sun rise north of east, reach its highest altitude for the year at noon, and set north of west as shown in the diagram. The sun's apparent path is at its longest, so the daylight period is longest. On December 21, which is the winter solstice, an observer in New York State would see the sun rise to the south of east, rise to its lowest altitude for the year at noon, and set to the south of west. The Sun's apparent path is at its shortest, so the daylight period is shortest. Thus, compared to the Sun's apparent path on June 21, the Sun's apparent path on December 21 at this location will be shorter, and the noon Sun will be lower in the sky.

WRONG CHOICES EXPLAINED:
(2) As explained above, on June 21 the Sun's altitude at noon is highest and the Sun's apparent path is longest. On December 21, the Sun's altitude at noon is lowest and the Sun's apparent path is shortest. From June 21 to December 21, the altitude of the Sun at noon and the length of the sun's apparent path decreases daily. Thus, compared to the Sun's apparent path on June 21, the Sun's apparent path on December 21 will be shorter, not longer.

(3) and (4) From June 21 to December 21, the altitude of the Sun at noon and the length of the sun's apparent path decreases daily. They do not remain the same.

42. **3** The star *Polaris* is located very close to Earth's north celestial pole, that is, directly over Earth's North Pole. Therefore, as Earth spins on its axis, *Polaris* remains in a fixed position in the sky relative to an observer in the Northern Hemisphere. Thus, observers in New York State always see *Polaris* due north of their position. By simple geometry, it can be shown that to an observer in the Northern Hemisphere, the altitude of *Polaris* is the same as the observer's latitude. (Refer to Figure 4.5 in *Let's Review: Earth Science*,

38. **4** Find the Specific Heats of Common Materials table in the *Reference Tables for Physical Setting/Earth Science*. Note that liquid water has a specific heat of 4.18 Joules/gram • C°. Note, too, that basalt and granite (common materials from which land is derived) have specific heats of 0.84 Joules/gram • C° and 0.79 Joules/gram • C°, respectively. This means that if 1 gram of water absorbs 4.18 Joules of heat, the temperature of the water will increase by 1°C. However, if 1 gram of basalt absorbs 4.18 Joules of heat, the temperature of the basalt will increase by about 5°C (4.18 J/g ÷ 0.84 Joule/g • °C = 4.98°C). Thus, when basalt and granite absorb the same amount of heat as does water, a land surface will become warmer than a water surface. Therefore, the unequal heating rates of India's land and water are caused by water having a higher specific heat than land.

WRONG CHOICES EXPLAINED:

(1) and (2) Find the Equations section in the *Reference Tables for Physical Setting/Earth Science*. Note the equation for the density of a substance:

$$\text{Density} = \frac{\text{mass}}{\text{volume}}.$$

The property of density quantifies the concentration of matter in a substance, not the rate at which the matter absorbs heat. To compare heating of land and water, one would have to compare how equal masses of land and water that were receiving the same amount of energy changed temperature. Specific heat is how much heat it takes to raise the temperature of a unit mass by a unit degree of temperature. Find the Specific Heats of Common Materials table in the *Reference Tables for Physical Setting/Earth Science*, and note that specific heat is measured in Joules/g • °C. Thus, while density is dependent on volume, specific heat is independent of volume. Therefore, the unequal heating rates of land and water are not caused by the difference in their densities.

(3) Find the Specific Heats of Common Materials table in the *Reference Tables for Physical Setting/Earth Science*. Note that liquid water has a specific heat of 4.18 Joules/gram • C° and that basalt and granite (common materials from which land is derived) have specific heats of 0.84 Joules/gram • C° and 0.79 Joules/gram • C°, respectively. Thus, land has a lower specific heat than water, not a higher specific heat than water.

39. **3** When humid air rises, the pressure confining the air decreases and the air expands. When air expands, it cools adiabatically. When the air is cooled to its dew point, water vapor in the air condenses into water droplets, forming clouds. Thus, the processes that lead to cloud formation when humid air rises over India are expansion, cooling to the dew point, and condensation.

PART B-1

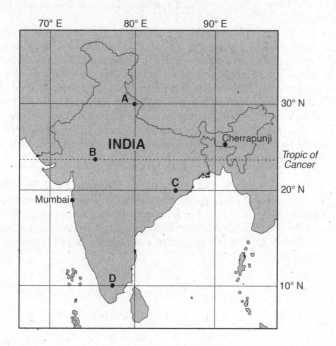

36. **2** When the Sun's rays strike at 90° to Earth's surface, an object will cast no shadow. Note that all of the locations shown are located in the Northern Hemisphere. On the first day of summer in the Northern Hemisphere, Earth is tilted 23.5° toward the Sun. On that day, the Sun's rays strike Earth's surface at 90° at latitude 23.5° N, which is the Tropic of Cancer. Thus, the map location at which no shadow would be cast by the vertical stick at solar noon on the first day of summer is location *B* on the Tropic of Cancer.

37. **1** According to the reading passage, during the summer a large low-pressure system forms over India. In a low-pressure system, the pressure gradient causes the air to move inward toward the low pressure center. At the same time, the Coriolis effect causes the moving air to be deflected to the right. The combination of these two motions results in winds that blow inward in a counterclockwise spiral as shown in map (1).

boundary along which a warm air mass is advancing against a cold air mass is called a warm front. Thus, the frontal boundary between Buffalo and Utica is a warm front.

Find the Key to Weather Map Symbols in the *Reference Tables for Physical Setting/Earth Science*. In the section labeled "Fronts," locate the symbols corresponding to a cold front and a warm front. The side of the cold front on which the symbols are drawn indicates the direction in which the cold air is advancing. The side of the warm front on which the symbols are drawn indicates the direction in which the warm air is advancing. Thus, the cold front between Detroit and Buffalo should be drawn with triangles on the side of the front facing Buffalo, and the warm front between Buffalo and Utica should be drawn with the half circles on the side of the front facing Utica. This is shown on weather map (1).

35. **4** Find the Tectonic Plates map in the *Reference Tables for Physical Setting/Earth Science*. Locate the San Andreas Fault along the west coast of North America. Note the bold black arrows pointing in opposite directions on either side of the fault showing that the plates are sliding horizontally past each other. Note that this pattern of plate motion is best represented by block diagram (4).

cial points, each called a focus (plural, foci). The distance from one focus to any point on the ellipse and back to the other focus is always the same. The closer together the foci, the more circular the ellipse is. The farther apart the foci, the flatter and more elongated the ellipse is. The flatness of an ellipse is called its eccentricity. Eccentricity is expressed as the ratio between the distance between the foci and the length of the major axis. A perfect circle has an eccentricity of 0; a straight line has an eccentricity of 1. As you can see from the diagram, the comet's orbit has a flatter, more elongated oval shape that is less circular than Earth's orbit. Thus, the comet's orbit is more eccentric than Earth's orbit. The flatter and more eccentric an ellipse, the farther apart the foci are. Therefore, compared to Earth's orbit, the comet's orbit is more elliptical and has a greater distance between foci than Earth's orbit.

33. **4** Find the Geologic History of New York State chart in the *Reference Tables for Physical Setting/Earth Science.* In the column labeled "Important Geologic Events in New York," locate the entries "Initial opening of the Iapetus Ocean," and "Dome-like uplift of the Adirondack region begins." From each entry, trace left to the time scale labeled "Million years ago" along the right-hand side of the column labeled "Epoch." Note that the initial opening of the Iapetus Ocean occurred close to 542 million years ago and that the dome-like uplift of the Adirondack region began close to 146 million years ago. In the column labeled "Era," locate the entries "Earliest stromatolites" and "Oceanic oxygen begins to enter the atmosphere." From each entry, trace left to the time scale labeled "Million years ago" along the left side of the column labeled "Eon." Note that the earliest stromatolites appeared about 3300 million years ago and oceanic oxygen began to enter the atmosphere about 2200 million years ago. Thus, the sequence of geologic events in the correct order, from oldest to most recent, is earliest stromatolites → oceanic oxygen begins to enter the atmosphere → initial opening of the Iapetus Ocean → dome-like uplift of the Adirondack region begins.

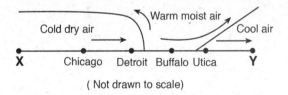

(Not drawn to scale)

34. **1** Note in the cross section that at the frontal boundary between Detroit and Buffalo, a cold, dry air mass to the west is advancing against a warm, moist air mass to the east. The boundary along which a cold air mass is advancing against a warm air mass is called a cold front. Thus, the frontal boundary between Detroit and Buffalo is a cold front. Now, note in the cross section that at the frontal boundary between Buffalo and Utica, a warm, moist air mass to the west is advancing against a cool air mass to the east. The

will sink downward (like the air near the cold window). Thus, diagram (1) represents convection, not conduction.

(2) Energy from the Sun must travel through a vacuum to reach Earth. Energy cannot be transmitted through a vacuum by conduction (molecular collisions) or convection (movements of fluids due to density differences) because both of these methods require a medium. Therefore, energy is transferred from the Sun to Earth mainly in the form of radiation consisting of electromagnetic waves. Thus, diagram (2) represents radiation, not conduction.

(4) Find the Inferred Properties of Earth's Interior chart in the *Reference Tables for Physical Setting/Earth Science*. In the upper portion of the diagram, locate the lines corresponding to the upper and lower boundaries of the asthenosphere. Follow these lines downward until they become dashed lines and intersect the bold line marked "Interior Temperature" in the "Temperature (°C)" graph. From these intersections, trace horizontally to the left to the temperature scale. Note that the interior temperature at the bottom of the asthenosphere is approximately 2600°C and the interior temperature at the top of the asthenosphere is about 1100°C.

Note that in diagram (4), the arrows show material rising to the top of the asthenosphere near the center of the diagram, spreading outward to either side, then sinking downward through the asthenosphere, and moving back toward the center along the upper surface of the stiffer mantle. When taken together, the arrows indicate a circular pattern of motion in which hot material at the bottom of the asthenosphere rises, spreads sideways, cools, and sinks back to the bottom of the asthenosphere, a process known as convection. Thus, diagram (4) represents convection, not conduction.

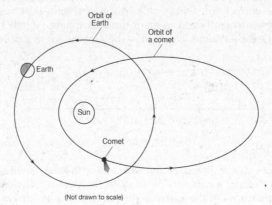

32. **4** An ellipse is an oval shape. It has a major axis that connects the two points farthest apart on the ellipse. It also has a minor axis that connects the two points closest together on the ellipse. The major axis contains two spe-

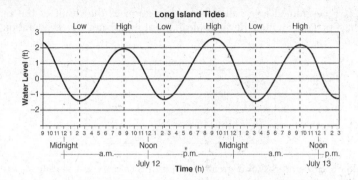

30. **2** Three high tides are shown on the graph. The first high tide occurred on June 12 at about 8:35 a.m. The second high tide occurred on June 12 at about 9:05 p.m.—a difference of about 12 hours 30 minutes. The third high tide occurred on June 13 at roughly 9:35 a.m.—a difference of about 12 hours 30 minutes. Based on the data, the next high tide occurred about 12 hours 30 minutes later at 10:00 p.m. on July 13.

31. **3** In the process of conduction, heat energy is transferred from one atom or molecule to another through collisions. Note that diagram (3) shows the center of a metal rod being heated by a flame. Arrows indicate that heat is moving away from the flame in both directions through the rod. The cutaway view shows that the metal rod is composed of closely spaced atoms that are vibrating. Heating the closely spaced atoms in a flame would cause them to vibrate more vigorously and collide with nearby atoms. During these collisions, the vibrating atoms would transfer some of their energy to their neighbors. These neighbors would, in turn, collide with their neighbors and so on down the line, transferring energy away from the source of heat. Thus, diagram (3), which shows heat transferred through a metal rod in both directions away from the flame by vibrating atoms, best represents heat transfer mainly by the process of conduction.

WRONG CHOICES EXPLAINED:
(1) The arrows in diagram (1) show a circular pattern of motion in which air is heated by the radiator, rises and spreads sideways, cools near the cold window, sinks back to the floor, and then spreads sideways back toward the radiator. Such a circular pattern of motion in a fluid (such as air) is called a convection current. Convection involves the transfer of heat due to density differences within a fluid. Heated fluids are less dense than cooler fluids. Thus, when a portion of a fluid is heated, it becomes less dense than the surrounding cooler fluid and rises upward (like the air over the radiator). As the heated fluid rises, it carries the heat energy within it upward. Conversely, cooling a fluid makes it denser than the surrounding fluid. The cooler fluid

(4) Fine-grained layers of volcanic lava form as the result of the rapid cooling of lava emitted during volcanic eruptions, not the evaporation of water.

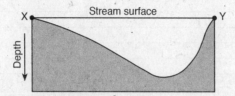

28. **4** Stream erosion is greatest where the velocity of the water is greatest. The greater the stream erosion, the greater the depth to which the streambed will be eroded. Thus, the depth of a streambed will be greatest where the velocity of the water is greatest. Along a straight section of a stream's channel, water velocity is greatest in the center of the stream farthest from the friction of the banks or bed. When the stream curves, or meanders, water velocity is greatest along the outside of the curve and lowest along the inside of the curve. The cross section shows that the streambed is deepest near point Y and is shallowest near point X. Thus, it can be inferred that the stream velocity and stream erosion are greater near point Y and least near point X. Therefore, point Y is located along the outside curve of a meander and point X is located along the inside curve as shown in map (4).

29. **4** As the velocity of wind or water decreases, the particle size that can be transported also decreases. Therefore, as wind and running water gradually decrease in velocity, larger, denser particles are deposited first, followed later by smaller, less-dense particles. This separation of particles during deposition is called sorting. Thus, when wind and running water gradually decrease in velocity, the transported sediments are deposited over a period of time and are sorted by size and density.

WRONG CHOICES EXPLAINED:
(1) and (2) The question states that the velocity of the wind or water decreases gradually. Gradually means slowly over a period of time, not all at once.

(3) As wind and running water gradually decrease in velocity, larger, denser particles are deposited first, followed later by smaller, less-dense particles. This separation of particles during deposition is called sorting. Thus, although the deposition would take place over a period of time, the particles deposited would be sorted, not unsorted.

of water at the center of the landscape feature into which the streams flow. Therefore, the landscape feature is not a glacial kettle lake.

(4) On a flat plain, the slope is gentle. Streams flow downhill slowly and typically form a dendritic stream drainage pattern. In a dendritic stream drainage pattern, there are numerous small streams that join together to form larger streams in a treelike pattern, not a pattern in which the streams radiate outward from the center as shown in the map.

27. **2** According to the map, Lake Bonneville has decreased greatly in size since the end of the last ice age. This means that Lake Bonneville lost a great deal of water. Since the end of the last ice age, global temperatures have increased. According to the map, ancient Lake Bonneville covered a large surface area in Utah and Nevada. Utah and Nevada currently have very arid climates. The combination of large surface area, increased temperatures, and an arid climate make it likely that the loss of water by Lake Bonneville was the result of increased evaporation. The high salt level in the Great Salt Lake remnant of Lake Bonneville further supports this inference. As the water in Lake Bonneville evaporated, the concentration of dissolved minerals increased. Eventually, the water became saturated and no more minerals were able to dissolve. At this point, as the water continued to evaporate, the minerals came out of solution and were precipitated in the lake. Mineral crystals that precipitate out of water settle to the bottom and form flat, horizontal layers called evaporites. Thus, material that was formerly on the bottom of Lake Bonneville and is most likely exposed on the land surface today is flat-lying evaporate deposits.

WRONG CHOICES EXPLAINED:

(1) Folded metamorphic bedrock forms when layers of rock are folded due to compression forces in Earth's crust, not the evaporation of water.

(2) Coal beds are sedimentary rock formed from compacted plant remains, not evaporation of water.

WRONG CHOICES EXPLAINED:

(1) Find the Scheme for Sedimentary Rock Identification in the *Reference Tables for Physical Setting/Earth Science*. In the column labeled "Rock Name," locate sandstone. Note that sandstone is an inorganic, land-derived sedimentary rock. Find the Rock Cycle in Earth's Crust diagram in the *Reference Tables for Physical Setting/Earth Science*. Note that the arrow leading to "Sedimentary Rock" is labeled "Deposition and Burial" and "Compaction and/or Cementation." Neither of these processes involves intense heat and pressure.

(3) and (4) It is given that the rock did not solidify from magma. Find the Rock Cycle in Earth's Crust diagram in the *Reference Tables for Physical Setting/Earth Science*. Note that the only arrow leading to igneous rock comes from magma and is labeled "Solidification." Thus, igneous rocks solidify from magma. Find the Scheme for Igneous Rock Identification in the *Reference Tables for Physical Setting/Earth Science*. Note that gabbro and rhyolite appear on the chart. Thus, gabbro and rhyolite cannot be the rock because they are igneous rocks that solidified from magma.

26. **3** Water flows from higher elevations to lower elevations. The streams in this stream drainage pattern radiate outward from the center. In other words, the streams are flowing outward from the center. Thus, the highest elevation in this landscape feature is in the center and decreases in all directions away from the center. That is, the landscape feature is a conical hill—a shape typically associated with a dome or a volcanic mountain. Therefore, of the choices, the landscape feature that would produce this stream drainage pattern is a volcanic mountain.

WRONG CHOICES EXPLAINED:

(1) A steep cliff would result in a rectangular stream drainage pattern in which streams would radiate outward from a central line representing the peak of the cliff, not a radial pattern as shown in the map.

(2) A glacial kettle lake forms in a circular depression created when a block of buried ice melts. In a depression, the center is at a lower elevation than the surrounding land. The streams would flow downhill toward the center of the depression and into the glacial kettle lake. The map shows no body

(4) Bedrock age alone does not determine how a landscape is formed, its structure, and how it is classified. For example, bedrock of many different ages may have been formed by the folding and faulting that creates mountains. Bedrock of many different ages may have been gently uplifted to form plateaus. The age of the bedrock does not determine the landscape's structure and how it is classified. Rather, the structure of the bedrock that resulted from the forces acting on the bedrock and caused the bedrock to be uplifted determine the landscape's structure and classification. Thus, bedrock age is not the factor most responsible for the difference in landscape classification of these two regions.

24. **2** At the melting edge of a glacier, sediments within the ice as well as those carried on top of the ice are released and drop to the ground. Particles of mixed shapes and sizes drop to the ground in a confused jumble, forming an unsorted deposit called glacial till. Drumlins are formed when the ice of a glacier slides over such previously deposited piles of sediment, causing them to become elongated into a teardrop shape. Thus, an elongated hill that is composed of unsorted sediments deposited by a glacier is called a drumlin.

WRONG CHOICES EXPLAINED:
(1) A delta is a large, flat, fan-shaped pile of sediment at the mouth of a stream. It is not composed of unsorted sediments deposited by a glacier.
(2) By definition, a sand dune is a mound of sand deposited by wind. It is not composed of unsorted sediments deposited by a glacier.
(3) An outwash plain is a flat plain next to the leading edge of a glacier formed by the deposition of sediments from meltwater streams. Sediments deposited by the running water in streams are typically sorted. Therefore, an outwash plain is a flat plain, not an elongated hill. An outwash plain is composed of sorted sediments deposited by streams, not unsorted sediments deposited by a glacier.

25. **2** Find the Rock Cycle in Earth's Crust diagram in the *Reference Tables for Physical Setting/Earth Science*. Note that the only arrow leading to igneous rock comes from magma and is labeled "Solidification." Thus, a rock that did not solidify from magma is not an igneous rock. Note that the arrow leading to "Sedimentary Rock" is labeled "Deposition and Burial" and "Compaction and/or Cementation." Also note that all of the arrows leading to the rectangle representing "Metamorphic Rock" are labeled "Heat and/or Pressure." It is given that the rock was subjected to intense heat and pressure; therefore, the rock in question is most likely a metamorphic rock. Find the Scheme for Metamorphic Rock Identification in the *Reference Tables for Physical Setting/Earth Science*. Locate the column labeled "Rock Name." Note that of the given choices, only schist is a metamorphic rock. Thus, the rock that was subjected to intense heat and pressure but did *not* solidify from magma is schist.

the line. As atom after atom is pulled out of place and then returns to its equilibrium position, the transverse wave moves through the solid. However in fluids (liquids or gases), atoms are far enough apart that they move freely past one another and have no equilibrium position. When an atom in a fluid is pulled out of place, there is no restoring force to pull it back to an equilibrium position. This ability for an atom in a fluid to move about freely prevents one atom from displacing its neighbor in a direction perpendicular to the direction of wave travel in fluids. Therefore, S-waves from an earthquake cannot be transmitted through liquids but are instead absorbed.

Find the Inferred Properties of Earth's Interior chart in the *Reference Tables for Physical Setting/Earth Science*, and locate the graph labeled "Temperature (°C)." Note the region in which the bold line labeled "Interior Temperature" is higher than the dotted line labeled "Melting Point." This indicates that the interior temperature of the rock is greater than its melting point and that the rock is therefore inferred to be liquid. Trace this region upward to the cross section, and note that this section corresponds to the region labeled "outer core." Thus, the seismic station recorded the P-waves, but no S-waves, from an earthquake because S-waves were absorbed by Earth's outer core.

23. **2** By definition, a plateau is a large, flat region of land elevated more than 150–300 meters above the surrounding land. Plateaus generally have an underlying structure made of horizontal layers of rock that were gently uplifted. By definition, mountains are parts of Earth's crust that project at least 300 meters above the surrounding land and have great relief, a restricted summit, and considerable bare-rock surface. Most mountains are formed by crustal motions that fold or fault rock and have an underlying bedrock structure composed of folded and faulted rock layers. Thus, the differences between mountains and plateaus are relief, elevation above the surrounding land, type of uplift, and underlying bedrock structure. Thus, of the factors given, the one most responsible for the difference in landscape classification of these two regions is bedrock structure.

WRONG CHOICES EXPLAINED:

(1) Landscapes are classified as mountains or plateaus mainly by their elevation, relief, and underlying bedrock structure, all of which are the result of the forces that formed the landscape by causing the land to be uplifted. Climate variations would affect the erosion of a landscape after it had already formed, not how the landscape formed, its structure, and how it is classified. Thus, climate variation is not the factor most responsible for the difference in landscape classification of these two regions.

(3) Although vegetation type may affect the rate at which a landscape erodes, it would be used to identify a location's climate, not how the landscape formed, its structure, and how it is classified.

19. **3** Find the Geologic History of New York State chart in the *Reference Tables for Physical Setting/Earth Science*. In the column labeled "Life on Earth," locate the statement "Earliest grasses." Follow this row left to the column labeled "Epoch," and note that "Earliest grasses" corresponds to "Oligocene." Thus, scientists infer that the earliest grasses first appeared on Earth during the Oligocene epoch.

20. **1** Find the Inferred Properties of Earth's Interior chart in the *Reference Tables for Physical Setting/Earth Science*. In the cross section at the top of the chart, locate the boundary between the stiffer mantle and the outer core. Trace the dashed line corresponding to this boundary vertically downward until it intersects with the bold line on the Pressure vs. Depth graph. From this intersection, trace left horizontally to the axis labeled "Pressure (million atmospheres)." Note the value—1.5 million atmospheres. Now continue tracing the dashed line corresponding to this boundary vertically downward until it intersects with the bold line labeled "Interior Temperature" on the Temperature vs. Depth graph. From this intersection, trace horizontally to the left to the axis labeled "Temperature (°C)." Note the value—4950°C. Thus, the inferred pressure and temperature at the boundary of Earth's stiffer mantle and outer core is 1.5 million atmospheres pressure and an interior temperature of 4950°C.

21. **3** It is given that the seismic station is 7600 kilometers from the epicenter of the earthquake. Find the Earthquake P-Wave and S-Wave Travel Time graph in the *Reference Tables for Physical Setting/Earth Science*. Locate 7600 kilometers (7.6×10^3 km) along the horizontal "Epicenter Distance ($\times 10^3$)" axis at the bottom of the graph. Trace vertically until you intersect the bold line labeled "P." Then trace horizontally to the left to the "Travel Time (min)" axis, and read the *P*-wave travel time—11 minutes. If the seismic *P*-wave takes 11 minutes to travel the 7600 kilometers from the epicenter to the seismic station and arrives at 2:25 p.m., the *P*-wave must have left the epicenter 11 minutes before 2:25 p.m., or at 2:14 p.m. Thus, the earthquake occurred at 2:14 p.m.

22. **1** *P*-waves are compression waves that compress and expand rock in the direction of wave travel. Solids, liquids, and gases are all compressible; therefore, *P*-waves can be transmitted through solids, liquids, and gases. *S*-waves, however, are transverse waves that twist rock back and forth, displacing rock atoms in a direction perpendicular to that of wave travel. *S*-waves can be transmitted through only solids. In solids, atoms are closely spaced and held in place by the forces exerted by neighboring atoms. Each atom has an equilibrium position where all the forces exerted on it by its neighbors are in balance. If an atom gets displaced, it tends to be pulled back to that equilibrium position by the forces exerted by neighboring atoms. When an atom is displaced, it pulls its neighbors out of place and so on down

impact event: levels of iridium higher than those found in Earth's crust but similar to those found in meteorites, shocked quartz grains, melted spherules, soot from the widespread forest fires ignited by the meteorite, and evidence of large waves such as would be caused by an ocean impact. The impact of an asteroid or a large meteorite would generate a massive shock wave and spew large amounts of dust high into the atmosphere. The resulting decrease in sunlight would cause global temperatures to decrease markedly. The effects on photosynthetic life and on all of the life-forms that depend on photosynthetic organisms for food would be devastating. Thus, many scientists infer that one cause of the mass extinction of dinosaurs and ammonoids that occurred approximately 65.5 million years ago was an asteroid impact that resulted in climate change.

WRONG CHOICES EXPLAINED:

(1) Tectonic plate subduction occurs when sinking convection currents in the mantle pull a plate downward into the mantle, where the plate melts and is destroyed. Find the Geologic History of New York State chart in the *Reference Tables for Physical Setting/Earth Science*. Locate the column labeled "Inferred Positions of Earth's Landmasses," and note the upper two maps labeled "59 million years ago" and "119 million years ago." Note that although the positions of the continents have changed, their size and shape has not. Thus, most of the continents have not been destroyed as they would have been if tectonic plate subduction of most of the continents had occurred.

(3) Find the Geologic History of New York State chart in the *Reference Tables for Physical Setting/Earth Science*. In the column labeled "Time Distribution of Fossils," locate the bold gray line labeled "Ammonoids." Note the lettered circle "G" indicating an index fossil. Locate "G" in the diagrams of index fossils at the bottom of the chart. Note that ammonoids are shelled organisms and most likely lived in a marine environment. Dinosaurs were terrestrial vertebrates. Therefore, it is unlikely that ammonoids and dinosaurs came into contact with one another. Furthermore, there are currently no known diseases that can be transmitted between such widely different species that are fatal to both species. Thus, it is unlikely that a disease spreading among many groups of organisms caused the extinction of dinosaurs and ammonoids 65.5 million years ago.

(4) Most earthquakes occur along plate boundaries because these boundaries are where the plates collide and scrape against each other. Catastrophic earthquake damage is generally confined to regions fairly close to the epicenter. So, large regions of Earth's surface would be unaffected by severe earthquake damage. Thus, it is unlikely that the mass extinction of ammonoids and dinosaurs was caused by severe earthquake damage from worldwide earthquakes.

diagram, locations *D* and *B* are located at equal distances south and north of the equator. Therefore, the angle of insolation at locations *D* and *B* will have decreased from 90° by an equal number of degrees. Thus, locations *D* and *B* have the same angle of insolation and will receive the same intensity of insolation.

16. **2** Find the Generalized Bedrock Geology of New York State map in the *Reference Tables for Physical Setting/Earth Science*, and locate the key labeled "Geologic Periods and Eras in New York State." In the key, note that rocks of Cambrian to Early Jurassic age include sandstones. Now find the Geologic History of New York State chart in the *Reference Tables for Physical Setting/Earth Science*. Locate the column labeled "NY Rock Record," and note that bedrocks found in New York State are indicated in black. Trace down this column to the longest black segment. From this segment, trace left to the column labeled "Period," and note that much of the bedrock in New York State dates from the Cambrian to the Devonian periods. Thus, most of the sandstone found in New York State was deposited during the Cambrian through Devonian periods. Now trace right to the column labeled "Important Geologic Events in New York." Note the entries corresponding to the different periods. The entry for the Cambrian states, "Widespread deposition over most of New York along edge of Iapetus Ocean." The entry for the Ordovician states in part, "Queenston delta forms." The entry for the Silurian states, "Salt and gypsum deposited in evaporite basins." The entry for the Devonian states in part, "Catskill delta forms." Consider that all of these entries indicate that deposition was most likely occurring in water. Thus, the sand in the sandstone found in New York State was most likely originally deposited in water. The principle of original horizontality states that sediments deposited in water form flat, horizontal layers. Therefore, most of the sand that makes up the sandstone found in New York State was originally deposited in horizontal layers.

17. **3** Note the position of New York State in North America shown in the diagram. Find the Geologic History of New York State chart in the *Reference Tables for Physical Setting/Earth Science*. Locate the column labeled "Inferred Positions of Earth's Landmasses." Locate the diagram in which the region of North America corresponding to New York State is located at the equator. Beneath that diagram, note that the caption says "359 million years ago." So, this New York State region was located at the equator 359 mya.

18. **2** Evidence suggests that an impact event was responsible for the extinction of many life-forms at the end of the Cretaceous period 65.5 million years ago. The dividing line between the Cretaceous and the Tertiary periods in the rock record is a thin layer of clay that has been identified in sediments worldwide. This boundary of clay contains numerous indications of a massive

the location showing the least difference between the air temperature and the dew point is the location with the greatest chance of precipitation. According to the table, the difference between the air temperature and the dew point is smallest at location D. Therefore, location D has the greatest chance of precipitation.

13. **2** Find the Electromagnetic Spectrum chart in the *Reference Tables for Physical Setting/Earth Science*. Note the arrows indicating that on the chart, wavelength decreases to the left and increases to the right. Thus, the electromagnetic radiation listed farthest to the left on the chart has the shortest wavelength. Of the choices given, gamma rays are farthest to the left on the chart, indicating that they have the shortest wavelength.

14. **1** Greenhouse gases absorb infrared radiation and trap heat in the atmosphere. Over time, the amount of energy received from the Sun should be about the same as the amount of energy radiated back into space, leaving the temperature of Earth's surface roughly constant. However, many gases exhibit these "greenhouse" properties, trapping heat and thus may be responsible for warming of the atmosphere. Some of these greenhouse gases occur in nature (water vapor, carbon dioxide, methane, ozone, and nitrous oxide), while others are exclusively human-made (like the chlorofluorocarbons used in aerosol sprays). Thus, greenhouse gases that may be responsible for global warming include carbon dioxide (CO_2), methane (CH_4), water vapor (H_2O gas), nitrous oxide (N_2O), ozone (O_3), and chlorofluorocarbons (CFCs.) Of the choices, only methane is a major greenhouse gas.

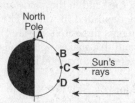

15. **4** Insolation is most intense when it strikes a surface at 90° (that is, perpendicular to a surface) because the insolation is concentrated in the smallest possible area. As the angle of insolation decreases from 90° toward 0°, the same amount of insolation is spread over larger and larger areas. Thus, as the angle of insolation decreases, the intensity of insolation decreases. However, if two locations have the *same* angle of insolation, they will receive the same intensity of insolation.

The diagram shows Earth in an equinox position. Earth is not tilted toward or away from the Sun, and the line between daylight and darkness passes directly through the poles. At the equinoxes, the Sun's rays strike Earth's surface at 90° at the equator. Moving either north or south of the equator will cause the angle of insolation to decrease. According to the

9. **3** Find the Key to Weather Map Symbols chart in the *Reference Tables for Physical Setting/Earth Science*, and refer to the "Station Model Explanation." According to the explanation, wind direction is indicated by a line extending out from the center pointing in the direction from which the wind is blowing. Note that the line extending outward from the center of the station model in the question points to the southeast. Wind speed is indicated by the number of feathers on this line. Whole feathers = 10 knots; half feathers = 5 knots. Since the line has 2 whole feathers, the wind speed is 20 knots. Thus, the weather station model shown in the question indicates that winds are coming from the southeast at 20 knots.

10. **3** Maritime (m) air masses form over water. Thus, they generally contain more humidity than continental (c) air masses that form over land. Tropical (T) air masses form near the equator and usually have higher air temperatures than polar (P) air masses that form near the poles. Thus, the type of air mass that most likely has high humidity and high temperature is a maritime tropical air mass. Find the Key to Weather Map Symbols in the *Reference Tables for Physical Setting/Earth Science*. In the section labeled "Air Masses," note that the symbol for a maritime tropical air mass is mT.

11. **1** It is given that the dry-bulb temperature is 16°C and that the wet-bulb temperature is 10°C. Thus, the difference between the wet-bulb and dry-bulb temperatures is 6°C. Find the Relative Humidity (%) chart in the *Reference Tables for Physical Setting/Earth Science*. Locate the column headed "6" in the "Difference Between Wet-Bulb and Dry-Bulb Temperatures (C°)" scale along the top of the chart. Find the "Dry-Bulb Temperature (°C)" scale along the left side of the chart. Locate the row labeled "16," trace to the right until it intersects the column headed "6," and note the value of 45. Thus, if the dry-bulb temperature is 16°C and the wet-bulb temperature is 10°C, the relative humidity is 45%.

Location	A	B	C	D
Air temperature (°F)	80	60	45	35
Dewpoint (°F)	60	43	35	33

12. **4** As the difference between the air temperature and the dew point decreases, the relative humidity increases. As the relative humidity increases, the likelihood of condensation that forms precipitation increases. Therefore,

Photosynthetic life-forms use energy from sunlight to convert carbon dioxide and water into carbohydrates (food) and oxygen. Therefore, once photosynthetic life evolved, these organisms released oxygen into the environment. However, evolving life-forms cannot account for the presence of other gases such as hydrogen, nitrogen, or ammonia.

(3) In the early stages of planetary formation, any gases left from the solar nebula would quickly have been swept away from the region surrounding Earth by the vigorous solar wind. With nearby space swept clean of gases, it is unlikely that Earth's early atmosphere contained carbon dioxide, sulfur dioxide, hydrogen, nitrogen, water vapor, methane, and ammonia because Earth's growing gravitational field attracted these gases from space.

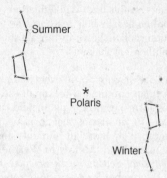

8. **3** The first day of summer (June 21) and the first day of winter (December 21) are 6 months apart. Therefore, the change in the apparent position of the Big Dipper between the first day of summer and the first day of winter is best explained by Earth revolving for 6 months.

WRONG CHOICES EXPLAINED:

(1) Earth rotates at a rate of 15°/h. Thus in 12 hours, the Big Dipper would rotate 180° (12 × 15 = 180) and would appear to change position relative to *Polaris* as shown. However, 12 hours after midnight on June 21 occurs at 12 noon on June 22, not the first day of winter 6 months later.

(2) In rotating for 1 day at a rate of 15°/h, the Big Dipper would move 360° around Polaris (15 × 24 = 360) and end up in roughly the same position relative to Polaris, not opposite its original position. Furthermore, 1 day after midnight on June 21 would occur at midnight on June 22, not at midnight on the first day of winter 6 months later.

(4) If Earth revolved for 1 year, it would be June 21 of the following year, not the first day of winter 6 months later.

6. **3** The best evidence that Earth spins on its axis was first demonstrated by the French astronomer Jean Foucault. He set up a pendulum by attaching a heavy weight to a long wire and allowing the pendulum to swing freely. Foucault knew that the path of a freely swinging pendulum would not change. He sealed the room so that no air movements would exert a force on the pendulum that would cause it to change direction. As the pendulum swung back and forth, it appeared to change direction slowly, passing over different lines on the floor until eventually coming full circle to its original position. Foucault concluded that the apparent shift in the direction of swing of the pendulum was due to the floor (Earth's surface) rotating beneath the pendulum. Thus, the pendulum movement and pattern of knocked-over pins at the museum is evidence of Earth's rotation on its axis.

WRONG CHOICES EXPLAINED:
(1) The shape of the surface beneath a freely swinging pendulum (flat or spherical) has no effect on the pendulum's direction of swing because there is no contact between the pendulum and the surface. Thus, a Foucault pendulum provides no evidence of Earth's shape.
(2) The swing of a Foucault pendulum is the result of Earth's gravity acting on the weight, not the Sun's gravity acting on Earth. Thus, the pendulum movement and pattern of knocked-over pins does not provide evidence of Earth's gravitational attraction to the Sun.
(3) The apparent change in the direction of swing of a Foucault pendulum is observed relative to Earth's surface, not the pendulum's position relative to the Sun. Thus, the pendulum movement and pattern of knocked-over pins does not provide evidence of Earth's nearly circular orbit around the Sun.

7. **4** Earth's early atmosphere most likely formed by outgassing, which is the release of water and trace gases trapped in rocks in the interior of the hot, young Earth. Even after Earth cooled and formed a solid crust, outgassing continued during volcanic eruptions. Analysis of gases emitted during volcanic eruptions shows the presence of carbon dioxide, sulfur dioxide, hydrogen, nitrogen, water vapor, methane, and ammonia. Thus, it is inferred that these gases were present in Earth's early atmosphere primarily because volcanic eruptions on Earth's surface released these gases from the interior.

WRONG CHOICES EXPLAINED:
(1) Find the Inferred Properties of Earth's Interior chart in the *Reference Tables for Physical Setting/Earth Science*. Note that Earth's core consists of iron and nickel. None of the gases listed in the question are radioactive decay products of iron or nickel. Therefore, it is unlikely that they were produced in Earth's core and released from Earth's surface.
(2) Earth's early atmosphere contained virtually no free oxygen. However, over time, life-forms evolved that were able to carry out photosynthesis.

a star emits energy relative to the Sun)" increases in value from the bottom to the top of the chart. Note that the horizontal scale along the bottom of the chart labeled "Surface Temperature (K)" increases from right to left. Thus, a star that is hotter than *Polaris* would be located to the left of *Polaris* on the chart. Additionally, a star that is less luminous than *Polaris* would be located lower down on the chart than *Polaris*. Of the choices, only *Sirius* is located both to the left of and lower on the chart than *Polaris*, indicating that *Sirius* is hotter and less luminous than *Polaris*.

4. **2** During the formation of the solar system, Earth and the other planets were very hot fluids. In this fluid state, materials of different densities were able to move under the influence of a planet's gravity. Denser materials have more mass per given volume than less-dense materials, and gravitational force is directly dependent on mass. Therefore, denser materials are more strongly attracted to a planet's center than are less-dense materials. This caused denser materials to sink toward a planet's center and less-dense materials to float toward a planet's surface, causing the planet's interior to separate into layers. Thus, Earth and the other planets of our solar system became layered as they were being formed because gravity caused more-dense materials to move toward the center of each planet.

WRONG CHOICES EXPLAINED:
(1) Less-dense materials have less mass per given volume than more-dense materials. Therefore, less-dense materials would not be pulled toward the center of a planet as strongly as would more-dense materials. Thus, gravity would cause more-dense materials to move toward the center of each planet, forcing less-dense materials to move upward toward the surface of each planet.
(3) and (4) Cooling a material, whatever the cooling rate, increases the material's density. Thus, gravity would cause this more-dense, cooler material to move toward the center of each planet, not stay at the surface.

5. **1** In order for water to seep into the ground, there must be openings, or pores, in the ground surface through which water can enter the ground. A substance through which water can flow is said to be permeable; a substance through which water cannot flow is impermeable. Thus, in order for water to seep into the ground, the ground must be permeable.
Water runs downhill more slowly on gentle slopes than on steep slopes. So on gentle slopes, there is more time for the water to seep into the ground before it runs off than on steep slopes. Therefore, in an area with gentle slopes, more water will seep into the ground and less will run off. Thus, a gentle slope and permeable surface will allow for the greatest amount of water to seep into the ground.

Answers Explained

PART A

1. **1** Find the Solar System Data chart in the *Reference Tables for Physical Setting/Earth Science*. In the column labeled "Celestial Object," locate "Earth." Trace right to the column labeled "Period of Revolution." Note that Earth's period of revolution is 365.26 days. One complete revolution consists of motion through 360 degrees. Thus, Earth completes one 360° revolution around the Sun in 365.26 days. This means the rate of revolution is about 1° per day (360°/365.26 days = 0.99°/day).

2. **2** Planetary winds tend to blow in a straight line from regions of high pressure toward regions of low pressure. As these winds are blowing, Earth is turning on its axis. This rotation causes the winds to appear to be turning toward the right in the Northern Hemisphere and toward the left in the Southern Hemisphere. This phenomenon is called the Coriolis effect. Thus, planetary winds in Earth's Northern Hemisphere generally curve to the right due to the Coriolis effect.

WRONG CHOICES EXPLAINED:
(1) The Doppler effect refers to the change in the perceived wavelength of sound or light emitted by a source that is moving toward or away from an observer, not the deflection of winds or currents.
(3) The deflection of winds and currents occurs because Earth rotates. If Earth's axis was not tilted but Earth still rotated, there would still be a Coriolis effect. Thus, planetary winds in the Northern Hemisphere are not deflected to the right due to the tilt of Earth's axis.
(4) Jet streams are relatively strong winds concentrated within a narrow band in the atmosphere, usually the upper troposphere. Jet streams are caused by a combination of Earth's rotation and strong temperature gradients that result in high wind speeds. Find the Planetary Wind and Moisture Belts in the Troposphere diagram in the *Reference Tables for Physical Setting/Earth Science*. Note that the polar front jet stream indicated by the symbol ⊗ (which blows in an easterly direction) just south of 60° N latitude. It is unlikely that such a narrow band, located so high in the troposphere and so far north, would be able to cause deflection of winds throughout the entire Northern Hemisphere.

3. **3** Find the Characteristics of Stars chart in the *Reference Tables for Physical Setting/Earth Science*, and locate *Polaris*. Note that the vertical scale along the left-hand side of the chart labeled "Luminosity (rate at which

Answers June 2016
Physical Setting/Earth Science

Answer Key

PART A

1. 1	8. 3	15. 4	22. 1	29. 4
2. 2	9. 3	16. 2	23. 2	30. 2
3. 3	10. 3	17. 3	24. 2	31. 3
4. 2	11. 1	18. 2	25. 2	32. 4
5. 1	12. 4	19. 3	26. 3	33. 4
6. 3	13. 2	20. 1	27. 2	34. 1
7. 4	14. 1	21. 3	28. 4	35. 4

PART B–1

36. 2	39. 3	42. 3	45. 4	48. 1
37. 1	40. 3	43. 2	46. 1	49. 3
38. 4	41. 1	44. 2	47. 4	50. 4

PART B–2 and **PART C.** *See* **Answers Explained**.

84 Explain why location A has a cooler average yearly air temperature than location B. [1]

_____

_____

85 The cross section below represents a mountain between locations C and D and the direction of prevailing winds.

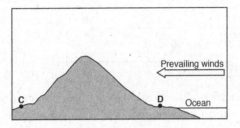

Explain why location D has a wetter climate than location C. [1]

_____

_____

82 The igneous rock gabbro is sometimes sold as "black granite." Compared to the density and composition of granite, describe how the density and composition of gabbro are different. [1]

Density of gabbro: _____

Composition of gabbro: _____

83 Identify *one* dimension stone mentioned in the passage that is composed primarily of calcite. [1]

_____

_____

Base your answers to questions 84 and 85 on the map of Australia below and on your knowledge of Earth science. Points *A* through *D* on the map represent locations on the continent.

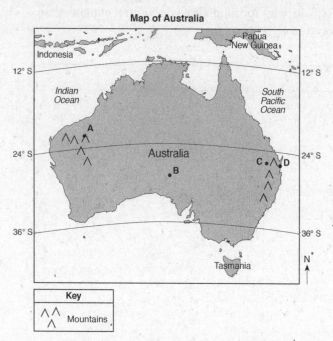

59 Identify the type of plate boundary where the earthquake occurred. [1]

_____

60 The diagram *below* represents an observer standing near the side of a building. Using the scale shown, place an **X** on the side of the building to show the maximum height of the tsunami that killed thousands of people as it engulfed towns on the northern coast of Japan. [1]

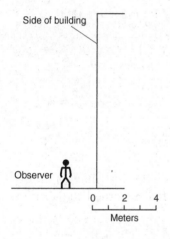

61 Identify the type of seismic wave traveling through Earth's crust that was the first to arrive at earthquake recording stations located in Japan. [1]

_____

62 Identify *one* safety precaution that residents of Maui could have undertaken in response to the tsunami warning. [1]

_____

Base your answers to questions 63 through 65 on the diagram below and on your knowledge of Earth science. The diagram represents a weather balloon as it rises from Earth's surface to 1000 meters (m). The air temperature and wet-bulb temperature values in degrees Celsius (°C) and the air pressure values in millibars (mb) are given for three altitudes.

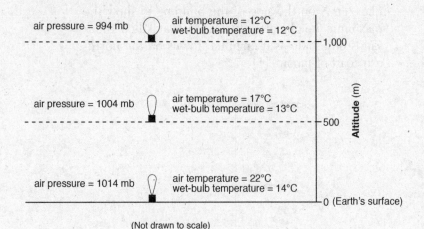

(Not drawn to scale)

63 Identify the names of the instruments carried by the weather balloon that recorded the air pressure and air temperature. [1]

Air pressure: _____

Air temperature: _____

64 Determine the dewpoint and the relative humidity of the air at Earth's surface. [1]

Dewpoint: _____ °C

Relative humidity: _____ %

65 A cloud is forming at 1000 meters. Identify the phase change that is occurring at 1000 meters to produce the cloud. [1]

_____

PART C
Answer all questions in this part.

Directions (66–85): Record your answers in the spaces provided. Some questions may require the use of the *2013 Edition Reference Tables for Physical Setting/Earth Science*.

Base your answers to questions 66 through 68 on the data table below and on your knowledge of Earth science. The data table shows five galaxies, *A* through *E*, their distances from Earth, and their recession velocities, the velocities at which they are moving away from Earth.

Galaxy Information

Galaxy	Galaxy's Distance from Earth (million light years)	Recession Velocity (km/s)
A	62	1210
B	978	15,000
C	1402	21,600
D	2510	39,300
E	3912	61,200

Note: One light year is the distance that light travels in one year.

66 State the general relationship between the galaxies' distances from Earth and their recession velocities. [1]

_____

_____

67 Another galaxy has a recession velocity of 30,000 kilometers per second. What is this galaxy's approximate distance from Earth in million light years if it follows the same pattern shown on the data table? [1]

_____ million light years

68 Identify the nuclear process that produces the energy released by stars within these galaxies. [1]

_____

_____

Base your answers to questions 69 through 71 on the information below and on your knowledge of Earth science.

A scientist found the bone of a mastodont. In the lab, the scientist found that 12.5% of the original radioactive C-14 still remained in the bone.

69 Identify the element formed when carbon-14 (^{14}C) undergoes radioactive decay. [1]

_____

70 Explain why ^{14}C was used to date the mastodont bone. [1]

_____

_____

71 Identify *one* important geologic event that occurred in New York State when mastodonts existed. [1]

_____

Base your answers to questions 72 through 75 on the diagram below and on your knowledge of Earth science. The diagram represents the Moon at four positions, A through D, in its orbit around Earth as viewed from above the North Pole (NP). The shaded parts of the Moon and Earth represent darkness.

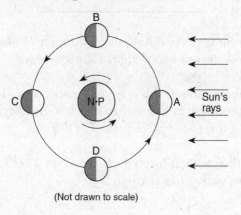

(Not drawn to scale)

72 The Moon phase shown below was seen by an observer in New York State.

On the diagram *above this question*, place an **X** on the Moon's orbit to indicate the Moon's position when this phase was observed. [1]

73 Calculate the number of days from the Moon phase at position *C* to the Moon phase at position *A* as seen from Earth. [1]

_____ d

74 Describe the effect on the heights of Earth's high and low tides when the Moon moves from position *D* to position *A*. [1]

Height of high tide: _____

Height of low tide: _____

75 Identify the celestial object in our solar system that has a period of rotation that is most similar to the period of rotation of Earth's Moon. [1]

_____

Base your answers to questions 76 through 78 on the map below and on your knowledge of Earth science. The map shows surface air temperatures for some locations in the United States on a day in November. The 20°F, 30°F, 40°F, and 70°F isotherms are shown. Points A, W, X, Y, and Z represent locations on Earth's surface. The air temperature at location A is shown.

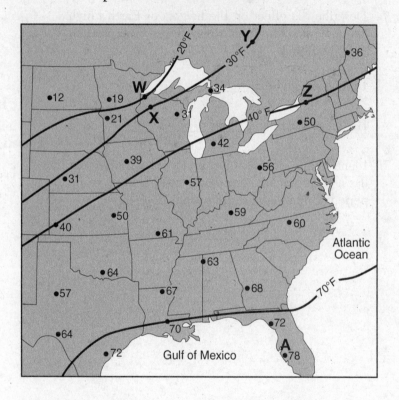

76 On the map *above*, draw both the 50°F and 60°F isotherms. Extend each isotherm to the edge of the map. [1]

77 Identify the air temperature at Watertown, New York. [1]

_____ °F

78 Describe the evidence shown on the map that indicates that the temperature gradient between locations W and X is greater than the temperature gradient between locations Y and Z. [1]

_____

_____

Base your answers to questions 79 through 82 on the passage, two diagrams, and table below and on your knowledge of Earth science. The passage describes a method used to mine gold and the diagrams represent two different views of a sluice box, which is used to separate gold from other sediments. The table shows the mineral characteristics of gold.

Gold Mining

A sluice box is used to remove gold pieces from other sediments in a stream. The box is placed in the stream to channel some of the water flow. Gold-bearing sediment is placed at the upper end of the box. The riffles in the bottom of the box are designed and positioned to create disruptions in the water flow. These disruptions cause dead zones in the current that allow the more dense gold to drop out of suspension and be deposited behind the riffles. Lighter material flows out of the box as tailings. Typically, particles of the mineral pyrite, which shares characteristics with gold, are deposited with gold particles in the sluice box. Since miners were fooled into thinking the nuggets of pyrite were gold, the name "fool's gold" is often applied to pyrite.

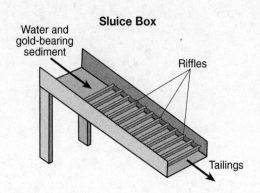

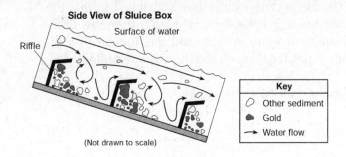

Mineral Characteristics of Gold						
Luster	Hardness	Dominant Form of Breakage	Color	Streak	Density g/cm³	Chemical Symbol
metallic	2.5 to 3	fracture	golden yellow	golden yellow	19.3	Au

79 Identify the characteristic of gold shown in the table that allows gold to be deposited behind the riffles, while other material flows out of the sluice box as tailings. [1]

_____

80 The velocity of the water leaving the sluice box was 90 centimeters per second (cm/s). State the diameter of the largest particle that could be found in the tailings. [1]

_____ **cm**

81 The angle of the sluice box is changed so that the box has a steeper slope. Describe the most likely change in water velocity and the amount of sediment passing through the sluice box as tailings. [1]

Water velocity: _____

Amount of sediment: _____

82 A gold nugget with a volume of 0.8 cubic centimeter (cm^3) was found in the sluice box. Calculate the mass of this gold nugget. [1]

_____ g

Base your answers to questions 83 through 85 on the data table below and on your knowledge of Earth science. The data table shows the average percentage of insolation from 2006 to 2012 that was reflected during the summer months by the ice sheet that covers a large portion of Greenland.

Data Table

Year	Average Insolation Reflected During the Summer (%)
2006	74.3
2007	72.8
2008	72.9
2009	71.8
2010	70.3
2011	70.1
2012	68.3

83 On the grid *below*, construct a line graph by plotting the average insolation reflected during the summer by the Greenland ice sheet from 2006 to 2012. Connect *all seven* plots with a line. [1]

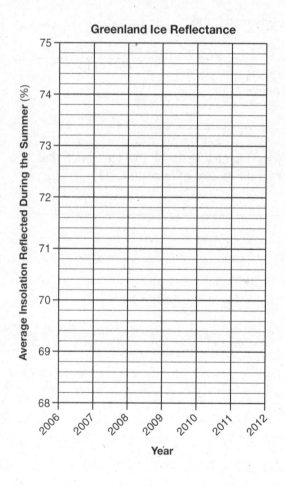

84 Describe the general trend for the average insolation reflected by the Greenland ice sheet from 2006 to 2012 and state what can be inferred about the change in size of the Greenland ice sheet during this time period. [1]

Insolation-reflected trend: _____

Inferred change in size: _____

85 Describe *one* characteristic of the ice sheet that makes it a good reflector of insolation. [1]

_____

Answers
August 2016
Physical Setting/Earth Science

Answer Key

PART A

1. 4	8. 4	15. 4	22. 3	29. 3
2. 2	9. 1	16. 2	23. 1	30. 3
3. 1	10. 4	17. 2	24. 2	31. 2
4. 2	13. 1	18. 3	25. 4	32. 4
5. 3	12. 1	19. 3	26. 2	33. 1
6. 2	13. 3	20. 1	27. 1	34. 4
7. 3	14. 4	21. 1	28. 2	35. 1

PART B–1

36. 2	39. 2	42. 3	45. 1	48. 3
37. 3	40. 1	43. 2	46. 3	49. 3
38. 1	41. 4	44. 4	47. 4	50. 4

PART B–2 and **PART C**. *See* **Answers Explained**.

Answers Explained

PART A

1. **4** Find the Characteristics of Stars chart in the *Reference Tables for Physical Setting/Earth Science*. Locate the circles labeled "Sun" and "Betelgeuse" on the chart. From the Sun, trace horizontally left to the axis labeled "Luminosity (Rate at which a star emits energy relative to the Sun)." Note that the Sun has a luminosity of 1. Trace vertically downward to the axis labeled "Surface Temperature (K)." Note that the surface temperature of the Sun is between 5,000 and 6,000 K. Repeat these steps for *Betelgeuse*. Note that the luminosity of *Betelgeuse* is roughly 100,000 and the surface temperature of *Betelgeuse* is between 3,000 and 4,000 K. Thus, compared to the Sun, *Betelgeuse* is more luminous and cooler.

2. **2** Observations indicate that the universe is expanding. If the motion of galaxies is traced back in time, there is a point about 13–14 billion years ago at which the galaxies were very close to each other. Thus, we have a model in which the universe started out with all its matter and energy in a very small volume and then expanded outward in all directions. This motion is similar to an explosion, hence the name for this model—the Big Bang Theory. According to this theory, the universe was initially smaller, denser, much hotter, and filled with a uniform glowing fog of hydrogen plasma. As the universe exploded outward in all directions from its origin point, it formed a sphere. Both the plasma and the radiation filling it spread out and grew cooler. Since an observer on Earth would be looking at the radiation from within the sphere of the expanding universe, that radiation should be coming at us from all directions, even from what may appear to be "empty space." Such radiation, observed coming from all directions in the universe, was detected in the mid-1960s by Arno Penzias and Robert Wilson at Bell Laboratories in New Jersey. They named this "cosmic background radiation." Thus, the existence of cosmic background radiation in space best supports the theory that the universe began with an explosion called the Big Bang.

WRONG CHOICES EXPLAINED:
(1) During the Big Bang, matter was thrown out uniformly in all directions. Thereafter, gravity caused matter to coalesce into clumps—galaxies. Within the galaxies, matter further coalesced to form disk-shaped protostars. Over a period of millions to tens of millions of years, protostars underwent further contraction and accretion that lead to the formation of a solar system containing planets. Furthermore, there is evidence that new stars and solar systems are currently forming in clouds of dust and gas formed by the explo-

sions of stars that occurred after the origin of the universe. Thus although the existence of planets is an expected outcome of the origin of the universe, it provides no evidence about either the origin or age of the universe.

(3) Star constellations are imaginary patterns of a group of stars as viewed from Earth. These patterns supposedly resemble the form of an animal, person, or thing. Since these patterns are imaginary, they provide no evidence about either the origin or age of the universe.

(4) The similarity of the composition of Earth and the Moon is a consequence of the formation of our solar system, which occurred long after the formation of the universe. Thus, the composition of Earth and the Moon provides no evidence about either the origin or age of the universe.

3. **1** Planetary winds tend to blow in a straight line from regions of high pressure toward regions of low pressure. As these winds are blowing, Earth is turning on its axis. This rotation causes the winds to appear to be turning toward the right in the Northern Hemisphere and toward the left in the Southern Hemisphere. This phenomenon is called the Coriolis effect. Thus, planetary winds generally curve to the right in Earth's Northern Hemisphere due to the Coriolis effect.

WRONG CHOICES EXPLAINED:

(2) The Doppler effect refers to the change in the perceived wavelength of sound or light emitted by a source that is moving toward or away from an observer, not the deflection of winds or currents.

(3) The deflection of winds and currents occurs because Earth rotates. If Earth's axis was not tilted but Earth still rotated, there would still be a Coriolis effect. Thus, planetary winds in the Northern Hemisphere are not deflected to the right due to the tilt of Earth's axis.

(4) The deflection of winds and currents occurs because Earth rotates. If Earth's gravity changed but Earth still rotated, there would still be a Coriolis effect. If Earth had no gravity, there would be no convection and thus no wind that could experience the Coriolis effect.

4. **2** It is given that 334 Joules of heat are released, not absorbed, by the process. Find the Properties of Water table in the *Reference Tables for Physical Setting/Earth Science*. Note that the heat energy released during freezing is 334 J/g. Therefore, the process that releases 334 joules of heat energy for each gram of water is freezing.

5. **3** The roots and stems of vegetation on a hillslope slow the rate at which water runs downhill, allowing more water to infiltrate so that there is less runoff. If the vegetation on a hillslope is removed, water can run downhill unimpeded and there is less time for the water to infiltrate. Therefore, less water will infiltrate and there will be more runoff. Thus as a result of the

removal of vegetation on a hillslope, infiltration will be less and runoff will be greater.

6. **2** When oceans are heated by solar radiation, the heated water warms the air over the oceans. The heating of the ocean water also causes an increase in evaporation rate. This increase in both the temperature and humidity of the air causes a decrease in air pressure. Decreasing air pressure further increases the evaporation of water from the ocean. This adds even more humidity to the air, which decreases the air pressure even further. The high temperature and moisture levels result in air with very low density. As this very low density air rises rapidly and cooler, denser air rushes in to replace it. Additionally, numerous convection cells and thunderstorms form. With very little wind shear, widespread thunderstorm activity can merge into a large updraft that is part of a huge convection cell. The large updraft further decreases the air pressure. Latent heat released when the moisture in the air condenses warms the air and also decreases the air pressure. These processes create a region of very low pressure. Winds begin to blow toward the low-pressure center as air moves in from surrounding areas of higher pressure. The Coriolis effect deflects the winds, and a cyclone forms. Heat is released when fresh moisture brought in by winds condenses, feeding the convection cell with new energy. As a result, the convection cell gets larger and stronger. When the winds in the cyclone exceed 119 km/hr, the phenomenon is called a hurricane. Thus, the ocean surface conditions responsible for the development of hurricanes are warm water temperatures and high evaporation rates.

7. **3** Earth's planetary wind and moisture belts are the result of the huge convection cells that form in the atmosphere because of unequal heating of Earth's surface. Since Earth is a sphere, insolation does not strike all points on Earth's surface at the same angle. The more direct the Sun's rays are, the more concentrated the insolation is and the greater the heating effect is. Thus, the location at which the Sun's vertical rays strike Earth's surface experiences the greatest heating effect. The location at which the Sun's vertical rays strike Earth's surface varies in an annual cycle with the seasons. For example, on June 21, Earth's Northern Hemisphere is tilted farthest toward the Sun and the Sun's vertical rays strike Earth's surface at 23.5° N latitude. At the equinoxes in March and September, Earth is tilted neither toward nor away from the Sun and the Sun's vertical rays strike Earth's surface at the equator. On December 21, Earth's Southern Hemisphere is tilted farthest toward the Sun and the Sun's vertical rays strike Earth's surface at 23.5° S latitude. These seasonal shifts in the location where the Sun's vertical rays strike Earth's surface cause seasonal shifts in the location of the centers of rising air that set the huge convection cells in Earth's atmosphere in motion. Thus, the seasonal shifts in Earth's planetary wind and moisture belts are due to changes in the latitude that receives the Sun's vertical rays.

WRONG CHOICES EXPLAINED:
(1) The seasonal shifts of Earth's planetary wind and moisture belts are shifts in latitude. Changes in distance between Earth and the Sun would not affect the latitude at which the Sun's vertical rays strike Earth's surface and thus cannot account for shifts in the positions of Earth's planetary wind and moisture belts.
(2) The amount of energy given off by the Sun changes with solar activity, such as solar flares or sunspots. Solar activity varies in cycles like the 11-year sunspot cycle and some longer cycles. The amount of energy given off by the Sun does not change on a seasonal basis and thus cannot account for shifts in the positions of Earth's planetary wind and moisture belts.
(4) The rate of Earth's rotation on its axis is changing—it is very slowly decreasing (slowing from once every 21 hours to once every 24 hours over the last 600 million years). Thus, Earth's rate of rotation does not change on a seasonal basis and cannot account for seasonal shifts in the positions of Earth's planetary wind and moisture belts.

8. **4** Clouds form when water vapor in the atmosphere changes back into tiny droplets of liquid water by the process of condensation or forms ice crystals by the process of deposition (sublimation). Condensation occurs when the air temperature cools to the dewpoint temperature. Deposition can occur if the dewpoint is below the freezing point of water. Processes in the atmosphere that may cause air to cool to the dewpoint include adiabatic cooling due to the expansion of air as it rises and cooling as air rises to higher altitudes where temperatures are cooler. Thus, the condition most likely to result in the formation of a cloud is warm, moist air rising.

WRONG CHOICES EXPLAINED:
(1) The rising of warm, moist air that results in cloud formation would result in an increase in wind speed as surrounding air moved in to replace the rising air. Thus, a decrease in wind speed would indicate that nearby air was not rising and would therefore not be likely to result in cloud formation.
(2) An increase in air pressure would be associated with cooler, drier air. If air was becoming cooler and drier, it would be less likely to result in cloud formation.
(3) As air sinks in the atmosphere, it is compressed and increases in temperature adiabatically. The increase in air temperature would favor an increase in evaporation and a decrease in the condensation that results in cloud formation. Thus, cool, moist air sinking is unlikely to result in the formation of a cloud.

9. **1** Since Earth is a sphere, insolation does not strike all points on the planet's surface at the same angle. Near the equator, insolation always strikes Earth's surface more directly than it does at the poles. Therefore, throughout the year, the insolation reaching the tropics is more concentrated (i.e., more

intense) than that reaching mid-latitude or polar regions. The more intense the insolation is, the more Earth's surface is warmed by the insolation. Thus, distance from the equator is typically associated with cooler temperatures.

Find the Selected Properties of Earth's Atmosphere graph in the *Reference Tables for Physical Setting/Earth Science*, and locate the "Temperature Zones" section. Note that as altitude increases in the troposphere, temperature decreases. Thus, an increase in elevation above sea level is associated with cooler temperatures. Therefore, the climate condition that generally results from both an increase in distance from the equator and an increase in elevation above sea level is cooler temperatures.

WRONG CHOICES EXPLAINED:

(2) As explained above, an increase in distance from the equator results in cooler temperatures. Cooler temperatures would result in cooler prevailing winds, not warmer prevailing winds.

(3) Find the Planetary Wind and Moisture Belts in the Troposphere diagram in the *Reference Tables for Physical Setting/Earth Science*. Note the wet zone at the equator and the dry zone at 30° N and 30° S. Thus, across more than half of Earth's surface, precipitation decreases with distance from the equator. Find the Selected Properties of Earth's Atmosphere graph in the *Reference Tables for Physical Setting/Earth Science*, and locate the section labeled "Water Vapor." Note that the concentration of water vapor decreases with altitude above sea level. The less moisture the air contains, the less likely that precipitation will occur. Thus, an increase in distance from the equator and an increase in elevation above sea level will typically result in decreased precipitation, not increased precipitation.

(4) Note that the question requires the climate condition to result from *both* distance from the equator and increase in elevation above sea level. An increase in distance from the equator will result in cooler temperatures and increased air pressure. Find the Selected Properties of Earth's Atmosphere graph in the *Reference Tables for Physical Setting/Earth Science*, and locate the "Atmospheric Pressure" section. Note that as altitude increases in the troposphere, pressure decreases. In other words, an increase in elevation above sea level results in decreased air pressure. Thus although distance from the equator results in increased air pressure, an increase in elevation above sea level results in decreased air pressure.

10. **4** Every object not at absolute zero radiates energy. The type of energy radiated depends on the temperature of the object. The Sun, which is at a high temperature, radiates energy of relatively short wavelengths such as ultraviolet and visible light rays. Earth's surface, which is at a much lower temperature, radiates longer wavelength infrared (heat) rays. Thus, most of the electromagnetic radiation given off by Earth's surface at night is infrared rays.

11. **1** Large volcanic eruptions spew ash and dust into the atmosphere, which are then carried around the globe by winds in the upper atmosphere. Ash and dust are tiny, solid particles of igneous rock. They block sunlight, thereby decreasing atmospheric transparency. When atmospheric transparency decreases, the amount of solar energy that reaches Earth's surface also decreases. Because it receives less solar energy, Earth's surface cools. As a result, surface air temperatures drop. Thus, after these eruptions, global temperatures are often cooler than normal because the atmosphere is less transparent.

12. **1** When sunlight strikes Earth's surface, some of it is reflected back toward space. However, some of that energy is absorbed and then reradiated as infrared radiation (heat). Over time, the amount of energy received from the Sun should be about the same as the amount of energy radiated back into space, leaving the temperature of Earth's surface roughly constant. However, many chemical compounds found in Earth's atmosphere act as greenhouse gases. These gases allow sunlight to enter the atmosphere and strike Earth's surface. However, the greenhouse gases absorb the infrared radiation reradiated by Earth's surface, trapping the heat in the atmosphere and resulting in global warming. Some of these greenhouse gases occur in nature (water vapor, carbon dioxide, methane, ozone, and nitrous oxide). Others are exclusively human-made (like the chlorofluorocarbons used in aerosol sprays). Thus, only list 1—carbon dioxide, methane, and water vapor—contains three major greenhouse gases found in Earth's atmosphere.

13. **3** Find the Generalized Bedrock Geology of New York State map in the *Reference Tables for Physical Setting/Earth Science*. In the key labeled "Geologic Periods and Eras in New York," note the statement "weakly consolidated to unconsolidated gravels, sands, and clays." Recall that gravels, sands, and clays are sediments. From this statement, trace horizontally to the left and note that it corresponds to "Cretaceous and Pleistocene (Epoch)." Now find the Geologic History of New York State chart in the *Reference Tables for Physical Setting/Earth Science*. Locate the column labeled "New York Rock Record," and note the symbol for sediment. Trace left from the two line segments corresponding to this symbol to the column labeled "Epoch." Note that they correspond to the Pleistocene and Late Cretaceous epochs. Thus, the New York State rock record consists of weakly consolidated to unconsolidated sediments during the Late Cretaceous.

14. **4** Find the Geologic History of New York State chart in the *Reference Tables for Physical Setting/Earth Science*. Locate *Valcouroceras* in the illustrations of index fossils printed along the bottom of the chart. Note that it is labeled with the circled letter *D*. Now locate the column labeled "Time Distribution of Fossils." Note the bold gray lines labeled with names and circled letters. The names are of types of fossil organisms. The circled letters are keyed to the illustrations of specific important fossils printed along the bottom of the chart. Note that the circled letter *D* lies on the line labeled

"Nautiloids." Thus, the New York State index fossil *Valcouroceras* is classified as a nautiloid.

15. **4** Find the Inferred Properties of Earth's Interior in the *Reference Tables for Physical Setting/Earth Science*. Find the region labeled "Asthenosphere (Plastic Mantle)," and locate the arrows beneath the Mid-Atlantic Ridge. Note that the arrows show material in the asthenosphere rising toward the top of the asthenosphere and spreading out to either side. Now look at the arrows beneath the Cascades. Note that they show material sinking downward to the bottom of the asthenosphere and moving back toward the Mid-Atlantic Ridge. Trace the dashed lines representing the top and bottom of the asthenosphere down to the lower graph labeled "Temperature (°C)." Note that temperatures are hotter at the bottom of the asthenosphere than they are at the top. When taken together, the arrows indicate a circular pattern of motion in which hot material at the bottom of the asthenosphere rises, spreads sideways, cools, and sinks back to the bottom of the asthenosphere, a process known as convection. Thus, the convection currents responsible for moving tectonic plates occur in the asthenosphere.

Florida

16. **2** Note the position of Florida in North America shown in the map. Find the Geologic History of New York State chart in the *Reference Tables for Physical Setting/Earth Science*. Locate the column labeled "Inferred Positions of Earth's Landmasses." Locate the diagram in which the region of North America corresponding to Florida is located at the equator. Beneath that diagram, note that the caption says "232 million years ago." From the tip of the gray arrow at the left edge of this diagram, trace horizontally left to the column labeled "Life on Earth." Note that 232 million years ago corresponds to the statement "earliest dinosaurs." Thus when Florida was located at the equator, the earliest dinosaurs appeared on Earth.

17. **2** The youngest crustal bedrock is formed when molten rock emerges from volcanoes and in the rift zones of the mid-ocean ridges and then hardens into bedrock. In fact, crustal bedrock is being formed at this very moment in volcanoes and mid-ocean ridges worldwide. Note that a volcano is not listed among the choices. As the tectonic plates diverge along a mid-ocean ridge, the bedrock fractures and new molten rock emerges, pushing

aside the bedrock. As this process is repeated, new bedrock is formed and the bedrock on either side of the mid-ocean ridges is pushed sideways. Thus, the most recently formed bedrock is found nearest a mid-ocean ridge and the oldest bedrock is found on either side farthest from the ridge. Therefore, the geologic feature composed of the youngest crustal bedrock is the Mid-Atlantic Ridge.

WRONG CHOICES EXPLAINED:
(1) Find the Tectonic Plates map in the *Reference Tables for Physical Setting/Earth Science*. Locate the Peru-Chile Trench along the west coast of South America. Note the symbol corresponding to the Peru-Chile Trench, and locate this symbol in the key at the bottom of the map. Note that this symbol represents a "Convergent Plate Boundary (subduction zone)" where one plate is being subducted beneath an overriding plate. Note that the arrows representing relative motion at the plate boundary show that the Nazca Plate is moving away from the East Pacific Ridge and toward the Peru-Chile Trench, where the Nazca Plate is subducted beneath the South American Plate. As explained above, the most recently formed bedrock is found nearest a mid-ocean ridge and the oldest bedrock is found on either side farthest from the ridge. Thus, the crustal bedrock being subducted at trenches, such as the Peru-Chile Trench, is older than the crustal bedrock at the mid-ocean ridges, such as the East Pacific Ridge or the Mid-Atlantic Ridge.

(3) Find the Generalized Landscape Regions of New York State map in the *Reference Tables for Physical Setting/Earth Science*. Locate the landscape region labeled "Adirondack Mountains." Now find the Generalized Bedrock Geology of New York State map in the *Reference Tables for Physical Setting/Earth Science,* and locate the region corresponding to the Adirondack Mountains. Note the symbols used to denote the rocks found in the Adirondack Mountains. Locate these symbols in the key labeled "Geologic Periods and Eras in New York." Note that most of the crustal bedrock in the Adirondack Mountains is Middle Proterozoic in age. Trace right and note the label "Intensely metamorphosed rocks (regional metamorphism about 1,000 m.y.a.)." Thus, the age of the crustal bedrock of the Adirondack Mountains is about 1,000 million years—much older than the crustal bedrock forming right now in the Mid-Atlantic Ridge.

(4) Find the Tectonic Plates map in the *Reference Tables for Physical Setting/Earth Science*. Locate the San Andreas Fault. Note that it lies wholly within the landmass of North America on the North American Plate. Find the Geologic History of New York State chart in the *Reference Tables for Physical Setting/Earth Science*. Locate the column labeled "Inferred Positions of Earth's Landmasses." In the diagrams, note that the North America landmass is shown in black and has existed for hundreds of millions of years. Thus, it is unlikely that the crustal bedrock at the San Andreas Fault is composed of the youngest crustal bedrock.

18. **3** Find the Earthquake *P*-Wave and *S*-Wave Travel Time graph in the *Earth Science Reference Tables*. Locate the point corresponding to 3200 kilometers along the horizontal axis at the bottom of the graph labeled "Epicenter Distance (× 10³ km)." From 3200 km, trace vertically until you intersect the bold line labeled *S*. From this intersection, trace horizontally to the left to the vertical axis labeled "Travel Time (min)." Read the *S*-wave travel time—about 10 minutes 40 seconds. Thus, the time needed for an *S*-wave to travel the 3200 km from the epicenter to the seismic station is 10 min 40 sec.

19. **3** If the climate becomes more humid, this landscape feature would experience greater rainfall. Water is important to many weathering processes. Many chemical weathering processes require water. Water is a reactant in hydration and carbonation, and it provides a medium in which acid reactions can occur. Humid climates also support increased biological activity ranging from plant action to production of humic acid as plant matter decomposes. Therefore, an increase in the amount of moisture present in the environment would result in more pronounced weathering. An increase in rainfall would also result in an increase in erosion. Abundant rainfall also promotes a protective layer of vegetation that protects soil from rapid runoff and erosion, leading to rounded slopes. Thus if the climate became more humid, this landscape feature would have more weathering and erosion, producing a more rounded landscape feature.

20. **1** It is given that this igneous rock contains only plagioclase feldspar, biotite, and amphibole. Find the Scheme for Igneous Rock Identification in the *Reference Tables for Physical Setting/Earth Science*. Locate the graph labeled "Mineral Composition (relative by volume)." Locate the section in which a vertical line drawn through the graph intersects only plagioclase feldspar, biotite, and amphibole. It is given that the mineral crystals in the rock range in size from 2 to 6 millimeters. Now find the column labeled "Crystal Size," and locate the row labeled "1 mm to 10 mm." Trace horizontally left in this row and vertically upward from the vertical line in the Mineral

Composition graph corresponding to rock consisting only of plagioclase feldspar, biotite, and amphibole until the two intersect. Note the igneous rock name at this intersection—diorite.

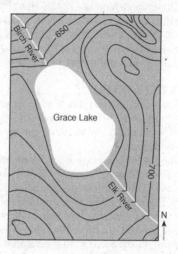

21. **1** Note the contour line labeled "700" that crosses the Elk River and the contour line labeled "650" that crosses Birch River. Water flows downhill. Therefore, it is logical to infer that the water in this lake and river system flows from the Elk River into Grace Lake and out of Grace Lake into the Birch River. Thus, the water in the Elk and Birch Rivers are flowing in the same general compass direction.

It is given that both rivers carry an equal volume of water. Therefore, volume is not a factor determining the stream velocity. However, water flows downhill more rapidly on steep slopes than it does on gentle slopes. On a topographic map, contour lines connect points of equal elevation. The distance between one contour line and the next indicates the distance over which the elevation changes. The farther apart the contour lines are on a map, the greater the distance one must travel in order to change elevation, that is, the gentler the slope is. Conversely, the closer together the contour lines are on a map, the shorter the distance one must travel in order to change elevation, that is, the steeper the slope is. Note that the contour lines are more closely spaced along the Birch River than they are along the Elk River. Thus, the slope down which the Birch River flows is steeper than the slope down which the Elk River flows. Therefore, equal volumes of water will flow faster down the Birch River than the Elk River. Thus, compared to the Elk River, the Birch River can best be described as flowing faster, and in the same general compass direction.

22. **3** Find the Properties of Common Minerals table in the *Reference Tables for Physical Setting/Earth Science.* In the column labeled "Mineral

Name," locate garnet. Trace left from garnet to the column labeled "Distinguishing Characteristics," and note the statement "often seen as red glassy grains in NYS metamorphic rocks." Find the Scheme for Metamorphic Rock Identification in the *Reference Tables for Physical Setting/Earth Science*. In the column labeled "Composition," locate the line segment labeled "garnet." From garnet, trace right to column labeled "Rock Name," and note that phyllite, schist, and gneiss contain garnet. Thus, of the choices given, the best source of the mineral garnet would be schist.

WRONG CHOICES EXPLAINED:
(1) Find the Scheme for Igneous Rock Identification in the *Reference Tables for Physical Setting/Earth Science*. Locate the graph labeled "Mineral Composition (relative by volume)." Note that igneous rocks, such as basalt, do not contain the mineral garnet.

(2) Find the Scheme for Sedimentary Rock Identification in the *Reference Tables for Physical Setting/Earth Science*. In the column labeled "Rock Name," locate limestone. From limestone, trace left to the column labeled "Composition." Note that limestone is composed of calcite, not garnet. Therefore, limestone would not be the best source of garnet.

(4) Find the Scheme for Metamorphic Rock Identification in the *Reference Tables for Physical Setting/Earth Science*. In the column labeled "Rock Name," locate slate. From slate, trace left to the column labeled "Composition." Note that slate is composed of mica and does not contain garnet.

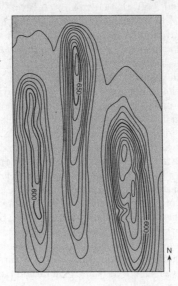

23. **1** At the melting edge of a glacier, sediments within the ice as well as those carried on top of the ice are released and drop to the ground. Particles

of mixed shapes and sizes drop to the ground in a confused jumble, forming an unsorted deposit called glacial till. Drumlins are formed when the ice of a glacier slides over such previously deposited piles of sediment, causing them to become elongated into a teardrop shape. The narrow end points in the direction that the glacier advanced. Thus, the shape of the drumlins in the topographic map indicates that the ice advanced from north to south. Thus, the direction of the advancing ice movement was from north to south and the most likely arrangement of the sediments in the drumlins is unsorted.

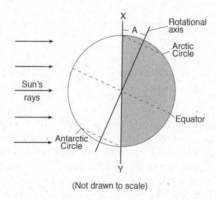

24. **2** By definition, Earth's rotational axis passes through Earth's North Pole, Earth's center, and Earth's South Pole. Earth's North Pole is located at 90° N latitude, and Earth's South Pole is located at 90° S latitude. Note that line XY passes through the Arctic Circle, Earth's center, and the Antarctic Circle. Find the Surface Ocean Currents map in the *Reference Tables for Physical Setting/Earth Science*. Locate the vertical axis along the right side of the map. Note that Earth's Arctic Circle is located at 66.5° N latitude and that Earth's Antarctic Circle is located at 66.5° S latitude. Therefore, letter A represents the angular distance measured from Earth's center between the North Pole at 90° N latitude and the Arctic Circle at 66.5° N latitude. Thus, the value represented by letter A is 23.5° (90° − 66.5° = 23.5°).

25. **4** Find the Scheme for Sedimentary Rock Identification in the *Reference Tables for Physical Setting/Earth Science*. In the column labeled "Grain Size," locate the entry "0.006 to 0.2 cm." Note that this grain size corresponds to sand. Trace the row for sand right to the column labeled "Rock Name," and note that a sedimentary rock composed of sand-sized grains is called sandstone. Continue tracing right to the column labeled "Map Symbol," and note the symbol for sandstone. Thus, the rock name and symbol used to represent a sedimentary rock that has a grain size of 0.006 to 0.2 centimeters is best shown in choice (4).

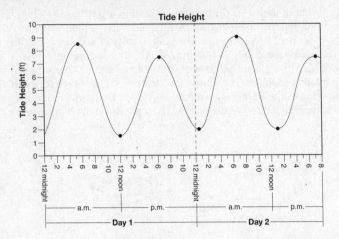

26. **2** Locate the first low tide on the graph. Trace vertically downward to the horizontal axis, and note that the first low tide occurred on Day 1 at 12:00 p.m. (noon). Note that the next low tide on the graph occurred on Day 2 at 12:30 a.m.—12 hours 30 minutes later. Note that the third low tide on the graph occurred at 1:00 p.m. on Day 2—12 hours 30 minutes after the second low tide. Therefore, the pattern is that a low tide occurs every 12 hours 30 minutes. Thus, if the pattern shown continues, the next low tide will occur 12 hours 30 minutes after 1:00 p.m. on Day 2, which is 1:30 a.m. on Day 3.

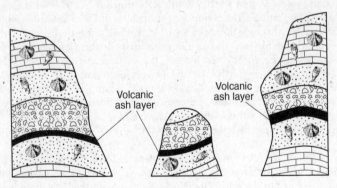

27. **1** Volcanic ash is ejected high into the atmosphere during a volcanic eruption. Once in the atmosphere, the ash is carried by global winds, quickly spreads out, and then settles to the ground, forming a volcanic ash layer over a wide area. After the volcanic eruption is over, the source of volcanic ash is cut off and deposition quickly tapers off. Thus, a volcanic ash layer is consid-

core. Find the Inferred Properties of Earth's Interior chart in the *Reference Tables for Physical Setting/Earth Science*. Note that the density of Earth's iron/nickel inner and outer cores ranges from 9.9–13.1 g/cm^3. Note, too, that Earth's rocky crust and mantle ranges in density from 2.7–5.6 g/cm^3. Thus, rock is less dense than iron/nickel and the density of both Earth and Mars increases toward their centers. Therefore, it is reasonable to infer that the interiors of terrestrial planets increase in density toward their center.

Now, note in the diagram of Jupiter that Jupiter's core is rock. Recall that rock ranges in density from 2.7–5.6 g/cm^3. Find the Solar System Data table in the *Reference Tables for Physical Setting/Earth Science*. In the column labeled "Celestial Object," locate Jupiter. From Jupiter, trace right to the column labeled "Density (g/cm^3)," and note that Jupiter has a density of 1.3 g/cm^3. If Jupiter's rock core has a density from 2.7–5.6 g/cm^3 and the entire planet has a density of 1.3 g/cm^3, then Jupiter's other layers must have densities much less than 2.7 g/cm^3 to average out to 1.3 g/cm^3. Thus, it is reasonable to infer that the interiors of Jovian planets also increase in density toward their centers. Therefore, the inference that best describes the interiors of the planets in our solar system is that both terrestrial and Jovian planets have layered interiors, with density increasing toward the center.

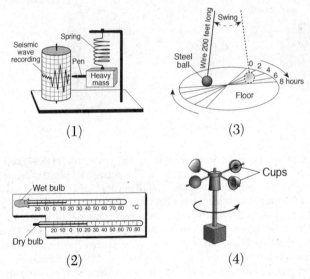

29. **3** In 1851, the French physicist Jean Foucault suspended a heavy iron ball on a long steel wire from the top of the dome of the Pantheon in Paris. As this pendulum swung back and forth, it appeared to change direction slowly, passing over different lines on the floor until eventually coming full circle to its original position after 24 hours. Since Foucault knew that the path of a freely swinging pendulum would not change on its own, he con-

ered a good time marker for correlating rocks because it was deposited over a large area in a short period of time when a volcano erupted.

WRONG CHOICES EXPLAINED:

(2) Fossils are generally helpful in determining the relative ages of rock layers. Thus, in and of itself, the lack of fossils in a volcanic ash layer would not explain why the ash layer was useful for correlating relative ages of the bedrock in the three outcrops.

(3) If the eruptions that produced the ash layer occurred over a long period of geologic time, the volcanic ash would not be associated with a specific time period. Therefore, the volcanic ash would therefore make a poor time marker for correlating the relative ages of bedrock.

(4) According to the cross sections, the rock symbols and fossils in the layers above and below the volcanic ash layer in each of the three outcrops are the same. Thus, the volcanic ash is found between two different layers of bedrock, not many different layers.

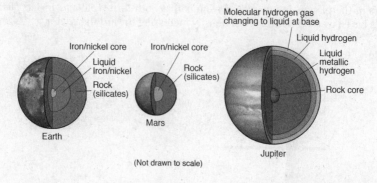

28. **2** The planets can be divided by mass and density into the terrestrial (Earthlike) and Jovian (Jupiter-like) planets. The terrestrial planets are small, dense, and rocky; composed mostly of metal silicates and iron; and include the four innermost planets Mercury, Venus, Earth, and Mars. The Jovian planets are large, low in density, gaseous, composed mainly of hydrogen and helium, and include the outer planets Jupiter, Saturn, Neptune, and Uranus. During the formation of the solar system, gravity caused Earth and the other planets to become layered according to the density of the materials of which they were composed. Gravity caused the denser materials to sink toward the planets' centers and the less dense materials to float toward the planets' surfaces. This separation of materials of different densities caused the planets to develop a layered internal structure. According to the diagrams, Earth, Mars, and Jupiter have a layered internal structure. Thus, it can be inferred that both terrestrial and Jovian planets have layered interiors. According to the diagrams, Earth and Mars have an iron/nickel core while Jupiter has a rock

cluded that the apparent shift in the direction of swing of the pendulum was due to the floor (Earth's surface) rotating beneath the pendulum. Thus, the device that provides evidence that Earth rotates on its axis is the pendulum shown in diagram 3.

WRONG CHOICES EXPLAINED:
(1) The device in diagram 1 is a seismograph. Seismographs detect motions of bedrock caused by seismic waves, not Earth's rotation.
(2) The device in diagram 2 is a psychrometer. A psychrometer is used to measure the dry-bulb and wet-bulb temperatures of the air so that the relative humidity and dewpoint may be determined using an appropriate chart.
(4) The device in diagram 4 is an anemometer. An anemometer has cups mounted on a shaft that cause the shaft to spin when wind blows against them. The faster the wind blows, the faster the cups spin the shaft and the higher the number that registers on the anemometer's scale. Thus, an anemometer is used to measure wind speed.

30. **3** By definition, a tornado is a violently rotating column of air in contact with both a cumulonimbus cloud base and the ground. A tornado is often (but not always) visible as a funnel cloud of condensation. Find the Key to Weather Map Symbols in the *Reference Tables for Physical Setting/Earth Science,* and locate the symbol for a tornado. Note that the symbol for a tornado corresponds to the symbol shown in choice (3).

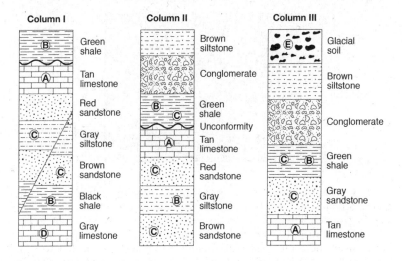

31. **2** It is given that the rock layers at these locations have not been overturned. The principle of superposition states that the bottom layer of a sedimentary series is the oldest, unless it was overturned or had older rock thrust

over it, because the bottom layer was deposited first. Similarly, in a sequence of rock layers, a rock layer is older than those above it and younger than those below it. Thus, it is reasonable to infer that in each of the cross sections, the youngest rock layer is the top layer because the top layer was deposited last. Note that in both column I and column II, the rock layer above the unconformity is a green shale containing fossil *B* and the rock layer below the unconformity is a tan limestone containing fossil *A*. Thus, we can correlate the green shale containing fossil *B* in columns I and II. Note that the green shale with fossil *B* is the uppermost, or youngest, layer in column I. However, note that the green shale containing fossil *B* in column II is overlaid by two more layers that were deposited on top of the green shale. Thus, the green shale is older than the conglomerate directly above it and the conglomerate is older than the brown siltstone directly above it. Therefore, the uppermost layer in column II, the brown siltstone, is the youngest. Now note the green shale containing fossils *B* and *C* in both columns II and III. In both columns, this green shale is overlaid by a conglomerate, which in turn is overlain by a brown siltstone. Thus, it is reasonable to correlate these three layers and infer that the brown siltstone in column III is the same age as the brown siltstone in column II. However, note that the brown siltstone in column III is overlaid by a layer of glacial soil containing fossil *E*. Since the glacial soil is on top of the brown siltstone, it can be inferred that the glacial soil is younger than the brown siltstone. Thus, the rock layer that is the youngest is the glacial soil containing fossil *E* in column III.

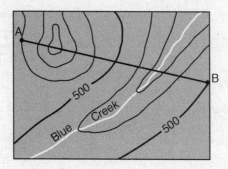

32. **4** Note the two contour lines labeled "500" on either side of Blue Creek. Streams flow in channels that are lower in elevation than the surrounding slopes. Thus, elevation decreases from the 500-foot contour lines toward Blue Creek. It is given that the contour interval is 20 feet. Thus, the elevations of the contour lines from point *B* to point *A* are as follows: Point *B* at 500 feet, 480 feet, 460 feet, Blue Creek, 460 feet, 480 feet, 500 feet, 520 feet, 540 feet, 560 feet, 560 feet, 540 feet, Point *A* at 520 feet. Therefore, the profile that best represents the topographic cross section along the line

from A to B will show point A on the profile at an elevation of 520 feet, will then slope upward to a hilltop more than 560 but less than 580 feet in elevation, then slope downward to less than 460 feet but more than 440 feet at Blue Creek, and then slope upward to 500 feet at point B, as shown in profile 4.

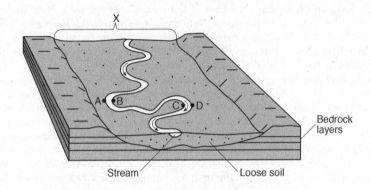

33. **1** The landscape feature labeled X is a broad, flat plain adjacent to the meandering stream. Such a landscape feature forms when a stream overflows or floods, spreads out over the valley floor where its velocity is slowed, and deposits sediments it is carrying. Therefore, such a broad, flat plain adjacent to a stream is called a flood plain.

WRONG CHOICES EXPLAINED:
(2) A sand bar is a long, narrow pile of sand deposited in open water, not a broad, flat plain adjacent to a meandering stream like feature X.
(3) A delta is a flat, fan-shaped mound of sediment deposited where a stream enters a body of standing water and slows down. It is not a broad, flat plain adjacent to a meandering stream like feature X.
(4) An escarpment is a long, clifflike ridge of land that formed by faulting or erosion and separates two relatively level areas of differing elevations. It is not a broad, flat plain adjacent to a meandering stream like feature X.

34. **4** Stream erosion is greatest where water velocity is greatest. Along a straight section of a stream's channel, velocity is greatest in the center of the stream farthest from the friction of the banks or bed. When the stream curves or meanders, velocity is greatest along the outside of the curve and lowest along the inside of the curve. Water velocity is greatest along the outside of the curve because the water is forced to cover a greater distance along the outside of the curve. Thus, erosion is most likely greatest at points D and A because they are both located at the outside of a curve in the stream where the water velocity is the greatest.

35. **1** Find the Generalized Landscape Regions of New York State map in the *Reference Tables for Physical Setting/Earth Science*. Locate "The Catskills," and note that they are part of the major geographic province labeled "Alleghany Plateau." Thus, New York State's Catskills are a plateau. Plateaus are large areas of flat land at high elevations. Plateaus generally have an underlying structure of horizontal layers of rock. Now find the Generalized Bedrock Geology of New York State map in the *Reference Tables for Physical Setting/Earth Science*. Locate the region corresponding to the Catskills, and note the map symbol for the surface bedrock. Locate this map symbol in the key labeled "Geologic Periods and Eras in New York," trace right, and note the statement "limestones, shales, sandstones, and conglomerates." Find the Scheme for Sedimentary Rock Identification in the *Reference Tables for Physical Setting/Earth Science*. Locate the column labeled "Rock Name." Note that limestone, shale, sandstone, and conglomerate are all sedimentary rocks. Thus, the chart that best describes the landscape category and the general bedrock structure, type, and composition of New York State's Catskills will indicate that they are a plateau with a horizontal underlying bedrock structure of sedimentary rock composed of limestone, shale, and sandstone as shown in chart 1.

PART B-1

36. **2** According to the reading passage, "Most asteroids follow orbits between 300 and 600 million kilometers from the Sun." Find the Solar System Data chart in the *Reference Tables for Physical Setting/Earth Science*. In the column labeled "Mean Distance from Sun (million km)," note that 300–600 million km lies between the values 227.9 and 778.4. Trace left from both 227.9 and 778.4 to the column labeled "Celestial Object." Note that 227.9 corresponds to Mars and that 778.4 corresponds to Jupiter. Thus, most asteroids are located between the planets Mars and Jupiter.

37. **3** According to the reading passage, "A meteor shower occurs, often filling the night sky with 'shooting star' trails as they burn up in the atmosphere 50 to 80 kilometers above Earth's surface." Find the Selected Properties of Earth's Atmosphere graph in the *Reference Tables for Physical Setting/Earth Science*. On the vertical axis labeled "Altitude," locate the range of 50 to 80 kilometers (note that kilometers are shown on the left side of the axis). Trace right from this range to the section labeled "Temperature Zones," and note that 50 to 80 kilometers corresponds to the mesosphere. Thus, the temperature zone of Earth's atmosphere in which most meteors burn up is the mesosphere.

38. **1** It is given that the extinction took place 65.5 million years ago. Find the Geologic History of New York State chart in the *Reference Tables for Physical Setting/Earth Science*. To the left of the column labeled "Life on Earth," locate 65.5 on the scale labeled "Million years ago." Trace right from 65.5 million years ago, and note the statement "Mass extinction of dinosaurs, ammonoids, and many land plants." Of these, only ammonoids are given as a choice. Thus, the impact event and global climate change that occurred 65.5 million years ago resulted in the extinction of ammonoids.

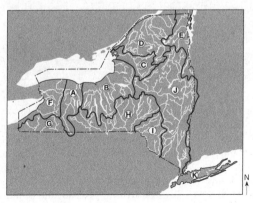

39. **2** Find the Generalized Bedrock Geology of New York State map in the *Reference Tables for Physical Setting/Earth Science*. Locate the

Susquehanna River near Binghamton. On the watershed map, locate the region corresponding to the Susquehanna River. Note that the Susquehanna River lies within the major watershed labeled H.

40. **1** Find the Generalized Bedrock Geology of New York State map in the *Reference Tables for Physical Setting/Earth Science*, and note the location of the Genesee River. Now find the Generalized Landscape Regions of New York State map in the *Reference Tables for Physical Setting/Earth Science*, and locate the region corresponding to the Genesee River. Note that the Genesee River extends from the Appalachian Plateau (Uplands) through the Alleghany Plateau to the Erie-Ontario Lowlands (Plains). By definition, a plateau is a large, flat region of land at high elevations (more than 150–300 meters above the surrounding land). Plains are large areas of land at low elevations, with an underlying structure of horizontal bedrock and very little relief. Water flows from higher elevations to lower elevations. Therefore, it is reasonable to infer that the Genesee River generally flows from the higher elevations of the Alleghany Plateau in the south toward the lower elevations in the Erie-Ontario Lowlands (Plains) in the north. In other words, the Genesee River in watershed A generally flows north.

41. **4** Find the Generalized Landscape Regions of New York State map in the *Reference Tables for Physical Setting/Earth Science*. Locate the region corresponding to watershed D. Note that watershed D extends across the landscape regions labeled "Adirondack Mountains" and "St. Lawrence Lowlands." Thus, the streams in watershed D flow over the Adirondack Mountains and the St. Lawrence Lowlands.

Mineral Chart

Mineral Name	Composition	Density (g/cm^3)	Hardness	Streak	Nonmetallic Luster	Common Colors
brucite	$Mg(OH)_2$	2.4	2.5-3	white	glassy to waxy	white
carnallite	$KMgCl_3 \cdot 6H_2O$	1.6	2.5	white	greasy	white
dolomite	$CaMg(CO_3)_2$	2.8	3.5-4	white	glassy to waxy	shades of pink
magnesite	$MgCO_3$	3.1	3.5-4.5	white	glassy	white
olivine	$(Fe,Mg)_2SiO_4$	3.3	6.5	white	glassy	green

42. **3** Find the Properties of Common Minerals table in the *Reference Tables for Physical Setting/Earth Science*. In the column labeled "Mineral Name," locate calcite. Trace left to the column labeled "Composition," and note that the composition of calcite is $CaCO_3$, which is calcium carbonate. Now locate the column labeled "Composition" on the Mineral Chart. Note that two formulas also indicate a carbonate composition (contain CO_3)— $CaMg(CO_3)_2$ and $MgCO_3$. Trace left from these formulas and note that they correspond to dolomite and magnesite. Thus, dolomite and magnesite have compositions most similar to calcite.

43. **2** Find the Properties of Common Minerals table in the *Reference Tables for Physical Setting/Earth Science*. In the column labeled "Mineral Name," locate both fluorite and amphibole. Trace left from these two minerals to the column labeled "Hardness." Note that fluorite has a hardness of 4 and amphibole has a hardness of 5.5. Hardness is a mineral's resistance to being scratched. On Mohs' Mineral Hardness Scale, ten minerals are arranged in order from the softest, which is talc (1), to the hardest, which is diamond (10). Each mineral on the scale will be scratched by those of higher number but will scratch all those of lower number. Thus, a mineral that will scratch fluorite but would not scratch the mineral amphibole will have a hardness that is greater than 4 but less than 5.5. Locate the column labeled "Hardness" on the Mineral Chart. Note that only the entry "3.5–4.5," contains a value that lies between 4 and 5.5. Trace left to the column labeled "Mineral Name," and note that this hardness value corresponds to magnesite. Thus, magnesite, which can have a hardness of up to 4.5, might scratch fluorite, but would not scratch the mineral amphibole.

44. **4** Streak is the color of the dust or powder left when a mineral is rubbed against a hard, rough surface such as an unglazed porcelain tile called a streak plate. In the Mineral Chart, find the column labeled "Streak." Note that all of the minerals on the chart have a white streak. In other words, all of the minerals are white in powdered form. Now find the column labeled "Common Colors." Note that two minerals have a common color that is not white—shades of pink and green. Trace left from these two colors to the column labeled "Mineral Name." Note that shades of pink corresponds to dolomite and that green corresponds to olivine. Thus, dolomite and olivine have common colors different from their color in powdered form. Of the choices given, only olivine has a common color different from its color in powdered form.

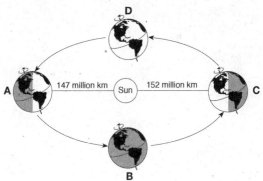

(Not drawn to scale)

45. **1** Gravitational attraction between any two bodies is greatest when the distance between their centers is smallest. It is given that "Earth is closest

to the Sun (perihelion) at position A." Therefore, the distance between the Sun and Earth is smallest when Earth is at position A. Thus, the gravitational attraction between the Sun and Earth will be greatest at position A.

46. **3** On the first day of summer in the Northern Hemisphere, Earth's axis of rotation is tilted farthest toward the Sun. The position in the diagram at which Earth's axis of rotation is tilted farthest toward the Sun is position C. Therefore, position C represents the first day of summer in the Northern Hemisphere, or June 21. Note the arrows indicating the direction in which Earth moves in its orbit. From position C, Earth moves approximately one-quarter revolution in its orbit to position D. Earth completes one revolution around the Sun in one year, or 12 months. Thus, Earth completes one-quarter of its revolution in 3 months (12 months ÷ 4 = 3 months). Therefore, Earth reaches position D 3 months after it was at position C. Three months after June 21 is September 21. On September 21, the Northern Hemisphere experiences the fall equinox, marking the first day of fall. Thus, the change in seasons that occurs in the Northern Hemisphere at position D is that summer is ending and fall is beginning in the Northern Hemisphere.

47. **4** As Earth revolves around the Sun, its axis of rotation remains parallel to itself. Therefore, Earth's north and south poles always point toward the same portions of the universe. The star *Polaris* is located almost directly in line with the north pole of Earth's axis of rotation. Therefore, the northern end of Earth's axis always points toward *Polaris*.

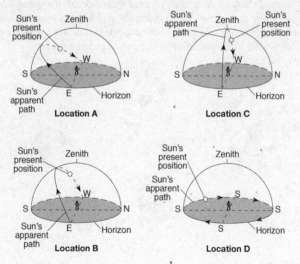

48. **3** Solar noon occurs when the Sun reaches its highest point in the sky for the day. At this moment, the Sun crosses an observer's meridian, which is

a semicircle connecting the North and South Poles that passes through the observer's zenith. Note that at locations A, B, and C, the Sun rises due east, crosses the observer's meridian at noon, and sets due west. The Sun rises due east and sets due west at all locations on Earth when Earth is in the equinox position. It is given that all four locations are shown on the same date. Therefore, all four locations are shown at an equinox. When Earth is at an equinox position, day and night are equal in length. Thus, an observer experiences 12 hours of daylight and 12 hours of darkness. If solar noon represents the midpoint of the Sun's path, then sunrise in the east would occur 6 hours earlier or at 6 a.m. and sunset would occur 6 hours later at 6 p.m. Note that at locations A, B, and C, the Sun's present position is about halfway between solar noon when it crosses the observer's meridian and sunset when it drops beneath the western horizon at 6 p.m. Note that at location D, the Sun's present position is about the same angular distance past the observer's meridian and represents the same time of day. Thus, the approximate time of day at each location is halfway between noon and 6 p.m., which is 3 p.m.

49. 3. The greatest intensity of insolation occurs where the Sun's rays strike perpendicular to Earth's surface. According to the diagrams, the Sun is at the observer's zenith at solar noon at location C. When the Sun is at an observer's zenith, it is directly overhead and the Sun's rays strike perpendicular to Earth's surface. Therefore, location C had the greatest intensity of insolation at solar noon.

50. 4 To the observer at location D, all directions are south of the observer. The only location at which this occurs is the North Pole.

PART B-2

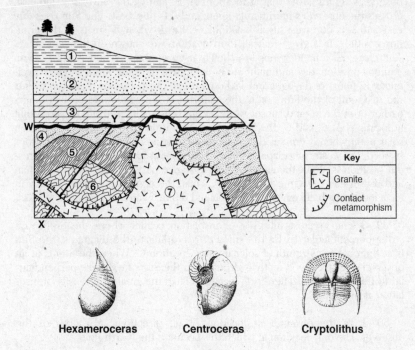

51. Note the symbols representing rock units 1, 2, and 3. Find the Scheme for Sedimentary Rock Identification in the *Reference Tables for Physical Setting/Earth Science*. In the column labeled "Map Symbol," locate the symbols for rock units 1, 2, and 3. From each symbol, trace left to the column labeled "Rock Name," and note that the symbols correspond to rock unit 1—shale, rock unit 2—sandstone, and rock unit 3—dolostone. Thus, all three rock units are sedimentary rocks. It is given that the rock layers at these locations have not been overturned. The principle of superposition states that the bottom layer of a sedimentary series is the oldest, unless it was overturned or had older rock thrust over it, because the bottom layer was deposited first. Similarly, in a sequence of rock layers, a rock layer is older than those above it and is younger than those below it. Thus, rock unit 1 is the youngest, rock unit 2 is older than rock unit 1, and rock unit 3 is older than both rock units 1 and 2. Find the Geologic History of New York State chart in the *Reference Tables for Physical Setting/Earth Science*. In the diagrams of index fossils at the bottom of the chart, locate E—*Hexameroceras*, F—*Centroceras*, and B—*Cryptolithus*. In the column labeled "Time Distribution of Fossils (including important fossils of New York)," locate

E, F, and B. From each letter, trace left to the time scale labeled "Million years ago." Note that fossil B, *Cryptolithus*, existed between 444 and 488 million years ago. Fossil E, *Hexameroceras*, existed between 416 and 444 million years ago. Fossil F, *Centroceras*, existed between 359 and 416 million years ago. Thus, from youngest to oldest, the fossils are *Centroceras*, *Hexameroceras*, and *Cryptolithus*. Therefore, rock unit 1, the youngest, would contain the youngest fossil—*Centroceras*. Rock unit 2, the next oldest, would contain the next oldest fossil—*Hexameroceras*. Rock unit 3, the oldest, would contain the oldest fossil— *Cryptolithus*.

One credit is allowed if *each of the three* **fossil names is in its correct row**.

Rock Unit	Fossil Name
1	*Centroceras*
2	*Hexameroceras*
3	*Cryptolithus*

Note: Credit is allowed if you list the correct sequence of letters corresponding to these fossils as shown on the Geological History of New York State chart of the *Reference Tables for Physical Setting/Earth Science*: (1) F, (2) E, (3) B.

52. Note the symbol corresponding to rock unit 6. It is given that rock unit 6 was metamorphosed. Find the Scheme for Metamorphic Rock Identification in the *Reference Tables for Physical Setting/Earth Science*. In the column labeled "Map Symbol," locate the symbol corresponding to rock unit 6. From this symbol, trace left to the column labeled "Comments," and note the statement "Metamorphism of limestone or dolostone." Find the Scheme for Sedimentary Rock Identification in the *Reference Tables for Physical Setting/Earth Science*. In the column labeled "Rock Name," locate limestone and dolostone. Thus, limestone and dolostone are sedimentary rocks that could have metamorphosed to form rock unit 6.

One credit is allowed for **limestone** *or* **dolostone**.

53. Note the symbol representing rock unit 3. Find the Scheme for Sedimentary Rock Identification in the *Reference Tables for Physical Setting/Earth Science*. In the column labeled "Map Symbol," locate the symbol for rock unit 3. Trace left to the column labeled "Rock Name," and note that rock unit 3 corresponds to dolostone. From dolostone, trace left to the column labeled "Composition." Note that dolostone is composed of dolomite. Find the Properties of Common Minerals table in the *Reference Tables for*

Physical Setting/Earth Science. In the column labeled "Mineral Name," locate dolomite. Trace left to the column labeled "Composition." Note the chemical formula for dolomite—$CaMg(CO_3)_2$. Thus, the most common mineral found in rock unit 3 (dolostone) is dolomite, and the chemical formula for dolomite is $CaMg(CO_3)_2$.

One credit is allowed for the chemical formula **$CaMg(CO_3)_2$**.

54. Layers of rock are generally deposited in an unbroken sequence. However, if forces within Earth cause rocks to be uplifted, deposition ceases. Weathering and erosion may wear away layers of rock before the land surface is low enough for another layer to be deposited. The uplift may cause rock layers to tilt or fold. The result is an unconformity, which is a break or gap in the sequence of a series of rock layers. Thus, the rocks above an unconformity are quite a bit younger than those below it. There are several types of unconformities. *Angular unconformities* form when rock layers are tilted or folded before being eroded. When new layers are deposited, they form horizontally and the layers below the unconformity are at an angle to those above it. *Disconformities* are irregular erosional surfaces between parallel layers of rock. Disconformities occur when deposition stops and layers are eroded but no tilting or folding occurs. These surfaces are not easy to discern and are often found when fossils of very different ages are discovered in adjacent layers. *Nonconformities* are places where sedimentary layers lie on top of igneous or metamorphic rocks and where the sedimentary layers are not metamorphosed in any way. According to the cross section, the rock units below the unconformity are tilted and folded. There is also a fault along line XY, and the layers to the right of the fault moved upward relative to the layers to the left of the fault. Also note that there is no contact metamorphism between rock unit 7 and rock unit 3 along WZ. Thus, the unconformity at WZ is most likely an angular unconformity resulting from tilting and folding due to uplift. When the uplifted land emerged, it was either exposed to weathering and erosion that wore down the land surface or the land then subsided until it was low enough for deposition to resume and bury the surface beneath new layers of sediment.

One credit is allowed if *both* processes are correct. Acceptable responses include but are not limited to:

- **uplift/emergence**
- **weathering**
- **erosion**
- **subsidence/submergence**
- **deposition/precipitation**
- **burial**

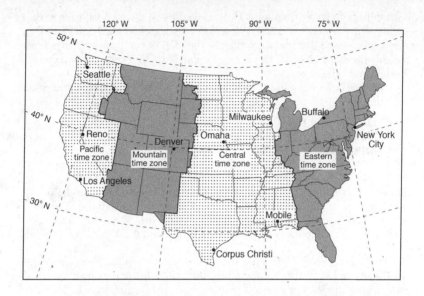

55. Lines of longitude run north-south. Find the Generalized Bedrock Geology of New York State map in the *Reference Tables for Physical Setting/Earth Science*. Find New York City, and note that it is located at approximately 74° W longitude. According to the given map, Reno, Nevada, is located at approximately 120° W longitude. Thus, the two cities are 46° of longitude apart (120° − 74° = 46°). Earth rotates 15° per hour. Therefore, time zones were set up around a series of meridians spaced at 15° intervals starting at the Prime Meridian (0° longitude). Since Earth rotates from west to east, for every time zone you travel west of the Prime Meridian, clocks are set 1 hour earlier. Note that New York City is located in the Eastern time zone and that Reno, Nevada, is located in the Pacific time zone. The Pacific time zone is 3 time zones to the west of the Eastern time zone. Therefore, the time difference between these two cities is 3 hours.

One credit is allowed if *both* the longitude difference and time difference are correct.

Longitude difference:

any value from 45° to 47°

Time difference:

3 hours

56. By simple geometry, it can be shown that to an observer in the Northern Hemisphere, the altitude of *Polaris* is the same as the observer's latitude. (Refer to Figure 4.5 in *Let's Review: Earth Science, The Physical*

Setting for a fuller explanation.) Thus, two cities on the map where measurements of the altitude of *Polaris* are within one degree of each other will be at roughly the same latitude. According to the map, several pairs of cities are at roughly the same latitude: Denver and Reno; New York City and Omaha; and Buffalo and Milwaukee.

One credit is allowed if *both* cities are correct. Acceptable responses include:

- **Denver and Reno**
- **New York City and Omaha**
- **Milwaukee and Buffalo**

57. As Earth rotates from west to east, the eastern part of the United States is carried from darkness into daylight (experiences sunrise) before the western part of the United States. Thus, the city labeled on the map where sunrise occurs first each day is the city located farthest east—New York City.

One credit is allowed for **New York City** *or* **New York** *or* **NYC**.

58. Observers on Earth experience solar noon when the Sun crosses their meridian, or longitude. Thus, each observer has his or her own time scale based upon his or her location. For example, while someone in New York City is experiencing solar noon, someone in Tokyo, Japan, is asleep in bed because it is 2 a.m. Earth rotates through 360° of longitude in 24 hours, or 15° of longitude per hour (360 ÷ 24 = 15). Thus, every 15° of longitude that separates two locations corresponds to a difference of 1 hour in solar time. Therefore, time zones were created and spaced at intervals of 15° of longitude so that adjoining time zones have a 1 hour time difference. Thus, the basis for our system of local time and time zones is Earth's 15° per hour rate of rotation.

One credit is allowed. Acceptable responses include but are not limited to:

- **rotation**
- **Earth rotates on its axis.**
- **a spinning Earth**

59. According to the reading passage, "The epicenter was located approximately 80 miles off Japan's eastern coast at the approximate coordinates of 38° N, 142° E." Find the Tectonic Plates map in the *Reference Tables for Physical Setting/Earth Science* (page 48). Locate the point corresponding to 38° N, 142° E, and note the plate boundary symbol at that point. According to the key at the bottom of the map, this symbol corresponds to a "Convergent plate boundary (subduction zone)." Convergent plate boundaries occur where adjacent plates are colliding.

One credit is allowed. Acceptable responses include but are not limited to:

- **convergent plate boundary**
- **subduction zone**
- **colliding plates**

60. According to the reading passage, the earthquake "generated a 7-meter-high tsunami." Use the scale in the diagram to mark off a distance corresponding to 7 meters on the straight edge of a piece of scrap paper. Now, mark off this distance vertically along the side of the building starting from the ground.

One credit is allowed if **the center of the X is within or touches the edge of the box below.**

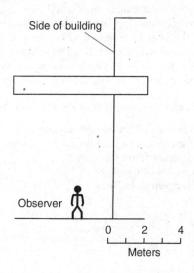

Note: Credit is allowed if a symbol other than an **X** is used.

61. When an earthquake occurs, all of the different types of seismic waves start moving outward from the focus at the same time. However, since they all travel at different speeds, they do not all arrive at a seismograph at the same time. The *P*-waves (compressional waves), which travel fastest, will arrive first, followed by the *S*-waves some time later.

One credit is allowed. Acceptable responses include but are not limited to:

- **P-wave**
- **primary wave/P**
- **compressional wave**

62. Tsunamis may be only a few meters high, but they have very long wavelengths and travel much more rapidly than ordinary ocean waves. Tsunamis have been clocked moving faster than 500 km/hr with wavelengths as great as 200 kilometers. When tsunamis reach shallow water, they are slowed by friction with the ocean bottom and begin to pile up as large waves. In bays and narrow channels, their high speed and long wavelength

may be funneled into huge breaking waves more than 20 m high. The force exerted by such a huge, fast-moving mass of water can do extensive damage and cause widespread coastal flooding. According to the Tsunami Travel Time map, it would take between 6 and 9 hours for the tsunami to reach Maui, Hawaii. Thus, when the residents of Maui received the tsunami warning, they probably had only a few hours to respond. In that time, they could have evacuated to higher elevations by moving inland away from the coast. If unable to evacuate before the tsunami struck, they could have moved to higher floors of buildings to try to get above the level of the flood waters. If on a boat, they could have moved the boat to deeper waters where the tsunami waves had not yet been pushed upward by shallow water to form large breaking waves. Finally, many coastal cities in areas prone to tsunamis have established emergency shelters in safe locations. Residents could have sought out these emergency shelters.

One credit is allowed. Acceptable responses include but are not limited to:

- **Evacuate to higher elevations/evacuate.**
- **Move to higher floors of buildings.**
- **Move inland, away from the coast.**
- **Move boats to deeper water.**
- **Seek out emergency shelters.**

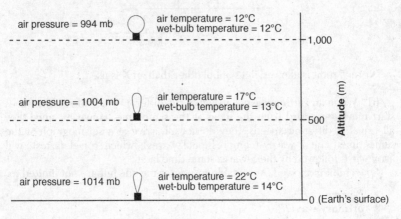

(Not drawn to scale)

63. The instrument used to measure air pressure is a barometer. The instrument used to measure air temperature is a thermometer. Thus, the names of the instruments carried by the weather balloon that recorded the air pressure and air temperature are a barometer and a thermometer.

One credit is allowed if *both* responses are correct.

Air pressure: barometer
Air temperature: thermometer

64. According to the diagram, the air temperature (dry-bulb temperature) at Earth's surface is 22°C and the wet-bulb temperature is 14°C. The difference between the wet-bulb and dry-bulb temperatures is 8°C (22° − 14° = 8°). Find the Dewpoint (°C) chart in the *Reference Tables for Physical Setting/Earth Science*. Find the horizontal axis at the top of the chart labeled "Difference Between Wet-Bulb and Dry-Bulb Temperatures (C°)," and locate the column labeled "8." Along the left side of the chart, find the vertical axis labeled "Dry-Bulb Temperature (C°)," and locate the row labeled "22." Where row 22 intersects column 8, read the dewpoint temperature, which is 8°C.

Now find the Relative Humidity (%) chart in the *Reference Tables for Physical Setting/Earth Science*. Find the horizontal axis at the top of the chart labeled "Difference Between Wet-Bulb and Dry-Bulb Temperatures (C°)," and locate the column labeled "8." Along the left side of the chart, find the vertical axis labeled "Dry-Bulb Temperature (C°)," and locate the row labeled "22." Where row 22 intersects column 8, read the % relative humidity, which is 40%.

One credit is allowed if *both* dewpoint and relative humidity are correct.

Dewpoint: 8°C
Relative humidity: 40%

65. Clouds form when water vapor (gas) in the atmosphere changes back into tiny droplets of liquid water by the process of condensation. Condensation occurs when the air temperature cools to the dewpoint temperature. Thus, the phase change occurring at 1000 meters to produce the cloud is condensation.

One credit is allowed. Acceptable responses include but are not limited to:

- **condensation**
- **water vapor changing to liquid water**
- **gas to liquid**

PART C

Galaxy Information

Galaxy	Galaxy's Distance from Earth (million light years)	Recession Velocity (km/s)
A	62	1210
B	978	15,000
C	1402	21,600
D	2510	39,300
E	3912	61,200

Note: One light year is the distance that light travels in one year.

66. According to the Galaxy Information table, both "Galaxy's Distance from Earth (million light years)" and "Recession Velocity (km/s)" increase from the top to the bottom of the table. Thus, as the galaxies' distance from Earth increases, their recession velocities also increase. In other words, galaxies that are closer to Earth are receding more slowly than galaxies that are farther from Earth. This type of relationship is called a direct relationship. In a direct relationship, when one value increases, the other also increases. Additionally in a direct relationship, when one value decreases, the other also decreases.

One credit is allowed. Acceptable responses include but are not limited to:

- **As the Earth-to-galaxy distance increases, the recession velocity increases.**
- **Galaxies closer to Earth are moving more slowly.**
- **direct relationship/positive relationship**

67. As explained earlier, the table shows a direct relationship between a galaxy's distance from Earth and its recessional velocity. A recession velocity of 30,000 km/s is roughly midway between the recessional velocities of galaxy C (21,600 km/s) and of galaxy D (39,300 km/s) shown on the table. To find the midpoint of these values, calculate $(21,600 + 39,300) \div 2 = 30,450)$. Therefore, it is reasonable to infer that the distance from Earth of a galaxy having a recession velocity of 30,000 km/s will be midway between the distance of galaxies C and D from Earth. On the Galaxy Information table, note that galaxy C is located 1402 million light years from Earth and that galaxy D is located 2510 million light years from Earth. Thus, the distance from Earth of a galaxy with a recession velocity of 30,000 km/s will be roughly midway between 1402 and 2510 million light years, or 1956 million light years $(1402 + 2510 \div 2 = 1956)$.

One credit is allowed for any value from **1800 to 2200 million light years.**

68. Within stars, the force of gravity is strong enough to overcome the force of repulsion between atomic nuclei, allowing the nuclei to combine in a process called nuclear fusion. By definition, nuclear fusion involves the combining of several atoms of a lighter element to form a single atom of a heavier element. The single heavier atom typically has less mass than the several lighter atoms from which it formed. The mass "missing" from the heavier atom is not lost. Instead, it is converted into energy according to Einstein's formula $E = mc^2$. This formula means that if mass is converted to energy, the amount of energy released, E, is equal to the mass, m, times the speed of light, c, squared. The speed of light squared is a very large number. Therefore, the conversion of even a small amount of mass to energy during nuclear fusion results in the release of a very large amount of energy. Thus, great amounts of energy are released in the core of a star as lighter elements combine and form heavier elements during the process of fusion. Therefore, the nuclear process that produces the energy released by stars within these galaxies is nuclear fusion.

One credit is allowed. Acceptable responses include but are not limited to:

- **fusion/nuclear fusion**
- **Light elements combine to form heavier elements.**

69. Find the Radioactive Decay Data table in the *Reference Tables for Physical Setting/Earth Science*. In the column labeled "Radioactive Isotope," find Carbon-14. Trace right to the column labeled "Disintegration," and note the reaction $^{14}C \rightarrow {}^{14}N$. Find the Average Chemical Composition of Earth's Crust, Hydrosphere, and Troposphere in the *Reference Tables for Physical Setting/Earth Science*. In the column labeled "Element (symbol)," locate Nitrogen (N). Thus, the element formed when carbon-14 (^{14}C) undergoes radioactive decay is an isotope of nitrogen—^{14}N, or nitrogen-14.

One credit is allowed. Acceptable responses include but are not limited to:

- ^{14}N
- **nitrogen-14**
- **nitrogen**

70. Carbon-14 is used to date only organic remains, i.e., remains of once-living things such as a mastodon. Carbon-14 is present in the environment and is incorporated into the bodies of living things as they carry out their life processes. When an organism dies, its life processes cease. The carbon-14 in its body that decays is no longer replenished from the environment. Therefore, the amount of carbon-14 present in remains decreases over time. This decrease can be used to measure the time elapsed since the organisms died.

Carbon-14 is most useful for dating relatively recent organic remains—that is, organic remains less than about 50,000 years old. Find the Radioactive Decay Data table in the *Reference Tables for Physical Setting/Earth Science*.

In the column labeled "Radioactive Isotope," find Carbon-14. Trace right to the column labeled "Half-life (years)," and note that the half-life of carbon-14 is 5.7×10^3 years, or 5700 years. After about 10 half-lives (57,000 years), too little of the original carbon-14 is left in the remains to be useful in radioactive dating. Therefore, any remains more than 50,000 years old could not be dated using carbon-14. However, it is given that 12.5% of the original radioactive C-14 remained in the mastodont bone. After 1 half-life, 50% of the original radioactive carbon-14 would remain. After 2 half-lives, 25% would remain. After 3 half-lives, 12.5% would remain. Thus, the age of the mastodont bone is 3 half-lives, or 17,100 years (5700 $\times$ 3 = 17,100.) Therefore, the mastodont bone is young enough (less than 50,000 years old) to be radioactively dated using carbon-14.

Therefore, ^{14}C was used to date the mastodont bone because the mastodont bone is an organic remain and is less than 50,000 years old.

One credit is allowed. Acceptable responses include but are not limited to:

- **C-14 is used to date recent organic remains.**
- **The mastodont bone is less than 50,000 years old.**
- **Carbon-14 has a short half-life.**
- **Carbon-14 decays at a predictable rate.**

71. Find the Geologic History of New York State chart in the *Reference Tables for Physical Setting/Earth Science*. In the column labeled "Life on Earth," locate the statement "Humans, mastodonts, mammoths." Trace right to the column labeled "Important Geologic Events in New York," and note the statement "Advance and retreat of last continental ice." Thus, one important geologic event that occurred in New York State when mastodonts existed was the advance and retreat of the last continental ice. This last major advance and retreat occurred at the end of the last ice age. It was responsible for the glaciation of much of New York State and the formation of many current features, such as Long Island and the Finger Lakes.

One credit is allowed. Acceptable responses include but are not limited to:

- **advance and retreat of last continental ice**
- **last ice age**
- **glaciation**
- **formation of Long Island**
- **formation of New York State Finger Lakes**

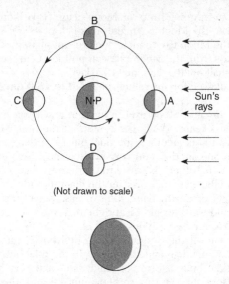

72. Note that the diagram shows the Moon's surface almost completely in darkness with a thin, crescent-shaped illuminated portion to the observer's right. This corresponds to the new crescent phase of the Moon. The new crescent phase occurs midway between the new Moon phase (position A in the answer diagram) and before the first-quarter phase (position B in the answer diagram). Thus, the **X** should be drawn along the Moon's orbit midway between positions A and B as shown below.

One credit is allowed if the **center of the X is located within or touches the bracket (clear band) below**.

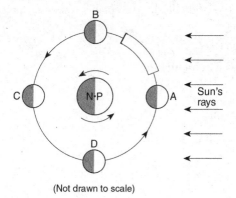

Note: Credit is allowed if a symbol other than an **X** is used.

73. Note that in traveling from the Moon phase at position *C* to the Moon phase at position *A*, the Moon completes one-half of its cycle of phases. The time it takes for the Moon to complete one cycle of phases is 29.5 days. Thus, the number of days from the Moon phase at position *C* to the Moon phase at position *A* is one-half of 29.5 days, or 14.75 days.
One credit is allowed for any value from **14.0 to 15.0 days**.

74. Both the Sun and Moon exert a force of gravity on Earth that causes tidal bulges in the oceans. When the centers of the Sun, Moon, and Earth are at right angles to one another, the tidal bulges of the Sun and Moon are at right angles to one another and very nearly cancel each other. The result is neap tides, in which there is very little difference between high and low tide. The arrows labeled "Sun's rays" in the diagram indicate that the Sun is located to the right of Earth. Thus, the two positions of the Moon in which the Sun, Moon, and Earth would be at right angles to one another and where neap tides would occur are positions *B* and *D*.

When the gravity of the Sun and Moon pull in the same direction, the tidal bulges of the Sun align and combine with the tidal bulges of the Moon. The combined tidal bulges form very high tides and very low tides, called spring tides because they "spring so high," not because they happen in the spring season. The arrows labeled "Sun's rays" in the diagram indicate that the Sun is located to the right of Earth. Thus, the two positions of the Moon in which the Sun, Moon, and Earth would align and where spring tides would occur are positions *A* and *C*. Therefore, when the Moon is at position *D*, a neap tide will occur and there will be very little difference between the heights of Earth's high and low tides. When the Moon is at position *A*, a spring tide will occur and there will be a great difference between the heights of Earth's high and low tides. Thus, when the Moon moves from position *D* to position *A*, the high tides will be higher and the low tides will be lower.

One credit is allowed for *both* a correct effect on high-tide height and a correct effect on low-tide height. Acceptable responses include but are not limited to:

Height of high tide:
- **High tides will be higher.**
- **higher**
- **increase**

Height of low tide:
- **Low tides will be lower.**
- **lower**
- **decrease**

75. Find the Solar System Data chart in the *Reference Tables for Physical Setting/Earth Science*. In the column labeled "Celestial Object," locate Earth's Moon. Trace right to the column labeled "Period of Rotation at Equator," and note that Earth's Moon rotates once every 27.3 days. Trace upward in this column, and note the most similar period of rotation—27 days. From 27 days, trace left to the column labeled "Celestial Object," and note that it is the Sun that has a period of rotation of 27 days. Thus, the celestial object in our solar system that has a period of rotation most similar to the period of rotation of Earth's Moon is the Sun.

One credit is allowed for the **Sun**.

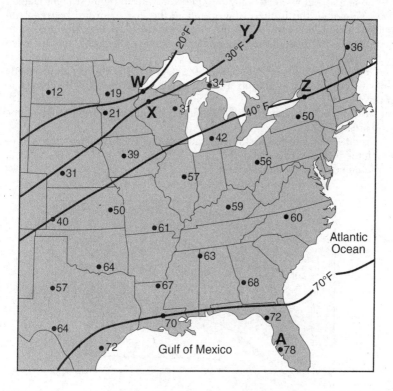

76. Isotherms are lines connecting points of equal temperature. On the map provided, construct the 50°F isotherm by drawing a smooth line connecting all points labeled 50. Not every point that is experiencing a temperature of 50°F has been labeled. However, between any locations, the temperature steadily changes from the temperature at one location to the

temperature at the other. Note how the 40°F isotherm is drawn between values that are greater than and less than 40°F. Use the same technique to draw the 50°F isotherm through regions where a temperature of exactly 50°F was not recorded. For example, a temperature of 42°F was recorded in the state of Michigan and a temperature of 56°F was recorded to the south in the state of Ohio. The 50°F isotherm would pass between these two points. Begin by drawing a line from the edge of the map south of the 40° isotherm to the point labeled "50" in New York State. Continue drawing the line westward between the points labeled "42" and "56." Then extend the line westward between the point labeled "57" and the 40° isotherm that is already drawn on the map. Continue extending the line westward to the next point labeled "50." Complete the isotherm by extending the line westward between the points labeled "40" and "57" to the edge of the map. Repeat this process to complete the 60°F isotherm. Remember to extend each isotherm to the edge of the map.

One credit is allowed if *both* the 50°F and 60°F isotherms are correctly drawn. If additional isotherms are drawn, all isotherms must be correct to receive credit.

Note: Credit is allowed if the isotherms extend only to the edge of the land area.

Credit is *not* allowed if student-drawn isotherms do *not* pass through or touch the 50 and 60 data points.

Example of a 1-credit response:

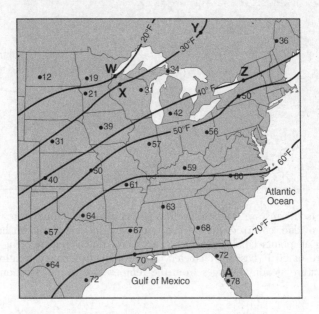

77. Find the Generalized Bedrock Geology of New York State map in the *Reference Tables for Physical Setting/Earth Science*. Locate Watertown, and note its position within New York State. On the isotherm map, note that the position corresponding to Watertown is labeled "Z" and lies directly on the 40°F isotherm. Therefore, the air temperature in Watertown is 40°F.
One credit is allowed for any value **from 39°F to 41°F**.

78. Isotherms are isolines connecting points of equal temperature. The more closely spaced the isotherms are on a field map, the greater the change in temperature is over a given distance. In other words, the more closely spaced the isotherms are, the steeper the temperature gradient is. Thus, the evidence shown on the map that indicates that the temperature gradient between locations W and X is greater than the temperature gradient between locations Y and Z is that the isotherms are closer together between locations W and X than they are between locations Y and Z.
One credit is allowed. Acceptable responses include but are not limited to:

- **The isotherms are closer together between locations W and X than they are between locations Y and Z.**
- **Temperatures between W and X show the same change over a shorter distance.**
- **The isotherms are farther apart between Y and Z.**
- **The isolines are closer together.**

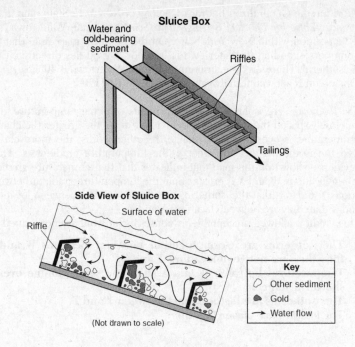

Mineral Characteristics of Gold						
Luster	Hardness	Dominant Form of Breakage	Color	Streak	Density g/cm^3	Chemical Symbol
metallic	2.5 to 3	fracture	golden yellow	golden yellow	19.3	Au

79. According to the reading passage, "The riffles in the bottom of the box are designed and positioned to create disruptions in the water flow. These disruptions cause dead zones in the current that allow the more dense gold to drop out of suspension and be deposited behind the riffles. Lighter material flows out of the box as tailings." Note in the Mineral Characteristics of Gold table that gold has a density of 19.3 g/cm^3. Find the Inferred Properties of Earth's Interior chart in the *Reference Tables for Physical Setting/Earth Science*. In the column labeled "Density (g/cm^3)" along the right side of the cross section, note that the density of Earth's granitic continental crust is 2.7 g/cm^3 and that the density of Earth's basaltic oceanic crust is 3.0 g/cm^3. Thus, gold has a much higher density than other rock materials that make up Earth's crust. Therefore, the characteristic of gold shown in the table that

allows gold to be deposited behind the riffles, while other material flows out of the sluice box as tailings, is gold's high density of 19.3 g/cm³.

One credit is allowed for **density** *or* **high density** *or* **19.3 g/cm³**.

80. It is given that the velocity of the water was 90 cm/s. Find the Relationship of Transported Particle Size to Water Velocity graph in the *Reference Tables for Physical Setting/Earth Science*. On the horizontal axis labeled "Stream Velocity (cm/s)," locate the line corresponding to 90 cm/s. Trace vertically upward until you intersect the black curve. From this intersection, trace horizontally to the left to the vertical axis labeled "Particle Diameter (cm)." Note the value corresponds to approximately 2.0 cm.

One credit is allowed for a value from **1.5 to 2.5 cm**.

81. In general, the steeper the slope down which water is flowing, the greater the water velocity is. The greater the water velocity is, the larger the particle sizes that can be transported by the water. Thus, if the angle of the sluice box is changed so that the box has a steeper slope, the water velocity will increase. The amount of sediment passing through the sluice box as tailings will also increase because larger particles that were previously left behind will now be carried out as tailings.

One credit is allowed if *both* responses are acceptable. Acceptable responses include but are not limited to:

Water velocity:

- **increases**
- **speeds up**
- **gets greater**
- **flows faster**

Amount of sediment:

- **increases**
- **becomes greater**
- **more sediment**
- **Less sediment is left behind in the sluice box.**

82. Find the Equations section in the *Reference Tables for Physical Setting/Earth Science*, and note the equation for density:

$$\text{Density} = \frac{\text{mass}}{\text{volume}}$$

According to the Mineral Characteristics of Gold table, the density of gold is 19.3 g/cm³. It is given that the volume of the gold nugget is 0.8 cubic centimeters. Substitute these values for density and volume in the equation for density, and solve for the mass of the gold nugget as shown below:

$$\text{Density} = \frac{\text{mass}}{\text{volume}}$$

$$\text{Density} \times \text{volume} = \text{mass}$$
$$19.3 \text{ g/cm}^3 \times 0.8 \text{ cm}^3 = \text{mass} = 15.44 \text{ g}$$

Thus, the mass of the gold nugget is 15.44 g.
One credit is allowed for **15.44 g** *or* **15.4 g** *or* **15 g**.

Data Table

Year	Average Insolation Reflected During the Summer (%)
2006	74.3
2007	72.8
2008	72.9
2009	71.8
2010	70.3
2011	70.1
2012	68.3

83. To construct a graph of the average insolation reflected during the summer by the Greenland ice sheet from 2006 to 2012, do the following. In the data table, note that in 2006, the average insolation reflected during the summer (%) was 74.3. On the graph provided on your answer sheet, locate the intersection of 2006 on the horizontal axis labeled "Year" and 74.3 on the vertical axis labeled "Average Insolation Reflected During the Summer (%)." Draw an **X** at this intersection. Repeat this process for each of the remaining years and their corresponding average insolation reflected during the summer based on information in the table. When all of the **X**s have been plotted, connect them with a smooth line. A graph with all points correctly plotted and connected with a smooth line is shown below.

Allow one credit if **the centers of *all seven* plots are within or touch the circles shown and are correctly connected with a line that passes within or touches each circle**.

Note: Credit is allowed if the line you drew does not pass through your plots but is still within or touches the circles.

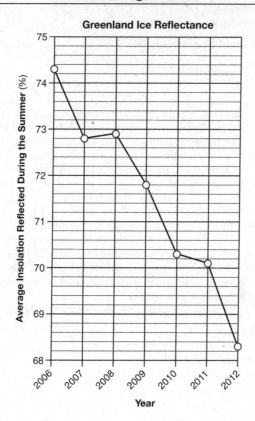

84. Note that with the exception of 2007 to 2008, the average insolation reflected in the summer by the Greenland ice sheet decreased every year from 2006 to 2012. Thus, from 2006 to 2012, the ice sheet reflectivity generally decreased. If the average reflectivity decreased, it can be inferred that there was less ice to reflect insolation. Thus, it can be inferred that the size of the Greenland ice sheet decreased during this time period, most likely because some of the ice melted.

One credit is allowed if *both* the insolation-reflected trend and the inferred change in size are correct. Acceptable responses include but are not limited to:

Insolation-reflected trend:

- **From 2006 to 2012, the ice sheet reflectivity generally decreased.**
- **became less**
- **lower**

Inferred size change:

- **The ice sheet became smaller.**
- **less**
- **shrunk or melted**
- **decreased**

85. The surfaces of ice sheets are typically smooth and light colored. Color is a fairly good indicator of whether a substance absorbs more light than it reflects. The lighter the color of a substance is, the more light is being reflected by it. The darker the color is, the more light is being absorbed. Snow and ice appear white and shiny, which indicates that snow and ice reflect more light than they absorb. Snow and ice also tend to form smooth, flat surfaces. If a surface is smooth, light is more likely to be immediately reflected away from the surface than if the surface is rough. The irregularities on a rough surface can cause some of the light to hit the surface several times before leaving. Each time the light strikes the surface, a little more of its energy is absorbed. Thus, a smooth, light-colored surface reflects more insolation than a rough, dark-colored surface. Therefore, the characteristics that make the Greenland ice sheet a good reflector of insolation are that ice and snow are light in color and smooth.

One credit is allowed. Acceptable responses include but are not limited to:

- **light in color**
- **smooth**
- **Ice and snow are white.**
- **shiny/glassy**

SELF-ANALYSIS CHART August 2016

Topic	Question Numbers (Total)	Wrong Answers (x)	Grade
Standards 1, 2, 6, and 7: Skills and Application			
Skills			
Standard 1 Analysis, Inquiry, and Design	1, 4, 7, 10, 13–16, 18, 20, 22, 25, 26, 30, 35–43, 49, 51–53, 55, 57–60, 64, 66–71, 75, 77–85		$\frac{100(49-x)}{49} = \%$
Standard 2 Information Systems	62		$\frac{100(1-x)}{1} = \%$
Standard 6 Interconnectedness, Common Themes	12, 16, 19–21, 23–35, 39–57, 59–61, 63, 65–67, 72–74, 76–79, 81, 83, 84		$\frac{100(54-x)}{54} = \%$
Standard 7 Interdisciplinary Problem Solving	62		$\frac{100(1-x)}{1} = \%$
Standard 4: The Physical Setting/Earth Science			
Astronomy			
The Solar System (MU 1.1a, b; 1.2d)	36, 38, 45, 46, 72, 73		$\frac{100(6-x)}{6} = \%$
Earth Motions and Their Effects (MU 1.1c, d, e, f, g, h, i)	3, 24, 26, 29, 47–50, 55–58, 74		$\frac{100(13-x)}{13} = \%$
Stellar Astronomy (MU 1.2b)	1, 68		$\frac{100(2-x)}{2} = \%$
Origin of Earth's Atmosphere, Hydrosphere, and Lithosphere (MU 1.2e, f, h)			
Theories of the Origin of the Universe and Solar System (MU 1.2a, c)	2, 28, 66, 67, 75		$\frac{100(5-x)}{5} = \%$

SELF-ANALYSIS CHART August 2016

Topic	Question Numbers (Total)	Wrong Answers (x)	Grade
Meteorology			
Energy Sources for Earth Systems (MU 2.1a, b)	37		$\dfrac{100(1-x)}{1} = \%$
Weather (MU 2.1c, d, e, f, g, h)	8, 30, 63–65, 76–78		$\dfrac{100(8-x)}{8} = \%$
Insolation and Seasonal Changes (MU 2.1i; 2.2a, b)	4, 6, 7, 10, 11, 83–85		$\dfrac{100(8-x)}{8} = \%$
The Water Cycle and Climates (MU 1.2g; 2.2c, d)	5, 9, 12		$\dfrac{100(3-x)}{3} = \%$
Geology			
Minerals and Rocks (MU 3.1a, b, c)	20, 22, 25, 42–44, 52, 53, 79, 82		$\dfrac{100(10-x)}{10} = \%$
Weathering, Erosion, and Deposition (MU 2.1s, t, u, v, w)	19, 23, 33, 34, 39, 40, 80, 81		$\dfrac{100(8-x)}{8} = \%$
Plate Tectonics and Earth's Interior (MU 2.1j, k, l, m, n, o)	15–18, 59–62		$\dfrac{100(8-x)}{8} = \%$
Geologic History (MU 1.2i, j)	13, 14, 27, 31, 51, 54, 69–71		$\dfrac{100(9-x)}{9} = \%$
Topographic Maps and Landscapes (MU 2.1p, q, r)	21, 32, 35, 41		$\dfrac{100(4-x)}{4} = \%$
ESRT			
Earth Science Reference Tables 2011 Edition	1, 4, 7, 10, 13–16, 18, 20, 22, 25, 30, 35–43, 51–53, 59, 64, 69–71, 75, 77, 80, 82		$\dfrac{100(34-x)}{34} = \%$

To further pinpoint your weak areas, use the Topic Outline in the front of the book.
MU = Major Understanding (see Topic Outline)

NOTES

NOTES

NOTES

NOTES

NOTES

It's finally here—
online Regents exams from the experts!

- Take complete practice tests by date, or choose questions by topic
- All questions answered with detailed explanations
- Instant test results let you know where you need the most practice

Online Regents exams are available in the following subjects:

- Biology–The Living Environment
- Chemistry–The Physical Setting
- Earth Science–The Physical Setting
- English (Common Core)
- Global History and Geography
- Algebra I (Common Core)
- Geometry (Common Core)
- Algebra II (Common Core)
- Physics–The Physical Setting
- U.S. History and Government

Getting started is a point and a click away!

Only $24.99 for a subscription…includes all of the subjects listed above!

Teacher and classroom rates also available.

For more subscription information and to try our demo visit our site at *www.barronsregents.com*.

Join the half-million students each year who rely on Barron's Redbooks for Regents Exams success

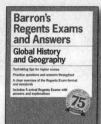

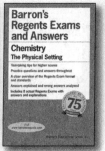

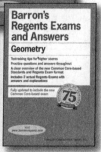

Algebra I
ISBN 978-1-4380-0665-9, $7.99, Can$9.50

Algebra II
ISBN 978-1-4380-0843-1, $7.99, Can$8.99

Biology—The Living Environment
ISBN 978-0-8120-3197-3, $7.99, Can$9.50

Chemistry—The Physical Setting
ISBN 978-0-8120-3163-8, $7.99, Can$9.50

Earth Science—The Physical Setting
ISBN 978-0-8120-3165-2, $7.99, Can$9.50

English
ISBN 978-0-8120-3191-1, $7.99, Can$9.50

Geometry
ISBN 978-1-4380-0763-2, $7.99, Can$9.50

Global History and Geography
ISBN 978-0-8120-4344-0, $7.99, Can$9.50

Physics—The Physical Setting
ISBN 978-0-8120-3349-6, $7.99, Can$9.50

U.S. History and Government
ISBN 978-0-8120-3344-1, $7.99, Can$9.50

Prices subject to change without notice.

Available at your local book store or visit **www.barronseduc.com**

Barron's Educational Series, Inc.
250 Wireless Blvd.
Hauppauge, NY 11788
Order toll-free: 1-800-645-3476
Order by fax: 1-631-434-3217

In Canada:
Georgetown Book Warehouse
34 Armstrong Ave.
Georgetown, Ontario L7G 4R9
Canadian orders: 1-800-247-7160
Order by fax: 1-800-887-1594

For extra Regents review visit our website: **www.barronsregents.com**

(#224) R8/16

Barron's Let's Review — Textbook companions to help you ace your Regents Exams

Let's Review: Algebra I
ISBN 978-1-4380-0604-8, $14.99, Can$16.99

Let's Review: Algebra II
ISBN 978-1-4380-0844-8, $14.99, Can$16.99

Let's Review: Biology—The Living Environment, 6th Ed.
ISBN 978-1-4380-0216-3, $13.99, Can$16.99

Let's Review: Chemistry—The Physical Setting, 5th Ed.
ISBN 978-0-7641-4719-7, $14.99, Can$16.99

Let's Review: Earth Science—The Physical Setting, 4th Ed.
ISBN 978-0-7641-4718-0, $14.99, Can$16.99

Let's Review: English, 5th Ed.
ISBN 978-1-4380-0626-0, $13.99, Can$16.99

Let's Review: Geometry
ISBN 978-1-4380-0702-1, $14.99, Can$17.99

Let's Review: Global History and Geography, 5th Ed.
ISBN 978-1-4380-0016-9, $14.99, Can$16.99

Let's Review: Physics, 5th Ed.
ISBN 978-1-4380-0630-7, $14.99, Can$17.99

Let's Review: U.S. History and Government, 5th Ed.
ISBN 978-1-4380-0018-3, $14.99, Can$16.99

Available at your local book store or visit www.barronseduc.com

Barron's Educational Series, Inc.
250 Wireless Blvd.
Hauppauge, N.Y. 11788
Order toll-free: 1-800-645-3476
Order by fax: 1-631-434-3217

In Canada:
Georgetown Book Warehouse
34 Armstrong Ave.
Georgetown, Ontario L7G 4R9
Canadian orders: 1-800-247-7160
Order by fax: 1-800-887-1594

For extra Regents review visit our website: **www.barronsregents.com**

Prices subject to change without notice.

#225 R8/16

Barron's Review Course Series

Let's Review:

Earth Science—
The Physical Setting

Fourth Edition

Edward J. Denecke, Jr., B.A., M.A.
Formerly, William H. Carr J.H.S. 194Q
Whitestone, New York

© Copyright 2016, 2015, 2014, 2013, 2012, 2011, 2010, 2009, 2008, 2007, 2006, 2005, 2004, 2003, 2002, 2001, 2000, 1995 by Barron's Educational Series, Inc.

All rights reserved.
No part of this book may be reproduced or distributed in any form or by any means without the written permission of the copyright owner.

All inquiries should be addressed to:
Barron's Educational Series, Inc.
250 Wireless Boulevard
Hauppauge, NY 11788
www.barronseduc.com

ISBN: 978-0-7641-4718-0

ISSN: 2164-7577

PRINTED IN THE UNITED STATES OF AMERICA

9 8 7 6 5 4 3 2

10% POST-CONSUMER WASTE
Paper contains a minimum of 10% post-consumer waste (PCW). Paper used in this book was derived from certified, sustainable forestlands.

TABLE OF CONTENTS

PREFACE   v

CAUTION! DON'T USE THIS BOOK UNTIL YOU READ THIS   vii

TOPIC ONE	**Astronomy**	**1**
UNIT ONE	**From a Geocentric to a Heliocentric Universe**	**1**
Chapter 1:	Early Astronomy and the Geocentric Model	1
Chapter 2:	The Development of the Heliocentric Model	18
Chapter 3:	Heliocentric Earth Motions and Their Effects	41
Chapter 4:	Earth's Coordinate System and Mapping	67
Chapter 5:	Our System of Time	98
UNIT TWO	**Modern Astronomy**	**125**
Chapter 6:	Tools of the Modern Astronomer	125
Chapter 7:	Stars, Their Origin and Evolution	139
Chapter 8:	The Solar System	163
Chapter 9:	Theories of the Origin of the Universe	190
UNIT THREE	**Earth's History**	**204**
Chapter 10:	The Origins of Earth and Its Moon	204
Chapter 11:	The Origin and Structure of Earth's Atmosphere	215
Chapter 12:	The Origin and Nature of Earth's Hydrosphere	228
Chapter 13:	The Origin and History of Life on Earth	253
TOPIC TWO	**Geology**	**291**
UNIT FOUR	**Earth Materials**	**291**
Chapter 14:	Minerals	291
Chapter 15:	Rocks	314
UNIT FIVE	**The Dynamic Earth**	**348**
Chapter 16:	Earthquakes and Earth's Interior	348
Chapter 17:	Volcanoes and Earth's Internal Heat	380
Chapter 18:	Plate Tectonics	395

Table of Contents

UNIT SIX   **Weathering, Erosion, and Deposition**   **430**

Chapter 19:	Weathering and Soil Formation	430
Chapter 20:	Erosion	451
Chapter 21:	Deposition	492

TOPIC THREE   **Meteorology**   **535**

UNIT SEVEN   **The Atmosphere, Weather, and Climate**   **535**

Chapter 22:	The Atmosphere	535
Chapter 23:	Weather	566
Chapter 24:	Climate	621

Appendix A: Answers to Review Questions   647

Appendix B: Reference Tables for Physical Setting/Earth Science   695

Glossary of Earth Science Terms   711

Regents Examination   729

Answers to Regents Examination   770

Index   771

Preface

To the Student:

"On the surface, islands may seem separate, but underneath they are all connected."

Earth science is not the study of isolated facts, but rather the development of a deep understanding, and appreciation of, the interconnectedness of Earth phenomena, processes, and systems. In earth science, the key to success is thorough understanding and the ability to demonstrate what you understand. This book is a concise text and review aid in which the author has tried to make earth science as understandable as possible to you, the student, by incorporating the following features:

- Explanations of concepts and understandings are detailed, yet simply and clearly stated. They are designed to help you grasp the "how" and "why" of an idea, rather than just stating the idea.
- Important terms are printed in **boldface type** where they are defined in the text.
- The illustrations are designed to make difficult ideas easier to understand. Many of the illustrations are similar to those used in Regents Examination questions in order to familiarize you with the types of diagrams you will be asked to analyze and interpret.
- Each unit ends with a wide range of review questions, including constructed response and extended constructed response questions.
- A glossary and a complete index make it easy for you to find the definition of a specific term and the pages in the book where a topic is covered.
- A full-length Regents examination provides you with the opportunity to assess your understanding and to test your earth science knowledge and skills before taking the Regents examination.

Earth science offers the challenge and excitement of new theories, new discoveries, and new problems to be solved. It is a science in which sweeping new theories are being tested and applied to puzzling new observations. It is a science in which revolutionary advances in human knowledge of Earth and the other planets of our solar system are being made almost daily. I hope that studying earth science will fill you with wonder and delight at the complexities of our planet Earth.

Preface

To the Teacher:

Let's Review: Earth Science is a concise text and review aid for courses based on the New York State Physical Setting/Earth Science Core Curriculum, a comprehensive course of study in earth science on the secondary level. However, the material in this book provides such comprehensive coverage of topics in earth science that it can be used as a review text to supplement virtually any secondary course in earth science taught in the United States, using any major textbook.

This edition reflects the content of the New York State Physical Setting/ Earth Science Core Curriculum. It is organized into three major topics: astronomy, geology, and meteorology. Each topic addresses a key idea of Standard 4: The Physical Setting Earth Science, and is divided into units based upon the performance indicators for that key idea. Each unit is subdivided into chapters that deal with groups of related major understandings underlying the performance indicator. Specific skills identified in Standards 1, 2, 6, and 7 are introduced with the appropriate major understanding. Figures and text follow the 2011 edition of the *Earth Science Reference Tables*.

The review questions in this edition have been chosen from previous Regents Examinations to reflect content consistent with the Physical Setting: Earth Science Core Curriculum and suitability for use with the reference tables. Constructed response and extended constructed response practice questions are included at the end of each chapter.

I wish to thank my wife, Gerry, for her infinite patience and my children, Meredith, Abigail, and Benjamin, for their loving support during the preparation of this manuscript.

CAUTION!
Don't Use This Book Until You Read This

You are taking earth science and want to get good grades on your exams. Well, before you even look at the earth science subject matter in this book, let's talk about how to use this book. Many students don't really know how to read a textbook or a review book. Suppose you get a homework assignment such as "Read pages 73–82 in the text, and answer questions 1–5 on page 83." What will you do? Typical students turn to page 73 and start reading, sentence by sentence, paragraph by paragraph. Once in a while they may stop to see how much more they have to read. When they finally reach the last sentence on page 82, they consider that they have "studied" the material and then turn to page 83 and try to answer the questions. This simple procedure doesn't require a lot of time or effort. But just reading the text once is a really weak approach to studying. So what is a better way to study?

Understand How the Text Is Organized

First, it is important to realize that textbooks are not written like novels. Textbooks are not meant to be read straight through; instead, they are structured to guide in-depth study. If you understand how a textbook is arranged, you can organize your reading into small, useful blocks.

Let's Review is based upon the New York State Physical Setting Earth Science Core Curriculum and is organized as follows:

ORGANIZATION

Physical Setting/Earth Science Core Curriculum	Let's Review: Earth Science
STANDARDS tell what you are expected to know and be able to do.	
KEY IDEAS tell you the really important ideas relating to the standard.	**TOPICS** are based upon the major disciplines in earth science—astronomy, geology, and meteorology—that are addressed in the key ideas of the standards.
PERFORMANCE INDICATORS describe what you should be able to do in order to show that you understand the key ideas.	**UNITS** are based upon the elements of the performance indicators for each key idea. For example, Unit 2: Modern Astronomy addresses Performance Indicator 1.2 – Describe current theories about the origin of the universe and solar system.

Caution!

MAJOR UNDERSTANDINGS list what you need to know in order to do the things described in the performance indicators.	**CHAPTERS** address groups of related major understandings underlying each performance indicator. For example, Chapter 13: The Origin and History of Life on Earth addresses Major Understandings 1.2h–j. **SECTIONS** break chapters into manageable reading blocks to help you better understand what you read. For example, Chapter 19: Weathering has such sections as Physical Weathering, Chemical Weathering, and The Products of Weathering.

Know How to Spell Success: S-Q-3R

Before you read, you need to realize that studying means not just reading, but also *thinking deeply* about what you have read. One good approach is to take notes while reading and then study your notes before a test. A better approach is to read each chapter subsection once, read it again and highlight the important points, and then study those points after you finish reading each chapter.

One of the best approaches to studying is called **S-Q-3R**, which stands for **S**urvey, **Q**uestion, **R**ead, **R**ecite, and **R**eview. Here's how it works:

1. You **Survey** the chapter by reading through the section and subsection headings to get a mental map of the material.

2. You ask **Questions** about the material by turning each heading into a question.

3. Then you **Read** the text, a subsection at a time, with the purpose of answering these questions.

4. Next, you **Recite** by jotting down brief notes about what you have read, making an outline or a graphic organizer, or writing a summary. (Note: In this case the word *recite* doesn't mean to "speak publicly"; it means "to re-cite, or cite again." To *cite* is to quote, or mention. Here, *recite* means to list or itemize important ideas you have read.)

5. Finally, you **Review** by rereading your notes and answering questions about the material. They may be questions that you pose to yourself or that have appeared on prior tests.

This five-step approach will take more time than just reading through, but the reward for the extra time you spend will be better grades. Therefore, plan to have enough time to study before you sit down for a session with this book.

Know What's on the Test

The key to success in preparing for any test is to know what will be expected of you so that you can review and practice beforehand. The New York State Physical Setting/Earth Science Regents Examination has four parts:

Part A—Multiple-Choice. In a multiple-choice question you are given several choices from which to select the one that best answers the questions or completes the statement. Many practice questions of this type from previous Regents Examinations are included at the end of each chapter of this book. Part A of the exam focuses on earth science content from Standard 4.

Part B—Multiple-Choice and Constructed Response. In a constructed response question there is no list of choices from which to select an answer; rather, you are required to provide the answer. Constructed response questions can test skills ranging from constructing graphs or topographic maps to formulating hypotheses, evaluating experimental designs, and drawing conclusions based upon data. Practice questions of this type are also included at the end of each chapter. In Part B you will be asked to demonstrate skills identified in Standards 1, 2, 6, and 7 in the context of earth science.

Part C—Extended Constructed Response. The constructed response questions require more time (5–10 minutes per item) and effort on your part to answer. Questions in Part C require you to apply your earth science knowledge and skills to real-world problems and applications. You may be asked to produce short essays, design controlled experiments, predict outcomes, or analyze the risks and benefits of various solutions to a problem.

Part D—Laboratory Performance Tasks. These tasks test your laboratory skills. You will take this part of the exam sometime during the 2 weeks before the written Regents. Laboratory performance tasks involve skills such as using instruments (e.g., rulers, external protractors, triple beam balances, graduated cylinders, stopwatches), observing properties of Earth materials, performing calculations, and collecting and analyzing data.

Note: The following description represents that information the State Education Department has stated may be shared with students before taking the performance part of the examination. You should be familiar with the skills being assessed because you have used them in laboratory activities throughout the year. However, you will not be allowed to practice the entire test or any of the individual stations before this performance component is administered.

Caution!

Station 1...*Mineral and Rock Identification*
The student determines the properties of a mineral and identifies that mineral using a flowchart. Then the student classifies two different rock samples and states a reason for each classification based on observed characteristics.

Station 2...*Locating an Epicenter*
The student determines the location of an earthquake epicenter using various types of data that were recorded at three seismic stations.

Station 3...*Constructing and Analyzing an Asteroid's Elliptical Orbit*
The student constructs a model of an asteroid's elliptical orbit and compares the eccentricity of the orbit with that of a given planet.

Extensive analyses of questions of all types can be found in the companion volume to this book, ***Barron's Regents Exams and Answers: Earth Science—The Physical Setting***. Together, these review books will help you prepare for the New York State Physical Setting/Earth Science Regents Examination by clearly explaining what you should know and be able to do in order to perform well on the exam and by providing you with practice questions from prior exams that are thoroughly explained.

Unit One: FROM A GEOCENTRIC TO A HELIOCENTRIC UNIVERSE

CHAPTER 1

EARLY ASTRONOMY AND THE GEOCENTRIC MODEL

> **KEY IDEAS** People have observed the stars for thousands of years, using them to find direction, note the passage of time, and express human values and traditions. To an observer on Earth, it appears that Earth stands still and everything else moves around it. Thus, in trying to make sense of how the universe works, it was logical for early astronomers to start with those apparent truths. To comprehend our modern view of the universe, it is helpful to begin by understanding these first attempts to explain the universe in terms of what can be seen from our vantage point on Earth. As technology has progressed, so has our understanding of celestial objects and events.

KEY OBJECTIVES
Upon completion of this chapter, you will be able to:

- Explain the meaning of the term celestial object.
- Compare and contrast "apparent" and "real" motion.
- Explain how the celestial sphere model of the sky accounts for the motions of celestial objects.
- Explain how Earth's rotation makes it appear that the Sun, the Moon, and the stars are moving around Earth once a day.
- Locate *Polaris* in the night sky.

OBSERVING THE SKY

If you kept a list of things observed in the sky, it might include birds, smoke, clouds, rainbows, halos, lightning, stars, the Moon, the Sun, and comets. One of the first ideas that might occur to you is that the sky has depth. Some things in it appear closer, and some appear farther away. Why? Perspective! From everyday experiences you know that closer objects block your view of more distant objects. For example, if you hold your hand in front of your

Figure 1.1 Motion in the Sky. (a) Photograph showing the crescent Moon and Venus setting. Exposures were made every 8 minutes, showing the changes in position of these celestial objects over time. Note the motion of the Moon relative to Venus.

Source: Horizons: *Exploring the Universe*, Michael A. Seeds, Wadsworth, 1987.

(b) A time exposure taken with a camera aimed at *Polaris* over Mauna Kea Observatory, latitude 20°. Note the circular star trails. Source: *Astronomy: The Cosmic Journey*, William K. Hartman, Wadsworth, 1987.

eyes, you cannot see a more distant tree. Therefore, if a bird flying by blocks your view of a cloud, you logically conclude that the bird is closer to you than the cloud. Then you see a cloud move "in front of" the Sun, and you conclude that the cloud is closer to you than the Sun. Or perhaps you see a solar eclipse, and you conclude that the Moon is closer to you than the Sun. In this way, all of the objects on your list could be put in order of distance from an observer.

Careful observation also leads you to realize that many of these closer objects or phenomena are associated with the atmosphere. You feel a wind and see it moving the clouds. You see a rainbow in the spray of a waterfall or in a distant rain shower and realize that it is caused by the interplay of sunlight and tiny droplets of water in the air. You see lightning flash between a cloud and the ground. In this way, you can classify the things seen in the sky into two groups: those that are part of, or occur in, the atmosphere, and those that are beyond the atmosphere.

Celestial Objects

Celestial objects are objects that can be seen in the sky that are not associated with Earth's atmosphere. The most numerous of the celestial objects are the stars. To an observer on Earth, stars are simply points of light that vary in size, brightness, and color. The Sun, the Moon, the planets, and comets are also examples of celestial objects. Clouds, rainbows, halos, and other phenomena seen in the sky that are part of, or occur in, Earth's atmosphere are *not* considered celestial objects.

Celestial Motion

If you observe celestial objects for even a short while, it is clear that they change position in the sky over time. You have probably noticed the Sun in different places in the sky at different times of the day. Thus, it seems that the Sun is "moving." Similarly, if you observe the Moon and stars carefully, you find that they, too, are seen in different places in the sky at different times of the night. Try going out on a moonlit night and noting the Moon's position at 7:00 P.M. If you go out again at 10:00 P.M., you'll notice that the Moon has changed position in the sky. The same is true of stars. When you observe the sky, you find that every celestial object changes position over time, or is in motion. See Figure 1.1 on pages 2 and 3.

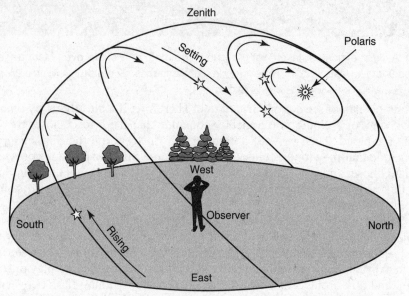

Figure 1.2 The Apparent Motion of Celestial Objects to an Observer in New York State.

If you keep track of this motion, you discover something curious. The motion of celestial objects is not random. They don't all move in different directions at different speeds. Instead, with very few exceptions, every single one of these thousands of objects appears to move in the same general direction—*from east to west*. And if you measure the rates at which all of these celestial objects are moving, you discover something even more curious—with few exceptions (such as the Moon), *they appear to move at the same rate!*

Careful records of this motion reveal that all celestial objects appear to move across the sky from east to west along a path that is an arc, or part of a circle. Since celestial objects appear to follow a circular path at a constant rate of 15 degrees per hour, or one complete circle every day (24 hr/day × 15°/hr = 360°/day), this motion is called **apparent daily motion**.

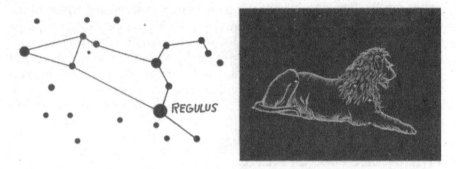

Figure 1.3 Constellations Are Imaginary Patterns of Stars. Source: *Astronomy: A Self-Teaching Guide*, 4th Ed., Dinah L. Moché, John Wiley, 1998.

In the Northern Hemisphere, all the circles formed by completing the arcs along which celestial objects move are centered very near the star Polaris. The apparent circular motion of celestial objects causes them to come into view from below the eastern horizon and to sink from view beneath the western horizon (that is, to rise in the east and set in the west). See Figure 1.2.

Early observers noted that the positions of celestial objects change in a daily and yearly cyclic pattern. They discovered that understanding these patterns of motion was very useful. Since the positions of celestial objects change with time and location, such changes can be used to *determine time* and to *find one's position on Earth*. Since the distribution of stars is random, these observers devised **constellations**, imaginary patterns of stars, to help them keep track of the changing positions of celestial objects. See Figure 1.3.

"Apparent" Versus "Real" Motion

So far, we have used the word *apparent* when referring to celestial motions because the motion of an object is always judged with respect to some other object or point. The idea of absolute motion or rest is misleading because there are several possible reasons why an object may *appear* to an observer to be moving. One possibility is that the observer is standing still and the *object* is moving. Another possibility is that the object is standing still and *the observer* is moving. A third possibility is that *both the observer and the object* are moving, but one is moving faster, or in a different direction, than the other. This is the case when you are in a car speeding down a highway;

as you look out of the car window, trees along the side of the road seem to whiz by. Of course, your brain tells you that the trees are rooted to the ground and that they only seem to whiz by because you are riding in a car. But to your eyes alone, you are not moving; the image of the trees is moving from one side of your window to the other. Now think about driving past a person walking along the sidewalk. The person is moving, but also seems to whiz by your window. Now think of a person sitting next to you in your car. To you, would that person look as though he or she was moving?

By now you should realize that the problem of determining which of the two is moving, the object or the observer, is not always easy to solve. If the signs that tell the body it is moving are removed, an observer may not realize that he or she is in motion. (Do you really feel as if you are moving at 400 miles per hour when watching a movie in an airplane cruising in level flight at that speed?) Without signs telling the observer's body that he or she is moving, any object seen changing position will be interpreted as a moving object by the observer.

THE CELESTIAL SPHERE

Early observers reasoned that when they looked at the sky they were standing still because their senses gave them no signs that they were moving. They *felt* as if they were standing still. Therefore, they interpreted the changing positions of celestial objects to mean that the *celestial objects* were moving. They visualized all celestial objects as revolving around a motionless Earth.

One effect of apparent daily motion is that the sky appears to move as if it were a single object. Here's a simple analogy. If you watch a truck with "Moving Van" painted on its side roll past you, the word "Moving" doesn't get closer to the word "Van" just because the truck is moving in that direction. All of the letters in the two words stay in a fixed pattern even though they are all moving because they are part of a single object—the sign. In much the same way, the stars in the sky stay in a fixed pattern even as you observe them moving through the sky.

It is not surprising, then, that early observers imagined that the sky *was* a single object—a huge dome. Since the "dome" of the sky was in motion, and new parts would come into view as others dropped out of sight, these observers imagined that the dome extended beyond the horizon. As they followed through on this model, they realized that, if the dome were extended far enough, it would form a hollow ball, or sphere, surrounding Earth. They imagined a huge "sky ball," or **celestial sphere**, slowly spinning around a motionless Earth. See Figure 1.4. To these observers the Sun, Moon, and stars were either holes in the celestial sphere or objects attached to it.

The "celestial sphere" was a nice model because it accounted for many observations. It explained why objects appeared, arced across the sky, disappeared, and then reappeared the next day. Imagine it as a ball tied to a rope and swung in a circle around your head. First the ball arcs across your line of sight as you swing it in front of you, next it disappears as it swings around behind you, and then it reappears as it swings around in front of you again. This model explained why all of the celestial objects moved in the same direction at the same speed. It also explained why the stars remained in fixed positions relative to one another. This Earth-centered, or **geocentric**, model of the universe was used successfully for thousands of years to explain most observations of celestial objects.

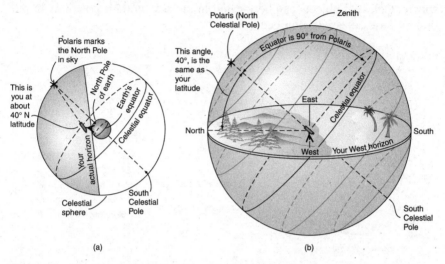

Figure 1.4 The Celestial Sphere, an Imaginary Sphere Surrounding Earth. The most you see at any one time is half of this sphere. Certain reference points on the celestial sphere are defined in relation to reference points on Earth. The celestial poles lie directly over Earth's poles; the celestial equator lies over Earth's equator midway between the celestial poles. Other points are defined by their positions in relation to the observer: the zenith is a point directly above the observer, the celestial meridian is the circle that runs through the celestial poles and the zenith. As Earth rotates from west to east, all objects in the sky appear to move from east to west, revolving around the north celestial pole. (a) View from a spot outside the celestial sphere. (b) Observer's view. Sources: *The Practical Astronomer*, Brian Jones, Simon and Schuster, 1990, and *Astronomy: A Self-Teaching Guide*, 4th Ed., Dinah L. Moche, John Wiley, 1998.

Even though we now know that the motion of celestial objects is due to Earth's rotation, it is still sometimes useful, when discussing objects in the sky, to think of them as part of a sphere surrounding Earth. The most that an observer would see at any one time would be half of this sphere; but we still refer to this imaginary half-sphere, or dome, visible over our heads as the celestial sphere. The circle formed by the intersection of the celestial sphere and the ground is called the **horizon**. The point on the celestial sphere that is right over an observer's head at any given time is the **zenith**. The imaginary circle that passes through the north and south points on the horizon and through the zenith is the **celestial meridian**.

A Simple Celestial Coordinate System

A useful coordinate system for locating objects on the celestial sphere can be set up by projecting Earth's Equator and poles onto the sky. As shown in Figure 1.5, Earth's Equator, North Pole, and South Pole correspond to a "celestial equator" and "north and south celestial poles" on the celestial sphere. Celestial objects can be located in the sky by their positions in relation to these celestial reference points.

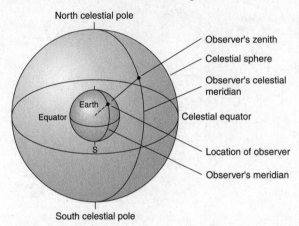

Figure 1.5 Projection of Earth's Latitude-Longitude System onto the Celestial Sphere. Source: *The Practical Astronomer*, Brian Jones, Simon and Schuster, 1990. (page 66).

The star Polaris is located very close to the north celestial pole, making it a convenient reference point for determining the north-south positions of celestial objects in the Northern Hemisphere. Polaris can be located by following the "pointer stars," Dubhe and Merak, in the bowl of the Big Dipper in the constellation Ursa Major. See Figure 1.6.

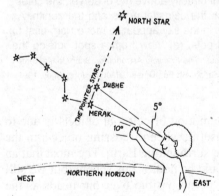

Figure 1.6 The "Pointer Stars," Dubhe and Merak, in the Bowl of the Big Dipper. Use these two stars to find the North Star, Polaris, and also to judge angular distances; they are about 5° apart. Another way to estimate angular distances is to hold a fist out at arm's length; the fist marks about 10°. Source: *Astronomy: A Self-Teaching Guide*, 4th Ed., Dinah L. Moché, John Wiley, 1998.

A convenient reference point for determining the east-west positions of objects on the celestial sphere is the Sun. Objects to the west of the Sun on the celestial sphere will "rise" before the Sun and "set" before it. Likewise, objects to the east of the Sun trail behind it and will "rise" after the Sun and "set" after it. See Figure 1.7.

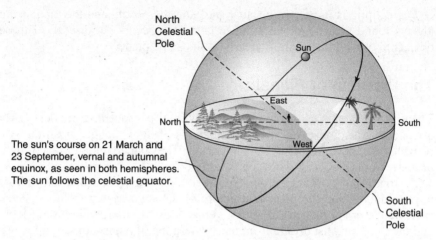

The sun's course on 21 March and 23 September, vernal and autumnal equinox, as seen in both hemispheres. The sun follows the celestial equator.

Figure 1.7 The Sun's Path on March 21 and September 23, the Vernal and Autumnal Equinox. The Sun follows the celestial equator. Source: *The Practical Astronomer*, Brian Jones, Simon and Schuster, 1990.

The Sun's Path

Each day, because of Earth's rotation, the Sun moves along an imaginary path on the celestial sphere. Over the course of a year, however, it also follows an imaginary path on the celestial sphere. As you can see in Figure 1.8, the apparent position of the Sun with respect to the background stars changes continuously as Earth orbits the Sun. When Earth has made one complete revolution in its orbit, the Sun will return to its starting point against the background stars. In other words, the Sun traces out a closed path on the celestial sphere once a year. The apparent path of the Sun through the stars on the celestial sphere over the course of the year is called the **ecliptic**. Since Earth's axis of rotation is tilted 23½° to the plane of its orbit, the ecliptic is tilted 23½° with respect to the celestial equator.

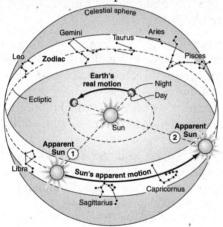

Figure 1.8 During Earth's Annual Journey Around the Sun, We View Stars from a Slightly Different Position from Day to Day. Thus, the Sun appears to travel around the celestial sphere during the course of a year along a path called the *ecliptic*. The part of the sky through which the Sun passes is known as the *zodiac*, and the Sun crosses the celestial equator at the vernal and autumnal equinoxes.

Early Astronomy and the Geocentric Model

The ecliptic is important because the Sun, the Moon, and the planets are always found near it. As we will see later, this occurs because all of these objects in our solar system lie nearly in the same plane.

The Problem of Planets

There were, however, some problems with the geocentric model. Early astronomers also observed that certain points of light changed position with respect to the background of stars in the sky. They called these points of light *planets*, from the Greek word for "wanderer."

Astronomers working before the invention of the telescope and before anyone understood the present structure of the solar system counted seven such wanderers or planets: Mercury, Venus, Mars, Jupiter, Saturn, the Moon, and the Sun. This list differs from our modern list of planets in several ways:

- Earth is missing, because no one realized that the points of light wandering in the sky and the Earth on which these observers stood were in any way alike.
- The Sun and the Moon were classified as planets because they wandered on the celestial sphere, just like Mars and Jupiter and the other planets.
- Uranus and Neptune are missing because they were not discovered until the telescope made them easily visible. Uranus, which is barely visible to the naked eye, was discovered in 1781. Neptune, which can't be seen at all without a telescope, was discovered in 1846.

Planets differ from stars in a number of ways. As already mentioned, the relative positions of stars on the celestial sphere are fixed, while planets move relative to the stars. Stars can be seen anywhere on the celestial sphere; planets are always found near the ecliptic (that imaginary yearly path of the Sun on the celestial sphere). Stars appear to "twinkle," but the brighter planets do not. Even through a telescope, stars appear as points of light, while the larger and nearer planets appear as disks.

These observed differences between planets and stars, particularly the "wandering" of planets on the celestial sphere, attracted a lot of attention from early astronomers. Their attempts to explain these differences ultimately led to the development of a new model of the universe.

MULTIPLE-CHOICE QUESTIONS

In each case, write the number of the word or expression that best answers the question or completes the statement.

1. Which of the following is *not* a celestial object?
   (1) the Sun
   (2) the Moon
   (3) a rainbow
   (4) a star

2. As viewed from Earth, most stars appear to move across the sky each night because
   (1) Earth revolves around the Sun
   (2) Earth rotates on its axis
   (3) stars orbit around Earth
   (4) stars revolve around the center of the galaxy

3. Which real motion causes the Sun to appear to rise in the east and set in the west?
   (1) the Sun's revolution
   (2) the Sun's rotation
   (3) Earth's revolution
   (4) Earth's rotation

4. An observer in New York State took a time-exposure photograph from 10 P.M. until midnight of the stars over the *northern* horizon. Which diagram best represents this photograph?

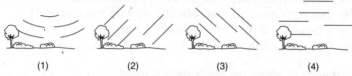

5. A camera was placed in an open field and pointed toward the northern sky. The lens of the camera was left open for a certain amount of time. The result is shown in the photograph below. The angle of the arc through which two of the stars appeared to move during this time exposure is shown.

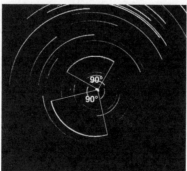

How many hours was the lens kept open to create the star trails in this photograph?
   (1) 12
   (2) 2
   (3) 6
   (4) 4

Early Astronomy and the Geocentric Model

6. How many degrees does the Sun appear to move across the sky in four hours?
   (1) 60°   (3) 15°
   (2) 45°   (4) 4°

Base your answers to questions 7 through 11 on your knowledge of Earth science and on the diagram, which represents observations of the apparent paths of the Sun in New York State on the dates indicated.

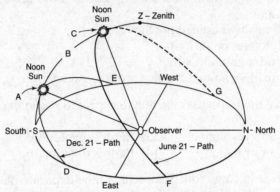

7. On the basis of the diagram, which statement is true?
   (1) The Sun passes through the zenith on December 21.
   (2) The Sun rises due east and sets due west on December 21.
   (3) The Sun passes through the zenith on June 21.
   (4) The Sun rises north of east and sets north of west on June 21.

8. Which statement about the Sun's path is true?
   (1) The Sun's path varies with the seasons.
   (2) The midpoint of the Sun's path is the zenith.
   (3) The angle of the Sun's path to the horizon is greatest on December 21.
   (4) The Sun's path on certain days of the year is shown by line *SZN*.

9. Which arc represents a part of the observer's horizon?
   (1) *DAE*   (3) *SBZN*
   (2) *FCG*   (4) *DSEG*

10. On which date will the noon sun be nearest to position *B*?
    (1) September 21   (3) December 21
    (2) November 21   (4) January 21

11. Which arc represents part of the observer's celestial meridian?
    (1) *SBZ*   (3) *SDF*
    (2) *DFN*   (4) *GCF*

12. The star located almost directly above Earth's North Pole is
    (1) Alpha Centauri   (3) Polaris
    (2) Betelgeuse   (4) Sirius

13. The constellation Pisces changes position during a night as shown in the diagram below.

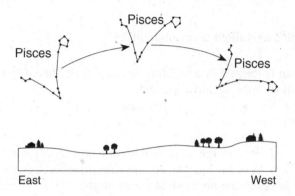

   Which motion is mainly responsible for this change in position?
   (1) revolution of Earth around the Sun
   (2) rotation of Earth on its axis
   (3) revolution of Pisces around the Sun
   (4) revolution of Pisces on its axis

14. The spinning of Earth on its axis causes the apparent rising and setting of the
    (1) Sun, only
    (2) Sun and the Moon, only
    (3) Moon and some stars, only
    (4) Sun, the Moon, and some stars

Base your answers to questions 15 and 16 on the diagram below, which shows sunlight entering a room through the same window at three different times on the same winter day.

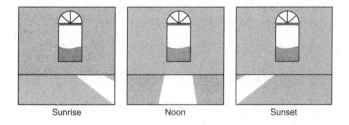

Early Astronomy and the Geocentric Model

15. The apparent change in the Sun's position shown in the diagram is best explained by
    (1) the Sun rotating at a rate of 15° per hour
    (2) Earth rotating at a rate of 15° per hour
    (3) the Sun's axis tilted at an angle of $23\frac{1}{2}°$
    (4) Earth's axis tilted at an angle of $23\frac{1}{2}°$

16. This room is located in a building in New York State. On which side of the building is the window located?
    (1) north (3) east
    (2) south (4) west

17. The dashed line on the map below shows a ship's route from Long Island, New York, to Florida. As the ship travels south, the star Polaris appears lower in the northern sky each night.

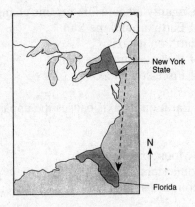

The best explanation for this observation is that Polaris
    (1) rises and sets at different locations each day
    (2) has an elliptical orbit around Earth
    (3) is located directly over Earth's Equator
    (4) is located directly over Earth's North Pole

18. Which diagram correctly shows the apparent motion of Polaris from sunset to midnight for an observer in northern Canada?

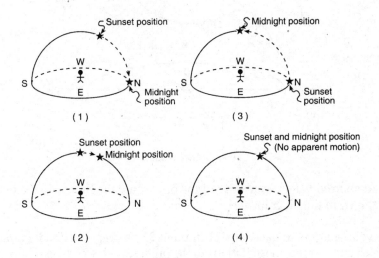

19. The diagram below represents the constellation Lyra.

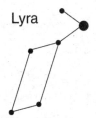

Which statement best explains why Lyra is visible to an observer in New York State at midnight in July but not visible at midnight in December?
(1) Earth spins on its axis.
(2) Earth orbits the Sun.
(3) Lyra spins on its axis.
(4) Lyra orbits Earth.

Early Astronomy and the Geocentric Model

CONSTRUCTED RESPONSE QUESTIONS

Base your answer to question 20 on the diagram below, which shows the Sun's apparent path as viewed by an observer in New York State on March 21.

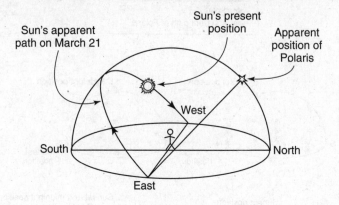

(Not drawn to scale)

20. At approximately what hour of the day would the Sun be in the position shown in the diagram? [1]

Base your answers to questions 21 through 23 on diagram 1 and on diagram 2, which show some constellations in the night sky viewed by a group of students. Diagram 1 below shows the positions of the constellations at 9:00 p.m. Diagram 2 shows their positions two hours later.

21. Circle Polaris on diagram 2. [1]

22. In which compass direction were the students facing? [1]

23. Describe the apparent direction of movement of the constellations Hercules and Perseus during the two hours between student observations. [1]

Base your answer to question 24 on the diagram below, which represents an observer's local reference lines on the celestial sphere. [3]

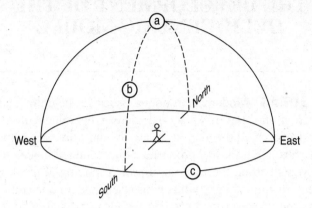

24. Identify the observer's zenith; celestial horizon; and celestial meridian.
    (a) _____ [1]
    (b) _____ [1]
    (c) _____ [1]

Extended Constructed Response Questions

25. Photographs of circumpolar stars, made with a time exposure, showed star trails having arcs of 60 degrees. How long had the film most likely been exposed? (Show all work.) [2]

26. Explain the difference between "real" motion and "apparent" motion. [3]

27. Explain why the stars on the celestial sphere appear to move during the night when you observe them from Earth. [1]

28. List three ways in which planets differ in appearance and/or behavior from stars. [3]

29. Explain why the North Star does *not* appear to change its position during the night. [2]

Chapter 2
THE DEVELOPMENT OF THE HELIOCENTRIC MODEL

> **KEY IDEAS** Modern astronomy traces its beginning to the publication in May 1543 by Nicolaus Copernicus of a new **heliocentric**, or Sun-centered, model of the universe. Although Aristarchus of Samos had proposed a Sun-centered model almost 1,800 years earlier, the idea that Earth is moving at great speed had been dismissed as obvious nonsense since no one could feel any motion. Copernicus discarded the idea of a stationary Earth and argued that Earth and the planets circle the Sun. His logical and mathematical arguments paved the way for further investigations. The shift from an Earth-centered to a Sun-centered model was revolutionary and has evolved into our current concept of the universe.

KEY OBJECTIVES
Upon completion of this chapter, you will be able to:

- Compare and contrast the geocentric and heliocentric models of the universe.
- Describe the investigations that led scientists to understand that most of the observed motions of celestial objects are the result of Earth's motion around the Sun.
- Explain how gravity influences the motions of celestial objects.
- Determine the gravitational force between two objects, given their masses and the distance between their centers.
- Analyze the relationships among a planet's distance from the Sun, gravitational force, period of revolution, and speed of revolution.

EARLY MODELS: ARISTOTLE AND PTOLEMY

Ancient Greek thinkers, particularly Aristotle, set a pattern of belief that persisted for 2,000 years—the universe had a large, stationary Earth at its center; and the Sun, the Moon, and the stars were arranged around Earth in a perfect sphere, with all of these bodies orbiting Earth in perfect circles at constant speeds. The Egyptian astronomer Ptolemy refined this concept into an elegant mathematical model of circular motions that enabled astronomers to predict the positions of celestial objects fairly accurately and could account for many of the "problem" observations that plagued Aristotle's model.

Aristotle's Geocentric Universe

Aristotle, a Greek philosopher who lived from 384 B.C. to 322 B.C., wrote about and taught many subjects, including history, philosophy, drama, poetry, and ethics. His wide-ranging knowledge and insight earned him a prominent place among the great thinkers of antiquity.

Aristotle's was a common sense view of the universe. He understood the celestial sphere model and its ability to explain most casual observations, such as the apparent movements of celestial bodies. As records of careful measurements were kept over time, however, some problems arose. The Sun doesn't follow the same path through the sky all year long. The Moon changes position relative to the stars from night to night. Five (actually, nine) "stars," out of the thousands seen in the sky, don't stay in fixed positions relative to the others, but "wander" around in the sky. These moving objects, as explained in Chapter 1, came to be called **planets**, from *planetes*, the Greek word for "wanderer." Aristotle realized that a one-sphere model couldn't explain these "problem" observations, so he revised the model.

Spheres Within Spheres

Aristotle reasoned that, if some objects move differently, they must be on different celestial spheres! Aristotle explained the "problem" observations by proposing a universe consisting of eight crystalline (i.e., transparent) spheres nesting one inside the other like a Chinese box puzzle, with Earth at the very center. The Sun, Moon, stars, and planets were fixed to the surface of separate spheres, which rotated around the unmoving Earth. All motions of the spheres were perfect circles. By having the spheres spinning at slightly different rates and at slightly different angles in relation to one another, most of the "problem" observations could be accounted for. Either the spheres moved because they were self-propelled, or, as was thought more likely, their motion was initiated by a supernatural being. See Figure 2.1.

Common Sense and Parallax

In Aristotle's model, Earth too was a sphere, the perfect shape, as could be seen when its shadow was visible against the Moon during an eclipse. Common sense indicated that Earth wasn't moving because no motion could be felt, but Aristotle believed there was other evidence as well. If Earth moved, objects falling in a straight line should fall to the side of points directly beneath them. Later astronomers argued against a moving Earth by citing a lack of stellar parallax. **Parallax** is the change in apparent position of closer objects in relation to farther objects due to a change in the position of the observer. If Earth was moving, the apparent positions of the stars should change as Earth (and the observer) moved. See Figure 2.2.

The Development of the Heliocentric Model

Figure 2.1 Aristotle's Model of the Universe. Crystalline spheres were nested one inside the other, with Earth at the center. The spheres and their attached stars and planets rotated around Earth. Source: *Horizons: Exploring the Universe*, Michael A. Seeds, Wadsworth, 1987.

According to Aristotle, the natural state of things on Earth was to be at rest. Natural motion on Earth was towards its center. Unlike the perfect circular motion of the spheres, the motion of objects on Earth was imperfect straight-line motion. The spheres were perfectly clear and were composed of ether, a substance that could not be changed or destroyed. The other four elements were earth, water, air, and fire. All objects were made of mixtures of these four elements and decayed as a result of being forced to move in unnatural directions.

Since Aristotle's universe has Earth at its center, it is called a **geocentric**, or Earth-centered, model of the universe.

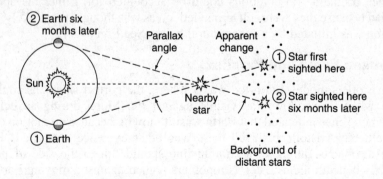

Figure 2.2 Stellar Parallax: the apparent Shift in Position of a Nearby Star Against the Background of More Distant Stars. Stars are so far away that the parallax angle was too small to be detected by early astronomers. Source: *Astronomy: A Self-Teaching Guide*, 4th Ed., Dinah L. Moché, John Wiley, 1998.

Ptolemy's Geocentric Model

There were some obvious problems with Aristotle's view of the universe. The most obvious was visible to the naked eye. There were times when the planets changed course in the sky; for example, at times Mars would stop and then move backwards, a phenomenon called **retrograde motion**. Since the crystal spheres of the Aristotelian universe could not stop or change direction, this observation could not be explained until the second century A.D., when Claudius Ptolemaeus, usually referred to as Ptolemy, proposed an ingenious theory.

Ptolemy, an Egyptian, lived and worked in the Greek settlement at Alexandria in about A.D. 140. There he studied mathematics and astronomy and developed a model of the universe based upon Aristotle's teachings. The details of his model are carefully spelled out in his great book, *Almagest*.

Explaining Retrograde Motion

Ptolemy's view was that each planet was fixed to a small sphere that was in turn fixed to a larger sphere. The smaller sphere and its attached planet turned at the same time that the larger sphere turned. As a result there could be times when, to an observer on Earth, the planet appeared to be moving backward. Ptolemy called the circular motions of the larger spheres *deferents* and the motions of the smaller spheres *epicycles*. He placed Earth's sphere off the center of its deferent. See Figure 2.3.

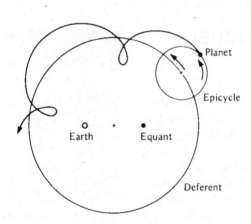

Figure 2.3 Ptolemy's Universe. Ptolemy added epicycles to Aristotle's model to explain retrograde motion and changes in apparent diameter. Source: *Horizons: Exploring the Universe*, Michael A. Seeds, Wadsworth, 1987.

With Ptolemy's ingenious modifications, Aristotle's universe could explain all casual, naked-eye observations of the universe. For 1,000 years astronomers studied and preserved Ptolemy's work, making no changes in his basic theory. It became part of the accepted thinking of the time. See Figure 2.4.

Problems with Predictions

At first, the Ptolemaic system was able to predict the motions of celestial objects with a fair degree of accuracy. However, as the centuries passed, the differences between what the Ptolemaic system predicted and what was actually observed grew so large they could not be ignored. At first, earlier

The Development of the Heliocentric Model

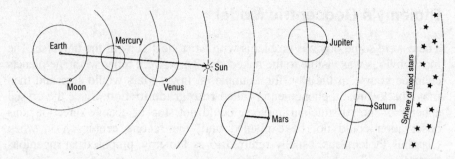

Figure 2.4 Ptolemy's Geocentric Model of the Universe. This model was accepted for well over 1,000 years. Source: *Horizons: Exploring the Universe*, Michael A. Seeds, Wadsworth, 1987.

astronomers blamed these discrepancies on poor instruments or inaccurate observations. Arabian and, later, European astronomers corrected the system, recalculated constants, and even added new epicycles. King Alfonso X of Castile paid for the last great correction of the Ptolemaic model. Ten years of observations and calculations were then published as the Alfonsine Tables. By the 1500s, however, the Alfonsine Tables were also inaccurate, often being off by as much as 2°, which is four times the angular diameter of the moon—a significant error.

THE HELIOCENTRIC MODEL

Copernicus

At about the same time that astronomers were struggling with the inaccurate Alfonsine Tables, there was a serious need for calendar reform. By the beginning of the 1500s, the Julian calendar was off by about 11 days. Easter, a major church holiday, was particularly hard to determine. Both the Hebrew calendar, which was based upon the Moon, and the Julian calendar, which was based upon the Sun, had to be used to calculate the phase of the moon, upon which the date of Easter depended. A secretary of Pope Sixtus IV asked Nicolaus Copernicus, a priest-mathematician from Poland (see Figure 2.5), to examine the problem of calendar reform.

Figure 2.5 Nicolaus Copernicus. Source: *Astronomy: The Cosmic Journey*, William K. Hartman, Wadsworth, 1987.

Copernicus recognized that any calendar reform would have to resolve the relationship between the Sun and the Moon. After much study of the problem, Copernicus proposed a mathematically elegant solution in which he suggested a **heliocentric**, or Sun-centered, universe with a moving Earth.

In 1514, he distributed a brief manuscript outlining his ideas, but was discreet because he recognized the potential dangers of questioning church dogma. Not until his death in 1543 was his full argument in favor of a Sun-centered system published. Even then, he avoided heresy charges by crediting classical Greek sources with the idea, thus implying that the concept did not originate with him.

In Copernicus' model of a heliocentric universe, the center of the universe was a point near the Sun. Earth orbited the Sun and spun once a day on its axis. See Figure 2.6.

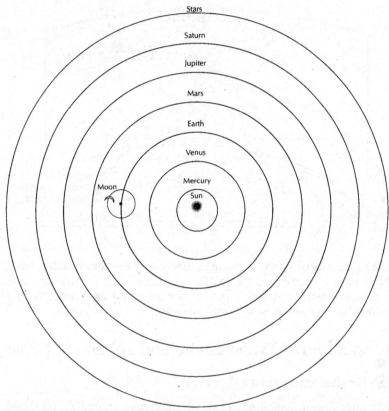

Figure 2.6 The Copernican Heliocentric Universe. Copernicus proposed a Sun-centered model in which all planets and stars moved in perfect circles around the Sun. Source: *Discovering Astronomy*, Robert D. Chapman. W. H. Freeman, 1978. Used with permission.

Copernicus explained the lack of stellar parallax by stating that stars are so distant that parallax is too small to measure. However, he avoided the problem of objects falling to the side of positions directly under them. Retrograde motion occurs because Earth moves faster in its orbit than do planets farther from the Sun. Earth and the other planets all move continuously in their orbits around the Sun, but Earth moves toward an outer planet in one part of its orbit and then passes it and moves away from it. However, planets moving in perfect circles around the Sun could not explain all of the observed details of their motions, and in the end Copernicus, too, resorted to epicycles and did no better at predicting the positions of celestial objects than Ptolemy.

The Development of the Heliocentric Model

While Copernicus' system was also erroneous, his idea that the universe was heliocentric, or Sun-centered, was correct and gradually gained acceptance. Probably the most important reasons why his theory was eventually accepted were the revolutionary mood of the world in his lifetime and the simple, forthright way in which his model explained retrograde motion. See Figure 2.7.

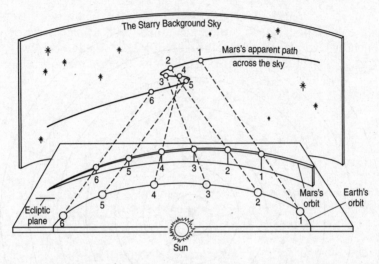

Figure 2.7 Copernicus' Simple, Forthright Explanation of Retrograde Motion. Both Earth and Mars move in a continuous path, but the inner planet (Earth) covers more of its orbit in the same time period, changing its point of view toward the outer planet (Mars). Source: *Astronomy: The Cosmic Journey*, William K. Hartman, Wadsworth, 1987. Used with permission.

Contributions of Tycho Brahe and Johannes Kepler

Tycho Brahe: Precision Observer

Shortly after Copernicus died, a Danish nobleman named Tycho Brahe (see Figure 2.8a) became interested in astronomy. After observing that the Alfonsine Tables were nearly a month off in predicting a conjunction of Jupiter and Saturn, and observing a "new star" produced by a supernova, that is, the explosion of a very large star, Tycho questioned the Ptolemaic system of a perfect, unchanging heaven in a small book he wrote. His book was widely read, and the King of Denmark gave him funds to build a world-class astronomical observatory with no telescopes, but many ingenious devices for precisely measuring celestial motions. When the King of Denmark died, Tycho fell out of favor and accepted a position as court astronomer to the Holy Roman Emperor in Prague, taking with him all of the data from the observatory in Denmark. In Prague, the emperor commissioned him to publish a revision of the Alfonsine Tables. Tycho hired several young mathematicians to help him with his task.

Johannes Kepler: Orbits Are Ellipses, Not Circles

One of Tycho's young assistants was Johannes Kepler (see Figure 2.8b). Shortly after beginning the project commissioned by the emperor, Tycho died unexpectedly. Before he died, however, he recommended Kepler to take over his position. As court astronomer, Kepler spent six years trying to work out the orbit of the planet Mars, using Ptolemy's system of the planet moving in a small circle that moved in a larger circle around the Sun. But no matter how hard he tried, he could not get the theoretical orbit to match the observed orbit. Finally Kepler realized that the orbit of Mars was elliptical, or oval, and that Mars moved at a speed that varied with its distance from the Sun.

Figure 2.8
(a) Tycho Brahe (1546–1601)
(b) Johannes Kepler (1571–1630)
Source: *Astronomy: The Cosmic Journey*, William K. Hartman, Wadsworth, 1987. Used with permission.

(a)   (b)

Kepler's Laws of Planetary Motion

After years of studying observations of celestial objects, Kepler made three important discoveries about the motions of planets as they revolve around the Sun.

1. **Each planet revolves around the Sun in an elliptical orbit with the Sun at one focus.**

    An **ellipse** has a major axis and a minor axis that are lines connecting the two points farthest apart and the two points closest together on the ellipse. It also contains two special points along the major axis, each called a **focus** (plural, foci). The distance from one focus to any point on the ellipse and back to the other focus is always the same.

    As a result it is very easy to draw an ellipse using two tacks and a loop of string. Press the tacks into a board, loop the string around them, and place a pencil in the loop on page 26. Keep the string taut; then, as you move the pencil, it will trace out the shape of an ellipse. See Figure 2.9.

    The closer together the foci, the more nearly circular the ellipse. The farther apart the foci, the flatter the ellipse. The flatness of an ellipse is called its **eccentricity**. Eccentricity is expressed as the ratio between the distance between the foci and the length of the major axis:

The Development of the Heliocentric Model

A perfect circle would have an eccentricity of 0; a straight line has an eccentricity of 1.

$$e = \frac{d}{l}$$

where e = eccentricity,
d = distance between foci, and
l = length of major axis.

Since the orbit of each planet is an ellipse, the distance from each planet to the Sun varies during its orbit. See Figure 2.10. For example, Earth's distance to the Sun varies from 147×10^6 kilometers on January 3 at its closest (perihelion) to 152×10^6 kilometers on July 6 at its farthest (aphelion). The difference between these distances, 5×10^6 kilometers, is the distance between the foci of Earth's elliptical orbit. This measurement is very small compared to the length of the major axis, 299×10^6 kilometers, so the eccentricity of Earth's orbit is very small (0.17), indicating that the orbit is very nearly a circle. Although many illustrations show Earth's orbit around the Sun in a perspective view that exaggerates its eccentricity, if viewed from directly overhead the orbit would appear very nearly circular. (In this connection, it is interesting to note that Ptolemy's system of circular orbits *almost* worked because Earth's orbit is *almost* a circle. However, that slight difference from a perfect circle was enough to throw Ptolemy's system into question over time.)

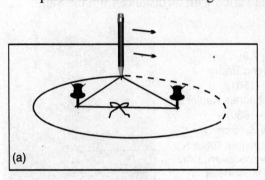
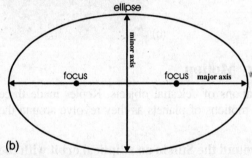

Figure 2.9 (a) The way to draw an ellipse (b) The main parts of an ellipse

As the distance between Earth and the Sun changes, the apparent diameter of the Sun changes in a cyclic manner. When Earth is closest to the Sun on January 4, the Sun has its greatest apparent diameter. On July 4, when the Sun is farthest away, it has its smallest apparent diameter.

2. **The planets do not move at a constant velocity.**
In Figure 2.11, the elliptical shape of a planet's orbit is exaggerated to show variation in velocity more clearly. The times the planet takes to

move from 1 to 2, from 3 to 4, and from 8 to 9 are all equal. However, if you look at the diagram carefully, you can see that the distance from 1 to 2 is less than the distance from 8 to 9 even though the distances were covered in the same time. This means that the planet is moving fastest when it is closest to the Sun and slowest when it is farthest from the Sun.

Kepler did not know *why* this was so; he only determined that the variation did occur. We now know that this cyclic changing velocity of the planets as they move around the Sun is due to changes in the gravitational force between a planet and the Sun as distance changes. As a planet approaches the Sun (distance decreases), gravitational force increases and, since it is acting in the same direction in which the planet is moving, causes the planet to move faster. As a planet moves away from the Sun (distance increases), gravitational force decreases and, since it is now acting opposite to the direction in which the planet is moving, causes the planet to slow down.

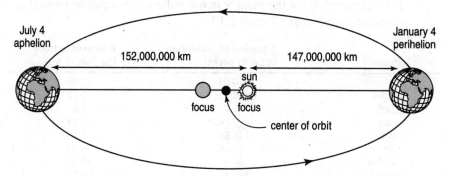

Figure 2.10 View of Earth's Elliptical Orbit with the Sun at One Focus. Earth is closest to the Sun at perihelion and farthest away at aphelion. Distances are approximate.

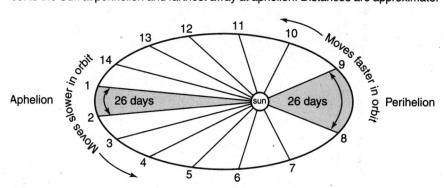

Figure 2.11 Kepler's Law of Equal Areas. A planet sweeps out equal areas of its elliptical orbit in equal periods of time. Since the planet travels less distance in the 26 days it takes to move from 1 to 2 on the ellipse above, it is traveling slower than when it moves from 8 to 9.

The Development of the Heliocentric Model

3. **There is a mathematical relationship between the time a planet takes to complete one revolution around the Sun and its average distance from the Sun.**

The time required for a planet to make one revolution is called its **period of revolution**. It was already known that the farther a planet is from the Sun, the longer its period of revolution. Kepler's careful analysis showed that there is a mathematical relationship between the two factors. See the accompanying table. A planet's period of revolution *squared* is proportional to its distance from the Sun *cubed*:

$$T^2 \propto R^3$$

If we measure the period of revolution in units of Earth-years, and let Earth's average distance from the Sun equal 1 unit of distance, called an *astronomical unit* (AU), the relationship simplifies. Then T^2 becomes $(1)^2$, R^3 becomes $(1)^3$, and, since $(1)^2 = 1$ and $(1)^3 = 1$:

$$T^2 = R^3$$

This relationship for the planets in our solar system can be plotted as a curve on a graph. See Figure 2.12.

Planet	T (period of revolution in Earth-years)	R (distance from the Sun in AU)
Mercury	0.24	0.39
Venus	0.62	0.72
Earth	1.0	1.0
Mars	1.88	1.52
Jupiter	11.86	5.23
Saturn	29.46	9.54
Uranus	84.0	19.18
Neptune	164.8	30.05

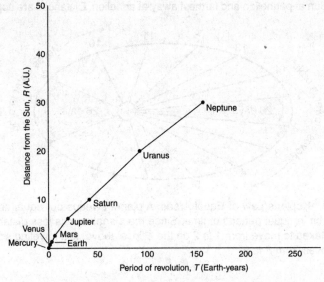

Figure 2.12 Relationship between Period of Revolution and Distance from the Sun

Galileo and Newton: Improving the Heliocentric Model

Galileo: Observations That Challenged Aristotle's Geocentric Universe

Shortly after Kepler's works were published, Galileo Galilei, an Italian astronomer and physicist (see Figure 2.13a), turned his telescope on the heavens and made several discoveries that further undermined the Ptolemaic system. His discovery of imperfections on the Moon's surface, spots on the surface of the Sun, and moons circling Jupiter challenged Aristotle's view of the heavens as perfect and unchanging. He saw Venus go through a full set of phases, which was not possible according to the Ptolemaic system of epicycles. Galileo also studied motion and solved Copernicus' falling-object problem. He argued that, if Earth is in motion, so are all objects on it. Therefore, an object that is dropped moves sideways at the same speed as Earth and falls on a spot directly beneath its point of release. Galileo became an outspoken champion of the heliocentric model.

Figure 2.13
(a) Galileo Galilei (1564–1642)
(b) Isaac Newton (1642–1727)
Source: *Astronomy: The Cosmic Journey*, William K. Hartman, Wadsworth, 1987. Used with permission.

(a)   (b)

Newton: Explaining Motion

Eleven months after Galileo died in 1642, Isaac Newton (see Figure 2.13b) was born in England. Newton brought the discoveries of Copernicus, Galileo, and Kepler together. Using a few key concepts (mass, momentum, acceleration, and force), three laws of motion (inertia, the dependence of acceleration on force and mass, and action and reaction), and the law of universal gravitation, Newton was able to explain both the motions of objects on Earth and the distant motions of celestial objects.

Newton's laws of motion made it possible to predict how an object would move if the forces acting on it were known. Newton thought about the forces that would be needed to keep a satellite moving in orbit around another object. In considering the Moon, Newton realized that the Moon would circle Earth only if some force pulled the Moon toward Earth's center; otherwise it would continue moving in a straight line off into space. Newton's

The Development of the Heliocentric Model

genius was to realize that the force that keeps the Moon in orbit around Earth is the same force that causes objects close to Earth (e.g., apples on a tree) to fall to the ground: gravity. He realized that the force of **gravity** is universal, that all objects are attracted to one another with a force that depends on the masses of the objects and their distance from each other. Newton expressed this relationship in a simple mathematical formula. See Figure 2.14.

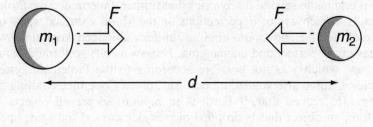

$$F = -G \frac{m_1 m_2}{d^2}$$

Figure 2.14 Newton's Law of Gravity. In this equation, F is the force of gravity acting between two masses, G is the gravitational constant, m_1 and m_2 are the masses, and d is the distance between them. Source: *Astronomy: A Self-Teaching Guide*, 4th Ed., Dinah L. Moché, John Wiley, 1998. Used with permission.

Gravity and Orbital Motion

How does gravity keep satellites moving in a curved orbit? Imagine that a cannonball is shot out of a cannon aimed horizontally. If there were no gravity, inertia would cause the cannonball to fly off horizontally until some force stopped it. However, with gravity pulling the cannonball downward toward Earth's center, as the cannonball is flying horizontally its path curves downward and eventually the cannonball strikes Earth's surface. If a more powerful charge is used in the cannon, the cannonball will travel farther horizontally before it strikes Earth. With a sufficiently powerful charge, the cannonball would travel far enough horizontally that, as its path curved downward because of gravity, Earth's surface would curve away because of its spherical shape, and the cannonball would never strike Earth's surface. Instead, gravity would cause the cannonball to fall downward at the same rate that Earth's surface curves away from it, and it would fall unendingly in a circular path around Earth—it would be in orbit. See Figure 2.15.

Similarly, as Newton explained, it is a combination of two forces that keeps the planets moving in their curved paths around the Sun. The combination of a planet's forward motion and its motion toward the Sun due to gravity results in circular motion—the planet's orbit around the Sun. See Figure 2.16.

Now let's return to our cannonball analogy. For an orbit just above Earth's surface, the cannonball would have to be shot out of the cannon at 7.9×10^3 meters per second, or about 18,000 miles per hour. If the cannonball was pro-

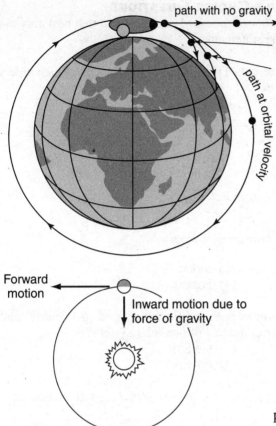

Figure 2.15 Orbital Velocity. As a fired cannonball travels through the air, it is drawn downward by gravity in a curved path until it strikes Earth's surface. If it is fired with more energy, it will travel farther in a curved path before crashing into Earth's surface. If it is fired with enough energy, its path will curve downward at the same rate that Earth curves away from its path, and it will go into orbit.

Figure 2.16 The Two Motions That Explain Why a Planet Travels in Orbit. Source: *Astronomy: A Self-Teaching Guide*, 4th Ed., Dinah L. Moché, John Wiley, 1998. Used with permission.

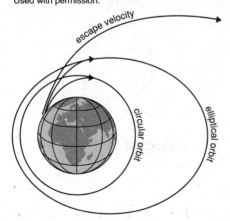

Figure 2.17 Circular Orbit, Elliptical Orbit, and Escape Velocity.

pelled at a higher speed, it would travel farther outward before being pulled back by gravity, and the orbit would be elliptical instead of circular. If the cannonball was shot out at a velocity equal to or greater than 11.2×10^3 meters per second, or about 25,000 miles per hour, it would be able to escape Earth's gravity and fly out of orbit. See Figure 2.17.

With Newton's explanation of the causes of motion, the heliocentric theory began to firmly displace the geocentric theory as the generally accepted model of the universe. Our modern view of planetary motions in the solar system is based upon the heliocentric model.

The Development of the Heliocentric Model

MULTIPLE-CHOICE QUESTIONS

For each case, write the number of the word or expression that best answers the question or completes the statement.

1. The diagram below represents a simple geocentric model. Which object is represented by the letter X?

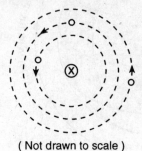

(Not drawn to scale)

(1) Earth (3) Moon
(2) Sun   (4) Polaris

2. Which object orbits Earth in both the Earth-centered (geocentric) and Sun-centered (heliocentric) models of our solar system?
   (1) the Moon   (3) the Sun
   (2) Venus   (4) Polaris

3. Which diagram best represents the motions of celestial objects in a heliocentric model?

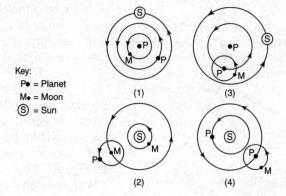

4. What is the exact shape of Earth's orbit around the Sun?
   (1) a slightly eccentric ellipse
   (2) a very eccentric ellipse
   (3) an oblate spheroid
   (4) a perfect circle

5. Which information about a nearby star must be known to determine its distance from an observer?
   (1) size
   (2) color
   (3) temperature
   (4) parallax

6. In each diagram below, the mass of the star is the same. In which diagram is the force of gravity greatest between the star and the planet shown?

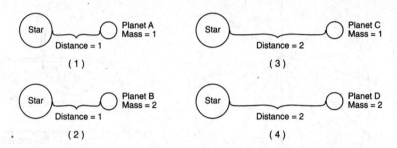

7. If the average distance between Earth and the Sun were doubled, what changes would occur in the Sun's gravitational pull on Earth and Earth's period of revolution?
   (1) The gravitational pull would decrease, and the period of revolution would increase.
   (2) The gravitational pull would decrease, and the period of revolution would decrease.
   (3) The gravitational pull would increase, and the period of revolution would increase.
   (4) The gravitational pull would increase, and the period of revolution would decrease.

8. The diagram shows Earth (E) in orbit about the Sun. If the gravitational force between Earth and the Sun were suddenly eliminated, toward which position would Earth then move?
   (1) 1
   (2) 2
   (3) 3
   (4) 4

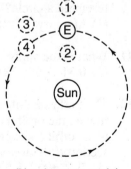

(Not drawn to scale)

The Development of the Heliocentric Model

Base your answers to questions 9 through 13 on your knowledge of Earth science, the *Earth Science Reference Tables*, and the diagrams, tables, and information below. Diagram I represents the orbit of an Earth satellite, and diagram II shows how to construct an elliptical orbit using two pins and a loop of string. The table shows the eccentricities of the orbits of the planets in the solar system.

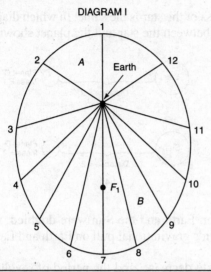

The satellite was at position 1 precisely at midnight on the first day. It arrived at position 2 the next midnight, at 3 the next, and so on.

Planet	Eccentricity of Orbit
Mercury	0.206
Venus	0.007
Earth	0.017
Mars	0.093
Jupiter	0.048
Saturn	0.056
Uranus	0.047
Neptune	0.008

9. At which position represented in diagram I would the gravitational attraction between the Earth and the satellite be greatest?
   (1) 1   (2) 7   (3) 3   (4) 11

10. According to the table, the orbit of which planet would most closely resemble a circle?
    (1) Mercury   (2) Venus   (3) Saturn   (4) Mars

11. What is the approximate eccentricity of the satellite's orbit?
    (1) 0.31   (2) 0.40   (3) 0.70   (4) 2.5

12. The Earth satellite takes 24 hours to move between each numbered position on the orbit. How does the orbital speed of the satellite in section *A* of its orbit (between positions 1 and 2) compare to its orbital speed in section *B* (between positions 8 and 9)?
    (1) It is moving faster in section *A* than in section *B*.
    (2) It is moving slower in section *A* than in section *B*.
    (3) Its speed in section *A* is equal to its speed in section *B*.
    (4) It is speeding up in section *A* and slowing down in section *B*.

Note that question 13 has only three choices.

13. If the pins in diagram II were placed closer together, the eccentricity of the ellipse being constructed would
(1) decrease   (2) increase   (3) remain the same

Base your answers to questions 14 through 16 on the diagram below, which represents the elliptical orbit of a planet traveling around a star. Points *A*, *B*, *C*, and *D* are four positions of this planet in its orbit.

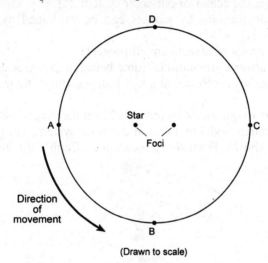

(Drawn to scale)

14. The calculated eccentricity of this orbit is approximately
(1) 0.1   (3) 0.3
(2) 0.2   (4) 0.4

15. The gravitational attraction between the star and the planet will be greatest at position
(1) *A*   (3) *C*
(2) *B*   (4) *D*

16. As the planet revolves in orbit from position A to position D, the orbital velocity will
(1) continually decrease
(2) continually increase
(3) decrease, then increase
(4) increase, then decrease

Constructed Response Questions

17. Listed below are statements of several theories and observations, some of which may have served as partial bases for Newton's Law of Universal Gravitation. For each statement, (a)–(e), write the number preceding the name of the scientist, chosen from the list below, who first made the observation or proposed the theory.

    (1) Brahe, (2) Newton, (3) Copernicus, (4) Ptolemy, (5) Galileo, (6) Aristotle, (7) Kepler

    (a) The Sun is the center of our solar system. [1]
    (b) Retrograde motions by planets can be explained by the use of epicycles. [1]
    (c) The true orbits of planets are ellipses. [1]
    (d) There is always gravitational force between two masses. [1]
    (e) Falling bodies accelerate at a rate independent of their masses. [1]

Base your answers to questions 18 through 21 on the diagram below, which shows the heliocentric model of a part of our solar system. The planets closest to the Sun are shown. Point B is a location on Earth's equator.

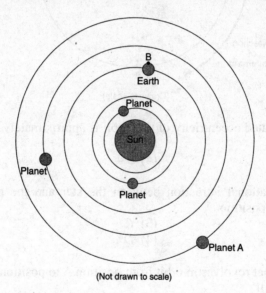

(Not drawn to scale)

18. State the name of planet A.

19. Explain why location B experiences both day and night in a 24-hour period.

20. On the graph below, draw a line to show the general relationship between a planet's distance from the Sun and the planet's period of revolution.

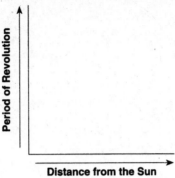

21. Identify one feature of the geocentric model of our solar system that differs from the heliocentric model shown.

Base your answers to questions 22 through 24 on the diagram below, which represents a model of Earth's orbit. Earth is closest to the Sun at one point in its orbit (perihelion) and farthest from the Sun at another point in its orbit (aphelion). The Sun and point B represent the foci of this orbit.

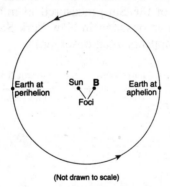

(Not drawn to scale)

22. Explain why Earth's orbit is considered to be elliptical.

23. Describe the change that takes place in the gravitational attraction between Earth and the Sun as Earth moves from perihelion to aphelion and back to perihelion during one year.

24. Describe how the shape of Earth's orbit would differ if the Sun and focus B were farther apart.

The Development of the Heliocentric Model

Base your answers to questions 25 and 26 on the diagram of the ellipse below.

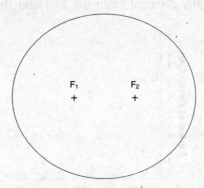

25. Calculate the eccentricity of the ellipse to the *nearest thousandth*. [1]

26. State how the eccentricity of the given ellipse compares to the eccentricity of the orbit of Mars. [1]

EXTENDED CONSTRUCTED RESPONSE QUESTIONS

Base your answers to questions 27 and 28 on the data table below, which lists the apparent diameter of the Sun, measured in minutes and seconds of a degree, as it appears to an observer in New York State. (Apparent diameter is how large an object appears to an observer.)

Apparent Diameter of the Sun During the Year

Date	Apparent Diameter (' = minutes " = seconds)
January 1	32'32"
February 10	32'25"
March 20	32'07"
April 20	31'50"
May 30	31'33"
June 30	31'28"
August 10	31'34"
September 20	31'51"
November 10	32'18"
December 30	32'32"

27. On the grid below, graph the data shown on the table by marking with a dot the apparent diameter of the Sun for *each* date listed and connecting the dots with a smooth, curved line. [2]

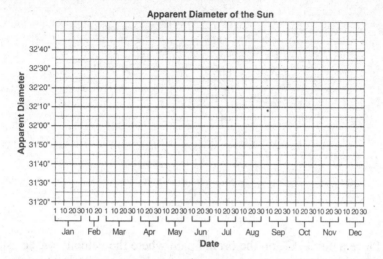

28. Explain why the apparent diameter of the Sun changes throughout the year as Earth revolves around the Sun. [1]

Base your answers to questions 29 and 30 on the diagram below, which shows Earth's orbit and the orbit of a comet within our solar system.

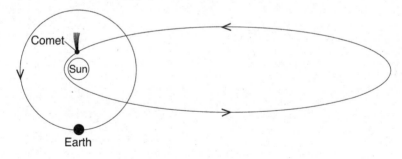

(Not drawn to scale)

29. Explain how this comet's orbit illustrates the heliocentric model of our solar system.

30. Explain why the time required for one revolution of the comet is more than the time required for one revolution of Earth.

The Development of the Heliocentric Model

Base your answers to questions 31 and 32 on the diagram below, which represents an asteroid's elliptical orbit around the Sun. The dashed line is the major axis of the ellipse.

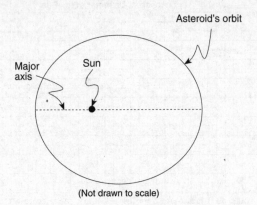

(Not drawn to scale)

31. Place a circle, **O**, on the orbital path where the velocity of the asteroid would be the least.

32. The Sun is located at one focal point of the orbit. Place an **X** on the diagram at the location of the second focal point.

CHAPTER 3

HELIOCENTRIC EARTH MOTIONS AND THEIR EFFECTS

> **KEY IDEAS** The shift from a geocentric to a heliocentric model of the universe changed the way most people saw themselves in relation to the physical universe. The heliocentric model required the apparently immobile Earth to spin completely around on its axis once a day and the universe to be far larger than anyone had imagined; worst of all, Earth lost its position at the center of the universe and became just one of nine planets that orbit the Sun, a typical star in a vast and ancient universe.
>
> However, the heliocentric model has greatly improved human understanding of complex phenomena, such as variations in day length, seasons, ocean tides, eclipses, phases of the Moon, the apparent motion of planets, and the annual traverse of the constellations.

KEY OBJECTIVES
Upon completion of this chapter, you will be able to:

- Explain how the Foucault pendulum and the Coriolis effect provide evidence of Earth's rotation.
- Describe Earth's rate of rotation, the orientation of Earth's axis of rotation with respect to the plane of its orbit, and the effects of Earth's changing position with regard to the Sun.
- Explain how seasonal changes in the apparent positions of constellations provide evidence of Earth's revolution.
- Interpret diagrams of Earth's orbit to determine how the Sun's apparent path through the sky and the relative position of the noon Sun vary with the seasons, and how the length of daylight varies throughout the year at different locations.
- Explain how the changing relative positions of Earth, the Moon, and the Sun account for the phases of the Moon, tides, and eclipses.

EARTH MOTIONS

Rotation

In the modern heliocentric model, Earth rotates at a rate of 15° per hour, or one complete rotation every 24 hours. **Rotation** is a motion in which every part of an object is moving in a circular path around a central line called the **axis of rotation**, like an ice skater spinning around a line through his or her body. Earth's axis of rotation is a line passing through the North Pole, Earth's center, and the South Pole. Earth's axis of rotation is almost directly aligned with the star Polaris and is tilted at an angle of 23½° from a perpendicular to the plane passing through the centers of Earth and the Sun. See Figure 3.1.

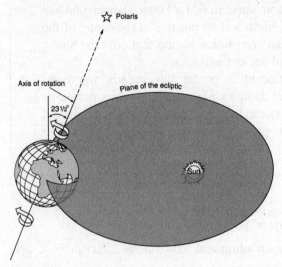

Figure 3.1 Rotation. Earth rotates from west to east once every 24 hours around an axis that runs through the poles.

Evidence of Rotation

The heliocentric theory requires Earth to rotate on its axis once every 24 hours. It is one thing to be able to successfully explain away a problem such as the lack of deflection of a falling object; it is quite another to actually *prove* Earth's rotation. The arguments in favor of rotation may be quite reasonable, but reasonableness is not scientific proof. A French physicist, Jean Foucault, and a French mathematician, Gaspard Coriolis, provided this proof.

The Foucault Pendulum
Jean Foucault used an ingeniously simple method to prove Earth's rotation. Foucault knew that gravity pulls a pendulum only toward Earth's center. It does not act laterally to change the plane of the pendulum's swing. Therefore, if set freely swinging in the absence of any lateral forces, a pendulum should swing in a fixed plane. Foucault further reasoned that the heavier he made the weight of the pendulum, the more force would be required to push it out of its plane of motion, making it unlikely that small breezes would deflect the pendulum. Finally, he realized that, if he made the pendulum really long, it would have a long period and would swing for many hours without needing a push to get it going again (which would introduce a possibility of exerting a lateral force). Foucault argued that, if Earth were motionless, such a pendulum

would swing in a plane whose direction would not change. However, if Earth rotated, it would rotate beneath the pendulum and would change position relative to the swinging pendulum. The result would be a pendulum whose plane of swing would appear to rotate relative to the ground in a direction opposite to that of Earth's rotation.

Foucault's idea is easiest to understand if you visualize a pendulum swinging over the North Pole. As the pendulum swings, Earth moves beneath it. In 24 hours the plane of the swinging pendulum will appear to make one rotation. At the Equator, however, the plane of the pendulum will not appear to change. In between, the period of rotation of swinging pendulums will vary from 24 hours at the poles to infinity at the Equator. At 40°N, a pendulum takes 37 hours to make one rotation. See Figure 3.2.

In 1851, Foucault suspended a freely swinging pendulum from the inside

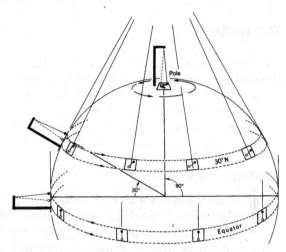

Figure 3.2 The Foucault Pendulum.

of the dome of the Pantheon building in Paris. The high dome allowed the use of a very long (60-meter) pendulum. Foucault fastened the pendulum's heavy weight to one side of the room with a thin cord and sealed all entrances to eliminate possible drafts that could exert a lateral force on the pendulum. He then burned the cord to set the pendulum swinging smoothly in a fixed plane. Through windows, observers were able to watch the pendulum rotate slowly in a direction opposite to that of Earth's rotation, thus proving that Earth rotates.

The Coriolis Effect

Gaspard Coriolis first described the behavior of objects moving in a rotating frame of reference. His work successfully predicted the behavior of objects moving near the rotating Earth. In a stationary system, fluids such as the atmosphere or the oceans move directly in a straight line from regions of high pressure to regions of low pressure. However, that behavior does not occur on Earth. Large-scale movements of both the atmosphere and the oceans follow curved paths.

Study the following diagrams in the *Reference Tables for Physical Setting/ Earth Science* (Appendix B): Surface Ocean Currents (page 698) and Planetary Wind and Moisture Belts in the Troposphere (page 708). Note that ocean currents form large circular patterns, called *gyres*, that flow clockwise in the Northern Hemisphere and counterclockwise in the Southern Hemisphere.

Note, too, that planetary winds follow paths that curve to the right in the Northern Hemisphere and to the left in the Southern Hemisphere. Such motions should not occur if Earth is stationary, but are precisely what would be expected if Earth is rotating. For a full description of why this occurs, refer to the section on the Coriolis effect in Chapter 22. Since the curving paths of ocean currents and weather systems would not occur on a stationary Earth, their existence is considered proof of Earth's rotation.

Revolution

As Earth rotates, it revolves around the Sun once every 365¼ days. **Revolution** is the motion of one body around another body in a path called an *orbit*. The revolving body is termed a **satellite** of the body it orbits. The body that a satellite orbits is its *primary*. Thus, Earth is a satellite of the Sun, its primary. The plane of Earth's orbit is called the *ecliptic* because eclipses occur when Earth, the Moon, and the Sun align in this plane.

Evidence of Revolution

Parallelism of Earth's Axis of Rotation

As stated previously, Earth's axis of rotation is tilted 23½° from a perpendicular to the plane of its orbit. Spinning like a top, Earth holds its axis fixed in space as it moves around the Sun. Another way to describe this is to say that Earth's axis of rotation at any point in Earth's orbit is parallel to its axis at any other point. As a result, the Northern Hemisphere is tipped toward the Sun in June and away from the Sun in December.

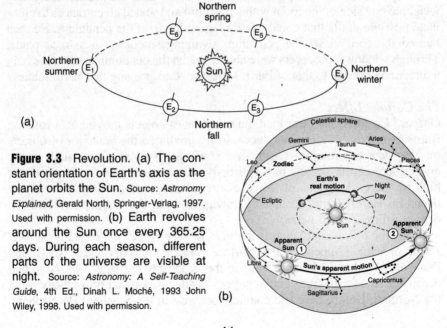

Figure 3.3 Revolution. (a) The constant orientation of Earth's axis as the planet orbits the Sun. Source: *Astronomy Explained*, Gerald North, Springer-Verlag, 1997. Used with permission. (b) Earth revolves around the Sun once every 365.25 days. During each season, different parts of the universe are visible at night. Source: *Astronomy: A Self-Teaching Guide*, 4th Ed., Dinah L. Moché, 1993 John Wiley, 1998. Used with permission.

Annual Traverse of the Constellations

As Earth revolves around the Sun, the side of Earth facing the Sun experiences day and the side facing away experiences night. Since the stars are visible only at night, the portion of the universe whose stars are visible to an observer on Earth varies cyclically as Earth revolves around the Sun. See Figure 3.3.

Precession

Precession is the very slow change in the direction of Earth's axis of rotation. Earth's axis sweeps around in a cone-shaped path like the wobbling of a spinning top. Almost 26,000 years are required for each sweep, causing Earth's North Pole to move slowly with respect to Polaris. Precession is caused by the gravitational pulls of the Sun and Moon. See Figure 3.4.

Since Earth's precession takes place over such a long period of time, it has little effect during a human lifetime. During your lifetime, Earth's axis will point slightly closer to Polaris. It will point closest to Polaris around the year A.D. 2100 and will then begin to move away from Polaris. In 13,000 years, Earth's axis will point at the star Vega instead of Polaris.

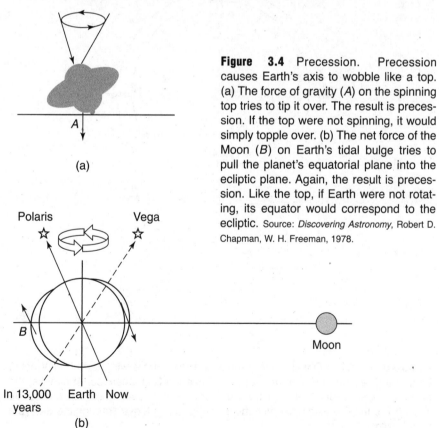

Figure 3.4 Precession. Precession causes Earth's axis to wobble like a top. (a) The force of gravity (*A*) on the spinning top tries to tip it over. The result is precession. If the top were not spinning, it would simply topple over. (b) The net force of the Moon (*B*) on Earth's tidal bulge tries to pull the planet's equatorial plane into the ecliptic plane. Again, the result is precession. Like the top, if Earth were not rotating, its equator would correspond to the ecliptic. Source: *Discovering Astronomy*, Robert D. Chapman, W. H. Freeman, 1978.

EFFECTS OF EARTH'S MOTIONS

The value of any model lies in its ability to explain past and current observations and to predict future observations. A revolving and rotating planet with its spin axis tilted at 23½° to a line perpendicular to its orbital plane fits all terrestrial and celestial observations.

The Sun's Path

The Sun's apparent path through the sky from sunrise to sunset is an arc, like the paths of all other celestial objects, and is the result of Earth's rotating on its axis. As Earth (and any observer on Earth's surface) rotates from west to east, the Sun appears to move from east to west. See Figure 3.5.

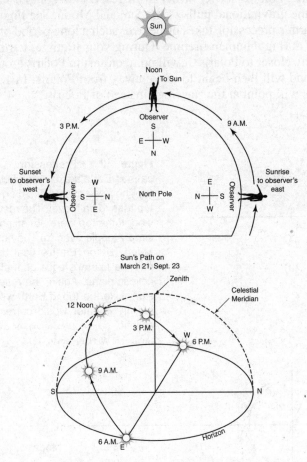

Figure 3.5 (a) As an Observer on Earth's Surface Rotates with Earth from West to East, the Direction in which the Observer Sees the Sun Changes. At sunrise, the observer sees the Sun to the east; at noon, toward the south; at sunset, toward the west. (b) To the Rotating Observer, the Sun Appears to Move Through the Sky from East to West.

At 6 A.M., an observer at sunrise would see the Sun to the east and low in the sky, at the horizon. At 9 A.M., Earth has rotated 45° from west to east and an observer would see the Sun to the southeast and higher up, above the horizon. At 12 noon, Earth has rotated another 45° from west to east and an observer would see the Sun to the south and at its highest point above the horizon, its noon position. At 3 P.M., Earth has rotated a further 45° and an observer would see the Sun to the southwest and lower above the horizon than at noon. At 6 P.M., Earth has rotated another 45° and an observer would see the Sun set to the west, low in the sky and at the horizon.

Note that, as Earth rotated through 180° from west to east, an observer saw the Sun move from east to southeast to south to southwest to west. The Sun's path through the sky began at sunrise at the eastern horizon at 6 A.M., arced upward, reaching a high point at noon, and then arced downward, ending at sunset at the western horizon at 6 P.M. What the observer saw as Earth rotated is summarized by path *B* in Figure 3.6.

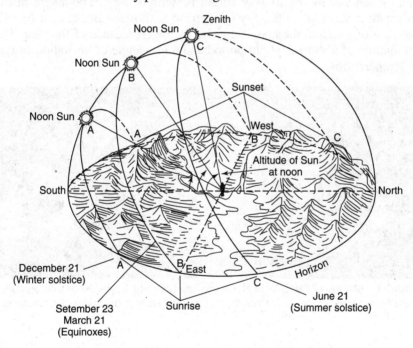

Figure 3.6 The Sun's Apparent Path. An observer at 42° N latitude would see the Sun move along different paths across the sky on different dates. Source: *Earth Science: A Study of a Changing Planet*, Daley, Higham, and Matthias, Prentice-Hall, 1986, and *The Story of Maps*, Lloyd Brown, 1977.

Changes in the Sun's Path

The length and the position of the Sun's path vary with the seasons and latitude. Sunlight that reaches Earth's surface is called **insolation**, short for *in*coming *sol*ar radi*ation*. The length of the Sun's path determines the length of time that insolation reaches Earth's surface, or the **duration of insolation**. The longer the Sun's path, the greater the duration of insolation, that is, the more hours of daylight. The position of the Sun's path determines the angle at which sunlight strikes Earth's surface, or **angle of insolation**. The closer to perpendicular the angle of insolation, the greater the intensity of the insolation, that is, the more it warms Earth's surface. The position of the Sun's path also determines the altitude of the Sun at noon. See Figure 3.6.

Changes with the Seasons

On different days of the year, an observer will see the Sun follow different paths through the sky. This is so because the Sun's position relative to an observer changes as Earth orbits the Sun. The points of sunrise and sunset vary, as does the altitude of the Sun at noon. See Figure 3.7.

In December, the position of sunrise to an observer in the United States is south of east and the position of sunset is south of west. The length of the Sun's path is shortest, so the daylight period is shortest. Because of Earth's tilted axis of rotation, the noon Sun is at its lowest altitude of the year. The combination of a short daylight period and a low angle of insolation causes low temperatures.

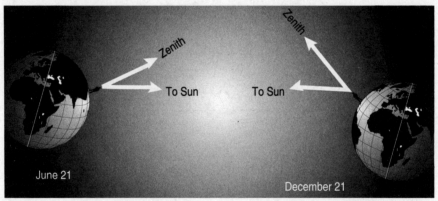

Figure 3.7 Altitude of the Sun at Noon on Two Different Dates. Note that on June 21 the Sun is seen closer to the zenith; that is, it is closer to being directly overhead.

In June, the Sun rises north of east, sets north of west, and rises to a higher altitude. The Sun's path is at its longest, so the daylight period is longest. The high altitude of the Sun results in a more direct angle of insolation. The combination of a long daylight period and a more direct angle of insolation results in high temperatures.

At the equinoxes in March and September, the Sun rises due east and sets due west. The length of the Sun's path results in exactly equal periods of daylight and darkness. The Sun's altitude at noon is halfway between its highest and lowest points. The length of day and the angle of insolation are intermediate between the two extremes; therefore, the temperatures are moderate.

From December to June, the Sun's path gets longer each day and the altitude of the Sun at noon increases. On June 21, the altitude of the Sun at noon stops increasing. Therefore, this date is called the **summer solstice**, meaning summer "Sun stop." From June to December, the Sun's path gets shorter each day and the altitude of the Sun at noon decreases. On December 21, the altitude of the Sun at noon stops decreasing. Therefore, this date is called the **winter solstice**, meaning winter "Sun stop." March 21 and September 21, when the daylight and darkness periods are equal, are called, respectively, the **spring equinox** and the **fall equinox**, meaning spring and fall "equal night." This cyclic pattern of change repeats in an annual cycle.

Changes with Latitude

Now consider what two observers at different latitudes will see on a given day as Earth rotates. As you can see in Figure 3.8, the observer near the Equator will see the Sun at a higher altitude at noon than the observer in New York State on the same day.

The altitude of the noon Sun at any location can be determined quite easily if you know the latitude of that location and the latitude at which the Sun is directly overhead on that day. To find the altitude of the Sun at noon, find the difference between the latitude of the location and the latitude at which the Sun is directly overhead. Then subtract this number of degrees from 90°.

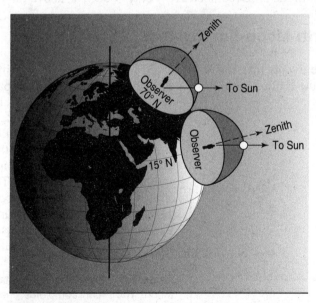

Figure 3.8 Altitude of the Sun at Different Latitudes. Observers at different latitudes will see the noon Sun at different altitudes on the same date.

Example:

On March 21, the Sun is directly overhead at the Equator, 0° latitude. On that day, what is the altitude of the Sun at noon in New York City?

$$\text{Latitude of NYC} - \text{latitude where Sun is directly overhead} = \text{difference in latitude}$$

$$41°\,N - 0° = 41°$$

$$90° - \text{difference in latitude} = \text{altitude of Sun at noon in NYC}$$

$$90° - 41° = 49°$$

The altitude of the Sun in New York City on March 21 is 49°.

Since the farthest Earth tips toward the Sun is 23½°, the Sun is never directly overhead north of 23½° N (Tropic of Cancer) or south of 23½° S (Tropic of Capricorn). There is always a difference between the latitude of a location in the continental United States and the latitude at which the Sun is directly overhead, and therefore the altitude of the Sun at noon is always less than 90° in the United States.

Because of New York's latitude, an observer in that State sees the noon Sun at its highest altitude on June 21, at about 73° above the horizon, and at its lowest altitude on December 21, at only about 23° above the horizon.

The Earth-Moon-Sun System

Few of the events that can be observed in the sky are as striking as those involving the Earth-Moon-Sun system. Phases of the Moon are probably the most familiar phenomena, but others involving this system include tides and eclipses of the Sun and Moon.

As Earth revolves around the Sun, the Moon circles Earth in an elliptical orbit with a period of 27.32 days. The Moon's orbit is tilted at an angle of about 5° from the plane of Earth's orbit around the Sun. The Moon moves rapidly in its orbit, covering 13° every day. As a result, each day

Figure 3.9 Revolution and Rotation of the Moon. The Moon completes one rotation in the same time it makes one revolution. Therefore, the same side of the Moon is always facing Earth.

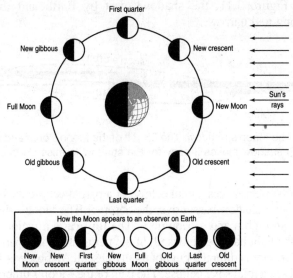

Figure 3.10 The Phases of the Moon.

its position against the backdrop of stars changes by 13°, or about 26 times its apparent diameter. The Moon also rotates on its axis once every 27.32 days. Thus, the same side of the Moon always faces Earth. See Figure 3.9.

Phases of the Moon

The Moon does not produce visible light of its own; the Moon is visible only because light from the Sun is reflected from its surface. An observer on Earth is able to see only that portion of the Moon that is illuminated by the Sun. As the Moon moves around Earth, different portions of the side of the Moon facing Earth are illuminated by sunlight, and the Moon passes through a cycle of **phases**. See Figure 3.10.

Although the Moon makes one revolution in 27.32 days, it takes 29.5 days to go through a complete cycle of phases. Why the extra two days? Let's start at new Moon, when the Moon is between Earth and the Sun. At the same time that the Moon revolves around Earth, Earth is revolving around the Sun at a rate of about 1° per day. Thus, by the time the Moon has completed one revolution, the Earth's position in relation to the Sun is not the same as it was when the Moon started its revolution. In 27 days the Earth has moved 27° in its orbit. Moving at about 13° per day, the Moon takes about 2 days to catch up to Earth and align with it and the Sun in a new Moon phase. The word *month* has its origin in "Moon-th," which referred to this 29.5-day cycle of phases.

Eclipses of the Sun and Moon

A **solar eclipse** occurs when the Moon passes directly between Earth and the Sun, casting a shadow on Earth and blocking our view of the Sun. Both the Moon and the Earth are illuminated by the Sun and cast shadows in space.

Heliocentric Earth Motions and Their Effects

As seen in Figure 3.11, the shadows cast by Earth and the Moon are extremely long and narrow.

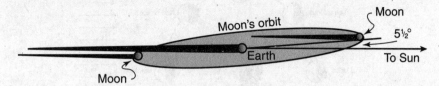

Figure 3.11 The Moon's Shadow. The 5½° tilt of the Moon's orbit and the small size of the Moon's shadow makes it easy for the shadow to miss Earth and not cause an eclipse.

Whether an observer sees a total eclipse or a partial eclipse depends on which part of the Moon's shadow passes over the observer. The Moon's shadow consists of two parts, an umbra and a penumbra. The **umbra** is a region of shadow in which all of the light has been blocked. In the **penumbra** only part of the light is blocked, so the light is dimmed but not totally absent. An observer in the Moon's umbra would see a total solar eclipse. The path of the Moon's umbra over Earth's surface is therefore called the **path of totality**. An observer in the penumbra would see a partial solar eclipse. An observer outside the Moon's shadow would see no eclipse. See Figure 3.12.

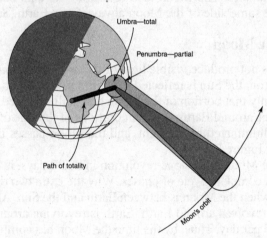

Figure 3.12 A Solar Eclipse. Observers in the umbra experience a total eclipse, while those in the penumbra experience a partial eclipse.

The umbra (i.e., the dark, inner part) of the Moon's shadow barely reaches Earth, and the small, circular shadow that the Moon casts on Earth's surface is never more than 269 kilometers in diameter. The 5° tilt of the Moon's orbit, together with the small size of the shadow that can reach Earth, makes it easy for the shadow to miss Earth at full and new Moon. Thus, total eclipses of the

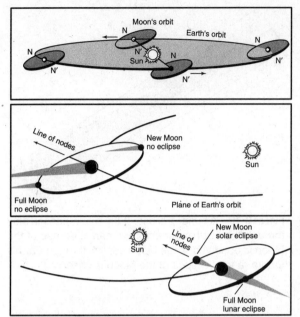

Figure 3.13 Eclipses of the Sun and Moon. Solar and lunar eclipses can occur when the Moon intersects the ecliptic in new-moon or full-moon position. If the Moon is above or below the ecliptic during these phases, no eclipse occurs. Source: *Horizons: Exploring the Universe*, by Michael A. Seeds, Wadsworth Publishing, 1987.

Sun are rare. For an eclipse to occur, the plane of the Moon's orbit must intersect the shadows being cast by both Earth and the Moon. See Figure 3.13.

Since the Moon's orbit is elliptical, the Moon's distance from Earth varies. If the Moon lines up directly with the Sun at a point in its orbit when the Moon is farthest away from Earth, the umbra of the Moon's shadow does not reach Earth's surface. If the umbra does not reach the surface, there is no total solar eclipse. However, a type of partial eclipse called an **annular eclipse** occurs under these conditions. In an annular eclipse the Moon is too far from Earth to completely block our view of the Sun. During the eclipse a narrow ring, or annulus, of light is visible around the edges of the Moon. See Figure 3.14.

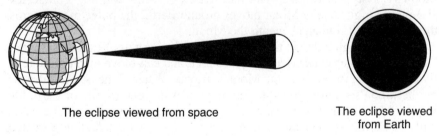

The eclipse viewed from space   The eclipse viewed from Earth

Figure 3.14 An Annular Eclipse and the Way It Appears When Viewed from Space and from Earth.

A **lunar eclipse** occurs when the full Moon moves through Earth's shadow. If the Moon moves into Earth's umbra, a total lunar eclipse is seen. If the Moon moves into Earth's penumbra, a partial lunar eclipse is seen. When the Moon is totally in Earth's umbra, it does not completely disappear from view. While no direct sunlight reaches the Moon, some light that is bent as it passes through Earth's atmosphere reaches the Moon. Since only the long waves of red light are bent far enough to reach the Moon, the Moon glows with a dull red color during a total lunar eclipse. During a partial lunar eclipse, however, the Moon is only partially dimmed as some of the Sun's light is blocked by Earth. Partial lunar eclipses are not very impressive. See Figure 3.15.

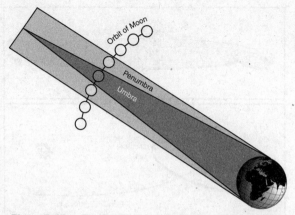

Figure 3.15 A Lunar Eclipse. If the moon passes only through the Earth's penumbra, a partial eclipse of the Moon is observed. If the Moon passes through Earth's umbra, a total eclipse of the Moon is observed.

Tides

Every day you feel the mutual attraction of gravity between Earth and your body pulling you downward with a force called your *weight*. But Earth's gravity is not the only gravity acting on you. While the Moon is farther away from you than Earth's center and has less mass than Earth, it still exerts a measurable force on you and everything else on Earth.

The part of Earth's surface that faces the Moon is about 6,000 kilometers closer to the Moon than is Earth's center. Therefore, the force of gravity the Moon exerts on this surface is stronger than the gravity exerted on Earth's center. Although Earth's surface is solid, it is not absolutely rigid. The Moon's gravity causes Earth's surface to flex outward, forming a bulge several inches high. As the Moon moves around Earth, the bulge moves across the surface as it remains beneath the Moon.

An inches-high bulge in the bedrock spread over half of Earth's surface is barely noticeable. Water is a fluid, however, and can move much more readily than rock in response to the Moon's gravity. Water in the oceans attracted by the Moon's gravity flows into a bulge of water on the side of Earth facing the Moon. A bulge of water also forms on Earth's far side. The Moon pulls on Earth's center more strongly than on Earth's far side, thus attracting Earth away from the oceans on the far side. The water flows into this space, creating a bulge. The water from the area between these bulges flows into them, creating a deep region and a shallow region in the ocean waters, or **tides**.

As Earth rotates on its axis, the positions of the tidal bulges remain fixed in line with the Moon. As the rotating Earth carries a location into a tidal bulge, the water deepens and the tide rises on the beach. As the location rotates out of the tidal bulge, the water becomes shallower and the tide falls. Since there are two bulges on opposite sides of Earth, the tide rises and falls twice a day.

The Sun also produces tidal bulges in Earth's surface and oceans. At new Moon and full Moon, the Sun's tidal bulges and the Moon's tidal bulges align with one another and combine. The result is very high and very low tides, called **spring tides** because they "spring so high," not because they happen during the spring season. See Figure 3.16a. Spring tides occur at every new and full Moon whatever the season. During the first- and third-quarter phases of the Moon, the tidal bulges of the Sun and Moon are at right angles to one another and very nearly cancel each other. The result is **neap tides**, in which there is very little difference between high and low tides. See Figure 3.16b.

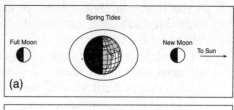

(a)

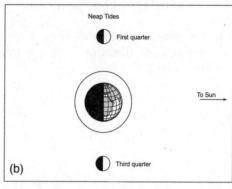

(b)

Figure 3.16 Spring and Neap Tides. (a) When the Sun and Moon pull in the same direction, their tidal forces combine and tidal bulges on Earth are larger. Spring tides occur at new Moon or full Moon. (b) When the Sun and Moon pull at right angles, their tidal forces do not combine and tidal bulges are much smaller. Neap tides occur at first- and third-quarter Moon.

MULTIPLE-CHOICE QUESTIONS

In each case, write the number of the word or expression that best answers the question or completes the statement.

1. Which observation provides the best evidence that Earth rotates?
   (1) The position of the planets among the stars changes during the year.
   (2) The location of the constellations in relationship to Polaris changes from month to month.
   (3) The length of the shadow cast by a flagpole at noontime changes from season to season.
   (4) The direction of swing of a freely swinging pendulum changes during the day.

Heliocentric Earth Motions and Their Effects

2. In the Northern Hemisphere, planetary winds blowing from north to south are deflected, or curved, toward the west. This deflection is caused by the
   (1) unequal heating of land and water surfaces
   (2) movement of low-pressure weather systems
   (3) orbiting of Earth around the Sun
   (4) spinning of Earth on its axis

3. The diagram below shows a heavy mass moving back and forth in a straight-line direction. The apparent direction of movement changes over time.

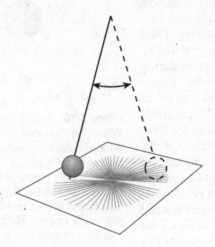

   This device provides evidence that
   (1) Earth rotates   (3) Earth revolves
   (2) Earth's axis is tilted   (4) Earth has a magnetic pole

4. In October, observers in New York State looking due south at the night sky would see a different group of constellations than they had seen in March. What is the best explanation for this change in the night sky?
   (1) Constellations revolve around Earth.
   (2) Constellations revolve around the Sun.
   (3) The Sun revolves around the center of our galaxy.
   (4) Earth revolves around the Sun.

5. If the tilt of Earth's axis were decreased from 23.5° to 15°, New York State's winters would become
   (1) warmer and summers would become cooler
   (2) warmer and summers would become warmer
   (3) cooler and summers would become cooler
   (4) cooler and summers would become warmer

6. A student in New York State looked toward the eastern horizon to observe sunrise at three different times during the year. The student drew the following diagram that shows the different positions of sunrise, A, B, and C, during this one-year period.

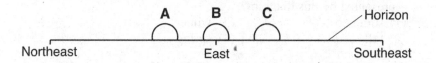

Which list correctly pairs the location of sunrise to the time of the year?
(1) A—June 21   B—March 21   C—December 21
(2) A—December 21 B—March 21   C—June 21
(3) A—March 21   B—June 21   C—December 21
(4) A—June 21   B—December 21   C—March 21

7. The apparent daily path of the Sun changes with the seasons because
(1) Earth's axis is tilted
(2) Earth's distance from the Sun changes
(3) the Sun revolves
(4) the Sun rotates

Base your answers to questions 8 through 10 on the diagram below, which shows the altitude and apparent position of the noontime Sun, as seen from various latitudes on Earth on a particular day of the year. Letters A through D represent locations on Earth's surface.

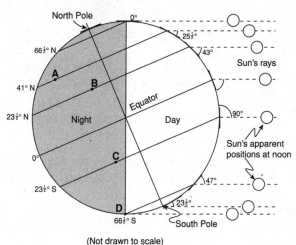

(Not drawn to scale)

8. Which lettered location will experience the *shortest* period of daylight during one Earth rotation on this day?
(1) A   (3) C
(2) B   (4) D

Heliocentric Earth Motions and Their Effects

9. What is the altitude of the noontime Sun at the Equator on this date?
   (1) 23½°
   (2) 43°
   (3) 66½°
   (4) 90°

10. Which season will begin at 41° N latitude, three months after the date represented by this diagram?
    (1) summer
    (2) fall
    (3) winter
    (4) spring

11. A cycle of Moon phases can be seen from Earth because the
    (1) Moon's distance from Earth changes at a predictable rate
    (2) Moon's axis is tilted
    (3) Moon spins on its axis
    (4) Moon revolves around Earth

Base your answers to questions 12 through 14 on the diagram below, which shows Earth in orbit around the Sun and the Moon in orbit around Earth. M_1, M_2, M_3, and M_4 indicate positions of the Moon in its orbit. Letter A indicates a location on Earth's surface.

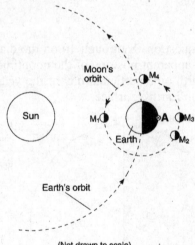

(Not drawn to scale)

12. An observer at location A on Earth views the Moon when it is at position M_3. Which phase of the Moon will the observer see?

(1) (2) (3) (4)

13. At which Moon position could a solar eclipse be seen from Earth?
    (1) M_1
    (2) M_2
    (3) M_3
    (4) M_4

14. An observer at location A noticed that the apparent size of the Moon varied slightly from month to month when the Moon was at position M_4 in its orbit. Which statement best explains this variation in the apparent size of the Moon?
    (1) The Moon expands in summer and contracts in winter.
    (2) The Moon shows complete cycles of phases throughout the year.
    (3) The Moon's period of rotation is equal to its period of revolution.
    (4) The Moon's distance from Earth varies in a cyclic manner.

15. The diagram below shows the relative positions of the Sun, the Moon, and Earth when an eclipse was observed from Earth. Positions A and B are locations on Earth's surface.

    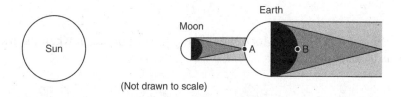

    (Not drawn to scale)

    Which statement correctly describes the type of eclipse that was occurring and the position on Earth where this eclipse was observed?
    (1) A lunar eclipse was observed from position A.
    (2) A lunar eclipse was observed from position B.
    (3) A solar eclipse was observed from position A.
    (4) A solar eclipse was observed from position B.

Base your answers to questions 16 through 19 on the diagram below, which shows Earth and the Moon in relation to the Sun. Positions A, B, C, and D show the Moon at specific locations in its orbit. Point **X** is a location on Earth's surface.

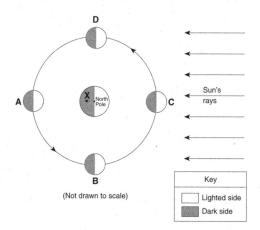

(Not drawn to scale)

Unit One GEO- TO HELIOCENTRIC

Heliocentric Earth Motions and Their Effects

16. What is the time of day at point **X**?
    (1) 6 A.M.
    (2) noon
    (3) 6 P.M.
    (4) midnight

17. On what date does the line separating day and night pass through Earth's North Pole, as shown in this diagram?
    (1) December 21
    (2) January 21
    (3) March 21
    (4) June 21

18. A solar eclipse might occur when the Moon is at location
    (1) *A*
    (2) *B*
    (3) *C*
    (4) *D*

19. Which phase of the Moon would be observed on Earth when the Moon is at location *A*?

(1) (2) (3) (4)

Base your answers to questions 20 and 21 on the graph below, which shows two days of tidal data from a coastal location in the northeastern United States.

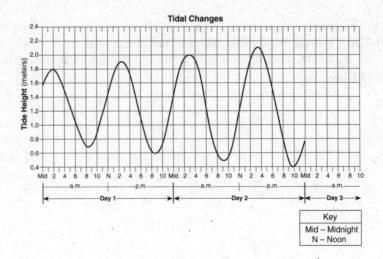

20. The change in the tides as shown on the graph is primarily the result of
    (1) Earth's rotation and the Moon's revolution
    (2) Earth's rotation and revolution
    (3) the Moon's rotation and Earth's revolution
    (4) the Moon's rotation and revolution

21. If the pattern shown continues, the most likely height and time for the first high tide on Day 3 would be
    (1) 2.2 meters at 4 A.M.
    (2) 2.3 meters at 4 A.M.
    (3) 2.2 meters at 5 A.M.
    (4) 2.3 meters at 5 A.M.

CONSTRUCTED RESPONSE QUESTIONS

Base your answers to questions 22 and 23 on the diagram below, which represents the sky above an observer in Elmira, New York. Angular distances above the horizon are indicated. The Sun's apparent path for December 21 is shown.

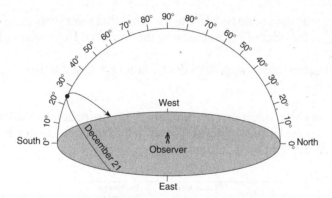

22. On March 21, the altitude of the noon Sun in Elmira is 48°. On the diagram, draw the Sun's apparent path for March 21 as it would appear to the observer. Be sure your path begins and ends at the correct positions on the horizon and indicates the correct altitude of the Sun at noon. [1]

23. On what date of the year does the maximum duration of insolation usually occur at Elmira? [1]

Base your answers to questions 24 and 25 on the diagram below, which shows Earth's orbit around the Sun as viewed from space. Earth is shown at eight different positions labeled *A* through *H*. Earth's North Pole, Arctic Circle, and equator have been labeled at position *C*. The arrows show the direction of orbital motion.

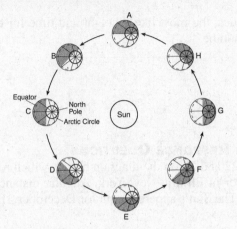

Season	Earth's Position
spring	
summer	
fall	
winter	

24. Complete the data table by placing the letter that represents the position of Earth at the start of *each* season in the Northern Hemisphere. [1]

25. Approximately how many days does Earth take to move from position *A* to position *C*? [1]

Base your answers to questions 26 through 29 on the data table below. The data table shows the latitude of several cities in the Northern Hemisphere and the duration of daylight on a particular day.

Data Table

City	Latitude (°N)	Duration of Daylight (hr)
Panama City, Panama	9	11.6
Mexico City, Mexico	19	11.0
Tampa, Florida	28	10.4
Memphis, Tennessee	35	9.8
Winnipeg, Canada	50	8.1
Churchill, Canada	59	6.3
Fairbanks, Alaska	65	3.7

26. On the grid, plot with an **X** the duration of daylight for each city shown in the data table. Connect your **X**s with a smooth, curved line. [1]

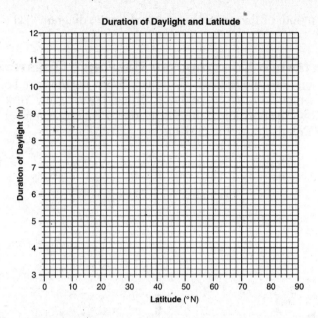

27. Based on the data table, state the relationship between latitude and the duration of daylight. [1]

28. Use your graph to determine the latitude at which the Sun sets 7 hours after it rises. [1]

29. The data were recorded for the first day of a certain season in the Northern Hemisphere. State the name of this season. [1]

Base your answers to questions 30 through 32 on the diagram below, which represents the Sun's rays striking Earth at a position in its orbit around the Sun.

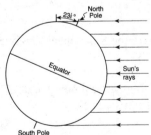

30. On the diagram above, neatly and accurately shade the area of Earth that is in darkness. [1]

31. On the diagram above, draw the line of latitude that is receiving the Sun's direct perpendicular rays on this date. [1]

32. What month of the year is represented by the diagram? [1]

EXTENDED CONSTRUCTED RESPONSE QUESTIONS

Base your answers to questions 33 through 37 on the diagram below, which represents a model of the sky above a vertical post in New York State. The diagram shows the position of the Sun at solar noon on September 23 and the position of Polaris above the horizon.

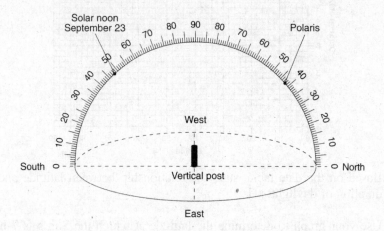

33. On the diagram, draw the apparent path of the Sun across the sky on September 23 from sunrise to sunset. [1]

34. On the diagram, draw the shadow of the vertical post as it would appear at solar noon on September 23. [1]

35. Place an **X** on the diagram above to indicate the altitude of the Sun at solar noon on June 21. [1]

36. How many degrees will the Sun appear to move across the sky from 1 P.M. to 3 P.M. on June 21? [1]

37. At which latitude is this vertical post located? Include the unit and compass direction in your answer. [1]

Base your answers to questions 38 through 41 on the passage below.

The Moon Is Moving Away While Earth's Rotation Slows

Tides on Earth are primarily caused by the gravitational force of the Moon acting on Earth's surface. The Moon causes two tidal bulges to occur on Earth: The direct tidal bulge occurs on the side facing the Moon, and the indirect tidal bulge occurs on the opposite side of Earth. Since Earth rotates, the bulges are swept forward along Earth's surface. This advancing bulge helps pull the Moon forward in its orbit, resulting in a larger orbital radius. The Moon is actually getting farther away from Earth, at a rate of approximately 3.8 centimeters per year.

The Moon's gravity is also pulling on the direct tidal bulge. This pulling on the bulge causes friction of ocean water against the ocean floor, slowing the rotation of Earth at a rate of 0.002 second per 100 years.

38. The diagram below shows the Moon and Earth in line with each other in space. On the diagram, place an **X** on Earth's surface to indicate where the direct tidal bulge is occurring. [1]

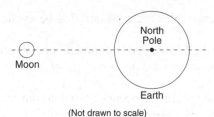

(Not drawn to scale)

39. Explain why the force of gravity between the Moon and Earth will decrease over time. [1]

40. In 100,000 years, the rotation of Earth will be slower by how many seconds? [1]

41. Explain why the Moon has a greater influence than the Sun on Earth's tides. [1]

Heliocentric Earth Motions and Their Effects

Base your answers to questions 42 through 45 on the diagram below, which shows the Moon's orbit around Earth.

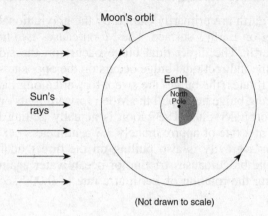

(Not drawn to scale)

42. On the diagram, place a small circle on the Moon's orbit at the new-Moon phase where none of the lighted portion of the Moon is visible from Earth. [1]

43. Explain why ocean tides are considered to be cyclic. [1]

44. How long does the Moon take to complete one revolution around Earth? Express your answer to the *nearest tenth of a day*. [1]

45. Explain why lunar eclipses occur only when the Moon and the Sun are on opposite sides of Earth. [1]

Chapter 4
EARTH'S COORDINATE SYSTEM AND MAPPING

> **KEY IDEAS** Very early on, a variety of evidence led astronomers to conclude that Earth is a sphere. However, if only the shape of Earth is known, celestial observation can be used only to determine relative position—for example, 60° south of the North Pole. But how far is that from other locations: 1,000 kilometers? 10,000 kilometers? To determine actual distances between places on Earth's surface, the planet's size must also be determined. Only then can true-to-scale maps be constructed.
>
> The size and shape of Earth can be readily determined by combining Earth-based measurements with simple observations of the sky. Observations show that Earth is a very slightly oblate sphere with an equatorial diameter of 12,757 kilometers.
>
> The ability to locate and map positions on Earth's surface is essential to a wide variety of human activities ranging from engineering to urban planning to national defense. Topographic maps represent the three-dimensional shape of Earth's surface on a two-dimensional surface. A field is a region of space with a measurable quantity at every point. Field maps can be drawn to represent any quantity that varies in a region of space.

KEY OBJECTIVES
Upon completion of this chapter, you will be able to:

- Describe Earth's shape and explain how it can be determined by simple observations.
- Explain how Earth's size can be calculated using Earth-based measurements and simple observations of the sky.
- Use the latitude and longitude coordinate system to locate points on Earth's surface.
- Construct field maps and calculate gradient within a field.
- Read and interpret a topographic map.

THE NATURE OF EARTH'S SHAPE AND SURFACE

In order to map Earth's surface, its size and shape must be known. As stated above, Earth's size and shape can be readily determined by combining Earth-based measurements with simple observations of the sky. Let's begin by considering what we currently know about Earth's shape, and then examine how we arrived at this knowledge.

Earth's Shape

Earth's shape is very nearly a perfect sphere. A perfect sphere has exactly the same diameter when measured in any direction. Actual measurements of Earth's dimensions deviate slightly from this ideal. The polar diameter, or diameter measured from the North Pole through the center of Earth to the South Pole, is 12,714 kilometers. The equatorial diameter, or diameter measured from a point at the Equator through the center of Earth to the Equator, is 12,756 kilometers. Thus, Earth's spherical shape "bulges" very slightly at the Equator and is very slightly "flattened" at the poles; this shape is called an **oblate** (flattened) **spheroid**.

Roundness

Earth's actual shape, however, is so close to being a perfect sphere that the eye cannot detect its oblateness. When viewed from space, Earth appears perfectly round. Any cross section of Earth looks like a perfect circle. To gain an idea of how close Earth is to being perfectly round, let's suppose that we made a scale model of Earth—a globe. If we used a scale of 1 centimeter = 1,000 kilometers, the globe would have a polar diameter of 12.714 centimeters and an equatorial diameter of 12.756 centimeters—a little bigger than a softball. The difference in diameters would be 0.042 centimeter, or less than half a millimeter! We would need a micrometer to measure this difference in diameters.

Look carefully at the circle in Figure 4.1. Its polar diameter is 4.238 centimeters, and its equatorial diameter is 4.252 centimeters. It was drawn exactly one-third the size of the globe described above so that it would fit on this page. Can you tell that it is not perfectly round?

Earth's oblateness is the result of forces produced by Earth's rotation on its axis. Just as a loose skirt will swirl outward if the wearer spins around, planet Earth "swirls"

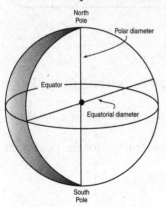

Figure 4.1 Schematic Diagram of Earth Showing Polar and Equatorial Diameters.

outward when it rotates. However, since Earth is much stiffer than a skirt, the distance it moves outward is much less.

Smoothness

In addition to being almost perfectly round, Earth is also very smooth. Compared with its diameter, the irregularities on Earth's surface (mountains and ocean basins) are relatively small. Mount Everest rises 8,848 meters, or about 8.8 kilometers, above sea level. That is about 7/10,000 of Earth's diameter (8.8 km/12,756 km). If you were to draw Mount Everest to scale on the circle in Figure 4.1, it would protrude from the surface less than 1/20 of a millimeter. Mark off a millimeter on a blank piece of paper, and then try to make dots small enough so that 20 will fit between the lines. If you put one of those dots on the outside of the circle, it would stick out of the circle as much as Mount Everest protrudes from Earth. As you can see, the dot barely affects the smoothness of the circle.

When you consider that most of the protrusions and indentations on the Earth's surface are much smaller than Mount Everest, you can see why Earth appears smooth. The only reason that Earth's mountains and valleys seem so large to us is that, compared to Earth, we are tiny. If you were to draw humans to scale on the model globe described in the preceding section, you would need a microscope to see them.

Evidence of Earth's Shape

Earth's true shape is a question that has captured the minds and imaginations of humans for thousands of years. Contrary to the popular belief that Columbus was among the first to believe that Earth is a sphere, Aristotle reached this conclusion in the third century B.C.

Ships and Eclipses

Aristotle based his conclusion upon simple observations of the sky and Earth-based measurements. This use of observation to support his ideas, rather than merely stating them as theories, places Aristotle among the earliest scientific thinkers. First, Aristotle observed that Earth's shadow during a lunar eclipse is definitely round. Although this could also happen if Earth was a cone or a cylinder, there was additional evidence supporting a spherical shape. Travelers reported that, as they went farther north or south, some stars appeared lower and lower in the sky. Furthermore, Aristotle noted that ships disappeared bow-first over the horizon, no matter in which direction they sailed. These observations could occur only if the ships were constantly moving along a curved surface that changed the angle at which they were viewed by an observer.

Today, a number of observations provide additional evidence of Earth's shape.

Photographs Taken from Space

The most direct evidence of Earth's shape consists of photographs taken from space. These photographs show that Earth is indeed spherical. When very precise photographs are taken, they can be analyzed and precise measurements of Earth's image in the photographs can be made. These measurements indicate that the Earth's polar and equatorial diameters are indeed slightly different, confirming that its shape is an oblate spheroid.

Observations of the Altitude of *Polaris* (the North Star)

Observations of Polaris provide us with information that can be interpreted, using simple geometry, to provide evidence of Earth's spherical shape and its oblateness. Polaris is a distant star that is located almost exactly over Earth's North Pole and is almost in line with its axis of rotation. Polaris is very far away from Earth, literally trillions of Earth-diameters distant. If it were observed from two points farthest apart on Earth's surface, the directions in which it was seen would vary by only a minute fraction of a degree. See Figure 4.2.

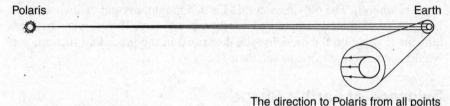

Figure 4.2 Direction to Polaris.

The direction to Polaris from all points on Earth is not measurably different.

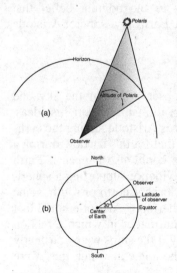

Figure 4.3 (a) Altitude of Polaris (b) Latitude of an Observer

The **altitude** of Polaris is the angle between the star and the horizon with the observer at the vertex, as shown in Figure 4.3a. The latitude of an observer is the angle of the observer north or south of the Equator; see Figure 4.3b.

If Earth were a flat disk, the altitude of Polaris would be almost exactly the same at all locations and at all times. See Figure 4.4a.

It is true that, to a fixed observer, as Earth rotates, the altitude of Polaris does not appear to change, as shown in Figure 4.4b. However, if the observer travels north or south of the original location, the altitude does change, a result that would be expected if Earth is spherical. See Figure 4.4c.

If Earth were a perfect sphere, the altitude of Polaris would be the same as an observer's latitude and the distance traveled to change the altitude of Polaris by 1° would always be the same. See Figure 4.5.

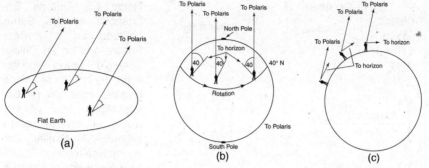

Figure 4.4 The Altitude of Polaris on a Flat and on a Spherical Earth. (a) If Earth were a flat disk, the altitude of Polaris would always be almost exactly the same at any location and at all times. (b) To a fixed observer on a spherical Earth, the altitude of Polaris does not appear to change as Earth rotates. (c) However, if the observer travels north or south, the altitude does change, a result that would be expected if Earth is spherical.

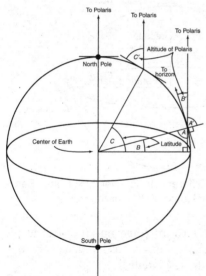

Figure 4.5 Relationship of the Altitude of Polaris to the Latitude of an Observer. By simple geometry it can be shown that the altitude of Polaris is the same as an observer's latitude. The sum of the angles of a triangle is equal to 180°. Similarly, since a straight line has a measure of 180°, <A + <B + 90° = 180°. Also, <A = <A' because they are alternate interior angles. Therefore, <B (latitude) = <B' (altitude of Polaris).

However, what is observed is that the altitude of Polaris is *not* always exactly equal to an observer's latitude, and the distance traveled to change the altitude of Polaris by 1° also varies. This is evidence that Earth is an oblate spheroid. See Figure 4.6.

Measurements of Gravity

Gravity is the force of attraction that exists between any two objects. It is proportional to the mass of the objects and inversely proportional to the square of the distance between their centers. Simply stated, the more massive the objects, and the closer their centers, the stronger the gravitational attraction between them.

The gravitational attraction between Earth and any object can be measured with a spring scale and is called the object's **weight**. If Earth were perfectly spherical, the distance between its center and any point on its surface would always be the same. Therefore, if we measured the weight of the same object at any point on the Earth's surface, the object's weight should

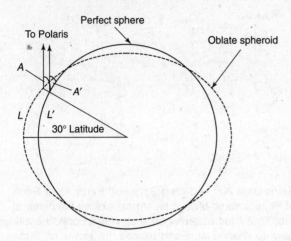

Figure 4.6 Altitude Differences at the Same Latitude on a Perfect Sphere and on an Oblate Spheroid. For the same latitude, the altitude of Polaris on a perfect sphere (*A'*) differs slightly from what it would be on an oblate spheroid (*A*). Similarly, for an identical change in latitude the distance covered on a perfect sphere (*L'*) would differ from the distance covered on an oblate spheroid (*L*).

always be the same. If, however, Earth is an oblate spheroid, then the distance between the center of an object and the center of Earth will be greater at the Equator than at the Poles. If the distance is greater, the gravitational attraction is weaker and the object will weigh less. Thus, if we measure the weight of the same object at the Equator and at the Poles, it will weigh less at the Equator and more near the Poles. Even taking into account the outward force on an object at the Equator due to Earth's rotation, this is exactly what is observed, as shown in Figure 4.7, and is further evidence of Earth's oblateness.

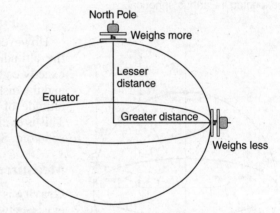

Figure 4.7 Differences in Weight of the Same Object at the Equator and near the Poles.

Determining Earth's Size

Modern measuring instruments based upon lasers allow us to measure Earth with great precision. However, Earth's size was estimated quite accurately more than 2,000 years ago. The principle is simple: if Earth is a sphere, then its circumference is a circle and all of the mathematical relationships between the various parts of a circle hold true for Earth.

Eratosthenes' Method

These mathematical relationships make it possible to determine Earth's circumference without having to measure the entire circumference directly. If it is a circle, then it consists of 360° and a 1° angle will intersect 1/360 of Earth's circumference. Therefore, if observers at two locations on a north-south line simultaneously determine the altitude of Polaris to differ by 1°, Earth's circumference must be 360 times the distance between those locations.

In 235 B.C., Eratosthenes used this technique but substituted the Sun for a star. Eratosthenes lived in Alexandria and had reliable reports that on June 21, the summer solstice, the Sun was directly overhead at midday in the city of Syene. Eratosthenes believed that Alexandria was due north of Syene. He also knew that a camel caravan traveling 100 stadia a day typically took 50 days to get to Syene. This meant that Syene was 50 × 100 or 5,000 stadia from Alexandria. The final piece of information he needed was the angle of the Sun at midday on June 21 in Alexandria. Using a gnomon, a vertical column erected perpendicular to the horizon, Eratosthenes determined the angle to be 7 degrees 12 seconds, or about 1/50 of 360°. Hence the distance between Syene and Alexandria was about 1/50 of Earth's circumference: 5,000 stadia × 50 = 250,000 stadia. The problem is that there is some dispute as to exactly how long a stadium, or stadion, a Greek unit of length, was. The likely figure is about 10 stadia to a mile, making Eratosthenes' estimate of Earth's circumference 25,000 miles (40,230 km) compared to the actual figure of 24,862 miles (40,010 km)—a remarkable achievement for his time. See Figure 4.8.

Once Earth's circumference is known, many other dimensions of Earth can be calculated using the formulas given below, where

C = circumference, d = diameter, r = radius, A = surface area, V = volume, and π (pi) = 3.1462:

$C = \pi d$, $d = 2r$, $A = 4\pi r^2$, $V = \frac{4}{3} \pi r^3$

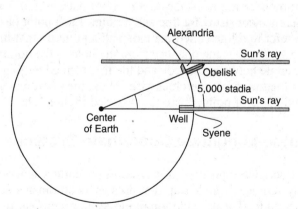

Figure 4.8 Eratosthenes' Method for Determining the Circumference of Earth.

MAPPING EARTH'S SURFACE

Maps

A **map** is a model of Earth's surface. Although models are different from the real thing, they are tools for learning about the situations or objects they represent. A map is meant to communicate a sense of place, of where one point is in relation to another. A map can be anything from a quick sketch showing a friend how to get to the park from school to an elaborate scale model of Earth complete with mountain ranges and ocean basins. The nature of a map depends on the purpose for which it was created.

Coordinate Systems

A **coordinate system** is a method of locating points by labeling them with numbers called coordinates. **Coordinates** are numbers measured with respect to a system of lines or some other fixed reference. The most commonly used coordinate system is the *Cartesian* system, in which two lines on a flat surface intersect at right angles. Each line is called an ***axis***, and the point where they intersect is the **origin**. The horizontal axis is usually referred to as the x-axis, and the vertical axis as the y-axis. Two coordinates describe any point on the flat surface, each defining one of the intersecting lines. For example, a fixed point can be located on a graph by describing the two lines $x = 4$ and $y = 3$. As shown in Figure 4.9a, these two lines intersect at a single point. By convention, coordinates in this system are written with the x value first and the y value second—(x,y), so the coordinates of the point shown are (4,3). The coordinate axes, in terms of which position is specified, form the **frame of reference** of a coordinate system.

Many coordinate systems can be devised to locate points on a surface. Another common example is the *polar* coordinate system, in which the frame of reference consists of a line and a point. Each point has coordinates (r,θ), which refer to a line called the **axis** and a point on that line called the **pole**. The r *coordinate* is the distance from the pole to the point. The θ *coordinate* is the angle between the axis and the line formed by joining the point and the pole measured counterclockwise. Thus, the coordinates of point P_1 are (4,30) and those of point P_2 are (3,240). See Figure 4.9b.

The Latitude-Longitude Coordinate System

In order to describe the position of any point on Earth's spherical surface, a coordinate system has been set up that uses two coordinates known as latitude and longitude. The latitude-longitude system consists of two sets of lines that cross each other at right angles. *Latitude* lines (called *parallels*) run in an east-west direction, and *longitude* lines (called *meridians*) run in a north-south direction. This type of system and the words *latitude* and *longi-*

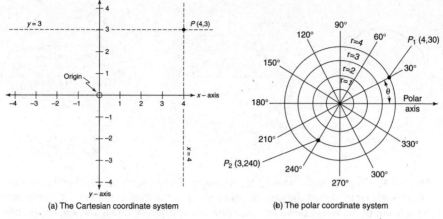

Figure 4.9 Two Common Coordinate Systems. (a) The Cartesian coordinate system (b) The polar coordinate system.

tude were already being used to describe such lines at the time of the Egyptian astronomer Claudius Ptolemy in A.D. 150.

Parallels

Since Earth is a sphere, east-west lines on its surface, such as the Equator, actually form circles. The circles formed by lines of latitude are called **parallels** because, if you drew a series of east-west lines, the circles formed would all be parallel to one another. See Figure 4.10a. Parallels lie in planes that are at right angles to Earth's axis of rotation. See Figure 4.10b.

To create a coordinate system on a globe or map, it is necessary to have a frame of reference, or fixed reference lines. The fixed reference line for latitude in this system is the **Equator**, an east-west line midway between the North and South Poles. Parallels are described by their angular distances north or south of the Equator as measured from the center of Earth. See Figure 4.10c.

Latitude is defined as *positive* for locations north of the Equator and *negative* for locations south of the Equator. Of course, latitude may also be given as **North (N)** or **South (S)** latitude. Thus, latitude is 0° at the Equator, and may be written as +90°, or 90°N, at the North Pole and –90°, or 90°S, at the South Pole. The latitude of New York City is about +41°30' or 41°30' N. Notice that the farther a location is from the Equator, the smaller the circle and the shorter its length. The east-west line at 60°N is only half as long as the Equator.

Great Circles and Meridians

Earth's axis and poles also provide the reference points for lines of longitude. To understand longitude, you must first grasp the concept of a great circle. A **great circle** is a line drawn around a sphere to form a circle whose

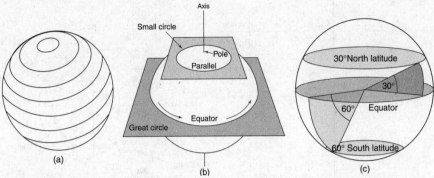

Figure 4.10 (a) Parallels are east-west lines on Earth's surface. (b) The Equator and all parallels lie in planes that are at right angles to the axis of rotation. (c) Latitude is a measure of the angle between the plane of the Equator and a line joining the center of Earth with a point on its surface.

plane passes through the center of the sphere. If the plane of a circle doesn't pass through the center of the sphere, a small circle is formed. See Figure 4.11. An infinite number of great circles can be drawn on a sphere. The Equator, however, is the *only* great circle whose plane is at right angles to Earth's axis.

One interesting property of great circles is that they all cut a sphere exactly in half (e.g., the Equator divides Earth into the Northern and Southern hemispheres). Another is that the arc of a great circle connecting any two points on a sphere's surface is the shortest distance between those points! This is why airliners travel along "great circle routes."

On Earth, any great circle that passes through both the North and the South poles is called a **meridian**. Earth's axis of rotation lies in the plane of every meridian. See Figure 4.12. **Meridians of longitude** go only halfway around Earth, from pole to pole. Although a meridian of longitude is actually only half of a meridian, it is often referred to simply as a meridian.

The **longitude** of a point is the angular distance between two meridians: a fixed reference meridian and the meridian passing though the point. It is measured from the center of Earth east or west along the Equator between the fixed reference meridian of longitude and the meridian of longitude passing through the point. See Figure 4.13.

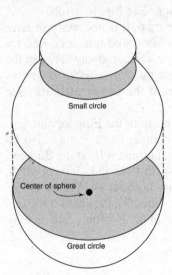

Figure 4.11 A Great Circle and a Small Circle.

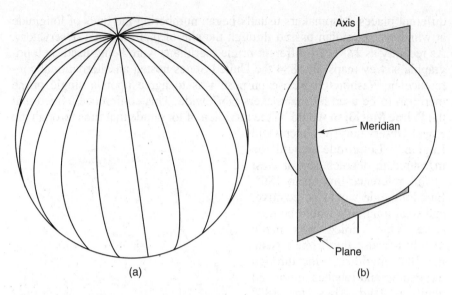

Figure 4.12 (a) Meridians are great circles that pass through both poles. (b) The plane of a meridian passes through Earth's axis.

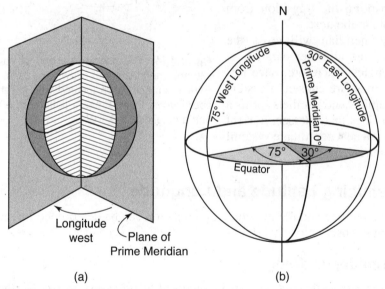

Figure 4.13 (a) Longitude can be thought of as the angle between the planes of two meridians. (b) Longitude is measured from the center of Earth as an angle east or west of the Prime Meridian.

Unlike the Equator, which is the *only* line halfway between the poles, the reference meridian, or *Prime Meridian*, is an arbitrarily chosen line and could be almost anywhere. Over the years, on different maps, it *has* been located in different places. Mapmakers usually began numbering the lines of longitude at whichever meridian passed through the site of their national observatory. As recently as 1881, 14 different prime meridians were being used on topographic survey maps. In 1884 the United States hosted an international conference in Washington whose purpose was to agree upon a single prime meridian to be used by mapmakers worldwide. This conference agreed that the **Prime Meridian** would be the meridian of longitude that runs through the Royal Observatory in Greenwich, England. Longitude would be measured in degrees east or west of this reference line up to 180°. East longitude would be positive, and west longitude would be negative. This choice was made mainly because the Prime's twin, the 180° meridian, runs through the Pacific and touches almost no habitable land. Thus, the 180° meridian is a good choice for an international date line, that is, a line marking the transition from one date to the next.

Any meridian will cross the Equator and all other parallels at right angles. Therefore, once reference lines are chosen, a system of meridians and parallels forms a grid of lines that intersect at right angles—a neat coordinate system! See Figure 4.14.

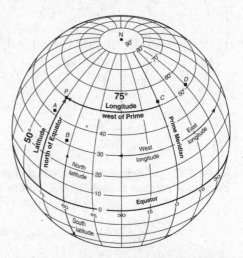

Figure 4.14 The geographic grid of parallels and meridians that allows any point on Earth's surface to be located with a set of coordinates. Point P has a latitude of 50°N and a longitude of 75°W.

Determining Latitude and Longitude

Both the latitude and the longitude of a location can be determined by making simple observations.

Determining Latitude

As described earlier and shown in Figure 4.5, observers in the Northern Hemisphere can determine their latitudes by measuring the altitude of *Polaris*. Observers in the Southern Hemisphere use the positions of other stars and make corrections for their deviation from Earth's axis of rotation.

Determining Longitude

The longitude of a particular location is determined by observing the time and the position of the Sun. Local noon is the time at which the Sun is at its highest altitude of the day. At that moment, a line drawn from the Sun to the center of Earth will pass through that location's meridian. A few moments later, Earth will have rotated from east to west and another meridian will align with the Sun and experience local noon. Thus, different longitudes will experience local noon at different times.

Since Earth makes one complete 360° rotation every 24 hours, the relationship between time and degrees rotated can be calculated. For example, in 1 hour Earth will rotate 360°/24, or 15°. Now, suppose you were at a location and the time was exactly local noon. One hour later, Earth would have rotated 15° and a location 15° of longitude away would be experiencing local noon. Two hours later, Earth would have rotated 30° and a location 30° of longitude away would experience local noon. By comparing the difference in time between local noons at any two locations, you can easily calculate their difference in longitude. Every hour of difference equals 15° of longitude, but this doesn't tell you a location's actual longitude, only its longitude relative to another point.

Suppose, however, that one of the times you knew was the time at which local noon occurred in *Greenwich*! Then you would know the difference in longitude at your location from *zero* longitude, and that would be your actual longitude. Therefore, in order to measure longitude at a particular location you need to know two times—local noon and the time of local noon in Greenwich. Each hour of difference between local time and Greenwich time equals 15° of longitude. Since Earth rotates from west to east, locations east of Greenwich will experience local noon earlier than Greenwich, while those west of Greenwich will experience local noon later than Greenwich. For example, if Old Forge, New York, experiences local noon about 5 hours later than Greenwich, England, its longitude is 5 × 15°, or 75°. Since local noon in Old Forge occurs after local noon in Greenwich, it is west longitude.

To measure longitude, then, you need a radio broadcast or a chronometer (a watch set to Greenwich time) to tell you what time it is in Greenwich when you experience local noon. You can then look up the time of local noon in Greenwich on that day in an astronomical table and, using the difference in times, calculate longitude. The need for such clocks drove the development of ever more precise timekeeping devices, culminating with today's atomic clocks.

Field Maps

A **field** is a region of space that has a measurable quantity at every point. Some examples of field quantities are temperature, pressure, magnetism, gravity, and elevation. Field maps can be used to represent any quantity that varies in a region of space.

Isolines

One way to represent field quantities on a two-dimensional field map is to use isolines. **Isolines** connect points of equal field values. For example, a

Earth's Coordinate System and Mapping

temperature field map shows lines connecting points of equal temperature, or isotherms. See Figure 4.15.

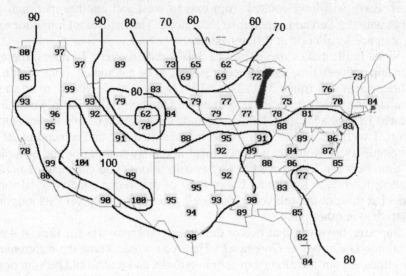

Figure 4.15 Temperature Field Map.

Gradient

Within a field, field values change as you move from place to place. The rate at which a field value changes is called its **gradient**. Gradient can be calculated as follows:

$$\text{Gradient} = \frac{\text{amount of change in field value}}{\text{distance through which change occurs}}$$

Example:
Figure 4.16 shows pollutant levels measured in Lake Meredith. Isolines connect points of equal pollutant concentration, measured in parts per billion (ppb). Abby's Island and Ben's Island are 2 kilometers apart, and the pollutant concentration between them changes from 30 to 50 parts per billion. The rate of change, or gradient, is:

$$\text{Gradient} = \frac{50 \text{ ppb} - 30 \text{ ppb}}{2 \text{ km}}$$

$$= \frac{20 \text{ ppb}}{2 \text{ km}}$$

$$= 10 \text{ ppb/km}$$

The gradient is 10 parts per billion per kilometer.

Isosurfaces

Field maps can also be represented in three dimensions using isosurfaces. Isolines show field values in a particular two-dimensional plane (e.g., air pressure at ground level). An **isosurface** is a three-dimensional surface on which every point has the same field value. Isosurfaces provide insights into the nature of a field that it may not be possible to gain by viewing only a two-dimensional map. See Figure 4.17.

Fields seldom remain unchanged over time. A field map shows a field at a particular point in time, the time at which its field values were measured. For example, the temperature field over the United States changes drastically from July to December. Also, Earth's magnetic field has changed position many times since Earth formed. Therefore, field maps need to be updated periodically in order to represent current field conditions. On weather maps, which represent rapidly changing fields, values are updated as often as once an hour.

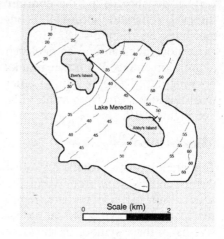

Figure 4.16 Meredith Lake Containing Abby's Island and Ben's Island.

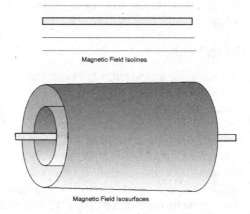

Figure 4.17 Magnetic Field Isolines Versus Isosurfaces.

Topographic Maps: Elevation Field Maps

Topographic maps are scale models of a part of Earth's surface that show its three-dimensional shape in two dimensions. A topographic map is actually a type of field in which the field value is elevation above sea level of points on Earth's surface.

Contour Lines

Isolines that connect points of equal elevation are called **contour lines** because they represent the shapes, or contours, of Earth's surface. A contour line shows the shape that would be formed if the land surface were sliced by a horizontal plane at a particular elevation above sea level. An unvarying vertical distance called the **contour interval** separates successive contour lines. See Figure 4.18.

Contour lines around an enclosed depression have **hachure marks** pointing into the depression in order to distinguish them from small hills. See Figure 4.19.

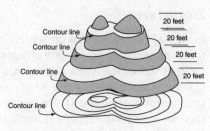

Figure 4.18 Three-Dimensional Mountain Sliced by Horizontal Plane, Showing Contour Line.

Map Symbols

Topographic maps provide a view of the ground as seen from vertically above. Surface features are represented by map symbols. An extensive list of these symbols is shown in Figure 4.20. Symbols are blue for bodies of water, black and red for human-made structures, and brown for contour lines and other relief symbols.

Map Scale

A *map scale* is the ratio between the distance shown on a map and the actual distance on the ground. For example, if the map scale is expressed as 1:100,000, then 1 unit of length on the map is equal to 100,000 of the

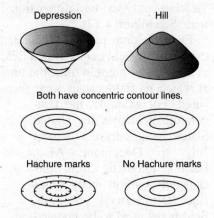

Figure 4.19 Depression Contours Versus Regular Contours.

same units on the ground. Both numbers in the ratio have the same units, but those units can be anything. For example, the map scale indicates that 1 centimeter on the map equals 100,000 centimeters on the ground. But it also means that 1 inch on the map equals 100,000 inches on the ground. On most topographic maps, a graphic scale such as the one shown in Figure 4.21 is used to represent the map scale. The graphic scale can be used as a ruler to measure distances on the map.

Map Direction

The convention used in most topographic maps is that the top of the map is north. However, since north is determined with a compass, and geographic north differs from magnetic north, most topographic maps include an arrow showing each. The map in Figure 4.22 shows magnetic north with an arrow labeled "MN" and geographic North with an arrow labeled "GN".

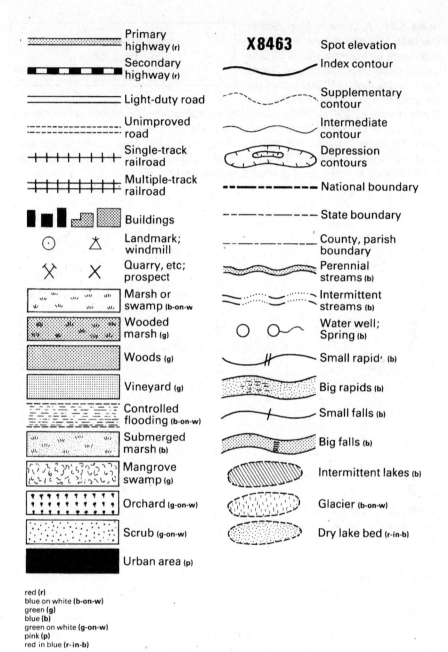

Figure 4.20 Selected Topographic Map Symbols.
Source: *Earth Science on File*. Copyright © 1988 by Diagram Visual Information Ltd.
Reprinted by permission of Facts on File, Inc.

Earth's Coordinate System and Mapping

Figure 4.21 A Graphic Map Scale. The ratio 1:62,500 means that 1 unit of distance on the map equals 62,500 units of distance on the ground. Thus, 1 inch on the map is 62,500 inches on the ground, which is slightly less than 1 mile (1 mil = 63,360 in.). One centimeter on the map equals 62,500 centimeters on the ground (0.625 kilometer). The graphic scale is printed so that each unit is precisely the correct size for the distance indicated. (In this case 1 mile on the scale measures 1 inch in length.) The first unit on the graphic scale is subdivided so that it can be used to make more precise measurements.

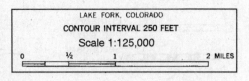

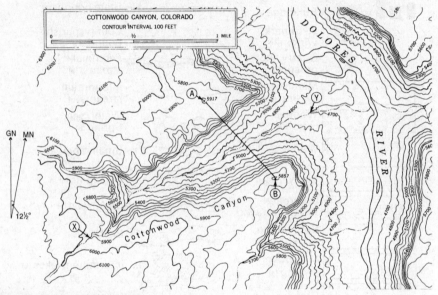

Figure 4.22 Topographic Map.

Source: *Exploring Earth Science*, Walter A. Thurber and Robert E. Kilburn, Allyn and Bacon, 1965. Used with permission of Prentice-Hall.

Map Profiles

It is often useful to construct a profile from a topographic map. A **map profile** shows what a cross section of land looks like between two points. Figure 4.23 shows how a profile can be constructed.

1. The two points between which the profile is to be drawn are chosen, and a line is drawn to connect them.
2. The edge of a piece of paper is placed along the line, and the edge of the paper is marked wherever it intersects a contour line.

Figure 4.23 Steps in Constructing a Map Profile. Source: *Exploring Earth Science*, Walter A. Thurber and Robert E. Kilburn, Allyn and Bacon, 1965. Used with permission of Prentice-Hall.

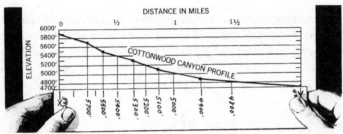

3. The paper is moved to a piece of graph paper on which a vertical scale has been marked to match the contour lines intersected. At each point where a contour line crosses the edge of the paper, a line of the appropriate height is drawn on the graph paper.
4. The endpoints of the lines are connected in a smooth line to form the finished profile.

Conventions Used in Topographic Maps

When topographic maps are drawn or read, the following conventions apply:

- All points on a contour line have the same elevation.
- Every fifth line, called an **index line**, is generally indicated in black ink or bold type and labeled with the line's elevation.
- All contour lines are closed, but they may run off the map.
- Two contour lines of different elevations may not cross each other.
- Contour lines may merge at a cliff or waterfall.
- The spacing of contour lines indicates the nature of the slope. The closer the contour lines are spaced, the steeper the slope. Even spacing indicates a uniform slope.
- Where contour lines cross a stream, they always form a V whose apex points up the valley.
- Where contour lines cross a ridge between valleys, they often form a V whose apex points down the valleys.

MULTIPLE-CHOICE QUESTIONS

In each case, write the number of the word or expression that best answers the question or completes the statement.

1. Which diagram most accurately shows the cross-sectional shape of Earth?

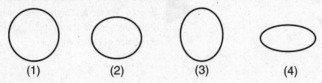

2. The diagrams below represent photographs of a large sailboat taken through telescopes over time as the boat sailed away from shore out to sea. The number above each diagram shows the magnification of the telescope lens.

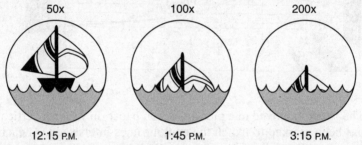

   Which statement best explains the apparent sinking of this sailboat?
   (1) The sailboat is moving around the curved surface of Earth.
   (2) The sailboat appears smaller as it moves farther away.
   (3) The change in density of the atmosphere is causing refraction of light rays.
   (4) The tide is causing an increase in the depth of the ocean.

3. At sea level, which location would be farthest from the center of Earth?
   (1) the North Pole   (3) 45° N
   (2) 23½° S   (4) the Equator

4. The best evidence of Earth's nearly spherical shape is obtained through
   (1) telescopic observations of other planets
   (2) photographs of Earth from an orbiting satellite
   (3) observations of the Sun's altitude made during the day
   (4) observations of the Moon made during lunar eclipses

5. Measurements taken with gravity meters would be most useful in providing information concerning Earth's
   (1) magnetic field intensity   (3) tidal range
   (2) shape   (4) distance from the Sun

6. Two students are located 800 kilometers apart on the same meridian of longitude. They measure the altitude of the Sun at noon on the same day. They can use these measurements to calculate
   (1) Earth's circumference
   (2) Earth's density
   (3) Earth's mass
   (4) the Sun's diameter

7. According to the Generalized Bedrock Geology of New York State map in the *Earth Science Reference Tables*, Triassic bedrock is found in New York State at approximately
   (1) 41° 05' N, 74° 00' W
   (2) 44° 05' N, 74° 00' W
   (3) 74° 00' N, 41° 05' W
   (4) 74° 00' N, 44° 05' W

8. What is the approximate location of the Canary Islands hot spot?
   (1) 32° S, 18° W
   (2) 32° S, 18° E
   (3) 32° N, 18° W
   (4) 32° N, 18° E

9. Polaris is used as a celestial reference point for Earth's latitude system because Polaris
   (1) always rises at sunset and sets at sunrise
   (2) is located over Earth's axis of rotation
   (3) can be seen from any place on Earth
   (4) is a very bright star

10. From Utica, New York, Polaris is observed at an altitude of approximately
    (1) 43°
    (2) 47°
    (3) 75°
    (4) 90°

11. The diagram below shows an observer on Earth viewing the star Polaris.

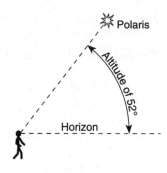

What is this observer's latitude?
(1) 38° N
(2) 38° S
(3) 52° N
(4) 52° S

Earth's Coordinate System and Mapping

The diagrams below illustrate systems that can be used to determine position on a sphere. Refer to them to answer questions 12 and 13.

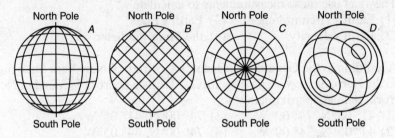

12. Systems of line like those illustrated above are called
    (1) latitude systems
    (2) coordinate systems
    (3) great circle systems
    (4) axis systems

13. Which of the systems illustrated is most like the latitude-longitude system used on Earth?
    (1) A
    (2) B
    (3) C
    (4) D

Base your answers to questions 14 and 15 on the map below, which shows the latitude and longitude of five observers, A, B, C, D, and E, on Earth.

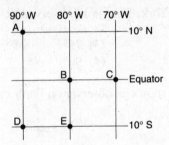

14. What is the altitude of Polaris (the North Star) above the northern horizon for observer A?
    (1) 0°
    (2) 10°
    (3) 80°
    (4) 90°

15. Which two observers would be experiencing the same apparent solar time?
    (1) A and C
    (2) B and C
    (3) B and E
    (4) D and E

16. The diagram below shows latitude measurements every 10 degrees and longitude measurements every 15 degrees.

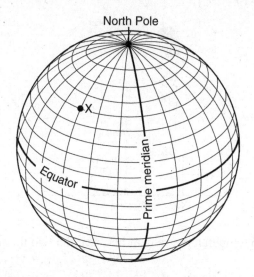

What is the latitude and longitude of point **X**?
(1) 40° S, 45° E
(2) 50° N, 45° W
(3) 60° S, 30° W
(4) 75° N, 30° E

17. When the time of day for a certain ship at sea is 12 noon, the time of day at the Prime Meridian (0° longitude) is 5 p.m. What is the ship's longitude?
(1) 45° W
(2) 45° E
(3) 75° W
(4) 75° E

18. Which statement is *not* true about fields?
(1) They do not change with time.
(2) They are often illustrated by the use of isolines.
(3) Any one place in a field has a measurable value at a specific time.
(4) Gradients indicate the degree of change from place to place in a field.

19. In the diagram below, the thermometer held 2 meters above the floor shows a temperature of 30°C. The thermometer on the floor shows a temperature of 24°C.

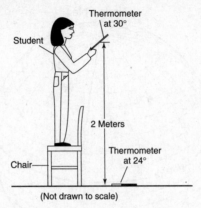

(Not drawn to scale)

What is the temperature gradient between the two thermometers?
(1) 6 C°/m
(2) 2 C°/m
(3) 3 C°/m
(4) 4 C°/m

20. A topographic map is a two-dimensional model that uses contour lines to represent points of equal
(1) barometric pressure
(2) temperature gradient
(3) elevation above sea level
(4) magnetic force

Base your answers to questions 21 through 24 on the topographic map below. Elevations are in feet. Points *A* and *B* are locations on the map.

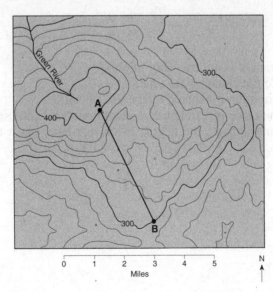

21. Toward which direction does the Green River flow?
    (1) northeast
    (2) northwest
    (3) southeast
    (4) southwest

22. What is the gradient along the straight line between points *A* and *B*?
    (1) 10 ft/mi
    (2) 20 ft/mi
    (3) 25 ft/mi
    (4) 35 ft/mi

23. Which graph best represents the profile along line *AB*?

    (1)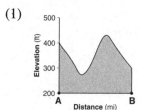

    (3)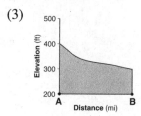

    (2)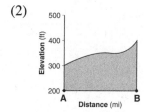

    (4)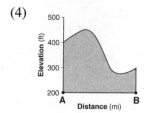

24. What evidence can be used to determine that the land surface in the northeast corner of the map is relatively flat?
    (1) a rapidly flowing river
    (2) a large region covered by water
    (3) the dark contour line labeled 300
    (4) the absence of many contour lines

CONSTRUCTED RESPONSE QUESTIONS

Base your answers to questions 25 through 27 on the diagram below, which represents a north polar view of Earth on a specific day of the year. Solar times at selected longitude lines are shown. Letter A represents a location on Earth's surface.

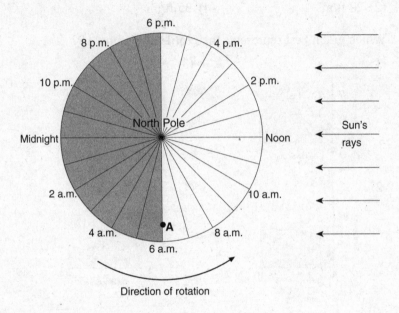

25. How many degrees apart are the longitude lines shown in the diagram? [1]

26. State the altitude of Polaris as seen by an observer at the North Pole. [1]

27. How many hours of daylight would an observer at location *A* experience on this day? [1]

Base your answers to questions 28 and 29 on the topographic map of an island shown below. Elevations are expressed in feet. Points A, B, C, and D are locations on the island. A triangulation point shows the highest elevation on the island.

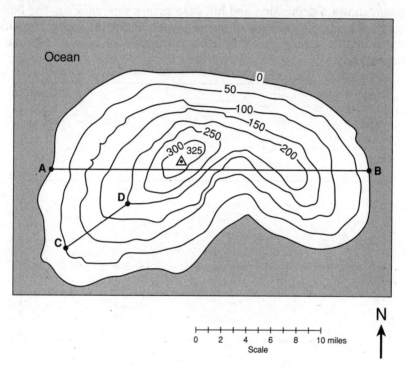

28. On the grid provided below, construct a topographic profile representing the cross-sectional view between point A and point B, following the directions below.
    (a) Plot the elevation of the land along line AB by marking, with a dot, the elevation of *each* point where a contour line is crossed by line AB. [2]
    (b) Connect the dots with a smooth, curved line to complete the topographic profile. [1]

Earth's Coordinate System and Mapping

29. What is the average gradient, in feet per mile, along the straight line from point C to point D? [1]

Base your answers to questions 30 and 31 on the diagrams below. The top diagram shows a depression and hill on a gently sloping area. The bottom diagram is a topographic map of the same area. Points A, X, and Y are locations on Earth's surface. A dashed line connects points X and Y. Elevation is indicated in feet.

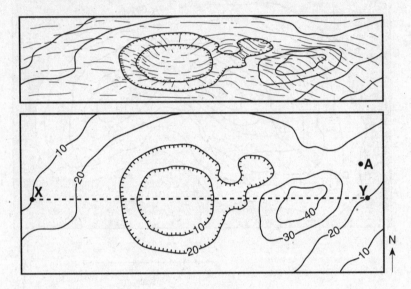

30. What is a possible elevation of point A? [1]

31. On the grid below, construct a topographic profile along line XY by plotting a point for the elevation of *each* contour line that crosses line XY. Points X and Y have already been plotted on the grid. Connect the points with a smooth, curved line to complete the profile. [2]

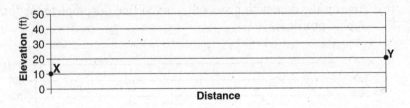

EXTENDED CONSTRUCTED RESPONSE QUESTIONS

Base your answers to questions 32 through 35 on the diagram provided below, which shows observations made by a sailor who left his ship and landed on a small deserted island on June 21. The diagram represents the apparent path of the Sun and the position of Polaris, as observed by the sailor on this island.

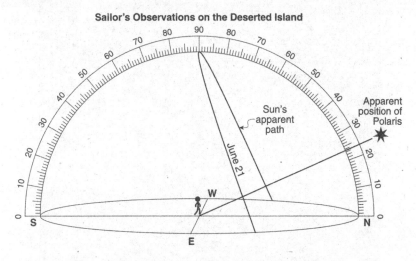

32. On the diagram above, draw an arrow on the June 21 path of the Sun to show the Sun's direction of apparent movement from sunrise to sunset. [1]

33. The sailor was still on the island on September 23. On the diagram above, draw the Sun's apparent path for September 23, as it would have appeared to the sailor. Be sure your September 23 path indicates the correct altitude of the noon Sun and begins and ends at the correct points on the horizon. [2]

34. Based on the sailor's observations, what is the latitude of this island? Include the units and the compass direction in your answer. [1]

35. The sailor observed a 1-hour difference between solar noon on the island and solar noon at his last measured longitude onboard his ship. How many degrees of longitude is the island from the sailor's last measured longitude onboard his ship? [1]

Earth's Coordinate System and Mapping

Base your answers to questions 36 through 39 on the map below, which shows partially drawn contour lines. **X**s indicate elevations in meters. Letters *A*, *B*, *C*, and *D* represent locations on the map.

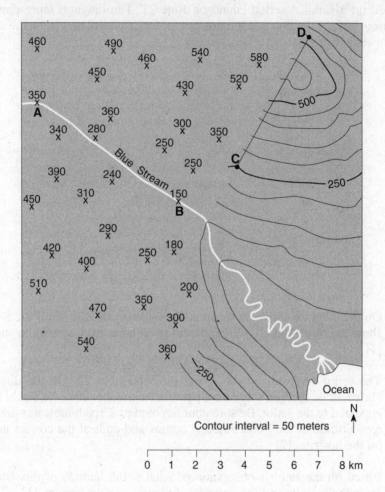

Contour interval = 50 meters

0 1 2 3 4 5 6 7 8 km

36. On the map above, complete the 250-meter contour line. [1]

37. On the portion of the map showing contour lines, place an **X** in an area where an elevation of 55 meters is located. [1]

38. Calculate the stream gradient from elevation *A* to elevation *B*. Label your answer with the correct units. [1]

39. On the grid below, construct a topographic profile along line *CD*. Plot with an **X** the elevation of *each* contour line that crosses line *CD*. Connect the **X**s from *C* to *D* with a smooth, curved line to complete the profile. Elevations *C* and *D* have already been plotted. [1]

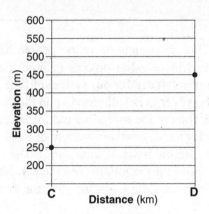

40. An island measures 10 kilometers from east to west and 8 kilometers from north to south. A single hill on the east side of the island has a maximum elevation of 57 meters and is steepest to the north. Draw a simple contour map to represent this island, using a distance scale of 1 centimeter = 1 kilometer and a contour interval of 10 meters. [4]

CHAPTER 5
OUR SYSTEM OF TIME

> **KEY IDEAS** For thousands of years, people have used the changing positions of celestial objects to note the passage of time. The frame of reference for our system of time has changed over the years; however, it is still based upon the changing positions of celestial objects due to Earth's motions. Earth's rotation provides a basis for the day and our system of local time; meridians of longitude are the basis of time zones. The revolution of the Moon is the basis for the month, and the revolution of Earth around the Sun provides the basis for the year and our current calendar system.

KEY OBJECTIVES
Upon completion of this chapter, you will be able to:

- Explain how Earth's rotation forms the basis for the day and our system of local time.
- Compare and contrast the solar day and the sidereal day.
- Explain the difference between a synodic month and a sidereal month.
- Describe how meridians of longitude are used to define time zones.
- Analyze the seasonal changes in the Sun's apparent path through the sky.

THE DAY

The seemingly endless cycle of daylight and darkness was probably the earliest timekeeping unit—the day. It is the most basic of all natural cycles. Few things influence our lives as deeply as the change from light to dark, from day to night, from being asleep and being awake. The problem with the day is where to begin and end it and how to divide it up.

The earliest way of expressing "time of day" was simply to point to a place in the sky where the Sun is found at that time. Here is where the Sun rises; over there is where it sets. Morning, noon, and afternoon can be easily determined simply by looking at which side of the sky the Sun is in. Selecting a specific landmark with which the Sun aligns, such as a hut, a tree, or a distant mountain peak, can further refine this system. Using such markers, it is not difficult to divide daytime into 6, 9, or even 12 divisions.

Dividing the Day into Hours

Timekeeping became a bit more precise with the simple invention of a stick stuck vertically into the ground. In sunlight, a vertical stick casts a shadow. As the Sun changes position, the shadow also changes position. The path of the Sun through the sky could be recorded as the line traced on the ground by the tip of the stick's shadow. This line could then be divided into equal units that, in turn, could be used to divide daytime into uniform time periods. The next small step was to realize that stars could also be sighted along the tip of a vertical stick and their motions tracked at night. In turn, these motions could then be used to divide nighttime into uniform time periods.

More than 3,000 years ago, the Egyptian priests of the Sun god Ra decided to divide the day into 12 hours of daytime and 12 hours of night. They chose 12 because they believed this number was mystical and symbolized the gods. The problem they faced was that most days did not have daylight and darkness periods of equal length. In New York State, for example, daylight varies from fewer than 9 hours in the winter to more than 15 hours in the summer. Consequently, dividing daylight into 12 equal time periods would yield a longer "hour" in the summer and a shorter "hour" in the winter. Eventually, the Egyptians discovered that there are only 2 days, the equinoxes, during which daylight and darkness are of equal length, so daylight and darkness were merged into a single 24-hour day with days and nights that vary in length. But the Egyptians still had to measure the length of that 24-hour day.

You might think that mea-

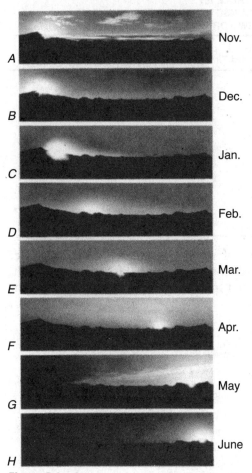

Figure 5.1 Changes in the Sunset Position along the Horizon During an 8-Month Period, Moving South from Position A to the Winter Solstice (B), Then North Again in C through the Spring Equinox (E) to the summer solstice (H).
Source: *Astronomy: The Cosmic Journey*, William K. Hartman, Wadsworth, 1987.

suring the length of a day is easy—simply pick a fixed point during the day, and then measure the time that elapses until that same point in the day is reached again. Then divide this time into 24 equal "hours." Great idea, but the question now becomes, At what point should you start measuring the day?

Suppose you decide to use sunrise to sunrise, or sunset to sunset, as your fixed time interval. You soon discover that the time of sunrise is not fixed. During part of the year the Sun rises a bit later each day, and during another part of the year it rises a bit earlier each day. See Table 5.1.

TABLE 5.1
TIME OF SUNRISE AND SUNSET DURING A TYPICAL YEAR AT MASSENA, NEW YORK (44°56'N, 74°51'W)

Date	Sunrise (A.M.) (Eastern Standard Time)	Sunset (P.M.) (Eastern Standard Time)
January 21	7:30	4:52
February 21	6:52	5.35
March 21	6:00	6:14
April 21	5:04	6:53
May 21	4:24	7:29
June 21	4:13	7:50
July 21	4:34	7:38
August 21	5:09	6:55
September 21	5:46	5:58
October 21	6:24	5:04
November 21	7:06	4:25
December 21	7:35	4:21

You also find that the positions of sunrise and sunset also change. See Figure 5.1 for changes in the sunset position over an 8-month period. If the times *and* the positions of sunrise and sunset vary, how, then, did people find a fixed point from which to measure the length of an entire day? The answer lies in the apparent path of the Sun. Look at Figure 5.2, which shows the Sun's apparent path through the sky on three different days of the year. Notice that, although the positions of sunrise and sunset and the length of the Sun's path vary, the midpoint of all three paths occurs when the Sun crosses the line running due north-south through the observer's zenith.

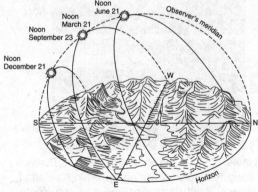

Figure 5.2 The Apparent Path of the Sun as Viewed by an Observer in New York State Varies from Season to Season, but the Sun Always Crosses the Observer's Meridian at Noon.

You recall from Chapter 4 that this imaginary line is the observer's meridian. Also notice that, when the Sun crosses the observer's meridian, it is at its highest point in the sky for that day and a vertical stick will cast the shortest shadow. The moment when the Sun is at its highest point in the sky is called **solar noon**. No matter how the number of daylight hours varies, solar noon always occurs exactly in the middle of the daylight period. Also, the Sun is always seen in the same direction at solar noon.

The Solar Day Versus the Sidereal Day

The position of solar noon can be marked using two fixed points, one on the horizon and one nearby, such that they line up with the Sun at solar noon. (Since the Sun crosses the observer's meridian at solar noon, a line connecting these markers will run due north-south; a shadow cast by a vertical stick will also lie along a north-south line—a handy way of determining direction without a compass.) Whenever the Sun aligns with these two markers, it is solar noon. The time Earth takes to rotate from one solar noon until the next is called a **solar day**; that is, a solar day is the time from when the Sun crosses an observer's meridian until the Sun next crosses the observer's meridian.

A similar technique can be used to track stars at night. The time from when a star crosses an observer's meridian until the same star next crosses the observer's meridian is called a **sidereal day**. However, a solar day and a sidereal day are not equal in length. How can *that* be?

The answer lies in the fact that stars are very much farther away from Earth than is the Sun. As a result, the direction to a star stays virtually the same throughout Earth's orbit, while the direction to the Sun changes drastically. Look at Figure 5.3. Even in this small diagram, you can see that from one place in Earth's orbit to another the direction to the star changes only slightly. Imagine how little it would change if the star was shown trillions of miles away! There would be almost no detectable difference in the direction to the star from any two places in Earth's orbit.

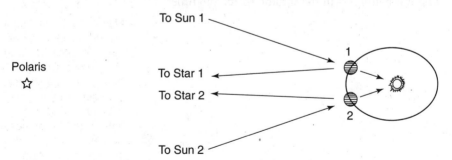

Figure 5.3 As Earth Orbits, the Change in the Direction to a Star Is Tiny Compared to the Change in Direction to the Sun.

Now look at Figure 5.4. Let's begin with Earth at position A when the Sun and a star are at an observer's meridian, and mark the direction to the Sun and to a star. When Earth has rotated exactly 360°, the star is back on the observer's meridian, marking the end of one sidereal day. But remember that, as Earth rotates on its axis, it also revolves around the Sun. Therefore, by the time an observer reaches position B, Earth's position is different too, and so is the direction to the Sun. Therefore, Earth must rotate *more* than 360° for the Sun to reach the observer's meridian and the next solar noon. Thus, a solar day is *more* than one Earth rotation long. As you can see, Earth must rotate an additional 1° per day (which takes about 4 minutes) in order for the Sun to reach the observer's meridian again. Thus, a solar day is about 4 minutes longer than a sidereal day.

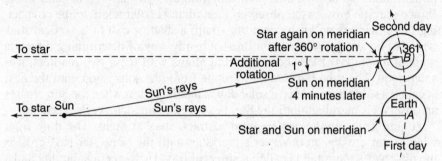

Figure 5.4 The Solar Day Is about 4 Minutes Longer than the Sidereal Day. (Angles are greatly exaggerated in this diagram.)

Which day is better to use as the basis for a timekeeping system? If the sidereal day is used, the problem shown in Figure 5.5 occurs. If we call the time at location M midnight when Earth is at position 1, then, by the time Earth has revolved to position 2, midnight would be occurring during the daytime! For that reason, the solar day is used as the basis of our timekeeping system. No matter where Earth is in its orbit, solar noon always occurs where it belongs—in the middle of the daytime.

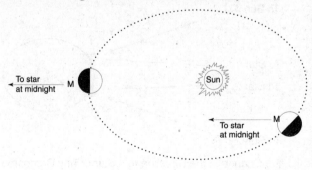

Figure 5.5 Midnight (12:00 A.M.) Sidereal Time at Location *M* at Two Times During the Year. Note that sidereal time gets "out of synch" with our periods of daylight and darkness because of Earth's changing position in relation to the Sun.

The Mean Solar Day

With the development of ever more accurate devices for measuring time, it became clear that the length of a solar day varies throughout the year. If we set a clock to 12:00 at solar noon one day, it will not read 12:00 at solar noon the next day. How can this be, since we do not detect the stars or the Sun speeding up or slowing down as they move through the sky because of Earth's rotation?

The answer lies in the relative motions of Earth and the Sun due to Earth's revolution. Although the rate at which Earth rotates remains steady, the rate at which it orbits the Sun doesn't. Remember that Johannes Kepler determined that Earth travels farther during a fixed time period in one part of its orbit than during another part. This, in turn, means that Earth is revolving faster during one part of its orbit and slower during another part.

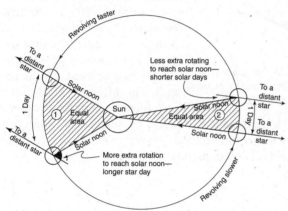

Figure 5.6 Varying Length of the Solar Day. Position 1: The faster rate of revolution causes a greater daily change in Earth's position, requiring more rotation to reach solar noon. The result is a longer solar day. Position 2: A slower rate of revolution results in less of a daily change in Earth's position; thus less rotation is required to reach solar noon and the solar day is shorter. (Angles are greatly exaggerated in this diagram.)

In Figure 5.6, the effect of this difference in orbital speed is shown. At position 1, Earth has to rotate farther between solar noons than it does in position 2. Thus, the solar day is longer at position 1 than at position 2. Making clocks that would speed up or slow down in time with the solar day is very difficult. Clocks are generally constructed to mark off absolutely uniform periods of time. As a result, if you were to set your watch to 12:00 at solar noon on one day, your watch would read earlier than 12:00 at solar noon the next day for part of the year, and later than 12:00 at solar noon for the remainder of the year. The solution has been to split the difference and use the *average* length of a solar day, or the **mean solar day**, as the basis for timekeeping. Therefore, clocks are constructed to complete a 24-hour cycle in a mean solar day.

Local Time

So far, we have talked about the day in a very general way, with solar noon occurring when the Sun crosses an observer's meridian. However, an observer's meridian depends on his or her longitude, meaning that each

Our System of Time

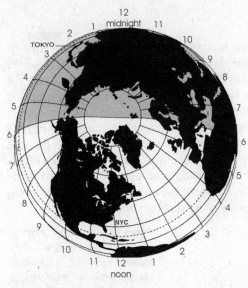

Figure 5.7 At Precisely the Same Instant, a Person in New York City Experiences Solar Noon While in Tokyo, Japan, the Time is 2:00 A.M.

observer has his or her own time scale based upon his or her location. A careful look at Figure 5.7 should make it clear that, while someone in New York City is experiencing solar noon, someone in Tokyo, Japan, is asleep in bed at 2 A.M. Our solar noon in New York (~75°W) occurs some 3 hours earlier than solar noon in California (~120°W) — a 45° difference in longitude — because Earth takes 24 hours to spin through 360°, rotating through 15° every hour.

In the past, every locality set its clocks to its own solar noon. Clocks in Syracuse would be set to a different local time from those in Buffalo. As soon as people began traveling long distances rapidly, by stagecoach, train, or steamboat, however, local timekeeping created severe problems for companies that wanted to set up arrival and departure timetables.

Time Zones

In 1883 and 1884, the countries of the world agreed to set up a series of time zones. In a **time zone** the same time is used throughout the zone, instead of the local time at each place within the zone.

Since Earth rotates 15° per hour, time zones were set up around a series of meridians spaced at 15° intervals starting at the Prime Meridian, which, you have learned, passes through Greenwich, England. Time zones officially cover 7½° on either side of these meridians, so places between longitude 7½° W and

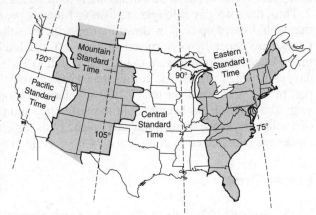

Figure 5.8 The Time Zones of the United States. Source: ESCP, *Investigating the Earth*, *Teachers Guide*, Part 1, p. 135, American Geological Institute, 1967.

7½° E all set their clocks to **Greenwich Mean Time** (GMT). The next zone to the west extends from longitude 7½° W to 22½° W, and all clocks in that zone are set 1 hour behind GMT. New York State falls in the time zone surrounding the 75° W meridian, which extends from 67½° W to 82½° W. This zone sets its clocks 5 hours behind GMT.

To avoid inconveniences, the boundaries are shifted where they pass over land so that small countries or whole states fall within the same time zone. Figure 5.8 shows how the time zone boundaries in the United States have been shifted to follow state lines.

The International Date Line

As you travel west of Greenwich, England, you pass through 12 time zones before reaching longitude 180°, ending up 12 hours *behind* GMT. Likewise, as you travel east of Greenwich, England, you also pass through 12 time zones to reach longitude 180°, but now you are 12 hours *ahead* of GMT. How can you be 12 hours behind and 12 hours ahead of GMT at the same location? The answer is that the international date line is located here, at 180° longitude, and marks the boundary between 2 consecutive days; as you cross the date line from east to west, you advance 1 day. The time of day doesn't change, only the date.

To understand how the date line works, consider that the change from one day to the next is made at midnight. Look at Figure 5.9a. At Greenwich, England, it is solar noon on Monday, and the international date line is at midnight. Point A is at 1:00 A.M. Monday, and point B is at 11:00 P.M. Monday. Now (see Figure 5.9b), Earth rotates for 1 hour, and Greenwich goes from noon

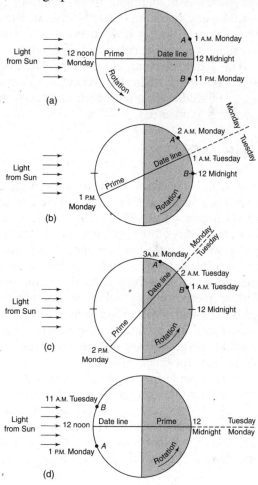

Figure 5.9 Understanding the International Date Line.

on Monday to 1:00 P.M. on Monday afternoon. Point *A* is now at 2:00 A.M. Monday and point *B* goes from 11:00 Monday to midnight, but the date line goes from midnight on Monday night to 1:00 P.M. on *Tuesday* morning. The date line is still 12 hours behind GMT (1:00 P.M. − 12 hr = 1:00 A.M.) and at the same time is 12 hours ahead of GMT (1:00 P.M. + 12 hr = 1:00 A.M.). But 12 hours behind 1:00 P.M. Monday is 1:00 A.M. Monday, while 12 hours *ahead* of 1:00 P.M. GMT is 1:00 A.M. on *Tuesday*. Thus, the date line marks the boundary between the two different days.

Now (see Figure 5.9c), Earth rotates yet another hour. Greenwich goes to 2:00 P.M. Monday, point *A* is at 3:00 A.M. Monday and the date line is at 2:00 A.M. Tuesday morning, but now point *B* is at 1:00 A.M. *Tuesday*. At this time it is Monday at point *A*, but Tuesday at point *B*. Let's continue this all the way to noon on Tuesday at the international date line (see Figure 5.9d). Now Greenwich is at midnight, finally entering Tuesday, and point *A* is at 1:00 P.M. Monday, but point *B* is at 11:00 A.M. Tuesday. The dividing line between Monday and Tuesday is the international date line. East of the date line it is Tuesday; west of the date line it is Monday.

This sequence continues until the date line once again rotates to the midnight position, when the boundary changes from Monday/Tuesday to Tuesday/Wednesday. With each successive rotation of Earth, the boundary advances another day.

THE MONTH

Just as the concept of a day arose from the natural cycle of light and darkness, the concept of a month originated in the natural cycle of lunar phases. (See Chapter 4.) The Moon, with its changes in shape, brightness, and position, has always fascinated humans. Even in the distant past, observers also perceived a connection between lunar phases and a natural human cycle—the menstrual cycle; the word *menstrual* means, literally, "recurring once a month" (and is derived from the Latin word *mensis* meaning "month"). Indeed, there is evidence that the Moon has been used from the very earliest days of human civilization as a measuring tool for time. A Cro-Magnon eagle bone found in Le Placard, France, with notches cut at regular intervals and marked into groups, suggests that the Moon was the first celestial object used by humans for timekeeping.

When people first began to use arithmetic to count time, it was not difficult to determine that the Moon went through its full cycle of phases in 29 or 30 days. By 430 B.C., the Greek astronomer Meton had determined that the new Moon occurred on the same day each year in a 19-year cycle. A hundred years later, another astronomer, Callippus of Cyzicus, had discovered that the 19-year cycle worked only if a day was dropped once every 76 years.

Synodic and Sidereal Months

We now know that we see the phases of the Moon because the Moon revolves around Earth in an elliptical orbit that is tilted at an angle of 5½° to the plane of the ecliptic. (Refer to Figure 3.11.) We also know that the Moon goes through its cycle of phases, from one new Moon until the next, in precisely 29.530588 mean solar days, or 29 days, 12 hours, 44 minutes, and 3 seconds. This 29½-day month, from one new Moon phase until the next, is called a **synodic month**.

Like a solar day, the synodic month is based upon the Moon's position in relation to the Sun. New Moon occurs when the Moon is aligned between the Sun and Earth. Because Earth moves in its own gigantic orbit around the Sun while the Moon is making its much smaller orbit of Earth, the Moon has to travel more than 360° around Earth to align again with the Sun. See Figure 5.10.

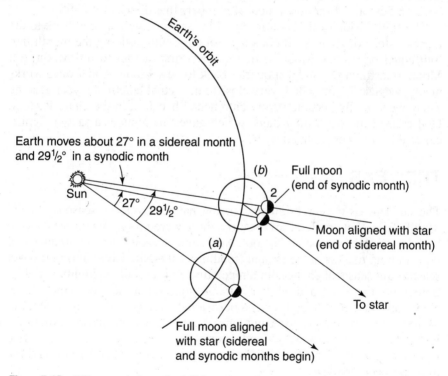

Figure 5.10 Difference in Length of Sidereal and Synodic Months. (a) The full Moon is observed to be near some bright star. (b) Roughly 27⅓ days later, at position 1, the Moon is again near the same bright star, completing a *sidereal month*. But Earth and the Moon have moved together around the Sun during that time. The Moon must revolve for 2⅙ more days before the Moon, Earth, and the Sun are again lined up so that the Moon is full again (position 2), completing a cycle of phases, or a **synodic month**. Thus, the time between full Moons, or the synodic month, is 27⅓ + 2⅙ = 29½ days.

If, instead of the Sun, we use alignment with a distant star, we find that the Moon completes a 360° orbit of Earth every 27.32166 mean solar days, or 27 days, 7 hours, 43 minutes, and 11 seconds. This 27⅓-day month, based upon alignment with a distant star and representing a 360° orbit of Earth, is called a **sidereal month**.

Since the phases of the Moon are so universally visible, the 29½-day synodic month has become the basis for all calendar months. However, lunar months will not fit neatly into a solar year without fractional division, nor will days fit exactly into lunar months. Since the solar year is 365.242199 mean solar days long, you can't get exactly a whole number of days into a solar year either. How, then, can the year be divided into months and weeks? A convenient week will have a number of days that will divide equally into 365. Yet, no matter how you look at it, the week is an artificial unit of time and could have been 5, 6, or even 10 days long. For example, there could have been a 73-week year made up of 5-day weeks for a total of 365 days, because 5 is one of the few practical numbers that divides into 365.

The subdivision of the month into 7-day weeks can be traced back to the Genesis story of creation. It coincides roughly with dividing the month into four equal parts: new Moon to first quarter, first quarter to full Moon, full Moon to last quarter, and last quarter back to new Moon. And 7 days works nicely because $7 \times 52 = 364$, very close to the actual length of a year. But the 7-day week really became firmly entrenched because, in the Genesis story, God created the world in 7 days, which gave this number a sacred significance that survives to this day.

THE YEAR

The concept of a year originated in the natural cycle of seasons. In the Middle East, that cradle of civilization, there were probably two seasons—a hot, dry summer season and a cool, wet winter season. The Egyptians had a built-in time marker in the annual flooding of the Nile River during the wet season. But seasonal changes in temperature and rainfall are highly variable. Sometimes the Nile floods earlier in the season, sometimes later. And who in New York State hasn't experienced occasional stretches of balmy weather in March, only to encounter freezing temperatures and a late spring snowstorm in April? Thus, a more reliable indicator of the time of year is needed. Here again, the observation of celestial objects provided a valuable tool, and the Sun takes center stage.

Seasons and the Sun's Path

Because of Earth's rotation, the Sun appears to move in an arc across the sky from east to west. The daily path that the Sun traces through the sky goes through a cycle of changes with the seasons. Also, although the Sun always rises to the east of an observer and sets to the west, the exact positions of sunrise and sunset on the horizon change cyclically with the seasons too. Three factors combine to produce these cyclic changes: Earth's *revolution* around the Sun, the *tilt* of Earth's axis of rotation, and the **parallelism of Earth's axis** of rotation (the position of the axis at any given time is always parallel to its position at any other time). Figure 5.11 shows how these factors result in important changes in Earth's position relative to the Sun on key dates in the seasonal cycle of changes.

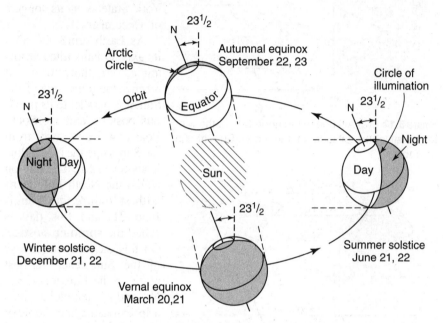

Figure 5.11 Earth's Changing Position in Relation to the Sun.

The Solstices

As Earth orbits the Sun, it also rotates on its axis. At one end of its axis is the North Pole, and at the other end the South Pole. Earth's axis of rotation is tilted at an angle of 23½°, so that during one part of its orbit the North Pole is leaning away from the Sun and the South Pole is leaning toward the Sun. The day on which the North Pole leans farthest *away from* the Sun is December 21, and this day is called the **winter solstice**. As you can see in Figure 5.12a, because of Earth's spherical shape only one parallel, the Tropic of Capricorn receives the Sun's direct rays (the direct rays strike Earth perpendicular to its surface).

On December 21, the direct rays of the Sun reach farthest south of the Equator—23½° S—and this parallel is given a special name, the Tropic of Capricorn. However, when the Sun's rays strike the Tropic of Capricorn at 90°, they are striking New York City, which is 64.5° of latitude away, at 90° − 64.5°,

Our System of Time

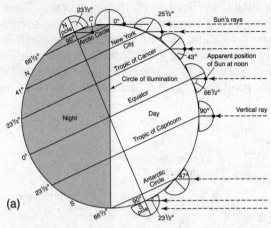

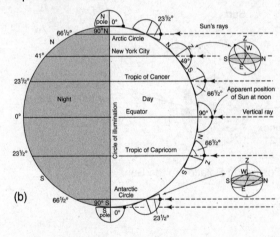

Figure 5.12 (a) Altitude of the noon Sun at the winter solstice (b) Altitude of the noon Sun at the equinoxes

or only 25.5°—the lowest angle of the year. At positions north of New York City, the Sun's rays strike at an even smaller angle. Therefore, the shadow cast by a stick at noon in New York State is at its longest on December 21.

As Earth orbits the Sun, its axis remains tilted at this angle, so that, when it reaches the *other* side of the Sun, it is parallel to its previous position and the South Pole now leans away from the Sun while the North Pole leans toward it. The day on which the North Pole leans farthest *toward* the Sun is June 21, and this day is called the **summer solstice**. On this day, the direct rays of the Sun reach farthest north of the Equator—23½° N—and this parallel is given a special name, the Tropic of Cancer. Locations north of the Tropic of Cancer never receive the Sun's direct rays, but on June 21 the rays are the most nearly direct of the year. Therefore, the shadow cast by a stick at noon in New York State is at its shortest, and the shadow falls due north.

The Equinoxes

Midway between these ends of Earth's orbit, however, neither the North nor the South Pole leans toward the Sun. To be sure, the poles are still tilted, but to the side, rather than toward or away from the Sun. In fact, at these midpoints, Earth's shadow cuts precisely through the poles. (Refer to the vernal and autumnal equinoxes shown in Figure 5.11.) In other words, the edge of Earth's shadow forms a great circle, cutting Earth precisely in half. And that great circle runs through the poles—it is a meridian! As Earth rotates, every point on its surface spends exactly one-half rotation in the light and one-half rotation in

Earth's shadow; day and night are equal in length. These midpoints occur on March 21 and September 23; and these dates are called, respectively, the spring and fall **equinoxes** (*equinox* is the Latin word for "equal night"). On these dates, the direct rays of the Sun strikes the Equator. See Figure 5.12b.

The Sun's Changing Path

At each of the positions shown in Figure 5.11, an observer on Earth's surface would see the Sun follow a different path as it rose, arced through the sky, and set. Figure 5.13 shows the changing path of the Sun to an observer at 42° N as Earth orbits the Sun.

Path *A* is the Sun's path on December 21, the winter solstice. Note that on this date the Sun rises at a point south of east on the horizon and sets south of west, and the arc traced by the Sun through the sky is the shortest. The shorter the Sun's daily path through the sky, the shorter the period of daylight, so December 21 is the shortest day of the year—less than 9 hours in New York State. The altitude of the Sun at noon is also at its lowest for the year, so sunlight strikes Earth's surface indirectly. Indirect sunlight is less intense than direct sunlight

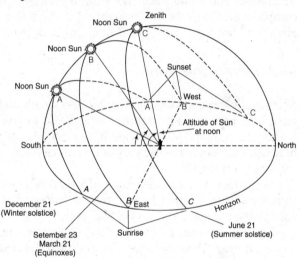

Figure 5.13 The Changing Path of the Sun to an Observer at 42° N.

and therefore has less of a warming effect. Indirect sunlight, coupled with a short daylight period, is the cause of the winter season's lower temperatures.

Path *B* is the Sun's path on March 21 and September 23, the equinoxes. After the winter solstice, the sunrise and sunset points begin to migrate northward a bit each day, and the arc traced by the Sun slowly lengthens. When the Sun reaches path *B* on March 21, sunrise is due east, sunset is due west, and the daylight period is exactly 12 hours long; this is the spring equinox. After the spring equinox, the points of sunrise and sunset keep migrating northward, and the Sun's path keeps lengthening, until June 21, the summer solstice.

Our System of Time

Path *C* is the Sun's path on June 21, the summer solstice. On this date, the Sun rises north of east and sets north of west on the horizon. The Sun's path is the longest of the year; therefore, the daylight period is also the longest of the year—more than 15 hours in New York State! The noon Sun also reaches its greatest altitude of the year, and sunlight strikes Earth's surface more directly and intensely than at any other time of the year. More direct, intense sunlight, coupled with longer daylight hours, is the cause of the summer season's higher temperatures.

After June 21, the points of sunrise and sunset begin to migrate southward each day, and the arc traced by the Sun shortens. By September 23, the Sun is back to path *B*. Once again the Sun rises due east and sets due west, and the daylight period is exactly 12 hours in length. After the fall equinox the Sun's path continues its steady southward migration. Each day it rises and sets farther south and traces a shorter arc through the sky until the cycle is complete on December 21, roughly 365 days later. Then the cycle begins anew, moving from path *A* to *B* to *C* and back through *B* to *A* again.

Measuring the Year

Such cyclic seasonal changes in the Sun's path and sunrise-sunset position have long been used by cultures worldwide as fixed points to measure the length of a year and to subdivide it. In principle, the average number of days between successive recurrences of an annual event will yield an estimate for the length of a year.

One method is to use the length of a shadow at noon to find the winter and summer solstices. The Chinese used the number of days between two such solstices to calculate the length of a year.

Between the tropics of Cancer and Capricorn, there is always at least one day each year when the Sun is directly overhead at noon and there are no shadows at noontime. The Incas of Peru used this fact to figure out the length of a year. They noted the days when the Sun could be seen at noon through a skinny vertical tube pointing at zenith, and counted the average number of days between recurrences.

Another method is to note the positions of sunrise and sunset on the horizon. At the solstices, the Sun rises and sets at its northernmost and southernmost positions. Of course, for this method to work, the exact same position must be used for observations and the exact point on the horizon where the sun rises and sets must be carefully marked. It is thought that the circular arrangement of the huge rocks at Stonehenge may have been used for this purpose.

Table 5.2 shows the length of the mean solar year determined by these methods in many different cultures over time—with an amazing degree of accuracy!

TABLE 5.2
ESTIMATED LENGTH OF THE MEAN SOLAR YEAR

Place	Date	Estimated Length of Year (mean solar days)	Error
Babylon	700 B.C.	365.24579	0.00344
Egypt	150 B.C.	365.2466	0.0043
China	8 B.C.	365.25016	0.0078
India	A.D. 476	366.2589	0.0167
Mexico	700	365.2420	−0.0001
Arabia	900	365.24056	−0.00170
Samarkand	1400	365.24223	0.00030
Europe	1500	365.24222	0.00034
China	1620	365.24219	−0.00003
Europe	1600	365.24222	−0.00003
Modern	2000	365.242199	

MULTIPLE-CHOICE QUESTIONS

In each case, write the number of the word or expression that best answers the question or completes the statement.

1. The length of an Earth day is determined by the time required for approximately one
   (1) Earth rotation
   (2) Earth revolution
   (3) Sun rotation
   (4) Sun revolution

2. The length of an Earth year is based on Earth's
   (1) rotation of 15°/hr
   (2) revolution of 15°/hr
   (3) rotation of approximately 1°/day
   (4) revolution of approximately 1°/day

3. The time required for the Moon to show a complete cycle of phases when viewed from Earth is approximately
   (1) 1 day
   (2) 1 week
   (3) 1 month
   (4) 1 year

4. On June 21, where will the Sun appear to rise for an observer located in New York State?
   (1) due west
   (2) due east
   (3) north of due east
   (4) south of due east

5. When does local solar noon occur for an observer in New York State?
   (1) when the clock reads 12 noon
   (2) when the Sun reaches its maximum altitude
   (3) when the Sun is directly overhead
   (4) when the Sun is on the Prime Meridian

6. A student in New York State looked toward the eastern horizon to observe sunrise at three different times during the year. The student drew the following diagram that shows the positions of sunrise, *A*, *B*, and *C*, during this one-year period.

   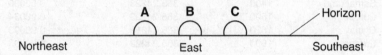

   Which list correctly pairs the location of sunrise to the time of the year?
   (1) *A*—June 21   *B*—March 21   *C*—December 21
   (2) *A*—December 21 *B*—March 21   *C*—June 21
   (3) *A*—March 21   *B*—June 21   *C*—December 21
   (4) *A*—June 21   *B*—December 21 *C*—March 21

7. The diagram below shows the latitude and longitude lines on Earth. Points *A* and *B* are locations on Earth's surface.

   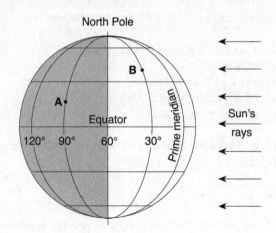

   If it is 4 A.M. at location *A*, what time is it at location *B*?
   (1) 10 A.M.   (3) 6 A.M.
   (2) 2 A.M.   (4) 8 A.M.

8. The diagram below shows the noontime shadows cast by a student and a tree. If the time is solar noon and the student is located in New York State, in what direction is the student facing?

(1) north
(2) south
(3) east
(4) west

Base your answers to questions 9 through 11 on the diagram below and on your knowledge of Earth science. The diagram shows a pin perpendicular to a card. The card was placed outdoors in the sunlight on a horizontal surface. The positions of the pin's shadow on the card were recorded several times on March 21 by an observer in New York State.

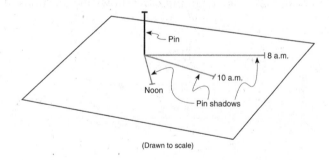

9. Which diagram best represents the length of the pin's shadow at 2 P.M. on March 21?

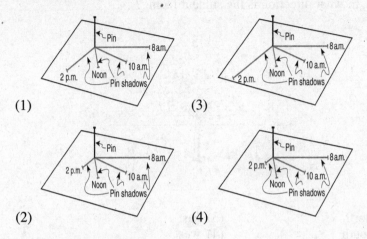

10. The changing location of the pin's shadow on March 21 is caused by
    (1) the Sun's rotation
    (2) the Sun's revolution
    (3) Earth's rotation
    (4) Earth's revolution

11. On June 21, the card and pin were placed in the same position as they were on March 21. The diagram below shows the positions of the pin's shadow.

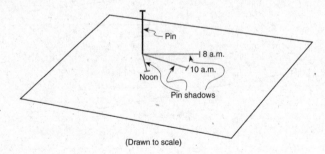

(Drawn to scale)

Which statement best explains the decreased length of each shadow on June 21?
(1) The Sun's apparent path varies with the seasons.
(2) The Sun's distance from Earth varies with the seasons.
(3) The intensity of insolation is lower on June 21.
(4) The duration of insolation is shorter on June 21.

Base your answers to questions 12 through 14 on the world map below. Letters A through D represent locations on Earth's surface.

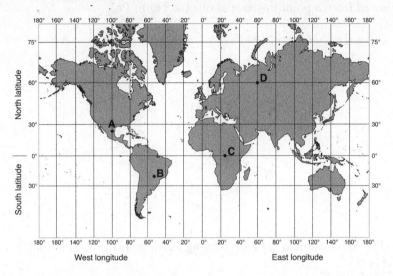

12. At which location could an observer *not* see Polaris in the night sky at any time during the year?
    (1) A   (3) C
    (2) B   (4) D

13. Which location receives 12 hours of daylight and 12 hours of darkness on June 21?
    (1) A   (3) C
    (2) B   (4) D

14. At which location on December 21 is the Sun directly overhead at solar noon?
    (1) A   (3) C
    (2) B   (4) D

15. How many times will the Sun's perpendicular rays cross Earth's Equator between March 1 of one year and March 1 of the next year?
    (1) 1   (3) 3
    (2) 2   (4) 4

Our System of Time

Base your answers to questions 16 through 20 on the diagram below, which represents Earth in relation to the Moon and Sun during the spring equinox as viewed from a point in space above the North Pole.

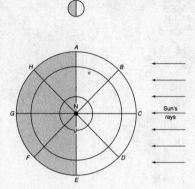

16. The Sun's vertical rays are striking Earth at this time at latitude
    (1) 0°
    (2) 23½° N
    (3) 23½° S
    (4) 66½° N

17. If it is 12 noon at point C, the time at F is
    (1) 3 A.M.   (3) 6 P.M.
    (2) 6 A.M.   (4) 9 P.M.

18. Sunrise on Earth is occurring at the point on the diagram labeled
    (1) A   (3) E
    (2) B   (4) G

19. If the Prime Meridian is at position B, the longitude at E is
    (1) 90° E   (3) 135° E
    (2) 90° W   (4) 135° W

20. In 7 days the Moon will have moved into a position nearly opposite the meridian indicated by letter
    (1) B   (3) G
    (2) E   (4) H

21. The diagram below shows the apparent daily path of the Sun, as viewed by an observer at a certain latitude on three different days of the year.

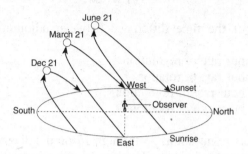

At which latitude were these apparent Sun paths most likely observed?

(1) 0°
(2) 23.5° N
(3) 43° N
(4) 66.5° N

22. What time is it in Greenwich, England (at 0° longitude), when it is noon in Massena, New York?
(1) 7 A.M.
(2) noon
(3) 5 P.M.
(4) 10 P.M.

Base your answers to questions 23 and 24 on the United States time zone map shown below. The dashed lines represent meridians (lines of longitude).

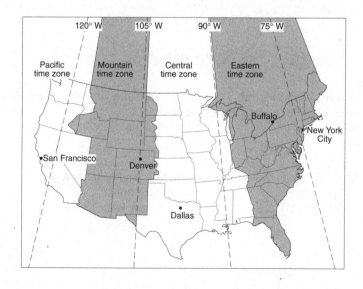

Our System of Time

23. If the time in Buffalo, New York, is 5 A.M., what time would it be in San Francisco, California?
    (1) 8 A.M.
    (2) 2 A.M.
    (3) 3 A.M.
    (4) 4 A.M.

24. The basis for the time difference between adjoining time zones is Earth's
    (1) 1° per hour rate of revolution
    (2) 1° per hour rate of rotation
    (3) 15° per hour rate of revolution
    (4) 15° per hour rate of rotation

Base your answers to questions 25 through 27 on the diagram below, which represents an exaggerated view of Earth revolving around the Sun. Letters A, B, C, and D represent Earth's location in its orbit on the first day of each of the four seasons.

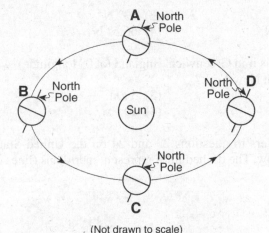

(Not drawn to scale)

25. Which location in Earth's orbit represents the first day of fall (autumn) for an observer in New York State?
    (1) A
    (2) B
    (3) C
    (4) D

26. Earth's rate of revolution around the Sun is approximately
    (1) 1° per day
    (2) 360° per day
    (3) 15° per hour
    (4) 23.5° per hour

27. Which observation provides the best evidence that Earth revolves around the Sun?
    (1) Stars seen from Earth appear to circle Polaris.
    (2) Earth's planetary winds are deflected by the Coriolis effect.
    (3) The change from high ocean tide to low ocean tide is a repeating pattern.
    (4) Different star constellations are seen from Earth at different times of the year.

CONSTRUCTED RESPONSE QUESTIONS

28. The diagram below shows the position of sunrise along the horizon for a period of time from September 10 until December 21 as seen by an observer near Binghamton, New York.

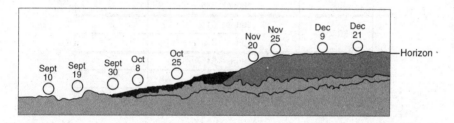

State one reason why the position of sunrise changes during this time period. [1]

Base your answers to questions 29 and 30 on the diagram below. The diagram shows a model of Earth's orbit around the Sun. Two motions of Earth are indicated. Distances to the Sun are given for two positions of Earth in its orbit.

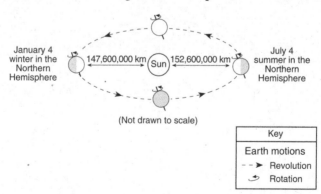

29. On the diagram above, place an **X** on Earth's orbit to indicate Earth's position on May 21. [1]

30. Explain why New York State experiences summer when Earth is at its greatest distance from the Sun. [1]

EXTENDED CONSTRUCTED RESPONSE QUESTIONS

Base your answers to questions 31 and 32 on the data table below, which shows the azimuths of sunrise and sunset on August 2 observed at four different latitudes. Azimuth is the compass direction measured, in degrees, along the horizon, starting from north.

Data Table

Latitude	Azimuths of Sunrise and Sunset	Letter Code
30° N	sunrise 69°	A
	sunset 291°	B
40° N	sunrise 66°	C
	sunset 294°	D
50° N	sunrise 61°	E
	sunset 299°	F
60° N	sunrise 51°	G
	sunset 309°	H

31. On the outer edge of the azimuth circle below, mark with an **X** the positions of sunrise and sunset for *each* latitude shown in the data table. Write the correct letter code beside each **X**. The positions of sunrise and sunset for 30° N have been plotted and labeled with letters *A* and *B*. [2]

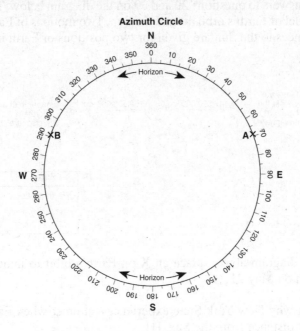

Azimuth Circle

32. State the relationship at sunrise between the latitude and the azimuth. [1]

Base your answers to questions 33 through 35 on the data table below. A student recorded the hours of daylight and the altitude of the Sun at noon on the twenty-first day of every month for one year in Buffalo, New York.

Data Table

Date	Hours of Daylight	Altitude of the Sun at Noon (°)
January 21	9.5	32.3
February 21	10.8	40.1
March 21	12.0	47.3
April 21	13.7	55.1
May 21	14.8	62.5
June 21	15.3	70.4
July 21	14.8	63.3
August 21	13.7	55.5
September 21	12.1	47.7
October 21	10.8	39.9
November 21	9.5	32.1
December 21	9.0	24.4

33. On the graph below, draw a line to represent the general relationship between the altitude of the Sun at noon and the number of hours of daylight throughout the year at Buffalo. [1]

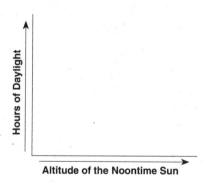

Our System of Time

34. The sky model diagram below shows the apparent path of the Sun on March 21 for an observer in Buffalo, New York. Draw a line to represent the apparent path of the Sun from sunrise to sunset at Buffalo on May 21. Be sure your path indicates the correct altitude of the noon Sun and begins and ends at the correct positions on the horizon. [2]

35. On the same sky model diagram, place an asterisk (*) at the apparent position of the North Star as seen from Buffalo. [1]

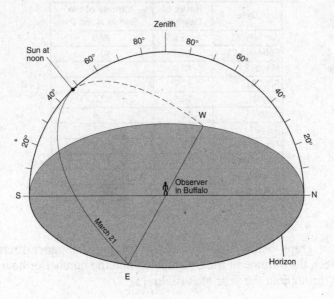

Unit Two: MODERN ASTRONOMY

CHAPTER 6
TOOLS OF THE MODERN ASTRONOMER

> **KEY IDEAS** Humans perceive the universe by the radiation it emits. Through much of recorded history, our understanding of the universe was based upon observations made with the naked eye. This was a great limitation, for our eyes perceive only a small fraction of the radiation emitted by the universe—radiation with wavelengths in the range called *visible light*. Technological advances have greatly extended the scope of human perception and have led to the observations upon which our current theories of the universe are based.

KEY OBJECTIVES
Upon completion of this chapter, you will be able to:

- Give examples of how progress in technology has increased our understanding of celestial objects.
- Explain how refracting and reflecting telescopes work and how they enhance celestial observations made with the unaided eye.
- Describe three types of spectra: continuous, absorption, and bright-line spectra.
- Explain how stellar spectra can be used to classify stars.
- Distinguish among stars, galaxies, and nebulae.

HOW HUMANS PERCEIVE THE UNIVERSE

When an observer "sees" a star or any other object, the lens in his eye has focused light emanating from that object to form an image on his retina. See Figure 6.1. The human retina is lined with specialized nerve cells that respond to light and send information to the brain, where it is interpreted. Nerve cells in the retina respond to two characteristics of light: intensity and wavelength. Intensity is interpreted as brightness; wavelength, as color.

Tools of the Modern Astronomer

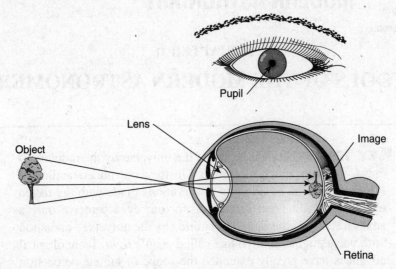

Figure 6.1 The Human Eye Forms an Image. Light from an object enters the human eye through a small opening called the pupil. The light then passes through a lens that focuses it into an image on the retina. The retina is composed of nerve cells that respond to light and send messages to the brain, where the information is interpreted and we "see" the object.

Limits of Human Perception

But the human eye has limitations. The light entering the eye is limited to what passes through the rather small lens. If the light intensity is too low, it doesn't trigger the nerve cells, and we don't "see" the light. If the intensity is too great, the nerves are overwhelmed and signal the pupil to become even smaller.

Also, the nerve cells respond to only a narrow range of wavelengths, commonly called *visible light*. Light of longer or shorter wavelengths doesn't trigger the nerve cells, and again we don't "see" it. Therefore, our eyes limit what we "see" of the universe to objects giving off light in a narrow range of wavelengths and brightness.

Another limitation is resolution. Think of the image sent to the brain as a series of dots, with each dot being the information from a single nerve cell in the retina. Although microscopic, nerve cells are huge compared to light waves. Many billions of light waves, containing lots of detail, can strike a single nerve cell, yet all the light that strikes that cell is perceived as a single dot—the detail is lost. That's the reason why planets look like stars to the naked eye. The image in the eye is too small and the light is too concentrated for us to perceive the planet as a disk. Instead, we see planets as bright dots. Most of the technological advances in observational astronomy have reduced these limitations.

Light and Lenses

The human eye contains a lens that forms images, and many of the instruments used by astronomers are simply improvements on this model. To begin, then, we need to understand how lenses focus images. A beam of light is bent, or **refracted**, whenever it passes at an angle from one substance (such as air) into another (such as glass). A **lens** is a curved piece of glass that bends, or refracts, light. Imagine a lens held vertically. The line through the center of the lens perpendicular to the plane of the lens is called the *optical axis*. A narrow beam of light parallel to the optical axis, shone through the lens, is bent so that it crosses the optical axis on the other side of the lens at a point called the lens's *focal point*. The distance from the center of the lens to the focal point is the lens's *focal length*. See Figure 6.2a.

If we shine a light beam through a lens so that the light appears to come from the focal point, it will leave the lens parallel to the optical axis, just the opposite of the result described above. See Figure 6.2b. If we shine a beam directly through the center of the lens, the lens does not bend the light. See Figure 6.2c.

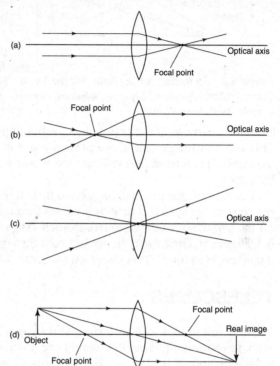

Figure 6.2 The Effect of a Lens on Light Beams. (a) Light beams entering a lens parallel to the optical axis are bent so that they cross at a focal point. (b) Light beams passing through the focal point are bent so that they emerge from the lens parallel to the optical axis. (c) Light beams that pass through the center of a thin lens are not bent. (d) A real image is formed by a lens as the three types of beams shown above meet. Source: *Discovering Astronomy*, Robert Chapman, W. H. Freeman, 1978.

We see an object by light that is emitted or reflected by the object. Whether it gives off its own light, or reflects light from another source, every point on the object sends out light in all directions. Let's place an object, for example, a flagpole, at some distance from a lens. See Figure 6.2d. A point at the tip of the flagpole reflects light in all directions. The light beams include one that is parallel to the optical axis, one that passes through the center of the lens, and one that passes through the focal point. As described

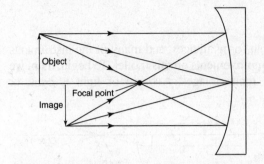

Figure 6.3 Formation of a Real Image by a Curved Mirror. Source: *Discovering Astronomy*, Robert Chapman, W. H. Freeman, 1978.

above, the lens bends these three light beams so that they all intersect at one point, or converge, on the other side of the lens. The result is the image we see of the point at the tip of the flagpole.

We can follow the same process with the light coming from the flag, the middle of the flagpole, the bottom of the flagpole, and every point in between. In this way, point by point, an image of the entire object is built up on the other side of the lens. This is a real image; and if we place a piece of paper at the position where the image has formed, we will see the image of the flagpole projected onto the paper.

A curved mirror can also be used to form a real image. Instead of the light converging because it is bent as it passes through a lens, the light is reflected off the surface of a curved mirror in such a way that it converges. See Figure 6.3. However, since the light doesn't pass through the mirror, but is reflected, it converges to form a real image on the same side of the mirror as the object.

TELESCOPES

What if you used a lens to form a real image, and then used a magnifying glass to examine it? Your eye would see an enlarged, close-up image. Such a combination of two lenses, one to form a real image and one to magnify the real image, is called a **refracting telescope**. See Figure 6.4a. If, on the other hand, a mirror is used to create the real image, which is then magnified, the instrument is a **reflecting telescope**. See Figure 6.4b. The lens or mirror that forms the real image in a telescope is called the ***objective***. The magnifier used to examine the image is the *eyepiece*.

The larger the lens or mirror in a telescope, the more light it gathers, and the larger, brighter, and more detailed the image. Light-gathering power is proportional to the area of the collecting surface, or to the square of the aperture (diameter) of the lens or mirror. A human eye has a lens with about a 5-millimeter diameter. The objective of the telescope at the Mount Palomar Observatory, with a diameter of 5 meters, has an aperture that is 1,000 times bigger than the lens in your eye. This means that the telescope on Mount Palomar has a light-gathering power that is $1,000^2$, or 1 million, times greater than that of your eye! Since telescopes can gather more light than the human eye, they allow us to see objects too dim to be visible with the naked eye, and under magnification we can see details in the images not otherwise visible. Some telescopes can even "see" light that is not visible to our eyes.

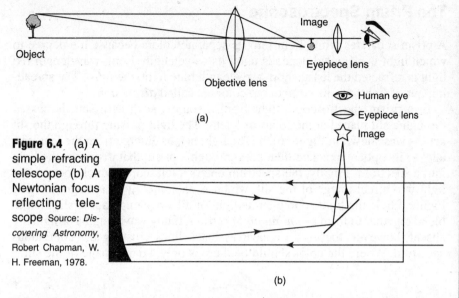

Figure 6.4 (a) A simple refracting telescope (b) A Newtonian focus reflecting telescope Source: *Discovering Astronomy*, Robert Chapman, W. H. Freeman, 1978.

They use sensors that detect electromagnetic radiation invisible to the human eye, and that convert it into images our eyes can see. All of these technological advances have added tremendously to our store of observational evidence about the universe.

SPECTROSCOPES

Another valuable tool of the modern astronomer is the spectroscope. In 1666, Isaac Newton demonstrated that white light is actually a combination of all colors. He passed light through a triangular piece of glass, or *prism*, and found that the prism spread the light out into a rainbow of colors, or **spectrum**. See Figure 6.5. The colors of the spectrum are, in order from longest to shortest wavelength, red, orange, yellow, green, blue, indigo, and violet. (This sequence can be remembered by using the name **Roy G. Biv**; each letter in the name is the first letter of a color, and the letters are in correct order.)

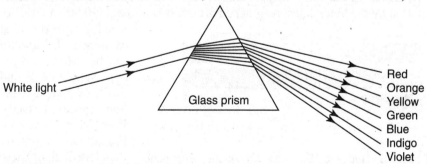

Figure 6.5 A Beam of White Light Entering a Prism Is Dispersed into a Spectrum.
Source: *Astronomy Explained*, Gerald North, Springer-Verlag, 1997.

The Prism Spectroscope

A prism separates white light into its separate colors because the degree to which light is refracted depends upon its wavelength. Long-wavelength red light is refracted the least; short-wavelength blue light, the most. The spreading out of light into its component colors is called *dispersion*.

In a prism spectroscope, light from a source, such as a star, is passed through a slit placed at the focus of a lens. The light passing through the slit acts as was shown in Figure 6.2b. The light beams emerge from the lens parallel to its optical axis and then pass through a prism that disperses the light into a spectrum. Finally, the spectrum enters a telescope, which focuses the light into a real image of the slit, which is examined with a magnifier. See Figure 6.6. If the light source emits light of all wavelengths, the slit images blend together to form a *continuous spectrum*. If any wavelengths of light are absent, however, these will appear as black lines crossing the continuous spectrum. Where the color should be there is only a dark image of the slit.

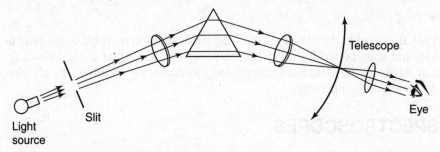

Figure 6.6 A Simple Prism Spectroscope. The telescope pivots around a point centered on the prism so that the viewer can examine all visible wavelengths. Source: *Discovering Astronomy*, Robert Chapman, W. H. Freeman, 1978.

When, in 1814, Joseph von Fraunhofer, a Bavarian optician and physicist, first used such a device to examine light from the Sun, he saw "an almost countless number of strong and weak vertical lines, which are, however, darker than the rest of the color image; some appeared to be almost perfectly black." Puzzled by the black lines, he mapped almost 600 of them (20,000 are recognized today). When he examined the spectra of the Moon, Mars, and Venus he found that they matched the Sun's spectrum exactly. From this discovery, Fraunhofer correctly concluded that these bodies do not produce their own light, but

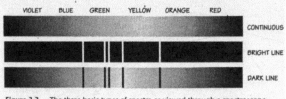

Figure 6.7 Types of Spectra: Continuous, Bright-line, Absorption. Source: *Astronomy: A Self-Teaching Guide*, 4th Ed., Dinah L. Moché, John Wiley, 1998.

instead reflect the light of the Sun. When, however, he turned the spectroscope on Sirius and other bright stars, the spectra told a different story. The pattern of black lines in each star's spectrum was different, and the spectra of all of the stars differed from the spectrum of the Sun. It appeared that these spectral lines were a kind of "fingerprint," each star having its own unique pattern of lines.

When a spectroscope is used to examine light from chemicals burned in the laboratory, a different kind of spectrum is observed. Only a few colors are present, and each color forms a distinct image of the slit, called a *spectral line*. The brightness of each colored spectral line is due to the amount of light of a particular wavelength that is being emitted by the gas, and the line's position in the spectrum is a function of its wavelength. Over time, it became clear that each element, when burned as a gas, produces a unique pattern of bright spectral lines, or a *bright-line spectrum*, which, like a fingerprint, identifies the element. Note in Figure 6.8 the differences in the spectral lines of sodium and hydrogen. This was a momentous discovery, and scientists soon used spectral analysis to identify the patterns of colored spectral lines for all known elements.

Around 1860, Gustav Kirchhoff, a German physicist, carried out a number of experiments that explained

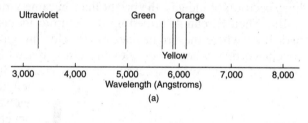

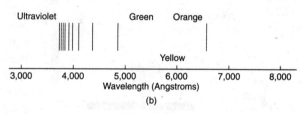

Figure 6.8 Bright-Line Spectra of (a) Sodium and (b) Hydrogen. The pattern of bright lines in every element's bright-line spectrum is unique; this spectral "fingerprint" can be used to identify elements. Source: *Discovering Astronomy*, Robert Chapman, W. H. Freeman, 1978.

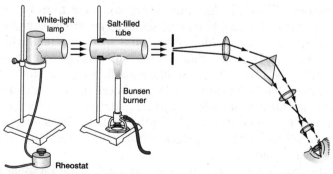

Figure 6.9 Types of Setups Used by Kirchhoff in Experiments that Led to His Discoveries Concerning Absorption Spectra. Source: *Discovering Astronomy*, Robert Chapman, W. H. Freeman, 1978.

the dark lines discovered by Fraunhofer. He knew that a glowing hot solid, liquid, or gas emits a continuous spectrum. He also knew that the glowing gas of an element forms a spectrum of bright lines. So Kirchhoff decided to view a glowing solid through a cloud of glowing gas to see what type of spectrum would result. He set up a very bright lamp whose filament was a glowing hot solid attached to a device like a dimmer switch so that he could vary the lamp's brightness. Then he passed light from the lamp (which produces a continuous spectrum) through a glowing gas (which produces a bright-line spectrum) and made an interesting discovery. When the lamp was turned low, he saw the bright lines of the glowing gas superimposed on the dim background of the lamp's continuous spectrum. When he slowly turned up the lamp so that its continuous spectrum was just as bright as the bright-line spectrum of the gas, the bright lines became virtually invisible.

But when Kirchhoff turned up the lamp to its greatest intensity, he saw dark lines where the bright lines of the glowing gas had been located! He concluded that, when a hot, glowing gas is placed in front of a still hotter source of a continuous spectrum, the gas absorbs light at the same wavelengths that it would emit if viewed alone, producing a spectrum with dark lines, called an *absorption spectrum*. In other words, the pattern of dark lines is identical to the pattern of colored spectral lines that would be produced by the elements in the gas! See Figure 6.10.

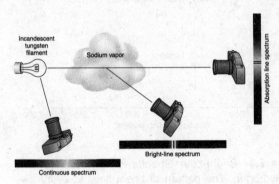

Figure 6.10 The Formation of Continuous, Bright-line, and Absorption Spectra. Source: *Discovering Astronomy*, Robert Chapman, W. H. Freeman, 1978.

Kirchhoff then applied these discoveries to astronomy. He concluded that light from the hot interior of the Sun passes through cooler gases in the Sun's outer atmosphere. Thus, the dark lines in the solar spectrum are the fingerprints of elements in the Sun! Kirchhoff set up an apparatus that allowed him to view, side by side, the dark lines in the Sun's spectrum and the bright lines produced by burning elements found on Earth. By matching lines in the two types of spectra, he identified many elements in the solar atmosphere that also are found on Earth, including hydrogen, sodium, iron, calcium, magnesium, nickel, and chromium. A few years later, during a solar eclipse, a spectrum matching no known element on Earth was found. Scientists called the new element helium, after the Greek word, *helios*, for Sun. Years later, helium was also found on Earth.

Since Kirchhoff's time the spectra of thousands upon thousands of stars have been examined. And although stars vary in composition, they consist of the same elements that are found on Earth and in our Sun.

The Spectrophotometer

In a **spectrophotometer**, a photometer (an electronic device that measures the brightness of light) is attached to a spectroscope and used to view starlight. A spectrophotometer works as follows. After the prism has dispersed the light, a narrow slit lets one color of light at a time pass through to a lens that focuses it onto the photometer. In this way the brightness of each wavelength of light in the spectrum can be very accurately measured. The analysis of the continuous, bright-line, and absorption spectra of light from stars and other celestial objects is a powerful tool, allowing scientists to identify the chemical elements present in them.

The Spectrograph

In 1872, when Henry Draper, a New York physician and amateur astronomer, used photographic plates to record the spectra of stars for later study, the spectroscope became a spectrograph. A **spectrograph** is a spectroscope equipped with a camera, a device that allows a star's spectrum to be photographed. Photographs of spectra can then be saved, analyzed, and categorized.

When Henry Draper died, his widow financed a program to photograph and classify stellar spectra at the Harvard College Observatory. In the 1890s, the Harvard project photographed the spectra of thousands upon thousands of stars. On the basis of similarity of spectral lines, the project director, Edward C. Pickering, devised a scheme for classifying stars with an alphabetical sequence of letters—A, B, C, D, etc.—that was

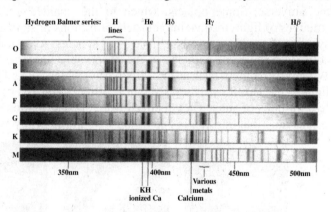

Figure 6.11 Representations of the Spectra of Stars from Each of the Major Spectral Classes. Source: *Astronomy: The Cosmic Journey*, William K. Hartman, Wadsworth, 1987.

used for a while. Between 1918 and 1924, however, the *Henry Draper Catalogue* of stellar spectra was published, and in it Annie Jump Cannon of the Harvard Observatory classified nearly 400,000 stellar spectra. Based upon her work, some of Pickering's classes were combined or dropped, and the sequence was reordered as O, B, A, F, G, K, M (which can be remembered using the sentence **O**h! **B**e **A** **F**ine **G**irl, **K**iss **M**e.) See Figure 6.11.

SOME OBSERVATIONS OF THE UNIVERSE MADE WITH INSTRUMENTS

Using instruments such as telescopes and spectroscopes, astronomers could see more of the universe, and obtain more detailed information about it, than ever before. Let's consider some of the observations made with these instruments that have led to current theories about the universe.

Stars

First, there are a LOT more stars in the universe than the 1,500 or so that can be seen on a really dark night with the naked eye. Since telescopes gather more light than do eyes, stars too dim to be seen with the eye alone *are* visible through a telescope. Patches of sky that look empty to the naked eye are revealed to be filled with stars when viewed through a telescope. As ever-larger telescopes have been built, ever-dimmer stars have been detected, and the number of observable stars has grown from thousands to hundreds of billions.

Galaxies

The higher resolution of telescopes has revealed much about the structure of the universe. Some objects, which to the unaided eye looked like dim smudges of light, resolve into clusters of billions of stars, called *galaxies*, in systems shaped like spirals, ellipses, and spheres. Astronomers have found a colossal number of galaxies in the observable universe. See Figure 6.12. When astronomers peer at the Milky Way, which looks like a pale band of light to the naked eye, they see that it actually consists of light from countless millions of stars. They realized from the distribution of stars that it is a galaxy, and that our Sun is one of the stars in the Milky Way galaxy.

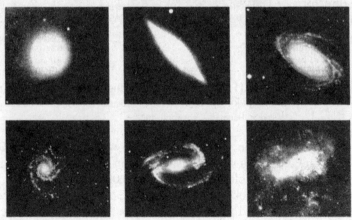

Figure 6.12 Galaxies Are Systems Containing Billions of Stars. The universe is filled with galaxies of all shapes and sizes. Source: *Astronomy: A Self-Teaching Guide*, 4th Ed., Dinah L. Moché, John Wiley, 1998.

Nebulae

Nebulae are clouds of gas and dust in the universe. The Orion nebula is a cloud of gas and dust lit by a group of stars within it. See Figure 6.13a. This nebula is dense enough to scatter starlight, and thus we can see it. By spectral analysis, the composition of the dust and gases in the cloud can be determined. When astronomers analyzed invisible radiation emitted by the cloud, they found that it extends out several times the area of the region we can see by visible light.

Sometimes dust is visible in a negative sort of way, as in the Horsehead nebula. There, a dust cloud is so thick that it blocks visible light, forming a black region (shaped like a horse's head) in front of the nebula. See Figure 6.13b. The dust actually absorbs starlight and reradiates it at longer wavelengths. To our eyes, it appears dark; to an infrared detector, it glows brightly.

(a)

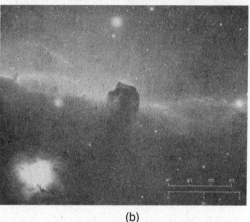

(b)

Figure 6.13 Nebulae. (a) Orion nebula. Source: *The Nature Company's Guides: Advanced Skywatching*, Robert D. Burnham et al., Time-Life Books, 1997. (b) Horsehead nebula. Source: *Astronomy: The Cosmic Journey*, William K. Hartman, Wadsworth, 1987.

Observations such as these indicate that there is a lot of "dark" matter in the universe, that is, matter that is not visible to the unaided eye.

MULTIPLE-CHOICE QUESTIONS

In each case, write the number of the word or expression that best answers the question or completes the statement.

1. Which instrument is used to study the compositions of stars?
   (1) sextant
   (2) spectroscope
   (3) seismograph
   (4) chronometer

2. An observer viewing the sky through a telescope sees a fuzzy, glowing region in the constellation Orion. The region has an irregular shape, and some stars seem to be shining through it. The observer is most likely viewing a
   (1) planet
   (2) comet
   (3) meteor
   (4) nebula

3. In a refracting telescope, the objective lens gathers light to form an image, and the eyepiece
   (1) magnifies the image formed by the objective
   (2) forms a spectrum
   (3) shields the eye from bright light
   (4) is the surface upon which the image forms

4. Compared to stars viewed with the unaided eye, stars viewed with telescopes appear
   (1) larger and brighter
   (2) larger and dimmer
   (3) smaller and brighter
   (4) smaller and dimmer

5. Telescopes reveal many more celestial objects than you can see with the unaided eye because telescopes
   (1) allow the observer to view a larger area of the sky
   (2) collect more light than an observer's eye can
   (3) convert ultraviolet rays into visible light rays
   (4) disperse light into a spectrum

6. The chemical composition of a star can be determined by examining its
   (1) celestial coordinates
   (2) absorption spectrum
   (3) distance from Earth
   (4) apparent brightness

7. When viewed by the unaided eye, the Milky Way looks like a pale band of light. When viewed through a telescope, the Milky Way
   (1) looks like a pale band of light
   (2) appears to be a glowing nebula
   (3) is seen to consist of billions of stars
   (4) appears to be the tail of a large comet

8. Two refracting telescopes, one with a 40-millimeter lens and another with a 160-millimeter lens, are used to observe the red star Betelgeuse in the constellation Orion. Compared to the image seen in the telescope with the 40-millimeter lens, the image seen in the telescope with the 160-millimeter lens will appear
   (1) brighter
   (2) dimmer
   (3) blue
   (4) smaller

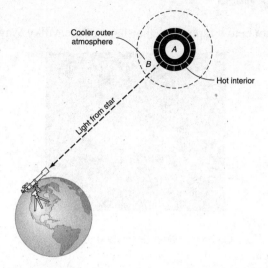

9. The diagram above shows light coming from a star's hot interior (A), passing through the star's cooler outer atmosphere (B), and traveling through space to enter a telescope fitted with a prism spectroscope. The star's spectrum is observed to contain many narrow black lines.

   Black lines are seen in the star's spectrum because
   (1) the star's gravity prevents some of the light from escaping
   (2) some wavelengths of light are absorbed by the star's outer atmosphere
   (3) some of the light coming from the star doesn't have enough energy to reach Earth
   (4) Earth's magnetic field blocks some of the star's visible light

10. Which of the following is *not* a way in which telescopes can aid human eyes?
    (1) magnification (the ability to make an image bigger)
    (2) light-gathering power (the ability to collect light)
    (3) resolution (the ability to resolve fine detail)
    (4) dispersion (the ability to separate light into its spectrum of colors)

11. In which list are celestial features correctly shown in order of increasing size?
    (1) galaxy, solar system, universe, planet
    (2) solar system, galaxy, planet, universe
    (3) planet, solar system, galaxy, universe
    (4) universe, galaxy, solar system, planet

12. Which of the following statements best describes the difference between a galaxy and a nebula?
    (1) A galaxy consists of stars; a nebula consists of dust and gas.
    (2) There are two types of nebula, but only one type of galaxy.
    (3) A galaxy always emits light; a nebula never emits light.
    (4) A galaxy consists of matter; a nebula consists of energy.

13. The diagram below represents the shape of the Milky Way Galaxy.

The Milky Way Galaxy is best described as
(1) elliptical
(2) irregular
(3) circular
(4) spiral

14. Which color of the visible spectrum has the *shortest* wavelength?
(1) violet
(2) blue
(3) yellow
(4) red

Constructed Response Questions

15. Compare the light-gathering powers of a 200-inch telescope and a 20-inch telescope. [2]

16. Describe three advantages of a telescope over the unaided eye. [3]

17. Explain the difference between the way in which a refracting telescope forms an image and the way in which a reflecting telescope forms an image. [2]

18. Construct a diagram showing why the image formed by a refracting telescope is inverted (upside down). [2]

19. Explain why some stars that are invisible to the unaided eye can be observed with a telescope. [1]

20. Compare and contrast continuous, absorption, and bright-line spectra. [3]

21. List the spectral classes of stars, and explain how stars are sorted into these classes. [2]

CHAPTER 7
STARS, THEIR ORIGIN AND EVOLUTION

KEY IDEAS The vast majority of observable objects in the universe are stars. A star is a hot, luminous, gaseous celestial body. Stars form when gravity causes clouds of matter to contract until nuclear fusion of light elements into heavier ones occurs. Fusion releases great amounts of energy over millions of years.

Stars vary in size, temperature, and age. The majority of stars, including the Sun, fall into the main sequence when plotted on the Hertzsprung-Russell diagram according to luminosity and spectral class. Stars undergo a series of changes as they age.

A galaxy is a system of stars, cosmic dust, and gas held together by gravitation. A galaxy typically contains billions of stars and may be thousands of light-years in diameter, and the universe contains billions of such galaxies. Our Sun is a medium-sized star within a spiral galaxy of stars known as the Milky Way.

KEY OBJECTIVES
Upon completion of this chapter, you will be able to:

- Describe the method by which distances to stars are determined using parallax.
- Explain the difference between apparent brightness and luminosity and the relationship of these qualities to distance.
- Describe the Hertzsprung-Russell diagram, and explain the relationship of a star's mass to its luminosity and temperature.
- List the three main steps leading to the birth of a star and the stages in the life cycle of a star such as our Sun.
- Compare and contrast what happens in the later stages of evolution for stars of large and small mass.
- Describe the origins of different chemical elements and the importance of supernovas to new generations of stars.

CHARACTERISTICS OF STARS

A *star* is a hot, luminous, gaseous celestial body. Stars are distant, blazing suns moving through space at different distances from Earth. We see stars by the light they emit. On a clear, dark night you can see about 2,000 stars with the unaided eye. With a telescope you can see billions. Stars differ from each other in size, temperature, and age.

Star Names

In order to keep track of observations of so many stars, each one must be identified in some unique way. Long ago, the brightest stars were given proper names by astronomers; today, modern astronomers use letters and numbers to identify hundreds of thousands of stars.

Many of the proper names of stars come from the Arabic and are usually descriptions of where the star appears in a constellation. For example, Orion, which represents a mighty warrior or hunter, is one of the most eye-catching winter constellations. The left shoulder of Orion is marked by the bright red star Betelgeuse (pronounced "beetle-juice"). This is a form of the Arabic name *Ibt al Jauzah*, meaning "the armpit of the central one." Rigel (pronounced "rye-jel") is from *Rijl Jauzah al Yusra*, meaning "left leg of the central one." But it is not practical to name, and to learn the names of, thousands upon thousands of stars. Clearly some other system had to be devised.

In 1603, Johann Bayer published a star atlas in which each star was named for the constellation in which it appeared and was assigned a Greek letter by brightness (α being the brightest, β the next brightest, and so on). As more and more stars were discovered, though, astronomers soon ran out of Greek letters. Sometime later, the English Astronomer Royal, John Flamsteed, published an atlas in which each star in a constellation was numbered, beginning with 1 for the brightest. Since then, many catalogs have been compiled that list the positions and characteristics of hundreds of thousands of stars.

Today, astronomers use only a handful of proper names—those of the very brightest stars. Then they use Bayer's Greek letter names; and in constellations such as Cygnus, with many bright stars, they use Flamsteed's numbers after Bayer's letters have been exhausted. In one of the more recent catalogs, fainter stars are referred to by their numbers.

The Distance to Stars *(not testable on Regents exam)*

The distances to stars that are relatively close to our solar system can be determined using simple trigonometry and a common optical effect. Hold a finger out at arm's length and alternately close your left and right eyes. The finger appears to shift back and forth against the background of more distant objects. This effect is known as *parallax*. Now hold your finger about 10 centimeters in front of your nose and repeat the experiment. Notice that the nearer the finger, the greater the parallax.

The same effect is seen when astronomers look at stars that are close to us on one date and then look at them 6 months later, when Earth has traveled halfway around its orbit. (Here the distance between your eyes in the finger experiment is replaced by the 300-million-kilometer diameter of Earth's orbit.) The stars appear to shift slightly in their positions relative to distant stars. Half of the total shift is the star's parallax angle or, simply, its parallax. See Figure 7.1.

The parallax of a star can be used to calculate its distance from Earth. As shown in Figure 7.1, the star and Earth at two positions in its orbit, E_1 and E_2, form an isosceles triangle. The angle at the star equals the total shift in position. Since the Sun is at the center of Earth's orbit, a line from the Sun to the star bisects this angle (which is why parallax is one-half the total shift) and also forms the right triangle star-Sun-E_1. Given the right angle, the parallax angle, and the fact that one side of the triangle is the Earth-Sun distance, simple trigonometry yields the distance to the star.

Stellar parallaxes are very small and are measured in seconds (") of arc. To get a sense of how small an angle that is, look at a protractor and note the size of 1 degree. Now consider that 1 degree = 60 minutes and 1 minute = 60 seconds. Thus, 1 second = 1/3,600 degree! The parallax of the star 61 Cygni is only 3/10 second—the size of a dime viewed from 12 kilometers! Imagine the precision of the instruments needed to measure such a tiny angle!

One **parsec** (from *par*allax *sec*ond) is the distance to an imaginary star that has a parallax of 1 second of arc. One parsec (abbreviated pc) equals about 31 trillion kilometers (19 trillion miles). Another, older unit of distance is the **light-year**, the distance light travels in 1 year. One parsec equals about 3.26 light-years. However, to simplify calculations, most astronomers use

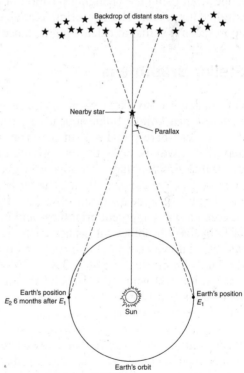

Figure 7.1 Stellar Parallax. A star's parallax is one-half the total angle of its apparent shift in position. Source: *Astronomy Explained*, Gerald North, Springer-Verlag, 1997.

parsecs for distances to stars in all their scientific work. To calculate the distance, in parsecs, to a star, they use this formula:

$$\text{Star's distance (pc)} = \frac{1}{\text{parallax (sec)}}$$

Thus, the distance to Proxima Centauri, the star with the largest parallax (0.76 sec), and therefore the closest, is

$$\text{Distance to Proxima Centauri} = \frac{1}{0.76 \text{ sec}} = 1.32 \text{ pc}$$

Stellar parallax decreases with distance. We can measure angles down to about 1/100 second, which corresponds to a distance of about 100 parsecs. For stars more than 100 parsecs away, the parallax angle becomes too small to be measured with any certainty, and more complex methods are needed. Although we will not describe them in detail, it is interesting to note that they range from using the changing position of our solar system over time as the baseline for parallax, to measuring relative velocities of clusters of stars based upon spectral shifts, to calculating distance based upon stellar brightness.

Stellar Brightness

Go outside on any clear night, and you will see that stars vary greatly in ***apparent brightness***. Some appear very bright in the sky, while others are so dim you can barely make them out. You might think that the brighter stars appear that way because they are giving off more light, but the explanation is not that simple. Stars may appear bright because they emit more light, *or* because they are closer to Earth. The farther away a source of light is located, the dimmer it appears to an observer. Therefore, astronomers distinguish between a star's apparent brightness and its ***luminosity***, or the actual amount of light the star shines into space each second. Since the Sun is the nearest and best known star, the luminosity of other stars is often stated in terms of the Sun's luminosity, which is 3.85×10^{28} watts (the equivalent of 3,850 billion trillion 100-watt light bulbs all shining together). The brightest stars are more than 1 million times as luminous as the Sun, while the dimmest are only 1/10,000 (0.0001) as luminous.

You cannot tell by just looking at stars in the sky which ones have the greatest luminosity. The farther away a source of light is located, the dimmer it appears to an observer. The reason is that the apparent brightness of a light source is measured by the amount of light energy that strikes each unit of area of the detector—be it your eye's retina or a photometer.

As light travels away from a source, it spreads out as shown in Figure 7.2. All of the light that passes through surface 1 also passes through surface 2 and then through surface 3. But note that surface 2 has *four times* the area of surface 1! Therefore, the amount of light that passes through each unit of area in surface 2 is *one-fourth* of the amount that passes through surface 1, and the brightness measured at surface 2 will be one-fourth of the brightness at surface 1. The observed brightness of a source falls off as the square of the distance from the source, a relationship known as the ***inverse-square law***. See Figure 7.2. Thus, if two stars have exactly the same luminosity, but one is twice as far away as the other, the distant one will appear ($\frac{1}{2}$)^2, or only one-fourth, as bright as the closer one because only one-fourth of the light reaches the eye. And at three times as far away, the distant star will appear only one-ninth as bright!

Over the centuries a scale to describe brightness has evolved. It began in the second century B.C., when the Greek astronomer Hipparchus classified stars into six brightness classes, or ***magnitudes***. Stars of the first magnitude were the very brightest in the sky, and stars of the sixth magnitude were just visible to the human eye. The number assigned to a star's brightness on this scale is called its ***apparent magnitude*** because it describes how bright a star *appears* to an observer on Earth.

The modern ***magnitude scale*** defines a first-magnitude star as exactly 100 times brighter than a sixth-magnitude star. Scientific studies show that this is consistent with the way our eyes respond to changes in brightness. What we see as a *linear* increase of one magnitude in brightness is precisely measured as a *geometric* increase in brightness of 2.5 times. If we measure the brightness of stars of different magnitudes with a photometer, we find that, when the magnitude increases by 1, the brightness increases 2.5 times. Thus, a difference of five magnitudes corresponds to a brightness difference of 100 times (2.5 × 2.5 × 2.5 × 2.5 × 2.5 = 100)! Stars that are 2.5 times brighter than first magnitude are zero-magnitude stars. The very brightest stars actually have negative magnitudes. For example, Sirius has magnitude –1.5, and the Sun has magnitude –27!

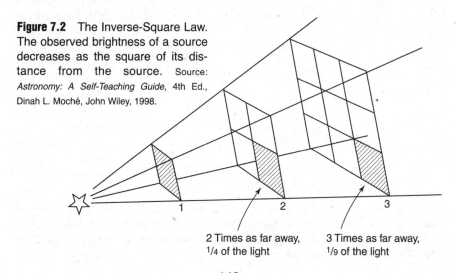

Figure 7.2 The Inverse-Square Law. The observed brightness of a source decreases as the square of its distance from the source. Source: *Astronomy: A Self-Teaching Guide*, 4th Ed., Dinah L. Moché, John Wiley, 1998.

To overcome this problem, astronomers devised a scale in which stars are assigned brightnesses based upon how bright they would appear if they were all located at the same distance from Earth. The *absolute magnitude* of a star is the magnitude it would have if it were moved from its actual location to a spot 10 parsecs from Earth. See Figure 7.3.

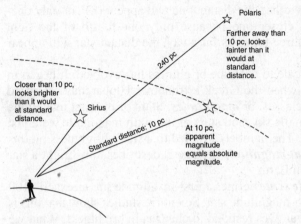

Figure 7.3 Absolute Magnitude and Apparent Magnitude. Source: *Astronomy: A Self-Teaching Guide*, 4th Ed., Dinah L. Moché, John Wiley, 1998.

Star Color and Temperature

If you look carefully at stars, you discover that many show definite shades of color. Betelgeuse is reddish. Another bright star, Arcturus, is a pale orange-red, while Vega, a prominent star in the summer sky, shows a definite blue tint. Why the different colors?

The temperature of a star is related to the average speed of the particles of which it is composed. We say "average" speed because some of the particles are moving faster and some are moving slower. The speed at which matter moves determines the wavelength of the energy it emits. The faster it is moving, the shorter the wavelength of the energy it gives off. Thus, some of the particles in a star emit short-wavelength radiation while others emit long-wavelength radiation. The result is that a star gives off radiation that is a mix of many different wavelengths.

A *radiation curve* shows the amount of radiation a hot object gives off at different wavelengths. The more radiation a star

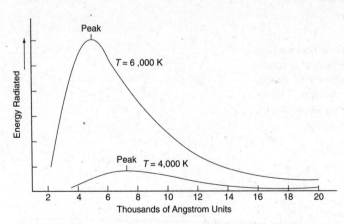

Figure 7.4 The Radiation Curves of a Theoretical Object at Two Different Temperatures. Note that, as temperature increases, the peak radiation shifts. Source: *Discovering Astronomy*, Robert Chapman, W. H. Freeman, 1978. P. 165.

emits at a particular wavelength, the higher the curve at that wavelength. The area under the curve indicates the total amount of energy being radiated by the star.

Figure 7.4 shows the radiation curves for a theoretical star at two different temperatures. As you can see, the higher the temperature, the shorter the peak wavelength of the energy emitted. Note, too, that the total amount of energy emitted by the star (the area under the curve) increases with temperature. The relationship between the total energy emitted by a star over all wavelengths and its temperature is directly proportional to its surface area. Thus, if we know the surface temperature of a star, we can calculate the amount of energy being emitted by each square meter of its surface, or its *luminosity*.

Figure 7.5 shows the actual radiation curve of the Sun and the radiation curve of a theoretical star at a temperature of 6,000 K. Note how closely the two curves match. For this reason, scientists believe the Sun has a surface temperature of about 6,000 K. Note, too,

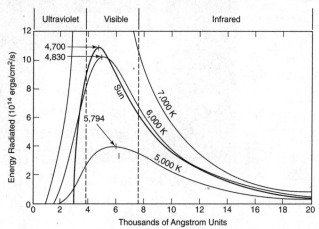

Figure 7.5 The Actual Radiation Curve of the Sun Closely Matches That of a Theoretical Object at 6,000 K. Source: *The Earth Sciences*, Arthur N. Strahler, Harper & Row, 1971.

that the Sun radiates most intensely in the yellow range, but remember that the *peak* wavelength emitted by a star is not the *only* wavelength it is emitting. Stars emit a mix of wavelengths, so we perceive their light as white tinted the color of the peak wavelength, or as pastel shades of color. Hence, the Sun appears white with a decidedly yellow hue.

The physical conditions in the surface of a star, such as the Sun, are such that theoretical models can be used to derive information about stars. By analyzing the amount of energy emitted at each wavelength in a star's spectrum, astronomers can determine the peak wavelength, and thus the temperature, of the star. Also, the pattern of spectral lines reveals which elements are in the star, and measurements of the intensities of various bright lines in the spectrum show how much of each element is present.

Shortly after the publication of the *Henry Draper Catalogue* of stellar spectra, it was recognized that the O, B, A, F, G, K, M sequence was a *temperature* sequence. The surface temperatures of blue O-type stars are high (around 30,000 K), while the red M-type stars are cool (temperatures around 3,000 K). The Sun, with a surface temperature of about 6,000 K, is a yellow G-type star. See Figure 7.6.

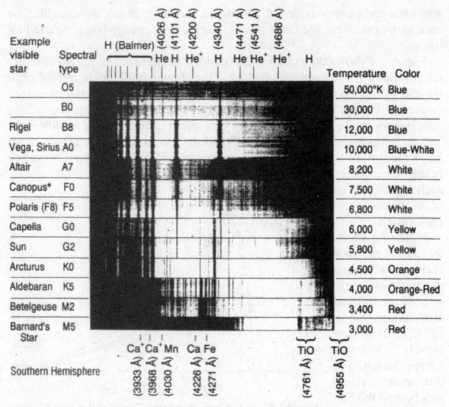

Figure 7.6 Spectra of Stars Representing the Seven Main Spectral Classes Arranged in Order of Decreasing Temperature. Source: *Astronomy: A Self-Teaching Guide*, 4th Ed., Dinah L. Moché, John Wiley, 1998.

The Hertzsprung-Russell Diagram

Early in the twentieth century, the relationship between the luminosities and temperatures of stars was discovered independently by two astronomers: Ejnar Hertzsprung of Denmark and Henry N. Russell of the United States. The ***Hertzsprung-Russell diagram*** (or H-R diagram) is a plot of luminosity versus surface temperature for the stars. See Figure 7.7. Every point on the H-R diagram represents a star. The star's temperature is read along the horizontal axis and its luminosity along the vertical axis. When several thousand stars are chosen at random and plotted on an H-R diagram, the dots do not scatter randomly over the graph. Instead, they fall into definite regions, forming patterns that show a meaningful relationship between a star's luminosity and its temperature.

Roughly 90 percent of all stars fall on a diagonal line across the diagram, called the ***main sequence***, which runs from the upper left (very hot, luminous blue supergiants) to the lower right (cool, dim red dwarfs). Why do stars follow this pattern? As was mentioned earlier, an increase in the temperature of a star results not only in a color shift in the radiation it emits, but also an increase in the total energy it gives off per unit of surface area. This increase in total energy emitted is seen as an increase in the star's brightness.

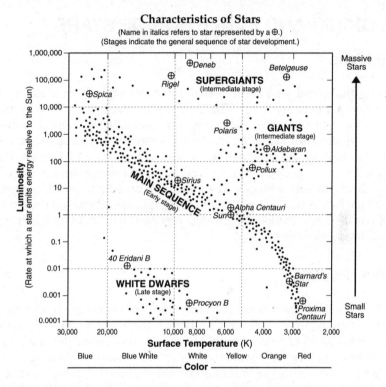

Figure 7.7 The Hertzsprung-Russell Diagram. When luminosity is plotted versus temperature, stars fall into distinct regions in patterns that represent the relationship between luminosity and temperature. Source: The State Education Department, *Earth Science Reference Tables*, 2011 Edition (Albany, New York; The University of the State of New York).

Most of the remaining 10 percent of stars fall above and to the right of the main sequence (cool, bright supergiants or red supergiants) or to the lower left (hot, dim white dwarfs). How can a star be cool yet bright, or hot yet dim? Consider, for example, 4,000°C stars. Since they all have the same surface temperature, they emit the same amount of energy per square meter of surface area. Thus, for a star above the main sequence, such as Aldebaran, to be more luminous than a red dwarf at the same temperature on the main sequence, it must have more square meters to radiate; in other words, it must be larger. Similarly, for 10,000°C stars that fall below the main sequence, such as white dwarfs, to be dimmer, they must be smaller than main sequence stars of the same temperature.

This inferred difference in sizes was confirmed after the first successful image of a star's disk was obtained in 1974. Using a technique called *speckle interferometry*, astronomers at Kitt Peak National Observatory measured the angular diameter of Betelgeuse to be 0.06 second of arc, making it 580 times larger than the Sun.

THE ORIGIN AND EVOLUTION OF STARS

Early Ideas

As information about stars accumulated, scientists made the first attempts to explain sunlight (and starlight). The British physicist Lord Kelvin hypothesized that the Sun formed by the collapse of interstellar gas and dust due to gravitational attraction, with the resultant compression causing the particles of matter to heat up. He thought that the Sun was simply radiating away this energy, just as any hot object radiates energy into its cooler surroundings, but his hypothesis ran into a time-scale problem. According to its size, temperature, and the rate at which it was emitting energy, the Sun would cool in tens of millions of years. Evidence indicates, however, that rocks on Earth are at least 10 times older than that.

Nuclear Fusion

By the 1930s, physicists began to understand the workings of the atomic nucleus. They realized that with sufficient force the repulsion between atomic nuclei could be overcome and the nuclei could combine in a process called *fusion*. In the Sun, which is composed mainly of hydrogen, four hydrogen nuclei (each a single proton) combine through a complex series of steps called the *proton-proton reaction*. See Figure 7.8a. During this process, two protons transform to neutrons and the most stable resulting nucleus is that of the isotope helium-4 (^{4}He). However, the resulting ^{4}He nucleus *has 0.7% less mass than the original four hydrogen nuclei*. The missing mass has been converted to energy according to Einstein's formula: $E = mc^2$. This formula says that, if mass, m, is converted to energy, the amount of energy created, E, is equal to m times the speed of light, c, squared. See Figure 7.8b.

If all of the hydrogen in the Sun were converted into helium, enough energy would be created to keep the Sun shining at its present rate for 100 billion years. However, as the ratio of hydrogen to helium in the Sun changes, so will its structure. After about 10 billion years the Sun will begin to undergo profound changes. But assuming that the Sun is about as old as the 4–5 billion-year-old Earth, it is only about halfway through its lifetime—so don't panic.

With star lifetimes measured in billions of years, it is not surprising that the stars we see today are pretty much the same as those observed by early astronomers. However, with all of written human history encompassing little more than 5,000 years, how can astronomers hope to learn anything about the

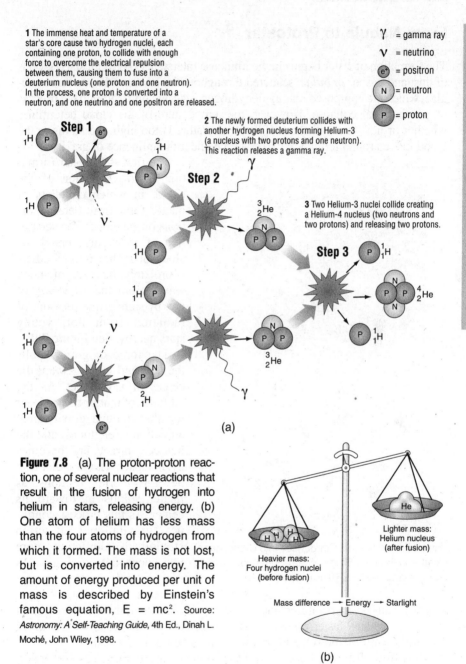

Figure 7.8 (a) The proton-proton reaction, one of several nuclear reactions that result in the fusion of hydrogen into helium in stars, releasing energy. (b) One atom of helium has less mass than the four atoms of hydrogen from which it formed. The mass is not lost, but is converted into energy. The amount of energy produced per unit of mass is described by Einstein's famous equation, $E = mc^2$. Source: *Astronomy: A Self-Teaching Guide*, 4th Ed., Dinah L. Moché, John Wiley, 1998.

birth, evolution, and death of a star? Luckily, stars did not all form at the same time, nor do they all go through their life cycle at the same rate. Therefore, it is possible to find stars at all stages of development in the universe and to piece together their life cycles from this evidence.

From Nebula to Protostar

The formation of a star begins in the immense interstellar (between stars) clouds of dust and gas, or **nebulae**, scattered throughout the universe. Inside these nebulae, which are composed mainly of hydrogen, stars are born through the process of gravitational collapse. However, the temperature of the cloud determines whether or not stars will form. If the temperature is too high, the atoms in the cloud move around too quickly to coalesce under the influence of gravity.

The first step in star formation is fragmentation of the cloud, in which cloudlets of matter form from denser portions of the cloud. The separation of an evenly spread out cloud of matter into cloudlets is probably the result of shock waves from the explosion of nearby stars or the pressure of radiation from hot, young stars nearby. See Figure 7.9.

The force of gravity pulls the dust and gas in toward the center of a cloudlet. As the particles of matter come closer together, mutual gravitational attraction strengthens, and the matter contracts and becomes even denser. The cloudlet also develops vortex motions, as matter spirals in toward its center. As this process continues, gravitational attraction strengthens, attracting more and more dust and gas. As more and more matter accretes, gravitational contraction causes temperatures and pressures to rise. Also, the cloudlet further contracts, rotates faster (as an ice skater spins faster as the arms are drawn inward toward the body), and becomes flattened out into a disk.

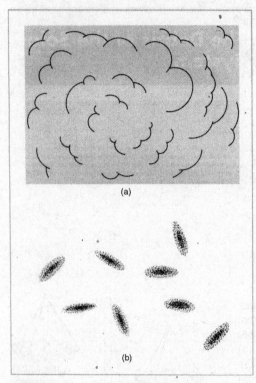

Figure 7.9 A Nebula Condenses into Cloudlets of Matter in the First Stage of Star Formation. Source: *Astronomy Explained*, Gerald North, Springer-Verlag, 1997.

Heat flows from the hot center of the condensing cloudlet to its cooler surface. At first, the heat generated in the condensing cloudlet is radiated away into the infinite reservoir of the universe around it. As the cloudlet thickens, however, the outer layers begin to absorb radiation. From this point on, the heat generated by its collapse can no longer escape the thickening cloudlet. Now the cloudlet heats up, and its internal pressure increases as the heat energy is shared among its gas and dust particles, causing their velocities to increase. Eventually, the energy of collisions between particles bouncing them apart balances the gravitational energy drawing them together. The rate of compression slows and stops, temperatures reach the point where the whole mass glows, and fragmentation ceases. The cloudlet has now become a hot, relatively dense, disk-shaped region called a ***protostar***. The protostar is approximately the size of a solar system, and its surface temperature is about 4,000 K. See Figure 7.10.

The balance between the outward pressure of very hot gases and the inward pull of gravity should prevent the collapse of a protostar. The interplay of magnetic fields and charged particles within the cloudlet, however, permits randomly formed clumps of matter poor in charged particles to collapse into the core, causing it to become increasingly dense and hot.

A Star Is Born: Mass Is Destiny

As the core of a protostar collapses, the dust and gas collide more and more vigorously, and in each collision some of the energy of motion is converted into heat. Temperatures at the center of the core reach tens of millions of degrees, and pressures rise to billions of atmospheres. When the temperature in the center of the protostar reaches 10 million K, nuclear fusion reactions are triggered, and a star is born. The energy liberated by the fusion of hydrogen into helium creates a tremendous, new outward pressure in the core. A new balance is then reached between the pressure of matter falling inward toward the core and the pressure of matter being blown outward by nuclear reactions.

If the core expands outward, it cools off, effectively stopping fusion. Too little expansion, on the other hand, causes more infalling of dust and gas, resulting in a denser, hotter core, a faster rate of fusion, and more outward pressure. This newly balanced core of fusing hydrogen is the astronomer's concept of a newly formed

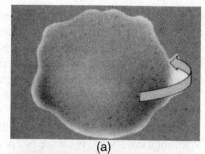
(a)

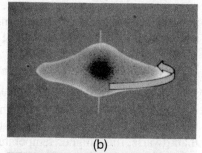
(b)

(c)

Figure 7.10 Three Stages in the Evolution of a Protostar. (a) An interstellar gas cloud begins to contract because of its own gravity. (b) A central condensation forms, and the cloud rotates faster and flattens. (c) The star forms in the cloud center, surrounded by a rotating disk of gas.
Source: *Astronomy: The Cosmic Journey*, William K. Hartman, Wadsworth, 1987.

star on the main sequence. What happens next depends upon the mass and composition of the protostar. Stars that have about the same mass and chemical composition go through the same stages of evolution in about the same time. High-mass stars evolve the fastest, while stars of very low mass take the longest time to evolve.

Red Dwarfs

Low-mass protostars have correspondingly low gravity. Their cores balance at a slow rate of fusion and low temperatures. With less than a third the mass of the Sun, these stars are smaller and cooler and therefore appear red rather than yellow; thus they are called red dwarfs. Red dwarfs fuse hydrogen at such a slow rate that they can remain stable for hundreds of millions of years. The smallest red dwarfs may last for trillions of years before all of their hydrogen is consumed.

Once a red dwarf has converted all of its hydrogen into helium, fusion ends. Without the outward pressure of energy released by fusion, the helium-rich star contracts and heats up; but, because of its small size, it never reaches the temperature needed to trigger the fusion of helium. The star slowly cools, dimming and contracting into an inert ball of gas known as a black dwarf.

Sun-Class Stars

Mid-sized stars such as the Sun have enough mass to escape the dead-end fate of a red dwarf. They are larger, have higher gravity, and, once fusion has begun, stabilize at a higher temperature. Although they contain more hydrogen than dwarfs, their higher temperature causes fusion to occur at a faster rate, exhausting their supply of hydrogen in less than 50 billion years. As the star's hydrogen is converted into helium, the core becomes mostly helium; eventually the star is composed of a helium core surrounded by a shell of hydrogen gas. Without fusion of hydrogen in the core to support it, inward pressure of gravity from the surrounding shell causes the helium core to contract. Heat from the contraction triggers fusion in the hydrogen remaining in the surrounding shell. This fusion, together with heat from the still contracting core, causes the shell to puff out to perhaps 100 times its former size. The surface cools because of the expansion and now appears red. The star has moved off the main sequence and is now a red giant. See Figure 7.11.

After about 100 million years, odd things begin to happen in the core of a red giant. As the density of the matter in the core increases, the matter changes its behavior and is said to be degenerate. Electrons can no longer move about freely and collide at random. Instead, they exert a damping effect on each other's motion. If heat is added to such a gas, the pressure actually decreases and the gas contracts! In the core of a red giant, degenerate matter contracts and heats up until it reaches the incredible temperature of 100,000,000 K, at which point helium begins to fuse into heavier elements, including carbon. In less than 100 million years this process ends, leaving a hot carbon core. But with insufficient gravity to fuse carbon, the star dies, its envelope of gases drifts off, and its hot, highly compressed core remains as a white dwarf.

Short-Lived Giants

Stars of at least three solar masses shine hot, bright, and blue. At the very upper end of the main sequence are hot, blue supergiants with 30 times the Sun's mass and 100,000 times its brightness. To shine 100,000 times brighter than the Sun, these stars must fuse hydrogen 100,000 times faster, but they have only 30 times the Sun's mass. Thus, a blue supergiant should last for a time period that is 100,000 ÷ 30, or roughly 3,000, times less than the period of time the Sun should exist. Thus, a typical blue supergiant exists for only about 3 million years.

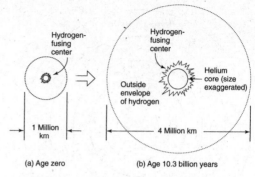

Figure 7.11 (a) At birth, a Sun-class star is mainly hydrogen, which fuses into helium, releasing energy. (b) After several billion years, a star consumes most of its hydrogen and develops a helium core. Heat from contraction of the helium core triggers fusion in the outer shell of hydrogen, causing the star to "puff up" into a red giant. Source: *Astronomy: A Self-Teaching Guide*, 4th Ed., Dinah L. Moché, John Wiley, 1998.

Such massive stars pass quickly through the same stages as a mid-sized star. They are so large, however, that even before all of their hydrogen is consumed, their helium core collapses; and contraction driven by their enormous gravity generates temperatures high enough to trigger the nuclear fusion of helium. Helium is converted into carbon, and eventually the core again collapses until temperatures needed to fuse carbon are reached. This cycle continues with carbon fusing to form neon and magnesium; then neon is converted into oxygen, oxygen into silicon and sulfur, and finally silicon into iron. Such massive stars, at the end of their lives, are layered like an onion. A thick hydrogen sheath encases successively thinner layers of helium, carbon, oxygen, and silicon shot through with magnesium, calcium, sulfur, and other elements. See Figure 7.12.

Iron, however, ends the process because the fusion of iron does not release energy, but instead requires energy to occur. When fusion ceases, the massive star's core collapses for the last time. The iron heart of the star crushes in on itself, and temperatures rise to 100 billion degrees while pressures get high enough so that all electrons and protons are squeezed together to form neutrons. Matter in the core then reaches a point at which it can no longer endure further compression. The repulsive force between nuclei overcomes the force of gravity, and the core rebounds like a tightly coiled spring. The decay of protons releases 99 percent of the energy of the explosion in a burst of exotic particles called **neutrinos**, which have little or no mass, have no charge, and can penetrate a solid object as if it were not there. In the wake of this neutrino burst, the recoil of the star's core hurls matter outward in a violent, explosive

Stars, Their Origin and Evolution

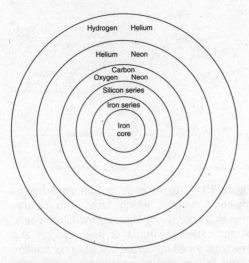

Figure 7.12 The "Onion-Layer" Structure That Develops in a Massive Star Prior to a Supernova. Massive stars undergo a succession of core collapses, each of which triggers fusion of heavier and heavier elements. The result is a star whose core consists of layers of elements produced by fusion. Source: *Astronomy Explained*, Gerald North, Springer-Verlag, 1997.

shock wave, or ***supernova***, that blasts through the surrounding layers.

As the explosion speeds through the layers of the core, it sets off fusion of nuclei and new elements are created. When the shock wave erupts from the core, the entire outer envelope of the star is blown away in a violent outburst, spewing energy and huge masses of newly created elements into space. For a time, the exploding star can outshine the rest of the stars in its galaxy. See Figure 7.13. After the outburst, all that remains is a small, superdense sphere composed almost entirely of neutrons—a ***neutron star***. A neutron star is so incredibly dense that the entire mass of the Sun may be packed into a sphere only a few kilometers across!

Figure 7.13 A Supernova Captured on Film. These two photographs were taken before and after the supernova, which can be seen as the intensely bright star to the lower right of the galaxy in the photograph to the right. Source: *Discovering Astronomy*, Robert Chapman, W. H. Freeman, 1978.

The elements ejected by a supernova are recycled. Incorporated into interstellar nebulae, they become part of a second and eventually a third generation of stars such as our Sun. In these younger, metal-enriched stars, fusion reactions occur that manufacture still other elements. Thus, all of the elements here on Earth can be traced to supernovas that occurred in the distant past.

If a very massive star (at least 10 solar masses) undergoes supernova, the remaining mass collapses beyond the neutron-star stage. The sphere left behind is so massive that irresistible gravitational contraction causes it to become so dense, with gravity so strong, that not even light can escape; this structure is called a ***black hole***. To an observer, the star simply disappears. See Figure 7.14.

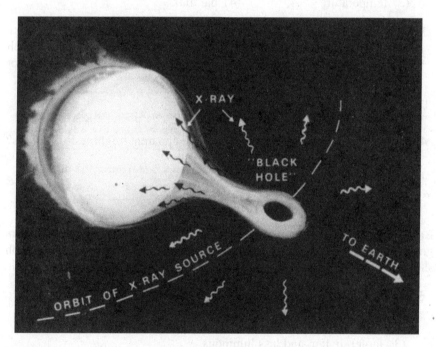

Figure 7.14 An Artist's Conception of a Black Hole. A stream of matter is ripped out of a star by the immense gravity of an orbiting black hole. X rays are emitted as the matter accelerates into the black hole. Source: *Astronomy: A Self-Teaching Guide*, 4th Ed., Dinah L. Moché, John Wiley, 1998.

Stars, Their Origin and Evolution

MULTIPLE-CHOICE QUESTIONS

In each case, write the word or expression that best answers the question or completes the statement.

1. A star differs from a planet in that a star
   (1) has a fixed orbit
   (2) is self-luminous
   (3) revolves about the Sun
   (4) shines by reflected light

2. Which information about a nearby star must be known to determine the distance of the star from Earth?
   (1) size
   (2) temperature
   (3) color
   (4) parallax

3. If three identical 100-watt light bulbs were placed at distances of 1 meter, 10 meters, and 100 meters, respectively, from an observer, each would seem to have a different brightness. Applied to stars, this concept is called
   (1) density
   (2) magnitude
   (3) volume
   (4) twinkling

4. Which factor does *not* directly affect the apparent brightness of Polaris?
   (1) its motion about the north celestial pole
   (2) its distance from Earth
   (3) its mass
   (4) its temperature

Base your answers to questions 5 through 11 on the Characteristics of Stars graph in the *Reference Tables for Physical Setting/Earth Science*. The graph shows the temperatures and luminosities of many stars observed from Earth.

5. Compared with our Sun, the star Betelgeuse is
   (1) smaller, hotter, and less luminous
   (2) smaller, cooler, and more luminous
   (3) larger, hotter, and less luminous
   (4) larger, cooler, and more luminous

6. Which list shows stars in order of increasing temperature?
   (1) Barnard's Star, Polaris, Sirius, Rigel
   (2) Aldebaran, the Sun, Rigel, Procyon B
   (3) Rigel, Polaris, Aldebaran, Barnard's Star
   (4) Procyon B, Alpha Centauri, Polaris, Betelgeuse

7. Which two stars are most similar in luminosity?
   (1) Betelgeuse and Barnard's Star
   (2) Procyon B and Proxima Centauri
   (3) Polaris and the Sun
   (4) Alpha Centauri and Sirius

8. The star Algol is estimated to have approximately the same luminosity as the star Aldebaran and approximately the same temperature as the star Rigel. Algol is best classified as a
   (1) main sequence star
   (2) red giant star
   (3) white dwarf star
   (4) red dwarf star

9. Which statement describes the general relationship between the temperature and the luminosity of main sequence stars?
   (1) As temperature decreases, luminosity increases.
   (2) As temperature decreases, luminosity remains the same.
   (3) As temperature increases, luminosity increases.
   (4) As temperature increases, luminosity remains the same.

10. Compared to the surface temperature and luminosity of massive stars in the main sequence, the smaller stars in the main sequence are
    (1) hotter and less luminous
    (2) hotter and more luminous
    (3) cooler and less luminous
    (4) cooler and more luminous

11. Compared to other groups of stars, the group that has relatively low luminosities and relatively low temperatures is the
    (1) Red Dwarfs
    (2) White Dwarfs
    (3) Red Giants
    (4) Blue Supergiants

12. The reaction below represents an energy-producing process.

    Hydrogen + Hydrogen → Helium + Energy
    (lighter element) (lighter element) (heavier element)

    The reaction represents how energy is produced
    (1) in the Sun by fusion
    (2) when water condenses in Earth's atmosphere
    (3) from the movement of crustal plates
    (4) during nuclear decay

Stars, Their Origin and Evolution

13. Which object forms by the contraction of a large sphere of gases causing the nuclear fusion of lighter elements into heavier elements?
    (1) comet
    (2) planet
    (3) star
    (4) moon

Base your answers to questions 14 and 15 on the diagram below, which represents possible stages in the life cycle of stars.

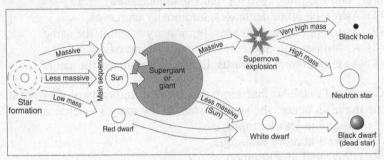

(Not drawn to scale)

14. Which star has the greatest probability of producing a supernova explosion?
    (1) Barnard's Star
    (2) Betelgeuse
    (3) Procyon B
    (4) the Sun

15. According to the diagram, a star like Earth's Sun will eventually
    (1) explode in a supernova
    (2) become a black hole
    (3) change into a white dwarf
    (4) become a neutron star

CONSTRUCTED RESPONSE QUESTIONS

Base your answers to questions 16 through 18 on the Characteristics of Stars graph in the *Reference Tables for Physical Setting/Earth Science*.

16. Describe the relationship between temperature and luminosity of main sequence stars. [1]

17. In which group of stars would a star with a temperature of 5000°C and a luminosity of approximately 100 times that of the Sun be classified? [1]

18. Complete the table below by identifying the color and classification of the star Procyon B. The data for the Sun have been completed as an example. [1]

Star	Color	Classification
Sun	yellow	main sequence
Procyon B		

Base your answers to questions 19 through 21 on the passage below and on your knowledge of stars and galaxies.

Stars

Stars can be classified according to their properties, such as diameter, mass, luminosity, and temperature. Some stars are so large that the orbits of the planets in our solar system would easily fit inside them.

Stars are grouped together in galaxies covering vast distances. Galaxies contain from 100 billion to over 300 billion stars. Astronomers have discovered billions of galaxies in the universe.

19. Arrange the terms *galaxy*, *star*, and *universe* in order from largest to smallest. [1]

20. Complete the table below by placing an **X** in the boxes that indicate the temperature and luminosity of each star compared to our Sun. [1]

Stars	Temperature		Luminosity	
	Hotter	Cooler	Brighter	Dimmer
Procyon B				
Barnard's Star				
Rigel				

21. The star Betelgeuse is farther from Earth than the star Aldebaran. Explain why Betelgeuse appears brighter or more luminous than Aldebaran. [1]

Base your answers to questions 22 through 24 on the graph below, which shows the early formation of main sequence stars of different masses (M). The arrows represent temperature and luminosity changes as each star becomes part of the main sequence. The time needed for each star to develop into a main sequence star is shown on the main sequence line.

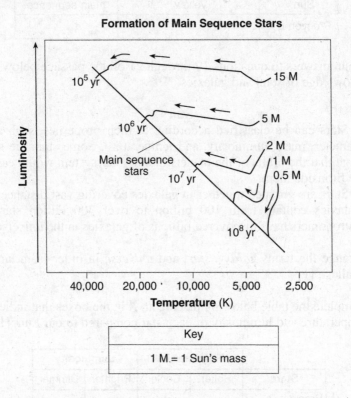

22. Describe the relationship between the original mass of a star and the length of time necessary for it to become a main sequence star. [1]

23. Describe the change in luminosity of a star that has an original mass of 0.5 M as it progresses to a main sequence star. [1]

24. Identify the force that causes the accumulation of matter that forms the stars. [1]

EXTENDED CONSTRUCTED RESPONSE QUESTIONS

Base your answers to questions 25 through 28 on your knowledge of Earth science and on the table below, which lists the seven brightest stars, numbered 1 through 7, in the constellation Orion. This constellation can be seen in the winter sky by an observer in New York State. The table shows the celestial coordinates for the seven numbered stars of Orion.

Location of the Seven Brightest Stars in Orion		
Star Number	Celestial Longitude (measured in hours)	Celestial Latitude (measured in degrees)
1	5.9	+7.4
2	5.4	+6.3
3	5.2	−8.2
4	5.8	−9.7
5	5.7	−1.9
6	5.6	−1.2
7	5.5	−0.3

25. On the grid provided below, graph the data shown in the table by following the steps below.
    (a) Mark with an **X** the position of *each* of the seven stars. Write the number of the plotted star beside each **X**. The first star has been plotted for you. [2]
    (b) Show the apparent shape of Orion by connecting the **X**s in the following order: 5 – 1 – 2 – 7 – 3 – 4 – 5 – 6 – 7 [1]

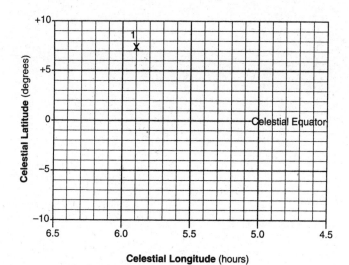

26. Star 1 plotted on the grid is the star Betelgeuse. Star 3 plotted on the grid is the star Rigel. How do the temperature and luminosity of Betelgeuse compare to the temperature and luminosity of Rigel? [1]

27. The seven stars of the constellation Orion that were plotted are located within our galaxy. Name the galaxy in which the plotted stars of Orion are located. [1]

28. State one reason why an observer in New York State can never observe the constellation Orion at midnight during July but can observe the constellation Orion at midnight during January. [1]

CHAPTER 8

THE SOLAR SYSTEM

Unit Two **MODERN ASTRONOMY**

KEY IDEAS Our solar system formed about 5 billion years ago from a giant cloud of gas and debris. The solar system consists of nine large planets, their satellites, and a variety of smaller objects, ranging from asteroids, meteors, and comets to tiny particles of dust and gas, all in orbit around a central star, our Sun.

During the formation of the solar system, gravity caused Earth and the other planets to become layered according to the density of the materials of which they were composed. The distance of each planet from the Sun was a key factor in determining the planet's characteristics. Powerful emissions from the Sun drove off most of the nearby gases, leaving behind the small, dense, rocky terrestrial planets. The more distant Jovian planets had sufficient gravity to retain much of the gas that surrounded them, and thus evolved into the large, low-density, gaseous planets.

Formation of the solar system left numerous smaller objects, such as asteroids, comets, and meteors, orbiting the Sun. From time to time, these objects collide with planets. Impact craters from such collisions have been identified in Earth's crust, and impact events have been correlated with mass extinctions and global climate change on Earth.

KEY OBJECTIVES
Upon completion of this chapter, you will be able to:

- List the members of the solar system.
- Describe the nebular theory of the formation of the solar system.
- Compare and contrast the characteristics of the nine large planets, distinguishing between the terrestrial planets and the Jovian planets.
- Describe the other components of our solar system, such as asteroids, meteors, and comets.
- Explain the evidence linking impact events with mass extinctions and global climate change on Earth.

THE SOLAR SYSTEM'S PLACE IN THE UNIVERSE

We currently know that Earth is one of several planets that orbit a star—the Sun. The Sun is millions of times closer to Earth than is any other star. Light travels at a finite speed of 300,000 kilometers per second. Light from the Sun reaches Earth in less than 10 *minutes*, but light from the next closest star takes several *years* to get to Earth. The light from very distant stars takes several billion years to reach Earth. When we look at distant stars, we see them as they were when the light we now see left each star. Therefore, when we look at distant stars, we look back in time.

The universe is so large that the distance light travels in a year, or a light-year, is used to measure its distances. A light-year is a distance of about 9½ trillion kilometers. Our fastest rockets would take thousands of years to reach the nearest star beyond the Sun. Compared with the vast distances between stars, the distances between the Sun and its planets are small. Most astronomers estimate the universe to be about 15 billion light-years in radius, and it is therefore thought to be about 15 billion years old.

Stars are not scattered evenly throughout the universe; gravity has drawn them together in huge clumps called galaxies. A **galaxy** is a system of hundreds of billions of stars. The universe contains many billions of galaxies, each, in turn, containing billions of stars. Our solar system is located near the edge of a disk-shaped galaxy of stars called the **Milky Way galaxy**, which gets its name from the faint white band of its stars that can be seen from Earth on a clear, dark night. The Milky Way galaxy contains more than 100 billion stars revolving in huge orbits around the center of the galaxy.

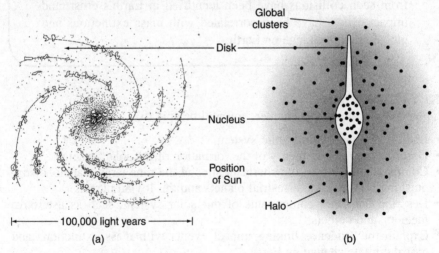

Figure 8.1 The Milky Way Galaxy. Viewed face on (a) and edge on (b), the shapes and locations of the nucleus, disk, and halo can be seen. Note the position of the Sun, about two-thirds of the way out from the nucleus in the Orion arm of the disk.

Sources: *Astronomy: The Cosmic Journey*, William K. Hartman, Wadsworth, 1987. *Horizons: Exploring the Universe*, Michael A. Seeds, Wadsworth, 1987.

THE ORIGIN AND EVOLUTION OF THE SOLAR SYSTEM

Our solar system consists of the Sun, the eight planets, fifty or so moons, and thousands of asteroids, meteors, and countless comets that orbit the Sun. On the whole, though, the solar system *is* the Sun, for the Sun contains 99.9 percent of all the mass of the whole system. It is unlikely that the origin of a family of objects (the planets, moons, etc.) representing such a tiny fraction of the solar system as a whole is not closely linked to the formation of the Sun. Any theory of the formation of the solar system has to account for observational evidence, such as the following:

1. More than 99 percent of the mass of the solar system is contained in the Sun.
2. All the planets move around the Sun in the same direction and in roughly the same plane.
3. All of the planets (except Venus and Uranus), spin in the same direction, close to the plane of the Sun's equator; and most of the moons also spin in the same direction as their planets, close to the plane of those planets' equators.
4. The planets can be divided by mass and density into the terrestrial and Jovian planets. The terrestrial planets are composed mostly of metal silicates and iron; the Jovian planets, mostly of hydrogen and helium.
5. The planets exhibit a fairly regular spacing of orbits.
6. All the solid bodies of the solar system that have been measured—Earth, the Moon, and meteorites—are roughly 4.6 billion years old, and formed within about 0.1 billion years of each other, that is, at about the same time.
7. Some meteorites, aside from their loss of volatile (easily vaporized) elements such as rare gases, are virtually identical in composition to the Sun.

The question is, Did the solar system form as part of the natural process of star formation or as some later event? As early as the eighteenth century, two theories were proposed to account for the formation of the solar system: the catastrophic theory and the nebular theory.

The *catastrophic theory* suggest that early in the Sun's life there was a collision or near collision between the Sun and a passing celestial body. Such an event would cause a solar upheaval, resulting in streamers and globs of gaseous solar material being thrown off into space. As this material condensed, it would form a great many small bodies, all revolving around the Sun. Some of these would be close enough to each other to aggregate by mutual gravitational attraction to form planets. Others would be left revolving around the Sun as asteroids, planetoids, and comets. However, the catastrophic theory requires an improbable event and does not adequately explain the even spacing of planets and their varying compositions.

The Solar System

In recent years, with the aid of computer modeling, the *nebular theory* has gained more widespread support. The modern **nebular theory** proposes that the solar system formed from an interstellar cloud of dust and gas, enriched in heavier elements from earlier supernovae, which condensed into a disk-shaped protostar. The source of planet formation was the disk of dust and gas surrounding the protostar. Turbulence in the dust and gas of the spinning disk caused it to fragment and sort itself out into concentric rings according to the mass and the speed at which the material in it was revolving. See Figure 8.2.

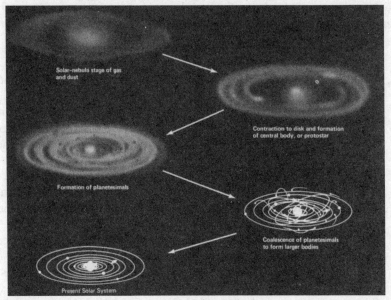

Figure 8.2 The Formation of the Solar System from a Solar Nebula to Our Present Solar System. Source: *Exploring the Cosmos*, Louis Berman and J. C. Evans, Little, Brown, 1986.

Most collisions between celestial bodies tend to cause material to fall inward and end up in the protostar. Some of the dust particles, however, stick together upon collision to form grains. This process of sticking, or accretion, continues, slowly forming larger and larger bodies called **planetesimals**. Computer simulations indicate that over a period of millions to tens of millions of years, Earth-sized bodies can be formed in this way. See Figure 8.3.

The nature of the dust that accretes depends upon its position in the disk, and it changes over time. The temperature of the disk decreases with movement outward from the core. Where temperatures are lower than 1,500 K, rocky and metallic grains can survive. This region, between 0.5 and 5 astronomical units from the Sun (recall that an AU, or astronomical unit, is the average distance between Earth and the Sun: about 150 million kilometers), is where the terrestrial planets are found today. Beyond about 5 astronomical units, the disk was cold enough so that water ice could condense and survive. Thus, today, in the outer solar system bodies made of rock and ice—the moons of the gas giants—are visible. The Jovian planets probably began forming with the accretion of rock and ice. But, as the body grew larger, it reached a point where it gravita-

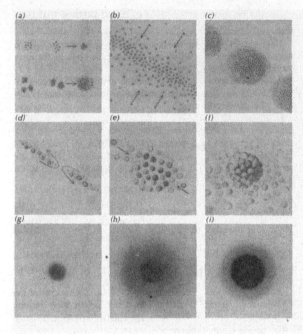

Figure 8.3 Planets Form by the Process of Accretion. Planets form because of the collision and sticking together of small grains (a) in the primordial solar system. The growing particles fall toward the plane of the original cloud (b), forming a loose disk of material. The disk breaks up into asteroid-sized bodies (c), which cluster together (d), collide (e), and coalesce (f) into planet-sized bodies (g). The planet-sized bodies have enough gravity to collect gas from the nebula (h). The result is a primordial planet (i). Source: *The Origin and Evolution of the Solar System*, A. G. W. Cameron, Scientific American, Inc., 1975. All rights reserved.

tionally attracted gas (mostly hydrogen) from the surrounding solar nebula. As the gas built up around the growing planet, the planet's gravity strengthened, attracting yet more gas, rock, and ice. As the giant planets formed, they produced disks of gas and dust of their own, out of which their moons formed. (The formation of Earth's Moon, which is not much smaller than Earth itself, must have occurred in a different way, which will be discussed later.) Computer simulations suggest that Jupiter and Saturn could have formed in this way in just a few million years. Uranus and Neptune, in the colder outer reaches of the disk, would have taken longer, perhaps 10 million years. As the planets drew material from the solar nebula, it dissipated and planet formation ceased as the planets literally ran out of gas and dust to draw upon.

That the terrestrial planets did not acquire huge gaseous envelopes may be due to a number of reasons. The surrounding gas may have been too hot—its molecules moving too vigorously—to form a stable envelope. The solar wind, or stream of charged particles and radiation emanating from the protostar, may have stripped the planets of their hydrogen envelopes as the solar nebula dissipated. Or perhaps the lack of water ice resulted in a slower rate of accretion, so that by the time the terrestrial planets were big enough to attract gas the solar nebula had already dissipated.

Beyond the orbit of Neptune, extending out to about 1,000 astronomical units, lies the **Kuiper belt**—a region of smaller orbiting objects left over from planet formation. Kuiper belt objects range in size from tiny grains to minor planets hundreds of kilometers in diameter. The dwarf planet Pluto and its moon Charon are believed to be Kuiper objects that strayed close enough to the Sun to be drawn into a planetary orbit. See Figure 8.4.

The Solar System

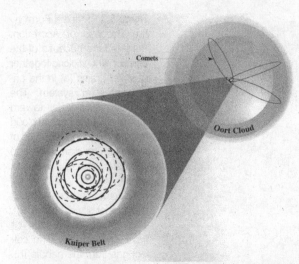

Figure 8.4 The Kuiper Belt. This is a region of objects surrounding the solar system and consisting of particles, ranging in size from dust to planetesimals, that are remnants of solar system formation. It extends from the orbit of Neptune out to about 1,000 AU. Source: *Earth: Evolution of a Habitable World*, Jonathan I. Lunine, Cambridge University Press, 1999.

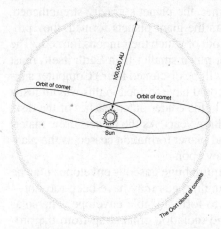

Figure 8.5 The Oort Cloud. A giant cloud of comets is believed to surround the solar system. Stars passing the Oort cloud tear comets from their orbits, sending some off into space and others into long-period orbits around the Sun. Source: *Discovering Astronomy*, Robert Chapman, W. H. Freeman, 1978.

Even farther out lies a huge, shell-like cloud of icy material called the **Oort cloud**, after the Dutch astronomer Jan Oort, who proposed its existence in 1950. The Oort cloud is thought to surround the solar system completely and extend to a distance of 100,000 astronomical units (more than 3,000 times the Sun–Neptune distance). See Figure 8.5. The very low temperatures in the Oort cloud allow gases such as methane, nitrogen, and carbon monoxide to form ices that accrete, along with water and dust, and preserve a record of the composition of the original solar nebula. It is believed that the clumps of icy material visible as comets originate within the Oort cloud. The gravitational influence of passing stars causes some of these orbiting ice balls to fall toward the inner reaches of the solar system. Drawn inward by the Sun's gravity, and perturbed by the gravity of major planets, some settle into short-term orbits; one example is Halley's comet, which orbits the Sun once every 76 years. Other, shorter period comets include Tempel 2 (5.3 years) and Encke's comet (3.3 years). Some comets, however, have longer periods, so long that astronomers cannot measure them and may be taken by surprise. One such comet, the Daylight comet of 1910, was probably the brightest seen in the twentieth century. See Figure 8.5.

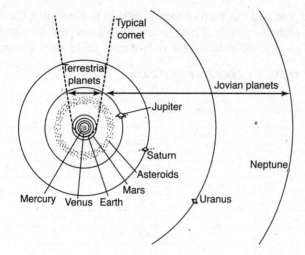

Figure 8.6 The Solar System. Note that in addition to the eight planets and their moons, the solar system includes the asteroid belt, dwarf planets such as Pluto, and comets that circle the Sun in highly eccentric orbits. *Source: Horizons: Exploring the Universe*, Michael A. Seeds, Wadsworth, 1987.

Although the disk-shaped protostars can be observed in molecular clouds such as the Orion nebula, planetary formation around stars has never been observed. Of the billions of stars in the universe, only ten or so planets orbiting other stars have been definitively identified. Thus, much of the current thinking about planetary formation is based upon computer modeling and is therefore subject to revision as new observational data are obtained.

THE STRUCTURE OF THE SOLAR SYSTEM

The solar system is defined as the Sun, the planets that orbit the Sun, the satellites of those planets, and many small interplanetary bodies such as asteroids and comets. See Figure 8.6. The **Sun** is a star, which is composed of gases and emits electromagnetic radiation produced by nuclear reactions in its interior. **Planets** are objects that are large enough that their gravity has pulled them into a round shape. Their gravity is also strong enough that they have swept up all nearby objects and debris and now orbit in a clear path around the Sun. **Satellites** are solid bodies that orbit planets.

Beginning at the center, the major planets of the solar system are Mercury, Venus, Earth, Mars, Jupiter, Saturn, Uranus, and Neptune. A common memory aid for this sequence is **My Very Educated Mother Just Served Us Nachos**. Some basic information about the bodies in the solar system is summarized in Table 8.1, the Solar System Data chart of the *Reference Tables for Physical Setting/Earth Science*. The major planets of the solar system are divided into two groups based upon their size and composition: the four innermost, small, dense **terrestrial planets** and the four large, much less dense outermost **Jovian planets.**

In addition to the eight major planets, several smaller dwarf planets orbit the Sun. **Dwarf planets** have enough gravity that they are rounded, or nearly round in shape. However, they have not swept up everything near their path and may orbit in a zone that still has many other objects in it. Currently, there are three known dwarf planets—Ceres, Pluto, and Eris. All are less than half the size of the planet Mercury and at best have only a trace of an atmosphere.

TABLE 8.1 SOLAR SYSTEM DATA

Celestial Object	Mean Distance from Sun (million km)	Period of Revolution (d=days) (y=years)	Period of Rotation at Equator	Eccentricity of Orbit	Equatorial Diameter (km)	Mass (Earth = 1)	Density (g/cm³)
SUN	—	—	27 d	—	1,392,000	333,000.00	1.4
MERCURY	57.9	88 d	59 d	0.206	4,879	0.06	5.4
VENUS	108.2	224.7 d	243 d	0.007	12,104	0.82	5.2
EARTH	149.6	365.26 d	23 h 56 min 4 s	0.017	12,756	1.00	5.5
MARS	227.9	687 d	24 h 37 min 23 s	0.093	6,794	0.11	3.9
JUPITER	778.4	11.9 y	9 h 50 min 30 s	0.048	142,984	317.83	1.3
SATURN	1,426.7	29.5 y	10 h 14 min	0.054	120,536	95.16	0.7
URANUS	2,871.0	84.0 y	17 h 14 min	0.047	51,118	14.54	1.3
NEPTUNE	4,498.3	164.8 y	16 h	0.009	49,528	17.15	1.8
EARTH'S MOON	149.6 (0.386 from Earth)	27.3 d	27.3 d	0.055	3,476	0.01	3.3

Source: The State Education Department, *Earth Science Reference Tables*, 2011 Edition (Albany, New York; The University of the State of New York).

The Terrestrial Planets

The four planets closest to the Sun are Mercury, Venus, Earth, and Mars. These "inner" planets are terrestrial; that is, they resemble Earth in size and rocky composition. They also have about the same density as Earth.

Mercury

Mercury is one of the five planets that can be seen from Earth with the unaided eye. Mercury is very difficult to spot, however, because it is never more than 28° from the Sun, whose glare usually hides it. Through a telescope from Earth, Mercury, like the Moon and Venus, can be seen to exhibit phases, but only dark, fuzzy features can be detected on its surface.

Mercury revolves around the Sun once every 88 days and rotates once every 58.65 days. Because it is the closest planet to the Sun, it receives the most intense radiation. The long rotation period causes the side of Mercury facing the Sun to receive nonstop sunlight for a long period of time. At the point where the Sun is directly overhead of Mercury, the surface temperature reaches 700 K—hot enough to melt lead. At the same time, the side facing away from the Sun is in darkness for long periods and cools to 100 K. Like the Moon, Mercury has no atmosphere because its small gravitational field was unable to hold on to any gases. The lack of an atmosphere allows debris from space to strike the surface of Mercury unhindered. A fly-by of Mercury by the Mariner 10 spacecraft showed a cratered surface remarkably like that of the Moon.

Venus

Venus is one of the brightest objects seen in the sky; only the Sun and the Moon are brighter. Since its orbit is larger than Mercury's, Venus may be seen as far as 47° from the Sun. This feature, together with its brightness, makes it one of the most obvious of all celestial objects.

Venus is almost identical in size to Earth. Like Mercury, Venus can be seen to go through a series of phases. However, Venus rotates in a direction opposite to that of Earth and most other planets.

A telescope can detect little about the surface of Venus because the planet's dense atmosphere, consisting mostly of carbon dioxide, hides the surface from view. This thick, cloudy atmosphere produces a greenhouse effect that traps heat; several Venera probes soft-landed on Venus by the Soviet Union recorded temperatures near 750 K and atmospheric pressures 90 times greater than Earth's. Spectroscopic studies indicate the presence of sulfuric acid, hydrochloric acid, and hydrofluoric acid. In February 1974, the Mariner 10 spacecraft came within 5,800 kilometers of Venus. As it flew by, cameras equipped with special filters and films recorded pictures by ultraviolet light. These photographs revealed details of Venus's atmosphere unseen in visible light, such as cloud motions indicating wind speeds of 100 meters per second (200 mph) in the upper atmosphere.

Earth

Earth is the third planet from the Sun and is the largest of the inner planets. Earth's atmosphere is rich in oxygen and nitrogen. From space, many details of Earth's surface can be seen; more than 70 percent is covered with liquid water. Earth has one natural satellite, the Moon.

Mars

Mars is the fourth planet from the Sun. In the sky, it appears as a reddish star. When Mars is viewed through a telescope, its reddish brown, desertlike surface and polar "ice caps" can be seen. Mars's axis is tilted 24° to its orbit (similar to Earth's 23½° tilt); but since its year is nearly twice as long as Earth's, its seasons are longer. The polar "ice caps," consisting mostly of carbon dioxide with some water ice, melt and refreeze as the seasons pass.

Beginning in 1965, a series of Mariner spacecraft passed Mars and photographed its surface, which is pockmarked with craters much like the Moon's. The photographs also revealed long, sinuous channels that look just like dry riverbeds. Most analysts think these channels were cut by liquid water, but have no idea where the water is now. The Viking probes that landed on Mars found a rocky surface with volcanoes (the largest is four times the size of Mount Everest) and vast, flat plains, but no traces of life. Mars atmosphere is very thin, with less than 1 percent of the surface pressure of Earth. It is composed mainly of carbon dioxide with traces of water vapor, argon, ozone, oxygen, carbon monoxide, and hydrogen.

Mars has two natural satellites, the moons Phobos and Deimos. These moons are tiny; Phobos is only 25 kilometers, and Deimos is just 15 kilometers, in diameter. However, they orbit much closer to the surface than our Moon does to Earth and would therefore appear large—about one-half the size of our Moon—to an observer on Mars. Phobos revolves around Mars in just under 8 hours, faster than the planet rotates, so Phobos's rising and setting are the result of its motion, not the rotation of Mars.

The Jovian Planets

All the more distant Jovian planets—Jupiter, Saturn, Uranus, and Neptune—are gas giants. Although they have more mass than the terrestrial planets, they are less dense. Gas giants have thick atmospheres (hence their name), composed largely of gaseous hydrogen compounds such as water (H_2O), methane (CH_4), and ammonia (NH_3) surrounding a small rocky or liquid core. Despite their large size, the gas giants rotate very rapidly, thereby causing a distinct equatorial bulge.

Jupiter

Jupiter is the most massive planet in the solar system, containing 70 percent of all the mass outside of the Sun (which is still only one-thousandth of the mass of the Sun). In spite of its large mass, Jupiter is less dense than Earth because its volume is 1,300 times that of Earth.

Through even a small telescope, Jupiter can be seen as a disk crossed by narrow, parallel bright and dark bands. Through more powerful telescopes, Jupiter's great red spot is visible. In 1973 and 1974, Pioneer 10 and 11 flew by Jupiter and took many pictures, which were sent back to Earth. They revealed Jupiter's surface in more detail than had ever been seen.

The visible features of Jupiter's disk are the tops of clouds in the deep atmosphere. A mottled appearance suggests the presence of convective cells. The current thinking is that heated gases from deep within the atmosphere rise and cool, causing clouds to condense and reflect sunlight, thereby forming bright spots. The clouds then spread out to the north and south, sink, and clear up, thus appearing darker. The rapid rotation of Jupiter causes these cloudy and clear regions to swirl together into parallel bands. The great red spot is thought to be the eye of an immense, hurricanelike storm in Jupiter's atmosphere.

Jupiter's interior structure was inferred from gravity measurements made by Pioneer spacecraft. The core is probably a small, rocky ball similar in composition to Earth. The rest of the planet is mostly hydrogen in two distinct layers: an inner layer of liquid hydrogen under such tremendous pressures that it acts like a metal, and an outer layer of liquid hydrogen that acts as hydrogen does on Earth. This outer layer changes gradually from pure hydrogen to hydrogen compounds such as water, ammonia, and ammonium hydrosulfide in the atmosphere.

Jupiter has many natural satellites because of its powerful gravitational field. Four bright moons of Jupiter were first discovered by Galileo. Over the years, ever more powerful telescopes were brought to bear on Jupiter, and by 1975 fourteen moons had been discovered. The four Galilean moons, Io, Europa, Ganymede, and Callisto, are all about the same size as Earth's Moon; the rest are much smaller. Voyager 1 discovered a faint ring around Jupiter, and photographs showed that one moon, Io, has active volcanoes that cover its surface with red and yellow sulfur compounds. Pictures also revealed that Europa has a smooth, ice-covered surface and that Ganymede and Callisto are covered with craters like those on Earth's Moon.

Saturn

Saturn resembles Jupiter in composition and structure, but its cloud markings show less contrast. The surface of Saturn consists mostly of bands of yellowish and tan clouds. The density of Saturn, 0.7 gram per cubic centimeter, is the lowest of any planet, and is, in fact, less than that of water. If a large enough body of water could be found, Saturn would float in it!

Saturn is best known for its ring system. When Galileo first saw Saturn through a telescope in 1610, he drew it as a blurry object with another blurry object on either side and thought it was a triple planet. In 1655, however, Christian Huygens, a Danish physicist and astronomer, discovered that a ring system surrounds the planet. The rings' dimensions are remarkable; they stretch for 274,000 kilometers from tip to tip but are barely 100 meters thick! In fact, the rings are so thin that observers from Earth lose sight of them when they appear edge-on as Earth passes even with their plane.

Saturn's rings are composed of countless particles ranging in size from that of a golf ball to that of a house. Spectroscopic studies have proved that these particles are frozen water or are covered by frozen water. The rings are separated by several large gaps and thousands of finer divisions. The reason that the gaps exist is poorly understood, but one theory is that gravitational effects are responsible. Where did the ring particles come from? Some possibilities are that the particles condensed from gas as Saturn formed, that they are fragments of a satellite that was blown apart by a collision with a comet or an asteroid, and that they are the remains of a comet or asteroid torn apart by tidal forces while passing very close to Saturn.

Like Jupiter, Saturn has an extensive system of natural satellites. It has at least 17 moons, including Titan, which is almost half the size of Earth, and many smaller moons clustered near the rings. Titan is so large it has an atmosphere of its own, consisting mainly of nitrogen with traces of ethane, acetylene, ethylene, and hydrogen cyanide.

Uranus

The English astronomer William Herschel first discovered Uranus in 1781. This planet is so far from Earth that it cannot be seen with the unaided eye and a telescope cannot resolve any markings on it; only a faint greenish color is visible. In 1986 Voyager 2 flew by Uranus and revealed that its atmosphere has an almost featureless blue haze overlying deeper clouds. Voyager 2 recorded a minimum temperature of 51 K (–368°F) and a composition matching that of Jupiter and Saturn. In 1977, evidence of rings around Uranus had been observed. Voyager 2 obtained the first clear pictures of those rings, showing them to be much narrower than those of Jupiter or Saturn.

Uranus's rotation, like that of Venus, is retrograde, and its axis of rotation, pointing toward the Sun, is almost in line with the plane of its orbit. Uranus has five major moons, as well as ten smaller moons only recently discovered by Voyager. The larger moons are composed of ice and black soil. All of the moons are heavily cratered, and cracks and canyons can be seen on several.

Neptune

The discovery of Uranus led astronomers to search for other planets. By 1800, observation of Uranus had revealed that it has irregular motions that run counter to Kepler's laws. Some scientists thought that this phenomenon was caused by a breakdown of gravitation at great distances from the Sun, but others guessed correctly that Uranus was being pulled from its theoretical orbit by an even further planet. Using laborious calculations, astronomers set out to predict where the new planet should be located in order to produce the effects observed on Uranus. In 1846, within 12 hours of beginning a search based upon the calculations of French astronomer Urbain Leverrier, two German astronomers discovered Neptune.

Neptune's atmosphere has been found to contain hydrogen, helium, and some methane, which gives the planet its bluish color. It also has faint cloud patterns resembling the bands of Jupiter and Saturn. Like the other gas giants, Neptune rotates rapidly and has a distinct equatorial bulge. The temperature of its atmosphere is 60 K, warmer than expected for a body so far from the Sun. Its high temperature suggests that Neptune may have an internal source of heat.

Neptune has eight known natural satellites. The largest moon, Triton, revolves around the planet in a direction opposite to that of all other satellites in the solar system. The second largest moon, Nereid, revolves in the normal direction, but its orbit is highly eccentric.

Interplanetary Bodies

In addition to the planets, many smaller bodies orbit the Sun and are therefore also considered part of the solar system. These objects include asteroids, comets, meteors, and meteoroids.

Asteroids

Asteroids are solid bodies having no atmosphere that orbit the Sun. They are tiny planets with well-determined orbits. More than 2,000 asteroids have been found in the solar system; most of them orbit in the gap between Mars and Jupiter. Some asteroids move in very elliptical orbits. The asteroid Icarus comes closer to the Sun than any other asteroid in the solar system.

The composition of asteroids has been studied by examining the wavelength of light that reflects from their surface. If asteroids were perfect mirrors, the light that reflects from them would be identical to the sunlight that strikes their surface. Minerals reflect light of different wavelengths in different ways, however, so that the light that reflects from an asteroid's surface differs from sunlight. By analyzing light reflected from asteroids, astronomers can infer their composition. There are two basic types of asteroids, those with surface reflections characteristic of metals and silicate minerals and those with reflections characteristic of carbonaceous chondrites (stony with a high carbon content).

Meteoroids

In addition to the asteroids between Mars and Jupiter, chunks of matter orbit the Sun in orbits that cross those of the planets. When one of these chunks of matter hits Earth's atmosphere at very high speed, it vaporizes because of friction with the air. As it streaks through the air and vaporizes, it emits light and to an observer on Earth appears as a streak of light. The chunk of matter is a **meteoroid**, the glowing object that streaks through the sky is called a **meteor**, and any of the matter that survives to strike the ground is a **meteorite**. Meteoroids vary in size from sand-sized particles to chunks weighing many tons. Meteor showers occur when Earth crosses the path of a clump or stream of meteoroids. See Figure 8.7.

Meteorites tell us a great deal about the solar system. There are three main types of meteorites: stony, stony-iron, and iron. Stony meteorites are like Earth's crust, whereas iron meteorites resemble its core. Stony-iron meteorites are like the iron-rich material of Earth's deep mantle. One explanation for this similarity between meteorites and portions of Earth is that a process called **gravitational differentiation** took place while the solar system formed. The premise is that the solar system originally had a uniform composition. As gravity drew clumps of the original matter together, however, the material separated into layers because lighter elements do not experience as strong a gravitational

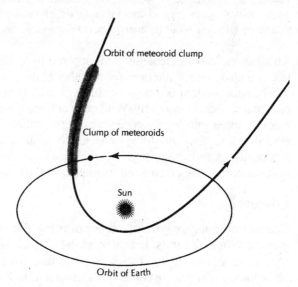

Figure 8.7 Meteor Shower. Source: *Discovering Astronomy*, Robert D. Chapman, W. H. Freeman, 1978. Used with permission.

The Solar System

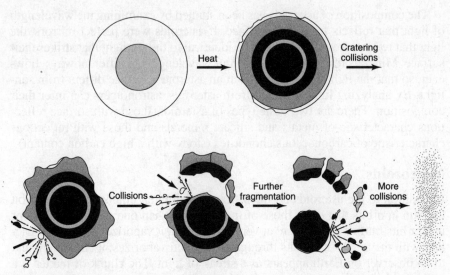

Figure 8.8 Production of Meteoroids. Collisions between large asteroids with layered interiors gave rise to the three types of meteorites found on Earth. Source: *Discovering Astronomy*, Robert D. Chapman, W. H. Freeman and Company, 1978. Used with permission.

attraction as heavier elements. This separation resulted in Earth's layered interior and also layered interiors for all other bodies in the solar system.

The solar system is thought to have originally contained many large asteroids that collided frequently. With each collision, material was broken off and ejected, some into highly eccentric orbits. If these large asteroids had a layered structure, they would be expected to break into fragments with different compositions reflecting the different layers. See Figure 8.8. For this reason, iron meteorites are considered examples of the composition of Earth's core.

Comets

Comets are another type of object orbiting the sun. Comets, like planets, move in elliptical orbits, but their orbital ellipses are highly elongated. Some comets vary from being 1 astronomical unit from the Sun at perihelion to more than 1,000 astronomical units at aphelion. Kepler's law of equal areas tells us that, although such a comet moves very fast while it is near the Sun, it must move very slowly when it is such a huge distance away. Therefore, some comets take as long as 2 million years to make one orbit of the Sun. Comets are generally divided into long-period and short-period comets. Short-period comets orbit the sun in 200 years or less; long-period comets require more than 200 years to complete an orbit.

Comets are thought to have a solid **nucleus** consisting of meteoroid particles embedded in ice. When the nucleus approaches the Sun and heats up, the ice sublimes into a cloud of gas, called a **coma**, around the nucleus. As the comet gets closer to the Sun, the coma increases in size as more gas sublimes, and particles emitted from the Sun collide with the gas molecules and push them out of the coma, forming a **tail**. Under the influence of this solar "wind," the tail always streams away from the Sun. It does *not* stream out along the path of the comet like a contrail out of a jet engine. See Figure 8.9.

Impact Events

From time to time, the orbits of asteroids, comets, and meteors are disturbed by collisions or by the gravity of other objects. Their new orbits may then place them on a collision path with a planet such as Earth. When a high-speed object (a comet, asteroid, or meteoroid) collides with a solid surface (a planet or moon), the collision creates a distinctive bowl-shaped hole called an **impact crater**.

The impact of the speeding object sends out shock waves that spread outward and downward into the ground, and the object's energy of motion is quickly changed to heat. The enormous pressures and heat of the shock wave

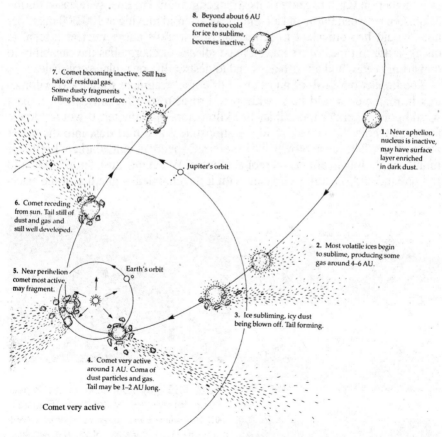

Figure 8.9 Changes That Take Place in a Comet as Its Orbit Carries It Past the Sun.
Source: *Astronomy: The Cosmic Journey*, William K. Hartman, Wadsworth, 1987.

shatter the ground. Close to the impact, rock is vaporized and melted. Farther away, it is pulverized. The shock waves also raise a rim around the crater and spew material off the sides. When the ground bounces back from the shock wave, it heaves up a central peak in the crater. See Figure 8.10a. These unique characteristics distinguish impact craters from other bowl-shaped holes, such as volcanic craters and sinkholes.

Photographs of planets and their moons reveal surfaces pock-marked with thousands upon thousands of craters. This evidence shows that impact events are not at all unusual, though they have decreased over time as gravity has swept most debris out of the inner solar system. On Earth, few impact craters have been found because they are quickly erased by water and geologic activity. Of the roughly 200 impact craters discovered on Earth, one of the largest and best known is Meteor Crater, near Flagstaff, Arizona. See Figure 8.10b.

The energy released by the impact of a high-speed object depends upon its mass and the speed of impact. Typical impact speeds on the Moon are about 40 kilometers per second, or about 100,000 miles per hour. Impact speeds would be even higher on Earth because of its stronger gravity. The energy released during a collision between Earth and a 1-kilometer asteroid moving at 100,000 miles per hour would be equivalent to setting off the world's entire nuclear arsenal at once—a dozen times over! Such a huge release of energy has the capability to transform oceans and atmospheres and to destroy life on a planetary scale.

The impact on Earth of an asteroid or comet fragment roughly 10 kilometers in diameter would have widespread disastrous effects. Such an impact would gouge a crater more than 100 kilometers in diameter. It would blow a temporary hole the same size in the atmosphere and hurl dust into the upper atmosphere. The dust blown into the stratosphere would darken Earth for months with drastic effects on global climate. When this dust settled, it would fall on land and sea alike, carrying with it the chemical signature of the aster-

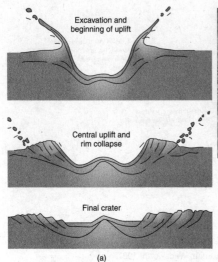

Figure 8.10 (a) Formation of an impact crater (b) Meteor crater near Flagstaff, Arizona Source: *Earth: Evolution of a Habitable World*, Jonathan I. Lunine, Cambridge University Press, 1999. Reprinted with the permission of Cambridge University Press.

oid. Rock near the site would be "shock-heated," and the shock wave and molten debris thrown out of the crater would knock down trees across thousands of kilometers of land. An ocean impact would create huge waves that would submerge land for hundreds of miles around the impact site.

There is much evidence that an impact event was responsible for the sudden extinction of many life-forms at the end of the Cretaceous period. At that time, about 65 million years ago, 15 percent of shallow-water *families* of organisms became extinct, including 80 percent of all shallow-water invertebrate species. The dinosaurs, too, disappeared around this time. The dividing line between the Cretaceous and the Tertiary, called the K/T boundary (K is the symbol commonly used by geologists for the Cretaceous), is a thin layer of clay that has been identified in sediments worldwide. K/T boundary sediments contain numerous indications of a massive impact event. The clay has an abundance of platinum-group elements—iridium, osmium, gold, platinum, etc.—that is more similar to that found in meteorites than that in Earth's crust. Iridium, in particular, is more abundant than in normal crustal rocks. Other features associated with an impact that are found in this thin boundary of clay include shocked quartz grains, melt spherules (droplets formed from molten rock), graphite and other evidence of burning, and evidence of large waves such as would be caused by an ocean impact.

Solar System Exploration

Exploration of the solar system has yielded information about the origin of Earth and the solar system. Exploration of other planets will help scientists to understand how or whether these planets can benefit humans. In the future, colonization of other planets may be possible and/or necessary. Asteroids may become a valuable source of metals and mineral resources.

MULTIPLE-CHOICE QUESTIONS

In each case, write the number of the word or expression that best answers the question or completes the statement.

1. The diagram below represents a side view of the Milky Way galaxy.

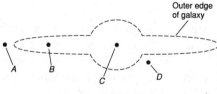

(Not drawn to scale)

At approximately which position is Earth's solar system?
(1) *A*   (3) *C*
(2) *B*   (4) *D*

The Solar System

2. According to the *Earth Science Reference Tables*, approximately how many years ago did the solar system originate?
   (1) 570,000,000
   (2) 1,000,000,000
   (3) 4,600,000,000
   (4) 10,000,000,000

3. Which celestial feature is largest in actual size?
   (1) the Moon
   (2) Jupiter
   (3) the Sun
   (4) the Milky Way

4. The Milky Way galaxy is best described as
   (1) a type of solar system
   (2) a constellation visible to everyone on Earth
   (3) a region in space between the orbits of Mars and Jupiter
   (4) a spiral-shaped formation composed of billions of stars

Base your answers to questions 5 and 6 on the diagram below, which shows an inferred sequence in which our solar system formed from a giant interstellar cloud of gas and debris. Stage *A* shows the collapse of the gas cloud, stage *B* shows its flattening, and stage *C* shows the sequence that led to the formation of planets.

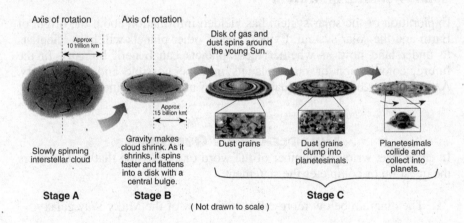

5. From stage *B* to stage *C*, the young Sun was created
   (1) when gravity caused the center of the cloud to contract
   (2) when gravity caused heavy dust particles to split apart
   (3) by outgassing from the spinning interstellar cloud
   (4) by outgassing from Earth's interior

6. After the young Sun formed, the disk of gas and dust
   (1) became spherical in shape
   (2) formed a central bulge
   (3) became larger in diameter
   (4) eventually formed into planets

7. What is the inferred age of our solar system, in millions of years?
   (1) 544
   (2) 1300
   (3) 4600
   (4) 10,000

8. Which pair of shaded circles bests represents the relative sizes of Earth and Venus when drawn to scale?

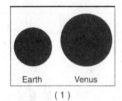
(1)
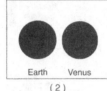
(2)

(3)

(4)

9. The diagram below shows cutaway views of the inferred interior layers of the planets Mercury and Venus.

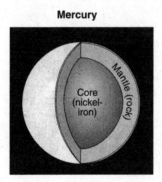
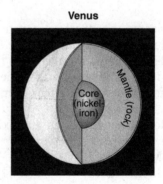

What is the reason for the development of the interior layers of these two planets?
   (1) Impact events added the mantle rock above the cores.
   (2) Heat from the Sun melted the surface rocks to form the mantles above the cores.
   (3) Gravity separated the cores and mantles due to their density differences.
   (4) Rapid heat loss caused the cores to solidify before the mantles.

The Solar System

10. Compared with the other planets in our solar system, Jupiter, Saturn, and Neptune have
    (1) shorter periods of rotation
    (2) shorter periods of revolution
    (3) greater eccentricities
    (4) greater densities

11. Which bar graph correctly shows the orbital eccentricity of the planets in our solar system?

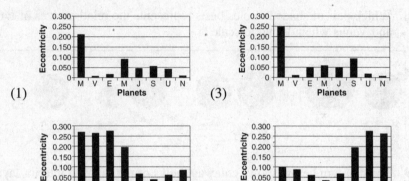

12. Compared with the Jovian planets in our solar system, Earth is
    (1) less dense and closer to the Sun
    (2) less dense and farther from the Sun
    (3) more dense and closer to the Sun
    (4) more dense and farther from the Sun

13. Which planet is located approximately ten times farther from the Sun than Earth is from the Sun?
    (1) Mars   (3) Saturn
    (2) Jupiter   (4) Uranus

14. Which object in our solar system has the greatest density?
    (1) Jupiter   (3) the Moon
    (2) Earth   (4) the Sun

Base your answer to question 15 on the diagram below. This diagram shows a portion of the solar system.

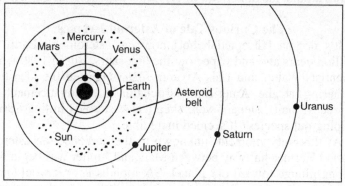

(Not drawn to scale)

15. What is the average distance, in millions of kilometers, from the Sun to the asteroid belt?
    (1) 129
    (2) 189
    (3) 503
    (4) 857

16. A person observes that a bright object streaks across the nighttime sky in a few seconds. What is this object most likely to be?
    (1) a comet
    (2) a meteor
    (3) an aurora
    (4) an orbiting satellite

17. The surface of Venus is much hotter than would be expected, considering its distance from the Sun. Which statement best explains this fact?
    (1) Venus has many active volcanoes.
    (2) Venus has a slow rate of rotation.
    (3) The clouds of Venus are highly reflective.
    (4) The atmosphere of Venus contains a high percentage of carbon dioxide.

Base your answers to questions 18 through 21 on the passage and diagram below. The diagram shows the orbits of the four inner planets and the asteroid Hermes around the Sun. Point *A* represents a position along Hermes' orbit.

The Curious Tale of Asteroid Hermes

It's dogma [accepted belief] now: an asteroid hit Earth 65 million years ago and wiped out the dinosaurs. But in 1980 when scientists Walter and Luis Alvarez first suggested the idea to a gathering at the American Association for Advancement of Sciences, their listeners were skeptical. Asteroids hitting Earth? Wiping out species? It seemed incredible.

At that very moment, unknown to the audience, an asteroid named Hermes halfway between Mars and Jupiter was beginning a long plunge toward our planet. Six months later it would pass 300,000 miles from Earth's orbit, only a little more than the distance to the Moon....

Hermes approaches Earth's orbit twice every 777 days. Usually our planet is far away when the orbit crossing happens, but in 1937, 1942, 1954, 1974 and 1986, Hermes came harrowingly [dangerously] close to Earth itself. We know about most of these encounters only because Lowell Observatory astronomer Brian Skiff rediscovered Hermes on Oct. 15, 2003. Astronomers around the world have been tracking it carefully ever since....

Excerpted from "The Curious Tale of Asteroid Hermes," Dr. Tony Phillips, Science @ NASA, November 3, 2003

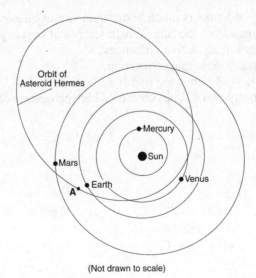

(Not drawn to scale)

18. When Hermes is located at position *A* and Earth is in the position shown in the diagram, the asteroid can be viewed from Earth at each of the following times except
    (1) sunrise
    (2) sunset
    (3) 12 noon
    (4) 12 midnight

19. How does the period of revolution of Hermes compare with the period of revolution of the planets shown in the diagram?
    (1) Hermes has a longer period of revolution than Mercury but a shorter period of revolution than Venus, Earth, and Mars.
    (2) Hermes has a shorter period of revolution than Mercury but a longer period of revolution than Venus, Earth, and Mars.
    (3) Hermes has a longer period of revolution than all of the planets shown.
    (4) Hermes has a shorter period of revolution than all of the planets shown.

20. Why is evidence of asteroids striking Earth so difficult to find?
    (1) Asteroids are made mostly of frozen water and gases and are vaporized on impact.
    (2) Asteroids are not large enough to leave impact craters.
    (3) Asteroids do not travel fast enough to create impact craters.
    (4) Weathering, erosion, and deposition on Earth have destroyed or buried most impact craters.

21. According to the diagram, as Hermes and the planets revolve around the Sun, Hermes appears to be a threat to collide with
    (1) Earth, only
    (2) Earth and Mars, only
    (3) Venus, Earth, and Mars, only
    (4) Mercury, Venus, Earth, and Mars

22. Large craters found on Earth support the hypothesis that impact events have caused
    (1) a decrease in the number of earthquakes and an increase in sea level
    (2) an increase in solar radiation and a decrease in Earth radiation
    (3) the red shift of light from distant stars and the blue shift of light from nearby stars
    (4) mass extinctions of life-forms and global climate changes

The Solar System

CONSTRUCTED RESPONSE QUESTIONS

Base your answers to questions 23 through 25 on the diagram below, which shows the relative diameter sizes of the planets compared with the radius of the Sun.

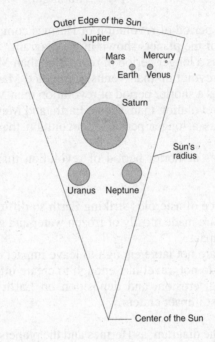

23. On the diagram above, circle only the terrestrial planets. [1]

24. On the diagram above, place an **X** on the planet with the lowest density. [1]

25. How many times larger is the diameter of the Sun than the diameter of Jupiter? [1]

Base your answers to questions 26 through 28 on the data table below, which shows the average distance from the Sun, the average surface temperature, and the average orbital velocity for each planet in our solar system.

Data Table

Planet	Average Distance from Sun (millions of km)	Average Surface Temperature (°C)	Average Orbital Velocity (km/sec)
Mercury	58	167	47.9
Venus	108	457	35.0
Earth	150	14	29.8
Mars	228	−55	24.1
Jupiter	778	−153	13.1
Saturn	1427	−185	9.7
Uranus	2869	−214	6.8
Neptune	4496	−225	5.4

26. State the relationship between the average distance from the Sun and the average surface temperature of the Jovian planets. [1]

27. Venus has an atmosphere composed mostly of carbon dioxide. Mercury has almost no atmosphere. Explain how the presence of the carbon dioxide in Venus's atmosphere causes the average surface temperature on Venus to be higher than the average surface temperature on Mercury. [1]

28. On the graph below, draw a line to indicate the general relationship between a planet's average distance from the Sun and its average orbital velocity. [1]

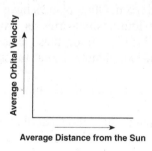

The Solar System

Base your answers to questions 29 and 30 on the data table below, which provides information about four of Jupiter's moons.

Data Table

Moons of Jupiter	Density (g/cm³)	Diameter (km)	Distance from Jupiter (km)
Io	3.5	3630	421,600
Europa	3.0	3138	670,900
Ganymede	1.9	5262	1,070,000
Callisto	1.9	4800	1,883,000

29. Identify the planet in our solar system that is closest in diameter to Callisto. [1]

30. In 1610, Galileo was the first person to observe, with the aid of a telescope, these four moons orbiting Jupiter. Explain why Galileo's observation of this motion did not support the geocentric model of our solar system. [1]

EXTENDED CONSTRUCTED RESPONSE QUESTIONS

Base your answers to questions 31 through 33 on the passage below.

Extrasolar Planets

Astronomers have discovered more than 400 planets outside of our solar system. The first extrasolar planet was detected in 1995 orbiting a star known as 51 Pegasi, which is similar in color and luminosity to our Sun. Astronomers can detect planets by identifying stars that move in response to the gravitational pull of planets revolving around them. Other planets have been discovered by finding stars whose luminosity varies as orbiting planets block outgoing starlight. Nearly all of these discovered planets are thought to be Jovian-like planets similar to Jupiter.

31. Other than Jupiter, identify *one* Jovian planet in our solar system. [1]

32. Compared with Jupiter, state how Earth's equatorial diameter and density are different. [1]

33. State the color and luminosity of 51 Pegasi. [1]

Base your answers to questions 34 through 37 on the graph below, which shows two conditions responsible for the formation and composition of some planets in our solar system. The distances of Earth and Neptune from the Sun, in astronomical units (AU), are shown beneath the horizontal axis. (1 AU = 149.6 million kilometers). The plotted line on this graph shows the relationship between a planet's distance from the Sun and the inferred temperature at its formation. The regions within the graph indicate the composition of planets formed within these zones.

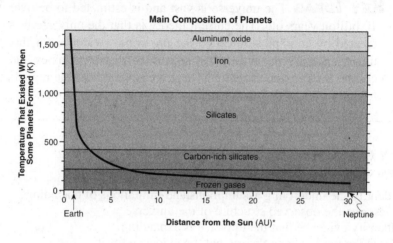

34. According to the graph, Neptune was mainly composed of which material at the time of its formation? [1]

35. Saturn is located 9.5 AU from the Sun. State the approximate temperature at which Saturn formed. [1]

36. State the relationship between a planet's distance from the Sun and the temperature at which that planet formed. [1]

37. What is Jupiter's distance from the Sun, in astronomical units? Express your answer to the *nearest tenth*. [1]

CHAPTER 9
THEORIES OF THE ORIGIN OF THE UNIVERSE

> **KEY IDEAS** The universe is vast and is estimated to be over 10 billion years old. The current theory is that the universe was created by an explosion called the *big bang*. Evidence for this theory includes the red-shifted spectra of distant galaxies and cosmic background radiation. There are several possible models for the future of the universe, depending upon its total mass.

KEY OBJECTIVES
Upon completion of this chapter, you will be able to:
- Identify the underlying assumptions and limitations of cosmology.
- Describe the observed structure of the universe.
- Identify evidence that the universe is expanding.
- Describe the big bang theory and the evidence for it.
- Compare and contrast the future of the universe according to the open and closed models of the universe.

WHAT IS THE UNIVERSE?

The universe, because of its seemingly orderly arrangement, is sometimes called the *cosmos*. **Cosmology** is the study of the nature of the universe as a whole. It is the scientific inquiry into the origin, evolution, and fate of the universe. By making assumptions that are not at odds with the observable universe, cosmologists create models, or theories, that attempt to describe the universe, its origin, and its future. The model is then used to make predictions until an observation is found that contradicts it, whereupon the model is modified or discarded in favor of a new model.

Astronomers define the universe in three different ways. The *observable* universe is all that we can see. It includes all the stars, gas clouds, galaxies, and other objects that we can detect by the radiation they emit. As new and more powerful instruments are developed to detect radiation, more objects can be perceived, and the observable universe grows. The *entire* universe is everything we can see, plus everything else that may be there. We can infer things about the entire universe only on the basis of what we know of the observable universe. We're not even sure that there is an entire universe! The *physical* universe is that part of the entire universe that can be described by the laws of physics. It extends a little beyond the observable universe because it includes objects that we cannot see, but that can be detected by their effects on objects we can see.

The Origin and Evolution of the Universe

Cosmologists assume that the laws of physics are identical throughout the universe and that the universe has the same appearance to all observers no matter where they are located. This assumption means that the universe cannot have an edge; if it did, it would appear different to an observer near the edge than to an observer near the center. The geometry of space must be such that all observers see themselves as being at the center. Cosmologists believe that the only motion that can occur in such geometry is expansion or contraction of the universe.

All this not only sounds mind-boggling; it actually is! The problem with thinking about the universe is that there are too few facts of which we can be certain. Therefore, we must make assumptions about the universe to make progress in cosmology. False assumptions, however, can lead us to nonsensical results! Consider a well-known example, Olber's paradox, shown in Figure 9.1.

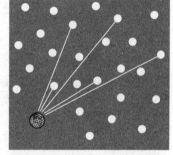

Figure 9.1 Olber's Paradox. If the universe is infinitely large, infinitely old, and uniformly filled with stars, any line of sight from Earth should intersect a star. This assumption predicts that nights should not be dark; instead, the sky should glow with starlight. Source: *Horizons: Exploring the Universe*, Michael A. Seeds, Wadsworth, 1987.

The discrepancy between what Olber's paradox theorizes and what we actually observe tells us that one of his assumptions is wrong. We now know that the light from many distant stars moving away from us is red-shifted into the infrared range, which is not visible to human eyes. Also, there are many stars that do not radiate energy in the range of visible light. If we can't see light, it can't make the sky glow. Furthermore, while the universe may be infinitely large, the observable universe is only 16–20 billion light-years in diameter. Light from distances beyond 20 billion light-years has not yet reached us.

The Structure of the Universe

The distribution of matter, such as stars, dust, and gas, throughout the universe is not uniform; there appears to be a high degree of clustering. On a very large scale, the universe seems to be made up of clusters of objects. Each of these objects is a cluster of smaller objects, which are clusters of yet smaller objects, which are clusters of even smaller ones. See Figure 9.2.

Theories of the Origin of the Universe

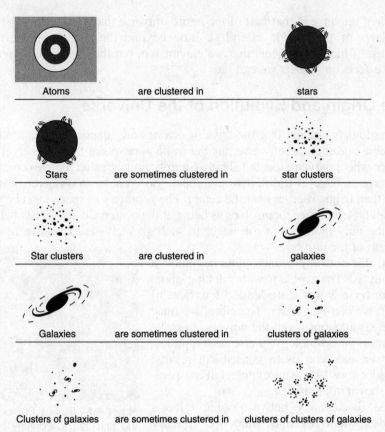

Figure 9.2 Clustering in the Universe. Source: *Concepts of the Universe*, Paul W. Hodge, McGraw-Hill, 1969.

The largest clusters detected so far are clusters of clusters of galaxies, called *superclusters*. Superclusters contain from 5 to 40 clusters of galaxies and are mind-boggling in size. An average supercluster is about 100 million light-years across; a **light-year**, you will recall, is the distance light travels in 1 year at the speed of 300,000 kilometers per second.

Clusters of galaxies are "smaller" objects—only a few million light-years across. Galaxies, which are clusters of stars, are even smaller—only about 100,000 light-years in diameter. But consider that an average galaxy contains about a million million stars, and the average star is 100 or so times the diameter of the Earth and millions of times its volume—not bad for a "small" object! Finally, each star is a cluster of atoms. And what are atoms but clusters of protons, neutrons, and electrons?

Despite the almost unimaginable number of stars, galaxies, and clouds of dust and gas that the universe contains, it is still mostly empty space. Vast distances separate the clusters of matter that can be observed.

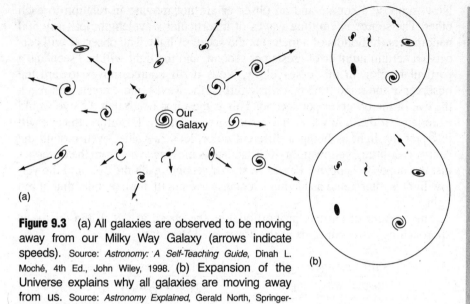

Figure 9.3 (a) All galaxies are observed to be moving away from our Milky Way Galaxy (arrows indicate speeds). Source: *Astronomy: A Self-Teaching Guide*, Dinah L. Moché, 4th Ed., John Wiley, 1998. (b) Expansion of the Universe explains why all galaxies are moving away from us. Source: *Astronomy Explained*, Gerald North, Springer-Verlag, 1997.

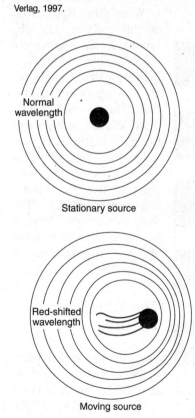

Figure 9.4 The Doppler Effect.

The Expanding Universe

Observations of objects in the universe indicate that it is expanding. Other galaxies are moving away from us, and the most distant galaxies are racing away the fastest. See Figure 9.3. The evidence of this motion away from Earth comes from the light reaching us from distant galaxies, which has a lower frequency than would be expected. The explanation for this shift lies in a phenomenon called the *Doppler effect*.

The **Doppler effect** is the shift of a spectrum line away from its normal wavelength caused by motion of the source toward or away from the observer. See Figure 9.4. If the source is approaching the observer, there is a blue-shift toward shorter wavelengths of light. If the source is moving away from the observer, there is a red-shift toward longer wavelengths of light.

Why does this occur? Light consists of electromagnetic waves. The human eye distinguishes between one wave and another by wavelength, that is, the distance between one crest of a wave and the next.

Theories of the Origin of the Universe

Now, suppose a source and an observer are not moving in relation to each other. The source is emitting waves of a particular wavelength; let's say 500 nanometers (billionths of a meter) at the speed of light. The observer will perceive a certain number of crests per second, and the light will be "seen" as a particular color, in this case yellow. Now, if the source moves toward the observer at the same time that it is emitting the waves, more crests will reach the eye of the observer per second. This is the same result that the eye would experience if the light waves had a shorter wavelength. Therefore, the eye will interpret the light as being a different color, let's say blue, even though the source is emitting yellow light. If the source is moving away from the observer, just the opposite happens. Fewer crests per second reach the eye, and the yellow light is interpreted as having a longer wavelength, for example, that of red light.

The amount of shift is proportional to the speed at which the source is approaching or receding from the observer. Therefore:

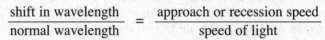

$$\frac{\text{shift in wavelength}}{\text{normal wavelength}} = \frac{\text{approach or recession speed}}{\text{speed of light}}$$

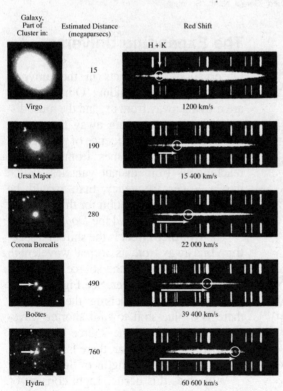

When the light from distant galaxies is observed, all the spectra are seen to be red-shifted, indicating that the galaxies are moving away from us. See Figure 9.5. Furthermore, the farther away the galaxy, the more it is red-shifted, indicating that distant galaxies are moving away from Earth faster than nearer galaxies. The only explanation for such an observation is that the universe is expanding in all directions.

The Big Bang Theory

Since the universe appears to be expanding, it must have been smaller at some time in the past. If the galaxies could be traced

Figure 9.5 Actual Photos of Red-Shifted Dark H and K Lines of Gaseous Calcium in the Spectra of Galaxies Moving Away from Earth at Different Velocities.

back in time, there would be a point at which they were all very close to one another. According to the observed expansion rate, this point occurred between 10 and 20 billion years ago. Thus, we have a model in which the universe started out with all of its matter in a small volume and then expanded outward in all directions, much like an explosion. For this reason, this model of the universe is called the **big bang theory**. See Figure 9.6.

(a) A region of the universe during the big bang

The big bang was, however, quite different from any explosion on Earth. An explosion on Earth can be observed from a distance; it comes from a specific point in space at a particular time. But the big bang did not occur at some pre-existing point in space. Instead, the entire universe, including space and time, originated in the big bang. Before it occurred, there was no place from which an observer could stand and watch the explosion, and there was no time as we know it. With the big bang, both space and time came into existence simultaneously. What we perceive as movement of galaxies away from us is due to the expansion of space, and what we perceive as the elapsing of time is the expansion of time. Thus, galaxies are not hurtling through space; the space itself is expanding and carrying the galaxies apart in the process.

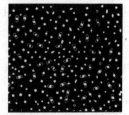

(b) A region of the universe now

After the big bang, energy and matter spread out as the universe expanded, so temperature and density dropped. The rapidly expanding matter cooled and fragmented into clouds of gas that coalesced into clusters of galaxies 10 to 20 billion years ago. When an observer looks at a distant galaxy today, she is not seeing it as it is now, but as it was long ago when the light now entering her eye first left the galaxy. The redshift in the light seen coming from distant galaxies is due to the expansion of the universe.

(c) The present universe as it appears from our galaxy

Figure 9.6 The Big Bang: An Artist's Conception of the Stages. (a) During the big bang, matter was thrown out uniformly in all directions. (b) Thereafter, gravity caused the matter to coalesce into clumps—galaxies. (c) Near us we see galaxies, farther away we see young galaxies (dots), and at a great distance we see radiation (arrows) coming from the hot clouds of the big bang explosion. Source: *Horizons: Exploring the Universe*, Michael A. Seeds, Wadsworth, 1987.

The most distant galaxies are so far away that the light reaching us today left them soon after the big bang occurred. Light from beyond these galaxies is so ancient it comes from the original hot clouds of the big bang. The radiation from the original big bang is even more red-shifted than the light from distant galaxies. In fact, it has such a large red-shift that the radiation reaching us from the original big bang arrives at Earth in the range of long-wave infrared or short microwave radio waves. When radiation is red-shifted, its wavelength makes it appear that the source is much cooler than it actually is. Thus, the radiation given off by the original big bang, when red-shifted, appears like radiation given off by an object with a temperature of 2.7 K (about 2.7° above absolute zero—very cold!). Since we are looking at the big bang from within the sphere of the expanding shock wave (see Figure 9.6c), that radiation should be coming at us from *all* directions, even from what may appear to be "empty space." Thus, space should be aglow with long-wavelength infrared and microwave radiation. Although predicted in the 1940s, this **cosmic background radiation** glow was not detected until the mid-1960s by Arno Penzias and Robert Wilson at Bell Laboratories in New Jersey. For their discovery of the big bang's primordial background radiation, they received the Nobel Prize in 1978.

The Future of the Universe

The future of the universe depends upon the total amount of matter in the universe. Gravity is a strong and pervasive force. If the density of matter in the universe is greater than a certain critical amount, gravity will be able to slow the expansion of the universe and slowly turn expansion into contraction. The problem is that measurements of the amount of matter that exists in the universe vary so much that none of the possibilities described below can be ruled out. One difficulty is that the matter we can see, that is, the matter that emits light, represents only a tiny fraction of all the matter in the universe. We can perceive matter only by the electromagnetic radiation it emits, and the amount and type of radiation emitted depends upon the temperature of the matter. The "dark" matter, which we cannot see because it doesn't emit electromagnetic radiation, will determine the fate of the universe.

The Open Universe

If the density of matter falls short of the critical amount, there won't be enough matter to exert the gravitational pull needed to stop the expansion. The universe will continue to expand and cool. Although stars will continue to form for a while, the amount of matter available for star formation will dwindle until no new stars form. Eventually, all of the mass in the universe will be contained in cold, dead planets and burned out stars. The universe will end as ever more widely scattered matter in a black void containing no light or heat.

If the universe has exactly the critical density of matter, expansion will continue, but will slow down at an ever-decreasing rate. In this case the fate of the universe will be similar to that of an ever-expanding universe, except that at some point it will stop moving entirely. See Figure 9.7.

(a) Beginning

(b) Billions of years ago

(c) Present

(d) Future

Figure 9.7 Stages of the Open Universe. Source: *Astronomy: A Self-Teaching Guide*, 4th Ed., Dinah L. Moché, John Wiley, 1998.

The Closed Universe

If the density of matter in the universe even slightly exceeds the critical density, then, as stated previously, the matter will exert enough gravitational attraction to stop the expansion and start a contraction. Ultimately, all of the matter would again collapse back into a small volume—the big crunch. And what then? One theory is that the universe is oscillating and the big crunch would give rise to another big bang. Another theory is that the big crunch would form an enormous black hole that would consume space and time, just as the big bang produced them. See Figure 9.8.

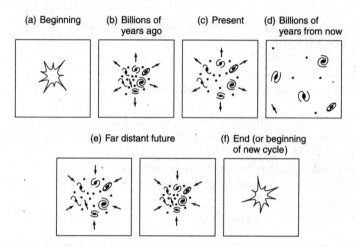

Figure 9.8 Stages of the Closed Universe. Source: *Astronomy: A Self-Teaching Guide*, 4th Ed., Dinah L. Moché, John Wiley, 1998.

Will we ever know which model is correct? Perhaps, but even then we will encounter new questions and continue to search for answers.

MULTIPLE-CHOICE QUESTIONS

In each case, write the number of the word or expression that best answers the question or completes the statement.

1. Observations of the universe indicate that the distribution of matter in the universe is
   (1) equally concentrated everywhere
   (2) concentrated in clusters of various sizes
   (3) most concentrated in the Milky Way galaxy
   (4) constantly increasing in concentration

2. The diagram below represents the bright-line spectrum for an element.

   Violet — Red

   The spectrum of the same element observed in the light from a distant star is shown below.

   Violet — Red

   The shift in the spectral lines indicates that the star is moving
   (1) toward Earth
   (2) away from Earth
   (3) in an elliptical orbit around the Sun
   (4) in a circular orbit around the Sun

3. Astronomers viewing light from distant galaxies observe a shift of spectral lines toward the red end of the visible spectrum. This shift provides evidence that
   (1) orbital velocities of stars are decreasing
   (2) Earth's atmosphere is warming
   (3) the Sun is cooling
   (4) the universe is expanding

4. In an expanding "open" universe, which of the following would be taking place?
   (1) Materials in the system would remain unchanged.
   (2) Galaxies would recede, leaving great open spaces.
   (3) Galaxies would come together, creating another big bang.
   (4) Galaxies moving away would be replaced by new material.

5. If the universe appears to be expanding, which conclusion is most logical concerning the universe?
   (1) The universe must have been smaller at some time in the past.
   (2) The universe must have been larger at some time in the past.
   (3) The density of the universe is increasing.
   (4) The temperature of the universe is increasing.

6. Background radiation detected in space is believed to be evidence that
   (1) the universe began with a primeval explosion
   (2) the universe is contracting
   (3) all matter in the universe is stationary
   (4) galaxies are evenly spaced throughout the universe

7. The big bang theory, describing the creation of the universe, is most directly supported by the
   (1) red shift of light from distant galaxies
   (2) presence of volcanoes on Earth
   (3) apparent shape of star constellations
   (4) presence of craters on Earth's Moon

Base your answers to questions 8 through 11 on the table below, which shows eight inferred stages describing the formation of the universe from its beginning to the present time.

Data Table

Stage	Description of the Universe	Average Temperature of the Universe (°C)	Time From the Beginning of Universe
1	the size of an atom	?	0 second
2	the size of a grapefruit	?	10^{-43} second
3	"hot soup" of electrons	10^{27}	10^{-32} second
4	Cooling allows protons and neutrons to form.	10^{13}	10^{-6} second
5	still too hot to allow the forming of atoms	10^{8}	3 minutes
6	Electrons combine with protons and neutrons, forming hydrogen and helium atoms. Light emission begins.	10,000	300,000 years
7	Hydrogen and helium form giant clouds (nebulae) that will become galaxies. First stars form.	−200	1 billion years
8	Galaxy clusters form and first stars die. Heavy elements are thrown into space, forming new stars and planets.	−270	13.7 billion years

8. How soon did protons and neutrons form after the beginning of the universe?
   (1) 10^{-43} second
   (2) 10^{-32} second
   (3) 10^{-6} second
   (4) 13.7 billion years

9. What is the most appropriate title for this table?
    (1) The Big Bang Theory
    (2) The Theory of Plate Tectonics
    (3) The Law of Superposition
    (4) The Laws of Planetary Motion

10. According to this table, the average temperature of the universe since stage 3 has
    (1) decreased, only
    (2) increased, only
    (3) remained the same
    (4) increased, then decreased

11. Between which two stages did our solar system form?
    (1) 1 and 3
    (2) 3 and 5
    (3) 6 and 7
    (4) 7 and 8

12. The explosion associated with the big bang theory and the formation of the universe is inferred to have occurred how many billion years ago?
    (1) less than 1
    (2) 2.5
    (3) 4.6
    (4) over 10

13. The observable universe is estimated to be roughly 16–20 billion years old. Which statement best describes why a galaxy located 25 billion light-years from Earth may not be visible to an observer on Earth?
    (1) Galaxies 25 billion light-years away would emit no visible light.
    (2) Light from beyond 20 billion light-years has not yet reached Earth.
    (3) Light from beyond 20 billion light-years passed our galaxy before Earth existed.
    (4) No galaxies are located farther than 5 billion light-years from Earth.

CONSTRUCTED RESPONSE QUESTIONS

Base your answers to questions 14 and 15 on the calendar model shown below of the inferred history of the universe and on your knowledge of Earth science. The 12-month time line begins with the big bang on January 1 and continues to the present time, which is represented by midnight on December 31. Several inferred events and the relative times of their occurrence have been placed in the appropriate locations on the time line.

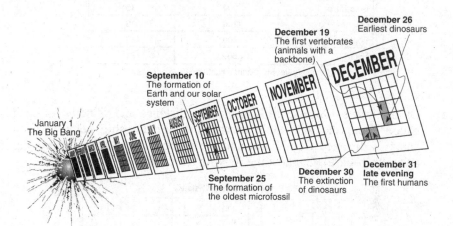

14. State one piece of evidence used by scientists to support the theory that the big bang event occurred. [1]

15. How many million years of Earth's geologic history elapsed between the event that occurred on September 10 and the event that occurred on September 25 in this model? [1]

16. State two observations successfully explained by the big bang theory. [2]

17. The critical density of the universe is the minimum average density of the matter required for the force of gravity to stop the expansion of the universe without reversing it. Infer the future of the universe for the following two conditions:
    (a) density of the universe > critical density
    (b) density of the universe < critical density

EXTENDED CONSTRUCTED RESPONSE QUESTIONS

Base your answers to questions 18 through 20 on the data table below, which shows some galaxies, their distances from Earth, and the velocities at which they are moving away from Earth.

Name of Galaxy	Distance (million light-years)	Velocity (thousand km/s)
Virgo	70	1.2
Ursa Major 1	900	15
Leo	1100	19
Bootes	2300	40
Hydra	3600	61

One light-year = distance light travels in one year

18. On the grid below, use an **X** to plot the distance and velocity for each galaxy from the data table to show the relationship between each galaxy's distance from Earth and the velocity at which it is moving away from Earth. Connect the **X**s with a smooth line. [1]

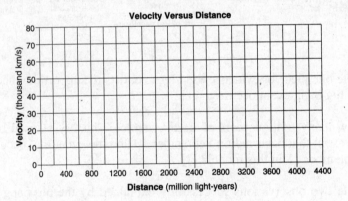

19. State the general relationship between a galaxy's distance from Earth and the velocity at which the galaxy is moving away from Earth. [1]

20. Another galaxy is traveling away from Earth at a velocity of 70 thousand kilometers per second. Estimate that galaxy's distance from Earth in millions of light-years. [1]

Base your answers to questions 21 and 22 on your knowledge of Earth science and on the newspaper article shown below, written by Paul Recer and printed in the *Times Union* on October 9, 1998.

Astronomers peer closer to big bang

WASHINGTON—The faintest and most distant objects ever sighted—galaxies of stars more than 12 billion light years away—have been detected by an infrared camera on the Hubble Space Telescope.

The sighting penetrates for the first time to within about one billion light years of the very beginning of the universe, astronomers said, and shows that even at that very early time there already were galaxies with huge families of stars.

"We are seeing farther than ever before," said Rodger I. Thompson, a University of Arizona astronomer and the principal researcher in the study.

Thompson and his team focused an infrared instrument on the Hubble on a narrow patch of the sky that had been previously photographed in visible light. The instrument detected about 100 galaxies that were not seen in the visible light and 10 of these were at extreme distance.

He said the galaxies are seen as they were when the universe was only about 5 percent of its present age. Astronomers generally believe the universe began with a massive explosion, called the "big bang," that occurred about 13 billion years ago.

Since the big bang, astronomers believe that galaxies are moving rapidly away from each other, spreading out and becoming more distant.

21. The big bang theory is widely believed by astronomers to explain the beginning of the universe. Why does the light from distant galaxies support the big bang theory? [1]

22. Compare the age of Earth and our solar system to the age of these distant galaxies of stars. [1]

Unit Three

EARTH'S HISTORY

CHAPTER 10

THE ORIGINS OF EARTH AND ITS MOON

> **KEY IDEAS** Earth formed as colliding matter accreted. A giant cloud of gas and debris contracted to form our solar system, with Earth reaching its present size about 4.5 billion years ago. The process of accretion heated the young Earth above the melting point of silicate minerals, creating a vast molten region extending downward from the surface. Within this molten region, denser materials sank toward the center and less dense materials floated to the surface, creating Earth's layered internal structure.
>
> Internal convection currents are responsible for Earth's magnetic field and the motion of crustal plates. Earth's Moon probably formed when collision of Earth with another large object caused a chunk of molten Earth material to spin off into a circular orbit.

KEY OBJECTIVES
Upon completion of this chapter, you will be able to:

- Describe the process of accretion by which Earth is thought to have formed.
- Explain how gravity caused Earth to become layered according to the densities of the materials of which it is composed.
- Describe the giant impact theory of the origin of the Moon.

THE ORIGIN OF EARTH

A Fiery Birth

Earth and the other inner, or terrestrial, planets accreted by collisions. When another body impacts a planet, its energy of motion (kinetic energy) is converted to heat. The larger and denser the planet, the stronger the gravitational force exerted on the incoming body, and the greater the energy of the impact. Therefore, of the terrestrial planets, Earth was heated the most, and Mercury the least. The temperature reached depends on the rate of impacts as compared to the rate at which heat is radiated away.

The heat produced by accretion caused Earth, Venus, and perhaps Mars to reach internal temperatures above the melting points of silicate minerals. Thus, the earliest Earth had a vast molten region, extending from the surface partway into the interior. Heat flowed from the hot surface toward the cooler center *and* outward to space. Shortly after accretion brought Earth to its present size, roughly 4.5 billion years ago, interior temperatures became high enough to partially melt the mixed solids of silicates, as well as iron. Once molten, elements that had been locked into crystalline structures were free to recombine with other, compatible elements.

The Layered Earth

Substances in the molten Earth then underwent ***differentiation***, a process by which denser materials (principally iron) sank toward the center, forming the core, while less dense materials (e.g., sodium and potassium ions) floated to the top. As a rock falling off a cliff releases gravitational potential energy as it falls, so the iron that fell toward the core released gravitational energy, creating extra heat. About 32 percent of the mass of our planet is iron, and iron is 50 percent denser than silicates, so the heat generated by formation of the core was substantial and fueled even more melting and differentiation. While this process was going on, about 4.5 billion years ago, an event that resulted in the formation of the Moon occurred, as discussed later in this chapter.

Eventually, Earth's gravity attracted most of the bodies in space surrounding it. With the surrounding space swept clean, the accretion rate fell off. In its molten state, Earth lost heat by convection. With little or no incoming energy from impacts, and removal of heat by convection, Earth cooled, forming a solid crust. The end result of differentiation and cooling was a planet with an internal structure that is layered like an onion, with the least dense materials at the surface and the densest ones at the center.

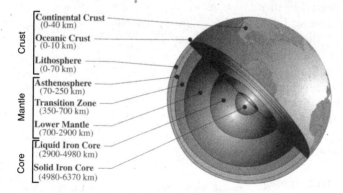

Figure 10.1 Cross Section Showing Earth's Layered Interior. Source: *Earth: Evolution of a Habitable World*, Jonathan I. Lunine, Cambridge University Press, 1999.

As you can see in Figure 10.1, there are three main zones within Earth: the crust, the mantle, and the core. The core is composed of iron, nickel, and cobalt and extends from the center of Earth out to a depth of 2,900 kilometers. It consists of two parts: a solid inner core and a fluid outer core. Over the core, out to a depth of about 70 kilometers, lies the mantle. The mantle is rather plastic in nature and is composed mainly of dark, dense silicates. It is subdivided into several layers based upon density, composition, and fluidity. Overlying the mantle is the solid crust, also subdivided by density and composition. Ranging from 0 to 70 kilometers in thickness, it is chemically diverse but contains a lot of lighter, less dense silicates than those in the mantle.

Earth also experiences radioactive heating. Among the elements present in the developing Earth were uranium, thorium, and potassium, each of which has radioactive isotopes. These elements also have large atomic radii and therefore floated to the top and became concentrated in crustal rocks, mainly granite. Decay of the radioactive isotopes in the crust and mantle generates roughly 30 trillion watts of power a year. Today, between radioactive decay and heat transferred to the mantle by the core, there is enough heating to soften mantle rock and allow convective flow to remove heat. The patterns of motion, however, are complex. Simple convection currents in the mantle are interrupted by plumes of hot material driven by heat from the core, which convects separately.

The Earth as a Magnet

The iron core was able to create a magnetic field. Magnetic fields in the solar nebula magnetized some of the rocks and iron grains that later accreted, forming Earth. Thus, some iron in the early Earth was magnetic. But iron is also a conductor of electricity. Convection in the molten core caused the electrically conductive iron to move in the presence of the magnetic field. This induced an electric current, which in turn created a stronger magnetic field. Further convection of the iron in this field induced an even larger current, thereby creating an even stronger magnetic field. Fueled by heat escaping from the core, this self-perpetuating process, called a *magnetic dynamo*, resulted in Earth's magnetic field. See Figure 10.2.

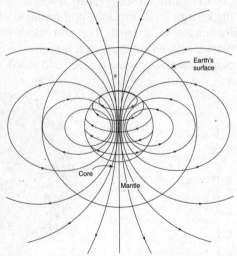

Earth's magnetic field, or *magnetosphere*, extends out into space about four Earth radii. Blasted by the solar wind's steady stream of high-energy charged

Figure 10.2 Earth's Magnetic Field. Convection motion of the molten iron core induces a magnetic field. Source: *The Earth Sciences,* Arthur N. Strahler, Harper & Row, 1971. Library of Congress Catalog Card Number 78-127335.

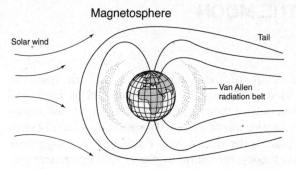

Figure 10.3 Earth's Magnetosphere and Van Allen Belts. The solar wind distorts Earth's magnetosphere. The Van Allen Belts are two doughnut-shaped regions of charged particles trapped by the magnetosphere that surround Earth. Source: *The American Heritage Dictionary of Science*, Robert K. Barnhardt, ed., Houghton Mifflin Company, 1986.

particles, the magnetosphere sticks out like a tail on the side of Earth facing away from the Sun. See Figure 10.3.

High-energy charged particles can be deadly. Penetrating organisms like a bullet, they can disrupt and destroy living tissue. Fortunately for life on Earth, moving charged particles have a magnetic field of their own, and many of the deadly particles of the solar wind are deflected by the magnetosphere. Also, some charged particles are trapped by Earth's magnetic field and move around rapidly inside two doughnut-shaped regions of the magnetosphere known as the Van Allen belts. See Figure 10.3.

High-energy charged particles moving through the atmosphere stimulate the atoms and ions of the atmosphere to radiate light, producing auroras. The Aurora Borealis, or northern lights, and Aurora Australis, or southern lights, are shimmering bands of light that shine in the night sky. Charged particles tend to be funneled down to Earth near the poles by the magnetosphere, and auroras are seen mainly in Earth's Arctic and Antarctic regions. See Figure 10.4.

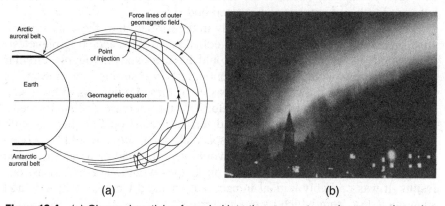

Figure 10.4 (a) Charged particles funneled into the upper atmosphere near the poles by lines of magnetic force surrounding Earth cause shining lights known as auroras. Source: *The Earth Sciences*, Arthur N. Strahler, Harper & Row, 1971. Library of Congress Catalog Card Number 78-127335. (b) An Aurora Borealis photographed during a winter night in Alaska. Source: *The Earth Sciences*, Arthur N. Strahler, Harper & Row, 1971. Library of Congress Catalog Card Number 78-127335.

THE ORIGIN OF THE MOON

Early Ideas

Until the mid-1980s, the origin of the Moon was a difficult issue for astronomers because the Moon is unusually large compared to Earth and moves in a circular orbit. A variety of theories, ranging from capture of a passing body by Earth's gravity, to ejection of a chunk of matter by a rapidly spinning Earth, to formation in place at the same time as Earth, had been suggested. Attempts to model the third theory, formation in place, with computers presents problems. The Moon's density is consistent with little or no iron content. Capture of the Moon, while possible, is unlikely, and capture into a circular orbit is even more unlikely. Finally, there are problems with the physical plausibility of a molten Earth spinning rapidly enough to throw off the Moon.

Apollo's Clues and a New Theory

Rock and dust samples brought back from the Moon by the Apollo missions in the 1980s virtually eliminated all of the early theories. Although the Moon's small size limits possible geologic activity, Moon rocks are more similar to Earth's mantle than to meteorites like those that would have accreted to form the Moon. Moon rocks are also richer in silicates and poorer in metals and volatile elements than Earth rocks. Simplistically put, Moon rocks could be made by heating rocks from Earth's mantle until they vaporize, and then recondensing the less volatile materials.

This geochemical mystery has prompted planetary scientists to consider a new theory—that the Moon is the result of a *giant impact*, a huge collision between Earth and another planet-sized body.

How likely is such an impact? Early in the formation of the solar system, conditions were right for this kind of collision. At first, accreting planetesimals would have been small and would have moved in roughly circular orbits. Collisions would have been gentle, largely resulting in accretion. As planets grew larger, however, close passes of these bodies would have sent them into elliptical orbits, which, in turn, would increase collision speeds. By the time the terrestrial planets had formed, catastrophic collisions with solar-system debris traveling in high-speed elliptical orbits would have been likely. Small bodies hitting big ones would vaporize and melt, adding to the larger body. Big bodies hitting other big bodies would have had disastrous results. It was probably a giant impact that knocked Uranus on its side and spun out a disk from which its moons formed.

The Giant Impact

Computer simulations show that, if a planet as big as Mars, or slightly bigger, struck Earth, a big chunk of Earth's mantle, but very little iron core, would have been spun off. Also, part of the material ejected would have entered a circular orbit, forming a ring around Earth. See Figure 10.5.

The rest of the ejected material would have reaccreted into Earth (computer simulations of the event indicate that the iron core of the planetesimal would have merged with Earth's in a matter of hours) or been lost into orbit around the Sun. The material ejected would have left a sizable hole, but the spinning of the still malleable Earth would quickly have restored the planet's normal shape.

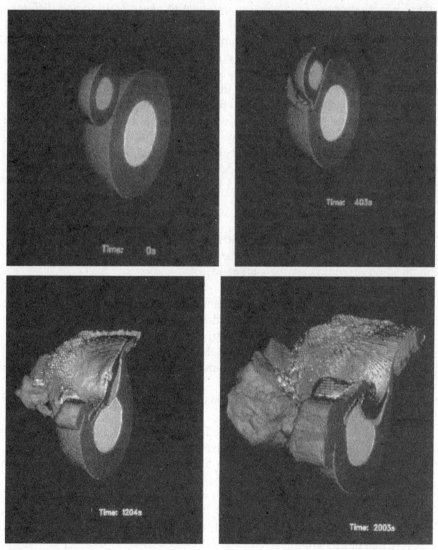

Figure 10.5 Computer Simulation of the Formation of the Moon by a Giant Impact.
Source: *Earth: Evolution of a Habitable World*, Jonathan I. Lunine, Cambridge University Press, 1999.

The Origins of Earth and Its Moon

Most of the material ejected into space would have been vaporized, with only the least volatile materials remaining solid. Much of the volatile material (water and less dense ions) would be lost. This hypothesis is consistent with the lower density minerals with which rocks from the lunar highlands are enriched. Since little of Earth's core was ejected, little iron would be present. Material in the ring then collected to form the Moon.

For the first billion years of its life, the newly formed Moon was bombarded by meteorites. They created great craters and melted the surface that is now the lunar crust. About a billion years into the Moon's life, accretion and radioactive decay produced enough heat to cause melting and differentiation in the upper 500 kilometers or so. Volcanoes erupted, pouring out huge floods of basalt, which now form the maria. By about 3 billion years ago, the Moon had cooled enough that volcanic activity ceased. With the exception of a few major impacts and small lava flows, the Moon has remained virtually unchanged since then.

When did this sequence of events take place? The oldest rocks recovered from the Moon date to 4.4–4.5 billion years ago, so it certainly didn't occur any later. This age also sets a time by which Earth's core must have formed. Formation of the core had to occur before the impact that formed the Moon, because the Moon is poor in iron. Most likely, the Moon formed early in Earth's history, close to or before 4.5 billion years ago.

MULTIPLE-CHOICE QUESTIONS

In each case, write the number of the word or expression that best answers the question or completes the statement.

1. Earth formed by accretion as a giant cloud of gas and debris contracted to form our solar system. During the process of accretion, the energy of motion of bodies that collided with Earth was converted into
   (1) magnetic force
   (2) heat energy
   (3) electrical force
   (4) matter

2. Of the terrestrial planets, Earth was heated the most by collisions during accretion because
   (1) collisions with Earth produced greater impact energy because of Earth's large mass and high density
   (2) Earth was closest to the Sun and received the most intense solar radiation
   (3) bodies that collided with Earth were hotter than bodies that collided with other planets
   (4) bodies that collided with Earth were heated by friction with Earth's magnetic field

3. Within the early Earth's vast molten region, substances underwent a process known as differentiation, during which
   (1) substances of low density rise to Earth's surface, while those of high density sink toward its center
   (2) substances of high density float to Earth's surface, while those of low density sink toward its center
   (3) substances of high and low density chemically combine to form uniformly dense substances
   (4) substances of high density form gases, while those of low density form solids

4. Compared to Earth's crust, Earth's core is believed to be
   (1) less dense, cooler, and composed of more iron
   (2) less dense, hotter, and composed of less iron
   (3) more dense, hotter, and composed of more iron
   (4) more dense, cooler, and composed of less iron

5. The density of Earth's crust is
   (1) less than the density of the outer core but greater than the density of the mantle
   (2) greater than the density of the outer core but less than the density of the mantle
   (3) less than the density of both the outer core and the mantle
   (4) greater than the density of both the outer core and the mantle

6. Earth's magnetic field is likely a result of
   (1) convection currents in Earth's mantle
   (2) convection currents in Earth's core
   (3) a high concentration of iron in Earth's crust
   (4) high-energy particles in the solar wind

7. Much of Earth's internal heat is produced by
   (1) decay of radioactive elements
   (2) fusion of hydrogen into helium
   (3) friction between Earth's surface and atmosphere
   (4) energy absorbed from sunlight

8. Analysis of rock and dust samples brought back from the Moon by the Apollo missions indicates that Moon rocks
   (1) have a composition more similar to that of Earth's mantle than to meteorites
   (2) contain elements never before identified on Earth
   (3) have a composition very different from that of Earth rocks
   (4) have a high concentration of iron, similar to that of Earth's core

The Origins of Earth and Its Moon

9. Which statement about the age of Earth rocks and the age of Moon rocks is true?
   (1) Moon rocks are much younger than Earth rocks.
   (2) Moon rocks are much older than Earth rocks.
   (3) Moon rocks are about the same age as Earth rocks.
   (4) Moon rocks cannot be dated because they contain no radioactive elements.

10. Which of the following models of the Moon's origin is currently considered most likely?
    (1) The Moon was spun off by a rapidly spinning, molten Earth.
    (2) The Moon was a passing body that was captured by Earth's gravity.
    (3) The Moon formed in place at the same time that Earth formed.
    (4) The Moon was ejected from a molten Earth by a giant impact.

11. Measurements of radioactivity in the rocks of Earth's crust currently place the age of Earth closest to
    (1) 100,000 years
    (2) 1.0 million years
    (3) 4.6 billion years
    (4) 16 billion years

12. Displays of auroras (northern lights) are believed to be caused by
    (1) radiation of infrared rays from Earth
    (2) reflection of the midnight Sun on northern snowfields
    (3) gases glowing under the impact of electrically charged particles
    (4) radioactive dust in the upper atmosphere

13. Craters and layers of basalt on the Moon's surface probably formed as a result of
    (1) meteorite impacts and the heating of surface rock due to accretion
    (2) recent volcanic eruptions
    (3) wrinkling of the Moon's molten surface as it cooled and contracted
    (4) bubbles of gas rising through the molten rock just beneath the Moon's surface

Base your answers to questions 14 and 15 on the data table below. The data table provides information about the Moon, based on current scientific theories.

Information About the Moon

Subject	Current Scientific Theories
Origin of the Moon	Formed from material thrown from a still-liquid Earth following the impact of a giant object 4.5 billion years ago
Craters	Largest craters resulted from an intense bombardment by rock objects around 3.9 billion years ago
Presence of water	Mostly dry, but water brought in by the impact of comets may be trapped in very cold places at the poles
Age of rocks in terrae highlands	Most are older than 4.1 billion years; highland anorthosites (igneous rocks composed almost totally of feldspar) are dated at 4.4 billion years
Age of rocks in maria plains	Varies widely from 2 billion to 4.3 billion years
Composition of terrae highlands	Wide variety of rock types, but all contain more aluminum than rocks of maria plains
Composition of maria plains	Wide variety of basalts
Composition of mantle	Varying amounts of mostly olivine and pyroxene

14. Which statement is supported by the information in the table?
    (1) The Moon was once a comet.
    (2) The Moon once had saltwater oceans.
    (3) Earth is 4.5 billion years older than the Moon.
    (4) Earth was liquid rock when the Moon was formed.

15. Which Moon feature is an impact structure?
    (1) crater   (3) terrae highland
    (2) maria plain   (4) mantle

16. In which parts of Earth's interior would melted or partially melted material be found?
    (1) stiffer mantle and inner core
    (2) stiffer mantle and outer core
    (3) crust and inner core
    (4) asthenosphere and outer core

CONSTRUCTED RESPONSE QUESTIONS

17. Explain the role of gravity in the process of differentiation that created Earth's layered internal structure. [2]

18. State one reason why, in light of current knowledge, it is unlikely that the Moon is a captured passing body. [2]

19. Construct a labeled diagram of Earth's magnetic field. [2]

EXTENDED CONSTRUCTED RESPONSE QUESTIONS

20. Base your answers to parts (a) through (c) on the *Earth Science Reference Tables* and the diagram below, which shows Earth's internal structure and its three main layers.

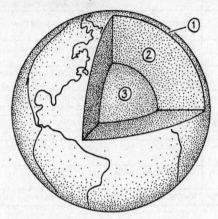

(a) Identify the three main layers of Earth's interior. [3]
(b) State the thickness of each of the three main layers of Earth's interior. [3]
(c) Describe the process by which Earth's layered internal structure is thought to have been formed. [2]

21. The photographs below show the Moon and Earth as viewed from space. It is inferred that Earth had many impact craters similar to those shown on the Moon.

(Not drawn to scale)

The Moon has many more impact craters visible on its surface than Earth has on its surface. State *two* reasons that Earth has so few visible impact craters. [2]

CHAPTER 11

THE ORIGIN AND STRUCTURE OF EARTH'S ATMOSPHERE

> **KEY IDEAS** Earth is surrounded by an atmosphere that extends several hundred kilometers out into space. Earth's early atmosphere formed as a result of the outgassing of water vapor, carbon dioxide, nitrogen, and other gases, in lesser amounts, from its interior. The evolution of life caused dramatic changes in the composition of Earth's atmosphere. Free oxygen did not form in the atmosphere until oxygen-producing organisms evolved. Earth's current atmosphere has a layered structure, with each layer differing in composition and temperature from other layers.

KEY OBJECTIVES
Upon completion of this chapter, you will be able to:

- Describe the process of outgassing that formed Earth's early atmosphere.
- Explain how the evolution of life led to the development of Earth's current oxygen-rich atmosphere.
- Compare and contrast Earth's early reducing atmosphere and current oxidizing atmosphere.
- Describe the layered structure of Earth's current atmosphere.

THE ORIGIN OF EARTH'S ATMOSPHERE

Earth's Lost Atmosphere

In an early stage, while still accreting and forming a core, Earth's atmosphere was a cloud of silicate vapor. Then, as accretion and core formation stopped, Earth cooled and the silicate vapor condensed, forming molten and solid rock. Any hydrogen and other gases left from the solar nebula would quickly have been swept away by the vigorous solar wind, leaving primordial Earth with no atmosphere.

Formation of Earth's Present Atmosphere

Outgassing

Where, then, did the gases of our present atmosphere originate? One likely source was **outgassing**, the release, from the interior of the hot, young Earth, of water and trace gases trapped in rocks. (Recall that the original matter from which Earth accreted was mostly bits of rock and water ice.) Compounds of hydrogen, carbon, oxygen, and nitrogen, such as carbon dioxide, ammonia, methane, and a large amount of water, worked their way to the surface of the molten Earth. Even after Earth had cooled and formed a solid crust, outgassing during volcanic eruptions would have continued. In addition to ash and lava, gases such as hydrogen sulfide, carbon dioxide, and water vapor are still being released during volcanic eruptions today. Table 11.1 compares, in terms of composition, the gases found in Earth's atmosphere and hydrosphere with those given off during volcanic eruptions and emitted by other sources such as hot springs, geysers, and fumaroles.

TABLE 11.1 COMPARISON OF THE COMPOSITIONS OF GASES FROM INTERNAL EARTH SOURCES WITH THOSE OF EARTH'S ATMOSPHERE AND HYDROSPHERE*

Gas	Some Gases of Earth's Hydrosphere and Atmosphere	Gases in Hot Springs, Fumaroles, and Geysers	Volcanic gases from Basaltic Lava of Mauna Loa and Kilauea
Water vapor, H_2O	92.8	99.4	57.8
Total carbon, as CO_2	5.1	0.33	23.5
Sulfur, S_2	0.13	0.03	12.6
Nitrogen, N_2	0.24	0.05	5.7
Argon, A	Trace	Trace	0.3
Chlorine, Cl_2	1.7	0.12	0.1
Fluorine, F_2	Trace	0.03	—
Hydrogen, H_2	0.07	0.05	0.04

*Data from W. W. Rubey (1952), *Geol. Soc. Amer., Bull.*, vol. 62, p. 1137. Figures in table represent percentages by weight.

Cometary Impacts

Another likely source of gases was comets and other bodies from farther out in the forming solar system. The outer solar system is cold enough to condense many volatile materials, and comets are rich in ices of water, carbon dioxide, carbon monoxide, ammonia, and organic compounds. Today, hundreds of Earth-masses of comets orbit the Sun. According to computer models, there were even more comets in the early solar system and they commonly intersected the orbits of Venus, Earth, and Mars. Collisions of comets with planets would have released the comets' icy gases into the atmospheres of the planets they struck. Hundreds of millions of years of cometary collisions could have added large amounts of material to Earth's atmosphere. As stated earlier, in addition to water, comets would have provided carbon dioxide, carbon monoxide, methane, ammonia, nitrogen, and other gases. Carbon dioxide was also released from rocks in Earth's mantle, so this gas probably dominated Earth's early atmosphere after water condensed out.

Virtually no molecular oxygen is found in comets, very little exists on Venus and Mars, and it was absent from Earth's early atmosphere. Evidence of the absence of oxygen comes from ancient rocks containing minerals that would have been unstable in an oxygen atmosphere.

Development of Earth's Atmosphere

There were a number of stages in the development of Earth's present, oxygen-rich atmosphere. Once liquid water had condensed on Earth's surface, the atmosphere began to change.

A Reducing Atmosphere

Carbon dioxide is very soluble in water, and was probably removed from the atmosphere by dissolving in the ocean and then combining with a substance such as calcium to form limestone. Oxygen began to be formed photochemically; water absorbs ultraviolet radiation from the Sun, and the water molecule (H_2O) breaks down into hydrogen gas and oxygen gas. The light hydrogen can escape from the atmosphere, so it does not recombine. The remaining oxygen forms molecular oxygen (O_2) and ozone (O_3). However, the weathering of minerals and reactions with volcanic gases bind the oxygen almost as quickly as it is formed, so that this element remained very scarce, less than one hundred-millionth of present atmospheric levels. Oxygen levels this low formed a **reducing atmosphere**, that is, an atmosphere in which elements and compounds that would combine with oxygen in today's atmosphere remained stable (e.g., iron would not form rust).

An Oxidizing Atmosphere

Once photosynthetic life (such as chlorophyll-containing life-forms) appeared and spread around the planet, however, there was a new source of oxygen. See Figure 11.1. As these organisms multiplied over time, photosynthesis steadily increased. Geological evidence suggests that between 2.2 and 2.0 billion years ago the oxygen level in the atmosphere skyrocketed to roughly one-hundredth of its present level—more than a millionfold increase. Oxygen levels grew high enough to form an **oxidizing atmosphere**, one in which oxidation of elements and compounds occurred but that was still too low to sustain aerobic respiration (the oxidation of food to produce energy). However, at this concentration, enough molecular oxygen, which strays into the upper atmosphere and is chemically changed into ozone by ultraviolet rays, is formed to sustain an ozone shield that blocks incoming ultraviolet rays.

The Origin and Structure of Earth's Atmosphere

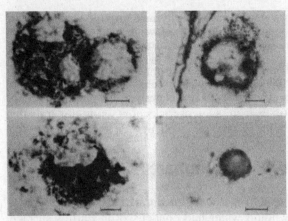

Figure 11.1 Possible Algaelike Fossils Found in Rocks More than 3.1 Billion Years Old. Source: *The Earth Sciences*, Arthur N. Strahler, Harper & Row, 1971.

As the oxygen level of the atmosphere continued to increase, and ozone shielded organisms from damaging ultraviolet rays, photosynthetic organisms proliferated. But these primitive photosynthesizers were anaerobic; their body metabolism did not use oxygen. In fact, anaerobic organisms cannot survive if oxygen levels rise too high. Oxygen is a waste product of anaerobic photosynthesizers, just as urea is a waste product of humans. Most organisms can tolerate waste products in moderate amounts, but high levels are toxic (which is why humans with damaged kidneys undergo dialysis to remove urea from their blood). At first this limited oxygen levels; as a population of anaerobic photosynthesizers would increase, oxygen levels would rise to toxic levels. This, in turn, would cause their population (and oxygen production) to decrease until oxygen levels were once again tolerable. By about 1.7 billion years ago, however, organisms had developed higher and higher oxygen tolerances, so production of this element surged again.

An Aerobic Atmosphere

Eventually, Earth formed an **aerobic atmosphere**, one with sufficient oxygen to support life-forms that metabolize oxygen. Over time, aerobic organisms (life-forms that use oxygen for body metabolism) evolved to take advantage of this oxygen-rich atmosphere. The number and vigor of photosynthetic life-forms (e.g., algae, green plants) increased. Photosynthesis has increased along with them, but is almost balanced by the consumption of oxygen by these new organisms during respiration. Photochemical oxygen production, as well as loss by weathering and volcanic activity, is now insignificant compared with photosynthesis and respiration. In this new balance, oxygen production slightly exceeds loss, and within the past billion years atmospheric oxygen has steadily risen to its current level. See Figure 11.2.

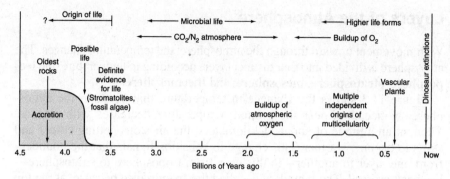

Figure 11.2 The Geologic and Climatological History of the Development of Earth's Atmosphere. Source: McKay, C.P., and Stoker, C.P. "The early environment and its evolution on Mars," in *Review of Geophysics* 27, 189–214, 1989, copyright © 1989 by the American Geophysical Union.

THE STRUCTURE OF EARTH'S ATMOSPHERE

The Composition of Air

The atmosphere, which now has a total mass of about 5,000 trillion tons, is held in place by Earth's gravity and extends several hundred kilometers into space. The **atmosphere** consists mostly of gases but also contains water, ice, dust, and other particles. This mixture of gases and other substances is called **air**.

Dry air is composed mainly of nitrogen and oxygen, with traces of other gases as well. On average, air today contains about 78 percent nitrogen and 21 percent oxygen; the remaining 1 percent consists mostly of argon, with traces of carbon dioxide and other gases. The water vapor content of air may range from 0 percent over deserts and polar ice caps to as much as 4 percent in a tropical jungle. See Table 11.2. Air also contains varying amounts of water vapor, dust, and chemicals released by industry and microorganisms.

TABLE 11.2 THE COMPOSITION OF AIR IN THE TROPOSPHERE

Gas	Percent by Volume in Dry Air
Nitrogen	78
Oxygen	21
Argon	Almost 1
Neon	Trace
Helium	Trace
Krypton	Trace
Xenon	Trace
Hydrogen	Trace
Ozone	Trace
Carbon dioxide	Trace
Nitrous oxide	Trace
Methane	Trace

Layers of the Atmosphere

With movement upward through the atmosphere, the temperature changes. The atmosphere is divided into four distinct layers according to temperature: the **troposphere**, **stratosphere**, **mesosphere**, and **thermosphere**.

Figure 11.3 shows the changes in temperature that occur in the atmosphere. Notice that, near the ground, temperature decreases with altitude. Then, at an altitude of about 11 kilometers, the air stops getting colder and starts to become warmer. This change in behavior signals that the boundary from one layer to another—in this case from troposphere to stratosphere—has been crossed. The boundary is called the **tropopause** because, at the top of the troposphere, the temperature stops decreasing, or pauses. You can see from Figure 11.3 that such a change happens twice more, at the stratopause and the mesopause.

These changes in temperature are the result of differences in the compositions of the different atmospheric layers. The upper stratosphere contains the gas ozone. Ozone absorbs ultraviolet light from the Sun, thereby warming the stratosphere. Once past the ozone, the temperature decreases in the mesosphere. In the thermosphere, gases and charged particles absorb ultraviolet light and other wavelengths of energy and turn them into heat energy. Once again, temperatures rise. In fact, the amount of energy absorbed by this outermost layer is so great that temperatures exceed the boiling point of water. Some gas molecules absorb so much energy they lose or gain electrons and become ions, that is, charged particles. These ions in the thermosphere form a layer, the ionosphere, that can reflect radio waves.

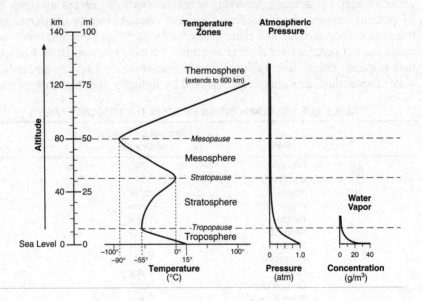

Figure 11.3 Selected Properties of Earth's Atmosphere. Source: The State Education Department, *Earth Science Reference Tables*, 2011. (Albany, New York; The University of the State of New York).

As you can see from the atmospheric pressure and water vapor sections of Figure 11.3, most of the air in the atmosphere and all of the water vapor is confined to the troposphere. Air exerts pressure in proportion to the amount of air that is present in the atmosphere. At the tropopause air exerts only one-tenth of the pressure that it does at sea level. This means that there is only about one-tenth as much air! At the stratopause there is only about one-thousandth of the air that is present at sea level, and by the mesopause less than one ten-thousandth the air. The atmosphere extends for hundreds of kilometers above the lithosphere and hydrosphere, but most of the air is confined to the 8–10 kilometers closest to their surface.

MULTIPLE-CHOICE QUESTIONS

In each case, write the number of the word or expression that best answers the question or completes the statement.

1. Scientists have inferred that Earth's original atmosphere was formed by the
   (1) outgassing from Earth's interior
   (2) erosion of Earth's surface
   (3) decay of microorganisms in Earth's oceans
   (4) radioactive decay of elements in Earth's core

2. Earth's early atmosphere formed during the Early Archean Era. Which gas was generally absent from the atmosphere at that time?
   (1) water vapor   (3) nitrogen
   (2) carbon dioxide   (4) oxygen

3. The diagram below shows four different chemical materials escaping from the interior of early Earth.

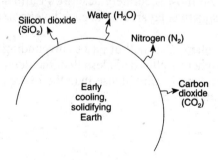

Which material contributed *least* to the early composition of the atmosphere?
   (1) SiO_2 (3) N_2
   (2) H_2O (4) CO_2

The Origin and Structure of Earth's Atmosphere

4. Most scientists believe Earth's Early Archean atmosphere was formed primarily by gases released from
   (1) stream erosion
   (2) chemical weathering
   (3) volcanic eruptions
   (4) plant transpiration

5. The gases in Earth's early atmosphere are inferred to have come primarily from
   (1) meteor showers
   (2) melting of glacial ice
   (3) volcanic eruptions
   (4) evaporation of seawater

6. Which event in Earth's history was dependent on the development of a certain type of life-form?
   (1) addition of free oxygen to Earth's atmosphere
   (2) formation of clastic sedimentary rocks
   (3) movement of tectonic plates
   (4) filling of the oceans by precipitation

7. Earth's atmosphere consists
   (1) mostly of oxygen, argon, carbon dioxide, and water vapor
   (2) entirely of ozone, nitrogen, and water vapor
   (3) of gases that cannot be compressed
   (4) of a mixture of gases, liquid droplets, and minute solid particles

8. The greatest mass of Earth's atmosphere is found in the
   (1) ionosphere
   (2) exosphere
   (3) troposphere
   (4) stratosphere

9. The atmosphere is bound to Earth by
   (1) magnetic fields
   (2) atmospheric pressure
   (3) the force of gravity
   (4) chemical bonding at the interface

10. Which statement most accurately describes Earth's atmosphere?
    (1) The atmosphere is layered, with each layer possessing distinct characteristics.
    (2) The atmosphere is a shell of gases surrounding most of Earth.
    (3) The atmosphere's altitude is less than the depth of the ocean.
    (4) The atmosphere is more dense than the hydrosphere but less dense than the lithosphere.

11. Which graph best shows the general relationship between altitude and temperature in the troposphere?

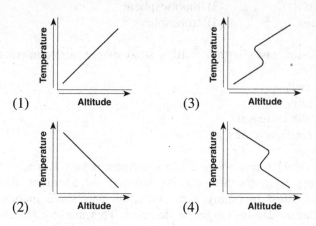

12. Which pie graph correctly shows the percentage of elements by volume in Earth's troposphere?

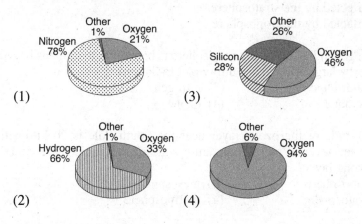

13. In which two temperature zones of the atmosphere does the temperature increase with increasing altitude?
    (1) troposphere and stratosphere
    (2) troposphere and mesosphere
    (3) stratosphere and thermosphere
    (4) mesosphere and thermosphere

14. If the base of a cloud is located at an altitude of 2 kilometers and the top of the cloud is located at an altitude of 8 kilometers, this cloud is located in the
    (1) troposphere, only
    (2) stratosphere, only
    (3) troposphere and stratosphere
    (4) stratosphere and mesosphere

15. At which position would the surrounding air temperature most likely be warmest?
    (1) mesosphere
    (2) stratosphere
    (3) thermosphere
    (4) troposphere

16. As the altitude increases within Earth's stratosphere, air temperature generally
    (1) decreases, only
    (2) increases, only
    (3) decreases, the increases
    (4) increases, then decreases

17. At an altitude of 95 miles above Earth's surface, nearly 100% of the incoming energy from the Sun can be detected. At 55 miles above Earth's surface, most incoming x-ray radiation and some incoming ultraviolet radiation can no longer be detected. This missing radiation was most likely
    (1) absorbed in the thermosphere
    (2) absorbed in the mesosphere
    (3) reflected by the stratosphere
    (4) reflected by the troposphere

18. Which gas in Earth's upper atmosphere is beneficial to humans because it absorbs large amounts of ultraviolet radiation?
    (1) water vapor
    (2) methane
    (3) nitrogen
    (4) ozone

19. The altitude of the ozone layer near the South Pole is 20 kilometers above sea level. Which temperature zone of the atmosphere contains this ozone layer?
    (1) troposphere
    (2) stratosphere
    (3) mesosphere
    (4) thermosphere

20. Scientists are concerned about the decrease in ozone in the upper atmosphere primarily because ozone protects life on Earth by absorbing certain wavelengths of
    (1) x-ray radiation
    (2) ultraviolet radiation
    (3) infrared radiation
    (4) microwave radiation

CONSTRUCTED RESPONSE QUESTIONS

Base your answers to questions 21 through 24 on the passage below.

Earth's Early Atmosphere

Early in Earth's history, the molten outer layers of Earth released gases to form an early atmosphere. Cooling and solidification of that molten surface formed the early lithosphere approximately 4.4 billion years ago. Around 3.3 billion years ago, photosynthetic organisms appeared on Earth and removed large amounts of carbon dioxide from the atmosphere, which allowed Earth to cool even faster. In addition, they introduced oxygen into Earth's atmosphere, as a by-product of photosynthesis. Much of the first oxygen that was produced reacted with natural Earth elements in the lithosphere, such as iron, and produced new varieties of rocks and minerals. Eventually, photosynthetic organisms produced enough oxygen so that it began to accumulate in Earth's atmosphere. About 450 million years ago, enough oxygen was in the atmosphere to allow for the development of an ozone layer 30 to 50 kilometers above Earth's surface. This layer was thick enough to protect organisms developing on land from the ultraviolet radiation from the Sun.

21. State *one* reason why the first rocks on Earth were most likely igneous in origin. [1]

22. Identify *one* mineral with a red-brown streak that formed when oxygen in Earth's early atmosphere combined with iron. [1]

23. Identify the temperature zone of the atmosphere in which the ozone layer developed. [1]

The Origin and Structure of Earth's Atmosphere

24. Complete the pie graph below to show the percent by volume of nitrogen and oxygen gases currently found in Earth's troposphere. Label each section of the graph with the name of the gas. The percentage of other gases is shown. [1]

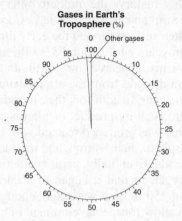

EXTENDED CONSTRUCTED RESPONSE QUESTIONS

Base your answers to questions 25 through 27 on the passage below and on your knowledge of Earth science.

Ozone in Earth's Atmosphere

Ozone is a special form of oxygen. Unlike the oxygen we breathe, which is composed of two atoms of oxygen, ozone is composed of three atoms of oxygen. A concentrated ozone layer between 10 and 30 miles above Earth's surface absorbs some of the harmful ultraviolet radiation coming from the Sun. The amount of ultraviolet light reaching Earth's surface is directly related to the angle of incoming solar radiation. The greater the Sun's angle of insolation, the greater the amount of ultraviolet light that reaches Earth's surface. If the ozone layer were completely destroyed, the ultraviolet light reaching Earth's surface would most likely increase human health problems, such as skin cancer and eye damage.

25. State the name of the temperature zone of Earth's atmosphere where the concentrated layer of ozone gas exists. [1]

26. Explain how the concentrated ozone layer above Earth's surface is beneficial to humans. [1]

27. Assuming clear atmospheric conditions, on what day of the year do people in New York State most likely receive the most ultraviolet radiation from the Sun? [1]

Base your answers to questions 28 and 29 on the table below and on your knowledge of Earth science. The table shows air temperatures and air pressures recorded by a weather balloon rising over Buffalo, New York.

Altitude Above Sea Level (m)	Air Temperature (°C)	Air Pressure (mb)
300	16.0	973
600	16.5	937
900	15.5	904
1,200	13.0	871
1,500	12.0	842
1,800	10.0	809
2,100	7.5	778
2,400	5.0	750
2,700	2.5	721

28. On the grid provided below, construct a graph of altitude above sea level and air temperature by following the directions below.
    (a) Plot an **X** for the air temperature recorded at *each* altitude shown on the table. [1]
    (b) Connect the **X**s with a solid line. [1]

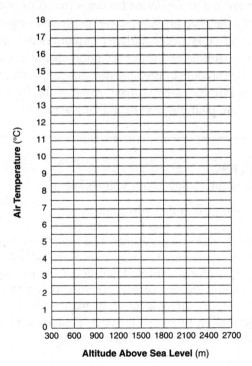

29. State the relationship shown in the table between altitude above sea level and air pressure recorded by the rising weather balloon. [1]

CHAPTER 12

THE ORIGIN AND NATURE OF EARTH'S HYDROSPHERE

KEY IDEAS Approximately 70 percent of Earth's surface is covered by a relatively thin layer of water. Earth's oceans formed as a result of precipitation over millions of years. The presence of early oceans is indicated by sedimentary rocks of marine origin, dating back about 4 billion years.

Earth has continuously been recycling water ever since the outgassing of water early in the planet's history. This constant recirculation of water at and near Earth's surface is described by the water (hydrologic) cycle. Water leaves Earth's surface and enters the atmosphere by evaporation or transpiration from plants. Water is returned from the atmosphere to Earth's surface by precipitation.

A portion of the precipitation becomes runoff over the land or infiltrates into the ground and is stored in the soil or groundwater below the water table. Soil capillarity influences these processes. Porosity, permeability, and water retention also affect runoff and infiltration. The amount of precipitation that seeps into the ground or runs off is influenced by climate, slope of the land, soil, rock type, vegetation, land use, and degree of saturation.

KEY OBJECTIVES
Upon completion of this chapter, you will be able to:

- Describe the origin of Earth's hydrosphere.
- Identify the forms in which water exists on Earth.
- Analyze the water cycle, and describe the transfer of water and energy into and out of the atmosphere at each step of the cycle.
- Explain why fresh water is a limited but renewable resource.
- Give examples of how factors such as porosity, permeability, water retention, and soil capillarity affect infiltration and runoff.
- Describe the formation of the water table and some methods by which groundwater is obtained for use at Earth's surface.

THE ORIGIN OF THE HYDROSPHERE

Earth is often referred to as "the water planet." Most people know that the oceans are full of water, and think water is commonplace because there is so much of it around. On a cosmic scale, however, water is not so common. Earth is the *only* known planet that has liquid water on its surface! This has resulted from a unique combination of Earth's temperature and the unusual properties of water. Water not only covers Earth's surface, but also makes life as we know it possible.

Let us now consider the origin and nature of Earth's oceans and the properties of water that make it such an unusual substance.

The Origin of Earth's Water

The origin of Earth's water is directly linked to the formation of Earth. Meteorites called Type I carbonaceous chondrites provide a clue to the probable origin of Earth's water. They contain 5 percent carbon compounds and 20 percent water tied up in mineral compounds. These meteorites are believed to be debris left over from the formation of the solar system. As such, they represent a type of material thought to have formed while the planets were condensing from a cloud of dust and hydrogen gas. See Table 12.1.

TABLE 12.1 COMPOSITION OF EARTH AND CHONDRITE METEORITES*

Rank	Element	Symbol	Earth Average (% by wt.)	Average of Chondrites (% by wt.)
1	Iron	Fe	34.6 ⎫	27.2 ⎫
2	Oxygen	O	29.5 ⎬ 92.0	33.2 ⎬ 91.8
3	Silicon	Si	15.2 ⎬	17.1 ⎬
4	Magnesium	Mg	12.7 ⎭	14.3 ⎭
5	Nickel	Ni	2.4	1.6
6	Sulfur	S	1.9	1.9
7	Calcium	Ca	1.1	1.3
8	Aluminum	Al	1.1	1.2
9	Sodium	Na	0.57	0.64
10	Chromium	Cr	0.26	0.29
11	Manganese	Mn	0.22	0.25
12	Cobalt	Co	0.13	0.09
13	Phosphorus	P	0.10	0.11
14	Potassium	K	0.07	0.08
15	Titanium	Tl	0.05	0.06

Source: *The Earth Sciences*, Arthur N. Strahler, Harper & Row, 1971.

*Data from *Principles of Geochemistry*, 3rd Ed., Brian Mason, John Wiley, 1966.

As Earth cooled and condensed from a cloud of dust and gas, hydrogen reacted with oxygen to form water, which in turn reacted with crystals of solids that were forming. The products of these reactions were silicate minerals that contain water and oxidized iron. Therefore, the original form in which water existed on Earth was as part of solid compounds—rocks. The common clay minerals found in soil are examples of this type of compound.

As these substances formed, they were drawn inward by gravity and heated. We know that, as rocks are heated above 500°C, the water in their molecules is released in a process called *degassing*. When vented to Earth's surface, the water is released into the atmosphere, a process called **outgassing**, and upon further cooling condenses to form liquid water. Much evidence supports the idea that much of Earth's water was released to the surface soon after it formed. Even today, however, the process continues on a small scale through degassing associated with volcanic eruptions.

From an Ocean of Gas to an Ocean of Water

Outgassing from Earth's interior and collisions of comets with Earth added large amounts of water vapor to the developing atmosphere. Consider that a typical comet contains about 10^{15} (one million billion) kilograms of water. Since all of Earth's oceans hold roughly 10^{21} (one trillion billion) kilograms of water, only a million or so comet collisions could have created Earth's entire water supply ($10^6 \times 10^{15} = 10^{21}$). This is not an unlikely number of collisions, considering the estimated number of existing comets, which are but a fraction of those that crowded the early solar system.

As Earth cooled, water vapor in the atmosphere would have condensed and fallen as torrential rains that ran off high places and filled low-lying basins. At first, striking a still warm Earth, water would vaporize quickly and return to the atmosphere. But as the planet continued to cool, liquid water would remain for longer and longer periods in the basins of Earth's surface. Eventually, the oceans and lakes were permanently filled with water that returned to the atmosphere only because of evaporation by the Sun, as is the case today.

This reservoir of liquid water blanketing Earth is called the **hydrosphere**. The hydrosphere, however, is very thin. If you dipped a basketball in water and then shook the water off, the thin film of water adhering to its surface would be a good model of the hydrosphere.

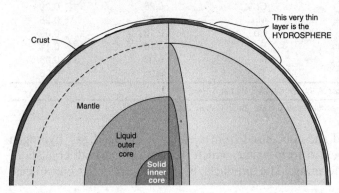

Figure 12.1 The Hydrosphere. The thicknesses of the crust and hydrosphere are greatly exaggerated in this diagram. The hydrosphere is actually much thinner than shown here. Source: *World Geomorphology*, E. M. Bridges, Cambridge University Press, 1990.

Outgassing of water during volcanic eruptions still occurs today, and so do collisions with icy comets. (Consider the 1994 collision of Comet Shoemaker-Levy with Jupiter.) As water continues to work its way to Earth's surface and to be carried in by comets, the oceans may gradually increase in volume until, in a few billion years, the hydrosphere covers the whole planet.

The Origin of Seawater

How did seawater become salty? The general assumption is that the ocean was originally a huge freshwater lake. Over the billions of years that the ocean has existed, soluble ions freed by chemical weathering have been carried into the ocean by rivers, creating the saline waters of today. We know, for example, that the extremely salty Dead Sea and Great Salt Lake have resulted because all types of substances entered them, but only water vapor left them. But at the current rate at which streams add salts to the ocean, it could have reached its present salinity in roughly 100 million years. This time span falls far short of the current estimates, based upon radioactive decay rates and other data, of 3.5–4 billion years for the age of the ocean. Why the discrepancy?

One answer is that the salinity of the ocean does not increase at a constant rate, and probably decreases from time to time. For example, the end of a glacial period would flood the oceans with melting ice, decreasing salinity worldwide. Shallow seas that become landlocked and then evaporate would remove substantial amounts of salt from the ocean. (Salt deposits near Moab, Utah, that formed when a vast inland sea evaporated are as much as 4,000 feet thick!)

Another answer is that the rate at which water has been added by outgassing has not been uniform or continuous. Thus, it appears that salinity, along with sea level, has risen and fallen with the freezing and melting of ice and the removal of huge amounts of salt by evaporation of landlocked seas.

Seawater Today

Seawater is a complex mixture of dissolved inorganic matter, dissolved organic matter, dissolved gases, and particulate matter. **Salinity** is the total amount of dissolved material in seawater. Salinity is measured in parts per thousand (‰) by weight in 1 kilogram of seawater.

Inorganic Matter

By weight, seawater is about 96.5 percent pure water and about 3.5 percent dissolved inorganic substances. See Figure 12.2. Six elements—chlorine, sodium, magnesium, sulfur, calcium, and potassium—make up more than 90 percent of the materials dissolved in seawater. The minor elements include strontium, bromine, and boron. The elements in seawater are almost always present as part of chemical compounds, most of which are salts. Although the salinity of seawater varies from 33 percent to 37 percent, depending upon where in the ocean the sample is taken, the ratio of the major dissolved substances remains about the same.

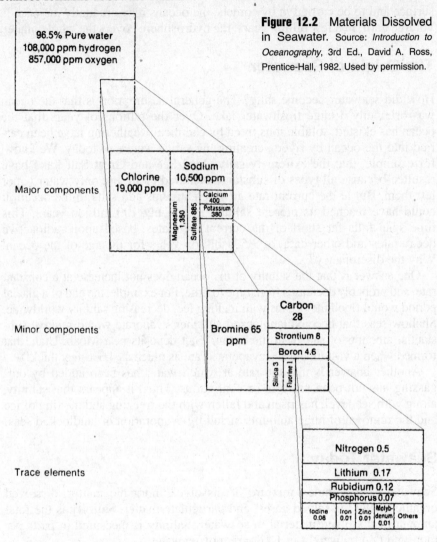

Figure 12.2 Materials Dissolved in Seawater. Source: *Introduction to Oceanography*, 3rd Ed., David A. Ross, Prentice-Hall, 1982. Used by permission.

Organic Matter

The dissolved organic matter in seawater comes mainly from wastes excreted by organisms and from the decaying bodies of dead organisms. It is usually present in small amounts (0–6 parts per million) and is highly variable. Dissolved organic matter includes organic carbon, carbohydrates, proteins, amino acids, organic acids, and vitamins. Organic compounds containing nitrogen and phosphorus oxidize to form nitrates and phosphates.

Dissolved Gases

Nitrogen, oxygen, and carbon dioxide are the major gases dissolved in seawater. Dissolved gases should not be confused with gaseous elements that are part of compounds. For example, water is a compound of hydrogen and oxygen, both of which are gases. However, the compound is not a gas. The oxygen that is part of the water molecule is not a dissolved gas; it is bound in the molecule and can't be "breathed in" and used for respiration. Other gases dissolved in seawater include helium, neon, argon, krypton, and xenon. With the exception of oxygen and carbon dioxide, the gases in seawater usually enter it from the atmosphere.

Seawater has two sources of oxygen: the atmosphere and the plants that live in the ocean. Oxygen gas dissolves directly into seawater from the atmosphere. Photosynthesis by plants living in the upper layers of the ocean (mainly phytoplankton) releases oxygen into seawater.

Particulates

Particulates are solid particles of material that is not dissolved in seawater, but is suspended or settling through it. Aside from living organisms, particulate matter consists mainly of fine-grained minerals and decaying organic material. Particulate matter in seawater varies greatly depending on ocean currents, winds, and nearness to rivers and streams.

WATER: A UNIQUE SUBSTANCE

Earth's hydrosphere is the result of a unique combination of Earth's temperature and the properties of water. Water has unusual properties that make it different from any other substance on Earth. These properties enable liquid water to exist on Earth's surface and are responsible for its chemical and physical behavior. If water had properties typical of a substance with its molecular size and weight, Earth would be a very different place indeed.

The Structure of Water

A water molecule consists of one relatively large oxygen atom combined with two smaller hydrogen atoms. The result is a somewhat lopsided molecule that creates an unequal distribution of electrical charges. The hydrogen side of the molecule is slightly positive; the oxygen side, slightly negative. This unusual structure causes water molecules to be attracted to one another and "stick" together. The attractions between molecules of liquid water are called **hydrogen bonds**. Hydrogen bonds are weak, but they make water behave differently from any other substance on Earth. See Figure 12.3.

Although weak, hydrogen bonds hold water molecules together more strongly than is the case in substances lacking such bonds. Thus, liquid water has a high **surface tension**; to push liquid water molecules apart and to

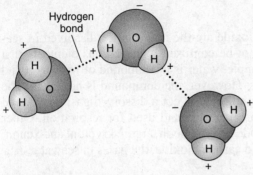

Figure 12.3 The Structure of Water Molecules Causes Hydrogen Bonding. The two ends of a water molecule have opposite electrical charges that attract each other as do opposite poles of a magnet. The weak attractions between the slightly positive end of one water molecule and the slightly negative end of another water molecule are known as hydrogen bonds. Source: *Marine Biology*, 3rd Ed., Peter Castro and Michael Huber, McGraw-Hill, 2000.

penetrate the surface of liquid water, a relatively strong force must be exerted to overcome the hydrogen bonds. Water striders, long-legged bugs that move about on the surface of water, spread their mass over several legs, so that no one leg exerts enough force to break through the surface.

The same "stickiness" that holds water molecules together also makes them adhere to the surfaces of other substances. For example, when water seeps through the ground, some of it "sticks" to the soil particles so that their surface stays "wet" even after most of the water has drained through them. The water that adheres to the surface of soil particles is an important source of water for plant life.

Water's "stickiness" is also responsible for *capillarity*, the tendency for water's surface to rise in a narrow opening. The water rises because its attraction to the walls of the opening pulls it upward, and surface tension drags the water surface between the walls upward along with it. Capillarity is an important process in the movement of water through porous rocks and soil.

The States of Water

Matter can exist in three different physical forms—liquid, gas, and solid—called the **states**, or **phases**, of matter. Water is the only substance on Earth that occurs naturally in all three phases of matter. Since the majority of the hydrosphere is in the liquid state, we'll begin our discussion with water as a liquid.

Liquid Water

In the liquid phase, molecules of water are in constant motion. Because of hydrogen bonds individual water molecules cluster together in clumps of two to eight molecules. Although weak, hydrogen bonds hold water molecules together more strongly than is the case in substances lacking such bonds.

The average speed of the molecules in a substance is reflected in its temperature; the faster the molecules are moving, the higher the temperature. The addition of heat to water makes its molecules move faster, and thus raises its temperature. However, some of the heat added to the water goes into breaking hydrogen bonds instead of promoting molecular motion. Therefore, a large amount of heat energy is needed to get water molecules to move faster, and thus increase the temperature of water. This heat energy is not lost; it is stored *in* the water. Thus, liquid water holds more heat than do substances without hydrogen bonds; and its ability to hold heat, or **specific heat capacity**, is high. In fact, water has one of the highest specific heat capacities of any naturally occurring substance.

Water Vapor

If a water molecule moves fast enough, it can break free of the molecules around it. This process, known as **evaporation**, marks the change from the liquid phase of water to the gaseous or vapor phase. See Figure 12.4. The molecules in water vapor are so fast moving, and move so much farther apart than in the liquid, that they are not held together by hydrogen bonds. As temperature rises, and liquid molecules move faster and faster, more and more escape, so the rate of evaporation increases. When the temperature reaches 100°C, nearly all of the molecules break free of their neighbors and try to enter the vapor state all at once; in other words, the water boils.

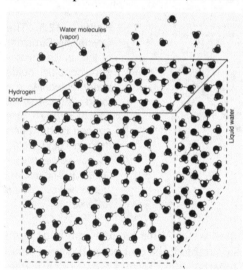

Figure 12.4 Evaporation Takes Place When Molecules Break Free of the Hydrogen Bonds of the Liquid and Enter the Gaseous State.
Source: *Marine Biology*, 3rd Ed., Peter Castro and Michael Huber, McGraw-Hill, 2000.

Once water begins to boil, added heat goes into breaking hydrogen bonds between molecules instead of increasing molecular speed. Therefore, while water is changing phase, it does not change temperature. The amount of heat required to evaporate a substance is called the **latent heat of vaporization**. The word *latent* means "hidden," and in this term refers to the added heat that is not evident as a temperature change. As long as there is boiling water to be evaporated, the temperature of the liquid remains constant at 100°C. Water absorbs a large amount of heat when it evaporates; in other words, it has a high latent heat of vaporization.

At temperatures lower than the boiling point, only the most energetic, fastest moving molecules can break free of the hydrogen bonds and become a gas. Since the fastest moving molecules leave the liquid, those that are left behind are slower moving. As a result, the average speed of the molecules in the liquid, and thus its temperature, decreases. This effect is known as **evaporative cooling**. Our bodies use evaporative cooling to regulate body temperature—when sweat evaporates, it cools the skin.

As liquid water cools, the molecules move slower, pack closer together, and take up less space. In other words, the volume of the water decreases. Since the same mass of water is occupying less volume, the water becomes denser. Thus, liquid water gets denser as it gets colder, but only down to a temperature of about 4°C.

Water as a Solid

As water continues to cool, the molecules move more slowly and rebound from one another more weakly. Eventually, the hydrogen bonds between the molecules suppress these motions and begin to hold the molecules in a fixed, three-dimensional pattern. This process, known as **freezing**, marks the change from the liquid phase to the solid phase. Solids in which atoms or molecules are arranged in such a regular, fixed pattern are called **crystals**. The solid, crystalline form of water is, of course, ice.

In ice, the hydrogen bonds of water hold the molecules in a hexagonal, or six-sided, pattern. As shown in Figure 12.5, the molecules of water are

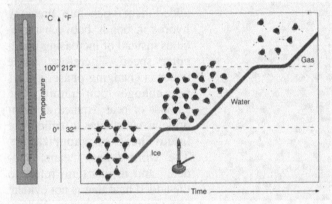

Figure 12.5 The Crystalline Structure of Ice. In ice, hydrogen bonding holds water molecules in a fixed structure in which they are spaced farther apart than in liquid water.
Source: *Marine Biology*, 3rd Ed., Peter Castro and Michael Huber, McGraw-Hill, 2000.

spaced farther apart in ice than in liquid water. What happens might be compared to putting up a tent. When disassembled, the tent poles fit into a small bag; but when they are connected together, they form a large, open framework. Similarly, when water freezes, it expands. Since the same mass of water occupies more volume in ice than in liquid water, ice is less dense than liquid water and therefore floats. It is extraordinarily unusual for the solid phase of a substance to be less dense than the liquid phase. This unique property of water has important consequences for aquatic life. As ice forms in a body of water, it floats and forms an insulating blanket that keeps the subsurface water from cooling rapidly in cold weather. Organisms can live in the liquid water below the surface ice.

If ice were denser than liquid water, it would sink as it formed and build up at the bottom. In colder climates, lakes and even some parts of the oceans would freeze solid. During the summer, only a thin surface layer would thaw, and these bodies of water would be much less hospitable to life.

A large amount of heat is required to melt ice. Water molecules in ice crystals vibrate, but they do not move around freely. For melting to occur, the hydrogen bonds holding ice crystals together must first be broken. As heat energy is added to ice, the molecules vibrate faster and eventually break the hydrogen bonds holding the crystal together; then the ice melts. The molecules then move around as well as vibrate. Since a lot of energy is needed to break the hydrogen bonds holding ice crystals together, ice melts at a much higher temperature than do substances that do not have hydrogen bonds. Similarly, since hydrogen bonds help pull water molecules together into crystals, water does not have to be cooled as much before freezing. Thus, water melts and freezes at a much higher temperature than do substances that lack hydrogen bonds. If water did not have hydrogen bonds, the change from solid to liquid (or liquid to solid) would occur at about −90°C instead of 0°C.

As ice crystals are melting, added heat goes into breaking hydrogen bonds instead of increasing molecular speed. Therefore, while water is changing phase it does not change temperature. The amount of heat required to melt a substance is called the **latent heat of fusion**. Here, again, the word *latent* means "hidden," and in this term refers to the added heat not evident as a temperature change. As long as there is ice left to be melted, the temperature of the ice-water mixture remains constant at 0°C.

Water: The Universal Solvent

Water is often called the universal solvent because it can dissolve more things than any other naturally occurring substance. Water is particularly good at dissolving molecules held together by the electrical attraction between oppositely charged particles. Common table salt, or sodium chloride, is such a molecule. It is composed of positively charged sodium (Na^+) and negatively charged chlorine (Cl^-). These sodium and chlorine ions have much stronger charges than the slightly charged ends of a water molecule. When a salt crystal is placed in water, the strong charges of the sodium and chlorine ions attract water molecules. Water's positive end is attracted to (or will attract) the negative chlorine; likewise, water's negative end will be attracted to the positive sodium. Thus, water molecules cluster around both ions in a salt molecule. As water molecules cluster around the ions, they cancel some of the attraction between the sodium and the chlorine, and the two move farther apart. Then more water molecules can come between them until the attraction is eliminated. Finally, the sodium and chlorine ions in the salt crystal pull apart, and the crystal **dissolves**. Once the sodium and chlorine ions separate, water molecules surround them completely, insulating them from each other and preventing them from recombining to form a crystal. See Figure 12.6.

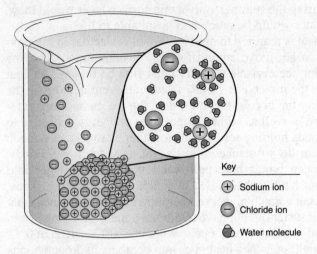

Figure 12.6 Water Molecules Cluster Around the Electrically Charged Sodium and Chlorine ions in a Salt Crystal. This weakens the bonds between the ions, causing them to separate, and the crystal dissolves. Source: *Marine Biology*, 3rd Ed., Peter Castro and Michael Huber, McGraw-Hill, 2000.

Key
(+) Sodium ion
(−) Chloride ion
Water molecule

THE WATER CYCLE

The *water cycle* is the process by which Earth's water moves into and out of the atmosphere. Let's begin by describing this process in the oceans that cover nearly three-quarters of Earth's surface.

1. Each day, insolation (*in*coming *sol*ar radi*ation*) reaching the surface of the oceans provides energy to the water molecules at the surface. As the molecules heat up, trillions of tons of water change from liquid to gas, or *evaporate*. The resulting water vapor enters the atmosphere, where it is circulated by winds and carried upward by rising warm air.
2. When air is cooled to its dewpoint by expansion during rising or contact with cooler surfaces, the water vapor changes from gas back to liquid, or *condenses*, forming tiny water droplets. These tiny water droplets suspended in air form clouds.
3. When the water droplets become too large to remain suspended, they fall back to the surface as *precipitation*. Precipitation may fall on the oceans or the land. Water that falls on the oceans has completed its cycle and may evaporate back into the atmosphere, beginning the entire process anew.
4. A number of things may happen to precipitation that falls on land. It may become *runoff*, water that flows downhill into rivers and streams that eventually carry it back into an ocean or other body of water. Or, it may seep into the ground to become **groundwater**, water that has infiltrated open spaces in the outer part of Earth's crust. Groundwater may move slowly back to the oceans through the ground, but most of it seeps into rivers and streams. It is then carried back to the oceans.
5. Some of the water may evaporate again to be carried farther over land, or be blown back over the oceans.
6. Once the water is back in the oceans, the cycle begins all over again.

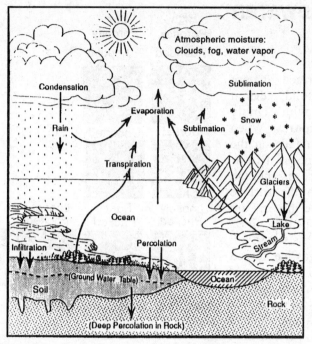

Figure 12.7 The Water Cycle. Source: Permission from Thomas McGuire.

There are many variations in the water cycle. Sometimes the water that evaporates over the oceans condenses there and falls directly back into the oceans as rain. Water falling on land may evaporate almost immediately, and in some cases precipitation evaporates before it ever reaches the ground.

Plants play a role in the water cycle as well. The roots of plants absorb water that has seeped into the soil. Then the water is transported to their leaves and released back into the atmosphere by a process called **transpiration**. In an area of abundant vegetation, such as a forest, transpiration returns more than one-third of precipitation back to the atmosphere as water vapor.

Other variations are possible, but all are part of the endless water cycle that receives its energy from the Sun. The water cycle continually renews our supply of fresh water. Each day an estimated 15 trillion liters of water in the form of rain or snow fall on the United States alone. Earth is a closed system, and water consumed today will eventually be recycled for use by a future generation.

In theory, Earth's water will never be exhausted. If, however, persistent, long-lasting pollutants enter the basins and ground from which Earth's water supply is drawn, water that has been cleansed by evaporation will be polluted as soon as it condenses and falls back into these polluted areas. Consider what will happen if the ground is contaminated with nuclear waste, which remains radioactive for tens or even hundreds of thousands of years. For the entire time the material remains radioactive, any water that infiltrates that ground will immediately become unusable.

GROUNDWATER

Porosity

Earth's surface is not completely solid; rather, it is filled with empty spaces, or **pores**. When precipitation falls on the surface, some of it **infiltrates**, or seeps through openings in the surface and into pore spaces. Pore spaces are usually filled with air unless water, oil, or natural gas has forced the air out. The amount of open space in a material is its **porosity**. Porosity is expressed as a percentage of the total volume of a given substance. For example, if half of the volume of a sample of rock or sediment is open space, the specimen is said to have a porosity of 50 percent. Earth materials differ greatly in porosity. Factors that affect porosity are shown in Figure 12.8. Loose sediments such as sand, gravel, and clay usually have the highest porosity.

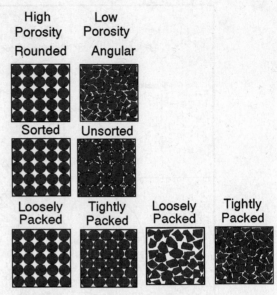

Figure 12.8 Factors That Affect the Porosity of a Material.

Porosity can be calculated by measuring the amount of water required to fill the pores in a substance and then setting up a ratio of the volume of water to the total volume of the sample being tested.

Example:
Water is poured into a 100-cubic-centimeter sample of sand. When 35 milliliters of water have been poured in, the water is just even with the surface of the sample. What is the porosity of the sand? [Note: 1 ml = 1 cm³]

$$\text{Porosity} = \frac{\text{volume of pore space}}{\text{total volume of sample}} \times 100\%$$

$$= \frac{35 \text{ cm}^3}{100 \text{ cm}^3} \times 100\% = 35\%$$

The porosity of the sand is 35%.

Permeability

Permeability is the rate at which water can infiltrate a material. A substance through which water passes rapidly is said to be **permeable**; a substance through which water cannot flow is **impermeable**. The permeability of a substance depends on two factors: the size of its pores and the degree to which they are interconnected. See Figure 12.9. Small pores constrict the flow of water; and if water cannot get from one pore to another, it cannot flow through a substance.

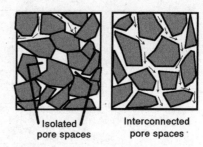

Figure 12.9 Isolated and Interconnected Pore Spaces. Permeable materials have interconnected pore spaces.

Knowing the permeability of a material is very useful. From it, the rate at which rainwater will sink into the ground can be determined, as well as the speed and direction in which sewage in a septic tank will flow. Also, the rate at which a well will refill with water after some has been pumped out can be determined.

The Water Table

When rain falls on land or when snow that has fallen melts, gravity pulls the water down into the ground through interconnected pore spaces. Some of the water clings to particles near the surface, forming a moist layer called the **soil moisture zone**. This is the layer from which plants obtain the water they need to survive. Most of the water that infiltrates continues to trickle downward through the ground until it reaches an impermeable substance. Then the water begins to fill all the pore spaces from the impermeable material upward. The region of permeable ground in which all of the pore spaces are filled with water is the **zone of saturation**.

The top of the zone of saturation, known as the **water table**, is the boundary between pores filled with water and pores filled with air. Just above the water table is a narrow region, called the **capillary fringe**, in which narrow pore spaces are filled with water drawn upward by capillary action. The area above the capillary fringe, where most of the pore spaces are filled with air, is called the **zone of aeration**. See Figure 12.10.

Groundwater can leave the ground in several ways. It can be drawn up by **capillary action** through tiny interconnected pore spaces and reach the soil moisture zone, where plant roots remove it. Or, instead, it may evaporate and slowly diffuse out of the ground through the zone of aeration, or it may seep through the ground sideways and slowly move out of the area. If the surface of the ground dips beneath the level of the water table, water can seep out and run off, forming streams, or collect in depressions, forming springs, ponds, or lakes. When the water table in a depression is just above the surface, a swamp may form. See Figure 12.11.

The Origin and Nature of Earth's Hydrosphere

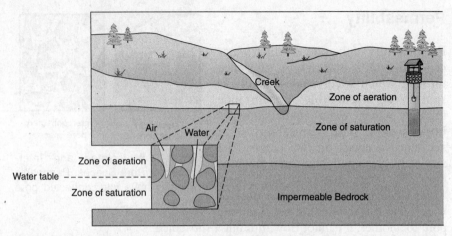

Figure 12.10 Zones of Soil Water and Ground Water. The water table is the boundary between the zone in which pores are filled with air and the zone in which pores are filled with water.

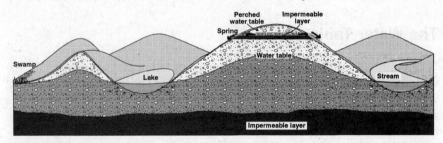

Figure 12.11 Variations in the Subsurface Position of the Water Table. The water table roughly follows the contours of Earth's surface. Where the surface dips beneath the water table, water seeps out of the ground and can collect in depressions, forming a body of water such as a lake or pond. If the water seeping out runs off, it may form a stream. Impermeable layers in hills can create a perched water table, which may seep out of a hillside as a spring.

Wells

Groundwater is an important source of fresh water. There is 37 times as much fresh water beneath the ground as there is on top of it in lakes, streams, glaciers, and other sources. Layers of permeable rock or loose sediments whose pore spaces are filled with water are called **aquifers**. Groundwater in aquifers can be tapped by digging a well, which is a hole sunk into the ground so that it penetrates the water table. Beneath the water table the well acts as a large pore space in which water accumulates. This water can then be pumped up to the surface through the hole.

Artesians

In artesian rock formations, groundwater can be tapped by wells and the water brought to the surface without pumping. In these rock formations an aquifer is sandwiched between two impermeable layers called *aquicludes*. The aquifer slopes downward away from the area where water seeps into it. The aquicludes act like the walls of a pipe, confining the water as it seeps downslope, so that water pressure builds up in the aquifer. If the upper aquiclude is pierced by a well, this pressure may push the water up to the surface or even higher. The amount of pressure in an artesian well depends on the difference in elevation between the well and the water table in the aquifer. See Figure 12.12. Artesian wells get their name from Artois, a former province of France, where the first well of this type was dug.

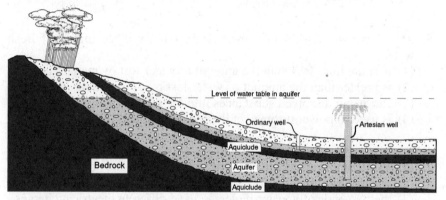

Figure 12.12 An Artesian Formation. The water rises in an artesian well because of hydrostatic pressure created by the weight of the water in the aquifer.

MULTIPLE-CHOICE QUESTIONS

In each case, write the number of the word or expression that best answers the question or completes the statement.

1. Large amounts of water vapor were added to Earth's developing atmosphere by
   (1) outgassing and collisions with comets
   (2) photosynthesis by early life-forms
   (3) dust and hydrogen gas
   (4) electrical discharges in the atmosphere

2. Approximately what fraction of Earth's surface do oceans now cover?
   (1) ¼   (3) ½
   (2) ⅓   (4) ¾

3. What is the major source of the dissolved minerals that affect the salinity of ocean water?
   (1) ice sheets
   (2) industrial pollutants
   (3) continental erosion
   (4) tropical storms

4. Which profile when drawn to true scale most accurately shows the relationship between the ocean width of 1,500 kilometers and a maximum depth of 8 kilometers?

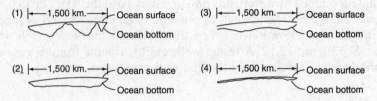

5. A lake without outlets may increase in salinity if the inflow of fresh water
   (1) is equal to or less than the amount of water lost by evaporation
   (2) is greater than the amount of water lost by evaporation
   (3) carries no dissolved substances into the lake
   (4) is used to power turbines in hydroelectric plants

6. Water is a good solvent because
   (1) a water molecule contains two hydrogen atoms and one oxygen atom
   (2) water is held together by strong molecular bonds
   (3) the two ends of a water molecule have different electrical charges
   (4) water is the most common compound on Earth

7. Liquid water can store more heat energy than an equal amount of any other naturally occurring substance because liquid water
   (1) covers 71% of Earth's surface
   (2) has its greatest density at 4°C
   (3) has the higher specific heat
   (4) can be changed into a solid or a gas

8. During which phase change will the greatest amount of energy be absorbed by 1 gram of water?
   (1) melting
   (2) freezing
   (3) evaporation
   (4) condensation

9. When 1 gram of liquid water at 0° Celsius freezes to form ice, how many total joules of heat are lost by the water?
   (1) 2.11
   (2) 4.18
   (3) 334
   (4) 2260

10. How many joules are required to evaporate 1 gram of boiling water?
    (1) 2.0   (3) 334
    (2) 2.11   (4) 2260

11. During some winters in the Finger Lakes region of New York State, the lake water remains unfrozen even though the land around the lakes is frozen and covered with snow. The primary cause of this difference is that water
    (1) gains heat during evaporation
    (2) is at a lower elevation
    (3) has a higher specific heat
    (4) reflects more radiation

12. The flowchart below shows part of Earth's water cycle. The question marks indicate a part of the flowchart that has been deliberately left blank.

    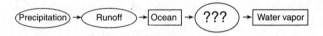

    (1) condensation   (3) evaporation
    (2) deposition   (4) infiltration

13. Most water vapor enters Earth's atmosphere by the processes of
    (1) condensation and precipitation
    (2) radiation and cementation
    (3) conduction and convection
    (4) evaporation and transpiration

Base your answers to questions 14 through 17 on the diagram of the water cycle below. Letter *A* represents a process in the water cycle. Points *X* and *Y* represent locations on Earth's surface.

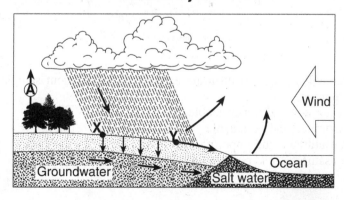

Water Cycle

14. The process represented by A is
    (1) precipitation
    (2) transpiration
    (3) condensation
    (4) saturation

15. Rainwater will enter the ground at X if the ground is
    (1) saturated and permeable
    (2) saturated and impermeable
    (3) unsaturated and permeable
    (4) unsaturated and impermeable

16. The amount of runoff at Y will increase as the
    (1) slope of the land decreases
    (2) porosity of the soil increases
    (3) evaporation rate exceeds the infiltration rate
    (4) precipitation rate exceeds the infiltration rate

17. Which process in the water cycle is directly responsible for cloud formation?
    (1) condensation
    (2) infiltration
    (3) precipitation
    (4) evaporation

Base your answers to questions 18 through 21 on the diagram below, which shows four tubes containing 500 milliliters of sediment labeled A, B, C, and D. Each tube contains well-sorted, loosely packed particles of uniform shape and size and is open at the top. The classification of the sediment in each tube is labeled.

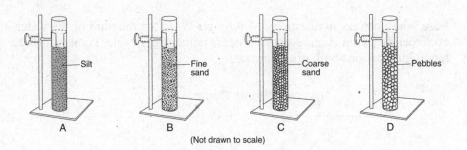

(Not drawn to scale)

18. Water will be able to infiltrate each of these sediment samples if the sediment is
    (1) saturated and impermeable
    (2) saturated and permeable
    (3) unsaturated and impermeable
    (4) unsaturated and permeable

19. Water was poured into each tube of sediment. The time it took for the water to infiltrate to the bottom was recorded, in seconds. Which data table best represents the recorded results?

Tubes	Infiltration Time (s)
A	5.2
B	3.4
C	2.8
D	2.3

(1)

Tubes	Infiltration Time (s)
A	2.4
B	2.9
C	3.6
D	3.8

(3)

Tubes	Infiltration Time (s)
A	3.2
B	3.3
C	3.2
D	3.3

(2)

Tubes	Infiltration Time (s)
A	3.0
B	5.8
C	6.1
D	2.8

(4)

20. Each tube is filled with water to the top of the sediments, and the tube is covered with a fine screen. The tubes are then tipped upside down so the water can drain. In which tube would the sediment retain the most water?
    (1) A
    (2) B
    (3) C
    (4) D

21. Which graph best represents the relationship between soil particle size and the rate at which water infiltrates permeable soil?

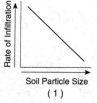
(1)

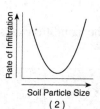
(2)

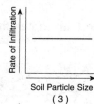
(3)

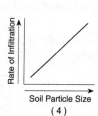
(4)

The Origin and Nature of Earth's Hydrosphere

Base your answers to questions 22 and 23 on the cross section below, which represents part of Earth's water cycle. Letters A, B, C, and D represent processes that occur during the cycle. The level of the water table and the extent of the zone of saturation are shown.

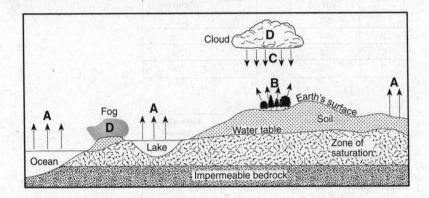

22. Which two letters represent processes in the water cycle that usually cause a lowering of the water table?
    (1) A and B (3) B and D
    (2) A and C (4) C and D

23. What are two water cycle processes not represented by arrows in this cross section?
    (1) transpiration and condensation
    (2) evaporation and melting
    (3) precipitation and freezing
    (4) runoff and infiltration

24. Which diagram best illustrates the condition of soil below the water table?

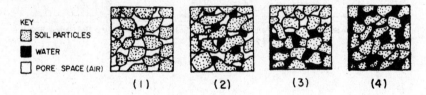

25. The water table usually rises when there is
    (1) a decrease in the amount of infiltration
    (2) a decrease in the amount of surface area covered by vegetation
    (3) an increase in the amount of precipitation
    (4) an increase in the slope of the land

26. Which soil conditions normally result in the greatest amount of runoff?
    (1) low permeability and gentle slope
    (2) low permeability and steep slope
    (3) high permeability and gentle slope
    (4) high permeability and steep slope

CONSTRUCTED RESPONSE QUESTIONS

Base your answers to questions 27 through 29 on the diagram below, which shows the changes in the molecular structure of water that occur as it changes temperature.

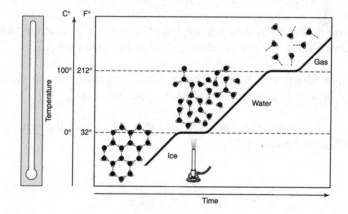

27. According to the diagram, in which phase would water have the greatest density? Explain your answer. [1]

28. What would happen to the spacing between molecules of water as a gas if the temperature was increased to 300°C? [1]

29. Describe how the volume of a particular sample of water changes as the sample changes from water into ice. [1]

The Origin and Nature of Earth's Hydrosphere

Base your answers to questions 30 through 32 on the diagram below, which shows Earth's water cycle. Numbers indicate the estimated volume of water, in millions of cubic kilometers, stored at any one time in the atmosphere, in the oceans, and on the continents. The yearly amount of water that moves in and out of each of these three portions of Earth is also indicated in millions of cubic kilometers.

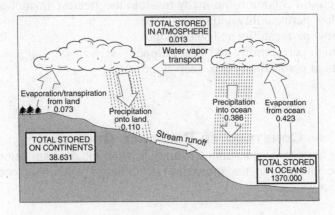

30. Calculate the total amount of water stored in the atmosphere, the oceans, and on the continents together at any one time. [1]

31. Explain why the yearly total precipitation over the oceans is greater than the yearly total precipitation over the continents. [1]

32. Describe *two* surface characteristics that will affect the rate of stream runoff into the ocean. [1]

EXTENDED CONSTRUCTED RESPONSE QUESTIONS

Base your answers to questions 33 through 37 on the cross section below, which shows the general pattern of water movement in the water cycle. Letter *X* represents a water cycle process.

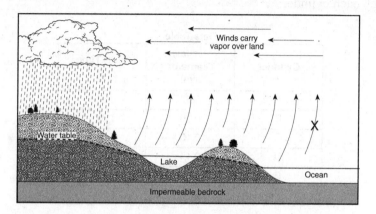

33. What process of the water cycle is represented by *X*? [1]

34. Describe the process of condensation. [1]

35. Describe *one* surface condition that would allow runoff to occur. [1]

36. Explain *one* role of plants in the water cycle. [1]

37. As the lake surface freezes in the winter, how many joules of heat are released by each gram of water? [1]

Base your answers to questions 38 through 40 on the data table below. Six identical cylinders, A through F, were filled with equal volumes of sorted spherical particles. The data table shows the particle diameters, in centimeters, and the amount of time, in seconds, for water to flow equal distances through each cylinder.

Data Table

Cylinder	Particle Diameter (cm)	Flow Time (s)
A	0.07	51
B	0.08	39
C	0.10	25
D	0.14	13
E	0.16	10
F	0.18	8

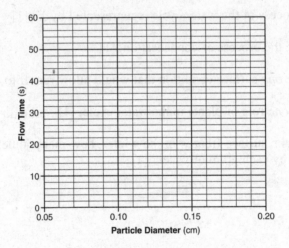

38. Use the information in the data table to construct a line graph. On the grid next to the data table, plot the data for the flow time for each of the particle sizes given in the data table. Connect the plotted data with a smooth, curved line. [1]

39. Determine the flow time in a cylinder containing particles with a diameter of 0.13 centimeter. [1]

40. State *one* reason why the water flows faster through the cylinders containing larger particles than through the cylinders containing smaller particles. [1]

Chapter 13
THE ORIGIN AND HISTORY OF LIFE ON EARTH

KEY IDEAS All of Earth's present-day life forms have evolved from common ancestors reaching back about three billion years to the simplest one-celled organisms. Before that time, simple molecules may have formed complex organic molecules that slowly evolved into cells capable of replicating themselves. The history of life on Earth has been pieced together from geological, anatomical, and molecular evidence. The geological evidence has come mainly from the sequence of changes seen in fossils found in successive layers of rock that have formed over more than a billion years. Human existence has been very brief compared to the expanse of geologic time.

Geologic history can be reconstructed by observing sequences of rock types and fossils to correlate bedrock at various locations. The characteristics of rocks indicate the processes by which they formed and the environments in which these processes took place. Fossils preserved in rocks provide information about past environmental conditions.

Geologists have divided Earth history into time units based upon the fossil record. Age relationships among bodies of rocks can be determined using principles of original horizontality, superposition, inclusions, cross-cutting relationships, contact metamorphism, and unconformities. The presence of volcanic ash layers, index fossils, and meteoritic debris can provide additional information. The regular rate of nuclear decay (half-life time period) of radioactive isotopes allows geologists to determine the absolute ages of materials found in some rocks.

KEY OBJECTIVES

Upon completion of this chapter, you will be able to:

- Identify the characteristics of life and two ideas about how it may have originated.
- Explain how a study of the fossil record shows that life-forms have evolved through geologic time.

- Describe how fossils provide evidence of past environments.
- Determine the relative ages of a series of rock layers and any igneous intrusions, faults, folds, or fossils they may contain, based upon the principle of superposition.
- Explain the significance of index fossils and volcanic ash deposits in correlating widely separated rock layers.
- Interpret the geologic time scale.
- Determine the absolute age of a rock, given the relative amounts of a radioisotope and decay product in the rock, and the half-life of the radioisotope.

THE GEOLOGIC RECORD OF LIFE'S HISTORY

Until the nineteenth century, nearly everyone believed that Earth was only a few thousand years old. Scientific evidence, though, indicates that some of the rocks on Earth's surface are several *billion* years old. Observations of patterns in rock layers and the location of various kinds of fossils allow inferences concerning the relative ages of rocks and the events that formed them. The absolute age of a rock can be determined from the relative amounts of a radioisotope and its decay products in the rock. The rock and fossil record reveals that life-forms originated early in Earth's history and that they and the environments in which they lived have changed over time.

How Can the Order in Which Geological Events Occurred Be Determined?

Much of what is known about Earth's history, and that of life, has been learned by studying rocks. Rock layers contain traces of events that occurred in the past and provide clues to the origins of the layers and the environments in which they formed. With careful logic, inferences can be drawn about Earth's past from data obtained from rocks. For example, the presence in a rock layer of sorted sand grains that are rounded implies that the grains were carried by running water. Also, the size of the grains may indicate how fast the water that deposited them was moving. If a rock layer contains shells, the type of shell may show that the rock formed in a lake, not an ocean. Shells can also provide clues to the depth and temperature of the water. Microscopic pollen grains in the rock could identify plants living around the lake when the rock formed. Still other rock layers may contain evidence of glaciers, deserts, or volcanic eruptions.

Such inferences are based upon an assumption proposed by James Hutton in 1795—uniformitarianism. **Uniformitarianism** is the idea that Earth's features, such as mountains, valleys, and rock layers, have formed gradually by processes still underway, not by instantaneous creation. It assumes that Earth has always behaved much as it does now; that water, for example, ran downhill in streams carrying sand and silt 10 million years ago just as it does today. This idea can be summarized in the statement "The present is the key to the past."

According to uniformitarianism, every rock layer contains a record of a short part of Earth's long history. By working out the age of each layer in a series, geologists can determine the order in which the layers formed and can arrange the events they represent in the order in which these events occurred. In this way a history of Earth can be pieced together.

There are two ways of expressing the ages of Earth materials or geologic events, relative age and actual age. **Relative age** is the age of a rock, fossil, geologic feature, or event relative to that of another rock, fossil, geologic feature, or event. Thus, relative age is not an exact age in years; it tells you only that one thing is older or younger than another. If you say that you are older than a first grader but younger than your parents, you are giving your relative age.

Absolute age is the age of a rock, fossil, geologic feature, or event given in units of time, usually years. If you state, "I am 15 years old," you are giving your actual age. The absolute age of Earth materials and events is usually determined by analyzing radioisotopes in objects.

Determining Relative Age

Two key ideas in determining relative age are the law of original horizontality and the principle of superposition. The **law of original horizontality** states that sediments deposited in water form flat, level layers parallel to Earth's surface.

The **principle of superposition** is the concept that the bottom layer of a series of sedimentary layers is the oldest, unless it has been overturned or older rock has been thrust over it. Each layer forms atop the one that is already there. (Layers of sediment do not form in midair, nor do they hover halfway between the ocean surface and the ocean bottom!) Therefore, each layer is younger than the one under it and older than the one on top of it. The relative ages of any two layers can be determined based upon which layer is on top and which is on the bottom. See Figure 13.1.

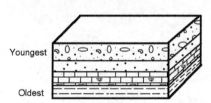

Figure 13.1 Series of Rock Layers Showing Oldest to Youngest.

The principle of superposition must be used with care, however, because there are events that can disturb the positions of rock layers. Forces within Earth may tilt, fold, or fault the layers. Older layers may be pushed on top of younger ones. In such cases, one must work out the original positions of the rock layers before applying the principle of superposition.

In general, a rock layer is older than any joint, fault, or fold that appears in it; the rock had to already exist in order to be folded or faulted. By unfolding or unfaulting the rock layers, one can determine the positions of the layers before they were disturbed, and can then apply the principle of superposition to determine their relative ages.

For example, let's look at Figure 13.2. If we were to judge only by the position of the rock layers in the core sample at (3), we would conclude that layer C is the youngest and A is the oldest because C is on top and A is on the bottom. If we look at the entire rock structure, however, we see that the three layers of rock have been folded and overturned. By unfolding the layers, we place the layers in their original position and see that layer A is the *youngest* while C is the oldest.

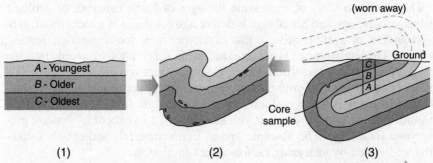

Figure 13.2 The Original Position of Rock Layers Must Be Determined Before Applying the Principle of Superposition. Originally horizontal rock layers (1), may be folded by compression (2), causing older rock layers to appear above younger rock layers (3). Source: *Macmillan Earth Science*, Eric Danielson and Edward J. Denecke Jr., Macmillan Publishing Co., 1989.

The same is true of faults, that is, fractures along which the rocks on either side have moved. If rock layers fracture and then move along that fracture, they are displaced. Any rock displaced by a fault is older than the fault. During faulting, underlying rock layers may be pushed up so that they are found on top of younger rock. See Figure 13.3. Again, to obtain true relative ages, one must work backward to the position the rock layers were in before the fault offset them.

Igneous intrusions or extrusions are often found associated with other types of rock. Igneous intrusions form when molten rock forces its way into preexisting rock, cools, and hardens. Thus, an intrusion is younger than any rock it cuts through. See Figure 13.4. Extrusions form during volcanic eruptions when molten rock flows out onto Earth's surface as lava and hardens, or is blown into the atmosphere and settles on the ground, forming a blanket of volcanic

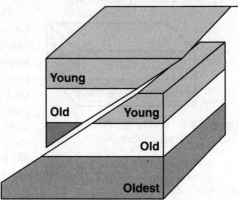

Figure 13.3 Fault, Showing Older Rocks Pushed Atop Younger Ones.

rock particles. Thus, extrusions are younger than the rocks beneath them, but older than any layer that may form over them. Therefore, if an igneous body is found in rock, one must first determine whether it is an intrusion or an extrusion before the relative ages of the layers can be established.

Unconformities

Layers of rock are generally deposited in an unbroken sequence. However, if forces within Earth uplift rocks, deposition ceases. Erosion may wear away many layers of rock before the land is low enough for another layer to be deposited. The result is an **unconformity**, a break or gap in the sequence of a series of rock layers. Thus, the rocks above an unconformity are quite a bit younger than those below it.

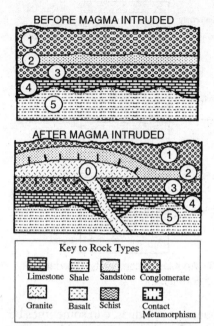

Figure 13.4 Five Rock Layers Before and After Intrusion of Magma. Layer O, which formed from the cooling magma, is younger than the five rock layers into which it was intruded.

There are three types of unconformity. See Figure 13.5. **Angular unconformities** form when rock layers are tilted or folded before being eroded. When new layers are deposited, they form horizontally and the layers below the unconformity are at an angle to those above it. **Disconformities** are irregular erosional surfaces between parallel layers of rock. Disconformities occur when deposition stops and layers are eroded, but no tilting or folding occurs. These surfaces are not easy to discern and are often found when fossils of very different ages are discovered in adjacent layers. **Nonconformities** are places where sedimentary layers lie on top of igneous or metamorphic rocks.

How Can Rocks and Geologic Events in One Place Be Matched with Those in Other Locations?

Correlation

With care and logic the relative ages of rock layers in any particular location can be worked out. However, this information provides only a small part of the overall picture of Earth's history, and unconformities may mean that even more details are missing. To broaden the scope of the picture, rocks in one place need to be correlated with rocks in other locations. The process of **correlation** involves determining that rock layers in different areas are the same age. In this way, rocks in one location may fill in gaps in the record in another location. Correlation also allows the relative ages of rocks in widely separated outcrops to be determined.

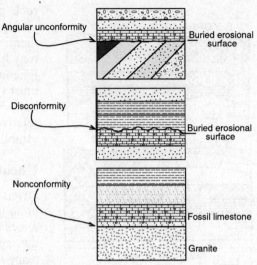

Figure 13.5 Angular Unconformity, Disconformity, and Nonconformity.

Walking the Outcrop

Rock layers can sometimes be followed from one location to another by walking the outcrop. See Figure 13.6. *Walking the outcrop* means physically tracing the layers from one place to another. In this way, rock layers in two different places, such as two adjacent mountains or ridges, can sometimes be correlated. Unfortunately, rock layers are rarely continuously visible for any distance. Most rocks are hidden beneath regolith, the loose fragments of rock and soil that blanket Earth's surface. Rock outcrops poke out here and there like islands. As a result, geologists are often faced with the task of correlating rocks in widely separated outcrops.

Matching Physical Characteristics

Rocks can sometimes be correlated on the basis of distinct similarities in physical characteristics such as composition, color, thickness, and fossil remains. Distinctive rocks or sequences of rock types can also be used to correlate rocks.

For example, Alfred Wegener used this type of correlation as part of his evidence for continental drift. He stated: "Igneous rocks in Brazil and Africa [have] no less than five parallels: 1) the older granite, 2) the younger granite,

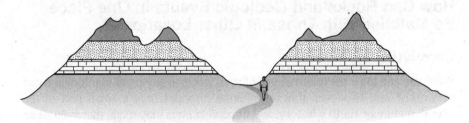

Figure 13.6 Walking the Outcrop.

3) alkali-rich rocks, 4) volcanic Jurassic rock and intrusive dolerite, and 5) kimberlite, alnoite, etc... The last of the rock groups (kimberlite, alnoite) is best known since both in Brazil and South Africa the beds yield the famous diamond finds. In both these regions the peculiar type of stratification known as 'pipes' occurs. There are white diamonds in Brazil in Minas Geraes State and in South Africa north of the Orange River only."

This type of correlation must be done with care, though, because different formations can look almost identical.

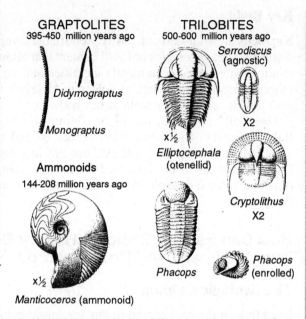

Figure 13.7 Graptolites, Trilobites, and Ammonoids Labeled with Ages. Source: The New York State Museum. Educational Leaflet #28, *Geology of New York, A Simplified Account*, Second Edition. Printed with permission of New York State Museum, Albany, N.Y.

Index Fossils

Index fossils are remains of organisms that had distinctive body features, were abundant, and had a broad, even worldwide range, yet existed only for a short period of time. The best index fossils include swimming or floating organisms that evolved rapidly and were distributed widely, such as graptolites, trilobites, and ammonoids. See Figure 13.7. Their distinctive body shapes and broad distribution make index fossils easy to find in widely separated rock layers (see Figure 13.8). Also, their short existence pinpoints the time period during which those rock layers were formed.

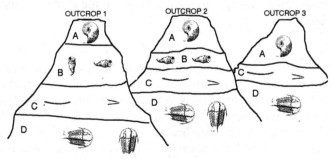

Figure 13.8 Correlation by Fossils of Strata in Several Locations.

Key Beds

Key beds are well-defined, easily identifiable layers or formations that have distinctive characteristics or fossil content that allows them to be used in correlation. Key beds can be readily identified and are not easily confused with other layers. Materials such as volcanic ash layers that are rapidly deposited over a wide area make excellent key beds.

One well-known key bed is the iridium-rich layer of rock discovered in Italy, Denmark, and many other places around the world. Iridium is an extremely rare element on Earth, but not in meteorites. The iridium-rich layer is thought to have formed 66 million years ago when a meteorite impact created a dust cloud that encircled Earth and then slowly settled to the ground.

How Can Earth's Geologic History Be Sequenced from the Fossil and Rock Record?

The Geologic Column

Even though the rock record in any one place is incomplete, correlation has made it possible to piece together a fairly complete record of Earth's geologic history by examining rocks in different locations. By the nineteenth century, geologists had correlated rocks worldwide into a single sequence called the **geologic column**. Rocks are still being added to the geologic column as more outcrops are mapped and described.

The Geologic Time Scale

Geologists then divided Earth's history into a sequence of time units called the **geologic time scale**. *Geologic time* is the entire time that Earth has existed. The geologic time scale divides geologic time into units and subunits based upon the fossil record. Geologic time is divided into eons, eons are subdivided into eras, eras are subdivided into periods, and periods are subdivided into epochs. At first, these time units were based only upon the relative ages of fossils in rocks of the geologic column, but radioisotope dating has enabled geologists to assign actual ages to the time units of the geologic time scale as shown in Figure 13.9.

Eons

Eons are the largest time units on the geologic time scale. The oldest is the *Archean* (Greek "ancient"); it covers the time from Earth's formation to the appearance of multicelled organisms. Archean rocks are the oldest known on Earth and contain microscopic fossils of single-celled, bacteria-like organisms. The *Proterozoic* (Greek "earlier life") is the next oldest eon; its rocks contain traces of multicelled organisms that had no preservable hard parts. The most recent eon is the *Phanerozoic* (Greek "visible life"), which provides an abundant fossil record of preserved hard parts.

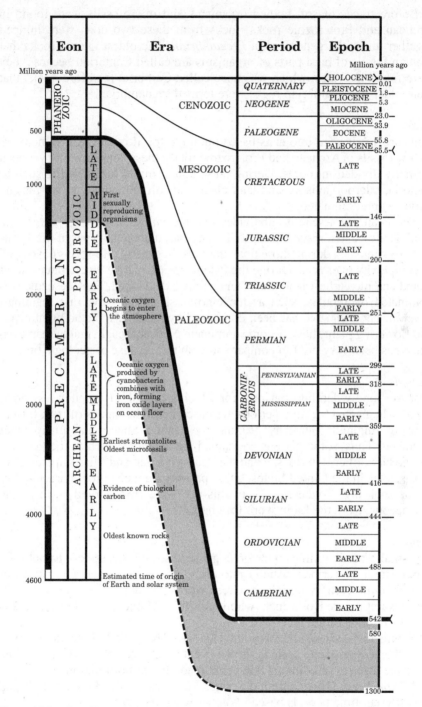

Figure 13.9 The Geologic Time Scale. Source: The State Education Department, *Earth Science Reference Tables*, 2011 ed. (Albany, New York; The University of the State of New York).

Before traces of soft-bodied organisms and microfossils were found in Archean and Proterozoic rocks, rocks from these two eons were lumped together under the general term *Precambrian*. The oldest known rocks that contain fossils of hard parts of organisms are called Cambrian because they were found in Wales, which was once called Cambria; therefore, rocks that were older than Cambrian rocks were termed Precambrian.

Eras

Eons are subdivided into **eras** based upon the fossil record. Since there are so few fossils in Archean and Proterozoic rocks, these eons have not yet been formally divided into eras. During these eons, simple forms of life such as bacteria, microorganisms, and later algae and soft animals (e.g., worms and jellyfish) predominated.

The Phanerozoic eon is subdivided into the *Paleozoic* (old life), *Mesozoic* (middle life), and *Cenozoic* (recent life) eras based upon the types of life-forms that predominated during those time intervals. In general, life-forms developed in complexity over time. During the Paleozoic era, marine invertebrates dominated and the earliest land plants and animals developed. The Mesozoic was dominated by reptiles, such as the dinosaurs, and saw the first mammals develop. The Cenozoic has been dominated by mammals, including humans, and flowering plants have become dominant. Notice, though, that human existence has been very brief in comparison with the expanse of geologic time.

Periods

Eras are subdivided into **periods**, for which terminology is much less organized. The names for time periods are based upon the names of rock formations in England, Germany, Russia, the United States, and many other countries. Thus, some periods are named for locations, such as the Permian for the city of Perm in Russia, and the Pennsylvanian and Mississippian after those states in the United States. Other periods are named for characteristics of the rock layers where rocks of this age were first studied, such as the Cretaceous, from the Latin word for chalk.

Epochs

The smallest unit of time on the geologic time scale is the **epoch**. **Epochs** are subdivisions of periods, usually into early, middle, and late. The names of epochs of the Tertiary Period are Greek words that denote degrees of recentness; examples are Holocene ("wholly recent"), Miocene ("less recent"), and Eocene ("dawn of the recent").

Figure 13.10 shows the Geologic History of New York State at a Glance. In it, the geologic time scale is shown. Next to, and correlated with, the time scale are a series of columns that contain additional information.

Life on Earth. This column describes the types of life that existed during the different time periods based upon the fossil record.

Rock Record in N.Y. This column shows the time periods for which there is a rock and fossil record in New York State. The thick black line indicates that rock or fossil evidence from that time period exists in New York State. If there is no line, there is no rock or fossil record for that time. These gaps reflect unconformities in the rocks of New York State.

Time Distribution of Fossils (Including Important Fossils of New York). This column contains bold black lines labeled with names and circled letters. The names identify types of fossil organisms. The circled letters are keyed to the illustrations of specific index fossils printed along the top of the chart, and are placed on the line to indicate the approximate time of existence of each specific index fossil. For example, find A at the bottom of the line labeled "Trilobites." Now look to the left, and note that organism A lived at the end of the Early epoch of the Cambrian period.

Important Geologic Events in New York. This column describes in more detail how the plate tectonic events shown in the preceding column affected New York State. It describes the origin of well-known features of New York State, such as the Palisades Sill, which was intruded during the late Triassic, and the Adirondacks, which were uplifted during the Pliocene.

Inferred Position of Earth's Landmasses. This column shows the changing positions of landmasses due to plate movements. North America is shown in black to highlight its movements. The latitude and longitude lines show the positions of landmasses in relation to the Equator and poles, as well as to each other. Notice that during the Devonian and Mississippian periods, North America was centered on the Equator, a location that explains how coal beds formed from tropical rain forests.

How Can the Absolute Age of a Rock Be Determined?

Early attempts at determining the age of Earth on the basis of uniformitarianism were inaccurate at best. For example, careful measurement of the accumulation of sediments in a shallow sea over objects of known age, such as shipwrecks, shows that about 15,000 to 30,000 years are needed to form a layer of sediment 1 meter thick. Layers of sedimentary rock exist that are more than 2,000 meters thick. If the sediment accumulated at a rate of 1 meter in 20,000 years, then the layers would have taken 2,000 meters × 20,000 years per meter or *40 million years to form*! If this method is applied to the entire geologic column, an estimate for the age of Earth can be obtained.

Another early method used to estimate Earth's age was based upon the salt content of the oceans. Salt is carried into the oceans by rivers, so scientists measured the amounts of salt in all the oceans. Then they measured the amounts of salt all the world's rivers are carrying into the oceans. Finally they calculated how long the rivers would take to carry in all the salt now in the oceans.

The Origin and History of Life on Earth

The Geologic History of New York State at a Glance Chart

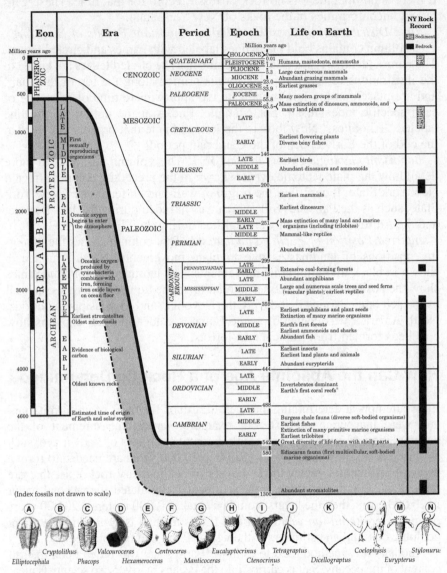

Figure 13.10 Geologic History of New York State. Source: The State Education Department, *Earth Science Reference Tables*, 2011 ed. (Albany, New York; The University of the State of New York).

Based on such attempts, the age of the Earth was estimated at several hundred million years. We now know that this figure is much too low. Dating by those early methods was not reliable because they depended too heavily on rates that vary widely from place to place and through time. What was needed was a way to measure time by a process that does not vary, a process that runs

Unit Three **EARTH'S HISTORY**

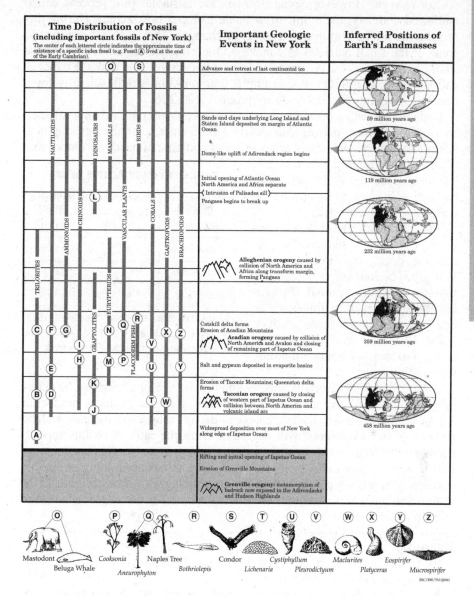

continuously through time and leaves a record with no gaps in it. The discovery of radioactivity by A. H. Becquerel in 1896 provided the needed process.

All elements are made up of tiny bits of matter called *atoms*. Most elements have a number of isotopes. **Isotopes** are varieties of the same element whose atoms differ slightly in mass. For example, most carbon atoms have a mass of twelve units. This isotope is called carbon-12. Some carbon atoms, however, have a mass of fourteen units. This isotope is carbon-14.

The most common isotopes of elements are stable, meaning their atoms do not change. However, some isotopes are unstable. In a process called **radioactive decay**, the atoms of unstable isotopes break apart. During this process the atoms go through a series of changes. They give off energy and some of the small particles that make up their nucleus. Finally, they form a stable isotope of a new element that is not radioactive. This new substance is called the decay product. For example, uranium-238, a radioactive isotope, decays slowly into the stable decay product lead-206.

Radioactive decay takes place at a steady, constant rate. It is not affected by outside factors such as changes in temperature, pressure, or chemical state. The minerals in certain types of rocks contain radioactive isotopes that start to decay when the rock forms. Therefore, the decay process may be used as a clock to determine the actual ages of these rocks.

The decay of a radioactive isotope occurs at a statistically predictable decay rate known as its half-life. The **half-life** of a radioisotope is the time required for one-half of the unstable radioisotope to change into a stable decay product. Table 13.1 lists the half-lives and decay products of four commonly used radioisotopes.

TABLE 13.1 RADIOACTIVE DECAY

RADIOACTIVE ISOTOPE	DISINTEGRATION	HALF-LIFE (years)
Carbon-14	$^{14}C \rightarrow {}^{14}N$	5.7×10^3
Potassium-40	$^{40}K \rightarrow {}^{40}Ar$, ^{40}Ca	1.3×10^9
Uranium-238	$^{238}U \rightarrow {}^{206}Pb$	4.5×10^9
Rubidium-87	$^{87}Rb \rightarrow {}^{87}Sr$	4.9×10^{10}

Source: The State Education Department, *Earth Science Reference Tables*, 2011 ed. (Albany, New York; The University of the State of New York).

If the half-life of a radioisotope is known, the age of a rock can be determined from the relative amounts of the radioisotope and its decay product in the rock. Every half-life, the percentage of decay product will increase and the percentage of radioisotope will decrease. Thus, the ratio of radioisotope to decay product indicates how many half-lives have gone by. This information, in turn, reveals how many years the decay process has been going on. In this manner, geologists are able to find the absolute ages of rocks. Figure 13.11 shows how the relative amounts of radioisotope and its decay product change as each half-life elapses.

Example:

A rock contains 50 grams of potassium-40 and 50 grams of its decay product argon-40. How old is the rock?

Since the entire decay product was originally radioactive isotope, the rock originally contained 100 grams of potassium-40 (the 50 g of the radioisotope that still exist plus the 50 g that decayed into argon-40). Thus, exactly one-half of the original radioisotope has decayed; the rock is one half-life old. From Table 13.1 we know that the half-life of potassium-40 is 1.3×10^9 years, or 1.3 billion years.

The rock is 1.3 billion years old.

There are two reasons why a certain isotope may be used to date a rock. It may be used because it is present in the minerals of that rock, or it may be used because its half-life is long or short. Isotopes with long half-lives are used to date very old rocks; those with short half-lives, to date younger objects. For example, potassium-40 can be used to date rocks between 100,000 and 4.6 billion years old. On the other hand, carbon-14 is best for dating objects between 100 and 50,000 years old. Table 13.2 gives data for six radioisotopes used in dating.

TABLE 13.2 SIX RADIOISOTOPES USED IN DATING*

Isotopes				
Parent	Decay Product	Half-life (years)	Dating Range (years)	Minerals or Other Materials That Can Be Dated
Uranium-238	Lead-206	4.5 billion	10 million–4.6 billion	Zircon
Uranium-235	Lead-207	710 billion	10 million–4.6 billion	uraninite
Uranium-232	Lead-208	14 billion	10 million–4.6 billion	
Potassium-40	Argon-40 Calcium-40	1.3 billion	50,000–4.6 billion	Muscovite, biotite, hornblende
Rubidium-87	Strontium-87	47 billion	10 million–4.6 billion	Muscovite, biotite, potassium feldspar
Carbon-14	Nitrogen-14	5,730 ± 30	100–70,000	Wood, peat, grain, charcoal, bone, tissue, cloth, shell, stalacites, glacier ice, ocean water

*From *The Dynamic Earth: An Introduction to Physical Geology*, Brian J. Skinner and Stephen C. Porter, copyright © 1992 by John Wiley & Sons, Inc. Reprinted by permission of John Wiley & Sons, Inc.

Carbon-14 is especially useful because it can be used to date the remains of living things. All living things contain carbon, and some of that carbon is carbon-14, which is present in air, water, and food. As long as an organism is alive, the amount of carbon-14 in its body remains constant. Whatever decays is quickly replaced from its surroundings. However, when the organism dies, the carbon-14 that decays is not replaced. The longer the organism has been dead, the less carbon-14 remains. Carbon-14 can be used to date fossils in which the

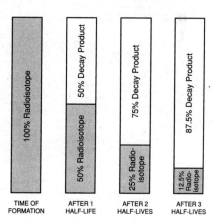

Figure 13.11 Ratios of Isotope and Decay Product after One, Two, and Three Half-Lives.

The Origin and History of Life on Earth

original material remains unchanged, such as wood, bones, and shells. However, this radioisotope has a short half-life. After about 50,000 years the amount of carbon-14 left in a fossil is too small to be measured. Therefore, other radioisotopes (e.g., potassium-40) must be used to date older rocks.

Example:

A wood branch is found buried beneath layers of sediment in a lake. The wood is found to contain one-fourth of the carbon-14 present in contemporary wood. How old is the layer of sediment in which the wood is found?

After one half-life, a sample would contain one-half the normal amount of carbon-14. After a second half-life, half of that would have decayed, leaving one-quarter the normal amount of carbon-14. Therefore, two half-lives have

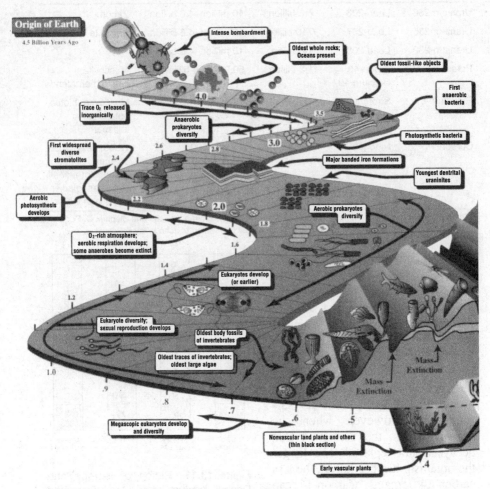

Figure 13.12 Schematic History of Life on Earth.
Source: *Earth: Evolution of a Habitable World*, Jonathan I. Lunine, Cambridge University Press, 1999.

elapsed since the tree died. Table 13.1 indicates that the half-life of carbon-14 is 5.7×10^3 years, or 5,700 years.

Since two half-lives have elapsed, the wood and therefore the layer of sediment are $5,700 \times 2$ or 11,400 years old.

How Can the Rock Record and Fossil Evidence Reveal Changes in Past Life and Ancient Environments?

Many thousands of layers of sedimentary rock contain evidence of the long history of Earth and the changing life-forms whose fossils are found in the rocks. Fossils show us that a great variety of plants and animals lived on Earth in the past. Fossils can be compared to one another and to living organ-

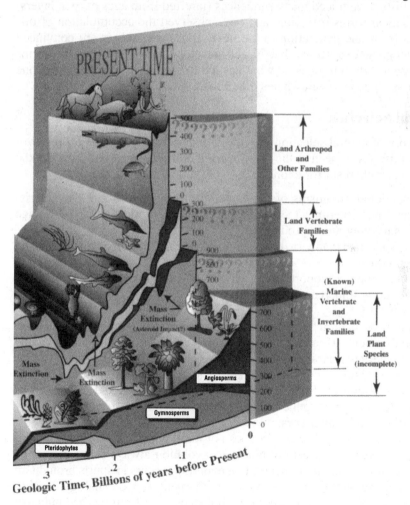

isms by their similarities and differences. Some fossil organisms are similar to existing organisms, but many are quite different. A study of the fossil record reveals that more recently deposited rock layers are more likely to contain fossils resembling existing life-forms and that most life-forms of the geologic past have become extinct.

Evolution

The history of life on Earth depicted by the fossil record is one of change. The fossil record shows that, since its beginning, many new life-forms have appeared and most old forms have disappeared. It reveals also that life-forms have changed gradually over time, or evolved, so that the current life-forms differ significantly from the earliest ones.

Scientists have traced many sequences (inferred from ages of rock layers) of anatomical forms over time and have observed the accumulation of differences from one generation to the next. These researchers are convinced that **biological evolution**, that is, the development of existing life-forms from earlier, different ones, is what has led to species as different from one another as algae are from whales. See Figure 13.12.

Natural Selection

The theory of natural selection provides a scientific explanation for the evolution of life-forms seen in the fossil record. **Natural selection** proposes the following mechanism for evolution:

1. Individual organisms of the same species have different characteristics, and sometimes these differences give one organism an advantage in surviving and reproducing.
2. Offspring that inherit the advantage are more likely to survive and reproduce.
3. Over time the proportion of individuals with the advantageous characteristic will increase.

In this way, natural selection leads to organisms that are well suited to their environments. Changes in the environment can affect the survival value of some inherited characteristics. Then the characteristics that are advantageous for survival may change, so changes in the environment can lead to changes in organisms. Small differences between parents and offspring can accumulate over many generations so that descendants may become very different from their ancestors.

Evolution by natural selection does not imply long-term progress toward a goal, or have a set direction. Nor does evolution always occur gradually. Much recent information indicates that evolution occurs in spurts brought on by sudden, large-scale changes in the environment. A good analogy for evolution is the growth of a hedge. Some branches exist from the beginning of the hedge's life with little or no change. Other branches grow and then die out. Still others grow a little, and then branch apart repeatedly. The end result is a complex network of large and small branches.

On the basis of the fossil record, life on Earth is thought to have begun about 4 billion years ago as simple, one-celled organisms. During the first 2 billion years, only these one-celled organisms existed. About 1 billion years ago, cells with nuclei began to develop. Since then, increasingly complex multicellular organisms have evolved.

One important reason to preserve life on Earth is that evolution builds on what exists. The more variety there is at the present time, the even greater variety can exist in the future. Human behavior that results in the extinction of organisms decreases the variety of life-forms on Earth and may have far-reaching effects in the future.

Ancient Environments

Rocks in which fossils are found provide evidence of ancient environments. Many fossils are clear indicators of particular types of environment. For example, fossil coral that is almost identical to existing corals can be found in some rocks. Corals that exist today can live only in shallow, tropical seas. Therefore, we can infer that rocks containing fossil coral existed in a similar

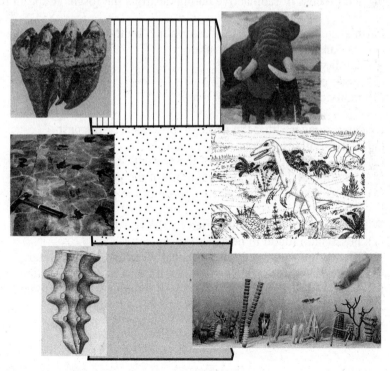

Figure 13.13 Fossil-Bearing Rock Layers with Illustrations of Environments Inferred from Fossil Evidence.

environment. Oak and birch trees inhabit moist, temperate climates. Abundant oak and birch pollen in a sedimentary layer implies that, when the rock layer formed, the climate was moist and temperate. The fossils in coal are the remains of tropical rain forest plants and animals, so layers of coal imply a tropical rain forest origin. Inferences of this type, together with the inferences that can be drawn from the composition of a given sedimentary layer, enable geologists to piece together a model of the environment that existed when that particular rock layer formed. See Figure 13.13.

MULTIPLE-CHOICE QUESTIONS

In each case, write the number of the word or expression that best answers the question or completes the statement.

1. It is difficult to state with certainty the way in which life on Earth originated because
   (1) Earth's original life-forms left no fossil record
   (2) no rocks date back to the time when life originated
   (3) the age of rocks formed at the time when life originated cannot be determined
   (4) the process of radioactive dating destroys the fossils found in rocks

2. Earth's age is most accurately determined by the
   (1) thickness of sedimentary strata
   (2) salinity of the oceans
   (3) study of fossils
   (4) radioactive dating of rock masses

3. Evidence suggests that the geologic processes of the past
   (1) were similar to those of the present
   (2) were different from those of the present
   (3) occurred at a faster rate than those of the present
   (4) occurred at a slower rate than those of the present

4. Which statement is best supported by the fossil record?
   (1) Fossils are found in nearly all rocks.
   (2) Fossils are found only in areas that were once under water.
   (3) Most early life-forms that left fossil remains are now extinct.
   (4) Most early life-forms that left fossil remains still exist today.

5. Unless a series of sedimentary rock layers has been overturned, the bottom rock layer usually
   (1) contains fossils
   (2) is the oldest
   (3) contains the greatest variety of minerals
   (4) has the finest texture

6. The cross section below shows a rock sequence that has not been overturned.

   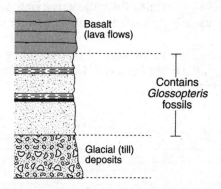

   Which event occurred last at this location?
   (1) Shale was deposited.
   (2) Glacial till was deposited.
   (3) Basaltic lava flows solidified.
   (4) *Glossopteris* flourished and then became extinct.

7. Older layers of rock may be found on top of younger layers of rock as a result of
   (1) weathering processes
   (2) igneous extrusions
   (3) joints in the rock layers
   (4) overturning of rock layers

8. Which feature in a rock layer is older than the rock layer?
   (1) igneous intrusions
   (2) mineral veins
   (3) rock fragments
   (4) faults

9. Unconformities (buried erosional surfaces) are good evidence that
   (1) many life-forms have become extinct
   (2) the earliest life-forms lived in the sea
   (3) part of the geologic rock record is missing
   (4) metamorphic rocks have formed from sedimentary rocks

10. The presence of brachiopod, nautiloid, and coral fossils in the surface bedrock of a certain area indicates the area was once covered by
    (1) tropical vegetation
    (2) glacial deposits
    (3) volcanic ash
    (4) ocean water

11. Why can layers of volcanic ash found between other rock layers often serve as good geologic time markers?
    (1) Volcanic ash usually occurs in narrow bands around volcanoes.
    (2) Volcanic ash usually contains index fossils.
    (3) Volcanic ash usually contains the radioactive isotope carbon-14.
    (4) Volcanic ash is usually deposited rapidly over a large area.

Base your answers to questions 12 through 14 on the geologic cross section below in which overturning has not occurred. Letters A through H represent rock layers.

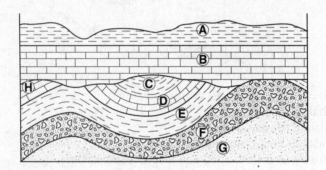

12. Which sequence of events most likely caused the unconformity shown at the bottom of rock layer B?
    (1) folding → uplift → erosion → deposition
    (2) intrusion → erosion → folding → uplift
    (3) erosion → folding → deposition → intrusion
    (4) deposition → uplift → erosion → folding

13. The folding of rock layers G through C was most likely caused by
    (1) erosion of overlying sediments
    (2) contact metamorphism
    (3) the collision of lithospheric plates
    (4) the extrusion of igneous rock

14. Which two letters represent bedrock of the same age?
    (1) A and E
    (2) B and D
    (3) F and G
    (4) D and H

15. The best method for the correlation of sedimentary rock layers several hundred kilometers apart is by comparing the
    (1) index fossils in the layers
    (2) layers by walking the outcrop
    (3) thickness of the rock layers
    (4) color of the rock layers

16. Which characteristic is most useful in correlating Devonian-age sedimentary bedrock in New York State with Devonian-age sedimentary bedrock in other parts of the world?
    (1) color
    (2) index fossils
    (3) rock types
    (4) particle size

Base your answers to questions 17 through 20 on the cross sections of three rock outcrops, *A*, *B*, and *C*. Line *XY* represents a fault. Overturning has not occurred in the rock outcrops.

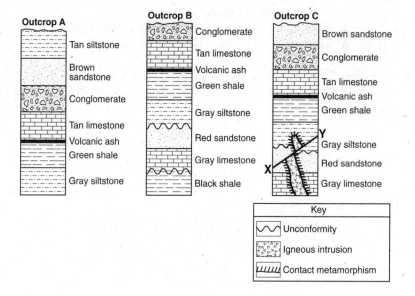

17. The volcanic ash layer is considered a good time marker for correlating rocks because the volcanic ash layer
    (1) has a dark color
    (2) can be dated using carbon-14
    (3) lacks fossils
    (4) was rapidly deposited over a wide area

18. Which sedimentary rock shown in the outcrops is the youngest?
    (1) black shale   (3) tan siltstone
    (2) conglomerate   (4) brown sandstone

19. What is the youngest geologic feature in the three bottom layers of outcrop *C*?
    (1) fault   (3) unconformity
    (2) igneous intrusion (4) zone of contact metamorphism

20. Which processes were primarily responsible for the formation of most of the rock in outcrop *A*?
    (1) melting and solidification
    (2) heating and compression
    (3) compaction and cementation
    (4) weathering and erosion

21. The division of Earth's geologic history into units of time called eons, eras, periods, and epochs is based on
    (1) absolute dating techniques   (3) climatic changes
    (2) fossil evidence   (4) seismic data

22. The cross sections below represent three widely separated outcrops of exposed bedrock. Letters *A*, *B*, *C*, and *D* represent fossils found in the rock layers.

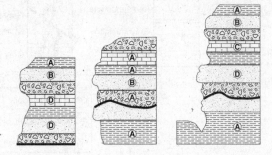

Which fossil appears to have the best characteristics of an index fossil?
(1) *A*   (3) *C*
(2) *B*   (4) *D*

Base your answers to questions 23 through 26 on the cross sections below, which represent two bedrock outcrops 15 kilometers apart. The rock layers have been numbered for identification, and some contain the index fossil remains shown.

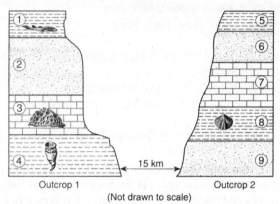

23. When these rocks were deposited as sediments, this area was most likely
    (1) under the ocean
    (2) a desert between high mountains
    (3) repeatedly covered by lava flows
    (4) glaciated several times

24. Both organisms that formed the fossils found in rock layers 3 and 4
    (1) lived during the same period of geologic time
    (2) lived in polar regions
    (3) are members of the same group of organisms
    (4) are still alive today

25. Evidence best indicates that rock layers 4 and 8 were deposited during the same geologic period because both layers
    (1) contain the same index fossil
    (2) are composed of glacial sediments
    (3) contain index fossils of the same age
    (4) are found in the same area

26. Which two types of organisms survived the mass extinction that occurred at the end of the Permian Period?
    (1) trilobites and nautiloids
    (2) corals and vascular plants
    (3) placoderm fish and graptolites
    (4) gastropods and eurypterids

The Origin and History of Life on Earth

27. One reason *Tetragraptus* is considered a good index fossil is that *Tetragraptus*
  (1) existed during a large part of the Paleozoic Era
  (2) has no living relatives found on Earth today
  (3) existed over a wide geographic area
  (4) has been found in New York State

28. According to the fossil record, which group of organisms has existed for the greatest length of time?
  (1) gastropods   (3) mammals
  (2) corals   (4) vascular plants

Base your answers to questions 29 through 31 on the passage below.

Fossils and the History of Earth's Rotation

Data from coral fossils support the hypothesis that Earth's rotation rate has been slowing down by about 2.5 seconds per 100,000 years. Scientists believe this is due to the frictional effects of ocean tides. This slowing rotation rate decreases the number of days in the year.

Scientists have discovered that corals produce a thin layer of shell every day, resulting in growth rings. These daily layers are separated by yearly ridges. The Devonian coral fossil, *Pleurodictyum*, has approximately 400 growth rings between each yearly ridge, which suggests that there were about 400 days in a year during the Devonian Period. Supporting this hypothesis, scientists have found coral from the Pennsylvanian Period that have about 390 growth rings per year, while present-day corals have about 365 growth rings per year.

29. Approximately how many fewer Earth days per year are there today than there were during the Devonian Period?
  (1) 10   (3) 35
  (2) 25   (4) 40

30. What inference can be made about the number of growth rings per year for a coral from the Permian Period and Ordovician Period compared with the number of growth rings per year for the Devonian coral, *Pleurodictyum*?
  (1) Ordovician coral would have fewer, but Permian coral would have more.
  (2) Ordovician coral would have more, but Permian coral would have fewer.
  (3) Both Ordovician and Permian coral would have fewer.
  (4) Both Ordovician and Permian coral would have more.

31. The evidence of the fossil *Pleurodictyum* found in surface bedrock in the Finger Lakes region of New York State suggests that this region was once
    (1) covered by a glacial ice sheet
    (2) covered by a warm, shallow sea
    (3) located in a desert area
    (4) located in a tropical rain forest

32. The graph below shows the radioactive decay of a 50-gram sample of a radioactive isotope.

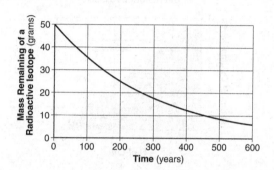

   According to the graph, what is the half-life of this isotope?
    (1) 100 years
    (2) 150 years
    (3) 200 years
    (4) 300 years

33. The graph below shows the rate of decay of the radioactive isotope K-40 into the decay products Ar-40 and Ca-40.

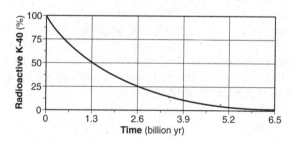

   Analysis of a basalt rock sample shows that 25% of its radioactive K-40 remained undecayed. How old is the basalt?
    (1) 1.3 billion years
    (2) 2.6 billion years
    (3) 3.9 billion years
    (4) 4.6 billion years

34. A whalebone that originally contained 200 grams of radioactive carbon-14 now contains 25 grams of carbon-14. How many carbon-14 half-lives have passed since this whale was alive?
    (1) 1   (3) 3
    (2) 2   (4) 4

Base your answers to questions 35 and 36 on the graph below, which shows the generalized rate of decay of radioactive isotopes over 5 half-lives.

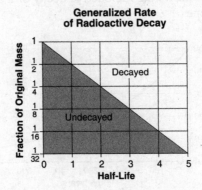

35. If the original mass of a radioactive isotope was 24 grams, how many grams would remain after 3 half-lives?
    (1) 12   (3) 3
    (2) 24   (4) 6

36. Which radioactive isotope takes the greatest amount of time to undergo the change shown on the graph?
    (1) carbon-14   (3) uranium-238
    (2) potassium-40   (4) rubidium-87

37. An igneous rock contains 10 grams of radioactive potassium-40 and a total of 10 grams of its decay products. During which geologic time interval was this rock most likely formed?
    (1) Middle Archean   (3) Middle Proterozoic
    (2) Late Archean   (4) Late Proterozoic

38. The graph below shows the extinction rate of organisms on Earth during the last 600 million years. Letters *A* through *D* represent mass extinctions.

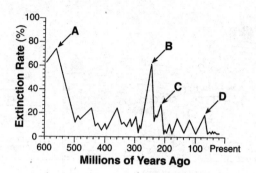

Which letter indicates when dinosaurs became extinct?
(1) *A*   (3) *C*
(2) *B*   (4) *D*

39. According to the fossil record, which sequence correctly represents the evolution of life on Earth?
(1) fish → amphibians → mammals → soft-bodied organisms
(2) fish → soft-bodied organisms → mammals → amphibians
(3) soft-bodied organisms → amphibians → fish → mammals
(4) soft-bodied organisms → fish → amphibians → mammals

40. Which time line most accurately indicates when this sequence of events in Earth's history occurred?

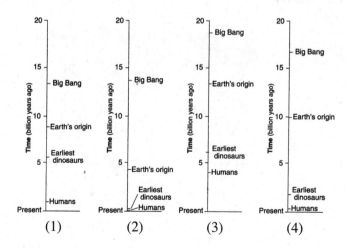

CONSTRUCTED RESPONSE QUESTIONS

Base your answers to questions 41 through 43 on the geologic cross section below. The rock layers have not been overturned.

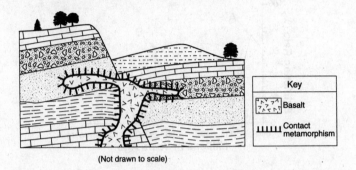

(Not drawn to scale)

41. The index fossil *Dicellograptus* was found in the shale layer. During which geologic time period did this shale layer form? [1]

42. Describe one piece of evidence from the cross section that supports the inference that the fault is older than the basalt intrusion. [1]

43. Explain why carbon-14 could not be used to determine the age of the *Dicellograptus* fossil. [1]

Base your answers to questions 44 through 47 on the diagrams below, which represent two bedrock outcrops, I and II, found several kilometers apart in New York State. Rock layers are lettered *A* through *F*. Drawings represent specific index fossils.

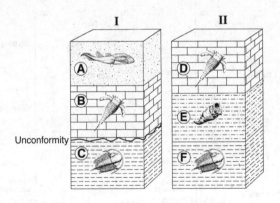

44. During which geologic time period was rock layer *C* deposited? [1]

45. Identify *two* processes that produced the unconformity in outcrop I. [1]

46. Describe *one* characteristic a fossil must have in order to be considered a good index fossil. [1]

47. Explain why carbon-14 *cannot* be used to find the geologic age of these index fossils. [1]

Base your answers to questions 48 through 50 on the table of index fossils shown below and on your knowledge of Earth.

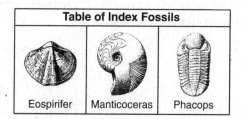

48. During what geologic time period did the oldest index fossil shown in this table exist? [1]

49. State *one* characteristic of a good index fossil. [1]

50. Complete the classification table below by filling in the general fossil group name for *each* index fossil. [1]

Fossil Classification

Index Fossil	Eospirifer	Manticoceras	Phacops
General Fossil Group			

51. The table below shows information about Earth's geologic history. Letter X represents information that has been omitted.

Period	Million Years Ago	Index Fossil Found in Bedrock	Important Geologic Event
Triassic	251 to 200	*Coelophysis*	X

Identify *one* important geologic event that occurred in New York State that could be placed in the box at X. [1]

Base your answers to questions 52 and 53 on the data table below, which shows the radioactive decay of carbon-14. The number of years required to complete four half-lives has been left blank.

Radioactive Decay of Carbon-14

Number of Half-Lives	Percentage of Original Carbon-14 Remaining	Time (years)
0	100	0
1	50	5700
2	25	11,400
3	12.5	17,100
4	6.3	
5	3.1	28,500
6	1.6	34,200

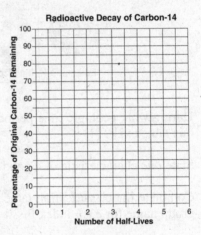

52. On the grid next to the data table, construct a graph that shows the radioactive decay of carbon-14 by plotting an **X** to show the percentage of original carbon-14 remaining after each half-life. Connect the **X**s with a smooth, curved line. [1]

53. How long does radioactive carbon-14 take to complete four half-lives? [1]

54. The cross section below shows part of Earth's crust. The objects in parentheses indicate materials found within each rock unit or deposit.

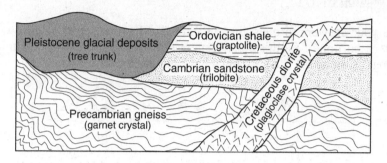

Which object in parentheses could be accurately dated using carbon-14? Explain your answer. [1]

Base your answers to questions 55 through 57 on the cross sections below, which show widely separated outcrops labeled I, II, and III. Index fossils are found in some of the rock layers in the three outcrops. In outcrop III, layers *A*, *B*, *C*, and *D* are labeled. Line *XY* represents an unconformity. Line *GH* represents a fault.

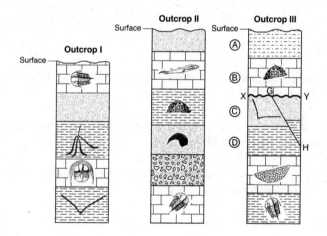

55. Describe one characteristic necessary for a fossil to be classified as an index fossil. [1]

56. On outcrop II above, place the symbol for an unconformity ◠◠ between the two rock layers where the Silurian-age bedrock has been removed by erosion. [1]

285

The Origin and History of Life on Earth

57. List in order, from oldest to youngest, the relative age of the four rock layers, *A*, *B*, *C*, and *D*, of fault *GH*, and of unconformity *XY* shown in outcrop III. [1]

Base your answers to questions 58 through 61 on the cross section which shows a portion of Earth's crust. Letters *A* through *J* represent rock units or geologic structures. The rock units have not been overturned.

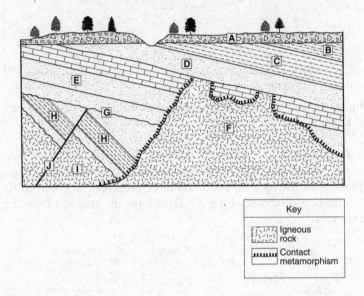

58. On the cross section above, draw a circle around the letter of the oldest rock unit shown. [1]

59. On the same cross section, place an **X** to indicate a location where the rock, marble, was formed. [1]

60. Describe one piece of evidence shown in the cross section that suggests rock unit *D* is younger than rock unit *F*. [1]

61. Explain why rock unit *H* is not one continuous layer. [1]

EXTENDED CONSTRUCTED RESPONSE QUESTIONS

Base your answers to questions 62 through 67 on the geologic cross section shown below. Rock units *A* through *H* are shown. Several rock units contain fossils. Rock unit *G* was formed in a zone of contact metamorphism.

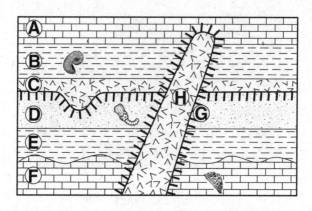

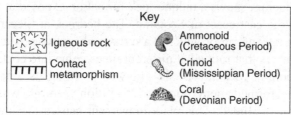

62. Place two **X**s on the cross section above to show the locations of two unconformities that formed at different times in geologic history. [1]

63. Identify two possible geologic periods during which the sediments that formed rock unit *E* could have been deposited. [1]

64. Describe the evidence shown in the cross section that indicates that rock unit *C* is younger than rock unit *D*. [1]

65. Identify the letter of the rock unit that was formed at the same time as igneous rock unit *H*. [1]

66. Identify one geologic period during which igneous intrusion *H* could have formed. [1]

67. Explain why the absolute age of the fossils shown in the cross section *cannot* be determined by using radioactive carbon-14. [1]

Base your answers to questions 68 through 71 on the passage below and on the diagram shown below.

Siccar Point

The diagram shows a unique rock formation exposed at Siccar Point on the east coast of Scotland. The bedrock at Siccar Point shows an unconformity, which is a surface where two separate sets of rock layers that formed at different times come into contact.

The bottom rock layers are graywacke, which is a form of sandstone, formed approximately 425 million years ago when tectonic plates collided. This plate movement caused the layers of graywacke to tilt into their present vertical orientation and eventually uplifted them above sea level to form mountains.

By about 345 million years ago, these mountains had been eroded to form a plain that submerged beneath the sea. More sediment was deposited on top of the vertical graywacke layers, eventually forming the nearly horizontal layers called the Old Red Sandstone.

68. On the diagram above, draw a dark, heavy line tracing the unconformity separating the graywacke from the Old Red Sandstone. [1]

69. During which geologic time period did the graywacke bedrock form? [1]

70. Describe the structural evidence shown by the bedrock at Siccar Point that led geologists to conclude that the graywacke was moved by converging tectonic plates. [1]

71. Identify *two* of the processes that produced the unconformity at Siccar Point. [1]

Base your answers to questions 72 through 76 on the passage and map below. Point *F* on the map shows the location where an unusual mammal fossil was found.

Fossil Jaw of Mammal Found in South America

Paleontologists working in Patagonia have found the tiny fossil jaw that may be the first evidence of early mammals in South America. The fossil, which measures less than a quarter-inch long, is believed to be from the middle or late Jurassic Period. Researchers said it suggests that mammals developed independently in the Southern Hemisphere.

The fossil, named *Asfaltomylos patagonicus*, was discovered in a shale formation in Patagonia. Dinosaurs were the dominant land animal at that time. Mammals were tiny and hunted insects in the dense tropical vegetation. The now-arid region also has yielded some remarkable dinosaur fossils from the same period in a vast ancient boneyard covering hundreds of square miles.

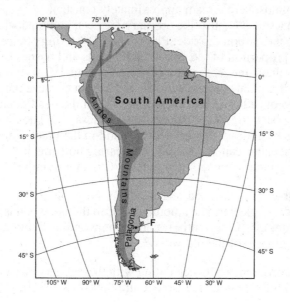

72. State the latitude and longitude of point *F*, to the nearest degree, where the fossil *Asfaltomylos patagonicus* was discovered. Include the correct units and compass directions in your answer. [1]

73. State the name of the dominant sediment particle that was compacted to form the shale in which this fossil was found. [1]

74. What other life-form first appeared on Earth during the geologic period when *Asfaltomylos patagonicus* existed? [1]

75. State one method used by geologists to determine the age of the bedrock in which this ancient mammal fossil was found. [1]

76. Explain how the uplift of the Andes Mountains changed eastern Patagonia's climate from a wet tropical forest at the time *Asfaltomylos patagonicus* lived to the arid conditions of today. [1]

Base your answers to questions 77 and 78 on the passage below and on your knowledge of Earth science.

Radiocarbon Dating

Radioactive carbon-14 (^{14}C), because of its short half-life, is used for the absolute dating of organic remains that are less than 70,000 years old. Carbon-14 is an isotope of carbon that is produced in Earth's upper atmosphere. High-energy cosmic rays from the Sun hit nitrogen-14 (^{14}N), producing radioactive ^{14}C. This ^{14}C is unstable and will eventually change back into ^{14}N through the process of radioactive decay. The proportions of ^{14}C and ordinary ^{12}C in Earth's atmosphere remain approximately constant.

Radioactive ^{14}C, just like ordinary ^{12}C, can combine with oxygen to make carbon dioxide. Plants use CO_2 during photosynthesis. The proportion of ^{14}C to ^{12}C in the cells and tissues of living plants is the same as the proportion of ^{14}C to ^{12}C in the atmosphere. After plants die, no new ^{14}C is taken in because there is no more photosynthesis. Meanwhile, the ^{14}C in the dead plant keeps changing back to ^{14}N, so there is less and less ^{14}C. The longer the plant has been dead, the less ^{14}C is found in the plant. The age of organic remains can be found by comparing how much ^{14}C is still in the organic remains to how much ^{14}C is in a living organism.

77. Radioactive ^{14}C was used to determine the geologic age of old wood preserved in a glacier. The amount of ^{14}C in the old wood is half the normal amount of ^{14}C currently found in the wood of living trees. What is the geologic age of the old wood? [1]

78. State *one* difference between dating with the radioactive isotope ^{14}C and dating with the radioactive isotope uranium-238 (^{238}U). [1]

Unit Four
EARTH MATERIALS

CHAPTER 14
MINERALS

KEY IDEAS The solid part of Earth is made up of rocks. Rocks, in turn, are composed of minerals, and most minerals are made of chemical elements. Earth consists of a great variety of minerals, whose properties depend on the history of how they were formed as well as the elements of which they are composed. Most properties of minerals can be explained in terms of the arrangement and properties of the atoms that compose them.

Minerals are made and remade on Earth's surface, in its oceans, and in the hot and high-pressure layers inside Earth. Observing and classifying minerals has helped us understand the great variety and complexity of Earth materials. Through the study of minerals, we have gained insight into Earth's historical development and its dynamics.

Minerals are important to us because many are sources of essential industrial materials, such as iron, copper, aluminum, and magnesium. The abundance of minerals ranges from almost unlimited to extremely rare. Many of the best sources, however, are depleted, making it more difficult and expensive to obtain those minerals. Also, the difficulty of extracting minerals from Earth's crust has important economic and environmental impacts. As limited resources, minerals must be used wisely.

KEY OBJECTIVES
Upon completion of this chapter, you will be able to:

- Explain how the physical properties of minerals are determined by their chemical compositions and crystal structures.
- Describe how minerals can be identified by well-defined physical and chemical properties, such as cleavage, fracture, color, specific gravity, hardness, streak, luster, crystal shape, and reaction with acid.
- Explain that chemical composition and physical properties determine how humans use minerals.
- Describe how minerals are formed inorganically by the process of crystallization as a result of specific environmental conditions. These include:

Minerals

- Cooling and solidification of magma.
- Precipitation from water caused by such processes as evaporation, chemical reactions, and temperature changes.
- Rearrangement of atoms in existing minerals subjected to conditions of high temperature and pressure.

WHAT IS A MINERAL?

If you walk outside and pick up any earth material—a rock, sand, soil, gravel, or mud—you will hold minerals in your hand. Nearly all rocks are composed of one or more substances called *minerals*. When rocks are broken down into smaller pieces such as pebbles or sand, those smaller pieces are composed of the same minerals that were in the rock.

Nearly every single thing we use is made of, or contains, minerals. Minerals are essential to the life processes of plants and animals. Industry is equally dependent upon an abundant supply of minerals as raw materials. Without minerals, none of the devices and structures that surround us in our daily lives could be made.

Some minerals are rare and highly prized for their characteristics, such as gold for its conductivity, resistance to corrosion, and high luster, and diamonds for their beauty and hardness. Some minerals are raw materials from which industries manufacture the products that are the basis of a nation's wealth. Hematite, an iron ore, is needed to make the steel from which products ranging from cars to skyscrapers are manufactured. As such, minerals are of strategic importance and wars have been fought to gain control over mineral resources. Some minerals, however, are so abundant that they have little commercial value.

The word *mineral* means different things to different people. To a nutritionist, minerals are things to be eaten, along with proteins, carbohydrates, and vitamins. To a jeweler, a mineral is a stone to be cut or polished. To a geologist, a **mineral** is a naturally occurring, inorganic compound with a fixed chemical composition and an orderly internal arrangement of atoms.

Defining Characteristics of Minerals

Although there are many different minerals, they all share certain characteristics.

Minerals are naturally occurring. A **naturally occurring** material is formed as result of natural processes in or on Earth. It is not manufactured in a factory or synthesized in a laboratory. For example, most diamonds are formed naturally in Earth and are minerals. Synthetic diamonds made in the laboratory are not minerals.

Minerals are inorganic matter. **Inorganic** substances are not alive, never were alive, and do not come from living things. Thus, amber (a tree resin in which insects are often found embedded) and the fossil fuels coal, petroleum, and natural gas are not true minerals. They were formed from organic substances, animal or vegetable material, that once lived on Earth.

A chemical symbol or formula can be written for a mineral. Minerals are either elements or compounds. Both elements and compounds have definite chemical and physical properties.

Elements are substances that cannot be broken down into simpler substances by ordinary chemical means. Ninety-two different elements have been found to occur naturally on Earth, each with distinctly different physical and chemical properties. Elements consist of particles called atoms, and all atoms of an element have the same properties. One- or two-letter symbols are used to represent the atoms of elements. For example, the symbol for the element oxygen is O and the symbol for the element silicon is Si.

Table 14.1 shows the relative abundances of different elements in Earth's crust, hydrosphere, and troposphere. Notice that only eight elements make up more than 98 percent of Earth's crust, the two most common elements being oxygen and silicon.

TABLE 14.1 AVERAGE CHEMICAL COMPOSITION OF EARTH'S CRUST, HYDROSPHERE, AND TROPHOSPHERE

ELEMENT (symbol)	CRUST		HYDROSPHERE	TROPOSPHERE
	Percent by mass	Percent by volume	Percent by volume	Percent by volume
Oxygen (O)	46.10	94.04	33.0	21.0
Silicon (Si)	28.20	0.88		
Aluminum (Al)	8.23	0.48		
Iron (Fe)	5.63	0.49		
Calcium (Ca)	4.15	1.18		
Sodium (Na)	2.36	1.11		
Magnesium (Mg)	2.33	0.33		
Potassium (K)	2.09	1.42		
Nitrogen (N)				78.0
Hydrogen (H)			66.0	
Other	0.91	0.07	1.0	1.0

Source: The State Education Department, Earth Science Reference Tables, 2011 ed. (Albany, New York; The University of the Sateof New York)

Although some minerals found in Earth's crust are pure elements, such as native copper and silver, the elements in Earth's crust rarely exist by themselves. Most are chemically combined with other elements as compounds. *Compounds* consist of molecules, that is, groups of atoms joined together in a definite proportion. For example, the mineral calcite is a compound of the elements calcium, carbon, and oxygen (see Figure 14.1a). Every calcite molecule contains one atom of calcium, one atom of carbon, and three atoms of oxygen. To show the chemical composition of a mineral, a formula can be written. The

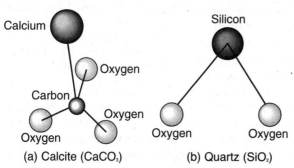

(a) Calcite ($CaCO_3$) (b) Quartz (SiO_2)

Figure 14.1 Molecular Models of Calcite and Quartz.

Minerals

chemical formula for calcite is $CaCO_3$. The mineral quartz consists of molecules each containing one atom of silicon and two atoms of oxygen (Figure 14.1b). The chemical formula for quartz is SiO_2.

Every compound has distinct properties of its own. Thus, a mineral, which is either an element or a compound, has both a definite composition and distinct physical and chemical properties. Table 14.2 shows the chemical names and formulas of a few common minerals. The last column of the Properties of Common Minerals chart in the *Earth Science Reference Tables* has a more extensive list of the chemical compositions of common minerals.

TABLE 14.2 CHEMICAL NAMES AND FORMULAS OF SOME COMMON MINERALS

Mineral	Chemical Name	Chemical Formula
Calcite	Calcium carbonate	$CaCO_3$
Galena	Lead sulfide	PbS
Gypsum	Calcium sulfate-water	$CaSO_2 \cdot 2H_2O$
Olivine (fosterite)	Magnesium silicate	Mg_2SiO_4
Potassium Feldspar	Potassium aluminum silicate	$KAlSi_3O_8$
Pyrite	Iron sulfide	FeS_2
Quartz	Silicon dioxide	SiO_2

Minerals have a crystalline form. The atoms or molecules of a mineral are the same throughout that mineral. When they are joined in fixed positions as a solid, a definite pattern is formed. A solid having a definite internal structural pattern is said to have a **crystalline** form. If the pattern is large enough to be seen with the unaided eye, the solid is called a **crystal**. The crystal form of a mineral determines its cleavage, or the way it splits or breaks, as well as many other properties. Mica, for example, splits into thin, flat sheets (see Figure 14.2).

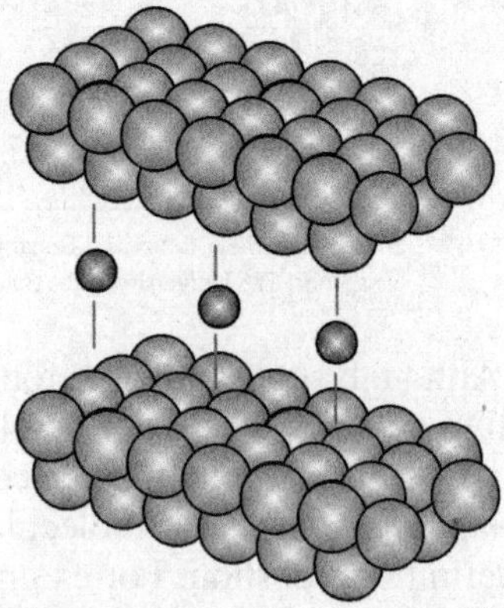

Identifying Minerals

Mineralogists have examined Earth materials and identified more than 2,000 minerals. Minerals can be identified on the basis of well-defined physical and chemical properties. Several important properties commonly used in mineral identification are color, luster, streak, hardness, and cleavage, parting, and fracture. Other properties used to identify minerals include specific gravity, radioactivity, luminescence, and chemical, thermal, electrical, and magnetic properties, as well as elasticity and strength.

Figure 14.2 The Crystalline Pattern of Mica. Mica consists of sheets of tightly bonded silicon and oxygen atoms held together weakly by metallic ions.

Since no single property can be used to identify all minerals, mineral identification is usually a process of elimination. As each property is observed, it becomes evident what a mineral is not, rather than what it is. Step by step, the possibilities are narrowed down until the identity of the mineral is determined.

Color

Color is often the first property noticed about a mineral. When observing color, it is important to use a fresh surface of a mineral since exposed surfaces are often discolored by weathering. Color alone is an unreliable property by which to identify a mineral. Although some minerals are always the same color (e.g., sulfur is yellow), others may occur in a variety of colors. Thus quartz is found in many colors, some of which are rare (these specimens are highly prized as semiprecious gems). For example, amethyst is purple quartz, citrine is yellow quartz, and rose quartz is pink quartz. Therefore, quartz cannot be identified by color alone. Another reason why color is unreliable is that many minerals have almost the same color. For example, calcite, quartz, and halite all occur in white and transparent varieties and can look strikingly alike.

Luster

The **luster** of a mineral is the way light reflects from its surface. Luster can be metallic or nonmetallic. Nonmetallic lusters can be described as glassy, brilliant, greasy or oily, waxy, silky, pearly, or earthy.

Related to the luster of a mineral are its transparency and iridescence, or the play of colors in its interior or exterior.

Streak

Streak is the color of the fine powder left when a mineral is rubbed against a hard, rough surface. A piece of unglazed porcelain, called a *streak plate*, is usually used to create a streak. Although the color of a mineral may vary, the streak of a particular mineral is always the same color. For example, fluorite may range in color from green to blue; yet its streak is always white. This characteristic makes streak a useful property for identifying a mineral. It must be remembered, though, that a mineral's streak is not always the same color as the mineral because powders reflect light in a different way than large crystals. Pyrite crystals have a brassy, yellow color, but their streak is greenish, or brownish black! Most minerals have colorless or white streaks; therefore, a colored streak is very useful in identifying a mineral.

Hardness

Hardness is a mineral's resistance to being scratched. Talc is so soft that it can be scratched by a fingernail. At the other extreme, diamond is so hard that no other mineral can scratch it. The hardness of a mineral is usually stated in terms of Moh's scale (see Table 14.3). On this scale, ten typical minerals are arranged in order from the softest to the hardest.

Minerals

TABLE 14.3 MOH'S SCALE

Mineral	Hardness	
Talc	1	SOFTEST
Gypsum	2	
Calcite	3	
Fluorite	4	
Apatite	5	
Orthoclase	6	
Quartz	7	
Topaz	8	
Corundum	9	
Diamond	10	HARDEST

To find the hardness of a mineral, you determine what minerals your sample can scratch and what minerals it cannot scratch. For example, tourmaline will scratch quartz or anything softer, but cannot scratch topaz or anything harder. The hardness of tourmaline, therefore, is between 7 and 8 (sometimes expressed as 7.5).

Cleavage, Parting, and Fracture

Some minerals break in ways that help to identify them. **Cleavage** (see Figure 14.3) is the tendency of a mineral to break parallel to atomic planes in its crystalline structure. *Parting* is the tendency to break along surfaces that follow a structural weakness caused by factors such as pressure, or along zones of different crystal types. These cleavage or parting surfaces often occur at very specific angles to one another and can be helpful in identifying a mineral.

Mineral	Cleavage	Appearance
Muscovite mica	Breaks in parallel sheets	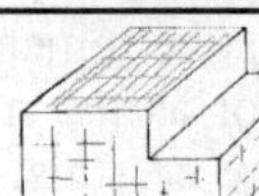
Pyroxene	Breaks along planes at 88° angle to one another in a prismatic shape	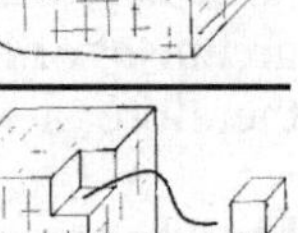
Halite	Breaks along planes at 90° to one another in a cubic shape	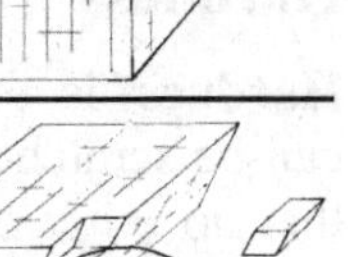
Calcite	Breaks along planes at a 75° angle to one another in a rhombohedral shape	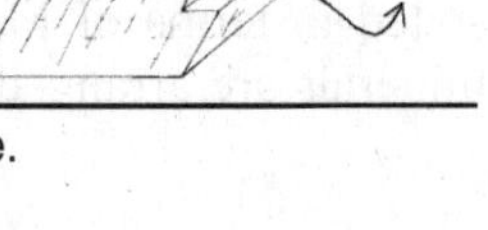

Figure 14.3 Some Common Minerals That Display Cleavage.

Some minerals have no planes of weakness, or are equally strong in all directions. They **fracture**; that is, they do not follow a particular direction when they break. Although fracture does not occur along smooth surfaces, it can display distinctive patterns that are very helpful in mineral identification (see Figure 14.4).

Patterns of Fracture	Description	Appearance
Conchoidal	Smooth, curved break that looks somewhat like the inside of a shell (e.g., quartz, olivine, gypsum)	
Fibrous or splintery	Fibers, as in asbestos Long, thin splinters (e.g., chrysotile, actinolite, tremolite, satin-spar calcite)	
Hackly	Jagged or sharp-edged surfaces (e.g, native metals such as copper, silver, iron, nickel)	
Uneven	Rough, irregular surfaces (e.g., sulfur, anhydrite, barite)	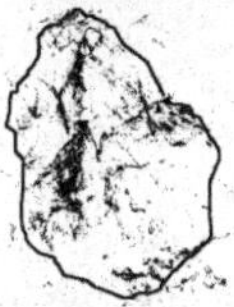

Figure 14.4 Some Common Minerals That Display Fracture.

Specific Gravity

Every mineral has a certain density. If you have samples that are about the same size, you can compare the densities of two different minerals. When you hold one in each hand, the denser mineral will feel heavier than the less dense one. A more exact way to determine the relative densities of different minerals is to compare them all to one standard. **Specific gravity** is the ratio of the density of a mineral to the density of water. It tells us how much heavier than water a mineral is. For example, the density of galena is 7.5 grams per cubic centimeter; the density of water is 1 gram per cubic centimeter. Galena is 7.5 times denser than water and therefore has a specific gravity of 7.5.

Most rocks have a specific gravity of 2.5 to 3.5. Most people have a feeling for how heavy a chunk of mineral of a given size should be. Sometimes, though, when you pick up a sample of a mineral and toss it in your hand, it feels heavier than you expected. This feeling of unexpected weight is termed *heft*. You can usually judge by a mineral's heft that its specific gravity is greater than 4.0.

Chemical Tests

Two chemical tests are commonly used to identify minerals in the field. The acid test uses hydrochloric acid. Some minerals bubble when a drop of hydrochloric acid dropped on the mineral reacts with it. Calcite bubbles vigorously; dolomite, slowly.

The second chemical test is the taste test. Halite tastes the same as table salt, which it is, though not purified. **The taste test should be used with caution and only at your teacher's direction.**

Special Properties

Special properties that are displayed by some minerals and can be used to distinguish them from other minerals are summarized in Table 14.4.

TABLE 14.4 SOME SPECIAL PROPERTIES OF MINERALS

Property	Definition or Description
Magnetism	Mechanical force of attraction associated with moving electricity (magnetite)
Luminescence	Emission of light by a substance that has received energy or electromagnetic radiation of a different wavelength from an external stimulus
includes:	
phosphorescence	Light emitted after stimulus is removed (scheelite autunite)
fluorescence	Light emitted while stimulus is applied (fluorite, willemite)
triboluminescence	Light emitted when pressure is applied (fluorite, lepidolite)
thermoluminescence	Light emitted when heated (chlorophane fluorite)
Piezoelectricity	Electricity emitted when pressure is applied (quartz)
Pyroelectricity	Electricity emitted when heat is applied (tourmaline)
Flame color	Color of the flame when a mineral is intensely heated in a flame
Double refraction	Double image produced when an object is viewed through the mineral (Iceland spar calcite)
Play of colors	Series of colors produced when the angle of the light shining on the mineral is changed (labradorite, opal)
Chatoyancy and asterism	Appearance of being made of tiny parallel fibers (satin spar gypsum, cat's-eye chrysoberyl)
	Light scattered in a pattern that looks like a three- or six-rayed star (star rubies and sapphires)

Using the Properties of Common Minerals Chart to Identify Minerals

As mentioned earlier, mineral identification is a process of elimination. As each property of a mineral is observed, possibilities are narrowed down until the identity of the mineral is determined. The Properties of Common Minerals chart in the *Earth Science Reference Tables* is arranged to assist you in this process. From left to right the columns list properties in the order of their usefulness in narrowing down the possible identity of a mineral.

Example:
Which mineral contains iron, has a metallic luster, is hard, and has the same color and streak?
(1) biotite mica   (2) galena   (3) potassium feldspar   (4) magnetite

First, rearrange the properties given in the question in the order in which they appear in the Properties of Common Minerals chart in the *Earth Science Reference Tables*: metallic luster, hard, same color and streak, contains iron. Now, referring to the chart, follow these steps:

- Since the mineral has a metallic luster, all of the minerals with nonmetallic luster are eliminated. This leaves graphite, galena, magnetite, pyrite, and the earthy-red form of hematite (earthy is a nonmetallic luster).
- Since the mineral is hard, graphite and galena are eliminated. This leaves magnetite, pyrite, and the metallic silver form of hematite.
- Since the mineral has the same color and streak, pyrite (brassy yellow color, green-black streak) and hematite (metallic silver color, red-brown streak) are eliminated. This leaves magnetite.
- Since the mineral contains iron, check to see that magnetite contains iron. The last column lists the composition of magnetite as Fe_3O_4; and since the chemical symbol for iron is Fe, magnetite does, indeed, contain iron.

Thus, the mineral in the question is magnetite, choice 4.

Minerals as the Building Blocks of Rocks

Of the thousands of known minerals, about 100 are so common that they make up more than 95 percent of Earth's crust. Not surprisingly, these minerals are generally compounds of the most abundant elements in the crust. Nearly every rock contains one or more of these so-called *rock-forming minerals* (see Table 14.5). Of the rock-forming minerals, fewer than 20 comprise most of the rocks you are likely to find.

Minerals

TABLE 14.5 SOME MAJOR ROCK-FORMING MINERALS

Mineral or Mineral Group	Composed of These Elements
Olivines	Iron, magnesium, silicon, oxygen
Pyroxenes	Calcium, iron, magnesium, silicon, oxygen
Amphiboles	Sodium, calcium, aluminum, silicon, oxygen
Micas	Potassium, aluminum, iron, magnesium, silicon, oxygen, hydrogen
Chlorite	Magnesium, iron, manganese, aluminum, silicon, oxygen, hydrogen
Kaolinite	Aluminum, silicon, oxygen, hydrogen
Feldspars	Potassium, sodium, calcium, aluminum, silicon, oxygen
Quartz	Silicon, oxygen
Nepheline	Sodium, potassium, aluminum, silicon, oxygen
Hematite	Iron, oxygen
Rutile	Titanium, oxygen
Magnetite	Iron, oxygen
Spinels	Magnesium, aluminum, oxygen
Pyrites	Iron, sulfur
Gypsum	Calcium, sulfur, oxygen
Calcite	Calcium, carbon, oxygen
Dolomite	Calcium, magnesium, carbon, oxygen
Apatite	Calcium, phosphorus, oxygen, hydrogen, fluorine, chlorine
Fluorite	Calcium, fluorine

HOW DO MINERALS DIFFER FROM EACH OTHER?

Minerals are grouped according to their chemical compositions. Major mineral groups include the silicates, sulfides, oxides, carbonates, and sulfates. Minor mineral groups include the nitrates, borates, chromates, phosphates, halides, hydroxides, and native elements.

Major Mineral Groups

Silicates

If we look at the chemical composition of Earth's crust (refer to Table 14.1), we see that oxygen is the most abundant element, by both mass and volume, and silicon is second, by mass. It is not surprising, then, that minerals composed mainly of silicon and oxygen are abundant and widespread. In all silicates, one ion (charged atom) of silicon is always joined to four ions of oxygen in the shape of a tiny **tetrahedron** (plural, tetrahedra), as shown in Figure 14.5. The bonds in this structure, which we will call the **silicon terahedron**, are very strong.

Silicon tetrahedra can join with positive ions of common elements in a wide variety of structures. Consider the next most abundant elements in Earth's crust after silicon and oxygen; aluminum, iron, calcium, sodium, magnesium, and potassium are all metals that form positive ions. When the positive ions of these elements join with negative silicon tetrahedra, a tremendous number of different silicates can be formed. Rings, chains, sheets, and frameworks are some of the different structures that silicon tetrahedra can form by joining with positive ions and sharing oxygen atoms.

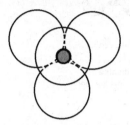 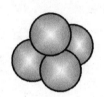 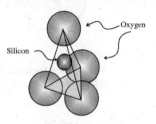

An overhead view, as though looking through the large oxygen ions at the smaller silicon ion between them.

The natural position of the silicon tetrahedon. Notice that the silicon ion is nestled among the oxygen ions and is hidden from view.

An exploded view, showing the positions of the silicon ion and oxygen ions.

Figure 14.5 Three Views of a Silicon Tetrahedron. The silicon ion has a +4 charge (Si^{4+}), and each oxygen ion has a –2 charge (O^{2-}), so the net charge of the silicon tetrahedron is –4 $(SiO4)^{4-}$.

In the complex structures shown in Figure 14.6, many other elements and groups of elements, such as extra oxygen, water, aluminum and other metal ions, and hydroxide ions, promote electrical and mechanical balance so that these structures are stable.

The physical properties of minerals depend on the arrangement and bonding of their atoms, as is very clearly seen in the structures formed by silicon tetrahedra. Look at a typical sheet silicate, mica, in Figure 14.6. The tetrahedra all share oxygens and are tightly bound together. The metal ions between the sheets, however, are isolated from each other and are less tightly bound. They form a plane of weakness in the structure—a weakness that manifests itself in the tendency of mica to break into thin sheets. Quartz, on the other hand, is a framework silicate. All of the tetrahedra are bound tightly to each other, and no place in the structure is weaker than any other place. Therefore, quartz does not break apart easily; its structure resists outside forces, so it is hard and strong. When it does break, there are no planes of weakness, and the break is random and uneven.

This relationship between the arrangement and bonding of atoms and the physical properties of minerals extends to all mineral compounds. Knowing the structure of a mineral helps us to understand why that mineral displays certain physical properties.

Silicate minerals are so abundant and widespread that almost every rock contains one or more. Although there are more groups of nonsilicate minerals than there are of silicates, nonsilicates make up only a very small part of Earth's crust, and very few are important rock-forming minerals. A brief overview of four nonsilicate minerals follows.

Minerals

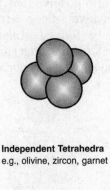

Independent Tetrahedra
e.g., olivine, zircon, garnet

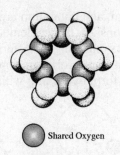

Shared Oxygen

Ring Silicates
e.g., tourmaline, beryl

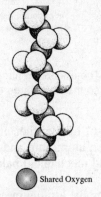

Shared Oxygen

Single-Chain Silicates
pyroxenes: e.g., enstatite, ferrosilite, augite, diopside, jadeite, spodumene

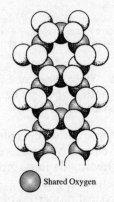

Shared Oxygen

Double-Chain Silicates
amphiboles: e.g., hornblendes, anthophyllite, tremolite, glaucophane

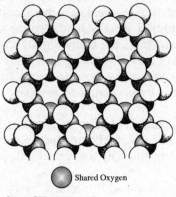

Shared Oxygen

Sheet Silicates
micas: e.g., muscovite, biotite, lepidolite
also, talc, chlorite, serpentine

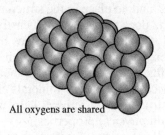

All oxygens are shared

Framework Silicates
e.g., feldspars, quartz, nephaline, sodalites

Figure 14.6 Some Structures That Form with Silicon Tetrahedra.

Sulfides

Sulfides are compounds in which one or more ions of sulfur are combined with a metallic ion. Sulfide minerals usually display a metallic luster, and many are important ores of metals. Some examples of sulfides are galena (PbS), pyrite (FeS), marcasite (FeS_2), cinnabar (HgS), sphalerite (ZnS), and stibnite (Sb_2S_3).

Oxides

Oxides are compounds in which oxygen is joined with ions of other elements, usually metals. Common oxides include hematite (Fe_2O_3), magnetite ($FeO\ Fe_2O_4$), corundum (Al_2O_3), ilmenite ($FeTiO_3$), chromite ($FeCr_2O_4$), and spinel ($MgAl_2O_4$). Magnetite is the only mineral attracted to a magnet. Iron oxides such as magnetite and hematite are mined to make iron and steel. Corundum is extremely hard and is widely used as an abrasive. When it forms with impurities, corundum produces red rubies and blue sapphires.

Carbonates and Sulfates

In oxides, oxygen forms a single ion and acts alone. In other minerals that contain oxygen, the oxygen ions join with other ions to form a group called a polyatomic ion. Two very common polyatomic ions are the **carbonate ion**, $(CO_3)^{2-}$, and the **sulfate ion**, $(SO_4)^{2-}$. Polyatomic ions are electrically charged and can combine with ions of the opposite charge to form mineral compounds.

When carbon dioxide dissolves in water, it often forms carbonate ions. These react with other ions in the water to form carbonate compounds that can precipitate out of the water or be left behind when the water evaporates. Some organisms take carbon dioxide from the environment and form, in their bodies as skeletons, compounds containing carbonate ions. In this way, carbonate minerals have accumulated into huge sedimentary rock formations and in some cases have been metamorphosed. Calcite (calcium carbonate) is the chief mineral in limestone and marble. Dolomite (calcium magnesium carbonate) is the chief mineral in dolostone.

Gypsum (hydrated calcium sulfate) is probably the best known sulfate mineral. It is mined for use in making wallboard, plaster, cement, fertilizer, and paint and as a filler in paper. Barite ($BaSO_4$) and celestite ($SrSO_4$) are important sources of barium and strontium, respectively, which are used in medicines and drilling.

Minor Mineral Groups

There are also other, less abundant ions that form minerals: nitrates (nitrogen and oxygen), borates (boron and oxygen), chromates (chromium and oxygen), phosphates (phosphorus and oxygen), and hydroxides (hydrogen and oxygen).

Halides are compounds of metallic ions with halogen elements such as chlorine, fluorine, and bromine. The only major rock-forming mineral in this group is halite (NaCl), a sedimentary mineral that forms when seawater evaporates.

Native elements are found as pure elements in Earth's crust. They include gold, silver, copper, and sulfur. These were among the first minerals to be mined by humans.

CONSERVATION OF EARTH'S RESOURCES

Minerals

The minerals in Earth are a treasure of almost unimaginable value. They help to fill many human needs, which, through time, have become greater. Minerals are used in making products ranging from steel to electric light bulbs. Unfortunately, minerals are a nonrenewable resource; that is, they can be used only once and then they are gone.

Imagine that every week you put a few cents of your allowance into a bank. For three years you do not take any money from the bank. Then you see a video game that you want, you open the bank, and you use the money to buy the game. In a few minutes, you have spent the money it took you three years to save. In the same way, it is easy to use up in a few years minerals that took millions of years to form. But while you will probably save more money in the future, mineral resources may not form again within your lifetime or even a thousand lifetimes. Every time a mineral is used, that much less remains. Thus, Earth's mineral resources will not last forever.

For this reason, people must conserve mineral resources, not just by saving these resources, but also by making sure that there is no waste in using them. Recycling and modernizing factories are parts of a good conservation policy. Another important part is research into better ways of production.

If conservation is practiced by everyone, Earth's mineral resources can be used to best advantage. Today there are recycling centers throughout the United States, and many communities have instituted mandatory recycling. Recycling centers process used aluminum cans, old bottles, scrap metals, and newspapers. These materials are then reused in the manufacture of aluminum, glass, steel, and paper products. You may have noticed on some cans a statement that they were made from recycled aluminum.

Fossil Fuels

At certain times in Earth's history, environmental conditions have enabled plant and marine organisms to grow at a rapid rate. Over millions of years, their bodies stored energy from sunlight that was converted into chemical bonds during photosynthesis. When these organisms completed their life cycles, their remains accumulated faster than decomposers could break them down and recycle them back into the environment. Over time, when these remains became buried by deposition, the pressure of overlying sediments and the increased temperatures at depth converted the layers of organic remains into coal beds and pools of petroleum. Since coal and petroleum are derived from organic remains and are used chiefly as fuels, they are called **fossil fuels**.

Fossil fuels are a vital source of energy and petrochemicals for the global economy. Transportation, urban development, industry, commerce, agriculture, and many other human activities depend on the amount and type of energy available. Most of the energy used today is obtained by burning fossil fuels. While coal was widely used in the past, petroleum and natural gas are now the fuels of choice because they are easier to obtain than coal, have many uses in industry, and are concentrated, easily portable sources of energy for cars, trucks, airplanes, and trains.

Unfortunately, the burning of fossil fuels releases into the atmosphere waste products that threaten the health of living things and have environmental risks associated with them. Current environmental conditions are not resulting in the large-scale accumulation of remains that can be converted into coal or petroleum. Therefore, for the foreseeable future, organic fossil fuels are a nonrenewable resource.

Global Implications of Resource Use and Distribution

The use and the distribution of Earth's resources have global political, financial, and social implications. Highly industrialized nations use more energy and mineral resources than less developed countries. The energy and natural resources we consume contribute to our high standard of living. This consumption, however, has led to a rapid depletion of Earth's natural resources and to mounting environmental crises, ranging from the devastation of marine life by oil-tanker spills to the breakdown of the ozone layer of the atmosphere by the chloroflourocarbons (CFCs) used in refrigerators and air conditioners.

As developing nations industrialize and their urban centers grow, they demand more energy and natural resources, thereby entering into competition with all of the other industrialized nations for these resources. Often this competition has resulted in friction between nations and even in war. If we are to live in peace and maintain our environment, we must develop ways to slow the depletion of natural resources and to use them more efficiently.

The depletion of energy and other natural resources can be slowed by decisions to use efficient technologies, to conserve, and to recycle. These decisions can be implemented at many levels ranging from the national down to the personal. Government can restrict low-priority uses of materials, such as the use of petrochemicals to manufacture purely ornamental packaging that adds to our solid-waste disposal problems. Government can also set standards of technical efficiency, for example, mileage and emission requirements for automobiles or the use of insulation in home construction. Cars that are built with more power than their function warrants waste energy. (If the speed limit

Minerals

is 55 mph, why build cars with the power to do 100 mph?) However, such decisions often involve trade-offs of cost and social values. For example, when other, less damaging compounds are substituted for CFCs, new compressors are required, so older units must be replaced before they have worn out. Reduced emissions depend on catalytic converters filled with toxic heavy metals. The development of wise policies for conserving and managing natural resources is one of the great challenges now facing all nations.

MULTIPLE-CHOICE QUESTIONS

In each case, write the number of the word or expression that best answers the question or completes the statement.

1. Which group of elements is listed in increasing order based on the percent by mass in Earth's crust?
   (1) aluminum, iron, calcium
   (2) aluminum, silicon, magnesium
   (3) magnesium, iron, aluminum
   (4) magnesium, silicon, calcium

2. Oxygen is the most abundant element by volume in Earth's
   (1) inner core   (3) hydrosphere
   (2) troposphere   (4) crust

3. The diagram below represents a part of the crystal structure of the mineral kaolinite.

   Structure of Kaolinite

   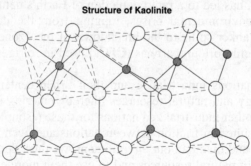

   An arrangement of atoms such as the one shown in the diagram determines a mineral's
   (1) age of formation   (3) physical properties
   (2) infiltration rate   (4) temperature of formation

4. The pie graph below shows the elements comprising Earth's crust in percent by mass.

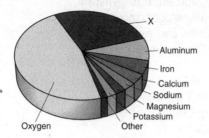

Which element is represented by the letter *X*?
(1) silicon   (3) nitrogen
(2) lead   (4) hydrogen

5. The diagrams below show the crystal shapes of two minerals.

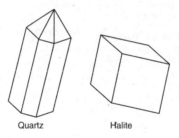

Quartz and halite have different crystal shapes primarily because
(1) light reflects from crystal surfaces
(2) energy is released during crystallization
(3) of impurities that produce surface variations
(4) of the internal arrangement of the atoms

6. A student created the table below by classifying six minerals into two groups, *A* and *B*, based on a single property.

Group A	Group B
olivine	pyrite
garnet	galena
calcite	graphite

Which property was used to classify these minerals?
(1) color   (3) chemical composition
(2) luster   (4) hardness

Minerals

7. Silicate minerals contain the elements silicon and oxygen. Which list contains only silicate minerals?
   (1) graphite, talc, and selenite gypsum
   (2) potassium feldspar, quartz, and amphibole
   (3) calcite, dolomite, and pyroxene
   (4) biotite mica, fluorite, and garnet

8. Which two minerals have cleavage planes at right angles?
   (1) biotite mica and muscovite mica
   (2) sulfur and amphibole
   (3) quartz and calcite
   (4) halite and pyroxene

Base your answers to questions 9 through 13 on the two tables below and on your knowledge of Earth science. Table 1 shows the composition, hardness, and average density of four minerals often used as gemstones. Table 2 lists the minerals in Moh's Scale of Hardness from 1 (softest) to 10 (hardest).

Table 1

Gemstone Mineral	Composition	Hardness	Average Density (g/cm³)
emerald	$Be_3Al_2(Si_6O_{18})$	7.5–8	2.7
sapphire	Al_2O_3	9	4.0
spinel	$MgAl_2O_4$	8	3.8
zircon	$ZrSiO_4$	7.5	4.7

KEY

Al = aluminum	O = oxygen
Be = beryllium	Si = silicon
Mg = magnesium	Zr = zirconium

Table 2

Moh's Scale of Hardness	
1	talc
2	gypsum
3	calcite
4	fluorite
5	apatite
6	feldspar
7	quartz
8	topaz
9	corundum
10	diamond

9. Part of a gemstone's value is based on the way the gemstone shines in reflected light. The way a mineral reflects light is described as the mineral's
   (1) fracture   (3) luster
   (2) hardness   (4) streak

10. Sapphire is a gemstone variety of which mineral on Moh's scale of hardness?
    (1) corundum   (3) fluorite
    (1) diamond   (3) topaz

11. If the mass of a spinel crystal is 9.5 grams, what is the volume of this spinel crystal?
    (1) 0.4 cm³   (3) 5.7 cm³
    (2) 2.5 cm³   (4) 36.1 cm³

12. The hardness and density of each gemstone is based primarily on the gemstone's
    (1) internal arrangement of atoms
    (2) geologic time of formation
    (3) oxygen content
    (4) natural abundance

13. Which gemstone minerals contain the two most abundant elements by mass in Earth's crust?
    (1) emerald and spinel   (3) sapphire and spinel
    (2) emerald and zircon   (4) sapphire and zircon

14. Which property is most useful in distinguishing pyroxene from amphibole?
    (1) sample size   (3) type of luster
    (2) hardness   (4) angles of cleavage

15. The mineral graphite is often used as
    (1) a lubricant   (3) a source of iron
    (2) an abrasive   (4) a cementing material

16. Which mineral has a metallic luster, a black streak, and is an ore of iron?
    (1) galena   (3) pyroxene
    (2) magnetite   (4) graphite

Base your answers to questions 17 and 18 on the diagram below, which shows the results of three different physical tests, A, B, and C, that were performed on a mineral.

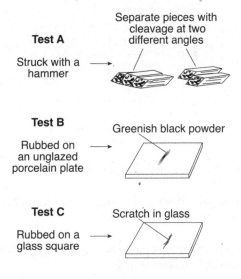

17. Which mineral was tested?
    (1) amphibole
    (2) quartz
    (3) galena
    (4) graphite

18. The luster of this mineral could be determined by
    (1) using an electronic balance
    (2) using a graduated cylinder
    (3) observing how light reflects from the surface of the mineral
    (4) observing what happens when acid is placed on the mineral

CONSTRUCTED RESPONSE QUESTIONS

Base your answers to questions 19 and 20 on the data table below, which shows the volume and mass of three different samples, A, B, and C, of the mineral pyrite.

	Pyrite	
Sample	Volume (cm^3)	Mass (g)
A	2.5	12.5
B	6.0	30.0
C	20.0	100.0

19. On the grid provided below, plot the data (volume and mass) for the three samples of pyrite and connect the points with a line. [2]

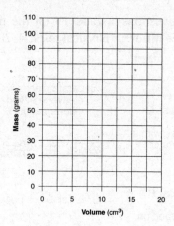

20. State the mass of a 10.0-cm^3 sample of pyrite. [1]

21. State one reason why color is frequently less valuable than cleavage in mineral identification. [1]

22. In one or more complete sentences, explain why minerals are considered a nonrenewable resource. [1]

Base your answers to questions 23 and 24 on the hardness of the minerals talc, quartz, halite, sulfur, and fluorite.

23. On the grid below, construct a bar graph to represent the hardness of these minerals. [1]

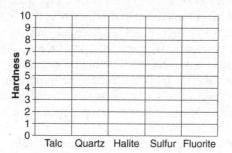

24. Which mineral shown on the grid would be the best abrasive? State one reason for your choice. [1]

Base your answers to questions 25 through 27 on the chart below, which shows some physical properties of minerals and the definitions of these properties. The letters A, B, and C indicate parts of the chart that have been left blank. Letter C represents the name of a mineral.

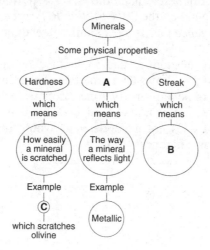

25. Which physical property of a mineral is represented by letter A? [1]

26. State the definition represented by letter B. [1]

27. Identify *one* mineral that could be represented by letter C. [1]

EXTENDED CONSTRUCTED RESPONSE QUESTIONS

Base your answers to questions 28 through 31 on the passage below.

Asbestos

Asbestos is a general name given to the fibrous varieties of six naturally occurring minerals used in commercial products. Most asbestos minerals are no longer mined due to the discovery during the 1970s that long-term exposure to high concentrations of their long, stiff fibers leads to health problems. Workers who produce or handle asbestos products are most at risk since inhaling high concentrations of airborne fibers allows the asbestos particles to become trapped in the workers' lungs. Chrysotile is a variety of asbestos that is still mined because it has short, soft, flexible fibers that do not pose the same health threat.

28. State one reason for the decline in global asbestos use after 1980. [1]

29. Chrysotile is found with other minerals in New York State mines located near 44° 30′ N, 74°W. In which New York State landscape region are these mines located? [1]

30. What determines the physical properties of minerals, such as the long, stiff fibers of some varieties of asbestos? [1]

31. The chemical formula for chrysotile is $Mg_3Si_2O_5(OH)_4$. State the name of the mineral found on the *2011 Edition Reference Tables for Physical Setting/Earth Science* that is most similar in chemical composition. [1]

Base your answers to questions 32 and 33 on the diagram below of a mineral classification scheme that shows the properties of certain minerals. Letters *A* through *G* represent mineral property zones. Zone *E* represents the presence of all three properties. For example, a mineral that is harder than glass, has a metallic luster, but does not have cleavage would be placed in zone *B*. Assume that glass has a hardness of 5.5.

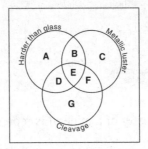

32. In which zone would the mineral potassium feldspar be placed? [1]

33. State the name of one mineral listed on the *Properties of Common Minerals Table* that could *not* be placed in any of the zones. [1]

Base your answers to questions 34 through 38 on the passage and cross section below, which explain how some precious gemstones form. The cross section shows a portion of the ancient Tethys Sea, once located between the Indian-Australian Plate and the Eurasian Plate.

Precious Gemstones

Some precious gemstones are a form of the mineral corundum, which has a hardness of 9. Corundum is a rare mineral made up of closely packed aluminum and oxygen atoms, and its formula is Al_2O_3. If small amounts of chromium replace some of the aluminum atoms in corundum, a bright-red gemstone called a ruby is produced. If traces of titanium and iron replace some aluminum atoms, deep-blue sapphires can be produced.

Most of the world's ruby deposits are found in metamorphic rock that is located along the southern slope of the Himalayas, where plate tectonics played a part in ruby formation. Around 50 million years ago, the Tethys Sea was located between what is now India and Eurasia. Much of the Tethys Sea bottom was composed of limestone that contained the elements needed to make these precious gemstones. The Tethys Sea closed up as the Indian-Australian Plate pushed under the Eurasian Plate, creating the Himalayan Mountains. The limestone rock lining the seafloor underwent metamorphism as it was pushed deep into Earth by the Indian-Australian Plate. For the next 40 to 45 million years, as the Himalayas rose, rubies, sapphires, and other gemstones continued to form.

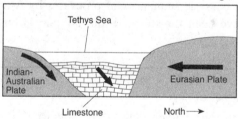

A Portion of the Tethys Sea 50 Million Years Ago

34. Which element replaces some of the aluminum atoms, causing the bright-red color of a ruby? [1]

35. State *one* physical property of rubies, other than a bright-red color, that makes them useful as gemstones in jewelry. [1]

36. Identify the metamorphic rock in which the rubies and sapphires that formed along the Himalayas are usually found. [1]

37. During which geologic epoch did the events shown in the cross section of the Tethys Sea occur? [1]

38. What type of tectonic plate boundary is shown in the cross section? [1]

CHAPTER 15
ROCKS

KEY IDEAS Rocks are the naturally formed, solid, coherent materials that comprise Earth. Most rocks are aggregates of minerals; that is, they are made of many grains stuck together. The grains may be mineral crystals or tiny bits of broken rock, or even solid parts of once-living organisms. The grains may be small or large, tightly or loosely bound together, and may all be made of one type of mineral or represent a variety of minerals.

Rocks are grouped into three "families" based upon how they were formed: igneous, metamorphic, and sedimentary. Igneous rocks form by the solidification of molten rock; metamorphic rocks form when heat, pressure, or chemical activity causes changes in existing rocks; and sedimentary rocks form by the chemical precipitation or the compaction and/or cementation of sediments. The composition and structure of a rock offer important clues as to its origin.

Of the three types of rock, no one type necessarily originates before another. Rock is constantly being created, destroyed, and altered in internal and external Earth processes. The schematic model of the possible sequences of processes by which any one rock may have been transformed into another is known as the rock cycle.

KEY OBJECTIVES
Upon completion of this chapter, you will be able to:

- Explain that rocks are usually composed of one or more minerals.
- Describe how rocks are classified by their origins, mineral contents, and textures.
- Explain how the characteristics of rocks indicate the processes by which they formed and the environments in which these processes took place.
- Describe how the properties of rocks determine their uses and also influence land usage by humans.

WHAT IS A ROCK?

A **rock** is a naturally formed, nonliving, firm and coherent mass of solid Earth material. A pile of sand is not a rock because the grains are not bound together, so the sand is not firm and coherent. A tree is not a rock even though it is solid, because a tree is living. But coal, which has been compressed into a coherent mass of dead leaves, twigs, and other plant parts, *is* a rock. Most rocks, though, are aggregates of minerals.

There are three large "families" of rocks, each defined by the processes that form them. **Igneous** rocks (from the Latin *ignis*, meaning "fire") form by the cooling of molten rock. **Sedimentary** rocks form when sediments are compacted and/or cemented together, or by the precipitation of dissolved minerals. *Sediments* are small bits of matter deposited by water, ice, or wind. They can be fragments of rocks, shells, or the remains of plants or animals. **Metamorphic** rocks are formed when already existing rock is changed by great heat, great pressure, or chemical action. Let us now consider how each type of rock forms, and how the way in which it forms determines its characteristics.

TYPES OF ROCK

Igneous Rocks

Formation of Igneous Rocks

Igneous rocks form by the crystallization of a molten mixture of minerals and dissolved gases. This molten mixture is called **magma** while it is beneath Earth's surface and **lava** when it escapes onto the surface. Most magmas can be thought of as fiery soups of fast-moving ions, that is, electrically charged atoms or molecules. As the magma cools, these ions slow down, are attracted to one another, and settle into stable structures—molecules of mineral compounds joined in a structural array, or mineral crystals. See Figure 15.1.

The formation of mineral crystals is a process, not an instantaneous event. As magma or lava cools, the amount of time available for crystal structures to form determines how large they can become. Thus, the size of the mineral crystals in an igneous rock provides a clue to the speed at which the rock crystallized, and that in turn is a clue to its origin.

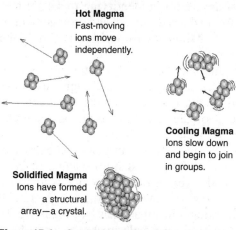

Figure 15.1 Cooling and Crystallization.

If cooling is very rapid, groups of atoms may solidify into a substance that lacks any distinctive structure—a glass. Such rapid cooling typically occurs at Earth's surface, where molten material comes into sudden contact with air or water at temperatures that are hundreds of degrees cooler than its temperature. Volcanic glasses, such as obsidian, can be observed forming when lava flows into the ocean and rapidly cools. From this type of observational data, we can infer the origin of glass found elsewhere, far from current volcanic activity or perhaps even buried beneath layers of other rock.

If cooling occurs more slowly, mineral crystals may begin to form, but may solidify while still microscopic in size. This, too, typically occurs at or near the surface of Earth. For example, the surface of lava, which may flow out of a fissure as a glowing liquid, can be observed to quickly cool and "skin over." This thin skin of rock shields the molten lava beneath it and allows the lava to cool more slowly. The lava inside may take days, weeks, or even months to solidify, thus allowing it to form crystals large enough to be seen with a microscope or magnifying glass.

Only very slow cooling allows the formation of crystals large enough to be visible to the unaided eye. Such slow cooling occurs underground, where the cooling magma is blanketed by the surrounding rock. Rock is a poor conductor of heat; and magma surrounded by rock, like hot coffee in a Thermos bottle, loses heat very slowly. Some underground pools of magma cool only a few degrees per century! Thousands of years may be required for the magma to totally crystallize. The result is large, visible crystals.

Characteristics of Igneous Rocks

Since most rocks are mixtures of minerals, rocks cannot be identified in the same way as minerals. A single rock may consist of minerals of several colors, hardnesses, and densities. For example, granite may contain white quartz with a hardness of 7, pink feldspar with a hardness of 6, and black mica with a hardness of 2–3. Thus, granite does not have a single characteristic color or hardness.

Two more general properties that are useful in identifying rocks are texture and mineral composition. **Texture** is the size, shape, and arrangement of the mineral crystals or grains in a rock. **Mineral composition** is simply the minerals that comprise the rock.

Igneous rocks are composed of randomly scattered, tightly interlocking mineral crystals. The main differences in the textures of igneous rocks involve the sizes and compositions of the mineral crystals. Most igneous rocks consist of a mixture of several of the following common minerals: quartz, feldspar, biotite, amphibole, pyroxene, and olivine.

Igneous rocks can be divided into two major types: intrusive and extrusive. Classification is based on whether the rocks formed from magma or lava. Magma and lava solidify into rock under very different conditions and thus produce rocks with different characteristics. Except for the volcanic glasses, all igneous rocks consist of tightly interlocking mineral crystals (see Figure 15.2). The texture of these crystals is a key to how a particular rock was formed.

Textures of Igneous Rocks

Intrusive Rocks

Figure 15.2 Tightly Interlocking Crystals of an Igneous Rock.

Magma, as previously defined, is molten rock inside Earth. Magma can move around by pushing its way into cracks or crevices in surrounding rock or by intruding into them. If the magma remains trapped underground and cools enough to solidify, an igneous rock will form. Igneous rock formed by the cooling of magma is called **intrusive**, or *plutonic*, rock.

As described earlier, magma generally cools slowly underground because it is blanketed by the surrounding rocks. Rocks are poor conductors of heat and do not allow the heat in the magma to escape very rapidly. When magma cools slowly, ions in the magma have time to align themselves in orderly structures called *crystals*. The slower the cooling, the more ions are able to align and the bigger the crystal. As a result, intrusive rocks generally have crystals large enough to be seen with the unaided eye. While you may not think a 1-millimeter crystal is large when you look at it, remember that millions upon millions of ions had to move into alignment with each other to reach that size. Rocks with large, visible crystals are said to have a coarse texture.

Granite, gabbro, and pegmatite are typical intrusive rocks.

Extrusive Rocks

When magma reaches the surface and pours out of Earth, it is called **lava**. Lava is usually pushed out, or extruded, from inside Earth in volcanoes or through large cracks in Earth's crust. Therefore, igneous rock formed from lava is called **extrusive**, or *volcanic*, rock.

Lava cools and solidifies into rock rapidly. When lava is extruded from a volcano or cracks in the crust, it is instantly exposed to an environment that may be 1,000°C colder than the lava's location just moments before. Since lava cools and solidifies rapidly, the ions in it have less time to align themselves into crystals, and the crystals that form are generally too small to be seen by the unaided eye. A magnifier or microscope may be needed to see the crystals in an extrusive rock. Rocks with crystals that can be seen only with magnification have a fine texture.

In some cases, magma cools so rapidly that no crystalline structures can form and the ions literally become frozen in place. When a rock of this type is examined with a microscope, no crystals can be seen. The physical characteristics of these unusual rocks are the same as those of glass, so they are called *volcanic glasses*. Their texture, in which no mineral crystals or grains are visible, is termed glassy. Obsidian is a common volcanic glass.

Sometimes, as lava is being extruded, gases dissolved in the lava form tiny bubbles. A glassy, lava froth forms similar to the froth that overflows from a champagne bottle after the cork is popped. In thick, viscous lava, the rock solidifies around the bubbles before the gas can completely escape, and a type of rock called *pumice* is formed. Since most of the glassy bubbles remain sealed, pumice will often float on water. In thinner, runnier lavas, the gas tends to bubble away rather than to form pumice. However, rapidly cooling surfaces may become pockmarked with small openings, or **vesicles**, made by the final escaping gas bubbles just as the lava solidified.

When lava is very forcibly extruded, as in an explosive volcanic eruption, a whole array of unusual rock materials form. These range from the thin, glassy strands ("Pele's hair") produced when very liquid lava is sprayed out, to large, streamlined globs ("volcanic bombs") that solidify while hurtling through the air. Such shapes are clues to the way in which the rock formed.

Basalt and rhyolite are typical extrusive rocks. The differences in the grain sizes of typical intrusive and extrusive rocks are shown in Figure 15.3.

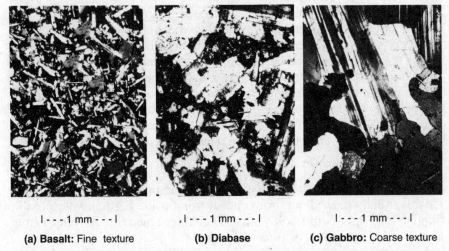

|---1 mm---| ,|---1 mm---| |---1 mm---|
(a) **Basalt:** Fine texture   (b) **Diabase**   (c) **Gabbro:** Coarse texture

Figure 15.3 Comparison of Grain Sizes in a Continuum from Extrusive (a) and Intrusive (b and c) Rocks. Basalt (a) forms at the surface. Diabase (b) may form as an intrusion near the surface as in a dike or sill. Gabbro (c) cools very slowly deep underground. Source: *Physical Geology*, 2nd Ed., Richard Foster Flint and Brain J. Skinner, John Wiley, 1974, 1977.

Porphyrys

In most igneous rocks the crystal grains are all roughly the same size, all large or all small. One type, however, has an unusual texture. In **porphyrys**, coarse mineral grains are embedded in a mass of fine mineral grains (see Figure 15.4). The isolated, large crystals are called *phenocrysts*; the surrounding fine-grained material is the *groundmass*. Porphyrys are thought to form in two stages: (1) a magma begins to solidify slowly at depth, and (2) the magma rises rapidly to the surface, where it finishes solidifying. The crystals formed at depth are large because the rate of cooling is slow. Nearer the surface, cooling is more rapid and the crystals formed are tiny.

Mineral Composition of Igneous Rocks

The mineral composition of an igneous rock is the product of the magma from which the rock solidifies. All common igneous rocks are mixtures of two or more of the following minerals: quartz, potassium feldspar, plagioclase feldspar, biotite mica, amphibole, pyroxene, and olivine. All are silicates, but they vary in composition, color, density, and other properties.

These silicate minerals are divided into two main groups: the **felsic** group, consisting of quartz and potassium and plagioclase feldspars, and the **mafic** group, consisting of silicates rich in magnesium and iron, such as olivine, pyroxene, amphibole, and biotite.

Figure 15.4 Porphyritic Texture—Large Phenocrysts of Plagioclase Feldspar in a Basalt Matrix.
Source: *Physical Geology*, 2nd Ed., Richard Foster Flint and Brian J. Skinner, John Wiley, 1974, 1977.

The word *felsic* comes from *fel*dspar and *si*lica (the name of the compound SiO$_2$, of which quartz is composed). *Mafic* comes from *ma*gnesium and the symbol *Fe* for iron. Felsic lavas have a thick, pudding-like consistency. As a result, they clog up fissures, allowing pressure to build up and leading to explosive eruptions. Mafic lavas have a thinner, syrupy consistency. They tend to flow more freely and to erupt as a flowing river of lava or spurt out of fissures like a fountain. Also, felsic minerals are lighter in color and lower in density than mafic minerals. Felsic rocks are common in the continents, while mafic rocks are more frequently found in the ocean basins.

The Scheme for Igneous Rock Identification

There are more than 1,500 different igneous rocks, and the system for classifying them is very complex. Petrologists are trying to simplify the system. Figure 15.5 can be used to make a very basic classification of an igneous rock.

This scheme has two main parts: an upper half, in which some of the most common igneous rocks are arranged by characteristics, and a lower half, a diagram that shows the percent mineral compositions of these rocks.

In the upper half, rocks such as granite and rhyolite, which are composed of light-colored minerals, are on the left. Rocks such as basalt and gabbro, which are composed of dark-colored minerals, are on the right. Since dark-colored minerals tend to be denser than light-colored minerals, the dark-colored rocks on the right are denser than the light-colored rocks on the left. Glassy extrusive rocks (e.g., obsidian and pumice) are at the top.

Fine-grained extrusive rocks (e.g., rhyolite and basalt) are in the middle. Coarse-grained, intrusive rocks (e.g., granite and gabbro) are on the bottom.

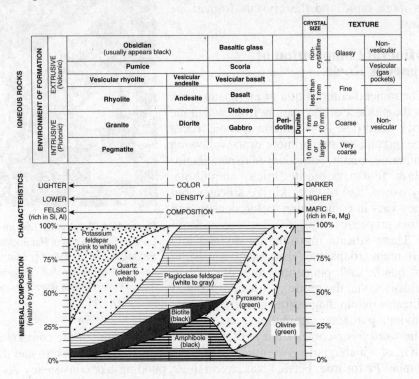

Figure 15.5 Scheme for Igneous Rock Identification. Source: The State Education Department, *Earth Science Reference Tables,* 2011 ed. (Albany, New York; The University of the State of New York).

In the lower half, the diagram shows the relative percents of the minerals of which the rocks are composed. As you can see, a particular type of rock can have a mineral composition that varies within a certain range. Look, for example, at the two different granites. Granite *A* consists of about 50% potassium feldspar, 30% quartz, 10% plagioclase feldspar, 7% biotite, and 3% amphibole. Granite *B* consists of only 10% potassium feldspar, 35% quartz, 25% plagioclase feldpsar, and 15% each of biotite and amphibole.

The following is a brief description of the rocks in Figure 15.5, starting at the top.

OBSIDIAN and BASALTIC GLASS form when lava cools so rapidly that crystals have no time to form before the lava solidifies. These rocks are very hard and brittle and break along shell-like depressions, forming sharp edges. Primitive peoples used them to make knives.

PUMICE hardens just as escaping gases whip the lava into a froth. The cooling rate is so fast that gases are trapped inside the frothy hardening rock. Thus, pumice is full of holes and looks somewhat like a rocky sponge. In fact, pumice is often used to scrub surfaces clean. Formerly, sailors used pumice stones to scrub the decks of wooden ships. Because of the holes, most pumice is so light it floats on water.

VESICULAR ROCKS form when gas escaping from lava bubbles away into the atmosphere rather than being trapped inside and forming pumice. Vesicles are small openings formed by escaping gas just as the lava flow hardens. In many flows the vesicles later become filled with calcite, quartz, or some other mineral deposited by heated groundwater. The type of lava determines which vesicular rock forms. For example, a rhyolite lava will form vesicular rhyolite. In scoria, the gases escape from the just-hardening lava as large bubbles, forming a texture full of cavities. Large bubbles are more likely to form in a thin, fluid basalt lava. Thus, scoria has a basaltic composition.

RHYOLITE contains the same minerals as granite, but it is fine grained and much rarer than granite. It is formed in lava flows and can be found in volcanic regions.

ANDESITE is named for the Andes Mountains, where it was first studied. Andesites are fine grained and usually gray or green in color. They have the same composition as diorites.

BASALT is the most abundant extrusive rock. Most lava flows form basalt, which is dark colored and fine grained. The ocean floor is mostly basalt. Sometimes basalt rock is crushed and used in road and railroad beds.

GRANITE consists mainly of quartz and feldspars. It is light colored and coarse grained and is thought to form by very slow cooling deep within Earth. Granite is the most abundant of all the igneous rocks. Most of the crust beneath the continents is made of granite. Granite is commonly used as a building stone and in making monuments.

DIORITE is coarse grained and contains about equal amounts of light- and dark-colored minerals, giving it a "salt and pepper" appearance.

GABBRO is a dark, coarse-grained rock. It is a common intrusive rock in New England and other mountain areas. It is sometimes used as a building stone.

PERIDOTITE is composed almost entirely of two minerals: olivine and pyroxene. It occurs widely, mostly in small plutonic bodies. It is a dark-colored rock with a high density—3.3 grams per cubic centimeter.

DUNITE is composed almost entirely of olivine. It is very rare and ranges in color from yellow to dark green, with a texture like hard brown sugar. It is named for the Dun Mountains of New Zealand, one of the few places it is found.

PEGMATITE has abnormally large crystals. It forms from a mixture of minerals that have relatively low melting temperatures, and are therefore the last to solidify from a cooling magma. It forms from a magma rich in silica and consists largely of quartz, potassium feldspar, and muscovite and biotite mica. Individual minerals grains several meters long have been found in some pegmatites, and it is mined for muscovite mica, which can sometimes be found in sheets up to a meter across.

Sedimentary Rocks

Sediments

All sedimentary rocks are composed of sediments. **Sediments** can be broadly defined as solid fragments of material that have been transported and then deposited by air, water, or ice. Most sediments are fragments of rock, but bits and pieces of plants and animals, or even molecules dissolved in water, can also be sediments.

When deposited, sediments typically form loose, unconsolidated layers on Earth's surface. These layers may then be transformed by physical and chemical changes into sedimentary rock.

Lithification

Lithification is the conversion of loose sediments into coherent, solid rock by processes such as cementation, compaction, dessication, and crystallization. This conversion may occur while the sediment is being deposited, soon after deposition, or long after deposition.

Cementation is the binding together of sediment particles by substances such as clay, carbonates, and hydrates of iron. These substances literally act as cement to hold particles together, in a solid mass, when they *crystallize* or otherwise accumulate between the loose sediment particles.

Compaction is the reduction in volume or thickness of a sediment layer due to the increasing weight of overlying sediments that are continually being deposited or to the pressure produced by movements of Earth's crust. Compaction usually results in a decrease in pore space within the sediments as they become more tightly packed. This, in turn, commonly leads to *desiccation*, the drying out or drying up of water, as it is forced out of the pore spaces during compaction. As the water dries up, minerals dissolved in the water crystallize and help to hold the rock together.

Figure 15.6 illustrates the compaction and cementation processes.

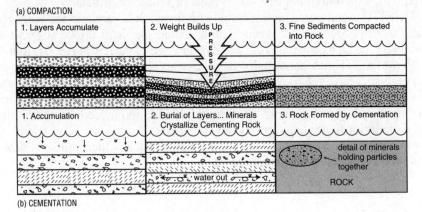

Figure 15.6 Compaction and Cementation.

Most sedimentary rocks form as a result of the compression and cementing of sediments. In most cases these sediments are rock fragments that range in size from tiny particles, such as clay and silt, to larger particles, such as sand and pebbles.

Some sedimentary rocks, however, result from the evaporation of seawater or from organic processes. Seawater has many substances dissolved in it; most are salts. When seawater evaporates, the salts are left behind. They form salt crystals, which settle out of the water in layers. Rock salt forms in this way as seawater evaporates.

Some organisms, such as corals, take substances dissolved in seawater into their bodies and form skeletons. When these organisms die, their skeletons settle to the ocean floor and can build up in layers. Biogenic limestones are formed in this way.

Plant materials that build up in layers can change drastically during compaction. They become desiccated, and the substances from which they were composed break down. Since carbon is the chief component of organic materials, it is often the end product of the lithification of plant parts. Coal typically forms in this way.

Characteristics of Sedimentary Rocks

Sedimentary rocks have three unique characteristics:

They *form at or near the surface of Earth at normal temperatures and pressures*. As a result, the environment in which sedimentary rocks form is quite different from the environments in which most other types of rock have their origins. Therefore, the minerals contained in sedimentary rocks are often quite different from those found in igneous or metamorphic rocks.

They *form layers*. Water is the chief transporter of sediments on Earth's surface. While being transported by water, most sediments become rounded. Sediments deposited in water settle into horizontal layers of rounded particles. This layering is preserved during lithification, so most sedimentary rocks usually contain rounded grains cemented in layers. These successive layers, known as strata, or beds, preserve evidence of past surface conditions. The individual layers can be easily distinguished because they usually differ in some way, such as color, thickness, size of particles, mineral composition, or degree of cementation.

The *surfaces of layers in sedimentary rock often contain distinctive features* identical to those found on modern sediments. Examples are ripple marks created by waves in shallow water, polygon-shaped cracks formed by fine sediment that was dried out by the Sun, footprints of walking animals, and the tracks and trails of crawling animals such as snails.

When sand and coarser sediments are moved along by air or water, they often form streamlined mounds such as dunes or ripples. Within these mounds, the layers of sediment are inclined in a down-wind or down-current direction. As successive layers of dunes or ripples pass over the surface, they leave behind layers of inclined sediments, called *cross-bedded sediments*, as shown in Figure 15.7.

They *contain fossils*. Plants and animals often live in areas where sediments are being deposited. The remains of these plants and animals are often incorporated into layers of accumulating sediment and into the sedimentary rocks that form from them.

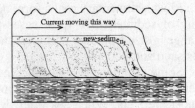

Figure 15.7 Diagram of Cross-Bedded Sediments.

Observations of current sediment layers and surfaces indicate that some characteristics enable us to infer which is the top, and which the bottom, of a series of sedimentary rock layers. This, in turn, allows us to infer the relative ages of the layers (bottom is oldest, and top is youngest). When a mixture of sediments settles in water, the larger particles reach the bottom first, and the smallest last. The result is graded bedding, as shown in Figure 15.8. When graded bedding is preserved in a sedimentary rock, we can infer that the original top was where the smaller particles are located. Similarly, cross-bedded sediments (Figure 15.8b) allow us to infer not only top and bottom, but also the direction in which the wind or water current was moving when the beds were deposited. If a rock layer contains ripple marks, mud cracks, footprints, or trails, we can infer the top of the layer from their position. Also, when certain shells settle in water (Figure 15.8c), they end up with one side facing upward most of the time, so their position in a sedimentary rock again allows us to infer top and bottom.

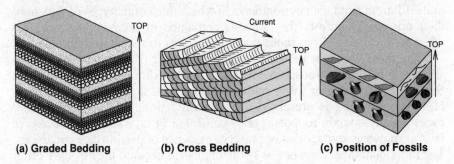

(a) Graded Bedding   (b) Cross Bedding   (c) Position of Fossils

Figure 15.8 Diagram of Graded Bedding, Cross-bedding, and Fossil Position All Showing the Top of a Series of Layers. (a) Graded Bedding. Within each bed, particles are sorted and form layers that decrease in size from the bottom to the top of the bed. (b) Cross Bedding. Particles are well sorted and form inclined layers within each bed. The layers in the bed are steeply sloping at the top and gently sloping at the bottom. (c) Position of Fossils. The hard parts of many organisms that become fossilized have a preferred orientation after settling through water. For example, clam shells settle more often with the cupped side of the shell facing down.

Types of Sediment Particles

There are three major types of sediment particles: clastic, organic or biogenic, and chemical.

Clastic sediments are rock or mineral fragments formed by the breakdown of rock due to weathering. The vast majority of sediments are of this type. These particles are grouped and named by size according to the Wentworth scale, as shown in Table 15.1.

TABLE 15.1 THE WENTWORTH SCALE

Particle-Size Range (diameter, cm)	Particle Name	Sediment Composed of That Particle
Greater than 25.6	Boulder	Boulder gravel
6.4–25.6	Cobble	Cobble gravel
0.2–6.4	Pebble	Pebble gravel
0.006–0.2	Sand	Sand
0.0004–0.006	Silt	Silt
Less than 0.0004	Clay	Clay

Organic, or **biogenic**, **sediments** are particles produced by the life activities of plants or animals. The most abundant organic sediment consists of the shells of aquatic animals. These may range from the large calcium carbonate shells of clams or snails to the microscopic silica skeletons of amoebas and diatoms. There are even some algae, which absorb calcium carbonate from seawater and incorporate it into their skeletons. When these organisms die, their shells and skeletons settle to the bottom and accumulate in layers of mud.

In warm seas, coral reefs form when huge populations of little animals called corals secrete calcium carbonate. Wave action smashes the reef into pieces, along with the shells of other animals living on the reef, producing a coarse sand or granule-sized sediment.

Organic sediments also include the woody tissue of plants (tree trunks, branches, twigs, and leaves), as well as spores and pollen. *Peat* is an organic sediment formed from plant material that accumulates in swamps, where stagnant water conditions prevent the oxidation of the plant material. Over time, buried peat may be transformed into coal.

The fatty or waxy tissues of land plants and the soft tissues of aquatic plants and animals do not form particles. However, this soft organic matter is usually mixed in with sediment particles and is considered to be the raw material from which petroleum forms underground.

Fossils in sedimentary rocks provide evidence of the environment in which they formed. The fossil of an organism that is clearly a fish, together with fossilized clam shells, indicates an aquatic environment. Pollen from citrus or palm trees in a rock is evidence of a warm terrestrial environment.

Chemical sediments consist of particles (crystals) that have crystallized out of solutions at or near Earth's surface. All the water on Earth's surface has some material dissolved in it. Seawater, for example, contains over 80 elements and hundreds of different compounds. A number of conditions can cause some of the substances dissolved in water to crystallize.

Let's use a simple analogy. Think of a solution as a container in which a substance is being held. A glass of juice is a good example. The solvent is like the glass, and the solute is like the juice. If there is more juice than can fit into the glass, the juice will overflow and spill to the ground. Now think of seawater. Water is the solvent, or glass; salts are the solute, or juice. Overflow is like crystallization; spilling to the ground is like precipitation.

In our example, there are two things that can change—the size of the glass and the amount of juice. Having more solvent is like having a bigger glass. Similarly, if more water is added, seawater can hold more salt. On the other hand, removing water is like having a smaller glass. If water is removed, less salt can be held.

Evaporation removes water from seawater. When the water in a solution evaporates, the solution becomes more and more concentrated. When it reaches saturation, the substance dissolved in the solution begins to crystallize out. Sediments made of particles that crystallized as a result of evaporation are called *evaporites*. Halite crystals are a common chemical sediment that forms along the edges of bodies of water undergoing extensive evaporation.

If you have ever tried to dissolve sugar in iced tea, you know that less sugar dissolves in cold tea than hot tea. The temperature of a solution affects the amount of material that can be dissolved in it. As the temperature of a solution falls, its ability to hold dissolved materials decreases. Cooling a solution is like making the glass in the preceding example smaller without removing any of the juice. The result is overflow. For example, when the mineral-laden water of a hot spring cools, calcite or even opal may precipitate from the water.

Particles may also crystallize out of a solution if more solute is added than the solution can hold. If you add juice to an already full glass, the glass will overflow. For example, aquatic plants and animals can increase the amount of carbon dioxide (CO_2) dissolved in the water around them and thereby cause calcium carbonate to crystallize out and sink to the bottom, or precipitate. Also, if a chemical reaction in a solution produces a substance that doesn't dissolve in water, that substance will also precipitate.

The Scheme for Sedimentary Rock Identification

Sedimentary rocks are classified as *fragmental*, *organic*, or *chemical*, depending on the sediments from which they lithified. As with igneous rocks, texture and mineral composition are the main properties used to classify and name sedimentary rocks, as shown in Figure 15.9. In sedimentary rocks, the mineral composition of the sediment particles tells us where the particles came from, and the texture provides clues about the processes that deposited the sediments.

In Figure 15.9, sedimentary rocks are divided into two main groups. In the upper part of the chart are "Inorganic Land-Derived Sedimentary Rocks," that is, rocks formed from clastic sediments. In the lower part are "Chemically and/or Organically Formed Sedimentary Rocks," that is, rocks formed from chemical or organic sediments.

Clastic sedimentary rocks are classified on the basis of grain size. Nonclastic rocks are identified primarily by composition and texture.

INORGANIC LAND-DERIVED SEDIMENTARY ROCKS					
TEXTURE	GRAIN SIZE	COMPOSITION	COMMENTS	ROCK NAME	MAP SYMBOL
Clastic (fragmental)	Pebbles, cobbles, and/or boulders embedded in sand, silt, and/or clay	Mostly quartz, feldspar, and clay minerals; may contain fragments of other rocks and minerals	Rounded fragments	Conglomerate	
			Angular fragments	Breccia	
	Sand (0.006 to 0.2 cm)		Fine to coarse	Sandstone	
	Silt (0.0004 to 0.006 cm)		Very fine grain	Siltstone	
	Clay (less than 0.0004 cm)		Compact; may split easily	Shale	
CHEMICALLY AND/OR ORGANICALLY FORMED SEDIMENTARY ROCKS					
TEXTURE	GRAIN SIZE	COMPOSITION	COMMENTS	ROCK NAME	MAP SYMBOL
Crystalline	Fine to coarse crystals	Halite	Crystals from chemical precipitates and evaporites	Rock salt	
		Gypsum		Rock gypsum	
		Dolomite		Dolostone	
Crystalline or bioclastic	Microscopic to very coarse	Calcite	Precipitates of biologic origin or cemented shell fragments	Limestone	
Bioclastic		Carbon	Compacted plant remains	Bituminous coal	

Figure 15.9 Scheme for Sedimentary Rock Identification. Source: The State Education Department, *Earth Science Reference Tables*, 2011 ed. (Albany, New York; The University of the State of New York).

Metamorphic Rocks

Formation of Metamorphic Rocks

Metamorphic rocks get their name from the Greek words *meta*, meaning "change," and *morphs*, meaning "form." A metamorphic rock has literally been changed from its original form. A rock changes when it is exposed to an environment significantly different from the one in which it originally formed. In general, three factors cause metamorphic changes in rock: heat, pressure, and chemical activity.

Heat

Most minerals expand when heated, causing the atoms to move apart, and the bonds that hold them together to be stretched and weakened. If a mineral is heated enough, all of the bonds break, and the mineral melts. If a rock experiences such heating, a magma forms. In metamorphism, however, heating breaks *some* but not all of the bonds in the minerals of the rock. *Some* atoms break loose, migrate through the rock, and join with other atoms to form new minerals—metamorphic minerals—in the rock. The result is to change both the chemical composition and the structure of the rock, that is, to change the form of the rock. Such extreme heating of rock may be caused by deep burial (temperature increases with depth) or by contact with hot materials, such as magma.

Pressure

The effect of pressure on minerals is opposite to that of heat. Pressure forces the atoms in the mineral much closer together. The resulting stress on the bonds causes some of them to break. This, in turn, enables the atoms to rearrange into a more compact structure. The result is a significant change—a denser, harder rock.

Pressure exerted on rock may be due to deep burial or to movements in Earth's crust. Pressure due to deep burial is equal in all directions, but crustal movements may produce pressures that are greater in one direction than another, thereby deforming or flattening rocks. The effects on the particles in a rock are different in these two cases, as shown in Figure 15.10.

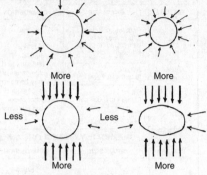

Figure 15.10 Equal Versus Directed Pressure on a Particle. Equal pressure (a) causes the particle to decrease in size, but the shape is unchanged. Unequal pressure (b) causes both a decrease in size and a change in shape.

Chemical Activity

A rock can also be changed when solutions rich in dissolved ions move through it. Such solutions are given off by cooling magmas; others come from metamorphism taking place deep underground. As these solutions move through pores in surrounding rock, they interact with minerals in the rock to form new minerals. Sometimes the water itself combines with minerals to form new ones. For example, olivine reacts with water to form talc or serpentine.

Metamorphism

The changes that occur in solid rock due to great temperature, great pressure, and chemical activity are known as **metamorphism**. Metamorphism occurs while rocks are still in the solid state. Changes produced by weathering, although they occur while rocks are solid, are not considered metamorphism. Neither are igneous processes such as the melting of rocks in high-temperature zones. True metamorphism occurs beneath Earth's surface at high temperatures and pressures. Under the conditions that produce metamorphism, ordinarily rigid, brittle rocks behave like soft modeling clay. The rocks can be bent and twisted without breaking. Rocks that originally had a clastic texture can recrystallize so that the metamorphic rock has a crystalline texture. Conversely, rocks that were originally crystalline can be crushed and broken into pieces and then relithified to form a clastic texture.

The preexisting rock from which a metamorphic rock forms is called the **parent rock**. By investigating the conditions that produce metamorphic changes, it is possible to infer the parent rock from the mineral compositions and structures of most metamorphic rocks.

Metamorphic rocks occur on a continuum from little alteration in the parent rock to major changes. Rarely are there sharp boundaries between parent rocks and metamorphic rocks; instead, gradations lead from one to the other. Limestone grades almost imperceptibly into marble. Shales grade into schists. Sandstones grade into quartzites. This characteristic has been helpful in determining the origins of certain metamorphic rocks. Metamorphic rocks are divided into groups based on the parent materials from which they were derived.

Types of Metamorphism

There are two major types of metamorphism: contact and regional. **Contact metamorphism** occurs when molten rock comes into contact with surrounding rocks. Heat given off by the cooling magma is conducted into the rock, where it causes changes in texture and mineral composition. Contact metamorphism is most intense at the contact between the magma and the surrounding rock, and decreases with distance from the magma until points are reached at which the effects of the heat are not felt. Generally, crystals are largest at the point of contact and become smaller with distance.

Regional metamorphism occurs over large areas and is generally associated with mountain building. Regional metamorphism takes its name from the large scale on which it occurs. For example, about 300 million years ago North America and Africa collided along a boundary between plates of Earth's crust. The result was the crushing of rocks from Alabama to New York and the formation of the Appalachian Mountains. During regional metamorphism, deeply buried sedimentary and igneous rocks are squeezed both by the weight of rock layers above and by movements of the crust, are intruded by igneous rocks, and are infused with chemically active fluids.

Characteristics of Metamorphic Rocks

When a rock is metamorphosed, nearly every characteristic can change, including texture and mineral composition. These changes can be so great that the metamorphic rock bears little, if any, resemblance to its original state. Therefore, to infer what a metamorphic rock once was, one needs to know what kinds of changes take place during metamorphism.

Changes in Texture

Crystalline Texture. During metamorphism the size, shape, and spacing of the grains (crystals or particles) in a rock are changed. Under both heat and pressure, the grains in a rock may partially melt and fuse together, forming larger grains (recrystallization); or, pressure may break brittle grains into smaller pieces, destroying the original texture of the rock. The heat and the pressure of metamorphism fuse any fragmental textured rock into a solid, crystalline mass. As a result, most metamorphic rocks have a crystalline texture.

Increased Density. Pressure from deep burial closes pore spaces in sedimentary rocks and openings in igneous rocks. Pressure also compresses grains so that they are smaller and closer together. The rock becomes more compact. Its mass is forced into a smaller volume, and its density increases. Thus, the igneous or sedimentary rock is changed—it becomes metamorphic rock.

Foliation and Banding. A rock is said to be foliated when its crystals are arranged in layers or bands along which the rock breaks easily. These layers may be microscopic or thick enough to be easily seen. *Foliation* results when mineral grains recrystallize or are flattened under pressure. Even pressure compresses the grains so that they are smaller and closer together, but retain their original shape. Uneven pressure not only compresses grains but also changes their shape. Grains that may have been randomly oriented will form parallel layers (foliation) or become aligned when deformed under pressure.

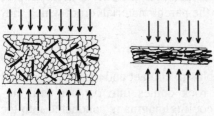

Compression causes randomly dispersed crystals to become oriented in a direction perpendicular to the pressure. The oriented crystals form thin sheets or layers called *foliation*.

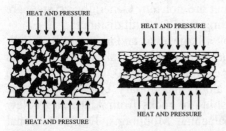

Migration of mineral crystals due to density differences produces *banding*—regions of light- and dark-colored minerals.

Figure 15.11 Random to Oriented Patterns Due to Pressure and Fusing of Grains, and Banding Caused by Density Separation.

Banding occurs when minerals of different densities recrystallize under pressure and separate into layers, like a mixture of oil and water. Since light-colored minerals tend to be less dense than dark ones, the layers that form are alternating light and dark.

Figure 15.11 shows the effects of pressure and recystallization on rocks.

Distortion of Layers. When sedimentary rocks are metamorphosed, the layers in the rock may be distorted. Softened by the heat and squeezed by the pressure, the once-flat layers become twisted and contorted. Such distortion is commonly seen in large-scale formations of metamorphic rock.

Changes in Mineral Composition

The changes described above can occur without altering the original mineral composition of the rock. When a rock is metamorphosed, however, the original minerals often disappear and new minerals form in their place. The underlying reason for this change is lack of stability. Minerals that are stable in one environment are not stable in another. Sugar is stable at room temperature; but when heated strongly, it becomes unstable and breaks down into carbon and water. Clay minerals, such as kaolinite, provide another good example. Kaolinite forms by the weathering of feldspar. It has an open, sheetlike molecular structure that is stable at the temperatures and pressures normally found at Earth's surface. Under the heat and pressure associated with deep burial, however, it cannot survive, and forms micas that have a more compact, sheetlike molecular structure.

Classification of Metamorphic Rocks

Metamorphic rocks are classified according to mineral composition and texture, including foliation and banding.

An interesting question associated with metamorphic rocks is, When has a rock changed enough to be called metamorphic? Rocks do not change instantaneously from one type to another; the changes occur along a continuum from little alteration to major changes. Generally, if a change can be recognized in a rock, the rock is considered metamorphic. Table 15.2 lists some common rocks and the rocks they can become when metamorphosed.

TABLE 15.2 DERIVATIONS OF COMMON METAMORPHIC ROCKS

Parent Rock	Metamorphic Rock
Shale	Compaction and reorientation → slate
	Slate → micas and chlorite form → phyllite
	Phyllite → grains recrystallize and get coarser → schist
	Schist → banding occurs → gneiss
Shale	Contact with magma → granofels
Sandstone	Quartzite
Limestone or dolostone	Marble
Basalt	Hydration reactions → greenschist
	Greenschist → dehydration → amphibolite
	Amphibolite → banding occurs → gneiss

The Scheme for Metamorphic Rock Identification

Metamorphic rocks are classified according to their textures (including foliation and banding) and compositions. In Figure 15.12 metamorphic rocks are divided into two main groups, foliated and nonfoliated, based on texture. Then, within each of these groups, the rocks are classified by grain size.

If you look at the column headed "Composition," you will see that the foliates share a common composition, while each of the nonfoliates has a distinctly different composition. The majority of foliated rocks are derived from clastic sedimentary rocks rich in clay minerals, such as shale and siltstone. The "Comments" column tell you what the parent rock was in each case.

Notice that slate, schist, and gneiss really form a continuum. Slate forms first as shale is slightly metamorphosed. The flakelike micas align because of compression and form surfaces along which the rock will split. With further metamorphosis, alignment of the mica becomes even more pronounced and phyllite is formed. As metamorphosis continues, chlorite changes to biotite, foliation becomes stronger, grains become coarser because of recrystallization, and a schist is formed. Still further metamorphism causes the crystals to grow even larger, alternating layers of light and dark minerals (bands) form as crystals migrate because of density differences, and a gneiss forms. The larger, interlocking crystals of a gneiss cause these rocks to break less regularly than schists, but they will often break completely across layers. In the highly metamorphic environment in which gneisses form, rock is plastic and will often be contorted into the weirdly twisted patterns seen in gneiss.

TEXTURE		GRAIN SIZE	COMPOSITION	TYPE OF METAMORPHISM	COMMENTS	ROCK NAME	MAP SYMBOL
FOLIATED	MINERAL ALIGNMENT	Fine	MICA / QUARTZ / FELDSPAR / AMPHIBOLE / GARNET / PYROXENE	Regional (Heat and pressure increases)	Low-grade metamorphism of shale	Slate	
		Fine to medium			Foliation surfaces shiny from microscopic mica crystals	Phyllite	
					Platy mica crystals visible from metamorphism of clay or feldspars	Schist	
	BANDING	Medium to coarse			High-grade metamorphism; mineral types segregated into bands	Gneiss	
NONFOLIATED		Fine	Carbon	Regional	Metamorphism of bituminous coal	Anthracite coal	
		Fine	Various minerals	Contact (heat)	Various rocks changed by heat from nearby magma/lava	Hornfels	
		Fine to coarse	Quartz	Regional or contact	Metamorphism of quartz sandstone	Quartzite	
			Calcite and/or dolomite		Metamorphism of limestone or dolostone	Marble	
		Coarse	Various minerals		Pebbles may be distorted or stretched	Metaconglomerate	

Figure 15.12 Scheme for Metamorphic Rock Identification. Source: The State Education Department, *Earth Science Reference Tables*, 2011 ed. (Albany, New York; The University of the State of New York).

WHAT IS THE ROCK CYCLE?

Where do rocks come from? You might say igneous rock comes from magma, sedimentary rock comes from sediments, and metamorphic rock comes from other rocks that have changed. But consider that magma is melted rock, sediments are broken rock, and there had to be an already-existing rock to be metamorphosed. Basically, then, rocks come from other rocks.

Rocks are constantly changing from one type to another in a never-ending **rock cycle**, shown in Figure 15.13. The outside circle shows the different forms in which rock matter can exist: magma, igneous rock, sediments, and so on. Arrows leading from one form to another are labeled with the changes that are taking place. Arrows inside the circle show some alternative paths in the rock cycle. For example, an igneous rock does not always break down into sediments. If buried, it may be metamorphosed; therefore, there is an arrow inside the circle leading from igneous rock to metamorphic rock.

Let's follow a rock through one possible cycle, beginning with magma. A magma solidifies and forms an igneous rock. The igneous rock is uplifted and exposed at the surface, where weathering breaks it down into sediments. Rain washes the sediments into a stream. The stream carries them to the ocean, where they are deposited and buried. Compaction and cementation produce a sedimentary rock. The sedimentary rock is exposed to heat and pressure, perhaps as two of the plates in Earth's crust collide, and a metamorphic rock is formed. If the metamorphic rock is forced deep enough into the crust, it melts and forms magma.

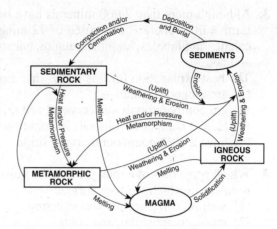

Figure 15.13 Rock Cycle in Earth's Crust. Source: The State Education Department, *Earth Science Reference Tables*, 2011 ed. (Albany, New York; The University of the State of New York).

Notice that we have come full circle back to where we started. This is an example of the rock cycle. It is, however, only one of many possible paths. Except for the stray meteorite that reaches the surface, Earth is essentially a closed system. Rock doesn't enter or leave, but is constantly recycling from one form to another.

The rock cycle diagram has no beginning and no end, but of course the rock cycle really did start somewhere. There is much evidence that Earth was originally totally molten. No solid rock existed; there was only magma. Therefore, it is thought that the original rocks formed as this magma cooled, and the rock cycle began with igneous rocks.

MULTIPLE-CHOICE QUESTIONS

In each case, write the number of the word or expression that best answers the question or completes the statement.

1. All rocks contain
   (1) minerals
   (2) sediments
   (3) intergrown crystals
   (4) fossils

2. Rocks are classified as igneous, sedimentary, or metamorphic based primarily on their
   (1) texture
   (2) crystal or grain size
   (3) method of formation
   (4) mineral composition

Rocks

3. Although more than 2,000 minerals have been identified, 90 percent of Earth's lithosphere is composed of 12 minerals: feldspar, quartz, mica, calcite, amphiboles, kaolinite, augite, garnet, magnetite, olivine, pyrite, and talc.
   The best explanation for this fact is that most rocks
   (1) are monomineralic
   (2) are composed only of recrystallized minerals
   (3) have a number of minerals in common
   (4) have a 10% nonmineral composition

4. Which properties of a rock provide the most information about the environment in which the rock formed?
   (1) texture, composition, and structure
   (2) mass, composition, and color
   (3) density, volume, and mass
   (4) color, texture, and volume

5. In which set are the rock drawings labeled with their correct rock types?

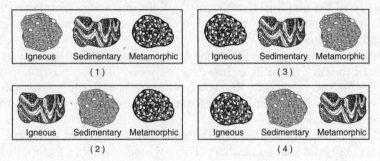

6. The photograph below shows an igneous rock.

   What is the origin and rate of formation of this rock?
   (1) plutonic with slow cooling
   (2) plutonic with rapid cooling
   (3) volcanic with slow cooling
   (4) volcanic with rapid cooling

7. Which igneous rock, when weathered, could produce sediment composed of the minerals potassium feldspar, quartz, and amphibole?
   (1) gabbro (3) andesite
   (2) granite (4) basalt

8. According to the *2011 Edition Reference Tables for Physical Setting/Earth Science*, generally as the percentage of felsic minerals in a rock increases, the rock's color will become
   (1) darker and its density will decrease
   (2) lighter and its density will increase
   (3) darker and its density will increase
   (4) lighter and its density will decrease

9. Which graph best represents the relative densities of three different types of igneous rock?

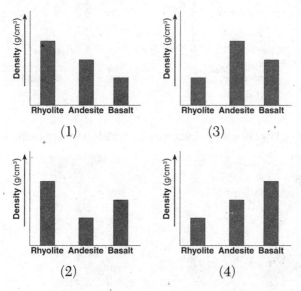

10. The flowchart below illustrates the change from melted rock to basalt.

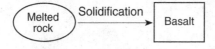

The solidification of the melted rock occurred
(1) slowly, resulting in fine-grained minerals
(2) slowly, resulting in coarse-grained minerals
(3) rapidly, resulting in coarse-grained minerals
(4) rapidly, resulting in fine-grained minerals

11. The graph below shows the relationship between the cooling time of magma and the size of the crystals produced.

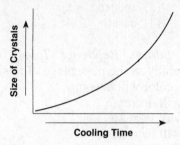

Which graph correctly shows the relative positions of the igneous rocks granite, rhyolite, and pumice?

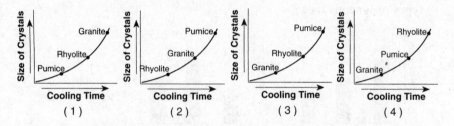

12. The photograph below shows the intergrown crystals of a pegmatite rock.

(Actual size)

Which characteristic provides the best evidence that this pegmatite solidified deep underground?
(1) low density
(2) light color
(3) felsic composition
(4) very coarse texture

13. The three statements below are observations of the same rock sample:
    • The rock has intergrown crystals from 2 to 3 millimeters in diameter.
    • The minerals in the rock are gray feldspar, green olivine, green pyroxene, and black amphibole.
    • There are no visible gas pockets in the rock.
    This rock sample is most likely
    (1) sandstone
    (2) gabbro
    (3) granite
    (4) phyllite

14. An extrusive igneous rock with a mineral composition of 35% quartz, 35% potassium feldspar, 15% plagioclase feldspar, 10% biotite, and 5% amphibole is called
    (1) rhyolite
    (2) granite
    (3) gabbro
    (4) basaltic glass

15. The diagram below shows magnified views of three stages of mineral crystal formation as molten material gradually cools.

    Which rock normally forms when minerals crystallize in these stages?
    (1) shale
    (2) gneiss
    (3) gabbro
    (4) breccia

16. Which igneous rock has a vesicular texture and contains the minerals potassium feldspar and quartz?
    (1) andesite
    (2) pegmatite
    (3) pumice
    (4) scoria

17. Which igneous rock is dark colored, cooled rapidly on Earth's surface, and is composed mainly of plagioclase feldspar, olivine, and pyroxene?
    (1) obsidian
    (2) rhyolite
    (3) gabbro
    (4) scoria

18. Which mineral can be found in all samples of rhyolite and andesite?
    (1) pyroxene
    (2) quartz
    (3) biotite
    (4) potassium feldspar

19. Where are Earth's sedimentary rocks generally found?
    (1) in regions of recent volcanic activity
    (2) deep within Earth's crust
    (3) along the midocean ridges
    (4) as a thin layer covering much of the continents

20. Rock layers showing ripple marks, cross-bedding, and fossil shells were formed
    (1) from solidification of molten material
    (2) from deposits left by a continental ice sheet
    (3) by high temperature and pressure
    (4) by deposition of sediments in a shallow sea

21. A student classified the rock below as sedimentary.

    Which observation about the rock best supports this classification?
    (1) The rock is composed of several minerals.
    (2) The rock has a vesicular texture.
    (3) The rock contains fragments of other rocks.
    (4) The rock shows distorted and stretched pebbles.

22. A student obtains a cup of quartz sand from a beach. A saltwater solution is poured into the sand and allowed to evaporate. The mineral residue from the saltwater solution cements the sand grains together, forming a material that is most similar in origin to
    (1) an extrusive igneous rock
    (2) an intrusive igneous rock
    (3) a clastic sedimentary rock
    (4) a foliated metamorphic rock

23. The diagram below shows a drill core of sediment that was taken from the bottom of a lake.

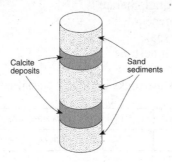

Which types of rock would most likely form from compaction and cementation of these sediments?
(1) sandstone and limestone
(2) shale and coal
(3) breccia and rock salt
(4) conglomerate and siltstone

24. Which rock is sedimentary in origin and formed as a result of chemical processes?
(1) granite
(2) shale
(3) breccia
(4) dolostone

25. Most sandstone bedrock is composed of sediment that was
(1) sorted by size and not layered
(2) sorted by size and layered
(3) unsorted and not layered
(4) unsorted and layered

Base your answers to questions 26 through 28 on the drawings of six sedimentary rocks labeled *A* through *F*.

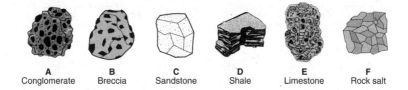

26. Most of the rocks shown were formed by
(1) volcanic eruptions and crystallization
(2) compaction and/or cementation
(3) heat and pressure
(4) melting and/or solidification

Rocks

27. Which two rocks are composed primarily of quartz, feldspar, and clay minerals?
    (1) rock salt and conglomerate
    (2) rock salt and breccia
    (3) sandstone and shale
    (4) sandstone and limestone

28. Which table shows the rocks correctly classified by texture?

Texture	clastic	bioclastic	crystalline
Rock	A, B, C, D	E	F

(1)

Texture	clastic	bioclastic	crystalline
Rock	A, C	B, E	D, F

(3)

Texture	clastic	bioclastic	crystalline
Rock	A, B, C	D	E, F

(2)

Texture	clastic	bioclastic	crystalline
Rock	A, B, F	E	C, D

(4)

29. Which two processes lead directly to the formation of both breccia and conglomerate?
    (1) melting and solidification
    (2) heat and pressure
    (3) compaction and cementation
    (4) evaporation and precipitation

30. Which diagram best shows the grain size of some common sedimentary rocks?

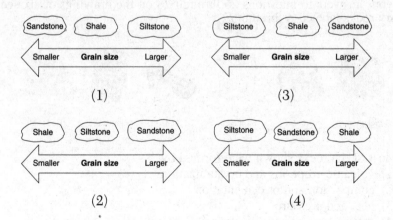

31. Metamorphic rocks result from the
    (1) erosion of rocks
    (2) recrystallization of rocks
    (3) cooling and solidification of molten magma
    (4) compression and cementation of soil particles

32. Which characteristics indicate that a rock has undergone metamorphic change?
    (1) The rock shows signs of being heavily weathered and forms the floor of a large valley.
    (2) The rock is composed of intergrown mineral crystals and shows signs of deformed fossils and structure.
    (3) The rock becomes less porous when exposed at the surface and is finely layered.
    (4) The rock contains a mixture of different-sized rounded grains or both felsic and mafic silicate minerals.

33. Which characteristic of rocks tends to increase as the rocks are metamorphosed?
    (1) density
    (2) volume
    (3) permeability
    (4) number of fossils present

34. Which two kinds of adjoining bedrock would most likely have a zone of contact metamorphism between them?
    (1) shale and conglomerate
    (2) shale and sandstone
    (3) limestone and sandstone
    (4) limestone and granite

35. Which statement is supported by information in the Rock Cycle in Earth's Crust diagram in the *Earth Science Reference Tables*?
    (1) Metamorphic rock results directly from melting and crystallization.
    (2) Sedimentary rock can be formed only from igneous rock.
    (3) Igneous rock always results from melting and solidification.
    (4) All sediments turn directly into sedimentary rock.

36. Which sequence of change in rock type occurs as shale is subjected to increasing heat and pressure?
    (1) shale → schist → phyllite → slate → gneiss
    (2) shale → slate → phyllite → schist → gneiss
    (3) shale → gneiss → phyllite → slate → schist
    (4) shale → gneiss → phyllite → schist → slate

37. Which rock is formed only by regional metamorphism?
    (1) slate
    (2) hornfels
    (3) dunite
    (4) marble

CONSTRUCTED RESPONSE QUESTIONS

Base your answers to questions 38 through 40 on the flowchart below and on your knowledge of Earth science. The flowchart shows the formation of some igneous rocks. The circled letters *A*, *B*, *C*, and *D* indicate parts of the flowchart that have not been labeled.

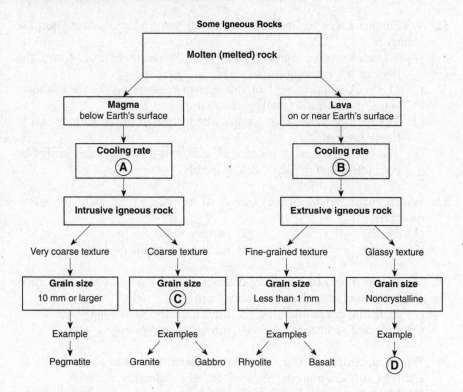

38. Contrast the rate of cooling at *A* that forms intrusive igneous rock with the rate of cooling at *B* that forms extrusive igneous rock. [1]

39. Give the numerical grain-size range that should be placed in the flowchart at *C*. Units must be included in your answer. [1]

40. State *one* igneous rock that could be placed in the flowchart at *D*. [1]

Base your answers to questions 41 through 43 on the photograph of a sample of gneiss below.

41. What observable characteristic could be used to identify this rock sample as gneiss? [1]

42. Identify *two* minerals found in gneiss that contain iron and magnesium. [1]

43. A dark-red mineral with a glassy luster was also observed in this gneiss sample. Identify the mineral and state *one* possible use for this mineral. [1]

Base your answers to questions 44 and 45 on the diagram below, which represents a part of the rock cycle. The igneous rock granite and the characteristics of sedimentary rock *X* and metamorphic rock *Y* are shown.

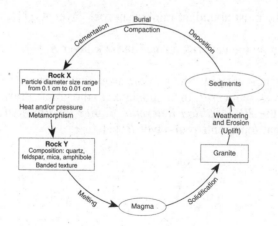

44. Identify sedimentary rock *X*. [1]

45. Identify metamorphic rock *Y*. [1]

Rocks

46. Complete the table below with descriptions of the observable characteristics used to identify granite. [1]

Characteristic of Granite	Description
Texture	
Color	
Density	

Base your answers to questions 47 through 49 on the block diagram below, which shows the landscape features of an area of Earth's crust. Two sedimentary rock layers, *A* and *B*, are labeled in the diagram. The rock symbol for layer *B* has been omitted.

47. Identify the most abundant mineral in rock layer *A*. [1]

48. Describe how the caverns formed in rock layer *A*. [1]

49. The graph below shows the particle sizes that compose the clastic sedimentary rock in layer *B*. In the box next to the graph, draw the map symbol from the *2011 Edition Reference Tables for Physical Setting/Earth Science* that represents rock layer *B*. [1]

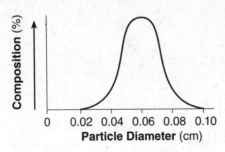

Base your answers to questions 50 through 52 on the data table below, which shows some characteristics of four rock samples, numbered 1 through 4. Some information has been left blank.

Data Table

Rock Sample Number	Composition	Grain Size	Texture	Rock Name
1	mostly clay minerals		clastic	shale
2	all mica	microscopic, fine	foliated with mineral alignment	
3	mica, quartz, feldspar, amphibole, garnet, pyroxene	medium to coarse	foliated with banding	gneiss
4	potassium feldspar, quartz, biotite, plagioclase feldspar, amphibole	5 mm		granite

50. State a possible grain size, in centimeters, for most of the particles found in sample 1. [1]

51. Write the rock name of sample 2. [1]

52. Write a term or phrase that correctly describes the texture of sample 4. [1]

Base your answers to questions 53 through 55 on the cross section below. The cross section shows a portion of Earth's crust. Letters *A*, *B*, *C*, and *D* represent rock units that have not been overturned.

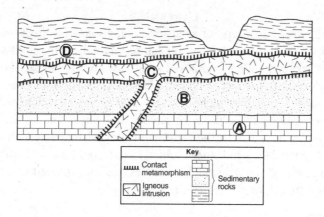

53. On the cross section, place an **X** where the metamorphic rock quartzite may be found. [1]

Rocks

54. Identify by name the most abundant mineral in rock unit *A*. [1]

55. State one piece of evidence shown in the cross section that indicates that rock unit *D* is older than igneous intrusion *C*. [1]

EXTENDED CONSTRUCTED RESPONSE QUESTIONS

Base your answers to questions 56 through 58 on the sequence of diagrams below, which show four stages in coal formation.

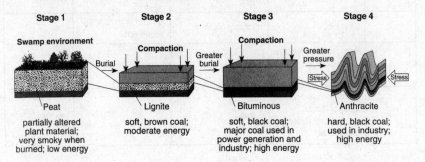

56. Which type of rock is forming above the coal material during stages 2 and 3? [1]

57. State the form of coal that normally has the highest density and explain why it has such a high density. [1]

58. Explain why coal deposits are *not* found in bedrock older than Silurian-age bedrock. [1]

Base your answers to questions 59 through 61 on the passage below.

Graywacke

Graywacke is a type of sandstone composed of a great variety of minerals. Unlike a "clean" sandstone where both the sand-sized grains and cement are composed mostly of quartz, graywacke is a "dirty" sandstone that can be composed of potassium feldspar, plagioclase feldspar, calcite, hornblende, and augite, as well as quartz.

Graywacke can be used for paving highways. The hard, massive bedrock is first drilled and then blasted into large chunks. Stone crushers grind these chunks into pebble-sized pieces. Truckloads of the graywacke pebbles are then hauled to plants where asphalt for paving is made.

59. State *one* difference in the mineral composition of a "clean" sandstone and a "dirty" sandstone. [1]

60. Identify *one* rock-forming process that must have occurred after the sediments were deposited to form graywacke. [1]

61. State *one negative* environmental impact a graywacke quarry could have on the area where it is located. [1]

Base your answers to questions 62 through 65 on the cross section below, which shows the bedrock structure of a portion of the lithosphere. Letters *A* through *D* represent locations in the lithosphere.

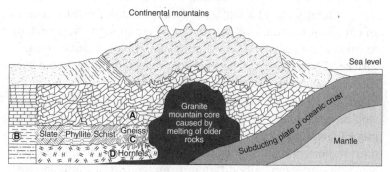

(Not drawn to scale)

62. Identify one of the most abundant minerals in the metamorphic rock at location *A*. [1]

63. Explain why the type of rock changes between locations *B* and *C*. [1]

64. Identify the grain size of the metamorphic rock at location *D*. [1]

65. Explain why the oceanic crust subducts beneath the continental crust when the two plates collide. [1]

Unit Five THE DYNAMIC EARTH

CHAPTER 16

EARTHQUAKES AND EARTH'S INTERIOR

KEY IDEAS With current technology, we cannot drill more than a few kilometers into the solid Earth. How, then, can we know anything of the internal structure of Earth? Earthquakes provide the key to inferring this internal structure. Earthquake waves literally travel around and through Earth, because rock is an excellent medium for conducting wave motion. Networks of seismometers, which detect earthquake waves, blanket Earth's surface. The records of earthquakes detected by seismometers can be analyzed to determine precisely when and where an earthquake occurred. By analyzing records of many earthquakes, as well as earthquake-wave behavior, scientists have been able to piece together a picture of Earth's internal structure. One of the most important results of such analysis has been the inference of a core that is chemically different from the rest of Earth and that is divided into a solid inner core and a liquid outer core.

KEY OBJECTIVES
Upon completion of this chapter, you will be able to:
- Describe the causes and effects of earthquakes.
- Explain how seismic records can be analyzed to locate the epicenter of an earthquake.
- Explain how earthquake waves can be used to create and refine a model of Earth's interior.
- Compare and contrast the composition, density, and phase of matter in the inner core, outer core, mantle, and crust.
- Identify regions of high earthquake activity, and relate them to the structure of Earth's crust.

EARTHQUAKES

An **earthquake** is a sudden trembling of the ground. Seismologists, scientists who study earthquakes, estimate that over one million quakes occur each year—about one every second! During most earthquakes, the shaking is so slight it is barely noticeable to the senses. During minor earthquakes, the ground feels a bit like a railway station when the train rumbles in. Major earthquakes cause violent shaking and lurching of the ground. The ground may split open, buildings may topple, pipes and power lines may be broken, and roadways may collapse. A destructive earthquake is a catastrophe, something to be planned for and guarded against.

Causes of Earthquakes

The major cause of earthquakes is **faulting**, the sudden movement of rock along planes of weakness, called faults, in Earth's crust. However, some earthquakes are caused by volcanic eruptions, and explosions set off by humans are responsible for a few. Nuclear testing is usually detected by the earthquakes it causes.

The **elastic rebound theory** explains the mechanism by which faulting causes earthquakes. At some places in Earth's crust, immense forces push or pull on the rock. Under this stress, rock can bend elastically, much as a wooden ruler can bend. However, if the stress increases, a point is reached where the rock can bend no farther without breaking; it is at its elastic limit. If stressed beyond this point, the rock snaps and the two broken edges whip back, or rebound. Great masses of rock suddenly scrape past each other, and the shock of this wrenching action jars the crust and sets an earthquake in motion. The point where the rock breaks is called the **focus** of an earthquake. The focus may be just beneath the surface or hundreds of kilometers down.

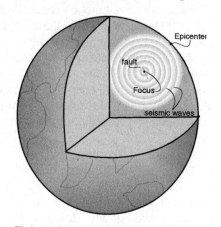

As the rock rebounds from this deformation, it pushes against surrounding rock, causing it to deform. The energy of this back and forth motion, or vibration, is transmitted as a series of deformations spreading out through the rock in all directions. The motion of a change in a medium is termed *wave motion*. The deformations of rock that spread out from the focus of an earthquake are therefore called *earthquake waves*, or **seismic waves**. Waves transport energy from one place to another through the motion of a change in the medium.

Figure 16.1 Fault, Focus, Epicenter, and Emanating Seismic Waves.

Seismic waves transport energy from the focus, through the surrounding rock, to Earth's surface. The earthquake is first felt at the **epicenter**, a point on the surface directly above the focus. See Figure 16.1.

Earthquake Waves

To understand the behavior of earthquake waves, it is necessary to review a few basic concepts relating to wave behavior in general.

Wave Behavior

A wave in a rope is a simple analogy that will help you visualize what is going on when seismic waves travel through Earth. It will also help you to understand some of the inferences that can be drawn from the behavior of seismic waves.

If you give one end of a stretched rope a quick shake, a bump, or wave, travels down the rope at some speed. See Figure 16.2. If the rope is uniform and completely flexible, the wave keeps the same shape as it moves down the rope. This is what happens in an earthquake, when faulting gives the crust a quick shake that produces seismic waves.

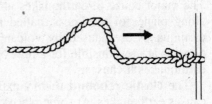

Figure 16.2 Wave Traveling Along a Rope.

As a wave travels down a rope, its forward part is moving upward and its rear part is moving downward. See Figure 16.3. The up and down motions of the string have kinetic energy. Work has to be done to produce the wave by pulling against the tension of the string, so the deformed string has potential energy. Thus, a wave contains both kinetic and potential energy as it travels along a rope. In like manner, seismic waves transfer energy as they travel through Earth. It is this energy that causes earthquake damage.

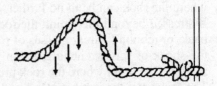

Figure 16.3 Movement of Rope as Wave Passes Through it.

The speed of the wave depends on the properties of the rope—how dense it is and how tightly it is stretched. Pulses move slowly down a loose, heavy rope because this rope has a lot of inertia and responds slowly to the forces acting on it. On the other hand,

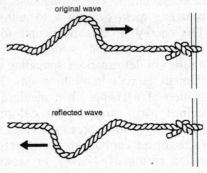

Figure 16.4 Wave Reflecting from a Fixed Support.

waves move rapidly down a taut, light rope because this rope has a greater tendency to straighten out. The same is true of seismic waves moving through Earth. The denser and more rigid the Earth material, the faster seismic waves travel through them.

When a wave reaches the end of the rope, it may be **reflected** and travel back toward its source. Depending on how the end is held, the reflected wave may be upright, be inverted, or disappear completely. If the end is fixed to a support (see Figure 16.4), the wave exerts a force on the support, which then exerts an equal but opposite force on the rope. This opposite force causes the rope to be displaced in the opposite direction from the original wave. An inverted wave forms and travels back along the rope in the opposite direction from the original wave. If the end is looped around a support so that it is connected but still free to move, an upright wave is reflected. If the end is held in a state somewhere between totally fixed and totally free, the wave disappears completely.

Now, consider what happens if two different ropes, a heavy (more dense) rope and a thin (less dense) rope are spliced together. See Figure 16.5. One end is tied to a support, and the other end is given a shake so that a wave is produced. When the wave reaches the splice, the wave passes from the thin rope to the heavy rope and is said to be transmitted. However, the transmission is not total because a reflected wave also appears at the splice and travels back down the thin rope. This type of wave behavior is what we see when seismic waves cross a boundary between two distinctly different regions in Earth's interior. This wave behavior was an important clue to Earth's layered internal structure.

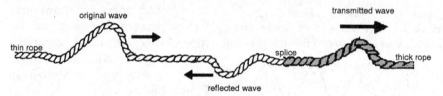

Figure 16.5 Transmitted and Reflected Waves at a Splice Between a Heavy and a Light Rope.

If one wave follows another in a rope, the result is a *periodic* wave. See Figure 16.6. In periodic waves a certain waveform, or shape, is repeated at regular intervals. Three quantities are often used to describe periodic waves: wave velocity, wavelength, and frequency. **Wave velocity** is the distance a wave moves each second; **wavelength** is the distance between adjacent crests or troughs; **frequency** is the number of waves that pass a given point in a second. Differences in wavelength, frequency, and the plane in which the waves oscillate are what distinguish different types of seismic waves.

If a periodic wave travels across a boundary between a more dense and a less dense medium at an angle, it is bent, or refracted. See Figure 16.7. This **refraction** occurs because, as the wave crosses the boundary, the front of the wave is moving at a different speed from the rear of the wave. The effect is somewhat like the result when the wheel at one end of an axle turns faster than the wheel at the other end — the path of the axle curves. Here, again, this is what we see when seismic waves cross a boundary in Earth's interior at an angle.

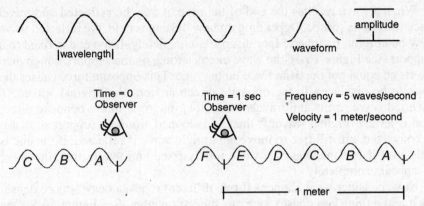

Figure 16.6 Periodic Wave, Showing Waveform, Wavelength, Wave Velocity, and Frequency.

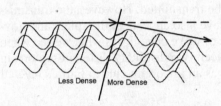

Figure 16.7 Periodic Wave Crossing a Boundary Between a More Dense and a Less Dense Medium and Being Refracted.

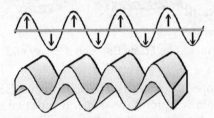

Figure 16.8 Transverse Waves, Showing Perpendicular Motion.

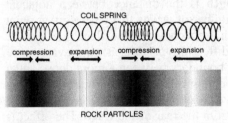

Figure 16.9 Longitudinal Waves in a Coil Spring and in Rock. As a longitudinal wave moves through a spring, its coils alternately compress and expand. In a rock, the individual particles are compressed and expanded.

Wave Types

In **transverse waves** the particles of the medium move back and forth in a direction perpendicular to the direction of wave motion. Waves in a stretched rope are transverse waves because any point on the rope moves side to side perpendicularly to the direction in which the wave is moving, as shown in Figure 16.8.

Longitudinal waves occur when the particles of the medium move back and forth in the same direction in which the waves travel. A good analogy is the movement of a compression down a coil spring when the end is pushed in and out. See Figure 16.9.

Seismic Waves

The complex motions that occur during faulting generate several types of periodic seismic waves. Some are transverse waves; others are longitudinal. Seismology took a giant step forward with the invention of the modern seismograph. A **seismograph** is a device that detects, measures, and records the motions of Earth that are associated with seismic waves.

A seismograph's operation is based upon the law of inertia: Objects that are at rest tend to remain at rest. A simple seismograph consists of a heavy, suspended weight that tends to remain at rest while Earth moves around it. A recording device (e.g., a pen) is attached to the weight. A recording medium, such as paper, is wound around a clockwork-driven drum mounted firmly in bedrock. When a series of seismic wave pulses causes the bedrock to move back and forth, the drum moves with it, but the heavy weight and its attached pen remain motionless for a long time. As the paper moves under the motionless pen, a line is drawn on the paper that records the back and forth motion of the bedrock, as shown in Figure 16.10a. The line recorded on paper by a seismograph is called a **seismogram** (see Figure 16.10b).

Modern seismographs use a laser beam reflected from a mirror on the weight instead of a pen. The laser beam exposes a track along photographic paper that is attached to the drum. When the bedrock vibrates, the track becomes a wavy line. In most earthquake recording stations, at least three seismographs are used, one to measure vibrations in each of three dimensions: x, y, and z (typically north-south, east-west, and up-down).

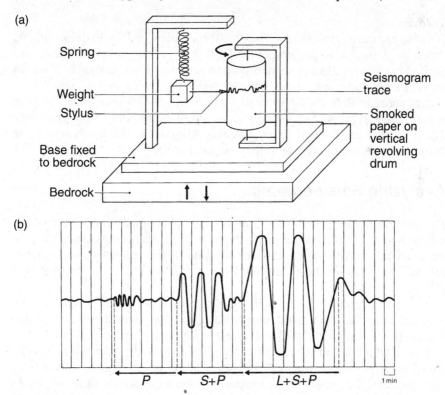

Figure 16.10 Simple Seismograph and Seismogram. P, S, and L represent P-, S-, and *Love* waves, respectively. Source: Earth Science on File. © Facts on File, Inc., 1988.

A careful study of seismograms reveals four different types of seismic waves. Two types, P-waves and S-waves, transport energy through the body of Earth and are therefore called **body waves**. Two other types, Love waves and Rayleigh waves, transport energy along the surface of Earth and are therefore called **surface waves**. Although surface waves play a major role in the damage caused by earthquakes, it is the body waves that reveal the structure of Earth's interior and make it possible to locate an earthquake's focus and epicenter. For this reason we will focus our attention on P-waves and S-waves.

P-waves

P-waves are longitudinal waves. They are alternating pulses of compression and expansion of rock in the same direction in which the wave travels. **P-waves** are the first (or **P**rimary) waves to reach a seismograph after an earthquake occurs because they travel fastest, at speeds of about 8 kilometers per second in the upper crust. P-waves can travel through solids, liquids, and gases because all three can be compressed and expanded. P-waves travel faster in these materials because these materials have stronger internal bonds; and, as occurs in a taut rope, the tendency to pull back into their original volume is greater. Since denser materials tend to be more rigid, P-wave speed increases in denser rock.

S-waves

S-waves are transverse waves. They twist rock back and forth, deforming its shape in a direction perpendicular to that of wave travel. **S-waves** (or **S**econdary waves) are the second type to arrive at a seismograph. S-waves arrive after P-waves because S-waves travel at a slower speed—about 4 kilometers per second in the upper crust. Like P-waves, S-waves travel faster in denser, more rigid rocks. S-waves can be transmitted only through solids. These waves cannot travel through liquids and gases because, when they are deformed, they do not return to their original shape.

Analyzing Seismograms

P-wave and S-wave recordings from a single event can be used to find the distance to the epicenter of an earthquake. With recordings from three or more seismic recording stations, the location of the epicenter can be determined. The basis for these determinations is the difference in the speeds of P-waves and S-waves.

When an earthquake occurs, all four types of seismic waves start moving outward from the focus at the same time. However, since they travel at different speeds, they do not all arrive at a seismograph at the same time. The P-waves, which travel the fastest, arrive first, followed by the S-waves some time later; the surface waves arrive last. The farther a seismograph is from the epicenter, the greater the difference between the arrival times of the P-waves and the S-waves.

For example, suppose *P*-waves travel 8 kilometers per second and *S*-waves travel 4 kilometers per second. The *P*-waves will arrive at a seismograph 80 kilometers away in 10 seconds. The *S*-waves will reach the same seismograph in 20 seconds. The arrival times at this first station will be 20 – 10 or 10 seconds *apart*. The same *P*-waves will reach a seismograph 200 kilometers away in 25 seconds, and the *S*-waves will arrive in 50 seconds. The arrival times at the farther station will be 50 – 25 or 25 seconds *apart*.

Determining Distance to the Epicenter of an Earthquake

For every distance that seismic waves travel, there is a corresponding difference in arrival times. This relationship between the difference in arrival time between *P*-waves and *S*-waves can be seen clearly in Figure 16.11. The travel time axis is marked off in minutes, and each minute is subdivided into three 20-second segments. The epicenter distance is marked off in thousands of kilometers ($10^3 = 1,000$) with each 1,000 kilometers subdivided into five 200-kilometer segments. The lines marked *S* and *P* show the times that *S*-waves and *P*-waves, respectively, take to travel a given distance. The vertical difference between the lines is the difference in arrival times at a given distance. It shows, for example, that *P*-waves travel 1,000 kilometers in 2 minutes while *S*-waves take about 4 minutes. The difference in arrival times is 2 minutes.

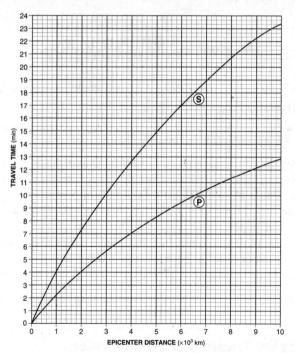

Figure 16.11 Earthquake *P*-wave and *S*-wave Travel Time. Source: The State Education Department, *Earth Science Reference Tables*, 2011 ed. (Albany, New York; The University of the State of New York).

Using Figure 16.11 and the relationship between difference in arrival times and distance, one can determine the distance between a seismograph and the epicenter of an earthquake as follows:

1. Determine the arrival times of the *P*-waves and the *S*-waves.
2. Calculate the *difference* in arrival times by subtracting the *P*-wave arrival time from the *S*-wave arrival time.
3. Using the travel time axis of the graph, mark off a distance along the edge of a piece of scrap paper equivalent to the *difference* in arrival times.
4. Keeping the edge of the paper vertical, find the point at which the marked-off difference in arrival times corresponds to the vertical distance between the *P*-wave line and the *S*-wave line.
5. Read the epicenter distance corresponding to that location on the graph.

Example:

The seismogram in Figure 16.12 is a record of an earthquake that was recorded by a seismograph. The seismogram does not tell how far away the earthquake occurred, but it does show when the *P*-waves and *S*-waves arrived. How far away from the seismograph was the epicenter of the earthquake?

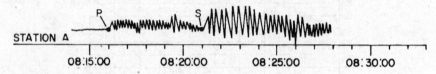

Figure 16.12 Seismogram.

To determine the epicenter, proceed as follows:

1. Note that the *P*-waves arrived at 08:16:00 and the *S*-waves arrived at 08:21:00.
2. Subtract: 08:21:00 − 08:16:00 = 00:05:00 or 5 min.
3. Using the travel time axis in Figure 16.11, mark off 5 min on the edge of a piece of paper.
4. Keeping the paper vertical, find the place where the marked-off portion of the paper just fits between the *S*-wave and *P*-wave lines.
5. Read the epicenter distance on the axis. It is 3,400 km.

The epicenter distance is 3,400 kilometers. These five steps are illustrated in Figure 16.13.

Determining the Location of the Epicenter

Knowing how far a seismograph station is from the epicenter does not pinpoint the location of the epicenter. In the example above, it could be 3,400 kilometers away in any direction. All of the points that are 3,400 kilometers from the seismograph in station A form a circle with a radius of 3,400 kilometers around that station. The epicenter is located somewhere on this circle. See Figure 16.14a.

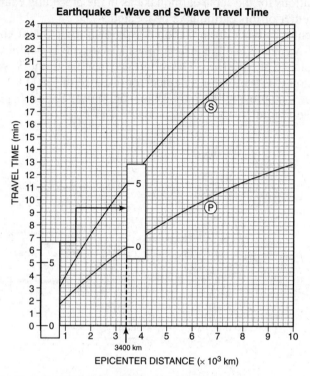

Figure 16.13 Five Steps in Determining Epicenter Distance.

However, if the same earthquake is 2,000 kilometers from a second station, station B, a circle can be drawn around that station too. The possible locations are now narrowed down to two points, as shown in Figure 16.14b. A third recording of the same earthquake, at station C, will eliminate one of these two points and pinpoint the location of the epicenter, as shown in Figure 16.14c.

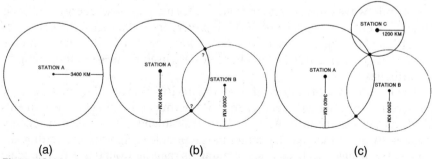

Figure 16.14 Information from Three Seismograph Stations Is Needed to Pinpoint the Location of the Earthquake Epicenter.

357

Example:

The seismograms shown in Figure 16.15 were recorded at three seismic recording stations for the same earthquake as in the preceding example. How can the epicenter of this earthquake be located?

The epicenter distance from station A is known to be 3,400 km. The difference in *P*-wave and *S*-wave arrival times at station B is 7 min, yielding an epicenter distance of 5,400 km. The difference in *P*-wave and *S*-wave arrival times at station C is 9 min, for an epicenter distance of 7,600 km.

Plotted on a map using the map scale, stations *A*, *B*, and *C* pinpoint the epicenter. See Figure 16.15.

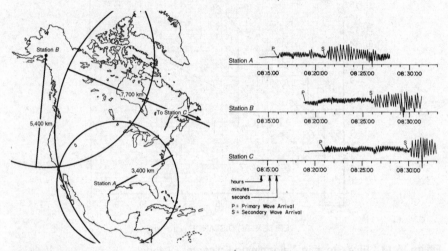

Figure 16.15 Locating the Epicenter. When the three seismograms above are analyzed, the distance from seismograph to the earthquake epicenter can be determined. Plotted on a map, only one place is the correct distance from all three stations.

Measuring Earthquakes

As seismic waves travel farther from the focus, they lose energy and their effects lessen. At a distance of 100 kilometers from the focus, the energy of an earthquake is only about 1/10,000 of what it was at 1 kilometer from the focus. The Richter and the Mercalli scales are used to measure the strengths of earthquakes.

The **Richter Magnitude Scale** (see Figure 16.16) is based upon the energy released by an earthquake, and energy is determined by direct measurements of the motions of the crust using seismic instruments. The greater the energy released by an earthquake, the larger the amplitude (height) of the earthquake waves. The Richter scale consists of numbers ranging from 0 to 8.6. The scale is logarithmic, meaning that the energy of a shock increases by powers of 10 in relation to the Richter magnitude numbers. In other words, each increase of

1 unit on the Richter scale corresponds to a tenfold increase in the amplitude (height) of the earthquake wave. Also, every tenfold increase in wave height corresponds to a hundredfold increase in energy! Thus, a magnitude-2 earthquake wave is 10 times higher than a magnitude-1 earthquake wave and releases 100 times as much energy.

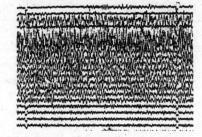

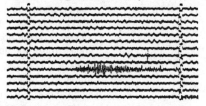

Figure 16.16 The Richter Magnitude Scale. This is based on the motion of bedrock as recorded by seismographs.

The actual destructiveness of an earthquake, though, is influenced also by factors other than the amount of energy released. For example, destructiveness depends also upon nearness to the epicenter and the nature of the subsurface materials. Therefore, a magnitude-5 earthquake may cause vastly different amounts of damage in two different places. Since humans are concerned about damage to their land and property, a scale of earthquake damage intensity was developed. The modified **Mercalli Intensity Scale** (see Table 16.1) is based upon observations of earthquake damage and is concerned chiefly with the impact of an earthquake on structures made by humans and on human activities. Engineers and city planners often use this scale in reaching decisions about structural and land-use issues in earthquake-prone areas.

Effects of Earthquakes

The effects of earthquakes depend mainly upon the strength of the seismic waves and the nature of the material through which they pass. Earthquake effects include ground shaking and failure, surface faulting, changes in wells and geysers, and tsunamis.

Ground shaking affects solid rock the least, and loose, water-soaked ground the most. The same shock that causes solid bedrock to sway slightly may cause intense jolting in nearby wet clay or landfill. Tapping a bowl of Jello is a good analogy. The solid bowl vibrates but moves hardly at all, while the Jello jiggles and quivers noticeably. In ground failure, strong vibrations cause loose or water-soaked ground to break apart and settle, forming cracks and fissures. On hillsides these fissures can trigger landslides, slumping, and mudflows.

Surface faulting occurs when movement along a fault causes the surface to be lifted, lowered, or shifted sideways. The result is displacement of surface features and structures.

TABLE 16.1 MODIFIED MERCALLI INTENSITY SCALE

Intensity Value	Description of Effects
I	Not felt.
II	Felt by persons at rest; felt on upper floors because of sway.
III	Felt indoors. Hanging objects swing. Feels like a passing truck.
IV	Hanging objects swing. Feels like heavy truck passing, or a jolt is felt. Standing cars rock. Windows and dishes rattle. Glasses clink. In the upper range of IV, wooden frames and walls creak.
V	Felt outdoors and direction can be estimated. Sleepers are awakened. Liquids are disturbed, some spill. Small, unstable objects upset. Doors swing open and close. Shutters and pictures move.
VI	Felt by all. Persons walk unsteadily, and many are frightened and run indoors. Windows, dishes, and glassware are broken. Furniture is moved or overturned. Pictures fall from walls. Weak plaster cracks. Trees and bushes visibly shaken or heard to rustle.
VII	Difficult to stand. Noticed by drivers of cars. Hanging objects quiver. Furniture is broken. Weak chimneys break at roof line. Fall of plaster, loose bricks, masonry. Waves on ponds and water muddied. Small slides and cave-ins of sand and gravel banks.
VIII	Steering of moving cars is affected. Partial collapse of masonry structures. Chimneys and smokestacks twist and fall. Frame houses move on foundation if not bolted down. Branches broken from trees. Wet ground and steep slopes crack.
IX	General panic. Weak masonry destroyed, stronger masonry cracks, is seriously damaged. Frame structures not bolted shift off foundations. Frames cracked. Reservoirs seriously damaged. Underground pipes broken. Conspicuous cracks in ground.
X	Masonry and frame structures destroyed along with their foundations. Some bridges destroyed. Serious damage to reservoirs, dikes, dams, and embankments. Large landslides. Water thrown out of lakes, canals, rivers, etc. Rails bent slightly.
XI	Rails bent greatly. Underground pipelines completely destroyed and out of service.
XII	Damage nearly total. Large rock masses displaced. Objects thrown into the air. Lines of sight and level are distorted.

Modified from: H.O. Wood and F. Neumann, "Modified Mercalli Intensity Scale of 1931," *Bulletin of the Seismological Society of America*, Vol. 21, No. 4, pp. 277–288.

As seismic waves travel through water-saturated rock or soil, water is compressed and forced through pore spaces. Like a hand alternately squeezing and releasing a soaked sponge in a bucket, seismic waves force water in and out of rock and soil surrounding wells, causing their water levels to fluctuate. Seismic waves have a similar effect on rock fissures, causing geysers to behave erratically.

Tsunamis are immense sea waves caused by earthquakes beneath the ocean floor or by undersea landslides. Tsunamis may be only a few meters high, but they have very long wavelengths and travel much more rapidly than ordinary ocean waves. Tsunamis have been clocked moving faster than 500 kilometers per hour with wavelengths as great as 200 kilometers. When tsunamis reach shallow water, they are slowed by friction with the ocean bottom. In bays and narrow channels, however, their high speed and long wavelength may be funneled into huge breaking waves more than 20 meters high. The force exerted by such a huge, fast-moving mass of water can do extensive damage.

Earthquakes are geologic hazards to humans, but anticipating the hazards earthquakes pose and preparing to meet them can minimize loss of lives and property. Figure 16.17 shows areas of various degrees of seismic risk in the United States.

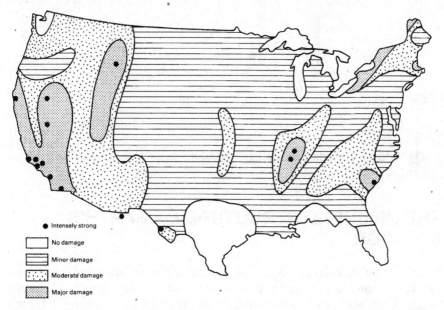

Figure 16.17 Seismic Risk Map of the United States. Source: Earth Science on FIle. © Facts on File, Inc., 1988.

Earthquake Zones

If the locations of earthquakes are plotted on a map of Earth, an interesting pattern emerges, as shown in Figure 16.18. Rather than being randomly spread over Earth, earthquakes occur in distinct, narrow zones. These zones include the mid-ocean ridges, the rim of the Pacific Ocean (called the **Ring of Fire** for its many active volcanoes), and the Mediteranean Belt. As you will discover in Chapter 18, the locations of earthquake epicenters reveal the locations of boundaries between the plates of Earth's crust.

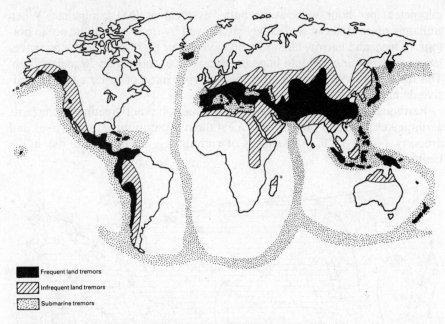

Figure 16.18 Worldwide Earthquake Distribution. Source: Earth Science on File. © Facts on File, Inc., 1988.

THE INFERRED STRUCTURE OF EARTH'S INTERIOR

Whenever an earthquake occurs, its seismic waves are recorded by seismographs at hundreds of seismic recording stations throughout the world. Analysis of these seismograms reveals much about the structure of Earth's interior.

As was discussed earlier, seismic waves are reflected from the surfaces of materials and refracted as they move across boundaries between substances of different densities. If Earth had a homogeneous composition and density, seismic waves would move through it in simple, straight-line paths with reflections only when the surface was reached. They would also travel directly from the earthquake's focus to all points on Earth's surface. Furthermore, if the location of an earthquake's epicenter is known, the arrival times of P-waves and S-waves at all stations should be easy to predict.

This is not the case. P-waves are absent from seismographs at certain distances from the epicenter and are more highly concentrated elsewhere. Predictions of arrival times based upon straight-line travel differ markedly from actual measurements. Moreover, the paths of seismic waves curve. This gradual bending is due to the gradual increase in density with depth, caused by increasing pressure. There are many more reflected waves than would be expected from a homogeneous Earth. There are also places on Earth's surface where *P*-waves or *S*-waves, or both, do not arrive directly from the focus of the earthquake. These facts led seismologists to conclude that Earth's interior is not homogeneous.

The Crust

In 1910, Andrija Mohorovicic discovered that two sets of *P*-waves and *S*-waves arrived at locations close to an earthquake's epicenter, but only one set arrived farther away. Mohorovicic explained this phenomenon by hypothesizing a boundary between rocks of different densities that would cause the waves to refract. As shown in Figure 16.19, stations close to the epicenter of an earthquake can receive both direct and refracted seismic waves. Farther away, however, waves cannot reach the station without crossing the boundary, so only refracted waves arrive.

This boundary, called the **Moho** in honor of Mohorovicic, marks the boundary between the outermost layer of Earth, the **crust**, and the **mantle** beneath it. The density of the crust varies from about 2.7 grams per cubic centimeter near the surface to about 3.1 grams per cubic centimeter near the Moho. The thickness of the crust varies, but is usually greatest beneath mountains and least beneath oceans.

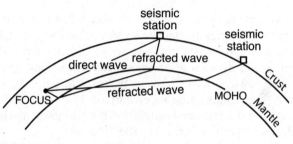

Figure 16.19 The Moho Boundary.

The Mantle and Core

Between the epicenter and an angle of 103° from the epicenter, *P*-waves and *S*-waves arrive directly from the focus, as would be expected. Beyond 103°, however, the *S*-waves *disappear*! This region is known as the **S-wave shadow zone**. Seismic stations in the *S*-wave shadow zone receive no *S*-waves from an earthquake.

Between 103° and 143° from the epicenter, the *P*-waves *also* disappear. But beyond 143°, the *P*-waves reappear. Therefore, the region between 103° and 143° from the epicenter is known as the **P-wave shadow zone**. Seismic stations located between 103° and 143° of an earthquake's epicenter fall within both the *S*-wave shadow zone and the *P*-wave shadow zone and record no seismic waves from the earthquake. Seismic stations beyond 143° in either direction from the epicenter record only *P*-waves. That this odd pattern of shadow zones can be explained by Earth's internal structure is shown in Figure 16.20.

Earthquakes and Earth's Interior

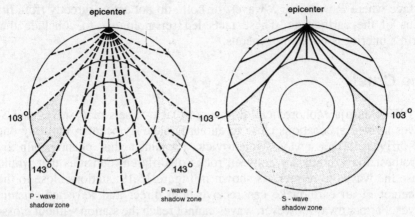

Figure 16.20 The *P*-Wave and *S*-Wave Shadow Zones.

A boundary at a depth of about 2,900 kilometers, between rocks of markedly different densities, produces a strong refraction of *P*-waves that cross it. Up to 103°, waves arrive directly from the epicenter. The strong refraction as soon as the boundary is encountered immediately angles the waves away to 143°, so no waves reach the region in between. Since *P*-waves continue to be received past 143°, they are able to pass through the material beneath the boundary. This boundary marks the bottom of the mantle; the layer beneath it is called the **core**.

The complete absence of *S*-waves beyond 103° indicates that they are unable to travel through the material beneath the mantle. Since transverse waves cannot travel through fluids, the disappearance of *S*-waves makes sense if the core is a fluid.

Careful analysis of refraction of *P*-waves that pass through the core and the presence of reflections from inside the core indicate that another boundary inside the core divides it into an **inner core** and an **outer core**. Only the outer core is thought to be a fluid.

Summary of Inferences about Earth's Interior

Figure 16.21 summarizes what has been inferred about Earth's interior based upon studies of seismic waves and other data. The diagram at the top shows a cross section of Earth with the interior layers labeled. Along the right edge, the density of each layer is shown. Notice that the mantle is divided into two regions, the asthenosphere, or plastic mantle, and the stiffer mantle. Notice, too, the sharp increase in density between the mantle and the core, which explains the strong refraction of *P*-waves. Directly beneath the cross section is a graph showing pressure in *millions of atmospheres* at different depths. The dotted lines indicate boundaries between crust, mantle, and outer and inner cores. On the bottom is a similar graph showing temperatures and melting points of rock at different depths. Notice that in the outer core the actual temperature is higher than the melting point of the rocks.

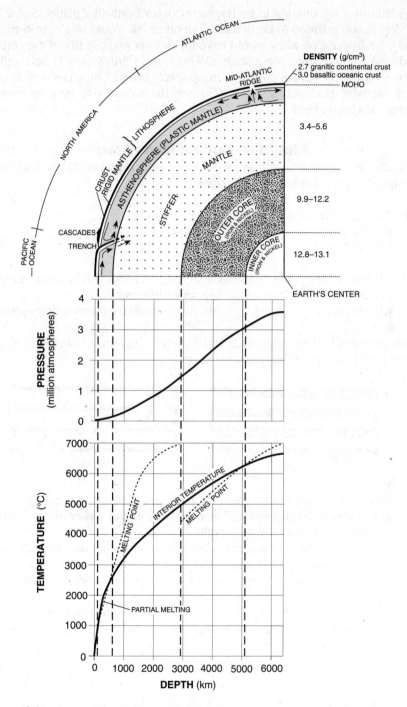

Figure 16.21 Inferred Properties of Earth's Interior. Source: The State Education Department, *Earth Science Reference Tables*, 2011 ed. (Albany, New York; The University of the State of New York).

Meteorites are thought to be fragments of an Earth-like planet. Some are similar in composition to Earth rocks, but others are composed of iron-nickel alloy. An iron-nickel alloy would have the density and rigidity of the core as deduced from seismic wave studies. Therefore, Earth's core is believed to consist of an iron-nickel alloy. An iron-nickel core in which convection currents transfer heat outward (see Chapter 10) would also help to explain Earth's magnetic field.

MULTIPLE-CHOICE QUESTIONS

In each case, write the number of the word or expression that best answers the question or completes the statement.

1. The most frequent cause of major earthquakes is
   (1) faulting   (3) landslides
   (2) folding   (4) submarine currents

2. The immediate result of a sudden slippage of rocks within Earth's crust is
   (1) isostasy   (3) an earthquake
   (2) erosion   (4) the formation of convection currents

Base your answers to questions 3 and 4 on the diagram below, which shows models of two types of earthquake waves.

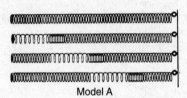
Model A

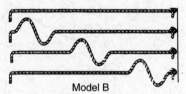
Model B

3. Model A best represents the motion of earthquake waves called
   (1) P-waves (compressional waves) that travel faster than S-waves (shear waves) shown in model B
   (2) P-waves (compressional waves) that travel slower than S-waves (shear waves) shown in model B
   (3) S-waves (shear waves) that travel faster than P-waves (compressional waves) shown in model B
   (4) S-waves (shear waves) that travel slower than P-waves (compressional waves) shown in model B

4. The difference in seismic station arrival times of the two waves represented by the models helps scientists determine the
   (1) amount of damage caused by an earthquake
   (2) intensity of an earthquake
   (3) distance to the epicenter of an earthquake
   (4) time of occurrence of the next earthquake

Base your answer to question 5 on the seismogram below. The seismogram was recorded at a seismic station and shows the arrival times of the first *P*-wave and *S*-wave from an earthquake.

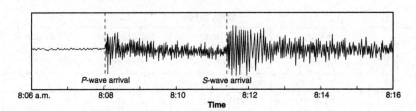

5. Which part of this seismogram is used to find the distance to the epicenter of the earthquake?
   (1) *P*-wave arrival time, only
   (2) *S*-wave arrival time, only
   (3) difference in the arrival time of the *P*-wave and *S*-wave
   (4) difference in the height of the *P*-wave and *S*-wave

6. How long would it take for the first *S*-wave to arrive at a seismic station 4,000 kilometers away from the epicenter of an earthquake?
   (1) 5 min 40 sec   (3) 12 min 40 sec
   (2) 7 min 0 sec   (4) 13 min 20 sec

7. Which statement best describes the relationship between the travel rates and travel times of earthquake *P*-waves and *S*-waves from the focus of an earthquake to a seismograph station?
   (1) *P*-waves travel at a slower rate and take less time.
   (2) *P*-waves travel at a faster rate and take less time.
   (3) *S*-waves travel at a slower rate and take less time.
   (4) *S*-waves travel at a faster rate and take less time.

8. A seismic station 4000 kilometers from the epicenter of an earthquake records the arrival time of the first *P*-wave at 10:00:00. At what time did the first *S*-wave arrive at this station?
   (1) 9:55:00   (3) 10:07:05
   (2) 10:05:40   (4) 10:12:40

9. A P-wave takes 8 minutes and 20 seconds to travel from the epicenter of an earthquake to a seismic station. Approximately how long will an S-wave take to travel from the epicenter of the same earthquake to this seismic station?
   (1) 6 min 40 sec
   (2) 9 min 40 sec
   (3) 15 min 00 sec
   (4) 19 min 00 sec

Base your answers to questions 10 through 12 on the data table below, which gives information collected at seismic stations W, X, Y, and Z for the same earthquake. Some of the data have been omitted.

Data Table

Seismic Station	P-Wave Arrival Time (h:min:s)	S-Wave Arrival Time (h:min:s)	Difference in Arrival Times (h:min:s)	Distance to Epicenter (km)
W	10:50:00	no S-waves arrived		
X	10:42:00	10:46:40		
Y	10:39:20		00:02:40	
Z	10:45:40			6200

10. Which seismic station was farthest from the earthquake epicenter?
    (1) W
    (2) X
    (3) Y
    (4) Z

11. What is the most probable reason for the absence of S-waves at station W?
    (1) S-waves were not generated at the epicenter.
    (2) S-waves cannot travel through liquids.
    (3) Station W was located on solid bedrock.
    (4) Station W was located on an island.

12. At what time did the S-wave arrive at station Y?
    (1) 10:36:40
    (2) 10:39:20
    (3) 10:42:00
    (4) 10:45:20

Base your answers to questions 13 and 14 on the earthquake seismogram below.

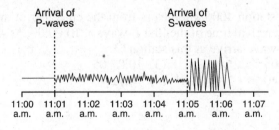

13. When did the first *P*-waves arrive as this seismic station?
    (1) 3 minutes after an earthquake occurred 2,600 km away
    (2) 5 minutes after an earthquake occurred 2,600 km away
    (3) 9 minutes after an earthquake occurred 3,500 km away
    (4) 11 minutes after an earthquake occurred 3,500 km away

14. How many additional seismic stations must report seismogram information in order to locate this earthquake?
    (1) one   (3) three
    (2) two   (4) four

Base your answers to questions 15 through 18 on the diagram and map below. The diagram shows three seismograms of the same earthquake recorded at three different seismic stations, *X*, *Y*, and *Z*. The distances from each seismic station to the earthquake epicenter have been drawn on the map.

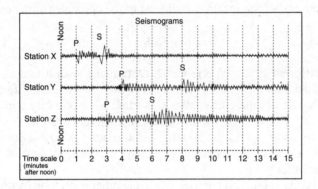

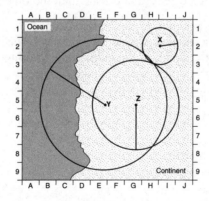

A coordinate system has been placed on the map to describe locations. The map scale has not been included.

15. Approximately how far away from station *Y* is the epicenter?
    (1) 1,300 km
    (2) 2,600 km
    (3) 3,900 km
    (4) 5,200 km

16. The *S*-waves from this earthquake that travel toward Earth's center will
    (1) be deflected by Earth's magnetic field
    (2) be totally reflected off the crust-mantle interface
    (3) be absorbed by the liquid outer core
    (4) reach the other side of Earth faster than those that travel around Earth in the crust

17. Seismic station *Z* is 1,700 kilometers from the epicenter. Approximately how long did the *P*-wave take to travel to station *Z*?
    (1) 1 min 50 sec
    (2) 2 min 50 sec
    (3) 3 min 30 sec
    (4) 6 min 30 sec

18. On the map, which location is closest to the epicenter of the earthquake?
    (1) *E*–5
    (2) *G*–1
    (3) *H*–3
    (4) *H*–8

19. Two different cities experience an earthquake with a magnitude of 5.5 on the Richter Magnitude Scale. However, on the Modified Mercalli Intensity Scale, the earthquake was rated V in one city and VII in the other city. The best explanation for this difference is that
    (1) one city is nearer the equator and the other is near a pole
    (2) the earthquake occurred at 8 P.M. in one city and at 4 A.M. in the other
    (3) one city is built on bedrock while the other is built on loose sediments
    (4) one city is in a drier climate zone than the other

20. An earthquake's magnitude can be determined by
    (1) analyzing the seismic waves recorded by a seismograph
    (2) calculating the depth of the earthquake faulting
    (3) calculating the time the earthquake occurred
    (4) comparing the speed of *P*-waves and *S*-waves

21. Scientists have inferred the structure of Earth's interior mainly by analyzing
    (1) the Moon's interior
    (2) the Moon's composition
    (3) Earth's surface features
    (4) Earth's seismic data

22. The theory that the outer core of Earth is composed of liquid material is best supported by
    (1) seismic studies that indicate that shear waves do not pass through the outer core
    (2) seismic studies that show that compressional waves can pass through the outer core
    (3) density studies that show that the outer core is slightly more dense than the inner core
    (4) gravity studies that indicate that gravitational strength is greatest within the core

Base your answers to questions 23 and 24 on the cross-sectional view of Earth below, which shows seismic waves traveling from the focus of an earthquake. Points *A* and *B* are locations on Earth's surface.

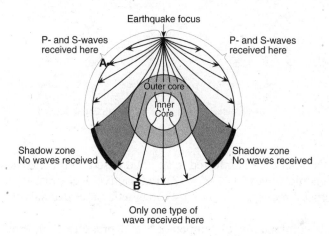

23. Which statement best explains why only one type of seismic wave was recorded at location *B*?
    (1) *S*-waves cannot travel through the liquid outer core.
    (2) *S*-waves cannot travel through the liquid inner core.
    (3) *P*-waves cannot travel through the solid outer core.
    (4) *P*-waves cannot travel through the solid inner core.

24. A seismic station located at point *A* is 5,400 kilometers away from the epicenter of the earthquake. If the arrival time for the *P*-wave at point *A* was 2:00 P.M., the arrival time for the *S*-wave at point *A* was approximately
    (1) 1:53 P.M.   (3) 2:09 P.M.
    (2) 2:07 P.M.   (4) 2:16 P.M.

25. Which graph best shows the inferred density of Earth's interior as depth increases from the upper mantle to the lower mantle?

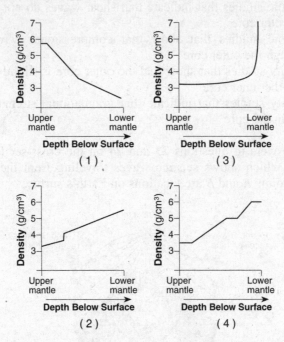

26. Earth's inner core is inferred to be solid based on the analysis of
    (1) seismic waves
    (2) crustal rocks
    (3) radioactive decay rates
    (4) magnetic pole reversals

27. Which combination of temperature and pressure is inferred to occur within Earth's stiffer mantle?
    (1) 3,500°C and 0.4 million atmospheres
    (2) 3,500°C and 2.0 million atmospheres
    (3) 5,500°C and 0.4 million atmospheres
    (4) 5,500°C and 2.0 million atmospheres

28. A huge undersea earthquake off the Alaskan coastline could produce a
    (1) tsunami
    (2) cyclone
    (3) hurricane
    (4) thunderstorm

29. Theories about the composition of Earth's core are supported by meteorites that are composed primarily of
    (1) oxygen and silicon
    (2) aluminum and iron
    (3) aluminum and oxygen
    (4) iron and nickel

Base your answers to questions 30 and 31 on the map below, which shows the risk of damage from seismic activity in the United States.

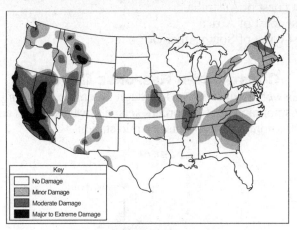

30. In the United States, most of the major damage expected from a future earthquake is predicted to occur near a
(1) divergent plate boundary, only
(2) convergent plate boundary, only
(3) mid-ocean ridge and a divergent plate boundary
(4) transform plate boundary and a hot spot

31. Which New York State location has the greatest risk of earthquake damage?
(1) Binghamton   (3) Plattsburgh
(2) Buffalo   (4) Elmira

32. The data table below shows the origin depths of all large-magnitude earthquakes over a 20-year period.

Data Table

Depth Below Surface (km)	Number of Earthquakes
0–33	27,788
34–100	17,585
101–300	7,329
301–700	3,167

According to these data, most of these earthquakes occurred within Earth's
(1) lithosphere   (3) stiffer mantle
(2) asthenosphere   (4) outer core

33. Which coastal area is most likely to experience a severe earthquake?
(1) east coast of North America
(2) east coast of Australia
(3) west coast of Africa
(4) west coast of South America

CONSTRUCTED RESPONSE QUESTIONS

Base your answers to questions 34 through 36 on the diagram below, which shows a seismograph that recorded seismic waves from an earthquake located 4,000 kilometers from this seismic station.

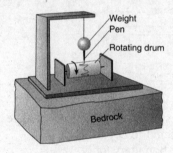

34. State *one* possible cause of the earthquake that resulted in the movement of the bedrock detected by this seismograph. [1]

35. Which type of seismic wave was recorded first on the rotating drum? [1]

36. How long does the first S-wave take to travel from the earthquake epicenter to this seismograph? [1]

Base your answers to questions 37 and 38 on the diagram below, which shows two seismogram tracings, at stations A and B, for the same earthquake. The arrival times of the P-waves and S-waves are indicated on each tracing.

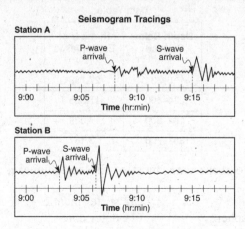

37. Explain how the seismic tracings recorded at station *A* and station *B* indicate that station *A* is farther from the earthquake epicenter than station *B*. [1]

38. Seismic station *A* is located 5,400 kilometers from the epicenter of the earthquake. How much time would it take for the first *S*-wave produced by this earthquake to reach seismic station *A*? [1]

Base your answers to questions 39 through 42 on the map below and on your knowledge of Earth science. The map shows the location of the epicenter, **X**, of an earthquake that occurred on April 20, 2002, about 29 kilometers southwest of Plattsburgh, New York.

39. State the latitude and longitude of this earthquake epicenter. Express your answers to the *nearest tenth of a degree* and include the compass directions. [1]

40. What is the *minimum* number of seismographic stations needed to locate the epicenter of an earthquake? [1]

41. Explain why this earthquake was most likely felt with greater intensity by people in Peru, New York, than by people in Lake Placid, New York. [1]

375

42. A seismic station located 1,800 kilometers from the epicenter recorded the *P*-wave and *S*-wave arrival times for this earthquake. What was the difference in the arrival time of the first *P*-wave and the first *S*-wave? [1]

EXTENDED CONSTRUCTED RESPONSE QUESTIONS

Base your answers to questions 43 through 46 on the information and map below and your knowledge of Earth science.

An earthquake occurred in the southwestern part of the United States. Mercalli scale intensities were plotted for selected locations on the map, as shown below. (As the numerical values of Mercalli ratings increase, the damaging effects of the earthquake waves also increase.)

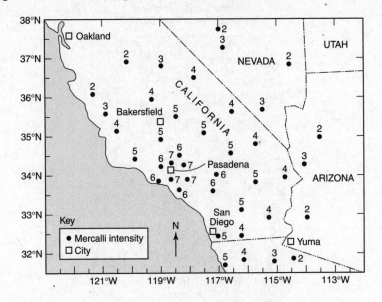

43. Using an interval of 2 Mercalli units and starting with an isoline representing 2 Mercalli units, draw an accurate isoline map of earthquake intensity. [4]

44. State the name of the city that is closest to the earthquake epicenter. [1]

45. State the latitude and longitude of Bakersfield. [2]

46. Using one or more complete sentences, identify the most likely cause of earthquakes that occur in the area shown on the map. [2]

Base your answers to questions 47 through 51 on the map and the modified Mercalli intensity scale below. The map shows modified Mercalli intensity scale damage zones resulting from a large earthquake that occurred in 1964. The earthquake's epicenter was near Anchorage, Alaska. The cities Kodiak and Anchorage are shown on the map. The Mercalli scale describes earthquake damage at Earth's surface.

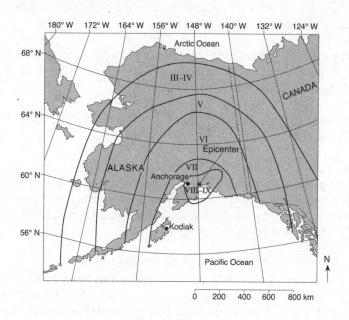

Modified Mercalli Intensity Scale

I	Instrumental: detected only by instruments	VII	Very strong: noticed by people in autos Damage to poor construction
II	Very feeble: noticed only by people at rest	VIII	Destructive: chimneys fall, much damage in substantial buildings, heavy furniture overturned
III	Slight: felt by people at rest Like passing of a truck	IX	Ruinous: great damage to substantial structures Ground cracked, pipes broken
IV	Moderate: generally perceptible by people in motion Loose objects disturbed	X	Disastrous: many buildings destroyed
V	Rather strong: dishes broken, bells rung, pendulum clocks stopped People awakened	XI	Very disastrous: few structures left standing
VI	Strong: felt by all, some people frightened Damage slight, some plaster cracked	XII	Catastrophic: total destruction

Earthquakes and Earth's Interior

47. Describe *one* type of damage that occurred in Anchorage but *not* in Kodiak. [1]

48. Write the names of the *two* converging tectonic plates that caused this earthquake. [1]

49. Explain why *S*-waves from this earthquake were *not* directly received on the opposite side of Earth. [1]

50. This earthquake produced a large ocean floor displacement. Identify *one* dangerous geologic event affecting Pacific Ocean shorelines as a result of this ocean floor displacement. [1]

51. Determine the latitude and longitude of this epicenter. Include the units and compass directions in your answer. [1]

Base your answers to questions 52 through 56 on the passage and the map below. The passage describes the New Madrid fault system. The numbers on the map show the predicted relative damage at various locations if a large earthquake occurs along the New Madrid fault system. The higher the number is, the greater the relative damage.

The New Madrid Fault System

The greatest earthquake risk area east of the Rocky Mountains is along the New Madrid fault system. The New Madrid fault system consists of a series of faults along a weak zone in the continental crust in the midwestern United States. Earthquakes occur in the Midwest less often than in California. When they do happen, though, the damage is spread over a wider area due to the underlying bedrock.

In 1811 and 1812, the New Madrid fault system experienced three major earthquakes. Large land areas sank, new lakes formed, the course of the Mississippi River changed, and 150,000 acres of forests were destroyed.

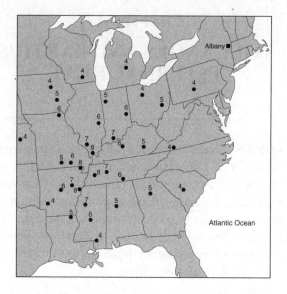

52. On the map, draw the 4, 6, and 8 isolines indicating relative damage. [1]

53. Using the predicted damage numbers, place an **X** on the map to indicate where the New Madrid fault system most likely exists. [1]

54. The distance between the New Madrid fault system and Albany, New York, is 1,800 kilometers. What was the time difference between the arrival of the first P-wave and the arrival of the first S-wave at Albany when the 1812 earthquake occurred? [1]

55. State *one* reason why earthquakes occur more frequently on the western coast of the United States than in the New Madrid region. [1]

56. An emergency management specialist near the New Madrid region is developing a plan that would help save lives and prevent property damage in the event of an earthquake. Describe *two* actions that should be included in the plan. [1]

CHAPTER 17

VOLCANOES AND EARTH'S INTERNAL HEAT

> **KEY IDEAS** Earth's internal heat engine is powered by heat from the decay of radioactive materials and residual heat from Earth's formation. Measurements of heat flow at and near Earth's surface indicate a steady increase in temperature with depth, reaching temperatures of 6,000°C or more at the core. Heat flowing outward from this fiery interior is largely responsible for a variety of phenomena ranging from Earth's magnetic field to crustal motion and volcanic activity.

KEY OBJECTIVES
Upon completion of this chapter, you will be able to:

- Explain that Earth has internal sources of energy that create heat.
- Explain how the transfer of heat energy within Earth's interior results in the formation of regions of different densities, and how these density differences lead to motion.
- Describe the origin of magma, and identify regions of frequent volcanic activity.
- Compare and contrast intrusive and extrusive volcanic activity.

EARTH'S INTERNAL HEAT ENGINE

Beneath its cool outer crust, Earth is glowing hot. Evidence for a hot interior comes from exploration of the crust, both above and below the surface. At the surface, we find numerous examples of heat escaping from Earth's interior: molten rock pours from volcanoes, and boiling hot water and steam rise from hot springs and jet out of geysers.

Further evidence comes from mines and deep wells. As we drill deeper into the crust, the temperature of the rock gets higher and higher. The deepest oil wells go down about 8 kilometers, and the temperature of the oil from these wells is 150°C. Measurements of rock temperature in mines and boreholes show an average temperature rise of 1°C per 30 meters; this increase in temperature with depth is known as the **geothermal gradient**. The rate of temperature increase, however, falls off rapidly with depth. See Figure 17.1. Although the rate of temperature increase in the mantle is lower, considering that Earth's center is more than 6,000 *kilometers* down, it is not unlikely that temperatures there exceed 6,000°C.

Heat flows continuously from Earth's fiery depths toward its cool surface. The flow of heat from Earth's interior is roughly 209 joules (50 calories) per square centimeter per year. This is extremely small compared to the amount of heat that Earth's surface receives from the Sun each year—an average 749,000 joules (179,000 calories) per square centimeter. Therefore, Earth's heat flow from its interior contributes little to its heat balance or to the powering of convection in the atmosphere or oceans. However, this heat does power convective motions within Earth's mantle and core.

Sources of Internal Heat

Original Heat

Some of the heat inside Earth was trapped when it was formed. As discussed in Chapter 10, the heat produced by accretion caused Earth to reach internal temperatures high enough to melt rock. Over time, the outer surface cooled, forming a solid crust. Solid rock is a poor conductor of heat; therefore, once the crust formed, the heat beneath

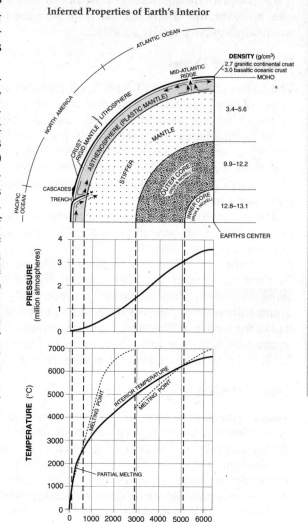

Figure 17.1 The Geothermal Gradient. The solid black line on the graph in the lower part of the diagram is the geothermal gradient, the actual change in temperature with increasing depth. The dotted line indicates the changing melting temperature of rock due to changes in composition and pressure. Source: The State Education Department, *Earth Science Reference Tables*, 2011 ed. (Albany, New York; The University of the State of New York).

it was trapped. Measurements of the rate at which Earth is losing heat indicate, however, that if this internal store were the only source of heat, Earth would already have totally solidified.

Radioactive Decay

The bodies that accreted to form Earth probably had a composition similar to that of stony meteorites known as chondrites. Among the elements found in chondrites are uranium, thorium, and potassium, each of which has radioactive isotopes. Since these elements have big ions, they don't fit into crystalline structures easily. When Earth's outer layers melted after accretion, these elements tended to stay in the molten magma and floated to the surface. As Earth cooled, they became concentrated in the rocks of the crust, particularly in granite.

The radioactive decay of these elements produces heat energy. It is estimated that the total mass of granite in the crust emits about 20 *trillion* watts of power per year. Volcanic rock derived from the mantle indicates that the concentration of radioactive elements in the mantle is lower. Nevertheless, the mantle is so much larger and more massive than the crust that it supplies about 10 trillion watts of power per year from radioactive decay alone. Thirty trillion watts per year is about 33 calories (140 joules), or more than half of the total heat flow from Earth's interior of 50 calories (209 joules) per square centimeter per year. Thus, more than half of the heat flowing out of Earth is due to radioactive decay within the crust and mantle.

Tidal Friction

Another possible source of ongoing heat production is tidal friction. Gravitational attraction between Earth and the Moon and between Earth and the Sun exert unequal forces on Earth that tend to deform it. The ever-changing configuration of the Earth-Moon-Sun system produces a rhythmic pattern of deformations—the tides. Not only Earth's oceans are affected; the atmosphere and solid earth show detectable tides as well. Measurements made in the 1970s showed that at middle latitudes a surface point on the crust is 30 centimeters farther from the center of Earth at high "earth tide" than at low "earth tide." In other words, tidal forces cause solid rock to bend upward, forming a tidal bulge. The tidal bending and unbending of rock layers in the crust releases heat, just as bending a wire coat hanger back and forth causes it to become hot to the touch.

Internal Heat Transfer

Heat tends to move from areas of high concentration to areas of low concentration. Simply stated, heat moves from materials at a higher temperature to materials at a lower temperature. Heat is transferred in three ways: conduction, radiation, and convection. See Figure 17.2.

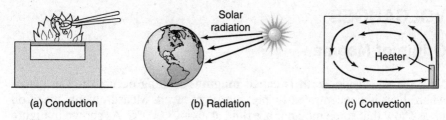

Figure 17.2 Mechanisms of Heat Transfer. Source: *Physics*, 2nd Ed., Arthur Beiser, Benjamin Cummings, 1978.

Conduction (Figure 17.2a) is the transfer of heat energy from one molecule to another through collision. Conduction involves actual contact between two materials. Molecules with a lot of heat energy are fast moving. When they collide with cooler, slower moving molecules, an energy transfer occurs. The faster moving molecule slows down, and the slower moving molecule speeds up. Think of a bowling ball hitting the pins at the end of the alley. The ball slows down when it hits the pins, but the pins speed up and go flying. Rock is a poor conductor of heat because its molecules are bound in crystalline structures and cannot move about freely. As a result, heat moves very slowly through rock by conduction.

Radiation (Figure 17.2b) is the transfer of energy by electromagnetic waves. As atoms and molecules vibrate, they send out waves of energy that travel through space. Within rock, however, surrounding atoms and molecules block and trap that energy. Thus, loss of Earth's internal heat by radiation occurs almost entirely at the surface.

Convection (Figure 17.2c) is the actual motion of a volume of hot fluid from one place to another. Heated fluids are less dense than cooler fluids. Thus, when a portion of a fluid is heated, it becomes less dense than the surrounding cooler fluid and rises upward. As the heated fluid rises, it carries upward the heat energy within it. The upward motion of the hot fluid displaces cold fluid in its path, thereby setting up convection currents.

Seismic evidence shows that the outer core of Earth is a fluid. Therefore, convection currents in the outer core can carry heat outward. Enough heating occurs in Earth's mantle to soften the rock and allow convection to remove the heat there as well. Thus, convection is occurring in both the core and the mantle. Plumes of hot material driven by heat from the core interrupt simple convection currents in the mantle. As a result, the nature of convection in the mantle is complicated. As these complex currents of hot material rise upward against the crust, they spread out and move sideways. As you will learn in Chapter 18, it is inferred that the force of these currents dragging along the base of the crust drives the motion of crustal plates.

VOLCANOES

Origin of Magma

Molten rock inside Earth is called **magma**. Melting occurs only in places where heat is concentrated or pressure is reduced. Measurements made on lavas show that some magmas are fluid at about 1,000°C. As shown in Figure 17.1, temperatures high enough to melt rock and form magma probably occur in the upper mantle at depths of 70–200 kilometers. This is the root zone of volcanoes.

Why is Earth's mantle not entirely molten? The answer is that pressure influences melting temperature. Although temperatures within the mantle are high enough to melt rock, the great pressures there hold crystals together and thus prevent the rock from turning into a liquid. For example, the mineral albite melts at 1,104°C at Earth's surface; but at a depth of 100 kilometers, where the pressure is 35,000 times greater, it melts at 1,440°C. Thus, the depth at which a rock melts depends on its composition and on the temperature and pressure at that depth.

The composition of continental crust and of oceanic crust is not the same. Continental crust is granitic, consisting of minerals with rather low melting points. Granitic magma can begin to form at depths of 30–60 kilometers. Ocean crust is basaltic, consisting of minerals with higher melting temperatures. Thus, basaltic magmas begin to form between depths of 100 and 350 kilometers. In either type of rock, some minerals melt at lower temperatures than others, so the magma that forms is a slush of hot, viscous liquid surrounding hot but still solid crystals. Geologists call this magma a "zone of partial melt." Until the liquid portion exceeds half of the melt, this magma behaves more like a solid than a liquid.

Once a body of magma has formed, it will start to rise because it is less dense than the surrounding rock. The rising magma works its way upward through small cracks and fissures. As magma rises into fissures, it widens them and can cause new ones to develop, resulting in earthquakes. Rising also decreases the pressure on the magma, allowing more minerals to remain molten at lower temperatures. Near Earth's surface, magmas can melt surrounding rock, forming underground pools of magma known as **magma chambers**. From these magma chambers, the molten rock may reach the surface through cracks and fissures to erupt, or may cool and solidify underground.

Basaltic magma is more fluid than granitic magma. Therefore, it rises more quickly; and, because of dropping pressure, its melting temperature falls faster than the magma is cooling. Thus, a lot of basaltic magma makes it all the way to Earth's surface as a fluid and erupts as lava. Granitic magma, on the other hand, is much more viscous, so its rate of ascent is slowed. It often cools faster than dropping pressure lowers its melting temperature. Therefore, most granitic magmas solidify underground rather than reaching the surface and forming lavas.

The Formation of Volcanoes

A **volcano** is both the opening in the crust (the vent) through which magma erupts and the mountain built by the erupted material. See Figure 17.3. Volcanoes form where cracks in Earth's crust lead to a magma chamber. Since liquid magma is less dense than the surrounding solid rocks and is under pressure, it rises toward the surface through these cracks. As magma rises, dissolved gases in it expand and are released, giving an upward boost to the magma. The closer to the surface, the less confining pressure there is to overcome and the faster the magma and gases move. Depending upon its viscosity, magma may pour out quietly onto the surface as a flood of molten rock, or spurt out explosively, showering the surrounding terrain with solid rock and globs of molten rock. Once magma emerges from the surface, it is called **lava**.

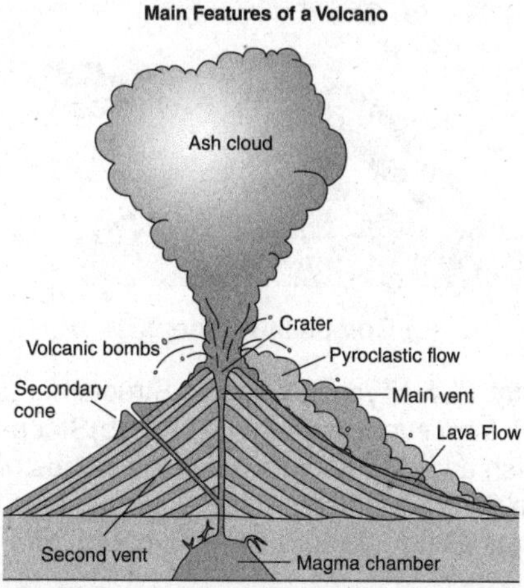

Figure 17.3 A Cross Section of a Volcano. Magma rises from chambers in the upper mantle until it exits the crust through an opening called the *vent*. Thereafter, it is called *lava*, and accumulates in a mound called a *volcano*. Source: *Painless Earth Science*, Edward J. Denecke, Jr., Barron's Educational Series, Inc., 2011.

Volcanic Structures

Erupted materials spread out in all directions around the vent, or central opening, of a volcano. With each new eruption, more material piles up around the vent and a cone-shaped mound forms. The shape of the cone depends upon the viscosity of the magma.

Fluid lava tends to flow quietly through the vent, spreading out in wide, thin sheets. The result is a flat, wide cone called a **shield volcano** (for its resemblance to a shield when viewed from overhead).

Thick, pasty lava containing a lot of dissolved gas does not flow easily through the vent, which tends to become clogged or completely blocked. As a result, gases cannot escape and pressure builds up until a violent explosion takes place. During an explosive eruption, rock and soil surrounding the vent are thrown upward, along with chunks and droplets of lava blown out of the vent. The lava cools as it hurtles through the air and is generally solid by the time it reaches the ground. Collectively, all of the fragments of solidified lava ejected during a volcanic eruption are called **tephra**. This "rain" of solid particles produces a steep, narrow cone called a **cinder cone volcano** (for the particles' resemblance to cinders).

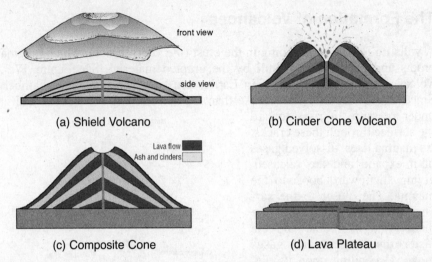

Figure 17.4 Typical Volcanic Structures. (a) Broad, gently sloping shield volcanoes form from successive lava flows. (b) Steep-sided cinder cones form from the buildup of ash and cinders. (c) Composite cones of intermediate slope form when periods of lava flows are interspersed with eruptions of ash and cinder. (d) Very fluid lava that erupts along a fissure spreads out to form a lava plateau.

Alternating explosive and quiet eruptions give rise to large, symmetrical **composite cones** made of alternating layers of solidified lava and volcanic rock particles such as ash and cinders.

In some cases, magma emerges through long, open cracks called **fissures**. Lava pouring out of fissures spreads out in wide, thin sheets. Instead of forming a cone as it would around a central vent, it forms a sheet, called a **lava plateau**, that blankets the surrounding land.

The four volcanic structures discussed above are illustrated in Figure 17.4.

Intrusive Activities and Structures

Some magma moves around within Earth but never reaches the surface. Underground flows of magma, or **intrusions**, move into and through the countless underground cracks in the rocks of Earth's crust. There, intrusions may solidify into rock structures called **plutons**. See Figure 17.5. Plutons are named for their shapes and include dikes, sills, laccoliths, stocks, and batholiths. **Dikes** are flat, slablike structures that cut across layers of rock like a wall. **Sills** have a similar shape, but lie parallel to layers of rock

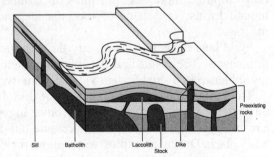

Figure 17.5 Some Types of Plutons.

(as a window sill is parallel to the ground). **Laccoliths** are large structures with flat bottoms and arched tops. The word laccolith means "lake-rock," and these structures resemble inverted lakes. Laccoliths form when magma is forced between rock layers faster than it can spread out, causing it to push the overlying layers of rock upward to form a dome mountain. Large, irregularly shaped plutons are called *stocks* if they cover less than 75 square kilometers, and *batholiths* if they cover a larger area.

Zones of Volcanic Activity

If the locations of active volcanoes are plotted on a map of Earth, the interesting pattern shown in Figure 17.6 emerges.

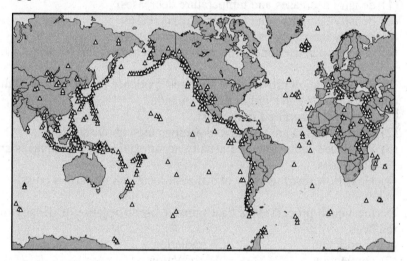

Figure 17.6 Earth's Active Volcanoes. Source: Adapted from image at *http://nmnhwww.si.edu/gvp/volcano/index.htm*, website of the Global Volcanism Program, National Museum of Natural History, E-121, Smithsonian Institution, Washington, D.C.

Active volcanoes are not distributed evenly on Earth. Rather, they occur in narrow, elongated zones in some areas and are completely absent from others. For example, active volcanoes surround the Pacific Ocean Basin, a zone known as the **Ring of Fire**, but none of the edges of continents bordering the Atlantic Ocean are volcanic. A narrow zone of volcanoes can also be found near the centers of most oceans, producing the vast chain of undersea mountains known as the mid-ocean ridge. A third zone runs along the northern edge of the Mediterranean Sea; known as the **Mediterranean Belt**, it includes historic volcanoes such as Mt. Etna and Mt. Vesuvius in Italy.

You may have noticed by now a remarkable similarity between the pattern of volcanic activity shown here, and the pattern of earthquake activity described in Chapter 16. As you will discover in Chapter 18, earthquake and volcano zones define the edges of the huge, platelike slabs of rock that make up Earth's crust.

Volcanoes and Earth's Internal Heat

MULTIPLE-CHOICE QUESTIONS

In each case, write the number of the word or expression that best answers the question or completes the statement.

1. The source of energy for the high temperatures found deep within Earth is
   (1) tidal friction
   (2) incoming solar radiation
   (3) decay of radioactive materials
   (4) meteorite bombardment of the Earth

2. What happens to the density and temperature of rock within Earth's interior as depth increases?
   (1) density decreases and temperature decreases
   (2) density decreases and temperature increases
   (3) density increases and temperature increases
   (4) density increases and temperature decreases

3. Which observation provides the strongest evidence for the inference that convection cells exist within Earth's mantle?
   (1) Sea level has varied in the past.
   (2) Marine fossils are found at elevations high above sea level.
   (3) Displaced rock strata are usually accompanied by earthquakes and volcanoes.
   (4) Heat-flow readings vary at different locations in Earth's crust.

4. During which process does heat transfer occur because of density differences?
   (1) conduction   (3) radiation
   (2) convection   (4) reflection

5. When Earth cools, most of the energy transferred from Earth's surface to space is transferred by the process of
   (1) conduction   (3) refraction
   (2) reflection   (4) radiation

6. The primary cause of convection currents in Earth's mantle is believed to be the
   (1) differences in densities of Earth materials
   (2) subsidence of the crust
   (3) occurrence of earthquakes
   (4) rotation of Earth

7. Which diagram best represents the transfer of heat by convection in a liquid?

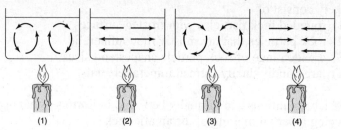

8. Which process transfers energy primarily by electromagnetic waves?
   (1) radiation
   (2) evaporation
   (3) conduction
   (4) convection

9. As depth beneath the surface increases, the temperature at which a particular mineral will melt
   (1) increases
   (2) decreases
   (3) remains the same
   (4) changes from Fahrenheit to Celsius degrees

10. According to the *Reference Tables for Physical Setting/Earth Science*, the melting point of materials in the outer core is
    (1) lower than their actual temperature
    (2) higher than their actual temperature
    (3) lower than that of materials in the crust
    (4) higher than that of materials in the inner core

11. The observed difference in density between continental crust and oceanic crust is most likely due to differences in their
    (1) composition
    (2) thickness
    (3) porosity
    (4) rate of cooling

12. Magma chambers tend to form in areas beneath Earth's surface where
    (1) heat builds up and pressure is reduced
    (2) pressure builds up and heat is reduced
    (3) pores in Earth are filled with groundwater
    (4) there are extensive deposits of coal, oil, or natural gas

13. Thick, pasty lava with a high gas content tends to erupt
    (1) quietly
    (2) explosively
    (3) from fissures in a broad sheet
    (4) from undersea vents

14. Rock ejected from a volcano during an explosive eruption will most likely consist of
    (1) rounded rock particles with frosted surfaces
    (2) rock particles that are mostly monomineralic
    (3) a wide range of rock particle shapes and sizes
    (4) large, individually formed mineral crystals

15. Which conditions can normally be found in Earth's asthenosphere, producing a partial melting of ultramafic rock?
    (1) temperature = 1,000°C; pressure = 10 million atmospheres
    (2) temperature = 2,000°C; pressure = 0.1 million atmospheres
    (3) temperature = 3,500°C; pressure = 0.5 million atmospheres
    (4) temperature = 6,000°C; pressure = 4 million atmospheres

16. Active volcanoes are most abundant along the
    (1) edges of tectonic plates
    (2) eastern coastline of continents
    (3) 23.5° N and 23.5° S parallels of latitude
    (4) equatorial ocean floor

17. The diagrams below represent four rock samples. Which rock was formed by rapid cooling in a volcanic lava flow? [The diagrams are not to scale.]

Bands of alternating light and dark minerals	Easily split layers of 0.0001-cm-diameter particles cemented together	Glassy black rock that breaks with a shell-shape fracture	Interlocking 0.5-cm-diameter crystals of various colors
(1)	(2)	(3)	(4)

18. A lava flow that has cooled and solidified into rock is observed to have a surface pockmarked by numerous spherical cavities. These cavities are most likely the result of
    (1) rainwater dissolving out water-soluble mineral crystals
    (2) the corrosive effects of sustained periods of acid rain
    (3) rounded pebbles pried out of the surface by frost action
    (4) gas bubbling out of the lava just as it hardened

19. An igneous intrusion loses heat to its surroundings primarily by
    (1) conduction   (3) radiation
    (2) convection   (4) absorption

20. Explosive volcanoes are usually characterized by
    (1) steeply sloped lava cones
    (2) steeply sloped cinder cones
    (3) gently sloped lava cones
    (4) gently sloped cinder cones

21. A flow of magma cools and hardens into rock before it reaches the surface. Which of the following structures most likely formed from the magma?
    (1) a cinder cone   (3) a geyser
    (2) a basalt plateau   (4) a dike

22. Which statement concerning a sill is true?
    (1) It may be vertical.
    (2) It must be vertical.
    (3) It may cut across sedimentary strata.
    (4) It must be horizontal.

23. Which statement applies equally to batholiths, laccoliths, dikes, and sills?
    (1) They create folded mountains.
    (2) They are composed of intrusive rocks.
    (3) They are flat sheets of metamorphic material.
    (4) They are composed largely of obsidian.

24. A study of rocks in an extinct volcano indicates a composition largely of cinders and other igneous fragments. This extinct volcano was probably formed as a result of
    (1) lava flows
    (2) igneous intrusions
    (3) explosive eruptions
    (4) magma solidifications

CONSTRUCTED RESPONSE QUESTIONS

25. State three types of evidence that support the statement "There is a tremendous amount of heat within Earth." [3]

26. In one or more complete sentences, explain the difference between magma and lava. [1]

27. Identify three types of volcanic structures. In one or more complete sentences, explain how each structure is formed. [3]

28. In one or more complete sentences, explain why the rock that forms from igneous intrusions typically consists of large mineral crystals. [1]

29. State two differences between a quiet eruption and an explosive eruption. [2]

EXTENDED CONSTRUCTED RESPONSE QUESTIONS

Base your answer to questions 30 through 34 on the diagram below, which represents a cross section of a portion of Earth's crust. Letters A through M identify rock layers, structures and boundaries within the cross section.

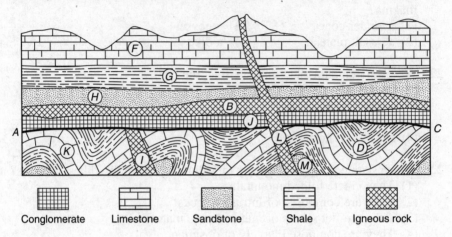

30. State the rock layer that best represents a dike. [1]

31. State the rock layer that best represents a sill. [1]

32. Place a circle around the rock structure that best represents a volcano. [1]

33. In one or more complete sentences, explain how the igneous intrusion labeled B most likely affected the rock adjacent to it in rock layer H. [1]

34. State *two* likely sources of the heat originally contained in the igneous rock that formed the intrusions in the cross section. [2]

Base your answers to questions 35 through 40 on the reading passage and maps below and on your knowledge of Earth science. The enlarged map shows the location of volcanoes in Colombia, South America.

Fire and Ice—and Sluggish Magma

On the night of November 13, 1985, Nevado del Ruiz, a 16,200-foot (4,938 meter) snowcapped volcano in northwestern Colombia, erupted. Snow melted, sending a wall of mud and water raging through towns as far as 50 kilometers away, and killing 25,000 people.

Long before disaster struck, Nevado del Ruiz was marked as a trouble spot. Like Mexico City, where an earthquake killed at least 7,000 people in October 1985, Nevado del Ruiz is located along the Ring of Fire. This ring of islands and the coastal lands along the edge of the Pacific Ocean are prone to volcanic eruptions and crustal movements.

The ring gets its turbulent characteristics from the motion of the tectonic plates under it. The perimeter of the Pacific, unlike that of the Atlantic, is located above active tectonic plates. Nevado del Ruiz happens to be located near the junction of four plate boundaries. In this area an enormous amount of heat is created, which melts the rock 100 to 200 kilometers below Earth's surface and creates magma.

Nevado del Ruiz hadn't had a major eruption for 400 years before this tragedy. The reason: sluggish magma. Unlike the runny, mafic magma that makes up the lava flows of oceanic volcanoes such as those in Hawaii, the magma at this type of subduction plate boundary tends to be sticky and slow moving, forming the rock andesite when it cools. This andesitic magma tends to plug up the opening of the volcano. It sits in a magma chamber underground with pressure continually building up. Suddenly, tiny cracks develop in Earth's crust, causing the pressure to drop. This causes the steam and other gases dissolved in the magma to violently expand, blowing the magma plug free. Huge amounts of ash and debris are sent flying, creating what is called an explosive eruption.

Oddly enough, the actual eruption of Nevado del Ruiz didn't cause most of the destruction. It was caused not by lava but by the towering walls of sliding mud created when large chunks of hot ash and pumice mixed with melted snow.

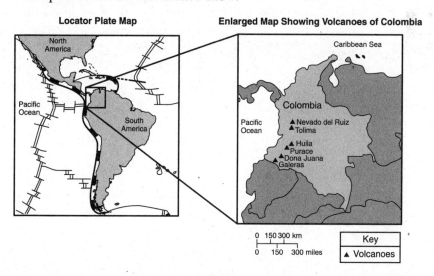

35. What are the names of the *four* tectonic plates located near the Nevado del Ruiz volcano? [1]

36. What caused most of the destruction associated with the eruption of Nevado del Ruiz? [1]

37. What caused the magma to expand, blowing the magma plug free? [1]

38. Vesicular texture is very common in igneous rocks formed during andesitic eruptions. Explain how this texture is formed. [1]

39. Why are eruptions of Nevado del Ruiz generally more explosive than most Hawaiian volcanic eruptions? [1]

40. Describe one emergency preparation that may reduce the loss of life from a future eruption of the Nevado del Ruiz volcano. [1]

CHAPTER 18
PLATE TECTONICS

KEY IDEAS There is much evidence that Earth is a dynamic geologic system whose crust is in nearly constant motion. The theory of plate tectonics is an excellent example of how our ideas about Earth change as new discoveries are made. Building upon earlier theories of continental drift, isostasy, and ocean floor spreading, plate tectonics was able to explain observations that could not be accounted for by these earlier theories.

According to the plate tectonics theory, Earth's crust, or lithosphere, consists of separate plates that rest on the more fluid asthenosphere and move slowly in relationship to one another, creating convergent, divergent, and transform plate boundaries. The outward transfer of Earth's internal heat drives convective circulation in the mantle that moves the lithospheric plates comprising Earth's surface. Plate boundaries are the sites of most earthquakes, volcanoes, and young mountain ranges.

Many of Earth's surface features, such as mid-ocean ridges/rifts, trenches/subduction zones/island arcs, mountain ranges (folded, faulted, and volcanic), hot spots, and the magnetic and age patterns in surface bedrock, are a consequence of forces associated with plate motion and interaction.

KEY OBJECTIVES
Upon completion of this chapter, you will be able to:

- Describe evidence of crustal movements.
- Explain how the theories of continental drift, isostasy, and ocean floor spreading were combined in a single unifying theory—the theory of plate tectonics.
- Explain how differences in density resulting from heat flow within Earth's interior cause movement of the lithospheric plates.
- Relate specific forms of crustal activity, such as earthquakes, volcanoes, and the deformation and metamorphism of rocks during the formation of mountains, to the various types of plate boundaries.
- Describe how many of the processes of the rock cycle are consequences of plate dynamics, including the production of magma, regional meta-

morphism, and the creation of major depositional basins through down-warping of Earth's crust.
- Explain how plate motions have resulted in global changes in geography, climate, and the patterns of organic evolution.

THE DYNAMIC CRUST

Earth's crust, or **lithosphere**, is almost constantly in motion. This motion may be abrupt, as in an earthquake, or so gradual that it is imperceptible to the senses. Movements of the lithosphere exert forces on rock, causing them to be deformed. Evidence that the lithosphere is moving ranges from direct observation of motion to inferences based upon displacement of structures and deformation of rock.

Evidence of Crustal Movements

Earthquakes are unmistakable evidence of crustal movement. During an earthquake, movement of the crust occurs along faults. Every one of the nearly 1 million earthquakes that occur annually involves movement of Earth's crust.

Volcanic eruptions also involve movements of the crust. As magma moves beneath and then out of a volcano, the crust around the volcano rises and sinks measurably, often causing the brittle crust to crack and form fissures. These breaks are direct evidence of movement.

Displaced structures are also evidence of crustal movement. When the crust moves, structures built on its surface move along with it.

Bench marks are permanent metal plaques set in the ground giving the exact locations and precisely determined elevations of points. See Figure 18.1a. Bench marks are used as references in geologic surveys and tidal observations. Measurements of bench-mark elevations for more than 100 years reveal that in large areas of the United States the ground is slowly moving upward or downward. See Figure 18.1b. Crustal motion can also be detected and measured using satellite laser technology.

Tilted or folded rock layers are also evidence of crustal movement. Sedimentary strata, lava flows, and tephra are originally deposited in horizontal layers. Therefore, where such materials are observed to be steeply tilted, or bent and folded, it can be inferred that they have been moved from their original positions.

Sedimentary rock layers at high elevations are further evidence of crustal movement. Most layers of sediments are deposited at or below sea level, and may be buried deeper before changing into sedimentary rock. In many places, however, layers of sedimentary rock are found high atop mountains or plateaus. Some are located as much as 2 kilometers above current sea level, far higher than any known body of water, past or present. From this it can be inferred that these rock layers have been uplifted.

Figure 18.1 (a) A United States Geological Survey Bench Mark. A bench mark is a small plaque, usually mounted in rock or on a concrete base, marked with the precise location and elevation of a point. Source: United States Geological Survey. (b) Changes in Bench Mark Elevations over a 100-Year Period. Source: *Physical Geology*, 2nd Ed., Richard Foster Flint and Brian J. Skinner, John Wiley, 1974, 1977.

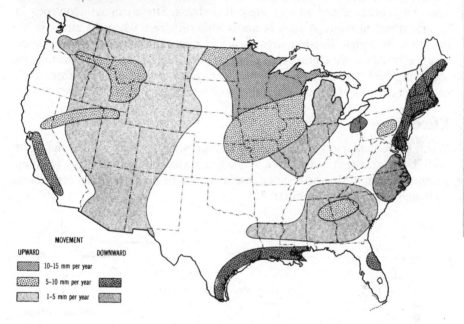

Fossils provide striking evidence of crustal movement. Fossils of marine life, such as corals and clams, found at mountaintops in the Himalayas are evidence of uplift. Fossils of terrestrial or shallow-water organisms beneath the deep ocean bottom indicate sinking, subsidence of the crust.

Thick layers of shallow-water sediments, as much as 15 kilometers thick in some places, are interesting evidence of crustal movement. The average depth of the ocean is 3.8 kilometers with several isolated points having depths of about 10 kilometers. How can 15 kilometers of sediment accumulate in a body of shallow water without filling it completely? One probable inference is that the crust beneath the sediment is sinking while the sediment is accumulating, so sediment layers build up while the water depth remains fairly constant.

Plate Tectonics

Causes and Effects of Crustal Movements

Causes

Unbalanced forces acting on Earth's crust cause crustal movement. Some of the forces acting on the crust are gravity (including gravitational attraction by the Sun and Moon) and forces produced by Earth's rotation, by the expansion and contraction of rock material due to heating and cooling, and by density currents in Earth's mantle.

Stress, Tension, Compression, and Shear

The many forces acting on rock are called **stress**. Stress can act upon rock in several different ways. A rock is under *uniform* stress when the stress in all directions is equal. **Tension** stresses act in opposite directions, pulling rock apart or stretching it. **Compression** stresses act toward each other, pushing or squeezing rock together. **Shear** stresses may act toward or away from each other, but they do so along different lines of action, causing rock to twist or tear.

Effects

When stress acts upon a rock, the rock **strains**, or changes in size, shape, or both. Uniform stress causes rock to change in size, but not in shape. Tension, compression, and shear stresses cause a change in shape and may also cause a change in size. See Figure 18.2.

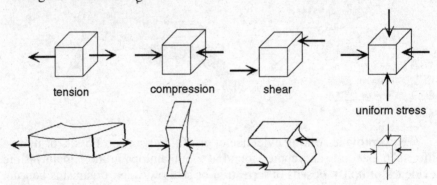

Figure 18.2 Changes in Rock Shape and/or Size over Time Due to Tension, Compression, Shear, and Uniform stress.

Deformation and Fracture

When rock is stressed, it goes through a series of changes. The first step is elastic deformation, in which the rock strains, but the change is not permanent. If the stress is removed, the rock will return to its original shape and size. The next step is ductile deformation, which begins when stress reaches a point called the **elastic limit**. At the elastic limit, the stress exceeds the strength of the rock's internal bonding and permanent changes occur. If stress is then removed, the deformation is permanent and the rock no longer returns to its original shape or size. Ductile behavior is similar to what happens when you squeeze or stretch modeling clay without breaking it. The final step is fracture, in which the stress actually causes the rock to break, or fracture. In general, high temperatures and pressures favor ductile behavior and make fracture less likely to occur. See Figure 18.3.

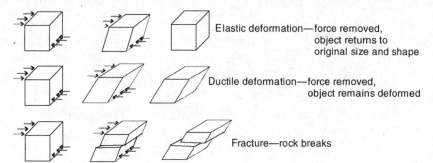

Figure 18.3 Elastic Deformation, Ductile Deformation, and Fracture.

Folds

Ductile deformation of layered rock forms bends or warps called **folds**. Folding is due to compression stresses. An upward-arched fold is an **anticline**; a downward, valleylike fold is a **syncline**. The sides of a fold are called its **limbs**. Folds range from simple **monoclines**, in which only one limb is bent, to **overturned** folds, in which both limbs are tilted in the same direction, to **recumbent** folds, which are bent back on themselves almost horizontally. See Figure 18.4.

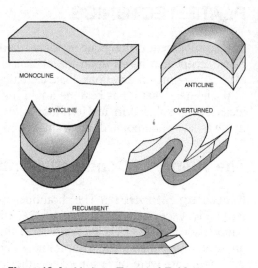

Figure 18.4 Various Types of Folds.

Joints and Faults

Joints and faults are fractures that form when rock is stressed until it breaks. **Joints** are fractures along which there has not been any movement of the rock. **Faults** are fractures along which the rock has moved and one side of the fracture is displaced relative to the other side. Faulting is always associated with earthquakes. A fault is classified by the angle of the fracture and the direction of relative movement along it. Five types of faults are illustrated in Figure 18.5.

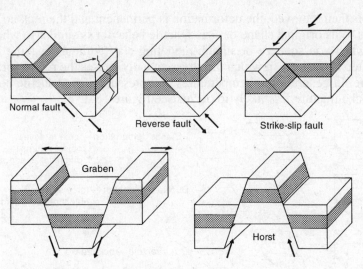

Figure 18.5 Various Types of Faults.

PLATE TECTONICS

The development of the plate tectonics theory is an example of how a theory gains acceptance based upon the evidence that supports it and its ability to explain puzzling observations. Plate tectonics makes sense of such a large range of phenomena that it has become a unifying principle in geology. Two earlier ideas—continental drift and seafloor spreading—were based upon evidence that was later incorporated into a single unifying theory—plate tectonics.

The Theory of Continental Drift: Supporting Evidence

Matching Shorelines The shorelines on both sides of the Atlantic match strikingly, especially the shapes of the Atlantic coasts of Africa and South America. The likeness is even clearer if the edges of the continental shelves are used rather than present shorelines. This match gave rise to the idea that the continents were once joined together but then drifted apart. At first the idea was rejected because it seemed ridiculous to think that anything as large as a continent could move around. In 1912, however, Alfred Wegener reintroduced the idea of **continental drift**—the concept that continents drift slowly across Earth's surface, sometimes colliding and sometimes breaking into pieces. Wegener and his colleagues presented striking evidence that the world's landmasses had once been joined together, but over time had broken apart into the continents of today and slowly drifted to their present positions. The original landmass was dubbed **Pangaea**, meaning "all land." See Figure 18.6.

Continental and Ocean Crust Statistical studies of elevations and depths indicate that there are two distinct levels for the world's surface—the continents and the ocean floors. These levels alternate and exist side by side with almost no transition between the two, suggesting a series of side-by-side blocks. Some blocks are continents; some are ocean floors. The compositions and thicknesses of the ocean floors and the continents differ. The ocean floors are thin and made of dense basalt, while the continents are thick and consist of less dense granite. See Figure 18.7.

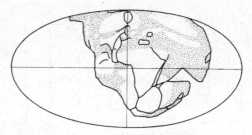

Figure 18.6 Pangaea as Proposed by Wegener. Source: Earth Science on File. © Facts on File, Inc., 1988.

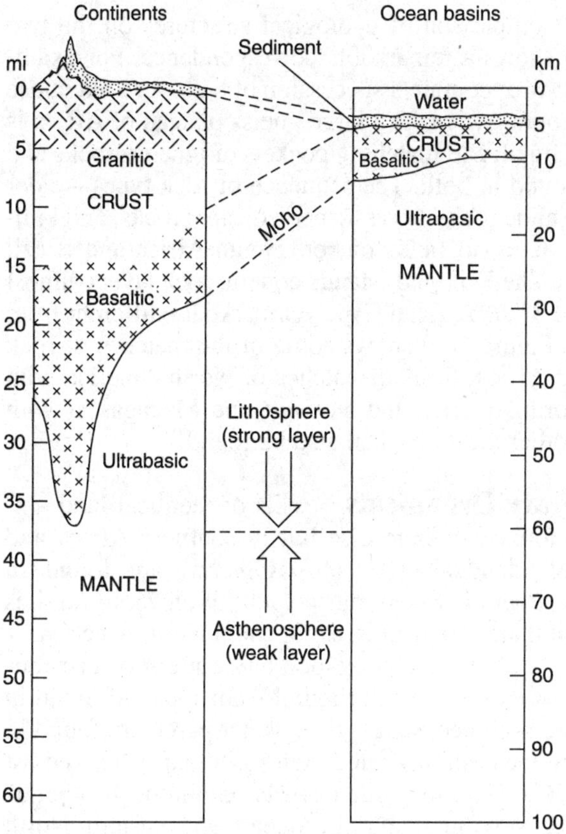

Wegener used a still earlier idea—isostasy—to explain the two levels of the crust. **Isostasy** stated that the crust was floating on hot, fluid rock in the mantle and offered two possible explanations for continents being higher than ocean floors: continents are higher because less dense materials float higher than more dense materials in the same fluid (Pratt's hypothesis), and continents are higher because thick objects float higher than thin objects in the same fluid (Airy's hypothesis). See Figure 18.8. The existence of blocks would explain how continents could "drift"—the blocks simply moved sideways.

Figure 18.7 Compositions and Thicknesses of Continents and Ocean Floors. Source: *The Earth Sciences*, Arthur N. Strahler, Harper & Row, 1971.

Unit Five **THE DYNAMIC EARTH**

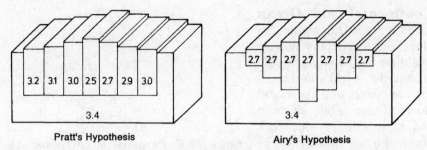

Figure 18.8 Isostasy. In both hypotheses, the rocks of the crust float on a denser (3.4 g/cm^3) mantle. In Pratt's hypothesis, mountains are the result of less dense rock (2.5 g/cm^3) floating higher than dense rock (3.2g/cm^3). In Airy's hypothesis, all the rocks of the crust have the same density (2.7 g/cm^3). Mountains occur where the crust is thicker, and ocean basins occur where it is thinner. Source: Earth Science on File. © Facts on File, Inc., 1988.

Geological Structures Comparison of geological structures on the two sides of the Atlantic Ocean shows a remarkable correspondence. For example, the Sierras near Buenos Aires contain a succession of beds very like those of the Cape Mountains in South Africa. The large gneiss plateau of Africa is strikingly similar to that of Brazil, and matching pockets of igneous rocks and sedimentary rocks can be found in both. The sequence of rock types—older granite, younger granite, alkaline rocks, Jurassic age volcanic rocks, and kimberlite—also matches. The diamond fields of both South Africa and Brazil occur in the kimberlite beds. The Falkland Islands contain rock layers almost indistinguishable from those of the African Cape yet markedly different from layers in nearby Patagonia. Figure 18.9 shows some of the matches of rock types in South America and Africa. Similar matches of North America with Europe, Greenland with North America and Norway, and Madagascar with Africa, as well as between other locations, can also be found.

Fossils and Contemporary Organisms Fossils of identical land animals (*Mesosauroidae*) are found nowhere else but in southern Africa and South America. Fossils of identical trees (*Glossopteris*) are found in Australia, India, and South America. Sixty-four percent of all fossil reptiles from the Carbonaceous Period are identical in Europe and North America.

Contemporary organisms also show a corresponding pattern of distribution. For example, *Lumbricus* earthworms are found from Japan to Spain in Eurasia, but only in the eastern United States. Freshwater perch are found in Europe and Asia, but only in the eastern United States. Similar patterns exist for pearl mussels, mud minnows, and garden snails. Common heather is found only in Europe and Newfoundland. European and eastern North American eels both spawn in the Sargasso Sea off the North American coast. Again, similar patterns can be found between other now-separated continents.

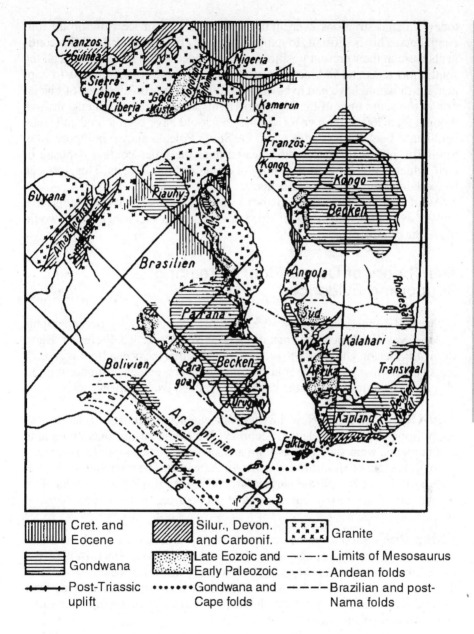

Figure 18.9 Evidence of Continental Drift. Alfred Wegener's correlation of rock type and surface features along the margins of Africa and South America. Source: *The Origins of Continents and Oceans*, Alfred Wegener, by permission of Dover Publications, Inc., 1966.

Past Climates There is also unmistakable evidence that 300 million years ago an ice sheet covered South America, southern Africa, India, and southern Australia. This ice sheet was huge, similar to the one covering Antarctica today. Glacial striations even indicate the direction of ice flow on all four continents. This is difficult to explain without continental drift. If the continents were in their present positions, the ice sheet would have covered all the southern oceans and in places crossed the Equator. In order for this to happen, Earth would have had to be very cold, yet there is no evidence of glaciation at the same time in the Northern Hemisphere. If the continents drifted, though, then 300 million years ago they all could have been joined and some could have been located adjacent to the South Pole, as shown in Figure 18.6. Similarly, places that are today located near the poles contain deposits of coal, which could have formed only in equatorial rain forests. This, too, can be explained by drifting continents.

All of the evidence presented above supports the idea that continents have moved over time. However, it does not suggest a mechanism that would move the continents.

The Theory of Ocean Floor Spreading: Supporting Evidence

In the 1950s the ocean floors were explored extensively by oceanographic research vessels. Much information was gathered and led Professor Harry Hess of Princeton University to propose the idea that the ocean floor is spreading sideways away from the mid-ocean ridges. The evidence supporting this idea suggests a mechanism that could have moved the continents apart.

Mid-Ocean Ridges New instruments enabled scientists to survey the ocean floor with unprecedented accuracy. The ocean basins, once thought to be flat plains, were shown to contain a chain of undersea mountains running down the center of almost every ocean in the world. The mountain chain was split by a deep rift, and the surface around it was riddled with faults. This rugged terrain suggested that movement is occurring on the ocean floors.

Young Rock Surveys of the ocean floors showed that the rocks beneath the oceans are generally younger than the rocks of the continents. Most ocean rocks are only a fraction of the age of some continental rocks. In addition, the age of the rocks on the ocean bottoms increases with distance away from the mid-ocean ridges, suggesting that rocks form at the ridges and then move sideways away from them.

Paleomagnetism Basalt, the major rock of the ocean floors, is rich in iron compounds. Magnetite, pyrrhotite, and hematite are strongly affected by magnetism. When lava is extruded in the mid-ocean rifts, crystals of minerals affected by magnetism align with Earth's magnetic field while the lava cools. When solidified, the igneous rock contains a record of Earth's magnetic field locked in its crystals. Studies of ancient rocks show that Earth's magnetic field has reversed many times in the past. The reversals seem to take place at irregular intervals ranging from 20,000 to several million years. Research ships carrying sensitive instruments that can measure small changes in magnetism have discovered that magnetism in rocks on the ocean floors reveals a striking pattern of strips of normal and reversed fields. See Figure 18.10. The fact that the pattern is identical on either side of the mid-ocean ridges also suggests that rocks form at the ridges and then move sideways away from them.

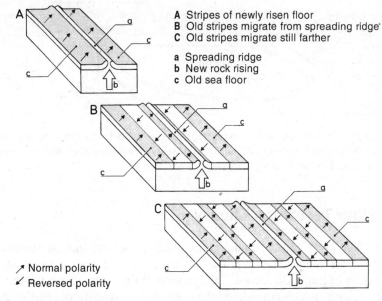

Figure 18.10 Paleomagnetism Patterns on the Ocean Floor. Source: *Earth Science on File.* Copyright © 1988 by Diagram Visual Information Ltd. Reprinted by permission of Facts on File, Inc.

The proposed mechanism for ocean floor spreading is as follows. Lava is extruded in the rift. The lava hardens to form new ocean floor with Earth's magnetic field "frozen" in its crystals. New lava erupts, splitting the just-formed rock and pushing it aside. The new lava hardens. As this process is repeated, new ocean floor is constantly being formed, and the floor on either side of the mid-ocean ridge is pushed sideways.

Trenches If new ocean floor is constantly being created along the ridges, Earth should be getting larger, but it is not. Therefore, ocean floor is being destroyed at the same rate at which it forms. But *where* is it being destroyed? The answer lies in the trenches. **Trenches** are deep crevices in the ocean floor where it bends downward sharply. Beneath the trenches, earthquakes occur frequently and the positions of their foci show a pattern of increasing depth with distance from one side of the trench. The plunging ocean floor and deepening pattern of earthquake foci beneath the trenches are consistent with a situation where sinking convection currents in the mantle are pulling the ocean floor downward into the mantle, where it melts and is destroyed in a process called **subduction**. See Figure 18.11. With this final piece of evidence the picture was complete, and scientists had not only an explanation for the features and formation of the ocean floor, but also a mechanism for continental drift. Shortly thereafter, continental drift and ocean floor spreading were unified into the theory of plate tectonics.

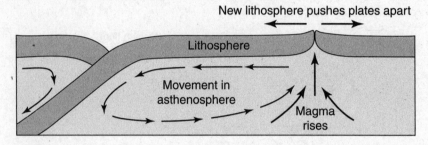

Figure 18.11 Ocean-floor is created along the ridges, but is destroyed as it plunges into the mantle (forming trenches) and melts. Source: *The Dynamic Earth*, Brian J. Skinner and Stephen C. Porter, John Wiley, 1989.

The Plate Tectonics Theory

By the 1960s, scientists had enough evidence to believe that continents and seafloors were indeed moving. The evidence showed not only motion, but also motion in many directions. This motion did not agree with the existing model of Earth, which considered the continents and seafloors part of a single, unbroken shell of rocky crust. Clearly, a solid shell cannot move in many different directions at once!

The Lithosphere and the Asthenosphere

As stated earlier, analysis of earthquake waves led to the discovery that Earth's interior consists of layers that have different properties. Based upon density differences, Earth can be divided into the inner and outer core, the mantle and the crust. If, instead of density, the *rigidity* of rock is considered, another pattern emerges. The crust and upper mantle to a depth of 100 kilometers is strong and rigid, forming a layer called the **lithosphere**. The region of the upper man-

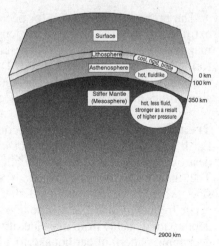

Figure 18.12 Lithosphere and Asthenosphere.

tle between 100 and 350 kilometers in depth is weak and behaves like a viscous fluid; this layer is called the **asthenosphere**. See Figure 18.12.

Scientists used the lithosphere/asthenosphere pattern to devise a new model of Earth's structure that fit the evidence. They combined aspects of continental drift, isostasy, and ocean floor spreading into a single unifying theory—**plate tectonics**. (*Tectonics* is the branch of geology dealing with the forces affecting the structure of Earth's crust.) In this new model, the lithosphere is seen, not as an unbroken solid, but fragmented and consisting of several huge pieces or **plates**—hence the name *plate* tectonics. The plates are envisioned as "floating" on a layer of fluid rock set in motion by huge convection currents in the mantle. Carried along like bobbing corks on these fiery currents of fluid rock, the plates collide with one another, jostle past one another, or drift apart. Continents and seafloors are in motion because they are part of these plates and move along with them, and it is the interaction between the edges of moving plates that explains all of Earth's features, as well as earthquakes, volcanic activity, and crustal movements.

In simple terms, then, plate tectonics embodies two key ideas: the rigid lithosphere consists of great slabs called plates; and these plates move sideways around Earth, sliding on the fluidlike asthenosphere. As the plates move laterally around Earth, their edges interact in one of three ways: they spread apart, they collide, or they slide past each other. Let us now consider these interactions and see how they account for a wide range of observations.

Interactions at Plate Boundaries

There are three types of plate boundaries. See Figure 18.13.

Divergent boundaries are places where adjacent plates are moving apart. Divergent boundaries create tension stresses, causing rock to fracture. Fractures result in earthquakes and open rifts through which magma can rise to the surface and solidify to form new lithosphere. The mid-ocean rifts are divergent boundaries, and the mid-ocean ridges are volcanic mountains formed by magma emerging from the rifts.

Convergent boundaries are places where adjacent plates are moving toward each other and colliding. When plate edges collide, compression and shear stresses cause rocks to fold, fracture, and move along faults. Plates consisting of ocean crust and plates of continental crust interact differently along divergent boundaries. Since ocean crust is denser than continental crust, ocean crust tends to plunge under continental crust when plates collide, forming a *subduction zone*. In subduction zones, the plunging plate is forced into the mantle and melts. If two plates carrying continental crust collide, they both crumple and fold. The Andes and Himalayan mountains formed along convergent boundaries.

Transform boundaries are places where plates move past each other in strike-slip motions. Shear stresses cause rock to fracture, forming numerous faults. The plates on either side of transform fault margins smash and rub against each other, as two ships in a near-collision grind their sides together.

Plate Tectonics

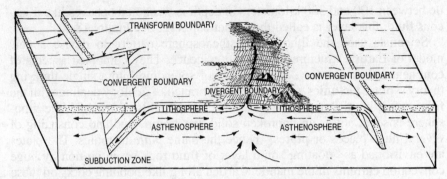

Figure 18.13 Convergent, Divergent, and Transform Boundaries.

The rock along these faults is intensely shattered, and the sliding motion of the plates causes many earthquakes. If the plates separate a little, some magma may "leak" through the boundary, causing small-scale volcanism. The San Andreas fault is part of a huge transform boundary along which the Pacific plate is moving past the North American plate.

Hot Spots Some volcanoes occur far from plate edges. For example, the

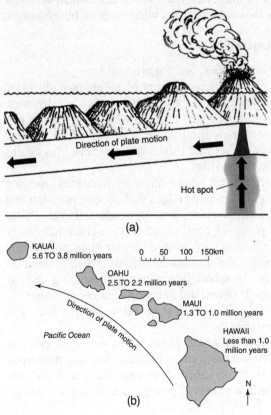

Figure 18.14 (a) Cross-section Showing a Series of Volcanoes Formed by a Plate Moving over a Hot Spot. The extinct volcanoes are eroded to nearly flat surfaces at sea level and are then submerged, forming flat-topped seamounts called *guyots*. Source: *Mountains of Fire*, Robert W. Decker and Barbara B. Decker, Cambridge University Press, 1991. (b) The Progressive Increase in Age of Hawaiian Islands with Distance Away from Currently Active Hawaiian Volcanoes. This supports the hot-spot hypothesis and indicates plate movement toward the northwest.

Hawaiian volcanoes are located 4,000 kilometers from the nearest plate edge. How does plate tectonics account for volcanic activity far from plate boundaries? To explain the unusual position of these volcanoes and relate them to plate tectonics, Canadian geophysicist J. Tuzo Wilson proposed the concept of hot spots.

According to this concept, there are long-lasting zones of rising hot magma, or **hot spots**, deep within Earth at several places beneath moving plates. Large batches of magma rise from these hot spots and work their way upward through the moving plate. The magma rises because it is more buoyant than the surrounding rock, wedges apart cracks in the plate, and erupts, forming a volcano. As the plate moves along, older volcanoes are carried away in the direction of plate motion and new ones form over the hot spot. See Figure 18.14a.

Wilson's idea explained something that had puzzled scientists for years—why the Hawaiian Islands have active volcanoes only at the southeastern edge of the chain. Furthermore, knowing the rate of plate motion, one should be able to predict the ages of the older volcanic islands. Radioactive dating of the rocks found on various Hawaiian Islands confirmed this hypothesis. See Figure 18.14b.

The geothermal activity—hot springs, geysers, and volcanic rocks—at Yellowstone National Park is the result of another hot spot. The Galapagos, the Azores, and the Society Islands are also examples of volcanic islands formed by hot spots. Thus, while plate tectonics explains the location of belts of volcanic activity, hot spot volcanoes reveal the direction and rate of plate motion.

Key Ideas of the Plate Tectonics Theory

The plate tectonics theory, then, consists of the following ideas.

1. Earth's crust consists of a series of rigid slabs of rock called lithospheric plates. Each plate is about 100 kilometers thick. Earth's surface consists of about six major plates and several smaller pieces. See Figure 18.15.
2. The plates rest on a relatively fluid layer of the upper mantle and lower crust called the asthenosphere. This fluid layer lets the plates move independently of the deeper rocks in the mantle. Boundaries between plates are regions of volcanic and earthquake activity because they are where the plates collide and scrape against each other. The interiors of plates are relatively quiet.
3. The type of crust atop each plate determines whether the plate carries an ocean floor, a continent, or both. The continents and ocean floors are merely passengers on the moving plates. When the plates change position, the continents and ocean floors move along with them. Past positions of the plates can be inferred from evidence such as fossils, paleomagnetism, and past climates.

Plate Tectonics

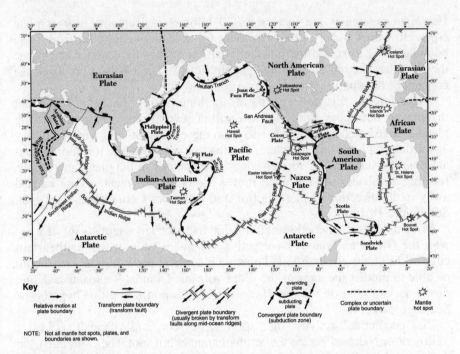

Figure 18.15 Plate Tectonics Diagram. Source: The State Education Department *Earth Science Reference Tables*, 2011 ed. (Albany, New York; The University of the State of New York).

4. There are three types of plate boundaries: convergent, divergent, and transform. Plates collide at convergent boundaries, separate at divergent boundaries, and slide past each other at transform boundaries. When an ocean plate collides with a continental plate, the denser ocean plate is pushed under the continental plate, or subducted. Most mountains form at divergent and convergent boundaries.

5. In subduction zones, ocean floor plates are consumed as they are pushed into the hot mantle and melt. As the plate plunges into the mantle, it remains rigid and produces deep-focus earthquakes. Friction between the subducted plate and the adjacent plate melts rock along the interface and produces volcanic activity directly above and parallel to the trench. For this reason most trenches are bordered by volcanic island arcs or volcanic mountain chains.

6. The creation of ocean floor in the mid-ocean rifts and its destruction in the trenches are in equilibrium. Therefore, Earth remains the same size.

7. Forces exist within Earth that are powerful enough to move the lithospheric plates. Two hypotheses have been proposed to explain plate motion: the plates are pushed by convection currents in the mantle, and they are pulled downward by gravity in subduction zones. Observations of heat flow support the concept of convection in the mantle. However, the scientific community is not yet unified in identifying the specific forces that propel the plates.

8. Convection in the mantle may produce "hot spots" over plumes of rising heat. As plates move over a hot spot, the magma works its way up through the plate and a volcano forms. Volcanic chains in the center of a plate, such as the Hawaiian Islands, are thought to have formed in this way.

Figure 18.16 is a simplified cross section of Earth showing the structures described above.

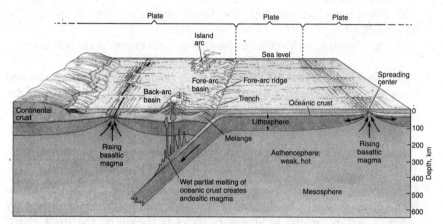

Figure 18.16 Cross Section of Earth Showing the Structures Found in Plate Tectonics. Source: *The Dynamic Earth*, 2nd Ed., Brian J. Skinner and Stephen C. Porter, John Wiley, 1992.

MULTIPLE-CHOICE QUESTIONS

In each case write the number of the word or expression that best answers the question or completes the statement.

1. The diagrams below show cross sections of exposed bedrock. Which cross section shows the *least* evidence of crustal movement?

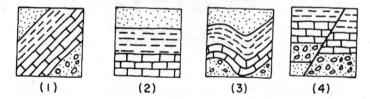

2. The best evidence of crustal movement is provided by
   (1) dinosaur tracks found in surface bedrock
   (2) marine fossils found on a mountaintop
   (3) weathered bedrock found at the bottom of a cliff
   (4) ripple marks found in sandy sediment

3. The Himalayan Mountains are located along a portion of the southern boundary of the Eurasian Plate. At the top of Mt. Everest (29,028 feet) in the Himalayan Mountains, climbers have found fossilized marine shells in the surface bedrock. From this observation, which statement is the best inference about the origin of the Himalayan Mountains?
   (1) The Himalayan Mountains were formed by volcanic activity.
   (2) Sea level has been lowered more than 29,000 feet since the shells were fossilized.
   (3) The bedrock containing the fossil shells is part of an uplifted seafloor.
   (4) The Himalayan Mountains formed at a divergent plate boundary.

4. Two geologic surveys of the same area, made 50 years apart, showed that the area had been uplifted 5 centimeters during the interval. If the rate of uplift remains constant, how many years will be required for this area to be uplifted a total of 70 centimeters?
   (1) 250   (3) 350
   (2) 500   (4) 700

5. The photograph below shows an escarpment (cliff) located in the western United States. The directions for north and south are indicated by arrows. A fault in the sedimentary rocks is shown on the front of the escarpment.

The photograph shows that the fault most likely formed
   (1) after the rock layers were deposited when the north side moved downward
   (2) after the rock layers were deposited when the north side moved upward
   (3) before the rock layers were deposited when the south side moved downward
   (4) before the rock layers were deposited when the south side moved upward

6. The diagrams below show four major types of fault motion occurring in Earth's crust. Which type of fault motion best matches the general pattern of crustal movement at California's San Andreas fault?

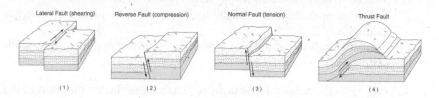

7. Folded sedimentary rock layers are usually caused by
   (1) deposition of sediments in folded layers
   (2) differences in sediment density during deposition
   (3) a rise in sea level after deposition
   (4) crustal movement occurring after deposition

8. Which statement best supports the theory that all the continents were once a single landmass?
   (1) Rocks of the ocean ridges are older than those of the adjacent seafloor.
   (2) Rock and fossil correlation can be made where the continents appear to fit together.
   (3) Marine fossils can be found at high elevations above sea level on all continents.
   (4) Great thicknesses of shallow-water sediments are found at interior locations on some continents.

9. The large coal fields found in Pennsylvania provide evidence that the climate of the northeastern United States was much warmer during the Carboniferous Period. This change in climate over time is best explained by the
   (1) movements of tectonic plates
   (2) effects of seasons
   (3) changes in the environment caused by humans
   (4) evolution of life

10. Compared with Earth's continental crust, Earth's oceanic crust is
    (1) thinner and more dense
    (2) thinner and less dense
    (3) thicker and more dense
    (4) thicker and less dense

Plate Tectonics

11. Which statement best supports the theory of continental drift?
   (1) Basaltic rock is found to be progressively younger at increasing distances from a mid-ocean ridge.
   (2) Marine fossils are often found in deep-well drill cores.
   (3) The present continents appear to fit together as pieces of a larger landmass.
   (4) Areas of shallow-water seas tend to accumulate sediment that gradually sinks.

12. According to the Inferred Positions of Earth Landmasses diagram in the *Earth Science Reference Tables*, on what other landmass would you most likely find fossil remains of the late Paleozoic reptile called Mesosaurus, shown below

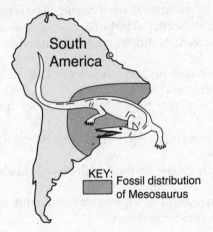

   (1) North America
   (2) Africa
   (3) Antarctica
   (4) Eurasia

13. Which observation about the Mid-Atlantic Ridge region provides the best evidence that the seafloor has been spreading for millions of years?
   (1) The bedrock of the ridge and nearby seafloor is igneous rock.
   (2) The ridge is the location of irregular volcanic eruptions.
   (3) Several faults cut across the ridge and nearby seafloor.
   (4) Seafloor bedrock is younger near the ridge and older farther away.

14. Igneous materials found along oceanic ridges contain magnetic iron particles that show reversal of magnetic orientation. This is evidence that
   (1) volcanic activity has occurred constantly throughout history
   (2) Earth's magnetic poles have exchanged positions
   (3) igneous materials are always formed beneath oceans
   (4) Earth's crust does not move

15. Alternating parallel bands of normal and reversed magnetic polarity are found in the basaltic bedrock on either side of the
    (1) Mid-Atlantic Ridge
    (2) Yellowstone Hot Spot
    (3) San Andreas Fault
    (4) Peru-Chile Trench

16. What is the primary force in the theory of ocean floor spreading?
    (1) Density differences in the mantle cause convection cells.
    (2) Planetary winds in the atmosphere blow against landmasses.
    (3) Earth rotates on its axis.
    (4) Earth revolves around the Sun.

17. In which set are the Earth processes thought to be most closely related to each other because they normally occur in the same zones?
    (1) mountain building, earthquakes, and volcanic activity
    (2) mountain building, shallow-water fossil formation, and rock weathering
    (3) volcanic activity, rock weathering, and deposition of sediments
    (4) earthquakes, shallow-water fossil formation, and shifting magnetic poles

18. According to the *Earth Science Reference Tables*, during which geologic period were the continents all part of one landmass, with North America and South America joined to Africa?
    (1) Tertiary
    (2) Cretaceous
    (3) Triassic
    (4) Ordovician

Base your answers to questions 19 through 21 on the diagram below. The diagram shows a model of the relationship between Earth's surface and its interior.

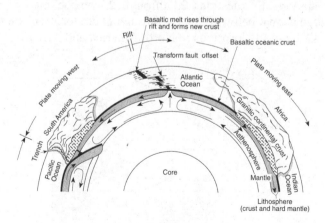

(Not drawn to scale)

Plate Tectonics

19. Mid-ocean ridges (rifts) normally form where tectonic plates are
    (1) converging
    (2) diverging
    (3) stationary
    (4) sliding past each other

20. The motion of the convection currents in the mantle beneath the Atlantic Ocean appears to be mainly making this ocean basin
    (1) deeper
    (2) shallower
    (3) wider
    (4) narrower

21. According to the diagram, the deep trench along the west coast of South America is caused by movement of the oceanic crust that is
    (1) sinking beneath the continental crust
    (2) uplifting over the continental crust
    (3) sinking at the Mid-Atlantic ridge
    (4) colliding with the Atlantic oceanic crust

22. Which diagram correctly shows how mantle convection currents are most likely moving beneath colliding lithospheric plates?

 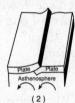  

Base your answers to questions 23 through 25 on the map of the Mid-Atlantic Ridge shown below. Points *A* through *D* are locations on the ocean floor. Line *XY* connects locations in North America and Africa.

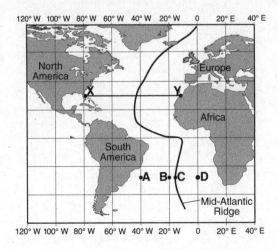

23. In which cross section do the arrows best show the convection occurring within the asthenosphere beneath line *XY*?

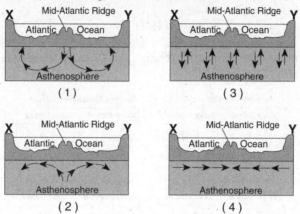

24. Samples of ocean-floor bedrock were collected at points *A*, *B*, *C*, and *D*. Which sequence shows the correct order of the age of the bedrock from oldest to youngest?
    (1) $D \to C \to B \to A$
    (2) $A \to D \to B \to C$
    (3) $C \to B \to D \to A$
    (4) $A \to B \to D \to C$

25. The boundary between which two tectonic plates is most similar geologically to the plate boundary at the Mid-Atlantic Ridge?
    (1) Eurasian and Indian-Australian
    (2) Cocos and Caribbean
    (3) Pacific and Nazca
    (4) Nazca and South American

Base your answers to questions 26 through 28 on the map below. The map shows the continents of Africa and South America, the ocean between them, and the ocean ridge and transform faults. Locations *A* and *D* are on the continents. Locations *B* and *C* are on the ocean floor.

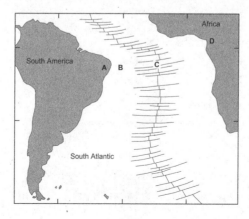

Plate Tectonics

26. The hottest crustal temperature measurements would most likely be found at location
    (1) A
    (2) B
    (3) C
    (4) D

27. Which table best shows the relative densities of the crustal bedrock at locations A, B, C, and D?

Relative Densities of Crust	
More Dense	Less Dense
A, B	C, D

(1)

Relative Densities of Crust	
More Dense	Less Dense
C, D	A, B

(3)

Relative Densities of Crust	
More Dense	Less Dense
B, C	A, D

(2)

Relative Densities of Crust	
More Dense	Less Dense
A, D	B, C

(4)

28. Which graph best shows the relative age of the ocean-floor bedrock from location B to location C?

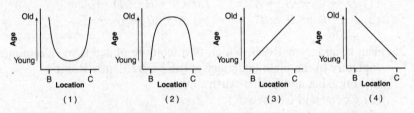

Base your answers to questions 29 and 30 on the map below, which shows Earth's Southern Hemisphere and the inferred tectonic movement of the continent of Australia over geologic time. The arrows between the dots show the relative movement of the center of the continent of Australia. The parallels of latitude from 0° to 90° south are labeled.

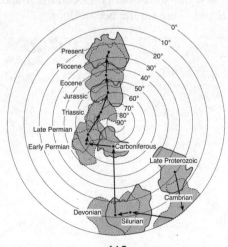

29. The geographic position of Australia on Earth's surface has been changing mainly because
    (1) the gravitational force of the Moon has been pulling on Earth's landmasses
    (2) heat energy has been creating convection currents in Earth's interior
    (3) Earth's rotation has spun Australia into different locations
    (4) the tilt of Earth's axis has changed several times

30. During which geologic time interval did Australia most likely have a warm, tropical climate because of its location?
    (1) Cambrian   (3) Late Permian
    (2) Carboniferous   (4) Eocene

31. Which map best indicates the probable locations of continents 100 million years from now if tectonic plate movement continues at its present rate and direction?

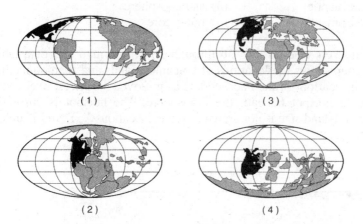

Base your answers to questions 32 through 35 on the cross section below, which shows the boundary between two lithospheric plates. Point *X* is a location in the continental lithosphere. The depth below Earth's surface is labeled in kilometers.

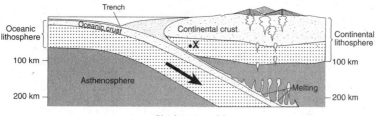

(Not drawn to scale)

419

Plate Tectonics

32. Between which two lithospheric plates could this boundary be located?
    (1) South American Plate and African Plate
    (2) Scotia Plate and Antarctic Plate
    (3) Nazca Plate and South American Plate
    (4) African Plate and Arabian Plate

33. Compared with the continental crust, the oceanic crust is
    (1) less dense and thinner   (3) more dense and thinner
    (2) less dense and thicker   (4) more dense and thicker

34. The temperature of the asthenosphere at the depth where melting first occurs is inferred to be approximately
    (1) 100°C   (3) 4,200°C
    (2) 1,300°C   (4) 5,000°C

35. Point *X* is located in which Earth layer?
    (1) rigid mantle   (3) asthenosphere
    (2) stiffer mantle   (4) outer core

Base your answers to questions 36 through 38 on the map and data table below. The map shows the locations of volcanic islands and seamounts that erupted on the seafloor of the Pacific Plate as it moved northwest over a stationary mantle hotspot beneath the lithosphere. The hotspot is currently under Kilauea. Island size is not drawn to scale. Locations *X*, *Y*, and *Z* are on Earth's surface.

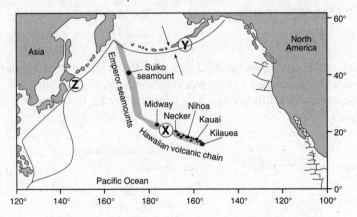

Map of Volcanic Features

Data Table
Age of Volcanic Features

Volcanic Feature	Distance from Kilauea (km)	Age (millions of years)
Kauai	545	5.6
Nihoa	800	6.9
Necker	1,070	10.4
Midway	2,450	16.2
Suiko seamount	4,950	41.0

36. Approximately how far has location X moved from its original location over the hotspot?
    (1) 3,600 km   (3) 1,800 km
    (2) 2,500 km   (4) 20 km

37. According to the data table, what is the approximate speed at which the island of Kauai has been moving away from the mantle hotspot, in kilometers per million years?
    (1) 1   (3) 100
    (2) 10   (4) 1,000

38. Which lithospheric plate boundary features are located at Y and Z?
    (1) trenches created by the subduction of the Pacific Plate
    (2) rift valleys created by seafloor spreading of the Pacific Plate
    (3) secondary plates created by volcanic activity within the Pacific Plate
    (4) mid-ocean ridges created by faulting below the Pacific Plate

39. When two tectonic plates collide, oceanic crust usually subducts beneath continental crust because oceanic crust is primarily composed of igneous rock that has
    (1) low density and is mafic   (3) high density and is mafic
    (2) low density and is felsic (4) high density and is felsic

40. At which plate boundary is one lithospheric plate sliding under another?
    (1) Nazca Plate and Antarctic Plate
    (2) Pacific Plate and Indian-Australian Plate
    (3) Indian-Australian Plate and Antarctic Plate
    (4) Nazca Plate and Pacific Plate

Plate Tectonics

41. The cross section below shows the direction of movement of an oceanic plate over a mantle hot spot, resulting in the formation of a chain of volcanoes labeled A, B, C, and D. The geologic age of volcano C is shown.

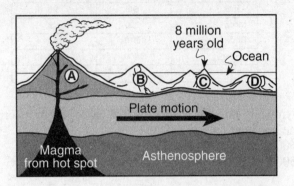

What are the most likely geologic ages of volcanoes B and D?
(1) B is 5 million years old, and D is 12 million years old.
(2) B is 2 million years old, and D is 6 million years old.
(3) B is 9 million years old, and D is 9 million years old.
(4) B is 10 million years old, and D is 4 million years old.

42. The Mariana Trench was most likely created by the
(1) convergence of the Pacific and Philippine Plates
(2) divergence of the Eurasian and Philippine Plates
(3) sliding of the Pacific Plate past the North American Plate
(4) movement of the Pacific Plate over the Hawaii Hot Spot

43. During which two geologic time periods did most of the surface bedrock of the Taconic Mountains form?
(1) Cambrian and Ordovician
(2) Pennsylvanian and Mississippian
(3) Triassic and Jurassic
(4) Silurian and Devonian

44. Which geologic event is inferred to have occurred most recently?
(1) collision between North America and Africa
(2) metamorphism of the bedrock of the Hudson Highlands
(3) formation of the Queenston delta
(4) initial opening of the Atlantic Ocean

CONSTRUCTED RESPONSE QUESTIONS

Base your answers to questions 45 through 48 on the world map below and on your knowledge of Earth science. The map shows major earthquakes and volcanic activity occurring from 1996 through 2000. Letter *A* represents a volcano on a crustal plate boundary.

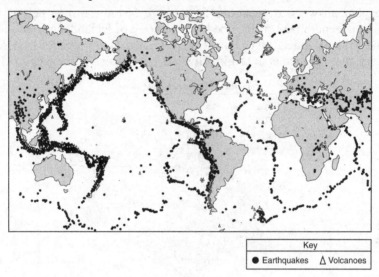

45. Place an **X** on the map to show the location of the Nazca Plate. [1]

46. Explain why most major earthquakes are found in specific zones instead of being randomly scattered across Earth's surface. [1]

47. Identify the source of the magma for the volcanic activity in Hawaii. [1]

48. Identify the type of plate movement responsible for the presence of the volcano at location *A*. [1]

Base your answers to questions 49 and 50 on the block diagram below. The diagram shows the tectonic plate boundary between Africa and North America 300 million years ago as these two continents united into a single landmass. The arrows at letters *A*, *B*, *C*, and *D* represent relative crustal movements. Letter *X* shows the eruption of a volcano at that time.

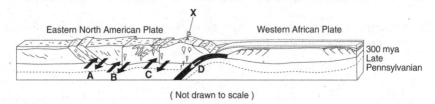

Plate Tectonics

49. Identify the type of tectonic plate motion represented by the arrow shown at *D*. [1]

50. Identify the type of tectonic motion represented by the arrows shown at *A*, *B*, and *C*. [1]

Base your answers to questions 51 through 55 on the map below, which shows the generalized surface bedrock geology of Iceland, an island located on the Mid-Atlantic Ridge. Points *A*, *B*, *C*, and *D* are locations on surface bedrock that is igneous in origin. Glaciers cover some surface bedrock.

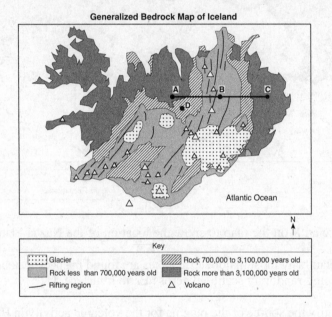

51. State the change in the relative ages of the surface bedrock along the line from *A* to *B* to *C*. [1]

52. According to the map, during which geologic era did the surface bedrock at location *D* form? [1]

53. Identify *one* fine-grained, highly mafic volcanic rock likely found as surface bedrock in Iceland. [1]

54. State the names of the *two* crustal plates that are diverging at Iceland. [1]

55. In addition to crustal plate divergence, what feature located in the mantle beneath Iceland may be causing Iceland's volcanic activity? [1]

EXTENDED CONSTRUCTED RESPONSE QUESTIONS

Base your answers to questions 56 through 58 on the map below, which is an enlargement of a portion of the *Tectonic Plates* map from the *Reference Tables for Physical Setting/Earth Science*. Points A and B are locations on different boundaries of the Arabian Plate.

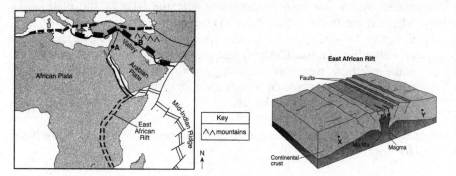

56. Identify the type of tectonic plate boundary located at point A. [1]

57. On the map shown, a valley is located south of point B and a mountain range north of point B. State the tectonic process that is creating these two land features. [1]

58. The block diagram next to the map represents Earth's surface and interior along the East African Rift. Draw *two* arrows, one through point X and one through point Y, to indicate the relative motion of each of these sections of the continental crust. [1]

Base your answers to questions 59 through 62 on the map and passage below. The map shows the outlines and ages of several calderas created as a result of volcanic activity over the last 16 million years as the North American Plate moved over the Yellowstone Hot Spot. A and B represent locations within the calderas.

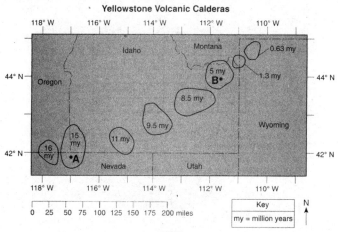

425

Plate Tectonics

The Yellowstone Hot Spot

The Yellowstone Hot Spot has interacted with the North American Plate, causing widespread outpourings of basalt that buried about 200,000 square miles under layers of lava flows that are a half mile or more thick. Some of the basaltic magma produced by the hot spot accumulates near the base of the plate, where it melts the crust above. The melted crust, in turn, rises closer to the surface to form large reservoirs of potentially explosive rhyolite magma. Catastrophic eruptions have partly emptied some of these reservoirs, causing their roofs to collapse. The resulting craters, some of which are more than 30 miles across, are known as volcanic calderas.

59. Describe the texture and color of the basalt produced by the Yellowstone Hot Spot. [1]

60. Identify *two* minerals found in the igneous rock that is produced from the explosive rhyolite magma. [1]

61. Based on the age pattern of the calderas shown on the map, in which compass direction has the North American Plate moved during the last 16 million years? [1]

62. Calculate, in miles per million years, the rate at which the North American Plate has moved over the Yellowstone Hot Spot between point *A* and point *B*. [1]

Base your answers to questions 63 through 65 on the map below, which shows the inferred position of Earth's landmasses at a particular time in Earth's history. The Taconic Mountains are shown near a subduction zone where they formed after the coast of Laurentia collided with a volcanic island arc, closing the western part of the Iapetus Ocean.

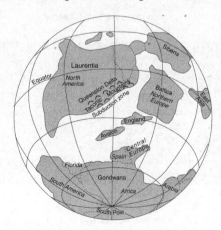

63. On the map, place an **X** to show the approximate location of the remaining part of the Iapetus Ocean. [1]

64. On the map, draw an arrow on the Laurentia landmass to show its direction of movement relative to the subduction zone. [1]

65. Identify the geologic time period represented by the map. [1]

Base your answers to questions 66 through 69 on the map and block diagram below. The map shows the location of North Island in New Zealand. The block diagram shows a portion of North Island. The Hikurangi Trench is shown forming at the edge of the Pacific Plate. Point *X* is at the boundary between the lithosphere and the asthenosphere.

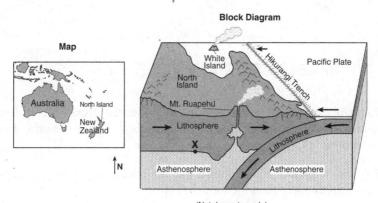

(Not drawn to scale)

66. State the approximate temperature at point *X*. [1]

67. On what tectonic plate are both North Island and White Island located? [1]

68. Describe the type of tectonic plate motion that formed the Hikurangi Trench. [1]

69. Describe *one* action that people on North Island should take if a tsunami warning is issued. [1]

Plate Tectonics

Base your answers to questions 70 through 73 on the world map shown below and on your knowledge of Earth science. Letters *A* through *H* represent locations on Earth's surface.

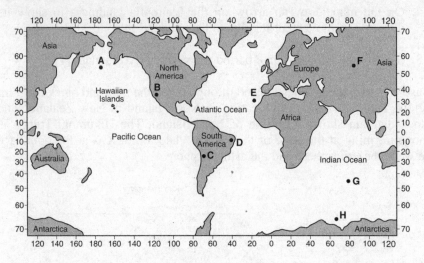

70. Explain why most earthquakes that occur in the crust beneath location *B* are shallower than most earthquakes that occur in the crust beneath location *C*. [1]

71. Explain why location *A* has a greater probability of experiencing a major earthquake than location *D*. [1]

72. Explain why a volcanic eruption is more likely to occur at location *E* than at location *F*. [1]

73. Explain why the geologic age of the oceanic bedrock increases from location *G* to location *H*. [1]

Base your answers to questions 74 through 78 on the passage and map shown below. The passage dscribes the Gakkel Ridge found at the bottom of the Arctic Ocean. The map shows the location of the Gakkel Ridge.

The Gakkel Ridge

In the summer of 2001, scientists aboard the U.S. Coast Guard icebreaker Healy visited one of the least explored places on Earth. The scientists studied the 1800-kilometer-long Gakkel Ridge at the bottom of the Arctic Ocean near the North Pole. The Gakkel Ridge is a section of the Arctic Mid-Ocean Ridge and extends from the northern end of Greenland across the Arctic Ocean floor toward Russia. At a depth of about 5 kilometers below the ocean surface, the Gakkel Ridge is one of the deepest mid-ocean ridges in the world. The ridge is believed to extend down to Earth's mantle, and the new seafloor being formed at the ridge is most likely composed of huge slabs of mantle rock. Bedrock samples taken from the seafloor at the ridge were determined to be the igneous rock peridotite.

The Gakkel Ridge is also the slowest-moving mid-ocean ridge. Some ridge systems, like the East Pacific Ridge, are rifting at a rate of about 20 centimeters per year. The Gakkel Ridge is rifting at an average rate of less than 1 centimeter per year. This slow rate of movement means that there is less volcanic activity along the Gakkel Ridge than along other ridge systems. However, heat from the underground magma slowly seeps up through cracks in the rocks of the ridge at structures scientists call hydrothermal (hot water) vents. During the 2001 cruise, a major hydrothermal vent was discovered at 87° N latitude 45° E longitude.

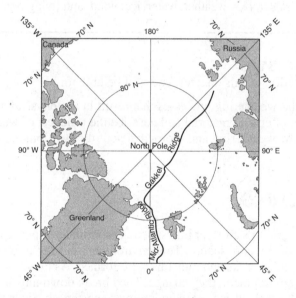

74. On the map, place an **X** on the location of the major hydrothermal vent described in the passage. [1]

75. Describe the relative motion of the two tectonic plates on either side of the Gakkel Ridge. [1]

76. The Gakkel Ridge is a boundary between which two tectonic plates? [1]

77. Identify *one* feature, other than hydrothermal vents, often found at mid-ocean ridges like the Gakkel Ridge that indicates heat from Earth's interior is escaping. [1]

78. State the *two* minerals that were most likely found in the igneous bedrock samples collected at the Gakkel Ridge. [1]

Unit Six: WEATHERING, EROSION, AND DEPOSITION

CHAPTER 19

WEATHERING AND SOIL FORMATION

> **KEY IDEAS** In preceding units, we explored the mechanisms by which the materials that comprise Earth's crust were formed. In this unit we will discuss the processes that shape the material exposed at the surface of Earth's crust; a dynamic, changing system of minerals, rocks, weather, water, ice, wind, and living organisms.

KEY OBJECTIVES
Upon completion of this chapter, you will be able to:

- Describe the weathering processes that result in the physical and chemical breakdown of crustal material and describe the products of weathering.
- Explain how weathering and biological activity over long periods of time create soils.

WEATHERING

The surface of Earth's crust, or lithosphere, is constantly exposed to the changing weather of the atmosphere. This environment is very different from the one in which most rocks and minerals were formed. As these materials adjust to their new environment, they change. They break down into smaller pieces, and chemical reactions change them into new substances. Since the processes that produce these changes result from exposure to the weather, they are called weathering. **Weathering** is the breakdown of rocks into smaller particles by natural processes. Whenever rocks are exposed to the air, water, and living things at or near Earth's surface, weathering occurs. Weathering processes are divided into two general types: physical and chemical.

Physical Weathering

Some of the changes caused by weathering involve form only. Weathering may break a large, solid mass of rock into loose fragments varying in size and shape but identical in composition to the original rock. Processes that break down rocks without changing their chemical compositions are called **physical weathering**.

Frost Action

Frost action is the result of an unusual property of water. Most materials expand when heated and contract when cooled. This is true of water except that, when water is cooled from 4°C to 0°C, it *expands*. It expands most when, at 0°C, it solidifies into ice; then its volume increases by 9%. The expansion of water as it cools and solidifies can exert huge forces on anything confining it—forces measuring tens of thousands of pounds per square inch!

Water, in the form of rain, melting snow, or condensation, seeps into any cracks or pores in rock. When the temperature drops below the freezing point of the water in these cracks and pores, the water changes into ice. The expanding ice exerts tremendous pressure against the confining rock. Acting like a wedge, it widens and extends the opening. Then, when the ice thaws, the water seeps deeper into the opening. When the water refreezes, the process is repeated. In this way, the alternate freezing and thawing of water, or **frost action**, breaks rocks apart (see Figure 19.1).

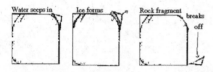

Figure 19.1 Frost Action.

Frost action is particularly effective where bedrock is directly exposed to the atmosphere, moisture is present, and the temperature fluctuates frequently above and below the freezing point of water. These conditions often exist during the winter in temperate climates such as New York State's, and can occur also on mountain tops and at high elevations in spring or fall. Daytime temperatures rise above freezing, causing snow and ice to melt, only to drop below freezing again at night, producing frost action.

Frost action on cliffs of bare rock breaks loose fragments that fall to the base of the cliff. When this takes place rapidly, a pile of fragments called a **talus slope** accumulates at the base of the cliff (see Figure 19.2).

The potholes that threaten motorists in many of our northern states are caused by frost action on exposed road surfaces. Farmers in New England have to clear their fields every spring of rocks and boulders pushed up out of the ground by frost action in soil. As you might expect, frost action is almost nonexistent in states such

Figure 19.2 A Talus Slope Forms at the Base of a Cliff. The angle of repose is the steepest slope at which the fragments remain stable.

as Florida and Hawaii. However, through much of the northern United States, it is probably the primary weathering process, and, throughout the world, probably the most significant physical weathering process.

Abrasion

Rocks can also be broken down by **abrasion**, that is, by rubbing against each other. Rock abrasion occurs mainly when fragments are being carried along by agents of erosion. A typical example occurs in streams. As the fragments are carried along by the water, they bounce off and rub against each other. This abrasion breaks smaller pieces off the surface, and the fragments become still smaller and also more rounded (see Figure 19.3). The fragments abrade the bedrock beneath the stream as well. Wind also weathers rock by abrasion. Anyone who has sat on a sandy beach on a windy day can attest to the abrasive power of wind-driven sand. Abrasion is a major physical weathering process.

Time Increases ⟶

Figure 19.3 Rounding of Particles Due to Abrasion.

Exfoliation

Exfoliation is the scaling off, or peeling, of successive shells from the surface of rocks. Exfoliation generally occurs in coarse-grained rocks that contain the mineral feldspar. Whenever the surface of such rocks becomes wet, moisture penetrates pores and crevices between the mineral grains and reacts with the feldspar. A chemical change occurs, producing a new substance: kaolin. This clay has a greater volume than the feldspar it replaces. The expansion pries loose the surrounding mineral grains, and a thin shell of surface material flakes away. (Note that this is a physical process caused by a chemical change.)

With successive wettings of the rock surface, the process is repeated. Since more surface is exposed at the corners of cracks or joints, these split off at the fastest rate. The result is a rounding in the shape of the rock as seen in Figure 19.4. Feldspar is a common rock-forming mineral; therefore, exfoliation is a significant weathering process.

Figure 19.4 Exfoliation.

Plant and Animal Action

Plants and animals interact with rock in a number of ways that cause the rock to break down into smaller pieces. When a rock develops cracks, small particles of rock and soil are washed into the cracks by rain or blown in by wind. If a seed finds its way into a crack, it can germinate and begin to grow. As the plant grows, it sends tiny rootlets deeper into the crack in search of water. The growing rootlets thicken and press against the sides of the crack. Like the expansion of ice in frost action, the growing roots widen and extend the crack (see Figure 19.5a). Eventually the rock is broken apart. The roots of tiny plants like mosses and lichens produce a rock-dissolving acid as they grow and decay, thereby further accelerating the breakdown of the rocks.

Although animals (with the exception of humans) do not directly attack rock, they contribute indirectly to its weathering in many ways. More than 100 years ago, Charles Darwin calculated that on 1 acre of land earthworms bring as much as 10 metric tons of particles to the surface each year. Exposed to the atmosphere again, these particles are subjected to further breakdown by weathering processes.

Ants (see Figure 19.5b), termites, woodchucks, moles and other burrowing animals also play a role in weathering. Furthermore, the burrows they create allow air and water to penetrate deeper beneath the surface and to weather the underlying bedrock.

Humans, however, have probably contributed more to the physical weathering of rock than any other single species. Roadcut building, rock quarrying, and strip mining are just a few examples of the human activities that break up rock. In addition, these activities expose vast quantities of fresh rock to other weathering processes.

The contribution of each individual organism to the weathering of Earth's crust may seem small. However, consider the number of living things on Earth and the length of time they have been at work. Taken collectively, the total amount of rock they have affected is tremendous.

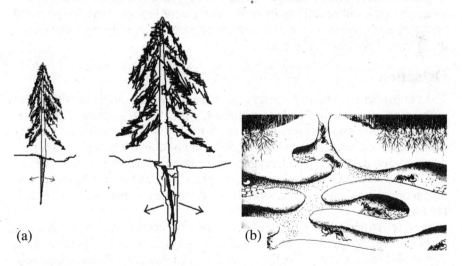

Figure 19.5 Plant (a) and Animal (b) Action on Rocks.

Changes in Temperature

Rocks are often exposed to large temperature changes. As rocks heat up during the day, they expand; as they cool off at night, they contract. You might expect this constant change to cause rocks to crack and break up, but experiments have indicated that this is not generally the case. Only extreme temperature changes, such as those resulting from forest and brush fires, cause rocks to crack or flake off at the surface.

Pressure Unloading

Some rocks form deep beneath Earth's surface. There they are under great pressure—millions of pounds per square inch. As the rocks form, stresses build up inside them that cannot be released because of the pressure. When forces within Earth bring these rocks to the surface, the pressure is reduced. The resultant expansion and release of stress cause the rocks to develop large cracks, or joints, at weak points in their structure—a phenomenon known as **pressure unloading**. Pressure unloading can also occur when glaciers melt away and the pressure exerted by the weight of the ice on underlying rock is released.

Chemical Weathering

Chemical weathering breaks down rocks by changing their chemical compositions. Most rocks form in an environment that is very different from that at the surface of Earth. Many of the substances in the atmosphere, for example, were not present in the environment where the rocks were formed. When the minerals in a rock are exposed to these substances, they may react with them to form new compounds with properties different from those of the original minerals. Such changes almost always weaken the structure of the rock, so that it either falls apart or is more easily broken down by physical weathering. Oxygen, water, and carbon dioxide are chiefly responsible for the chemical weathering of rocks.

Oxidation

The atmosphere is about 21 percent oxygen. Oxygen combines with many substances in a reaction called **oxidation**, which is an important chemical weathering process. Oxygen reacts most readily with minerals containing iron, such as magnetite, pyrite, amphibole, and biotite. Oxidation of iron produces compounds of iron and oxygen, iron oxides, of which hematite (Fe_2O_3) and magnetite (Fe_3O_4) are common examples. Since the iron in hematite and magnetite can be extracted economically, these compounds are also considered iron ores.

If water is present during oxidation, another reaction can occur; a compound of iron, oxygen, and water called *goethite* can form. Goethite has a yellowish brown color. If goethite is dehydrated, hematite forms. Many reddish or yellow-brown soils and weathered rocks get their color from the hematite or goethite they contain.

Oxidation of iron oxide in the presence of water to form goethite:
4FeO   +   2H$_2$O   +   O$_2$ →   4(FeO-OH)
iron oxide   water   oxygen   goethite

Dehydration of goethite to form hematite:
2(FeO-OH)   →   Fe$_2$O$_3$ +   H$_2$O
goethite   hematite   water

How does oxidation break down or weather rocks? When oxygen combines with iron, the chemical bonds between the iron and other substances in the rock are broken, thereby weakening its structure. Compare the strength of a rusty nail with that of a new one. The effect on rock is similar. In addition to iron, other elements in rock, such as aluminum and silicon, can combine with oxygen. The effects are much the same: an oxide is formed, the structure is weakened, and the rock disintegrates.

Hydration, Hydrolysis, and Solution

Most of Earth's surface is covered with water. Trillions of gallons fall on Earth as rain every day. With such abundance it is not surprising that water is an important agent of chemical weathering.

When water combines with another substance, the reaction is called **hydration**. For example, the hydration of anhydrite forms gypsum.

Hydration of anhydrite to form gypsum:
CaSO$_4$ +   2H$_2$O   →   CaSO$_4$ • 2H$_2$O
anhydrite   water   gypsum

Water can also form hydrogen ions (H+) and hydroxide ions (OH–). When these ions replace the ions of a mineral, the reaction is called **hydrolysis**. Feldspar, amphibole, and biotite are common minerals that can undergo hydrolysis, which causes them to swell and crumble. The product is a powder of clay minerals that are insoluble in water. For example, the hydrolysis of feldspar forms kaolinite, a clay mineral.

Hydrolysis of feldspar to form kaolinite:
4 KAlSi$_3$O$_8$ +   4 H$^+$ +   2H$_2$O   →   Al$_4$Si$_4$O$_{10}$(OH)$_8$ +   8SiO$_2$
K-feldspar   hydrogen   water   kaolinite   silica
                   ions

(Note that the symbol K does not appear on the right. The potassium is released as a positive ion into the soil water.)

Water also weathers rock simply by dissolving it. This process is called **solution**. So many materials dissolve, at least to some extent, in water that water is often called the *universal solvent*. Halite (rock salt) and gypsum are good examples of water-soluble minerals. Water will slowly dissolve these minerals out of a rock, thereby exposing the surrounding minerals to further weathering. In some cases the rock's structure may be so weakened by the empty spaces created that the rock just crumbles. In solution, the dissolved minerals may react with each other to form new substances. If these substances are insoluble in water, they precipitate out.

Carbonation

Carbon dioxide (CO_2) is a colorless, odorless gas that comprises 0.03–0.04 percent of Earth's atmosphere. The chemical combining of carbon dioxide with another substance is called **carbonation**. As a gas, carbon dioxide has almost no effect on rocks. When carbon dioxide comes in contact with water, though, the following carbonation reaction takes place:

Production of carbonic acid by solution of carbon dioxide:
$$CO_2 + H_2O \rightarrow H_2CO_3$$
carbon dioxide   water   carbonic acid

Although carbonic acid is a fairly weak acid, it attacks common rock minerals. Carbonic acid reacts readily with minerals containing the elements sodium, potassium, magnesium, and calcium. The compounds formed when these elements react with carbon dioxide are called *carbonates*.

Carbonic acid is most destructive to the mineral calcite, which is completely dissolved by carbonic acid. Rocks, such as limestones, that are almost entirely composed of calcite are totally dissolved away by carbonic acid in rainwater and groundwater. Spectacular caverns can form as groundwater containing carbonic acid seeps through bedrock composed of calcite and eats huge holes in it (see Figure 19.6).

Figure 19.6 Carbonation Caverns Form as Rainwater Containing Carbonic Acid Seeps into Cracks in Limestone.

Other Chemical Factors

In addition to carbonic acid, other naturally occurring acids attack rocks and minerals. Some of these acids are produced by lightning or the decay of organic material; others, as waste products of certain plants and animals. These acids dissolve in rainwater as it falls through the atmosphere and seeps through the soil. Upon reaching bedrock, they dissolve some rocks and cause others to crumble.

Primitive plants such as lichens can grow on bare rock surfaces. They have limited nutrient needs and can obtain these directly from the rock. Lichens grow when the rock surface is wet and lie dormant when it is dry. Secretions from the lichens eat away at the rock's surface, dissolving out mineral nutrients and loosening mineral particles. These mineral particles accumulate in rock crevices, along with dust from the atmosphere and debris from dead lichens. Seeds and spores may then take hold in these tiny pockets of soil and grow, further increasing the amount of chemical and physical weathering the rock undergoes.

An increasingly important source of acids that weather rock is human activity. Factories, homes, and automobiles release vast quantities of waste gases and other pollutants into the atmosphere. Many, such as the oxides of nitrogen and sulfur, react with water to form very strong, reactive acids.

In some areas surrounding large cities and industrial complexes, the concentration of these acids has reached alarming levels. Rainfall in these areas contains so much acid that it is called **acid rain**. Acid rain accelerates the weathering of rock and the breakdown of structures built by humans, and it damages plant and animal life.

Factors Affecting Weathering

How long does it take for a rock to be broken down by weathering processes? The answer is complex since many factors influence the rate at which a rock will weather. Among the more important factors affecting weathering are climate, particle size, exposure, mineral composition, and time.

Climate

Climate, which is the single most important factor affecting weathering, is the average condition of the atmosphere in a region over a long period of time. It is typically expressed in terms of two factors: temperature and precipitation (see Figure 19.7). Both factors influence the type and the rate of weathering. Warm climates favor chemical weathering; cold climates, physical weathering, principally frost action. In both cases, the more moisture present, the more pronounced the weathering.

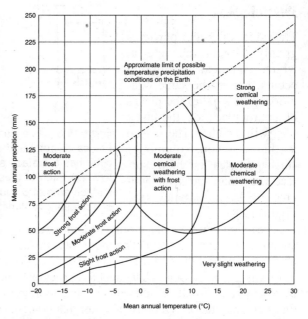

Figure 19.7 The Effect of Climate on Weathering Processes. Physical weathering is dominant in areas where rainfall and temperature are both low. High temperature and precipitation favor chemical weathering.

Weathering and Soil Formation

Chemical reactions tend to occur at a faster rate as temperature increases. Many of these reactions require water, which is a reactant in hydration and carbonation and provides a medium in which acid reactions can occur. Hot, moist climates also support increased biological activity, ranging from burrowing to the production of humic acid as plant matter decomposes. Thus, a hot, moist climate is the ideal environment for rapid chemical weathering.

In cold climates, physical weathering by frost action predominates. It is most effective in cold climates that alternately freeze and thaw. Here again, precipitation is a key factor. Without water, ice cannot form and frost action cannot occur. A cold, moist climate will therefore experience strong frost action.

Temperate climates, such as New York's, combine aspects of both extremes. Hot, moist summers followed by moist winters with many freeze-thaw cycles result in an environment in which rocks weather rapidly. A classic example of the destructiveness of this climate is its effect on Cleopatra's Needle, a granite obelisk with hieroglyphics cut into its surface that was moved from Egypt to Central Park in 1880. In Egypt's hot, dry climate the obelisk stood almost unchanged for 3,000 years. In the moist, changing climate of New York City, however, it began to rapidly deteriorate. Today, its surface cracked, worn, and discolored, the hieroglyphics almost unreadable, it stands as mute testimony to the effect of climate on weathering.

Particle Size

The size of rock particles greatly affects the rate at which chemical weathering occurs. Under the same conditions, the smaller the pieces of a particular rock, the faster they will weather. The reason is that a given volume of small particles has more surface area than the same volume of large particles (see Figure 19.8).

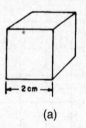
(a)

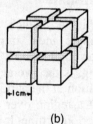
(b)

Figure 19.8 As a Particle Is Broken into Smaller Pieces, Its Total Surface Area Increases. (a) This cube has a surface area of 24 square centimeters. (b) Each of these eight small cubes has a surface area of 6 square centimeters, for a total of 48 square centimeters.

A reaction can take place only when the rock comes into contact with the chemical weathering agent. The more rock surface exposed, the more rock is able to react, and the faster the rate of the reaction.

Size is also a factor in what happens to rock particles after they are produced by weathering. Small particles may be transported to a new location that has a different climate, or to a body of water that contains a variety of reagents.

Exposure

Exposure determines the degree to which a rock comes into contact with weathering agents. Soil, ice, and vegetation can cover a rock and thereby decrease its contact with weathering agents. Rocks thus protected tend to weather more slowly than those completely exposed at the surface. Also, the weathered surface of a rock can itself shield fresh material underneath, thereby slowing the rate of weathering. Another factor that affects exposure is the slope of the land. On steep slopes, loose materials move downhill because of gravity or are carried downhill by erosion, thus continually exposing fresh rock.

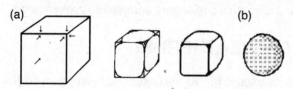

Figure 19.9 Results of Exposure. (a) Corners and edges weather fastest, resulting in a rounded particle (b).

Exposure also has its effects on each individual particle. As you can see in Figure 19.9a, the corners of a cube are exposed to weathering on three sides, the edges on two sides, and the faces on one side. The net effect is that the corners of a particle weather away most quickly and the faces most slowly. Over time the result is a spherical particle, as shown in Figure 19.9b.

Mineral Composition

The mineral composition of a rock determines its physical and chemical properties and thus its susceptibility to weathering. Mineral composition is of greatest importance in chemical weathering. Rocks composed of minerals that react readily with acids, water, or oxygen will weather more rapidly than those composed of less reactive minerals. For example, limestone, which is mostly calcite, is dissolved by even mildly acidic rainwater, while granite, which is mostly silicates, is almost unaffected.

Mineral composition also affects physical weathering (see Table 19.1). Softer rocks will abrade more readily than harder rocks. Solid, crystalline rocks have fewer openings into which water can penetrate than rocks composed of cemented-together particles.

TABLE 19.1 RELATIVE RESISTANCES OF THE MAJOR ROCK-FORMING MINERALS TO WEATHERING

Mineral	Resistance to Weathering
Olivine	Least resistance
Pyroxene	
Amphibole	
Biotite	↓
Plagioclase feldspars	
Muscovite	
Orthoclase	
Quartz	Most resistant

Time

Weathering is a slow process. The 3.5-billion-year-old gneisses in Greenland are scarcely weathered. Even easily weathered rocks may take hundreds of years to be completely broken down. The longer a rock is exposed to weathering processes, however, the more it deteriorates and disintegrates. Eventually, all rocks exposed at Earth's surface are completely broken down.

THE PRODUCTS OF WEATHERING

Since Earth formed, the rocks exposed at its surface have been attacked by forces that disintegrate them. Whether these forces are physical or chemical, the result is the same—solid rock is broken into fragments. These fragments range from the tiniest particle dissolved in water to the largest boulders. The fragments produced form sediments and soils.

Sediments

The fragments or particles of rock produced by weathering are called **sediments**. Sediments are named according to their sizes (Table 19.2).

TABLE 19.2 SEDIMENT SIZES AND NAMES

Particle Diameter (cm)	Name
<0.0004	Clay
0.0004–0.006	Silt
0.006–0.2	Sand
0.2–6.4	Pebbles
6.4–25.6	Cobbles
>25.6	Boulders

Soils

Soil is the accumulation of loose, weathered material that covers much of the land surface of Earth. Soil varies in depth, composition, age, color, and texture. Although its chief component is weathered rock, a true soil also contains water, air, bacteria, and decayed plant and animal material (humus).

The rock from which a soil forms is called the **parent material**. Soil that forms directly from the bedrock beneath it is **residual soil** (see Figure 19.10). If the soil forms from material that was transported to the location by erosion, it is **transported soil**. Transported soil may have a different mineral composition from the underlying bedrock.

As a soil forms, the processes of weathering and plant growth develop recognizable layers, or horizons, in the soil. These horizons differ in structure, composition, color, and texture. A soil that has been forming long enough to have developed distinct horizons is called a **mature soil**. In **immature soils**, horizons are indistinct or altogether lacking.

A **soil profile** is a cross section of a soil from surface to bedrock. Figure 19.11 shows the profile of a typical mature soil with a description of each horizon.

Parent material greatly influences the type of soil that forms, especially when the soil is just starting to form or is immature. Over a long period of time, however, climate has a greater influence on the type of soil that forms. Whatever the parent material, the profiles of mature soils that formed in similar climates are very much alike.

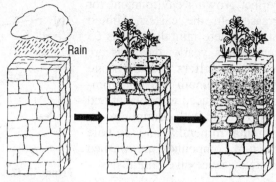

Figure 19.10 The Development of a Residual Soil. Weathering opens bedrock to air, water, and dust. Plant and animal life accelerate the breakdown of rock and allow air and water to penetrate ever more deeply. Eventually a deep layer of soil develops atop the bedrock.

In the United States, two general types of soils have developed (see Figure 19.12). In the western half of the country, where rainfall is less than 63 centimeters yearly, the soils that formed are mainly a type called **pedocals**. Pedocals are rich in calcium compounds and tend to be slightly alkaline, a

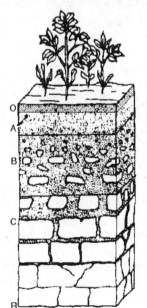

Horizon O (organic horizon)
The material in this horizon is commonly called topsoil and supports plant life. This horizon contains fresh to partly decomposed organic matter. Its color varies from dark brown to black.

Horizon A (leached horizon)
The top of this horizon is highly decomposed organic matter mixed with minerals. Its particles, exposed since the soil began to form, are sand sized or smaller. In a process called *leaching*, soluble minerals and tiny clay particles are carried down to lower layers as water seeps downward through this layer. Leaching is one of the processes that cause horizons to form. This horizon ranges from brown to gray in color.

Horizon B (accumulation horizon)
Materials leached out of horizon A are deposited here. This horizon is richer in clay, has less organic material, and has more and larger particles of bedrock than horizon A. As a result it is less fertile. The clay in this horizon gives it a clumpy or blocky consistency compared to loose, sandy horizon A. The leached materials deposited here (clays and iron oxides) give this horizon a reddish brown or tan color.

Horizon C (partly weathered horizon)
This horizon consists of partly weathered bedrock. Sometimes referred to as subsoil, it is the cracked, broken surface of the bedrock. It marks the final transition from topsoil to unaltered bedrock or parent material.

Parent material (R)
Unaltered bedrock

Figure 19.11 Profile of a Mature Soil.

perfect growing environment for grasses. In the eastern United States, where rainfall exceeds 63 centimeters yearly, the soils are mainly **pedalfers**. Pedalfers are rich in aluminum and iron compounds produced when water and oxygen react with common rock-forming minerals, and soluble calcium compounds are washed away by successive rainfalls.

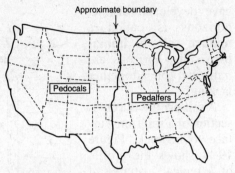

Figure 19.12 Two General Types of Soil in the United States.

MULTIPLE-CHOICE QUESTIONS

In each case, write the number of the word or expression that best answers the question or completes the statement.

1. Which is the best example of physical weathering?
    (1) the cracking of rock caused by the freezing and thawing of water
    (2) the transportation of sediment in a stream
    (3) the reaction of limestone with acid rainwater
    (4) the formation of a sandbar along the side of a stream

2. The diagram below shows granite bedrock with cracks. Water has seeped into the cracks and frozen. The arrows represent the directions in which the cracks have widened due to weathering.

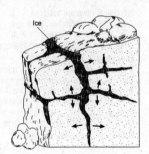

Which statement best describes the physical weathering shown by the diagram?
    (1) Enlargement of the cracks occurs because water expands when it freezes.
    (2) This type of weathering occurs only in bedrock composed of granite.
    (3) The cracks become wider because of chemical reactions between water and the rock.
    (4) This type of weathering is common in regions of primarily warm and humid climates.

3. At high elevations in New York State, which is the most common form of physical weathering?
   (1) abrasion of rocks by wind
   (2) alternate freezing and melting of water
   (3) dissolving minerals into solution
   (4) oxidation by oxygen in the atmosphere

4. Which weathering process tends to form spherical boulders?
   (1) carbonation
   (2) exfoliation
   (3) frost action
   (4) oxidation

5. Which characteristic would most likely remain constant when a limestone cobble is subjected to extensive abrasion?
   (1) shape
   (2) mass
   (3) volume
   (4) composition

6. The diagram below represents a naturally occurring geologic process.

   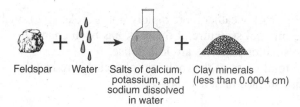

   Feldspar   Water   Salts of calcium, potassium, and sodium dissolved in water   Clay minerals (less than 0.0004 cm)

   Which process is best illustrated by the diagram?
   (1) cementation
   (2) erosion
   (3) metamorphism
   (4) weathering

7. Which event is an example of chemical weathering?
   (1) rocks falling off the face of a steep cliff
   (2) feldspar in granite being crushed into clay-sized particles
   (3) water freezing in cracks in a roadside outcrop
   (4) acid rain reacting with limestone bedrock

8. Landscapes will undergo the most chemical weathering if the climate is
   (1) cool and dry
   (2) cool and wet
   (3) warm and dry
   (4) warm and wet

9. Which factor has the greatest influence on the weathering rate of Earth's surface bedrock?
   (1) local air pressure
   (2) angle of insolation
   (3) age of the bedrock
   (4) regional climate

10. Four samples of the same material with identical composition and mass were cut as shown in the diagrams below. When the samples are subjected to the same chemical weathering, which sample will weather at the fastest rate?

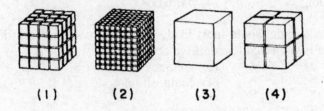

11. Why will a rock weather more rapidly if it is broken into smaller particles?
    (1) The mineral structure of the rock has been changed.
    (2) The smaller particles are less dense.
    (3) The total mass of the rock and particles is reduced.
    (4) More surface area is exposed.

12. The diagram below shows four mineral samples, each having approximately the same mass.

Quartz   Amphibole   Pyroxene   Galena

If all four samples are placed together in a closed, dry container and shaken vigorously for 10 minutes, which mineral sample would experience the most abrasion?
   (1) quartz
   (2) amphibole
   (3) pyroxene
   (4) galena

Base your answers to questions 13 through 16 on the graph below, which shows the effect that average yearly precipitation and temperature have on the type of weathering that will occur in a particular region.

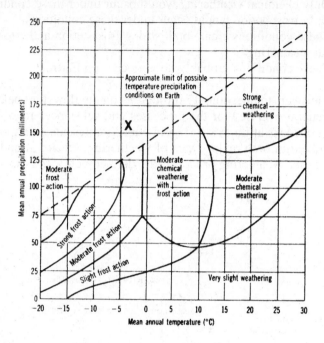

13. Which type of weathering is most common where the average yearly temperature is 5°C and the average yearly precipitation is 45 cm?
    (1) moderate chemical weathering
    (2) very slight weathering
    (3) moderate chemical weathering with frost action
    (4) slight frost action

14. The amount of chemical weathering will increase if
    (1) air temperature decreases and precipitation decreases
    (2) air temperature decreases and precipitation increases
    (3) air temperature increases and precipitation decreases
    (4) air temperature increases and precipitation increases

15. Why is no frost action shown for locations with a mean annual temperature greater than 13°C?
    (1) Very little freezing takes place at these locations.
    (2) Large amounts of evaporation take place at these locations.
    (3) Very little precipitation falls at these locations.
    (4) Very large amounts of precipitation fall at these locations.

Weathering and Soil Formation

16. No particular type of weathering or frost action is given for the temperature and precipitation values at the location represented by the letter X. Why is this the case?
    (1) Only chemical weathering would occur under these conditions.
    (2) Only frost action would occur under these conditions.
    (3) These conditions create both strong frost action and strong chemical weathering.
    (4) These conditions probably do not occur on Earth.

Base your answers to questions 17 through 19 on the flowchart below, which shows a general overview of the processes and substances involved in the weathering of rocks at Earth's surface. Letter X represents an important substance involved in both major types of weathering, labeled A and B on the flowchart. Some weathering processes are defined below the flowchart.

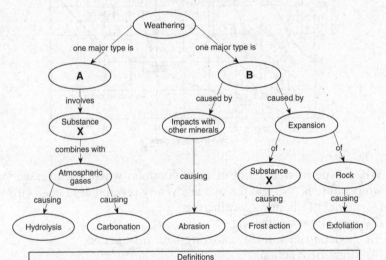

17. Which term best identifies the type of weathering represented by A?
    (1) physical   (3) chemical
    (2) biological   (4) glacial

18. Which substance is represented by X on both sides of the flowchart?
    (1) potassium feldspar   (3) hydrochloric acid
    (2) air   (4) water

19. Which weathering process is most common in a hot, dry environment?
(1) abrasion
(2) carbonation
(3) frost action
(4) hydrolysis

20. A variety of soil types are found in New York State primarily because areas of the state differ in their
(1) amounts of insolation
(2) distances from the ocean
(3) underlying bedrock and sediments
(4) amounts and types of human activities

21. The cross section below shows layers of soil.

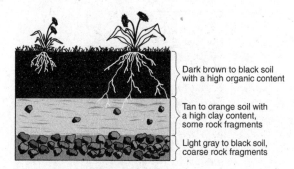

Which two processes produced the layer of dark brown to black soil?
(1) melting and solidification of magma
(2) erosion and uplifting
(3) weathering and biologic activity
(4) compaction and cementation

Constructed Response Questions

Base your answers to questions 22 through 24 on the cross section below, which shows limestone bedrock with caves.

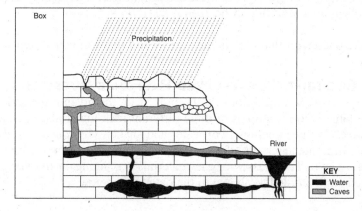

22. In the empty box on the left side of the cross section, draw a horizontal line to indicate the level of the water table. [1]

23. The precipitation in this area is becoming more acidic. Explain why acid rain weathers limestone bedrock. [1]

24. Identify one source of pollution caused by human activity that contributes to the precipitation becoming more acidic. [1]

Base your answers to questions 25 through 27 on the diagram below, which shows the soil profile formed in an area of granite bedrock. Four different soil horizons, A, B, C, and D, are shown.

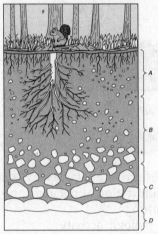

25. Identify the soil horizon containing the highest percentage of organic material. [1]

26. Soil horizon B is rich in clay and soluble minerals.
    (a) State the process by which this soil horizon has become enriched in these materials. [1]
    (b) In one or more complete sentences, compare the fertility of soil in this layer with that of horizon A. [1]

27. State one factor that affects the type of soil that forms in a region. [1]

EXTENDED CONSTRUCTED RESPONSE QUESTIONS

Base your answers to questions 28 through 31 on the given information and the data table below. Samples of three different rock materials, A, B, and C, were placed in three containers of water and shaken vigorously for 20 minutes. At 5-minute intervals, the contents of each container were strained through a sieve. The masses of the materials remaining in the sieve were measured and recorded as shown in the data table below.

MASSES OF MATERIAL REMAINING IN SIEVE

Shaking Time (min)	Rock Material A (gr)	Rock Material B (gr)	Rock Material C (gr)
0	25.0	25.0	25.0
5	24.5	20.0	17.5
10	24.0	18.5	12.5
15	23.5	17.0	7.5
20	23.5	12.5	5.0

28. Using the information in the data table, construct a line graph on the grid provided below, following the directions in parts (a) through (c) below.
    (a) Plot the data for rock sample *A* for the 20 minutes of the investigation. Surround each point with a small circle, and connect the points. [1]
    (b) Plot the data for rock sample *B* for the 20 minutes of the investigation. Surround each point with a small triangle, and connect the points. [1]
    (c) Plot the data for rock sample *C* for the 20 minutes of the investigation. Surround each point with a small square, and connect the points. [1]

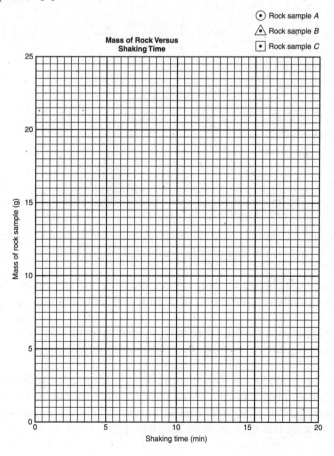

Weathering and Soil Formation

29. Using one or more complete sentences, state the most likely reason for differences in the weathering rates of the three rock materials. [2]

30. Using the directions in parts (a) through (c) below, calculate the average rate of change in the mass of rock material C for the 20 minutes of shaking.
    (a) Write the equation for the rate of change in mass. [1]
    (b) Substitute data into the equation. [1]
    (c) Calculate the rate of change in mass, and label your answer with proper units. [1]

31. Using one or more complete sentences, describe the most likely appearance of the corners and edges of rock material C at the end of the 20 minutes. [2]

Base your answers to questions 32 through 35 on the diagram below, which shows igneous rock that has undergone mainly physical weathering into sand and mainly chemical weathering into clay.

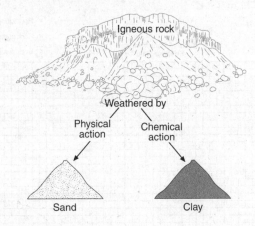

32. Compare the particle size of the physically weathered fragments to the particle size of the chemically weathered fragments. [1]

33. Describe the change in temperature and moisture conditions that would cause an increase in the rate of chemical weathering into clay. [1]

34. If the igneous rock is a layer of vesicular andesite, identify *three* types of mineral grains that could be found in the sand. [1]

35. A family wants to use rock materials as flooring in the entrance of their new house. They have narrowed their choice to granite or marble. Which of these rocks is more resistant to the physical wear of foot traffic and explain why this rock is more resistant. [2]

CHAPTER 20
EROSION

> **KEY IDEAS** Once weathering has broken rock down into smaller particles, the natural agents of erosion—generally driven by gravity—remove, transport, and deposit them. The natural agents of erosion include streams, ocean currents, and wave action (moving water), glaciers (moving ice), wind (moving air) and mass movements. Each of these agents of erosion produces distinctive changes in the material that it transports and creates characteristic surface features and landscapes. In certain erosional situations, destruction of property, personal injury, and loss of life can be reduced by effective emergency preparedness.

KEY OBJECTIVES
Upon completion of this chapter, you will be able to:

- Explain how natural agents of erosion, generally driven by gravity, remove, transport, and deposit weathered rock particles.
- Compare and contrast the distinctive changes each agent of erosion makes in the material it transports, as well as the characteristic surface features and landscapes each produces.
- Recognize the factors that affect erosion and relate them to landforms produced by erosion.
- Cite examples of the impacts of various types of erosion on human activities.

EROSION

What happens to the sediments produced by weathering? In most cases they are moved, often great distances from where they originated. Any process that moves sediments from one place to another on Earth's surface is called **erosion**. As you read these words, erosion is changing Earth in countless ways. Waves crashing against shores are scouring away sand, reshaping the coastline. In deserts, hot winds are moving towering dunes of sand grain by grain. On cold, barren mountaintops fragments of rock broken loose by weathering are tumbling to the ground. Glaciers creeping downhill are tearing huge boulders from the ground and carrying them along. Streams, from tiny trickles to raging torrents, are carving their way into the crust.

Erosion

Together, weathering and erosion wear away Earth's crust. Weathering breaks down solid rock; erosion carries away the pieces. In this way fresh rock is exposed to weathering, and the cycle repeats itself. Sometimes erosion is rapid and unmistakeable, as in a landslide. Usually, though, the process is gradual, and a long time passes before it is evident that erosion is taking place.

Evidence of Erosion

Any sediment moved from its source is evidence of erosion. At times, this evidence is striking. The owner of a house returning after a mudslide knows that the house was not built with one end hanging over the edge of a cliff. The lack of soil under the house is evidence of erosion. In much the same way, valleys and canyons are evidence of erosion. The Grand Canyon is one of the most spectacular examples of erosion in the United States. Imagine how much sediment had to be carried away to form it!

Quite often, though, the evidence of erosion is subtle. A boulder sitting on bedrock may not immediately seem to indicate erosion. If the boulder is granite and the bedrock is limestone, however, the boulder certainly wasn't derived from the bedrock; it had to be carried there from some other source. Similarly, layers of sediment can often be found overlying bedrock that has an entirely different composition. This, too, is evidence that erosion has occurred. Since the sediment did not come from the bedrock, it must have been carried there from some other source.

AGENTS OF EROSION

Erosion moves sediments. To move anything, a force is needed. The primary driving force of erosion is **gravity**. Gravity can move sediments by acting on them directly. On cliffs and steep slopes, sediments broken loose by weathering move downhill under the direct influence of gravity. Gravity can also move sediments by acting on them indirectly, through **agents of erosion**. For example, water runs downhill under the direct influence of gravity. The running water, in turn, can exert a force on sediments in its path, causing them to move. Thus the running water is an agent of erosion. Some other agents of erosion are waves, currents, winds, and glaciers.

An agent of erosion, together with the driving force (usually gravity) that sets in motion the agent that picks up and transports sediment, comprises an **erosional system**. Erosional systems are like natural conveyor belts, moving sediments from higher to lower elevations.

Mass Wasting

Mass wasting is the downhill movement of sediments under the direct influence of gravity. On most slopes, some kind of downhill movement of sediments is proceeding all the time because, on a slope, gravity acts as if it has two parts, or components. One part, which we will call the normal force (F_N), pulls downward perpendicular to the surface. The other part, which we will call the downhill force (F_D), pulls downhill parallel to the surface. The downhill force gives rise to a frictional force (F_F) that opposes the motion of the object. The frictional force depends on two factors: the normal force holding the two surfaces together and the nature of the two surfaces. The more tightly the two surfaces are pressed together by the normal force, the greater the frictional force between the two. (This explains why an empty box is easier to push across a floor than a similar box filled with something heavy.) As long as the frictional force is greater than the downhill force, the object will remain stationary.

As the slope gets steeper, more of the gravity acts in a downhill direction and less in a direction perpendicular to the surface (i.e., the downhill force increases and the normal force decreases). When the downhill force is greater than the frictional force, the object moves downhill. See Figure 20.1.

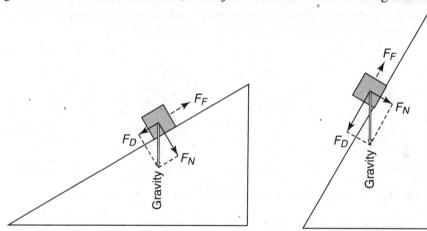

Figure 20.1 Forces Acting on an Object Resting on a Slope. In the diagram, F_N = normal force, F_D = downhill force, and F_F = frictional force. As a slope steepens, the downhill force increases while the normal force and frictional force decrease. When the downhill force exceeds the frictional force, the object moves downhill. Source: *Environmental Geology*, Carla W. Montgomery, McGraw-Hill, 1997.

The steepest slope angle at which a particular sediment remains stable is called its **angle of repose** (see Figure 20.2). The size, shape, and density of a rock particle affect its angle of repose. Thus, sand, gravel, and clay have different angles of repose. When their angles of repose are exceeded, sediments move downhill and mass wasting occurs. Any perceptible downslope move-

Erosion

ment of rock, soil, or a mixture of the two is commonly referred to as a **landslide**. This is a very general term, however, and may involve many different types of movement of many different types of material. Depending on the slope and the angles of repose of the sediments, mass-wasting processes may be rapid or slow.

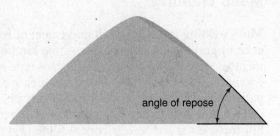

Figure 20.2 Angles of Repose in a Pile of Dry Sand.

Rapid Mass Wasting

Rockfalls (see Figure 20.3a) occur when rock fragments broken loose by weathering fall from cliffs or bounce by leaps down steep slopes. Rockfalls are the most rapid of all mass-wasting processes. "Fallen Rock Zone" is a common sign near roadcuts on many highways or in mountainous areas where rockfalls occur frequently. In populated areas or on heavily traveled roads, rockfalls can be very dangerous and have resulted in loss of life. Some localities have spent much money cutting back the rocky slopes of roadcuts or covering them with steel cable nets to prevent rockfalls.

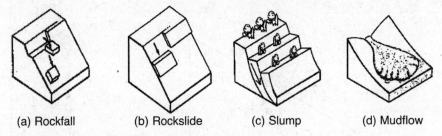

(a) Rockfall   (b) Rockslide   (c) Slump   (d) Mudflow

Figure 20.3 Four Rapid Mass-Wasting Processes.

Rockslides (see Figure 20.3b) occur on less steep slopes when rock masses or debris slide downhill. Most rockslides are triggered by heavy rains or earthquakes. Water acts as a lubricant between particles. The shock of an earthquake knocks particles apart, decreasing the friction between them and the underlying surface. The downhill component of gravity is then greater than the friction holding the particles in place, and the particles move downhill. Rockslides can move enormous amounts of material, often millions of cubic meters. In 1959, twenty-seven people were killed when an earthquake caused an entire mountainside to slide into the Madison River gorge in Montana.

Slump (see Figure 20.3c) is a mass-wasting process in which a huge mass of bedrock or soil slides downward from a cliff in one piece. The mass, or slump block, slides along a curved plane of weakness as shown in the diagram. The slump block rotates and comes to rest with its upper surface tilting toward the cliff. Slump is common where ocean waves or streams undercut cliffs.

A **mudflow** (see Figure 20.3d) is the rapid, downhill flow of a fluid mixture of rock, soil, and water. Mudflows generally occur after heavy rains saturate soil that has no protective covering of vegetation, and usually occur in semiarid regions or on slopes denuded by construction or lumbering. The rain mixes with the soil to form a thin mud, which flows downhill, thickening as more soil and debris are picked up and increasing in speed. Since the mudflow is much heavier and thicker than water, the impact when it hits something in its path is devastating. Entire houses, cars, and trees have been swept away by mudflows. A mudflow comes to rest when it reaches the bottom of the slope or when it has picked up enough material so that it thickens to a point where it can no longer move.

Slow Mass Wasting

In humid regions, slopes are usually covered with vegetation. The vegetation protects the surface from the impact of raindrops, and its root system holds the soil together, inhibiting downhill movement. However, mass wasting occurs even on slopes covered with vegetation, albeit slowly.

On vegetated slopes, rainwater entering the soil loosens the particles and makes them slippery, decreasing their angles of repose. In this situation, an earthflow may occur. In an **earthflow** (see Figure 20.4), a shallow layer of soil and vegetation, saturated with water, slowly slides downhill. An earthflow may take several hours to ooze its way down a slope. Earthflows are a common cause of road and rail blockages. While they are usually not life threatening since they move at a snail's pace, they can cause considerable property damage.

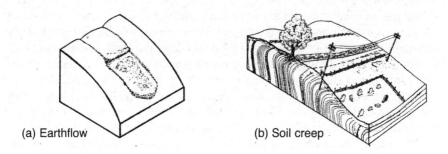

(a) Earthflow   (b) Soil creep

Figure 20.4 Two Slow Mass-Wasting Processes.

The slowest of all mass-wasting processes is soil creep. **Soil creep** (see Figure 20.4b) is the invisibly slow, downhill movement of soil, carrying vegetation with it. The tilting of old poles, fenceposts, or tombstones and bulging or broken retaining walls are all evidence of creep.

Any disturbance of the soil on a slope causes creep. Freezing and thawing, wetting and drying, and trampling or burrowing by animals are just a

Erosion

few of the things that can disturb the soil on a slope. When disturbed, the soil particles shift and settle. As they do so, gravity moves them downhill. Creep is common in regions where the soil alternately freezes and thaws frequently.

Large objects on or in a creeping slope are carried along by the moving soil. Boulders that have crept down a hillside may accumulate at the base of the slope, forming a boulder field. Trees growing on creeping slopes are often misshapen. The tree grows straight up, but the soil that its roots are in slowly creeps downhill.

Mass wasting is usually only the first step in the eroding of Earth. It delivers sediments to the base of slopes. There, agents of erosion, such as streams, can pick up the sediments and carry them farther.

Water in Motion: Runoff and Streams

Raindrops and Runoff

Raindrops may fall for thousands of feet before they hit the ground, and the force of falling raindrops can move sediments. When raindrops strike the ground, their impact causes a geyserlike splashing of loose soil. In a process called **splash erosion**, each raindrop impact lifts some soil and drops it in a new position. On one field observed during a rainstorm, splash erosion moved 91 metric tons of soil per acre.

Runoff is precipitation that does not evaporate or sink into the ground. Under the influence of gravity, runoff flows downhill. On even the gentlest slopes, runoff may flow downhill in a thin sheet called **overland flow**. Overland flow exerts a dragging force on the ground. Depending on its speed, it can exert a force great enough to move sediments ranging from fine clay to coarse sand or gravel. Erosion by overland flow is called **sheet erosion**.

Vegetation greatly decreases both splash and sheet erosion. The leaves and stems of plants absorb the impact of falling raindrops, and the network of plant roots anchors the soil in place. Plant stems also block and slow the downhill movement of water, thereby decreasing the force it exerts on sediments in its path. In arid regions or on slopes denuded of vegetation, however, splash and sheet erosion can remove vast quantities of loose soil.

Splash erosion and sheet erosion are serious problems on farms. They carry away the rich topsoil essential for healthy plant growth. Contour plowing, terracing, strip cropping, and crop rotation help lessen soil erosion.

On steep, unprotected slopes, erosion by runoff can become intense. Numerous tiny grooves, or rills, form as the water runs downhill, carrying away sediment. If erosion continues, the rills may widen and deepen into a gully. Gullies act as funnels for runoff. They concentrate the force of the flowing water, thus increasing the rate of erosion.

Streams

A **stream** is any body of flowing water that moves downhill under gravity, in a relatively narrow but clearly defined channel, at least part of the year. A stream may be small enough to step across or so wide that the other side is barely visible. Streams are the most important agent of erosion because they affect more of Earth's surface than any other agent. Most of the sediment carried downhill by runoff ends up in streams.

Stream Basics

The long trough-like depression that is normally occupied by the water in a stream is called the **stream channel**. The bottom of the channel is called the **stream bed**, and each side is called the **stream bank**. The water carried in a stream channel, or **streamflow**, comes mainly from two sources: runoff and ground-water seepage into the stream, known as **base flow**. A stream's rate of flow over a particular period of time is called stream **discharge** and is usually expressed in cubic meters per second. Stream discharge depends on the volume and velocity of the flow. Where the moving water in a stream comes in contact with the air or the ground, friction causes the water to slow down. As a result of this frictional drag, stream velocity is at a maximum at the center of the channel near the surface and a minimum near the bed and banks. Finally, flow is not always contained within the stream channel. During periods of high stream discharge a stream may overflow its banks, or **flood**.

A stream begins at its **source**, usually in mountains or hills where rain water or snowmelt collects and drains into low-lying areas between slopes, forming tiny streams. As these tiny streams flow downhill along natural passageways or depressions in the surface to lower and lower levels, they cut into the surface, forming a narrow trench, or gully. Once formed, the gully provides a pathway for all later runoff. Gullies either grow larger when they collect more water and become stream channels themselves or meet streams and add to the water already in the stream. When two streams meet, they merge, and the smaller stream is known as a **tributary**. A stream grows larger as it collects water from more tributaries. The water in most streams ends up in a lake, sea, or ocean. The point at which the stream enters these bodies of water is called the **mouth** of the stream. The change in elevation from a stream's source to its mouth, expressed in degrees or as a distance ratio $\left(\frac{\text{change in elevation}}{\text{distance}}\right)$, is called the stream **gradient**. Figure 20.5 shows a typical stream and its parts.

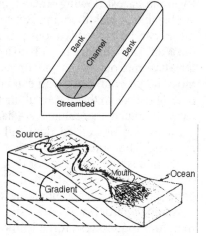

Figure 20.5 A Stream and Its Parts.

Stream Transport

Water flowing downhill can be very powerful. Have you ever seen whitewater rafters shooting a rapids? The water flowing downhill exerts quite a force! Imagine trying to paddle upstream through the rapids. It would be

Erosion

almost impossible because the force of the flowing water is so great. Now imagine a grain of sand or a pebble in the path of such a stream. Any loose sediments are likely to be carried away.

Material carried by a stream is called its **load**. A stream's load consists mainly of sediments, which can vary in amount and type, depending on the speed and volume of the stream. The steeper the slope, or gradient, of a stream's channel, the faster the water flows. Water can carry more and larger sediments as its speed increases. The graph in Figure 20.6(a) shows the sediment water carries at different speeds. Notice that as water velocity increases, the size of the particles it can carry increases. Of course, if water is traveling fast enough to carry pebbles, it is also traveling fast enough to carry any particles smaller than pebbles, too. Thus, the faster the streamflow, the more types of sediment the water can carry and the greater the stream's load. At any speed, the more water there is in a stream, the more load it can carry.

Streams transport different sediments in different ways. See Figure 20.6(b). Large sediments, such as pebbles, cobbles, and boulders, are heavy. Even the fastest flowing streams may not be able to pick them up. But sediments can be moved without picking them up! Do you have to pick up a car that has run out of gas in order to move it? Of course not; you can push or drag the car so that it rolls along the road. In much the same way, large particles can be carried downhill by streams. The force of the flowing water in a stream can push or drag large particles downhill by rolling or sliding in a motion called **traction**. If the sediment particles move in a series of bounces, hops, or leaps along the streambed, the motion is called **saltation**.

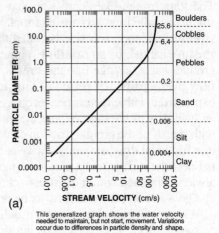

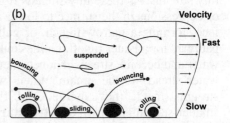

Figure 20.6 (a) Relationship of Transported Particle Size to Water Velocity. Source: The State Education Department, *Earth Science Reference Tables*, 2011 ed. (Albany, New York; The University of the State of New York). (b) Methods of Particle Transport.

Smaller particles, such as silt, sand, and clay, are lighter in weight and therefore require less force to move. Although they are heavier than water, the flowing water can pick up these small sediments; and, whenever they begin to sink, the force of the turbulent water pushes them back up again—they are carried downhill **suspended** in the water.

Some sediments dissolve in water. Water that has combined with carbon dioxide or other gases in the atmosphere to become acidic (see Chemical

Weathering, page 434) is especially effective at dissolving sediments. Sediments that are dissolved are carried downhill **in solution**.

Stream Erosion

Streams not only transport sediments but also change the sediments they carry, wear away the surfaces over which they flow, and create new sediments. Pure water flowing over rock wears it away very little, but water carrying sediment acts like a cutting tool. When the sediments being carried hit rock in the stream channel, chips of rock break off. Rolling and sliding pebbles, cobbles, and boulders collide and knock chips off each other. They crush and grind up smaller particles between them. The process of crushing, grinding, and wearing away rock by the impact of sediments is called **abrasion**. Abrasion wears away a stream's channel and causes the sediments carried by the stream to become rounded.

By abrasion, streams can cut through bedrock over which they flow. The potholes shown in Figure 20.7 were carved out of bedrock by sediments in a swiftly flowing stream. The waterfall formed where a stream flowed over hard bedrock onto soft bedrock. The soft bedrock wore away faster, lowering the streambed, and a waterfall was formed. Prolonged erosion can carve deep canyons and valleys into Earth's surface. The Grand Canyon is an amazing example of what sediment-laden water can do.

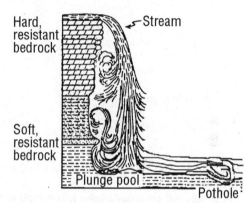

Figure 20.7 Waterfall and Pothole. The falling water creates turbulence that swirls sediments against the streambed, forming potholes.

A Stream's Life Cycle

As a stream cuts into the surface and carries away sediment, its channel becomes wider, deeper, and longer. Over time the shape and behavior of the stream change as the surface over which it flows is changed.

Youth. Newly formed streams generally flow down steep slopes. The water flows quickly, often forming rapids. See Figure 20.8a. Over time, the stream wears a long, deep groove, or valley, in the surface. Stream-cut valleys typically have steep sides that form a distinctive **V-shape**. A young stream valley is shown in Figure 20.8b.

Maturity. As a stream erodes the slope over which it flows, the slope becomes less steep. As a result, the water flows more slowly and sediment accumulates at the bottom of the valley, making it flat. As mass wasting and runoff wear away the sides of the valley, the V-shape becomes wider and less

Erosion

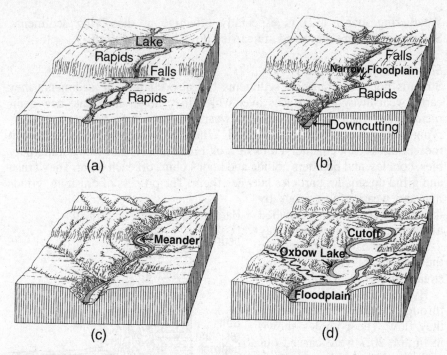

Figure 20.8 The Three Stages in the Life Cycle of a Stream. (a) The relief of the land is steep, and the stream has lakes, rapids, and falls. (b) Continued downcutting of the streambed eliminates falls, and cutting back of the valley walls allows a narrow floodplain to form. (c) Further downcutting and erosion of the valley walls enlarges the floodplain and allows the stream to meander widely between the valley walls. (d) Shifting meanders result in cutoffs and oxbow lakes. Source: *The Earth Sciences*, Arthur N. Strahler, Harper & Row, 1971.

steep. The slower moving water flows around obstacles instead of tumbling over them, and the stream's path forms curving loops called **meanders**. Meanders reflect the way in which a stream minimizes resistance to flow and spreads energy as evenly as possible along its course. Velocity is lowest along the bed and banks of streams because it is there that water encounters the most friction, and therefore the flow is reduced. Along a straight channel segment, water moves the fastest in mid-channel, near the surface. But as water moves around a bend, the zone of high velocity swings to the outside of the channel because a centrifugal force operates in the direction of the outside of the curve. As water rushes past the outer part of the bend, sediment is continuously eroded from the riverbank and is swept downstream. Along the inside of the bend, the slower-moving water deposits sediments. With continuous erosion at the outer banks and deposition at the inner banks, the bending of the meander becomes more pronounced. When runoff increases suddenly, as after a torrential downpour or sudden spring thaw, the stream overflows. Spreading out over the valley floor, the stream slows down and deposits sediments to form a broad, flat **floodplain** adjacent to the stream. Figure 20.8c shows a typical stream at the mature stage.

Old Age. Eventually, the slopes around a stream are worn away almost completely. A flat lowland of gently rolling hills, called a **peneplain**, is formed. Since the slope of the land is very gentle, the water in the stream flows slowly. Its meanders become highly curved and winding and at times may actually loop over one another. During floods, water may gush over the land between meanders, forming a **cutoff**. If the new channel cuts deep enough, part of the meander may become isolated, forming an **oxbow lake**. Figure 20.8d shows a typical stream at the old stage.

Whatever the stage of development, the water in most streams eventually reaches the oceans.

Water in Motion: the Oceans

The ocean never rests. Anyone who has swum in, sailed on, or just sat on the beach beside the ocean knows that it is constantly moving. Waves, currents, and tides all move ocean waters; and, once in motion, ocean water can transport sediments and erode the land.

Ocean Currents

Ocean currents are mainly the result of two interactions with the atmosphere: heating of the ocean and wind. The ocean is able to store more heat than the atmosphere or the land because of water's high specific heat. Differences in the intensity of solar radiation received at the Equator and near the poles results in an uneven distribution of heat in the oceans. The oceans contain more heat near the Equator than near the poles.

Convection Currents
The heat in the oceans near the Equator is transferred toward the poles by convection, or the movement of seawater due to density differences. Dense, cold water formed near the poles sinks and slowly flows toward the Equator. This cold water is oxygen-rich; without it, bottom waters would quickly become oxygen depleted by the oxidation of organic matter that sinks through it. As the cold water moves toward the Equator, it displaces warmer water upward, causing **upwelling**. Upwelling brings nutrients and oxygen to the surface, thereby supporting a rich growth of plants and animals.

Wind-Driven Currents
Uneven heating of Earth's surface and the air above it creates similar circulations in the atmosphere. As warm air at the Equator rises, air from adjacent areas gets sucked in to replace it, creating wind. Winds do not move straight toward the Equator, but curve because of the Coriolis effect. Thus, the winds near the Equator, called the **trade winds**, approach the Equator at about a 45° angle. Over the ocean, with little to block their path, the trade winds are among the steadiest winds on Earth. At middle latitudes, the more variable

prevailing westerlies move in the opposite direction. At high latitudes are the most variable winds, the **polar easterlies**. See Figure 20-9.

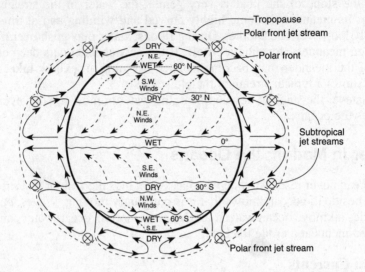

Figure 20.9 Planetary Wind and Moisture Belts in the Troposphere. The drawing shows the locations of the belts near the time of an equinox. The locations shift somewhat with the changing latitude of the Sun's vertical ray. In the Northern Hemisphere, the belts shift northward in summer and southward in winter. Source: The State Education Department, *Earth Science Reference Tables*, 2011 ed. (Albany, New York; The University of the State of New York).

Winds exert a force on the ocean surface and produce wind-driven currents. All of the major surface ocean currents are driven by the wind and follow roughly the pattern of surface winds. When pushed by wind, however, the surface water does not move in the same direction as the wind but, because of the Coriolis effect, curves off at about a 45° angle. As a result, the surface currents driven by the trade winds do not approach the Equator at the same 45° angle as the winds that are pushing them, but are themselves bent 45° from the wind direction. Thus, throughout Earth, the trade winds form westward-moving currents, known as **equatorial currents**, that are parallel to the Equator.

In the Atlantic and Pacific oceans, these westward-moving equatorial currents are obstructed by landmasses and are deflected north and south. These deflected currents, which run north and south along the western boundary of the oceans, are among the largest and strongest ocean currents. The Gulf Stream, a western boundary current in the Atlantic Ocean, runs north along the eastern coast of the United States and transports more than 100 times the output of all the rivers in the world.

When the western boundary currents reach the midlatitudes, the prevailing winds, which blow from the west, push them back east across the ocean. The result is a large, circular pattern of motion called a **gyre**. The North and South Atlantic and Pacific, as well as the Indian Ocean, all have large-scale gyres. In addition, the northern polar regions of the Atlantic and Pacific have smaller gyres. See Figure 20.10.

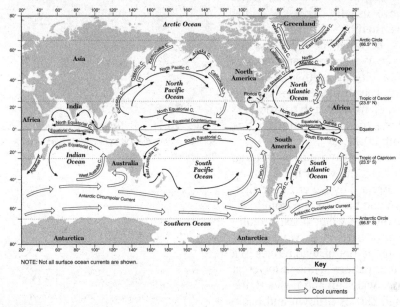

Figure 20.10 Surface Ocean Currents. Source: The State Education Department, *Earth Science Reference Tables*, 2011 ed. (Albany, New York; The University of the State of New York).

One unfortunate effect of the large-scale gyres is that they carry pollutants from the continental coastlines far out into the oceans. For example, many of the largest cities in the United States are located along the east coast. The sewage, garbage, and industrial waste from these cities have long been dumped into the ocean. The Gulf Stream then carries these pollutants far out into the mid-Atlantic. These practices may have had little impact when populations were small, but today they represent serious threats to the health of ocean life and the future well-being of our oceans.

Waves

Wind-Generated Waves

Wind not only drives surface currents but also creates waves. Wind forms waves on the surface of the ocean by transferring energy from the moving air to the water. The transfer of energy is not completely understood, but appears to involve pressure differences caused by the deflection of wind over the wave (much as air flowing over a wing provides lift). It also seems to be related to a resonating movement of the surface caused by wind pushing against the surface in turbulent conditions (like a hand pushing down repeatedly on the surface of the water). The result in either case is a transverse wave that moves through the surface water of the ocean.

The highest part of a wave is called the **crest**; the lowest part, the **trough**. The size of an ocean wave is usually expressed as the **wave height**—the vertical distance between the crest and the trough. Wave crests and troughs can be closer together or farther apart. The distance from crest to crest, or from trough to trough is called the **wavelength**. See Figure 20.11.

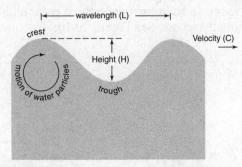

Figure 20.11 A Typical Transverse Wave. Source: *An Introduction to Oceanography*, David A. Ross, Prentice-Hall, 1970.

When a wave moves through the ocean, the water is carried up and forward by the crest and then down and back by the trough. As a result, the water doesn't really get anywhere as a wave passes through it, but instead moves in a circular pattern. Though the crest carries water up and forward, the trough carries it down and back, so the water ends up right where it started.

Near the surface, the movement of water particles is almost

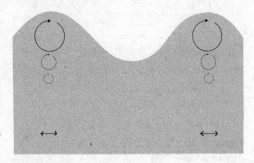

Figure 20.12 Movement of a Wave Through Seawater.

perfectly circular, but the diameter of the circular motion decreases quickly with depth. This decline is due to a loss of energy during transfer to deeper water. At a depth of one-fourth the wavelength, the circle is only one-fifth of its diameter near the surface. At greater depths the movement is little more than a back and forth rocking motion. Below 100 meters, little if any motion is felt from most wind-generated waves.

The height and the period of wind-generated waves are affected by three factors: the speed of the wind, the length of time it blows, and the *fetch*, or distance of water over which it blows. The higher the wind speed and the longer the wind blows, the higher the wave height. Fetch is important in determining wavelength. The longer the fetch, the longer the wavelength.

Wind-generated waves can be divided into three categories: seas, swells, and surf.

Seas are waves directly affected by the wind. While the wind is blowing, it pushes the wave crests up into sharp peaks and stretches out the troughs. The surface of the ocean is often a confused jumble of seas of varying heights moving in different directions, rather than a set of regular waves moving in one direction. This occurs because the ocean's surface at any one point is affected by a mixture of waves coming in from many different places, generated by winds blowing in different directions, at different speeds, and for different lengths of time. When the ocean is choppy, we are seeing a mixture of seas from different sources.

Swells are waves with smoothly rounded crests and troughs. Swells form as seas leave the area of turbulent winds that formed them because waves with a long wavelength travel faster than those with a short wavelength (velocity = wavelength/period). Thus, over time, the long waves outdistance the short ones and the waves sort out into regions of waves of similar wavelengths. The waves in these uniform patterns are called *swells*. As waves travel farther from their source, their height decreases but their length remains constant. Waves showing such patterns can travel across entire oceans.

Surf occurs near shore when a wave breaks. The difference between surf, or breaking waves, and seas or swells is that the water in breaking waves is not moving in a circular pattern but is actually moving toward the shore. As a result, a lot of energy is directed at the beach.

Tsunamis

Tsunamis are waves caused by undersea movements of the crust due to earthquakes, slumping, or volcanic eruptions. The sudden up-and-down movement of the bottom creates a wave that has a long wavelength and travels at high speed. In deep water, tsunamis may have wavelengths up to 700 kilometers and travel at speeds over 350 kilometers per hour, yet may be only several centimeters high. When tsunamis reach a coastline, though, they form huge waves that break against the coast and cause great destruction.

Coastal Processes

Breaking Waves

Breaking waves, or surf, forms as follows. When waves enter shallow water, the bottom of the wave comes into contact with the ground at a depth equal to one-half the wavelength. This contact slows the bottom of the wave and forces the wave upward, while the top continues at its original speed. When the top of the wave outpaces the rest of the wave, it is no longer supported and falls over. See Figure 20.13.

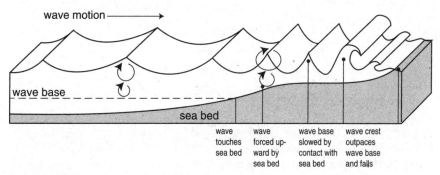

Figure 20.13 Mechanics of a Breaking Wave. Source: *Earth Science on File.* © Facts on File, Inc., 1988.

Erosion

Wave Refraction

Waves that enter shallow water at an angle to the beach are refracted; their direction of travel is changed. This refraction occurs because one part of the wave reaches shallow water and slows down while the rest of the wave is still in deep water and moving faster. Like a rolling log whose one end hits a tree, causing the whole log to swing around, the faster moving end of the wave swings around when the end in shallow water slows down.

On irregularly shaped coastlines, waves converge on headlands and diverge at depressions in the shoreline, as shown in Figure 20.14. The results are erosion of the headland by the force of the waves directed against it and the deposition of sediments in the calmer zone of divergence in inlets. Over time, the irregularly shaped shoreline is straightened out.

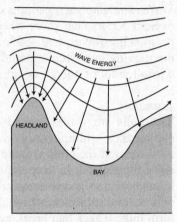

Figure 20.14 Wave Refraction at Headlands and Depressions.

Longshore Currents

When waves break, the water is carried into the surf zone and travels toward the beach. The water that washes up against the beach runs back along the incline under the surf zone. When waves approach the beach at an angle, the water moves upward at an angle and falls straight back. The result is a movement of water parallel to the beach called a **longshore current**. See Figure 20.15.

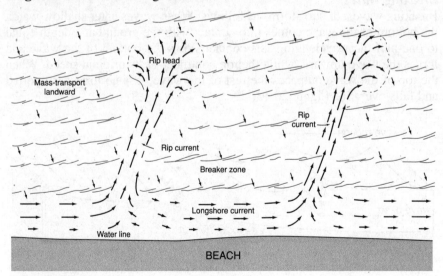

Figure 20.15 Formation of a Longshore Current. Source: *The Earth Sciences*, Arthur N. Strahler, Harper & Row, 1971.

The longshore current and the up-and-back motion of water washing along the beach move sand along the shore. The net result is to transport sediment along the beach and erode it inland. In populated areas, people try to impede this erosion by building structures in the surf zone to block the longshore current and trap beach sediments from being washed away. See Figure 20.16. These attempts are only moderately successful in the long run. Also, the structures often cause increased erosion down-current as the longshore current carries away sediment but brings none in from up-current.

Shoreline Erosion

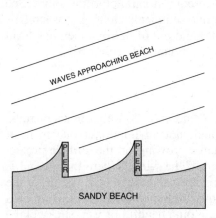

Figure 20.16 Jetties or Groins Built to Trap Sediment.

Ocean water is constantly in motion. Waves, currents, and tides are examples of the movement of ocean water. Moving ocean water can transport sediments and erode the land.

Waves

Waves can be very powerful. A single wave can send tons of water crashing against a shore. The force of waves can break up rocks on the shore. Loose fragments are stirred up and carried along in the turbulence of breaking waves.

Wave erosion produces a number of shoreline features, some of which are shown in Figure 20.17. When waves erode steep, rocky shores, they remove materials from the base of the slopes. The result is a **wave-cut notch** that undercuts the slope. Overhanging rocks and soil break away, leaving a steep-sided **sea cliff**. When wave erosion penetrates deeply into a cliff, a large hole or **sea**

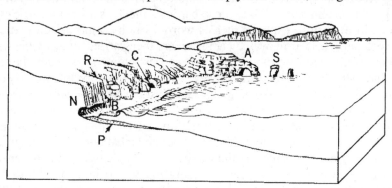

Figure 20.17 Shoreline Features Formed by Wave Erosion of a Rocky Coast. P—platform of abraded sediments, N—wave-cut notch at the base of a cliff, R—crevice eroded back into the rock, B—beach of sediments eroded out of the cliff, C—sea cave, A—arch, S—sea stack. Source: *The Earth Sciences*, Arthur N. Strahler, Harper & Row, 1971.

467

cave may form. As weaker rock is worn away from the cliff, pillars of resistant rock, called **stacks**, may be left behind. As the erosion continues, fragments broken from the cliff are ground up by abrasion in the turbulent waves. This buildup of material forms a flat platform, or **sea terrace**, at the base of the cliff.

Currents

Currents are movements of water. On gently sloping, sandy shores waves mainly shift sediment around. When waves strike perpendicularly to the shore, sediments are moved in and out with the waves. Waves that strike the shore at angles other than 90°, however, are reflected at an angle and interfere with each other, forming a longshore current parallel to the shore. See Figure 20.15. Longshore currents can move sediments along just as streams do. Along some coasts, tides flowing into and out of narrow openings form swift currents. These, too, can move large amounts of sediments.

Wind Erosion

Wind that blows fast enough can pick up and carry loose particles of rock. Wind is also an agent of erosion, though not an important one in most regions. Where rainfall is abundant, plants growing on the surface protect underlying sediments from the full force of the wind. Where the ground is moist, sediment particles tend to stick together and are more difficult for the wind to dislodge and move. (Think of the difference between blowing on a handful of dry, powdery clay and blowing on a handful of wet, sticky clay.) In arid regions, however, there is little plant cover, and the soil, loose sediments, and weathered bedrock are dry and exposed. Under such conditions, wind becomes an important agent of erosion.

Deflation and Abrasion

Wind erodes the land in two ways: deflation and abrasion. In **deflation**, the wind picks up and carries away loose particles much as a stream carries sediments. In **abrasion**, exposed rock is worn away by wind-driven particles. The rock particles worn away by abrasion are then carried away by the wind.

Wind is like a stream of air, and wind carries sediments in much the same way that a stream of water does. However, winds are usually able to carry only small particles, such as sand, silt, clay, and dust, because the winds' medium, air, is not very dense. It exerts less force on a particle than does a denser medium, such as water, moving at the same speed. Imagine the difference between being hit by a balloon filled with air and a balloon filled with water!

Wind carries very small particles, such as clay and dust, in suspension. The finest particles may be carried many meters up into the atmosphere. When volcanoes erupt, dust and ash are often spewed high into the atmosphere, where winds carry them for hundreds or even thousands of miles before they finally settle to the ground. Some particles of dust and ash may reach the jet streams of the stratosphere and remain suspended for years after they are picked up by these fast-moving winds.

Wind carries larger particles closer to the ground. They slide, roll, and bounce, skipping along the surface. As they do so, they collide with exposed bedrock or other particles. In the process of abrasion, these collisions scratch the surface of the exposed rock and cause tiny chips to break off. Like a sandblaster, the wind-borne particles cut and polish the rock surfaces they impact. Over time, rock exposed to wind-driven particles is worn away. The surfaces of both the exposed rock and the particles take on a frosted appearance because of the many tiny scratches inflicted on them.

The amount and the type of erosion caused by wind depend on three factors: the speed of the wind, the size of the particles it carries, and the length of time it blows. The faster the wind blows, the larger the particles it can transport. The larger the particles carried by the wind, the greater the abrasion of exposed surfaces. The longer the wind blows, the more deflation and abrasion that can occur.

Features Produced by Wind Erosion

As deflation removes layer after layer of loose surface materials, a shallow depression, known as a **deflation hollow**, is formed. The semiarid Great Plains of the midwestern United States are dotted with tens of thousands of deflation hollows. In wet years they are covered by grass; and when rainfall is unusually high, they may fill with runoff, forming shallow lakes. In dry years, however, the grass dies off and the wind continues to deflate the bare soil.

When deflation lowers the land surface to the level of the water table, erosion is stopped. The exposed groundwater holds the soil particles together and enables plants to grow. The plants' roots help hold the loose soil in place, and their leaves and stems block the wind, ending any further erosion. A patch of vegetation surrounded by arid wasteland is called an **oasis**.

Deflation usually removes only fine particles. If the soil is composed of a mixture of particle sizes, the larger particles, which are too heavy for the wind to move, are left behind. These larger particles shield the finer particles beneath them from deflation. As exposed fine particles are removed, coarse particles form an increasing percentage of the particles left at the surface. Eventually, a continuous layer of pebbles, gravel, and other coarse particles, called a **desert pavement**, may form, as shown in Figure 20.18.

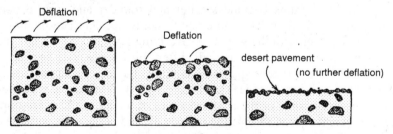

Figure 20.18 Formation of a Desert Pavement. As deflation removes finer sediments, larger sediments that are left behind eventually form a continuous layer called a desert pavement, which blocks further deflation. Source: *Physical Geology*, 2nd Ed., Richard Foster Flint and Brian J. Skinner, John Wiley, 1977.

Erosion

Abrasion by wind-blown particles produces spectacular features in arid regions. Caves, arches, and bridges may be cut into exposed bedrock. Unusually shaped rock formations may result when rocks of different hardnesses are exposed to abrasion simultaneously.

Loose rock particles, too heavy to be transported by the wind, can also be abraded by wind-blown particles. Over time, they develop flat surfaces on the side facing the prevailing winds (see Figure 20.19). These unusual particles are called **ventifacts**, from the Latin word *ventus*, meaning "wind."

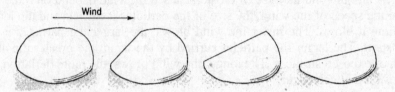

Figure 20.19 Ventifact Formation. Abrasion of the exposed surface of a rock fragment by wind-born particles forms a flat-faced ventifact. Source: Physical Geology, 2nd Ed., by Richard Foster Flint and Brian J. Skinner, John Wiley, 1977.

Glaciers

A **glacier** is a large mass of ice, formed on land by the compaction and recrystallization of snow, that is moving downhill or outward under the force of gravity. Modern glaciers are found only at high elevations and in polar regions where snowfall exceeds melting over an extended period of time.

Formation of Glaciers

For a glacier to form, the average amount of incoming snowfall must be greater than the average amount lost by melting or evaporation year after year.

When incoming snowfall exceeds snow loss, the net amount of snow remaining on the ground increases each year. Freshly fallen snow is a fluffy mass of delicate ice crystals that consists mostly of pockets of trapped air and therefore has a low density. The situation quickly changes, however, as snow piles up on the ground and sits there for weeks or months. Because of evaporation, the ice crystals become smaller and rounder and form a denser, **granular snow**. As more and more snow builds up, alternate melting and refreezing, together with the weight of overlying layers, compact the granular snow into an even denser mass called **firn**. Firn is usually a year or more old and has little or no pore spaces. Eventually, the ice crystals in firn grow together into a solid mass of ice (see Figure 20.20). When snow is converted to ice in this way, some of the dust and gases of the atmosphere are trapped in the ice. Scientists can sample the trapped air and compare it with the current atmosphere to determine what, if any, changes have occurred.

When ice has accumulated to a thickness such that its own weight deforms the ice mass and it moves slowly under pressure, a glacier has formed. Usually this stage is reached when the ice is 100 meters or more thick.

Movement of Ice

Under pressure, ice behaves like a very thick fluid. The ice *flows* downhill over Earth's surface, like thick syrup poured on a tilted plate. If there is just a little syrup, it spreads out into a thin layer and moves very slowly. The rate of flow of syrup can be increased in two ways: by adding more syrup, so that the layer gets thicker and flows more rapidly, and by increasing the tilt of the plate. In much the same way, glaciers flow faster where the glacial ice is thickest and where the slope is steepest.

Variations in the rate of flow of ice cause a glacier to thicken and also to thin out. Where the slope is steep, the ice thins out because it is flowing downslope faster than it is being replenished. The rapid downhill flow exerts a tension on the ice that causes cracks called **crevasses** to form. On extreme slopes the ice may break apart into blocks that cascade downslope in a rock fall. On gentler slopes the ice slows down. The ice from upstream pushes against the slower moving ice, and the compression causes the glacier to bulge and thicken.

These variations in the rate of flow of ice are illustrated in Figure 20.21.

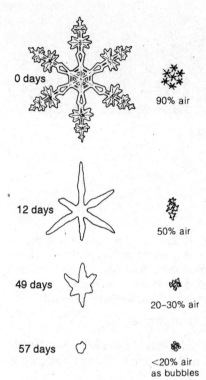

Figure 20.20 Transformation of Snow Crystals into Glacial Ice. Because of greater surface area, a snowflake's intricate edges melt and evaporate faster than its center. Over time, the snowflake becomes a rounded, compact pellet of ice. Sources: *Earth*, 4th Ed., Frank Press and Raymond Seiver, W.H. Freeman, 1986.

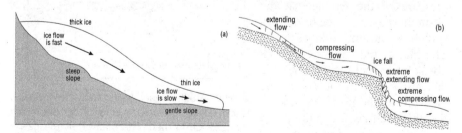

Figure 20.21 Glacial Flow Rates. (a) Ice flows fastest on steep slopes and in places where the ice is thickest. (b) On steep slopes, the ice thins out and may even split, forming crevasses, because it is flowing downhill faster than it is being replenished from farther upstream. On gentle slopes, the ice thickens as ice from upstream piles up behind slower moving ice. Source: *Living Ice: Understanding Glaciers and Glaciation*, Robert P. Sharp, Cambridge University Press, 1988.

Erosion

Because the ice in a glacier is flowing, the glacier moves in a unique way. Glaciers do *not* move over Earth's surface like a brick sliding down a board. A brick slides down an incline along its base as a single unit; nothing moves around *inside* the brick. In ice, however, both sliding, along the base of the glacier, and flowing, throughout the body of the ice, occur. See Figure 20.22 for a diagrammatic comparison of the two types of motion.

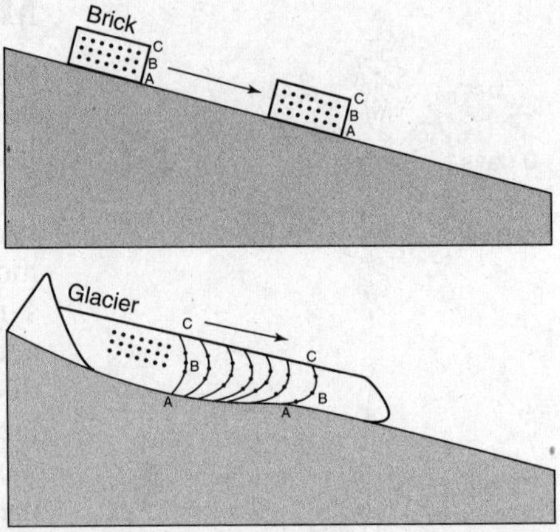

The amount of a glacier's movement that is due to the base sliding along the ground, as compared to the amount caused by internal flow, depends on the temperature of the glacier. Temperature affects the behavior of ice in two key ways:

Figure 20.22 A Brick and a Glacier in Motion. A glacier both slides along its base and flows like a fluid. The relative positions of particles inside the body of the ice change as the glacier flows downhill.

- Ice that is near its melting point is less rigid and is therefore more easily deformed than ice that is well below its melting point. For example, ice at 0°C deforms 10 times faster than ice at –22°C.
- Increasing the pressure on matter causes an increase in temperature. Pressure exerted on ice that is near its melting point may produce enough warming to result in melting.

Warm glaciers, whose ice is near its melting point, slip along their bases more readily than cold glaciers. When warm glaciers encounter an obstacle, the pressure of the ice against the obstacle causes the ice to deform and flow around the obstacle. The glacier may also bypass the obstacle by melting on one side and refreezing on the other. The pressure on the upstream side of the obstacle may cause the ice to melt and then refreeze on the downstream side, where pressure is lower. Meltwater produced by pressure can also fill in cavities and act as a lubricant, allowing the ice to move over the ground more easily. See Figure 20.23.

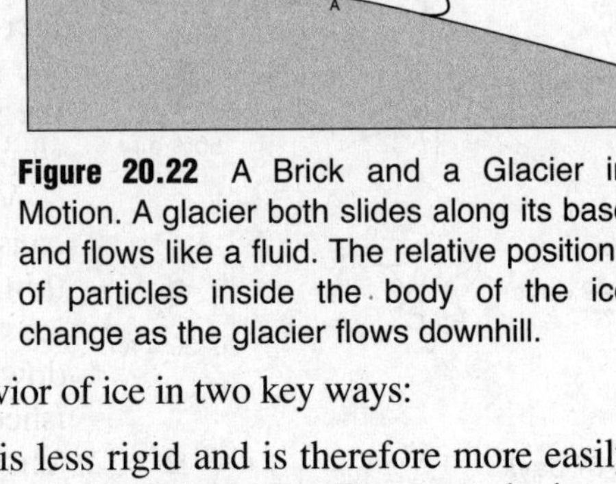

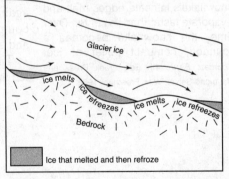

Figure 20.23 Basal Meltwater. High pressure on the upstream side of an obstacle causes ice to melt. The meltwater refreezes on the downstream side of the obstacle, where pressures are lower. Source: Adapted from *Glaciers and Landscape: A Geomorphological Approach*, David E. Sugden and Brian S. John, John Wiley, 1976.

In cold glaciers moving over bedrock, almost none of the movement is due to slipping of the base along the ground. The bond between the ice and the rock to which it is frozen is greater than the internal strength of the ice. Therefore, under pressure, the ice will deform before it breaks loose from the bedrock.

Types of Glaciers

There are two main types of glaciers: valley glaciers and the much larger continental glaciers. *Valley glaciers* are constrained by topography, and their form and flow are strongly influenced by the shape of the land. *Continental glaciers* submerge the land to the extent that the size and shape of the glacier, rather than the shape of the land, control glacial form and flow.

Valley Glaciers

Valley glaciers (see Figure 20.24) form between mountain slopes at high elevations. They start as snowfields that accumulate in bowl-shaped hollows called **cirques**. As the snowfield deepens and ice forms from firn, the ice flows out of the cirque and down the valley, confined by the surrounding slopes.

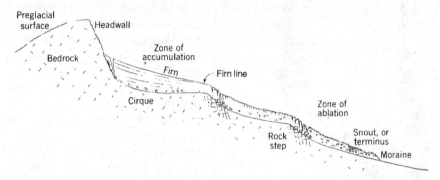

Figure 20.24 Side View of a Typical Valley Glacier. Snow accumulates in a bowl-shaped cirque and is compacted into firn. The firn recrystallizes into ice and moves downhill; the firn line marks the beginning of the glacial ice. When the glacier reaches warmer temperatures, ablation consumes the ice. The glacier ends in a thin snout edged by a pile of debris called a *moraine*. Source: *The Earth Sciences*, 2nd Ed., Arthur N. Strahler, Harper & Row, 1971.

As the glacier moves down the valley, it may meet other glaciers from adjacent valleys and the glaciers may merge, forming a larger glacier. Unlike streams, which mix when they merge, glaciers do not mix but rather "weld" together and flow downslope side by side. A large valley glacier may have been fed by dozens of smaller tributary glaciers flowing from high mountain valleys.

Valley glaciers end at a point where there is a balance between ablation and accumulation. This point can be a sensitive indicator of changes in climate. A general warming trend will cause the end of the glacier to melt back up the valley, or retreat. A cooling trend will cause the end of the glacier to move down the valley, or advance.

Erosion

When a glacier ends on land, streams of meltwater flowing from the glacier often form a river. Large blocks of ice that fall off the face of the glacier may be buried by sediment. A glacier that ends in a body of water forms steep cliffs of ice. Chunks of ice that break off and fall into the water will float since ice is less dense than liquid water. These floating masses of ice, or icebergs, then drift with the currents and gradually melt.

Continental Glaciers

Continental glaciers form at high latitudes where temperatures are always cold. These glaciers are immense masses of ice that spread over the entire land surface, rather than being confined to valleys. Two continental glaciers exist today: one covers Greenland, and the other Antarctica. The Greenland glacier (see Figure 20.25) covers more than 1.7 million square kilometers and is 3.2 kilometers thick at its center. The Antarctic glacier covers more than 12 million square kilometers (about 1.5 times the size of the 48 mainland U.S. states) and is 4 kilometers thick in places. It even extends over the ocean along the margins of the continent of Antarctica, forming several ice shelves.

A continental glacier has a lenslike shape; it is thickest at the center and thinner at the edges. The thick central area is where snow accumulates in a huge firn field and slowly turns to ice. The ice then flows outward in all directions under the pressure of its own weight. The huge weight of a continental glacier actually depresses the surface of the land, and the topography of the land is extensively changed by the movement of the glacier over its surface.

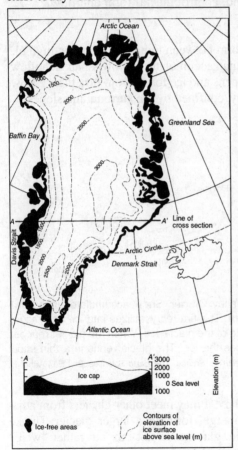

Figure 20.25 The Greenland Glacier. Contour lines show the elevations of the ice surface in meters, and a cross-sectional view reveals the lenslike shape of the glacier. Source: *Earth*, 4th Ed., Frank Press and Raymond Seiver, W.H. Freeman, 1986.

Processes of Glacial Erosion

Glaciers erode the surfaces over which they move by abrasion and plucking. **Abrasion** is the wearing, grinding, scraping, or rubbing away of rock surfaces by friction.

Plucking is the process by which rock fragments are loosened, picked up, and carried away by glaciers.

Abrasion
Abrasion is accomplished mainly by fragments of rock frozen in the ice at the bottom of a glacier. These rub and scrape against the underlying bedrock as the base of the glacier slips downslope. Sharp, angular fragments of hard materials such as quartz are very effective abraders. Clay is a good polishing material but doesn't abrade the bedrock very much.

Plucking
Plucking occurs in several ways. The water that forms when pressure on the ice causes melting can seep into cracks and pores in the bedrock. When this water refreezes, it shatters the bedrock and the broken pieces are picked up and carried away by the glacier. Plucking that occurs in this way is probably the most important because it goes on continuously over much of the glacier's bed as it moves along.

Plucking can also occur when glaciers move over bedrock that is broken into large pieces by jointing. Joints are fractures in rock and usually occur in three planes, forming large, rectangular blocks. These blocks freeze to the base of the glacier and are pushed out of the ground and carried along as the ice moves downslope.

A third type of plucking may occur where the glacier is thick and the bedrock is brittle. In such places, the pressure exerted by a glacier may be great enough to crush, shear off, or shatter brittle bedrock, forming fragments that are carried away by the glacier.

Features Resulting from Glacial Erosion

The processes of abrasion and plucking leave behind unmistakable evidence of a glacier's passage over bedrock. Numerous features, ranging from scratched and polished rock surfaces to basins excavated by plucking to streamlined hills of solid rock, are formed by glacial erosion.

Glacial Polish
Fine materials such as clays and silt that are frozen into the ice at the base of a glacier rub against bedrock as the glacier moves. These particles smooth the rock surface as sandpaper smooths wood. The smooth rock surfaces produced by this type of abrasion are said to be **glacially polished**.

Glacial Striations
Glacial striations (see Figure 20.26) are small scratches in the surface of the bedrock. They are created when fragments frozen into the bottom of a glacier scrape against the bedrock as the glacier slides over the bedrock. On smooth surfaces the striations are usually straight and parallel, but on uneven surfaces with protrusions the striations may curve because the ice deformed as it flowed around these obstacles. A striation may also thicken or thin out, depending on

Erosion

Figure 20.26 Glacial Striations and the Particles That Made Them. Striations vary in size and shape, depending on the nature of the rock particles that formed them and the way that the particles were dragged over the bedrock.

what happened to the abrading particle as it scraped against the bedrock. The long axis of a glacial striation indicates the direction of glacial movement.

Glacial Grooves

Glacial grooves are long, deep, U-shaped depressions in bedrock with smooth bottoms and sides and rounded edges. Most grooves are 10–20 centimeters deep, twice as wide, and tens of meters long. Really large grooves, however, can be more than 1 meter deep and hundreds of meters long. Grooves are shaped by abrasion and often form along fractures, soft layers, or other weak zones in the bedrock. Once started, a groove acts as a channel for further ice flow and concentrates the abrasive force of the glacier in this channel.

Rock Drumlins

Drumlins are hills of loose rock and soil molded into a streamlined shape by glaciers moving over them. Rock drumlins are large outcrops of rock that have been abraded into a streamlined shape by glaciers. Rock drumlins are large, often measuring 100 meters in length.

U-Shaped Valleys

Glaciers move from high elevations to low elevations along the easiest paths. Those paths are usually stream valleys (see Figure 20.27a) that were eroded before the glacier formed. Stream valleys in mountainous regions tend to be steep-sided and V-shaped and to follow a narrow, twisting path between adjacent slopes (see Figure 20.28a). When a glacier moves down a stream valley, it alters the valley's shape. The ice does not meander back and forth as readily as water, so it often breaks through obstacles to form a straighter path. The ice also cuts back the walls of the valley, making them steeper. The result is a wider, straighter, steeper-sided, **U-shaped valley** (Figure 20.27b).

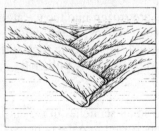 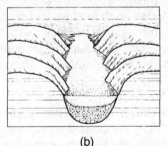

(a)   (b)

Figure 20.27 (a) A V-shaped Stream Valley and (b) a U-shaped Glacial Valley. Source: *Geology of New York: A Simplified Account*, Y. Isachsen et al., eds., New York State Museum Geological Survey, The State Education Department, The University of the State of New York, 1991.

Hanging Valleys

A **hanging valley** (Figure 20.28c) is a valley of a tributary glacier whose floor is above, or hanging over, the floor of the main glacier into which it feeds. The difference in the levels of the valley floors can be hundreds of meters; and if a stream flows down the tributary valley, it may form a spectacular waterfall as it plunges to the floor of the main valley. Hanging valleys may be formed if the main glacier cuts back the valley of a tributary, or if it simply erodes faster and deeper than the tributary.

Cirques, Arêtes, and Horns

A **cirque** is the open, half-bowl-shaped hollow high on a mountainside at the head of a valley glacier. Cirques form when snowfields on mountain slopes melt and refreeze during spring and fall. The repeated freezing and thawing breaks apart the rocks. The rock fragments creep downslope or, when broken into small enough pieces, are carried away by meltwater. In this way, the snowfield creates a hollow in the land on which it rests. As the hollow grows, so does the depth of the snowfield, eventually becoming large and deep enough to form a glacier. Once formed, the glacier plucks rock from the base of the cirque, further deepening it.

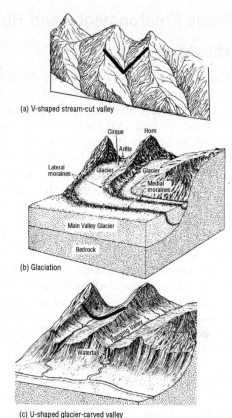

Figure 20.28 Landscape Before, During, and After Glaciation. (a) Valley before glaciation. (b) Glacial erosion forms cirques in hollows between slopes. Erosion in adjacent cirques forms sharp ridges called arêtes and converts rounded peaks into sharp horns. Eroded material carried along the ice margins forms moraines. (c) Hanging valleys form where tributary glaciers flowed into the main glacier. Source: *Earth*, 4th Ed., Frank Press and Raymond Seiver, W.H. Freeman, 1986.

Where cirques from adjacent slopes meet, a narrow, knife-edged ridge called an **arête** is formed between them. Where three or four cirques on the flanks of a mountain peak meet, the central peak takes on a pyramidal shape called a *horn*.

These three features—cirque, arête, and horn—are shown in Figure 20.28b.

Some Environmental and Human Impacts of Erosion

Mass Wasting

The sudden failure of slopes usually involves some triggering event. It can be the vibrations of an earthquake, the addition of water by intense or prolonged rainfall, an increase in the steepness of the slope (due to natural processes or human activities) or the loss of stabilizing vegetation.

The movement of rock material down a slope by mass-wasting processes can have a striking impact on the environment and on human activities. For example, a stream that is cutting a valley can create unstable slopes. The resulting landslides into the valley can dam up the stream, creating a natural reservoir that floods surrounding land. In 1983, a landslide in Utah blocked Spanish Fork Canyon, creating a large lake that cut off a transcontinental railway line and a major highway. Debris flows can bury or sweep away large areas of vegetation, leaving massive scars on the landscape.

Unstable slopes are a hazard to human structures built on or below them. Every year, rockfalls along highway cuts cause motorist fatalities. Housing developments built atop slopes are frequently undermined and destroyed by mass wasting. However, with careful planning, building regulations, and zoning laws, and the use of stabilization techniques, the impact of mass wasting on human activities can be reduced.

Hazardous slopes can be stabilized by changing the steepness of the slope, reducing the weight acting on the slope, removing water from the rocks and soil, and planting vegetation. Some commonly used stabilization techniques include (1) the use of retaining devices (steel nets, concrete walls, rock bolts), (2) the removal of water by installing drainage systems or by covering the slope with an impermeable material, (3) grading the slope to reduce its steepness, and (4) building walls to divert flows. All of these preventive measures, however, are costly. The best strategy is to avoid building on unstable slopes in the first place.

Streams

Erosion by moving water affects human lives in many ways. Overland flow may wash away the topsoil your family just bought and spread on the lawn. Splash and sheet erosion of topsoil on farmland decreases fertility and lowers crop yields, resulting in higher prices at the supermarket. Erosion and flooding by streams leave thousands homeless annually. Nevertheless, there are benefits to living near running water, such as fertile soils, access to water, food from marine life, energy, waste absorption and dispersal, and possibly a transportation route.

Flooding is *the* major stream hazard, causing loss of life and structural damage. Flooding is a natural response to an unusually high input of water over a short time, but human activities can influence flooding in many ways. Paving surfaces or constructing buildings changes runoff patterns and infiltration rates. Changes to river channels and dam construction alter a stream's flood patterns.

The primary impacts of flooding are due to actual contact with the flowing water. They include death by drowning, damage to structures, crop loss, and erosion of flooded areas. Floods also impact humans and the environment indirectly by interrupting services, changing river channels, and destroying wildlife habitat.

Regions at risk of flooding and the degree of risk each faces can be determined if accurate maps and records of past floods have been kept. These allow predictions to be made based upon the frequency and extent of past flooding. However, records may not go back long enough to permit precise predictions about rare, severe floods.

The recurrent hazard of flooding has prompted efforts to reduce risk in flood-prone areas by limiting the human population there. Some communities now have floodplain zoning, which prohibits new construction. The Federal Emergency Management Agency (FEMA) requires that, if the owner of a flood-damaged property gets more than 50 percent of its value in compensation, the structure must be demolished and the site rezoned to prohibit future construction. Many lenders place restrictions on mortgages for properties in floodplains, allowing only appropriate activities such as recreation or certain types of farming to be carried out there. In some places there are calls for the buyout of all properties in flood-prone areas and the return of the land to its original wetland status. For urban areas that already exist in floodplains, however, flood hazard will continue to be an issue.

Waves and Currents

Erosion is one of the most important problems in coastal zones, mainly because 85 percent of shorelines is privately owned and rising sea levels are eroding all shorelines. Erosion is most pronounced on beaches and coastal cliffs, where smaller sediments are easily moved by waves and currents. It tends to be less of a problem on rocky coastlines or beaches made of pebbles, which are harder to move.

There are really only three ways to fight erosion of shoreline areas: dredge sediments to fill in where erosion has occurred, dam rivers or build coastal structures such as jetties and breakwaters to decrease the force of eroding waves and currents, and develop coastal dune areas as barriers to inland erosion.

Unfortunately, some of these strategies have proved ineffective. For example, *beach replenishment* involves replacing eroded sand with sand brought in from someplace else. This is an expensive undertaking considering the huge amount of sand needed. Also, most beach replenishment doesn't last very long—40 percent of all replenished beaches are gone in less than 2 years, less than half last for 2–5 years, and only 12 percent last more than 5 years. Beach stabilization is expensive and, as you can see, usually ineffective in the long term. Furthermore, it often changes the nature of the shoreline, and this can have negative effects on the organisms that inhabit it.

Of course, the best solution would be to recognize that coastal erosion is a natural process and allow it to happen without interference. Perhaps the coastal zone is not an environment that is favorable for private ownership.

Erosion

Wind

Wind erosion presents humans with many problems. In arid regions, poles carrying electric power and telephone lines have been worn through by abrasion, causing them to fall and interrupt service. In times of drought, the precious topsoil of farmland stripped of vegetation must be protected from deflation. Vehicles used in arid regions require special features to keep abrasive particles out of their moving parts and to prevent particles suspended in the air from clogging fuel, air, and cooling lines.

Awareness of wind erosion and its causes enables humans to guard against it. Farmers plant windbreaks of drought-resistant trees. In many arid regions, off-road vehicles, such as dirt bikes and dune buggies, are banned because they destroy the fragile vegetation that protects the land from wind erosion.

Glaciers

Glaciers are responsible for much of North America's topography. Thus, many familiar landforms are the products of glacial erosion. Glacial erosion sculpted the rugged peaks of the Adirondacks and created the beautiful fjord known as the Hudson River Valley. The Great Lakes occupy basins carved out and deepened by glaciers and were once filled with glacial meltwater. Erosion by meltwater from the last glacial period formed most of the Mississippi River's drainage basin.

The modern-day burning of fossil fuels has increased the amount of carbon dioxide in the air. The result has been greenhouse-effect heating that has begun to melting Earth's remaining glaciers and ice sheets; this in turn has led to a rise in sea level and to coastal flooding. The global melting of ice may have other consequences, such as changes in global weather patterns that could have serious effects on agriculture.

MULTIPLE-CHOICE QUESTIONS

In each case, write the number of the word or expression that best answers the question or completes the statement.

1. Which event is the best example of erosion?
   (1) breaking apart of shale as a result of water freezing in a crack
   (2) dissolving of rock particles on a limestone gravestone by acid rain
   (3) rolling of a pebble along the bottom of a stream
   (4) crumbling of bedrock in one area to form soil

2. For which movement of Earth materials is gravity *not* the main force?
   (1) sediments flowing in a river
   (2) boulders carried by a glacier
   (3) snow tumbling in an avalanche
   (4) moisture evaporating from an ocean

3. Unsorted, angular, rough-surfaced cobbles and boulders are found at the base of a cliff. What most likely transported these cobbles and boulders?
   (1) running water
   (2) wind
   (3) gravity
   (4) ocean currents

4. The diagrams below represent four different examples of one process that transports sediments.

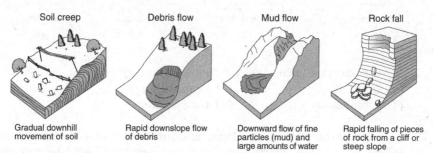

   Which process is shown in these diagrams?
   (1) chemical weathering
   (2) wind action
   (3) mass movement
   (4) rock abrasion

5. On Earth, which agent of erosion is responsible for moving the largest amount of material?
   (1) groundwater
   (2) glaciers
   (3) running water
   (4) wind

6. Compared with an area of Earth's surface with gentle slopes, an area with steeper slopes most likely has
   (1) less infiltration and more runoff
   (2) less infiltration and less runoff
   (3) more infiltration and more runoff
   (4) more infiltration and less runoff

7. During a dry summer, the flow of most large New York State streams generally
   (1) continues because some groundwater seeps into the streams
   (2) increases due to greater surface runoff
   (3) remains unchanged due to transpiration from grasses, shrubs, and trees
   (4) stops completely because no water runs off into the streams

Erosion

8. Which graph best represents the relationship between the discharge of a stream and the velocity of stream flow?

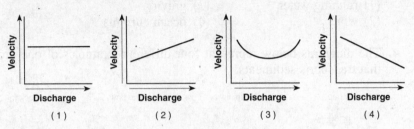

9. A river transports material by suspension, rolling, and
   (1) solution
   (2) sublimation
   (3) evaporation
   (4) transpiration

10. Four quartz samples of equal size and shape were placed in a stream. Which of the four quartz samples below has most likely been transported farthest in the stream?

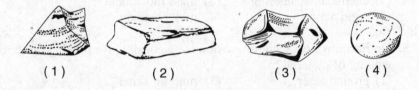

11. Which graph best represents the relationship between the slope of a river and the particle size that can be transported by that river?

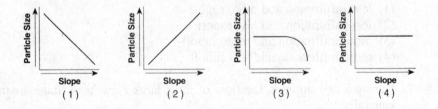

12. What is the largest sediment that can be transported by a stream that has a velocity of 125 cm/sec?
    (1) cobbles
    (2) pebbles
    (3) sand
    (4) clay

13. According to the *2011 Edition Reference Tables for Physical Setting/Earth Science*, which stream velocity transports cobbles but not boulders?
    (1) 50 cm/s
    (2) 100 cm/s
    (3) 200 cm/s
    (4) 400 cm/s

14. The diagram below shows points A, B, C, and D on a meandering stream.

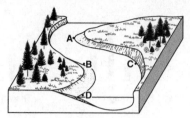

At which point is erosion greatest?
(1) A
(2) B
(3) C
(4) D

15. The map below shows the path of a river. The arrow shows the direction the river is flowing. Letters A and B identify the banks of the river.

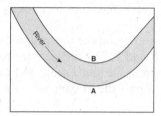

The water depth is greater near bank A than bank B because the water velocity near bank A is
(1) faster, causing deposition to occur
(2) faster, causing erosion to occur
(3) slower, causing deposition to occur
(4) slower, causing erosion to occur

Base your answers to questions 16 and 17 on the diagrams below. Diagrams A, B, and C represent three different river valleys.

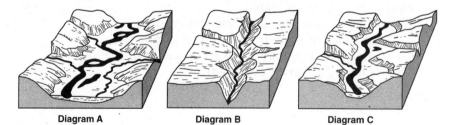

Diagram A   Diagram B   Diagram C

Erosion

16. Which bar graph best represents the relative gradients of the main rivers shown in diagrams A, B, and C?

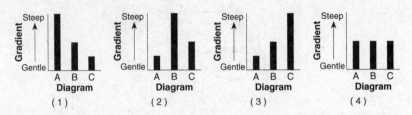

17. Most sediments found on the floodplain shown in diagram A are likely to be
    (1) angular and weathered from underlying bedrock
    (2) angular and weathered from bedrock upstream
    (3) rounded and weathered from underlying bedrock
    (4) rounded and weathered from bedrock upstream

18. The four streams shown on the topographic maps below have the same volume between X and Y. The distance from X to Y is also the same. All the maps are drawn to the same scale and have the same contour interval. Which map shows the stream with the greatest velocity between points X and Y?

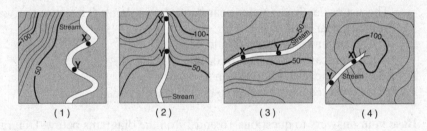

19. Which interaction between the atmosphere and the hydrosphere causes most surface ocean currents?
    (1) cooling of rising air above the ocean surface
    (2) evaporation of water from the ocean surface
    (3) friction from planetary winds on the ocean surface
    (4) seismic waves on the ocean surface

20. The arrows labeled A through D on the map below show the general paths of abandoned boats that have floated across the Atlantic Ocean.

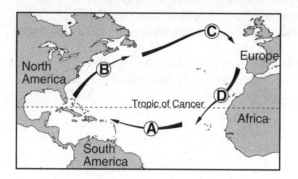

Which sequence of ocean currents was responsible for the movement of these boats?
(1) South Equatorial → Gulf Stream → Labrador → Benguela
(2) South Equatorial → Australia → West Wind Drift → Peru
(3) North Equatorial → Koroshio → North Pacific → California
(4) North Equatorial → Gulf Stream → North Atlantic → Canaries

21. Which natural agent of erosion is mainly responsible for the formation of the barrier islands along the southern coast of Long Island, New York?
(1) mass movement
(2) running water
(3) prevailing winds
(4) ocean waves

22. The block diagram below shows a part of the eastern coastline of North America. Points A, B, and C are reference points along the coast.

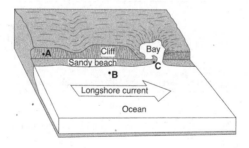

Which list best represents the primary process occurring along the coastline at points A, B, and C?
(1) A—folding; B—subduction; C—crosscutting
(2) A—weathering; B—erosion; C—deposition
(3) A—faulting; B—conduction; C—mass movement
(4) A—precipitation; B—infiltration; C—evaporation

Erosion

23. Which rock material is most likely to be transported by wind?
    (1) large boulders with sets of parallel scratches
    (2) jagged cobbles consisting of intergrown crystals
    (3) irregularly shaped pebbles that contain fossils
    (4) rounded sand grains that have a frosted appearance

24. The cross section below shows the movement of wind-driven sand particles that strike a partly exposed basalt cobble located at the surface of a windy desert.

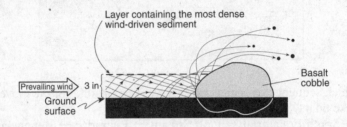

Which cross section best represents the appearance of this cobble after many years of exposure to the wind-driven sand?

25. The diagram below shows sand particles being moved by wind.

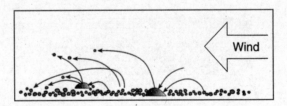

At which Earth surface locations is this process usually the most dominant type of erosion?
    (1) deserts and beaches
    (2) deltas and floodplains
    (3) glaciers and moraines
    (4) mountain peaks and escarpments

486

26. The photograph below shows a valley.

Which agent of erosion most likely produced this valley's shape?
(1) wave action
(2) moving ice
(3) blowing wind
(4) flowing water

27. U-shaped valleys and parallel grooves in bedrock are characteristics of erosion by
(1) mass movement
(2) wave action
(3) running water
(4) glacial ice

28. The photograph below shows a large boulder of metamorphic rock in a field in the Allegheny Plateau region of New York State.

The boulder was most likely moved to this location by
(1) glacial ice
(2) prevailing wind
(3) streamflow
(4) volcanic action

29. Which landscape surface resulted primarily from erosion by glaciers?

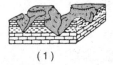   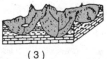   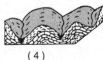
(1)   (2)   (3)   (4)

487

Erosion

CONSTRUCTED RESPONSE QUESTIONS

Base your answers to questions 30 through 33 on the cross section and block diagram below. The cross section shows an enlarged view of the stream shown in the block diagram. The sediments in the cross section are shown at 60% of actual size. Arrows show the movement of particles in the stream. The block diagram represents a region of Earth's surface and the bedrock beneath the region.

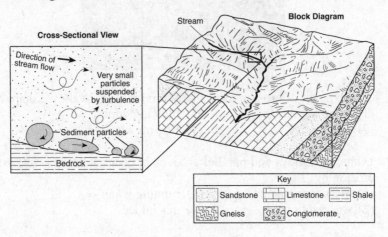

30. After measuring the size of each sediment particle, identify the name of the largest particle shown on the stream bottom in the cross section. [1]

31. What process is responsible for producing the rounded shape of the particles shown on the stream bottom in the cross section? [1]

32. Identify the type of rock shown in the block diagram that appears to be the most easily eroded. [1]

33. How does the shape of a valley eroded by a glacier differ from the shape of the valley shown in the block diagram? [1]

Base your answers to questions 34 through 36 on the block diagrams below, which show three types of streams with equal volumes.

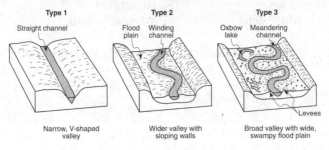

34. Explain how the differences between the type 1 and type 3 stream channels indicate that the average velocities of the streams are different. [1]

35. Explain why the outside of the curve of a meandering channel experiences more erosion than the inside of the curve. [1]

36. Explain how the cobbles and pebbles that were transported by these streams became smooth and rounded in shape. [1]

Base your answers to questions 37 through 40 on the map below, which shows the different lobes (sections) of the Laurentide Ice Sheet, the last continental ice sheet that covered most of New York State. The arrows show the direction that the ice lobes flowed. The terminal moraine shows the maximum advance of this ice sheet.

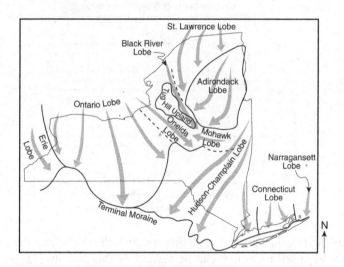

37. During which geologic epoch did the Laurentide Ice Sheet advance over New York State? [1]

38. Describe the arrangement of rock material in the sediments that were directly deposited by the glacier. [1]

39. According to the map, toward which compass direction did the ice lobe flow over the Catskills? [1]

40. What evidence might be found on surface bedrock of the Catskills that would indicate the direction of ice flow in this region? [1]

Extended Constructed Response Questions

Base your answers to questions 41 through 43 on the data table below and on your knowledge of Earth science. The data table shows the average monthly discharge, in cubic feet per second, for a stream in New York State.

Data Table

Month	Jan	Feb	Mar	Apr	May	Jun	Jul	Aug	Sept	Oct	Nov	Dec
Discharge (ft^3/sec)	48	52	59	66	62	70	72	59	55	42	47	53

41. On the grid below, plot with an **X** the average stream discharge for *each* month shown in the data table. Connect the **X**s with a line. [1]

Average Monthly Stream Discharge

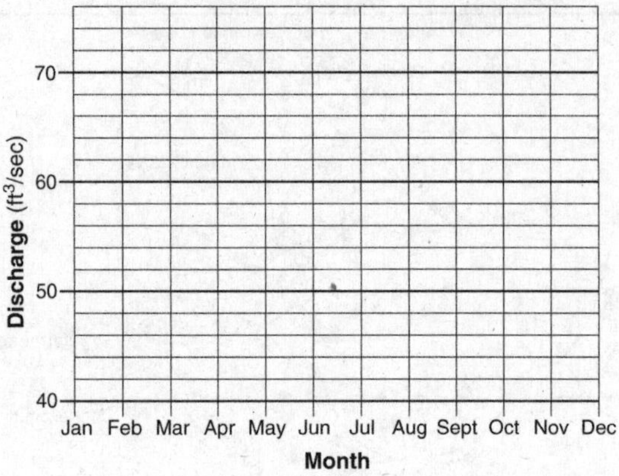

42. State the relationship between this stream's discharge and the amount of suspended sediment that can be carried by this stream. [1]

43. Explain *one* possible reason why this stream's discharge in April is usually greater than this stream's discharge in January. [1]

44. Complete the table below, by listing *three* agents of erosion and identifying *one* characteristic surface feature formed by *each* agent of erosion. [2]

Agent of Erosion	Surface Feature Formed
(1)	
(2)	
(3)	

Base your answers to questions 45 through 48 on the map and cross section of the Finger Lakes Region shown below and on your knowledge of Earth science.

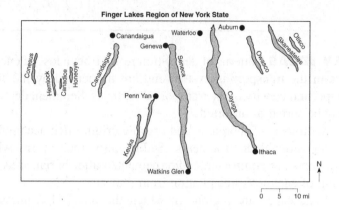

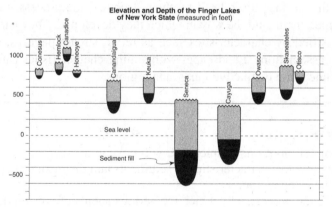

45. According to the cross section, how thick from top to bottom is the sediment fill in Seneca Lake? [1]

46. State *one* possible explanation for the north-south orientation of the Finger Lakes. [1]

47. During some winters, a few of the Finger Lakes remain unfrozen even though the land around the lakes is frozen. Explain how the specific heat of water can cause these lakes to remain unfrozen. [1]

48. Identify *two* processes that normally occur to form the type of surface bedrock found in the Finger Lakes Region. [1]

Chapter 21
DEPOSITION

> **KEY IDEAS** Patterns of deposition result from a loss of energy within the transporting system and are influenced by the size, shape, and density of the transported particles. Sediment deposits may be sorted or unsorted.
>
> Sediments of inorganic and organic origin often accumulate in depositional environments. Sedimentary rocks form when sediments are compacted and/or cemented after burial or as the result of chemical precipitation from seawater.
>
> Landscapes are regions in which the physical features of Earth's surface are related by their structure, the processes that formed them, and the way in which they developed. They can be classified by characteristics such as relief, stream patterns, and soil associations. Landscapes with similar characteristics can be grouped into distinctive landscape regions.

KEY OBJECTIVES
Upon completion of this chapter, you will be able to:
- Explain how the characteristics of sediments and the media carrying them affect their deposition.
- Describe the processes by which the various agents of erosion deposit the sediments they are transporting.
- Describe some of the common landforms resulting from erosion and deposition by mass movements, moving water, glaciers, and wind.
- Identify and describe landscape regions.

DEPOSITION

At some point, all agents of erosion begin to deposit sediment. **Deposition** is the process by which transported sediment is dropped in a new place. The same agents that erode sediment also deposit it. Several factors influence when deposition will begin.

Speed of Medium

The **medium** is the substance that carries sediment. In streams the medium is water, in winds it is air, and in glaciers it is ice. Generally, the slower a medium is moving, the less its carrying power. If the carrying power drops below the level needed to transport a particle, that particle will drop to the ground. Less carrying power results in deposition.

You can see this for yourself. Imagine you are flying a kite and the wind slows down. The slower the wind blows, the less carrying power it has. The kite will begin to drop. If the wind stops blowing, the kite will settle to the ground. Deposition occurs in much the same way. As the medium slows, sediment begins to settle out.

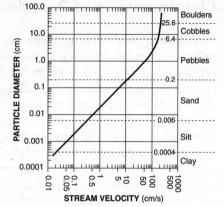

Figure 21.1 Relationship of Transported Particle Size to Water Velocity. This generalized graph shows the water velocity needed to maintain, but not start, movement. Variations occur because of differences in particle density and shape. Source: The State Education Department, *Earth Science Reference Tables*, 2011 ed. (Albany, New York; The University of the State of New York).

Figure 21.1 shows the sediment that water carries at different speeds. Notice that water carries more sediment sizes when it is moving fast. As water slows down, some sediment is deposited. Large, dense particles will settle first, and light, small particles last. The result is a separation of different particles into layers.

The separation of particles during deposition is called **sorting**. Figure 21.2 shows the sorting that occurs when a stream flows into a larger body of water, such as a lake or an ocean. As the stream flows into the larger body, it slows down. Coarse gravel and pebbles settle first. Farther from the shore, the movement of the water is slower and sand settles out. Still farther from shore, in very slow moving water, the finest particles of silt and clay settle. Sorted layers of particles result.

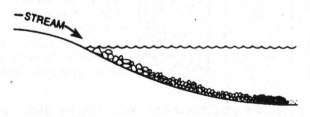

Figure 21.2 Horizontal Sorting of Particles Carried by a Stream. As a stream decreases in velocity, the heavier particles settle first; the lighter particles are carried farther downstream before settling.

Deposition

Characteristics of Particles

Particles may vary in size, shape, and density. Each of these characteristics influences the rate at which settling occurs. To learn how particle size influences settling rate, look at Figure 21.3a. Notice that, when all the rock particles are deposited, sorting has occurred. How can that be? The answer is that some particles must settle before others. The particles on the bottom settled fastest, and those on top slowest. Since the larger particles are on the bottom, they settled fastest. The smaller ones, on top, settled slowest.

In general, the larger the particle, the faster it will settle. Very small particles, such as fine clay, take a long time to settle. The muddy appearance of many rivers and lakes is due, in part, to particles that have not yet settled. To see how particle shape affects settling rate, look at Figure 21.3b. Notice that, when all the particles are deposited in water, the bottom layer consists of round particles. These particles must have settled first. Also notice in the figure a top layer of flat particles. These particles must have settled last.

In general, round particles settle faster than flat ones. To understand how particle shape affects settling rate, imagine dropping a crumbled paper and a flat paper to the ground at the same time. The crumbled paper, like a rounded particle, lands faster. Friction between the large surface area of the flat paper and the air slows that paper's fall. A flat shape in rock particles has the same effect in water. Friction between the flat particle and the water will decrease the rate at which the particle settles.

Finally, to see how the density of a particle affects settling, look at Figure 21.3c. Notice, once again, that sorting has taken place. The densest particles form the bottom layer; the least dense particles, the top layer. Thus it can be concluded that the densest particles settled fastest.

To summarize, several factors affect the deposition of sediment. The speed of the medium determines when particles will be deposited. The size, shape, and density of a particle determine the rate at which it will settle. Differences in the settling rates of sediments result in sorting during deposition.

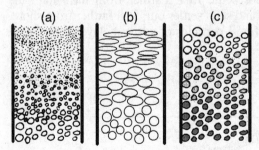

(a) All other factors being the same, large particles settle faster than smaller ones.
(b) All other factors being the same, particles with rounded shapes settle faster than flat particles.
(c) All other factors being the same, particles composed of denser materials settle faster than those composed of less dense materials.

Figure 21.3 Vertical Sorting of Different Particles in a Column. When sediments settle in a quiet medium, they form a series of layers. The fastest settling particles reach the bottom first and form the lowest layer. They are followed by slower settling particles, which form the next layer; the slowest settling particles form the top layer.

Bedding of Sediment

As a result of sorting during deposition, sediment forms layers consisting of different types of particles. Sediment deposited in a quiet body of water will usually form layers similar to those shown in Figure 21.4. Each layer of sediment is called a **bed**, and each bed usually represents a period of deposition. Beds of sediment are commonly found in sedimentary rock.

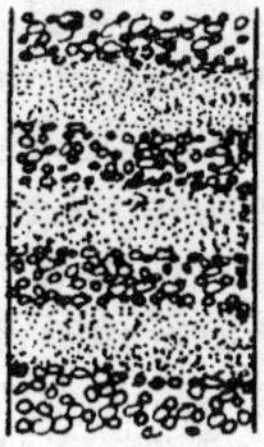

Hurricane, September
No rainfall, August
Summer storm, July
Periodic gentle rain, May-June
Heavy rains, May
No rainfall, April
Heavy rainfall and spring thaw, March

Figure 21.4 Bedding of Sediment. Rapid deposition results in **graded bedding**, a vertical sorting. Each sequence represents a depositional event.

DEPOSITION BY MASS MOVEMENTS

Deposition by Gravity

During mass movements, sediment is pulled downhill by gravity and is deposited, when it stops moving, at the base of a slope. Sediment deposited by gravity does not settle through a medium, as does sediment deposited by wind or water. Thus, little, if any, sorting takes place. Most deposits resulting from mass movements are unsorted and do not show any distinct layering. In **unsorted deposits**, all types of particles are arranged without any order, as shown in Figure 21.5.

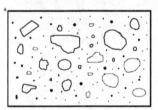

Figure 21.5 Unsorted Deposit. Unsorted sediment contains rock fragments of all sizes randomly mixed together.

Deposits by Rapid Mass Movements

Rapid mass movements deposit sediment. As deposition takes place, landforms develop. These landforms are evidence that deposition has occurred.

Rockfalls
During rockfalls and rockslides, fragments slide downhill until they reach the base of a slope, forming a talus. A **talus** is a pile of broken rock that builds up at the base of a cliff or steep slope. Boulders may roll down a talus and onto the flat ground at the base of the slope. As the talus weathers, soil may form on the surface. Then, if the talus is not being constantly covered by new rock fragments, plants may take root and grow. See Figure 21.6a.

Deposition

Mudflow

During a mudflow, mud moves downhill until it reaches the base of a slope. The mud then spreads out in a thin, wide sheet. See Figure 21.6b. Huge boulders carried by the mudflow may be left to stand isolated on the gently rolling land.

Deposits by Slow Mass Movements

Slow mass movements also deposit sediment. Earthflows, slump, and soil creep all produce sediment deposition from which landforms may develop.

Earthflow

An earthflow deposits a thick pile of sediment. See Figure 21.6c. Earthflow deposits are generally thicker than deposits produced by mudflows. The difference between the two can be demonstrated by spilling a jar of thick honey and a jar of thin tomato sauce next to each other. The thick honey, like an earthflow, produces a small, thick deposit. The thinner sauce, like a mudflow, produces a thin spread. Compare Figures 21.6b and 21.6c.

Slump

Slump forms a deposit that looks like an apron spread on the ground. Notice in Figure 21.6d the steplike appearance of the deposit. The steps are the tops of blocks of material that broke loose from the cliff. Very often, earthflows begin as slump.

Soil Creep

As stated earlier, soil creep is a very slow process. Material deposited by soil creep is often hard to recognize because it is usually covered with plants, and looks like part of the hill on which it formed. Only a person trained to recognize deposition will perceive the series of ripples at the base of the gentle slope as evidence of deposition by soil creep. See Figure 21.6e.

Effects of Mass Movements

Deposition by mass movement along roadsides can cause serious problems. Poorly planned roads

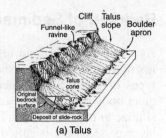
(a) Talus

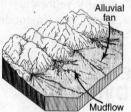
(b) Mudflow

(c) Earthflow

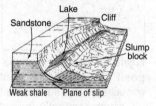
(d) Slump

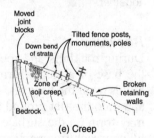
(e) Creep

Figure 21.6 Deposits Formed by Mass Movements. Source: *The Earth Sciences*, Arthur N. Strahler, Harper & Row, 1971.

may be blocked by deposits. In fact, material deposited by slides, flows, and slump may make a road impassable. Interstate 40, near Rockwood, Tennessee, is one such road; one stretch has been blocked more than 20 times by major slides. Clearing such roads costs taxpayers millions of dollars each year. In addition, these poorly planned roads pose dangers to motorists.

DEPOSITION BY MOVING WATER

More than 70 percent of Earth's surface is covered by water. Therefore, it is not surprising that most deposition occurs in water. In this section, you will learn how sediment carried by moving water is deposited and how deposition changes the shape of the land.

Deposition by Streams

Rivers and other streams carry large amounts of sediment. Water from the Colorado River is used to irrigate crops; canals carry the water to the fields. In just one of these canals, up to 5,000 metric tons of sediment has to be removed from the water every day.

Streams deposit sediment when they slow down or decrease in volume. Such slowdown or decrease in volume occurs when the slope decreases, runoff and seepage from groundwater decrease, or the stream enters a standing body of water, such as a lake or an ocean. Deposition by streams results in layers that are sorted according to particle size, shape, and density. In addition, these deposited sediments tend to be smoothed and rounded as a result of abrasion while being carried in the stream.

Many landforms develop as a result of stream deposition. These landforms are found in almost all parts of the United States.

Oxbow Lakes

In a preceding chapter you learned how meanders are formed. As meanders become more curved, their ends move closer together. Eventually, the stream erodes its way through the land, separating the meanders (see Figure 21.7). Most of the water then flows through this steeper, straighter shortcut. The slower moving water in the old meander deposits its load. When deposition blocks off the entrance to the old meander and it becomes separated from the stream, an oxbow lake is formed.

Floodplains and Levees

Because of its increased speed and volume, a stream in flood carries much more than the normal amount of sediment. Also, when a stream is in flood, it overflows its banks. Outside the channel, the water is slower moving and shallower. As soon as the flood water leaves the channel, it begins to deposit its sediment load over the surrounding land to form a **floodplain**. The larger

Deposition

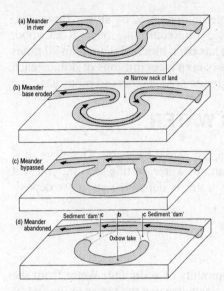

Figure 21.7 Formation of an Oxbow Lake. Water flows faster along the outside curve of a meander than it flows along the inside curve because it has farther to travel in order to make the turn. The faster moving water erodes the bank on the outside curve until just a narrow neck of land (a) separates the loops of the meander. When the meanders erode through the narrow neck, most of the stream flow bypasses the meander. Because less water flows through the meander, water velocity decreases and deposition builds sediment "dams" (c) that separate the meander from the stream, forming an oxbow lake (b). Source: *Earth Science on File.* © Facts on File, Inc., 1988.

particles settle out first, building a **levee**, that is, a low, thick, ridgelike deposit along the banks of both sides of the stream (see Figure 21.8).

Floodplains make good farmland; sediment deposited after each flood renews the soil. For this reason, farmers have been able to raise crops on the floodplains of the Nile River for over 4,000 years!

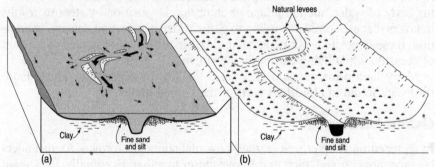

Figure 21.8 Formation of a Floodplain (a) and Levee (b). During a major flood, the stream's floodplain becomes a lake. Water in the main channel flows rapidly. However, water escaping the channel immediately slows down and deposits fine sand and silt, forming a natural levee. As the water continues to spread out and slow down, clay is deposited on the surrounding lower land. When the flood ends, the levees remain as low ridges along the sides of the channel and the floodplain is covered with wet clay and swampy areas.
Source: *Physical Geology*, 2nd Ed., Richard Foster Flint and Brian J. Skinner, John Wiley, 1977.

Deltas and Alluvial Fans

When a stream flows into a quiet body of water such as an inland sea, ocean, or lake, it stops moving. The point at which this occurs is called the **mouth** of the stream. Here most of the stream's sediment is deposited. The resulting

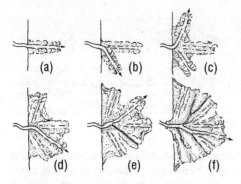

Figure 21.9 Stages (a)–(f) in the Formation of a Delta. A delta is formed as a stream flows into a standing body of water such as a lake or an ocean. Source: *The Earth Sciences*, by Arthur N. Strahler, Harper & Row, 1971.

landform is a **delta**, as seen in Figure 21.9. A delta is a large, flat, fan-shaped pile of sediment at the mouth of a stream. It is composed of sand, silt, and clay and is shaped like a triangle. It gets its name from the Greek letter delta, which is written as Δ.

When streams flowing down steep mountain valleys reach flat, open land, they slow down and deposit much of their coarser load. The resulting landform is an **alluvial fan**, that is, a large mound of coarse sediment deposited by a stream onto open, flat land (see Figure 21.10).

Although an alluvial fan is similar in appearance to a delta, there are some differences. An alluvial fan has steeper sides and is made of coarser sediment than a delta.

Deposition by Oceans

Oceans also deposit sediment. Whenever waves and currents slow down, they deposit their loads. Deposition by waves and currents builds new landforms that change the shape of shorelines.

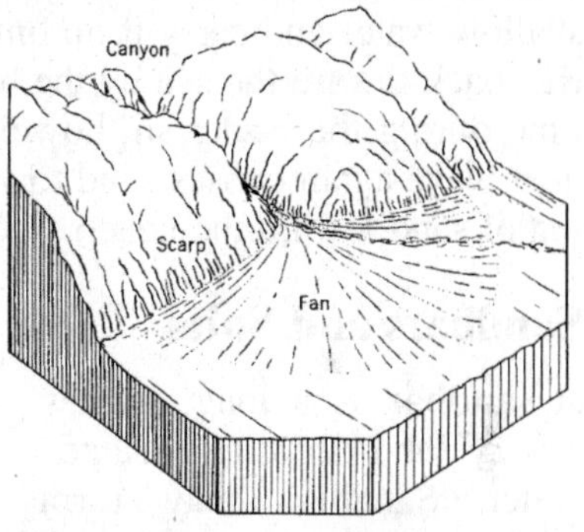

Beaches

A **beach** is a deposit of sediment along the shoreline of a body of water. Beaches are formed as waves wash up against the land and slow down. Beaches constitute an almost continuous fringe around the continents and most islands. Although most people think of a beach as a strip of fine, white sand along the seashore, beaches may consist of any earth material from fine clay to boulders.

Figure 21.10 Formation of an Alluvial Fan. An alluvial fan forms where a stream flows from a steep gradient onto a gentle gradient. Source: *The Earth Sciences*, Arthur N. Strahler, Harper & Row, 1971.

The nature of beach sediments depends on the source material and the weathering processes at work in the area. Beaches are shaped by waves and are changed daily and monthly by tides. Figure 21.11 shows the general features of a beach. **Berms** are flat areas formed by wave action. The breaking wave loosens sediments upslope and drags them back toward the water as the wave recedes.

Deposition

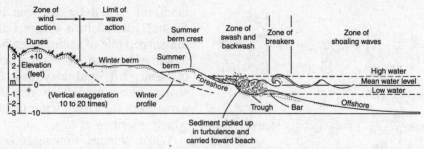

Figure 21.11 Beach Features. The usual direction in which breaking waves carry sand is toward the beach. This sand, together with sediments eroded from coastal cliffs, builds a flat area, or a berm. Source: *The Earth Sciences*, Arthur N. Strahler, Harper & Row, 1971.

Figure 21.12 shows the changing shoreline conditions in winter and summer. During stormy winter months, waves tend to be high and to have short periods. The sand picked up by one wave does not have a chance to settle to the bottom before the next wave sets the sand in motion again. As a result, the sand stays in suspension and is carried away from the beach by the backrush of water until it reaches deeper, calmer water and can settle, forming an offshore sand bar. See Figure 21.12a.

In summer months, the weather is calmer and the waves that reach the shore tend to be swells of long periods. The breaking waves pick up sand in shallow water and carry it up onto the beach. Most of the sand is then carried back toward the sea by the backrush of water along the beach. There is time during the backrush, however, for some particles to settle before the next wave again carries sand toward the beach. The overall result is a shifting of sand toward the beach. See Figure 21.12b.

Sandbars and Spits

A **sandbar** is a long, narrow pile of sand deposited in open water. Sandbars may form where receding waves wash beach sediments into deeper, quieter waters and also where the shoreline curves away from a longshore current. As the current curves away from the shoreline, it flows into deeper, quieter water and deposits its sediments. Most bars formed in this way are attached at one end to the mainland and are called **spits**. Sandbars that connect an island to the mainland are called **tombolos**. See Figure 21.13.

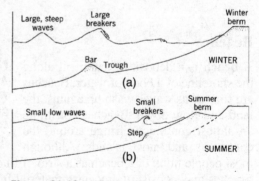

Figure 21.12 Winter (a) and Summer (b) Beach Profiles. High waves of short period generally occur in winter months and drag sediments offshore, forming a sandbar. In summer, waves generally have long periods and move sediment onto the beach. Source: *The Earth Sciences*, Arthur N. Strahler, Harper & Row, 1971.

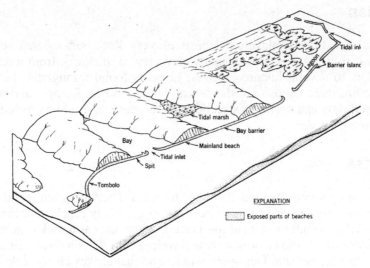

Figure 21.13 Some Coastal Landforms Created by Deposition of Sediments by Ocean Waves and Currents. Source: *Physical Geology*, 2nd Ed., Richard Foster Flint and Brian J. Skinner, John Wiley, 1977.

Barrier Bars and Lagoons

Eventually, waves deposit enough sand on a bar so that it is above water. Large waves continue to add material, and in time the bar is built up well above sea level. Such large bars are known as **barrier bars**. Barrier bars may completely block off a bay, forming a **tidal marsh**, or may parallel the coast, broken here and there by tidal inlets into elongated **barrier islands**. The calm, protected bodies of water between barrier bars and the mainland are called **lagoons**. See Figure 21.13.

Problems Associated with Deposition by Moving Water

Deposition by moving water can create problems in harbors, which must be deep and have clear openings to the sea to enable large ships to pass through. Occasionally, sandbars and spits block a harbor's entrance. In addition, sediment from streams and rivers deposited in harbors can create areas of shallow water. Ships are then in danger of running aground. To prevent this from happening, sediment must be removed from harbors by dredging.

DEPOSITION BY WIND

Wind is moving air. When a wind slows or stops moving, the particles it is carrying settle to the ground and are deposited. Wind deposits generally contain fine sediments and result in the formation of certain surface features.

Deposition

Loess

Loess is a thick, unlayered deposit of very fine, buff-colored sediment deposited by wind. Loess deposits may vary in thickness from a few centimeters to several hundred meters. Loess is found throughout the world. There are large deposits in the central United States, Europe, and parts of China. Loess is a source of excellent soil because of its ability to hold large amounts of water.

Dunes

Dunes are mounds of sand deposited by wind. They are often found in barren, desert regions and can also occur on large, sandy beaches. A dune often forms when windblown sand meets an obstacle such as a rock or a bush.

Figure 21.14 shows how a dune develops. On the windward side of the dune, that is, the side facing the wind, sand that strikes an obstacle falls to the ground. In time, the sand forms a mound that slopes gently up toward the tip of the obstacle. Wind then pushes sand grains up this slope to the crest, or top, of the dune.

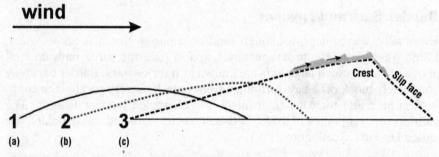

Figure 21.14 Formation of a Dune. (a) Wind-blown sand accumulates in low mounds. (b) Steady winds carry sand up the side of the dune, and it rolls over the crest onto the sheltered side. (c) As sand is picked up on the gently sloping side facing the wind and deposited on the steep slip face, the entire dune moves.

On the leeward side of the dune, that is, the side sheltered from the wind, winds are very slow. As a result, the sand grains rolling over the crest fall to the ground because the winds can no longer support them. A deposit of sand forms, creating a steep-sided mound. As sand continues to be deposited, this mound may become unstable. Indeed, sand may slip or slide downhill. For this reason, the steep, leeward side of a dune is called the **slip face**.

As winds blow sand from the windward side of a dune and deposit it on the leeward side, the entire dune moves in the direction in which the wind is blowing. In this way, dunes travel along a desert. Some desert dunes may move from 10 to 25 meters in a year. Moving dunes have been known to engulf forests, farmlands, and buildings.

Sand dunes have a variety of heights, shapes, and patterns. Dunes may reach heights of 30–100 meters, depending on wind speed and the amount of sand in the area. Three common shapes and patterns of sand dunes are shown in Figure 21.15. The type of dune that forms in an area depends on the prevailing wind patterns and the amount of sand.

In the United States, the dunes of our coastal areas are an important line of defense for inland properties from storm waters. From 1936 to 1940, the National Park Service built nearly 1,000 kilometers of fencing along the coast of North Carolina. The fencing formed an obstacle to the windblown sand, trapping it and thereby forming dunes. These dunes were then planted with grass, shrubs, and over two million trees! Today, the resulting barrier of dunes protects inland regions from serious damage by hurricanes.

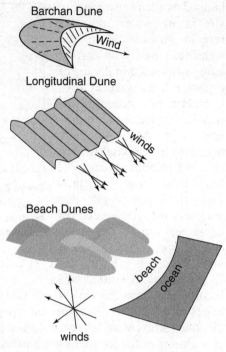

Figure 21.15 Common Shapes and Patterns of Sand Dunes.

DEPOSITION BY GLACIERS

The way in which glaciers transport and deposit sediments is very different from that of any other agent of erosion. To fully understand glacial deposition, let us first consider how glaciers transport their sediment load.

Glacial Transport

Glaciers transport materials suspended in the ice, along the sides and bottom of the ice, on the surface of the ice, and in the area directly alongside the ice. Although it is tempting to picture a glacier as a huge block of ice pushing a pile of debris ahead of it, most glaciers transport very little material by "bulldozing" ahead of the ice. Most of the material carried by a glacier is suspended in the ice.

One of the characteristics that affect the ability of any medium to carry materials in suspension is the viscosity, or resistance to flow, of the medium. The higher the viscosity, the greater is the medium's ability to hold materials in suspension. Ice is millions of times more viscous than liquid water and can carry just about anything in suspension. A boulder that couldn't be

Deposition

budged by the fastest moving stream of water, no less remain suspended in the water, can easily be carried along suspended in the ice of a glacier. See Figure 21.16.

Debris can become suspended in glacial ice in several ways. Many glaciers originate in snowfields high in mountainous regions. Landslides and rockfalls from the steep slopes that surround the glacier deposit debris on top of the snow. When more snow accumulates, the debris is buried and eventually encased in the ice that forms. Also, rock protrusions around which a glacier flows shed fragments that become incorporated into the ice. Where two glaciers merge, the debris along the edges of the glaciers becomes incorporated into the body of the larger glacier that forms. Material also becomes incorporated into the bottom and sides of a glacier as the ice melts and refreezes around loose rock fragments. The 10- to 100-centimeter layer along the base of a glacier is filled with suspended debris.

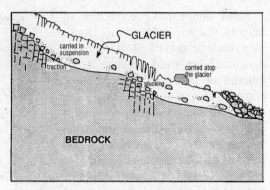

Figure 21.16 Transport of Sediment by a Glacier. A glacier carries sediments on its surface, suspended in the ice, and dragged along its sides and bottom.

Material along the bottom of the glacier is carried along by traction. In **traction**, the particles are not encased in the ice but are dragged along, slipping, sliding, and rolling beneath the moving ice. Material along the sides of the glacier is also moved by traction, and rock fragments may slide or roll off the ice onto ground next to the glacier. Meltwater-soaked debris with the consistency of wet concrete also flows off the sides of the glacier.

Most material carried on the top of a glacier was originally suspended in the ice but has been exposed as melting occurred. Thick debris, however, shields the ice beneath it from melting and results in high spots on the ice surface. Blocks of rock that fall onto a glacier from surrounding slopes can be carried for great distances, conveyor-belt style, atop the glacier.

Types of Glacial Deposits

Part of a glacier's load is carried frozen within the ice or scattered on its surface. The rest is pushed up in piles around the edges of the glacier and carried along as the glacier moves over Earth's surface. When the glacier reaches a warm region, or the climate changes and becomes warm enough, the ice begins to melt. As the glacier melts, the sediments it was carrying are deposited. These deposits may be sorted or unsorted, and each type is formed differently.

Unsorted Deposits

At the melting edge of the glacier, sediments within the ice, as well as those carried on top of it, are released. Particles of all shapes and sizes fall to the ground in a confused jumble, forming an unsorted deposit. In addition, as the ice melts back, piles of sediment that were pushed up around the edges of the glacier are left behind. These sediments, too, are unsorted. They were tumbled about and mixed as the ice scraped them from the ground and pushed them along. See Figure 21.17a.

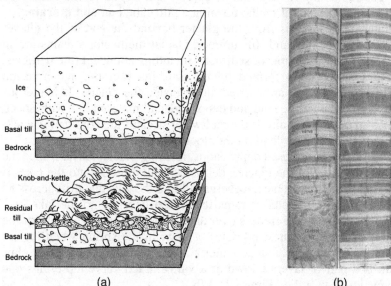

(a) (b)

Figure 21.17 Glacial Deposits. (a) Glacial till. As glacial ice melts, the sediments are released and drop to the ground in a confused jumble, forming an unsorted deposit. (b) Sorted varved clays from a glacial metlwater lake deposit. Meltwater streams may sort glacial sediments; or, as in this photograph, the sediments may settle out in sorted layers at the bottom of lakes fed by meltwater. Source: *The Earth Sciences*, Arthur N. Strahler, Harper & Row, 1971.

Sorted Deposits

Some glacial sediment is deposited into streams that originated from a melting glacier. This material is sorted by the running water and then deposited in layers. Fine sediments carried into lakes by meltwater lakes may settle out in sorted layers on the lake bed. See Figure 21.17b.

How Glaciers Deposit Sediments

Some deposition takes place beneath the ice. The base of a warm glacier is continuously melting because of heat radiating from the ground and heat created by friction with the ground. As particles in the base of the glacier are freed by melting, some are rolled or dragged along until they become lodged

Deposition

in the ground. Refreezing of large particles, which jut out more than finer ones, also leaves behind deposits enriched in fine particles.

Materials on top of a glacier are deposited in one of two ways: either meltwater and debris slides carry them off the glacier, or they are set down on the ground as the ice beneath them melts. When meltwater deposits these materials, they are well sorted and somewhat rounded. Materials that slide or fall off the glacier, however, are jumbled masses of large and small angular fragments. By the time a glacier reaches its end, it has already undergone much melting. The remaining material on its surface is dropped off the end of the glacier as the ice beneath it melts, forming a pile called an **end moraine**.

Meltwater carries debris from the glacier beyond the end of the glacier and deposits it over the land. In summer, glacial meltwater streams are in flood and carry huge amounts of sediment, including abundant fine particles. These fine sediments are formed by the crushing pressure and abrasion between particles as they are carried along inside the glacier. If numerous streams spread out over the land and deposit glacial material in a wide sheet, the result is an **outwash plain**. If the meltwater streams are confined in a valley, they may deposit their load on the floor of the valley.

Meltwater often fills closed depressions, forming glacial lakes. If the lake forms along the edge of the glacier, debris can fall directly into the lake. If the water undercuts the ice face, icebergs fall into the lake and drift across its surface. As an iceberg melts, it rains **ice-rafted** debris onto the lake bed. During the winter, the lake freezes over and deposition ceases except for the settling out of the very finest particles of clay. In the spring when the ice thaws, meltwater carries in coarser material, which is deposited atop the clay. The result is a pair of layers known as a **varve**. Each varve represents one seasonal cycle—a year. See Figure 21.17b.

Landforms Resulting from Glacial Deposition

Glacial drift is any material deposited as a result of glacial activity, including material deposited by meltwater and debris slides. **Glacial till** is material deposited directly from ice. The main difference between the two is sorting; drift may be sorted because of the action of meltwater, whereas till is unsorted.

When glacial drift and till are deposited, they form a number of distinctive landscape features, as shown in Figure 21.18.

Moraines

Moraines are deposits of till. A moraine that forms a thin, widespread layer of till is called a **ground moraine**. Ground moraines form gently rolling hills and valleys. Piles of till deposited around the side of a glacier are called **lateral moraines**. Deposits of till at the front of the glacier are **terminal moraines**. Terminal and lateral moraines form long, parallel ridges. The ridges mark the boundaries where a glacier once existed. See Figure 21.19.

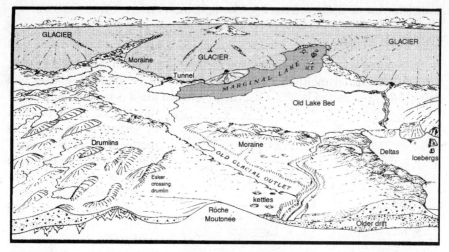

Figure 21.18 Landforms Resulting from Glacial Deposits. Source: Educational Leaflet #28, The New York State Museum, Albany, New York.

Drumlins

Drumlins are long, low mounds of till that have a rounded, teardrop shape, as seen in Figure 21.18. By looking at the figure, you can see the direction in which the glacier moved. The rounded part of the drumlin points in the direction from which the glacier advanced. Drumlins are thought to be molded by the ice of a glacier as it slides over previously deposited piles of sediment.

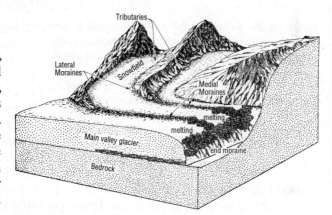

Figure 21.19 How Moraines Form. Lateral moraines consist of materials carried along the sides of a glacier. When two glaciers merge, their lateral moraines are welded together between the glaciers, forming a medial moraine. End moraines form where debris melts out of the ice at the end of the glacier. Source: *Earth*, 4th Ed., Frank Press and Raymond Seiver, W.H. Freeman, 1986.

Erratics

Erratics are large, isolated boulders deposited by a glacier. Many erratics are more than 3 meters in diameter and weigh thousands of metric tons. Such boulders are much too large to be carried by wind or water, but glaciers often transport them far from their sources. When the ice melts, the large boulder

Outwash Plains

At the leading edge of a glacier, some melting is almost always taking place. Streams of glacial meltwater carry sediments out beyond the glacier and deposit them in sorted layers. Over time, a broad plain is built up in front of the glacier by deposition of sediments from glacial meltwater. Such a landform is called an **outwash plain** because the sediments it is made of were "washed out" beyond the glacier by meltwater streams.

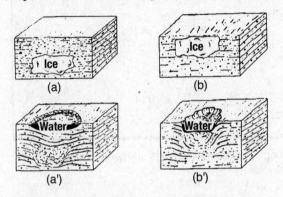

Figure 21.20 Formation of a Kettle. Kettles form when buried blocks of ice melt and the overlying sediments collapse, forming a hole. The melting of deeply buried blocks of ice forms rounder and shallower kettles. In (a) and (a'), the depression formed by melting a deeply buried ice block is seen, whereas (b) and (b') shows the depression formed by melting a shallowly buried ice block. Source: *Living Ice: Understanding Glaciers and Glaciation*, Robert P. Sharp, Cambridge University Press, 1988.

Kettles and Kettle Lakes

A **kettle** is a pit found in a glacial deposit. There are several steps in the formation of a kettle. First, a chunk of glacial ice breaks loose. Next, sediment from the glacier covers the chunk of ice. After a while, the ice melts, the overlying sediment sinks, and the kettle is formed. See Figure 21.20. A kettle may later fill with glacial meltwater, rainwater, or groundwater, forming a **kettle lake**.

Glacial Lakes

In addition to kettle lakes, glaciers may form lakes in two other ways. A glacier may gouge out a large depression in Earth's surface that may fill with water. Alternatively, a **glacial lake** may form when ice or moraines dam the course of a stream; the flow of water backs up behind the moraine, creating a lake. Long, narrow glacial lakes, such as the ones found in central New York, are known as **finger lakes**.

Ice-dammed lakes can cause disastrous floods when the ice dam melts and releases the lake water. There is much evidence that catastrophic floods caused by ice damming occurred repeatedly during the last ice age, 10,000–20,000 years ago.

Glacial lakes impact society in many ways. They form some of this country's most spectacular scenery. In addition to being beautiful, glacial lakes have both recreational and commercial uses. New York's Finger Lakes are used extensively for fishing, boating, and swimming. The Great Lakes, a system of five huge glacial lakes, are the largest repositories of fresh water on Earth. Native Americans and early explorers used them as key transportation routes. Later, they served as major shipping routes for getting the bountiful harvests of our country's heartland to major population centers back east. Large cities, such as Chicago and Detroit, grew from port cities along this route. The building of the Erie Canal created the link needed for a water route from the Great Lakes to New York City, and catapulted New York City from a small coastal port to the major international center of trade it is today.

LANDSCAPE DEVELOPMENT

Throughout this unit, you have learned about geologic processes that change the surface. Crustal movements cause the surface to be uplifted in some places and to subside in others. Volcanic activity injects new rock beneath the surface in some places while bringing new rock to the surface in others. Weathering and erosion reshape the land by wearing away rock exposed at the surface and transporting sediments downslope. When these sediments are deposited, characteristic features are formed. Together, all of these geologic processes create the **topography** of the land, that is, the landforms that shape the surface of Earth. **Landforms** are physical features of Earth's surface that have characteristic shapes and are produced by natural processes. Landforms include both major forms, such as mountains, plateaus, and plains, and minor forms, such as hills, valleys, drumlins, dunes, and many others.

Characteristics of Landscapes

A **landscape** is region in which the landforms are related by their structures, the processes that formed them, and the ways in which they developed. Landscapes are the product of the interaction of geological processes, climate, and human activities on the land over a long period of time. Landscapes can be classified by characteristics that can be readily observed and measured, such as relief, stream patterns, and soil associations.

Relief

Relief refers to the physical shape or general unevenness of a part of Earth's surface, such as variations in slope or elevation. It is expressed in terms of the vertical difference in elevation between the highest and lowest points in a given region. A region showing great variations in elevation has "high relief," and a region showing little variation in elevation has "low relief." Regions with high relief typically have steep slopes; those with low relief, gentle slopes. Most landscapes can be classified as mountains, plateaus, or plains based upon their reliefs.

Deposition

Mountains are parts of Earth's crust that project at least 300 meters above the surrounding land. A mountain generally has steep sides, a restricted summit area, and considerable bare-rock surface. Most mountains are formed by crustal movements. Two common types are fold mountains and fault-block

(a) (b)

Figure 21.21 Mountains. (a) Fold mountains. Source: *The Earth Sciences*, Arthur N. Strahler, Harper & Row, 1971. (b) Fault-block mountains. Source: *Macmillan Earth Science*, Eric Danielson and Edward J. Denecke, Jr., Macmillan, 1989.

mountains. Fold mountains form when layers of rock are folded by compression, wrinkling Earth's crust. Fault-block mountains form when tension forces cause blocks of crust to move along faults. On one side of the fault, rocks rise; on the other side, rocks subside. See Figure 21.21.

Plateaus are large areas of flat land at high elevations. A plateau generally has an underlying structure made of horizontal layers of rock that were gently uplifted. Plateaus are often found next to mountain ranges and were probably raised by the same forces that formed the mountains but were not faulted or folded as greatly as the rocks of the mountains. Although made of horizontal rock layers, the surface of a plateau is often not level. Streams and other agents of erosion can cut deep valleys and canyons into the surface of the plateau.

Plains are large areas of flat land at low elevations. Plains are usually formed by the deposition of sediments in horizontal layers at or below sea level. Thus, the underlying structure of a plain consists of horizontal layers of rock. Many plains have undergone uplifting, which has raised their surfaces above sea level, but have not been uplifted as much as plateaus.

Stream Patterns

Stream patterns refer to the patterns formed by the system of streams in an area as they flow down slopes and join with other streams. A stream that flows into and joins another stream is called a **tributary**. The area from which precipitation drains into a stream or system of streams is called a **drainage basin**. See Figure 21.22. Higher areas, called **divides**, separate drainage basins. Streams from neighboring drainage basins may join into larger streams, which, in turn, may join with others to form even larger streams. Together, all of the streams in an area and their tributaries form a **stream drainage system**. Drainage systems often show distinct patterns that reflect the geology of the region. Figure 21.23 shows some of the more common stream drainage patterns that develop on different geological structures.

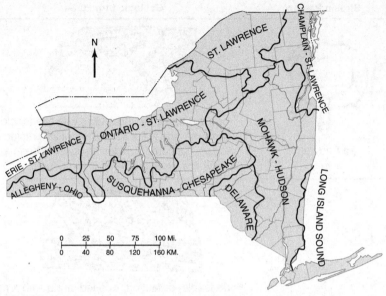

Figure 21.22 Drainage Basins of New York State.

Stream Pattern	Geologic Structures
 (a) Dendritic	 Form where underlying rocks have no pronounced differences in erosion rate, so stream flow is equal in all directions.
(b) Radial	Form on rounded surfaces, such as conical volcanoes or dome mountains, where stream flow away from a central high point.

Deposition

Stream Pattern	Geologic Structures
 (c) Annular	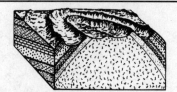 Form around dome mountains with tilted rock layers on their flanks. Differential erosion causes concentric ridges to form. Streams flow in low areas between these ridges.
 (d) Rectangular	 Form is folded, faulted, or tilted strata with long strips of rock of unequal resistance to erosion. Major streams flow along belts of weak rock, making many right-angle turns.
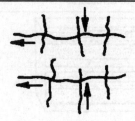 (e) Trellis	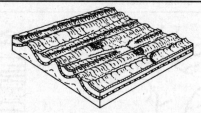 Form in tilted, folded, or faulted strata with long strips of rock of unequal resistance to erosion. Major streams run through long valleys, following belts of weak rock between parallel ridges of stronger rock. Tributary streams enter major streams at right-angles.

Figure 21.23 Stream Drainage Patterns. (a) Dendritic (b) Radial (c) Annular (d) Rectangular (e) Trellis.

Soil Associations

Soil associations are groups of two or more soils, occurring together in a characteristic pattern in a given geographical area. Soils are grouped according to characteristics such as composition, organic content, particle size and shape, porosity and permeability, and maturity (i.e., the degree to which the soil has developed horizons). Soil associations form patterns that reflect the soil parent material, underlying bedrock, slope, vegetative cover, and plant litter in a region.

Climate and Landscapes

Climate refers to the characteristic weather of a region over an extended period of time, particularly the region's precipitation and temperature patterns. Temperature and precipitation affect the development of a landscape by influencing its relief, stream patterns, and soil associations.

Climate influences relief by controlling such factors as the type and rate of weathering, the amount and duration of the runoff that removes and transports weathered materials, and the amount, type, and distribution of plant cover.

The development of a landscape's relief depends on the balance between weathering and the removal of weathered materials by erosion. Temperature and precipitation strongly influence the intensity of weathering. Hot, arid regions and cold, arid regions undergo minimal weathering. Chemical weathering dominates in hot, humid regions, while physical weathering (mainly frost action during colder months) dominates in cold, humid regions. Thus, bedrock tends to be broken down at a faster rate in humid climates than in dry regions. Since running water is the predominant agent of erosion on Earth, precipitation is essential to the removal and transport of weathered materials. Thus, precipitation influences the rate at which the relief of the landscape changes. Over the long term, the more abundant the precipitation, the more rapidly the landscape's relief changes.

Climate also influences relief by controlling the amount, type, and distribution of plant cover. In humid climates, abundant rainfall promotes a protective cover of vegetation that protects the soil from rapid erosion by runoff and produces gently rounded slopes. In arid climates, with little protective vegetation, erosion by rapid runoff leads to steep slopes and exposed bedrock.

You may be surprised to learn that, while arid regions may erode more slowly than humid regions in the long term, the lack of plant cover in arid regions allows the exposed soil to be very rapidly eroded during the few periods of heavy rainfall that do occur. Thus, some of the most rapid erosion rates occur in deserts after a storm! However, over the long term, the brief periods of intense erosion in arid regions are outweighed by the prolonged periods of more moderate erosion in humid regions.

Climate influences stream patterns by controlling such factors as the amount and duration of precipitation, differences between the windward and leeward sides of mountains, and the relationship between precipitation and evaporation. For example, in arid climates, smaller streams tend to remain dry for most of the year, while humid climates tend to have more permanent streams. Thus, humid climates often have a denser network of tributary streams than arid climates. The leeward sides of mountains tend to have less annual precipitation than the windward sides. Thus, the network of streams may vary from one side of the mountains to another.

Soil associations are strongly influenced by climate. Arid climates tend to have soils that are thinner and contain less organic matter than those in humid climates. Since physical weathering is dominant, soil particles in arid climates tend to be larger and more angular than those in humid climates.

Deposition

The more abundant rainfall in humid climates results in greater infiltration, which, in turn, speeds up the rate at which soil horizons form. Thus, humid climates develop mature soils much more rapidly than arid climates. Moisture and higher temperatures also favor chemical weathering processes, which also speed the breakdown of weathered materials and the development of soil horizons.

Landscape Regions

Landscapes with similar relief, stream patterns, and soil associations can be grouped into distinctive **landscape regions**. Landscape regions typically have distinctive physical features that set them apart from one another. On a large scale, such as the continental United States, regions with similar landscape regions are grouped into **physiographic provinces** that are classified mainly as mountains, plateaus, and plains. See Figure 21.24.

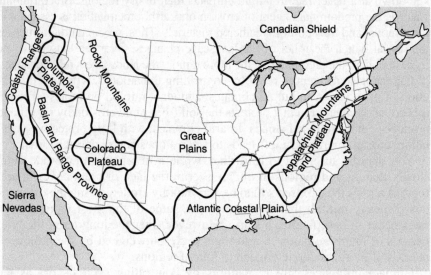

Figure 21.24 Physiographic Provinces of the United States. Source: *UPCO's Review of Earth Science*, Robert B. Sigda, United Publishing Co., 1995.

New York State is divided into a variety of smaller landscape regions based on relief, stream patterns, and the type and structure of the underlying bedrock. See Figure 21.25. The accompanying table gives a brief description of each of the landscape regions in New York State.

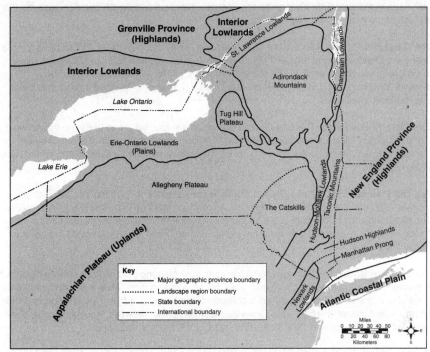

Figure 21.25 Generalized Landscape Regions of New York State. Source: The State Education Department, *Earth Science Reference Tables*, 2011 ed. (Albany, New York; The University of the State of New York).

LANDSCAPE REGIONS OF NEW YORK STATE

Adirondack Mountains. A circular region that is part of the Grenville Province, a large belt of deeply eroded metamorphic bedrock of Proterozoic age. Erosion has stripped away most of the overlying rock, and it is criss-crossed with faults. The resistant rocks of the Adirondacks eroded more slowly than the rocks of the St. Lawrence and Champlain lowlands around them. The Adirondacks have long, straight valleys that formed along faults, gently curved ridges of resistant rock, and a radial drainage pattern.

St. Lawrence and Champlain Lowlands. The St. Lawrence and Champlain lowlands are part of the Interior Lowlands that extend west through the Great Plains. They consist of layers of sedimentary rock that once covered the Adirondacks but were later eroded away. They now form the low-lying plains occupied by the St. Lawrence River and Lake Champlain.

Taconic Mountains. The Taconic Mountains are a region of intensely folded and faulted metamorphic rocks that were thrust into that area from the east by the collision between a volcanic island arc and the North American continent about 450 million years ago. Ridges and valleys generally run north-south because softer rocks have worn away to produce valleys, while more resistant rocks form the ridges.

Allegheny Plateau. The Allegheny Plateau forms the northern end of the Appalachian Plateau to the southwest. It consists of flat-lying layers of sedimentary rock that were deposited in a warm, shallow sea that covered much of New York State during the Late Silurian and Devonian. Later the entire area was uplifted to form the plateau. The plateau sur-

Deposition

face rises to the east until it becomes the Catskills. Streams and their tributaries have cut the surface of the plateau into hilly uplands.

The Catskills. The Catskills are the deeply eroded remains of a huge, apronlike delta formed during the Devonian from sediments that eroded from the ancient Acadian Mountains along the shore of the warm, shallow sea that covered much of New York State. Since then, streams have cut deeply into the sedimentary layers, forming steep-sided valleys that separate rounded "mountains."

Erie-Ontario Lowlands. The Erie-Ontario Lowlands are low, flat areas to the north and west of the Alleghany Plateau, and separated from it by escarpments, or cliffs. Though they are plains, the Erie-Ontario Lowlands are covered with hills of unsorted glacial till and sorted meltwater deposits.

Tug Hill Plateau. East of Lake Ontario, the land rises steadily to form the Tug Hill Plateau, a relatively flat, rocky area separated from the Adirondack Mountains by the Black River Valley. The plateau stands in the path of winter storms that pick up moisture from Lake Ontario. Annual snowfall averages 20 feet, so the plateau is one of the snowiest regions east of the Rocky Mountains.

Hudson Highlands. The Hudson Highlands consist of metamorphic rocks that were originally deposited as sedimentary and volcanic rocks 1.3 billion years ago. They were then metamorphosed into gneiss and marble during a collision of continents 1.1 billion years ago that thrust up an ancient mountain chain—the Grenville Mountains. The Grenville Mountains have since been eroded flat, exposing their roots—the rocks of the Hudson Highlands.

Hudson-Mohawk Lowlands. This region covers most of the Hudson River Valley and the Mohawk River Valley. The bedrock consists of relatively soft sedimentary rocks that are easily eroded. The surrounding highlands are made of more resistant rocks. These lowlands provide the nation's only natural, navigable waterway through the Appalachian Mountains and serve as an important transportation route between the Atlantic and the Great Lakes.

Newark Lowlands. The Newark Lowlands form a gently rolling surface broken by ridges, between the Hudson Highlands and the Manhattan Prong. Formed on layers of igneous and sedimentary rock dating from the Age of Dinosaurs (Triassic-Jurassic), the surface of these lowlands is broken by ridges of igneous rock that is more resistant to erosion. The Palisades Sill intruded into the rocks of the Newark Lowlands 195 million years ago.

Manhattan Prong. The Manhattan Prong consists of metamorphic rock that was folded and faulted by the same collision that formed the Taconic Mountains and was later eroded by glacial ice into a landscape of rolling hills and valleys. Resistant gneiss, schist, and quartzite form the hills, and less resistant marble underlies the valleys.

Atlantic Coastal Plain. The Atlantic Coastal Plain is a flat, low-lying area that slopes toward the Atlantic Ocean. In New York State, it consists of Staten Island and Long Island, both of which are important residential areas. Long Island is composed mainly of glacial sediments. Two terminal moraines form its hilly northern edge while a broad, flat glacial outwash plain stretches from the terminal moraine to the Atlantic Ocean. Long offshore barrier islands with broad beaches make Long Island a popular recreational area.

MULTIPLE-CHOICE QUESTIONS

In each case, write the number of the word or expression that best answers the question or completes the statement.

1. Why do the particles carried by a river settle to the bottom as the river enters the ocean?
   (1) The density of the ocean water is greater than the density of the river water.
   (2) The kinetic energy of the particles increases as the particles enter the ocean.
   (3) The velocity of the river water decreases as it enters the ocean.
   (4) The large particles have a greater surface area than the small particles.

2. A stream flowing at a velocity of 250 centimeters per second is transporting sediment particles ranging in size from clay to cobbles. Which transported particles will be deposited by the stream if its velocity decreases to 100 centimeters per second?
   (1) cobbles, only
   (2) cobbles and some pebbles, only
   (3) cobbles, pebbles, and some sand, only
   (4) cobbles, pebbles, sand, silt, and clay

3. The largest particles that a stream deposits as it enters a pond are 8 centimeters in diameter. The minimum velocity of the stream is approximately
   (1) 100 cm/sec   (3) 300 cm/sec
   (2) 200 cm/sec   (4) 400 cm/sec

4. A stream is transporting the particles W, X, Y, and Z, shown below.

Density = 3.8 g/mL   Density = 3.8 g/mL   Density = 2.4 g/mL   Density = 2.4 g/mL

Which particle will most likely settle to the bottom first as the velocity of this stream decreases?
   (1) W   (3) Y
   (2) X   (4) Z

5. When small particles settle through water faster than large particles, the small particles are probably
   (1) lighter   (3) better sorted
   (2) flatter   (4) more dense

Deposition

Base your answers to questions 6 through 8 on the diagram below, which shows a coastal region in which the land slopes toward the ocean. Point X is near the top of the hill, point Y is at the base of the hill, and point Z is a location at sea level. The same type of surface bedrock underlies this entire region. A stream flows from point X through point Y to point Z. This stream is not shown in the diagram.

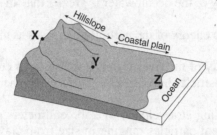

6. Which diagram best shows the most probable path of the stream flowing from point X to point Z?

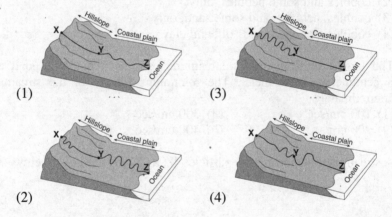

7. Compared with the stream velocity between point X and point Y, the stream velocity between point Y and point Z is most likely
   (1) greater since the slope of the land decreases
   (2) greater since the slope of the land increases
   (3) less since the slope of the land decreases
   (4) less since the slope of the land increases

518

8. Which cross section best shows the pattern of sediments deposited by the stream as it enters the ocean near point Z?

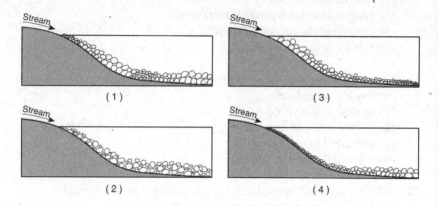

Base your answers to questions 9 through 12 on the diagram below. The arrows show the direction in which sediment is being transported along the shoreline. A barrier beach has formed, creating a lagoon (a shallow body of water in which sediments are being deposited). The eroded headlands are composed of diorite bedrock. A groin has recently been constructed. Groins are wall-like structures built into the water perpendicular to the shoreline to trap beach sand.

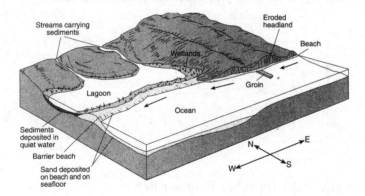

9. The groin structure will change the pattern of deposition along the shoreline, initially causing the beach to become
   (1) wider on the western side of the groin
   (2) wider on the eastern side of the groin
   (3) narrower on both sides of the groin
   (4) wider on both sides of the groin

Deposition

10. Which two minerals are most likely found in the beach sand that was eroded from the headlands?
    (1) quartz and olivine
    (2) plagioclase feldspar and amphibole
    (3) potassium feldspar and biotite
    (4) pyroxene and calcite

11. The sediments that have been deposited by streams flowing into the lagoon are most likely
    (1) sorted and layered
    (2) sorted and not layered
    (3) unsorted and layered
    (4) unsorted and not layered

12. Which event will most likely occur during a heavy rainfall?
    (1) Less sediment will be carried by the streams.
    (2) An increase in sea level will cause more sediment to be deposited along the shoreline.
    (3) The shoreline will experience a greater range in tides.
    (4) The discharge from the streams into the lagoon will increase.

Base your answers to questions 13 through 16 on the contour map below, which shows a hill formed by glacial deposition near Rochester, New York. Letters *A* through *E* are reference points. Elevations are in feet.

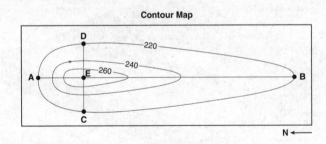

Contour Map

13. This glacial deposit is best identified as
    (1) a U-shaped valley   (3) a drumlin
    (2) a sand dune   (4) an outwash plain

14. Which description best compares the gradients of this hill?
    (1) *AE* and *EB* have the same gradient.
    (2) *AE* has a steeper gradient than *EB*.
    (3) *CE* has a steeper gradient than *ED*.
    (4) *CE* and *AE* have the same gradient.

15. Which set of characteristics most likely describes the sediment in this glacial deposit?
    (1) sorted and layered
    (2) sorted and not layered
    (3) unsorted and not layered
    (4) unsorted and layered

16. The hill shown on this map is found in which New York State landscape region?
    (1) Adirondack Mountains
    (2) Catskills
    (3) Atlantic Coastal Plain
    (4) Erie-Ontario Lowlands

Base your answers to questions 17 through 19 on the map of Long Island, New York. *AB*, *CD*, *EF*, and *GH* are reference lines on the map.

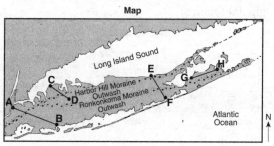

17. Which agent of erosion transported the sediments that formed the moraines shown on the map?
    (1) water
    (2) wind
    (3) ice
    (4) mass movement

18. The cross section below represents the sediments beneath the land surface along one of the reference lines shown on the map.

Along which reference line was the cross section taken?
    (1) *AB*
    (2) *CD*
    (3) *EF*
    (4) *GH*

19. A major difference between sediments in the outwash and sediments in the moraines is that the sediments deposited in the outwash are
    (1) larger
    (2) sorted
    (3) more angular
    (4) older

Deposition

20. What will be the most probable arrangement of rock particles deposited directly by a glacier?
(1) sorted and layered
(2) sorted and not layered
(3) unsorted and layered
(4) unsorted and not layered

21. Which statement presents the best evidence that a boulder-sized rock is an erratic?
(1) The boulder has a rounded shape.
(2) The boulder is larger than surrounding rocks.
(3) The boulder differs in composition from the underlying bedrock.
(4) The boulder is located near potholes.

22. The cross sections below show a three-stage sequence in the development of a glacial feature.

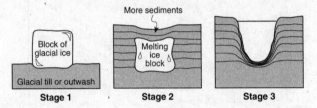

Which glacial feature has formed by the end of stage 3?
(1) kettle lake
(2) finger lake
(3) drumlin
(4) parallel scratches

Base your answers to questions 23 through 25 on the block diagram below, which shows some of the landscape features formed as the most recent continental glacier melted and retreated across western New York State.

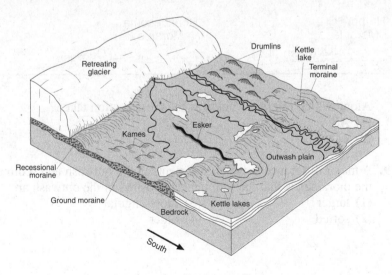

23. During which geologic epoch did this glacier retreat from New York State?
    (1) Pleistocene
    (2) Eocene
    (3) Late Pennsylvanian
    (4) Early Mississippian

24. The moraines pictured in the block diagram were deposited directly by the glacier. The sediments within these moraines are most likely
    (1) sorted by size and layered
    (2) sorted by size and unlayered
    (3) unsorted by size and layered
    (4) unsorted by size and unlayered

25. The shape of elongated hills labeled drumlins is most useful in determining the
    (1) age of the glacier
    (2) direction of glacial movement
    (3) thickness of the glacial ice
    (4) rate of glacial movement

26. The particles in a sand dune deposit are small, are very well sorted, and have surface pits that give them a frosted appearance. This deposit was most likely transported by
    (1) ocean currents
    (2) glacial ice
    (3) gravity
    (4) wind

27. The photograph below shows farm buildings partially buried in silt.

    Which erosional agent most likely piled the silt against these buildings?
    (1) glacial ice
    (2) ocean waves
    (3) wind
    (4) mass movement

Deposition

28. New York State landscape regions are identified and classified primarily by their
    (1) surface topography and bedrock structure
    (2) existing vegetation and type of weather
    (3) latitude and longitude
    (4) chemical weathering rate and nearness to large bodies of water

29. The table below describes the characteristics of three landscape regions, A, B, and C, found in the United States.

Landscape	Bedrock	Elevation/Slopes	Streams
A	Faulted and folded gneiss and schist	High elevation Steep slopes	High velocity Rapids
B	Layers of sandstone and shale	Low elevation Gentle slopes	Low velocity Meanders
C	Thick horizontal layers of basalt	Medium elevation Steep to gentle slopes	High to low velocity Rapids and meanders

    Which list best identifies landscapes A, B, and C?
    (1) A—mountain, B—plain, C—plateau
    (2) A—plain, B—plateau, C—mountain
    (3) A—plateau, B—mountain, C—plain
    (4) A—plain, B—mountain, C—plateau

30. Which sequence shows the order in which landscape regions are crossed as an airplane flies in a straight course from Albany, New York, to Massena, New York?
    (1) plateau→plain→mountain
    (2) plateau→mountain→plain
    (3) plain→mountain→plain
    (4) mountain→plain→plateau

31. The Catskills landscape region is classified as a plateau primarily because the region has
    (1) V-shaped valleys
    (2) jagged hilltops
    (3) horizontal bedrock structure
    (4) folded metamorphic rock

32. In which New York State landscape region is most of the surface bedrock composed of metamorphic rock?
    (1) Adirondacks
    (2) Catskills
    (3) Erie-Ontario Lowlands
    (4) Newark Lowlands

33. The block diagrams below, labeled A, B, and C, show the relative elevation and rock structure of three different landscape regions.

    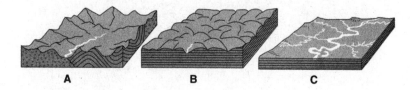

    Which set correctly identifies the landscape region shown in each block diagram?
    (1) A—mountain, B—plateau, C—plain
    (2) A—mountain, B—plain, C—plateau
    (3) A—plateau, B—mountain, C—plain
    (4) A—plateau, B—plain, C—mountain

34. Which two landscape regions in New York State have the oldest surface bedrock?
    (1) Allegheny Plateau and Newark Lowlands
    (2) Tug Hill Plateau and Erie-Ontario Lowlands
    (3) Taconic Mountains and the Catskills
    (4) Adirondack Mountains and Hudson Highlands

35. Which two New York State landscape regions are formed mostly of surface bedrock that is approximately the same geologic age?
    (1) Manhattan Prong and Atlantic Coastal Plain
    (2) Erie-Ontario Lowlands and Adirondack Mountains
    (3) Adirondack Mountains and Allegheny Plateau
    (4) Tug Hill Plateau and St. Lawrence Lowlands

36. The entire area drained by a river and its tributaries is called a
    (1) delta   (3) valley
    (2) watershed   (4) floodplain

Deposition

Base your answers to questions 37 through 39 on the map below, which shows watershed regions of New York State.

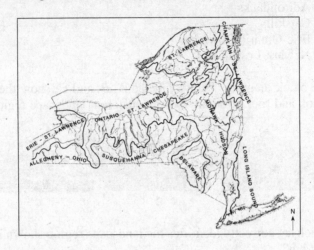

37. On which type of landscape region are both the Susquehanna-Chesapeake and the Delaware watersheds located?
    (1) plain
    (2) plateau
    (3) mountain
    (4) lowland

38. In which watershed is the Genesee River located?
    (1) Ontario-St. Lawrence
    (2) Susquehanna-Chesapeake
    (3) Mohawk-Hudson
    (4) Delaware

39. Most of the surface bedrock of the Ontario–St. Lawrence watershed was formed during which geologic time periods?
    (1) Precambrian and Cambrian
    (2) Ordovician, Silurian, and Devonian
    (3) Mississippian, Pennsylvanian, and Permian
    (4) Triassic, Jurassic, and Cretaceous

Base your answers to questions 40 through 42 on the map below, which shows the drainage basin of the Mississippi River system. Several rivers that flow into the Mississippi River are labeled. The arrow at location *X* shows where the Mississippi River enters the Gulf of Mexico.

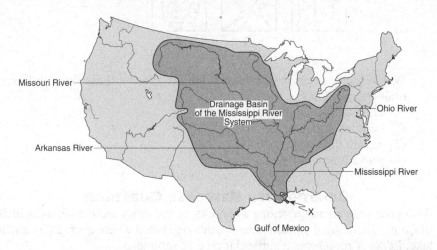

40. The entire land area drained by the Mississippi River system is referred to as a
   (1) levee
   (2) watershed
   (3) meander belt
   (4) floodplain

41. Sediments deposited at location *X* by the Mississippi River most likely have which characteristics?
   (1) angular fragments arranged as mixtures
   (2) rock particles arranged in sorted beds
   (3) rocks with parallel scratches and grooves
   (4) high-density minerals with hexagonal crystals

42. The structure formed by the deposition of sediments at location *X* is best described as a
   (1) moraine
   (2) tributary
   (3) delta
   (4) drumlin

Deposition

43. The maps below labeled A, B, and C show three different stream drainage patterns.

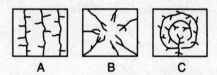

Which factor is primarily responsible for causing these three different drainage patterns?
(1) amount of precipitation
(2) bedrock structure
(3) stream discharge
(4) prevailing winds

CONSTRUCTED RESPONSE QUESTIONS

Base your answers to questions 44 and 45 on the cross section below, which illustrates the normal pattern of sediments deposited where a stream enters a lake. Letter X represents a particular type of sediment.

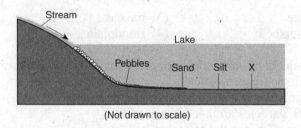

(Not drawn to scale)

44. Briefly explain why deposition of sediment usually occurs where a stream enters a lake. [1]

45. Name the type of sediment most likely represented by letter X. [1]

46. New York State's Adirondacks are classified as a mountain landscape region. Describe one bedrock characteristic and one land surface characteristic that were used to classify the Adirondacks as a mountain landscape region. [2]

Base your answers to questions 47 through 50 on maps *A*, *B*, and *C* below, which show evidence that much of New York State was once covered by a glacial ice sheet. Map *A* shows the location of the Finger Lakes Region in New York State. The boxed areas on map *A* were enlarged to create maps *B* and *C*. Map *B* shows a portion of a drumlin field near Oswego, New York. Map *C* shows the locations of glacial moraines and outwash plains on Long Island, New York.

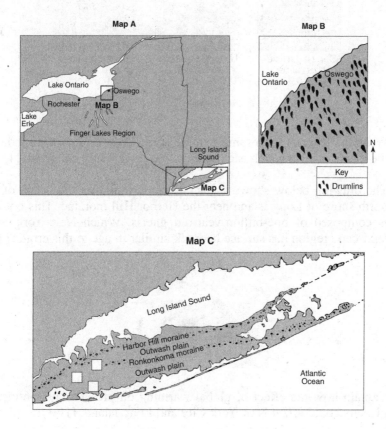

47. The arrangement of the drumlins on map *B* indicates that a large ice sheet advanced across New York State in which compass direction? [1]

Deposition

48. The diagrams below represent three sediment samples labeled *X*, *Y*, and *Z*. These samples were collected from three locations marked with empty boxes (□) on map *C*.

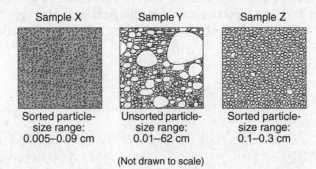

Sample X — Sorted particle-size range: 0.005–0.09 cm

Sample Y — Unsorted particle-size range: 0.01–62 cm

Sample Z — Sorted particle-size range: 0.1–0.3 cm

(Not drawn to scale)

Write the letter of each sample in the correct box on map *C* to indicate the location from which each sample was most likely collected. [1]

49. The drawing below shows a glacial erratic found on the beach of the north shore of Long Island near the Harbor Hill moraine. This boulder is composed of one-billion-year-old gneiss. Which New York State landscape region has surface bedrock similar in age to this erratic? [1]

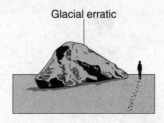

Glacial erratic

50. Explain how the effect of global warming on present-day continental glaciers could affect New York City and Long Island. [1]

Base your answers to questions 51 through 53 on the map below, which shows the generalized surface bedrock for a portion of New York State that appears in the *2011 Edition Reference Tables for Physical Setting/Earth Science*.

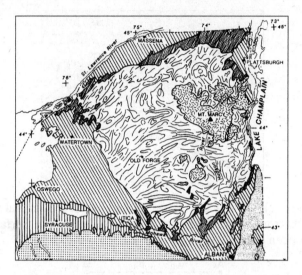

51. Place an *X* on the map to represent a location in the Tug Hill Plateau landscape region. [1]

52. State the longitude of Mt. Marcy, New York, to the nearest degree. The units and compass direction must be included in your answer. [1]

53. Identify the geologic age and name of the surface metamorphic bedrock found at Mt. Marcy. [1]

Deposition

EXTENDED CONSTRUCTED RESPONSE QUESTIONS

Base your answers to questions 54 through 58 on the map below, which shows a meandering stream as it enters a lake. Points A through D represent locations in the stream.

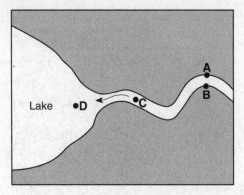

54. In the box below, draw a cross-sectional view of the general shape of the stream bottom between points *A* and *B*. The water surface line has already been drawn. [1]

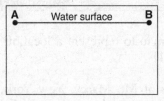

55. State the relationship between stream velocity and the size of the sediment the stream can carry. [1]

56. Describe how the size and shape of most pebbles change when the pebbles are transported in a stream over a great distance. [1]

57. The stream velocity at point *C* is 100 centimeters per second. The stream velocity at point *D* is 40 centimeters per second. Identify one sediment particle most likely being deposited between points *C* and *D*. [1]

58. Deposition is affected by particle density. On the grid below, draw a line to show the relationship between particle density and settling rate. [1]

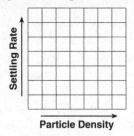

Base your answers to questions 59 through 62 on the topographic map below, which shows an area of the Saranac River just west of Plattsburgh, New York. Points A and B are locations in the river.

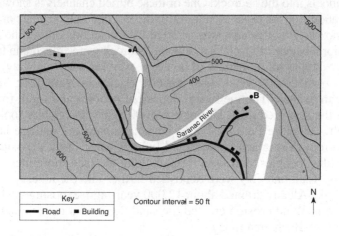

59. In this region of the Saranac River, the land area that is lower in elevation than 450 feet is a floodplain. On the map, draw a diagonal-line pattern, ▨, to indicate the entire floodplain area. [1]

60. Describe how the contour lines shown on the map indicate that the Saranac River flows from point A to point B. [1]

61. Why is erosion of the stream bank more likely at point A than at point B? [1]

62. Identify one emergency preparedness activity that people living in the floodplain area can take to protect themselves and their property from possible flooding. [1]

Base your answers to questions 63 through 67 on the passage and the cross section below. The passage describes the geologic history of the Pine Bush region near Albany, New York. The cross section shows the bedrock and overlying sediment along a southwest to northeast diagonal line through a portion of this area. Location A shows an ancient buried stream channel, and location B shows a large sand dune.

The Pine Bush Region

The Pine Bush region, just northwest of Albany, New York, is a 40-square-mile area of sand dunes and wetlands covered by pitch pine trees and scrub oak bushes. During the Ordovician Period, this area was covered by a large sea. Layers of mud and

Deposition

sand deposited in this sea were compressed into shale and sandstone bedrock.

During most of the Cenozoic Era, running water eroded stream channels into the bedrock. One of these buried channels is shown at location A in the cross section. Over the last one million years of the Cenozoic Era, this area was affected by glaciation. During the last major advance of glacial ice, soil and bedrock were eroded and later deposited as till (a mixture of boulders, pebbles, sand, and clay).

About 20,000 years ago, the last glacier in New York State began to melt. The meltwater deposited pebbles and sand, forming the stratified drift. During the 5000 years this glacier took to melt, the entire Pine Bush area became submerged under a large 350-foot-deep glacial lake called Lake Albany. Delta deposits of cobbles, pebbles, and sand formed along the lake shorelines, and beds of silt and clay were deposited farther into the lake.

Lake Albany drained about 12,000 years ago, exposing the lake bottom. Wind erosion created the sand dunes that cover much of the Pine Bush area today.

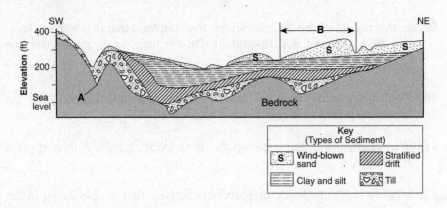

63. According to the passage, how old is the bedrock shown in the cross section? [1]

64. What evidence shown at location A suggests that the channel in the bedrock was eroded by running water? [1]

65. List, from oldest to youngest, the four types of sediment shown above the bedrock in the cross section. [1]

66. Explain why the till layer is composed of unsorted sediment. [1]

67. How does the shape of the sand dune at location B provide evidence that the prevailing winds that formed this dune were blowing from the southwest? [1]

Unit Seven: THE ATMOSPHERE, WEATHER, AND CLIMATE

Chapter 22

THE ATMOSPHERE

> **KEY IDEAS** The atmosphere can be thought of as a huge heat engine that converts heat energy into mechanical energy. Earth's atmospheric heat engine is powered primarily by solar energy and is influenced by gravity.
>
> When the various electromagnetic waves that emanate from the Sun strike the atmosphere, only a small percentage are directly absorbed, especially by gases such as ozone, carbon dioxide, and water vapor. Clouds and Earth's surface reflect some energy back into space, and Earth's surface absorbs some. The energy absorbed is then transferred between Earth's surface and the atmosphere by conduction, convection, radiation, and evaporation.
>
> The transfer of heat energy from solar radiation, and of water vapor, into and out of the atmosphere causes regions of different densities to form. The rise or fall of regions of different densities due to the action of gravitational force produces atmospheric circulation, which is affected by Earth's rotation. Atmospheric circulation distributes solar energy over the whole Earth. The interaction of these processes results in the complex atmospheric occurrence known as weather.

KEY OBJECTIVES
Upon completion of this unit, you will be able to:

- Explain why the Sun emits different types of electromagnetic radiation.
- Compare and contrast the different forms of electromagnetic radiation shown in the *Reference Tables for Physical Setting/Earth Science*.
- Describe the mechanisms by which solar energy is absorbed by the atmosphere and the factors that determine the amount of solar energy an area receives.
- Explain how density differences in the atmosphere cause atmospheric circulation.
- Explain how Earth's rotation influences atmospheric circulation (a phenomenon known as the Coriolis effect), and interpret a diagram of the planetary wind and pressure belts.
- Explain how energy can be stored or released during phase changes.

The Atmosphere

SOLAR RADIATION: THE ATMOSPHERE'S PRIMARY ENERGY SOURCE

The Sun is the atmosphere's major source of energy. Energy from the Sun reaches Earth in the form of electromagnetic waves.

Energy from Electromagnetic Waves

A magnet can make a compass needle move from a distance because the magnet is surrounded by an invisible magnetic field. If you move the magnet, the magnetic field will move and the moving magnetic field, in turn, will cause the compass needle to move. In much the same way, a statically charged balloon held near your head will attract your hair from a distance because the balloon is surrounded by an invisible electric field. If you move the balloon, the invisible electric field moves with it and you can feel the effect of the moving field on your hair. Such fields extend outward infinitely in all directions from their sources, but they weaken with distance from these sources (see Figure 22.1).

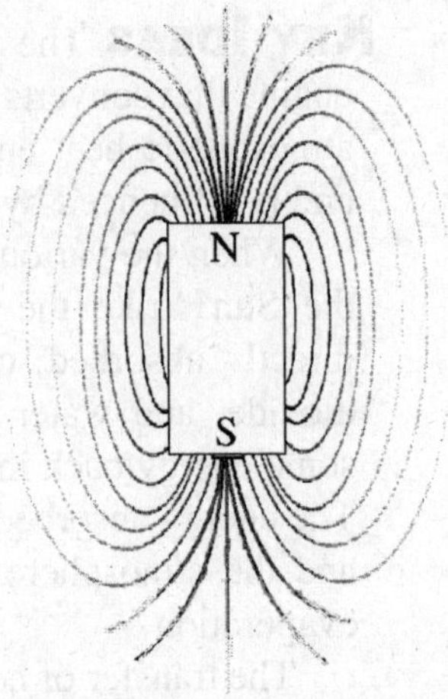

All matter is composed of atoms. Every atom consists of electrically charged particles such as protons and electrons, which are surrounded by an electric field. Whenever a charged particle moves, a magnetic force is produced and the particle is then also surrounded by a magnetic field. These two fields—an electric field and a magnetic field—exist simultaneously around all electrically charged particles that are moving. Together, they are called an **electromagnetic field**, and they extend outward infinitely in all directions around the particles.

Figure 22.1 An Invisible Magnetic Field Surrounding a Magnet. The lines represent magnetic force; closely spaced, the lines mean stronger magnetic force. Note that the strength of the magnetic field decreases with distance.

When a particle moves back and forth, its electromagnetic field moves with it (see Figure 22.2a). Like a ripple in a pond, this movement spreads out through the field in the form of a transform wave. The way that the wave moves through the electromagnetic field depends on the way in which the particle moves. The faster the particle oscillates, the shorter the wavelength, or distance between successive "ripples" in the electromagnetic field. Every frequency of oscillation produces a wave of a different wavelength (see Figure 22.2b).

A moving wave contains energy. It can exert forces on matter with which it interacts. Since an electromagnetic field does not require a medium to exist, it can extend through space. Disturbances in electromagnetic fields

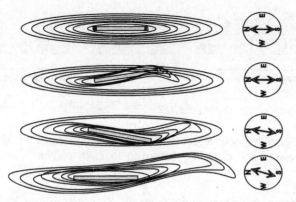

(a) As a magnet moves, its magnetic field moves too. Like a ripple in a pond, the movement of the field spreads outward in all directions at a rate of 3×10^8 meters per second—the speed of light. When the moving field reaches *another* field (such as the one around a compass needle), it exerts a force of attraction or repulsion that can cause motion of the field and even of the object that produced the field. A similar disturbance would travel through the electric field around an electrically charged particle in motion.

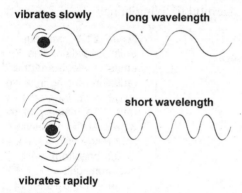

(b) Electromagnetic waves produced by particles vibrating at different rates. The higher the temperature, the faster the particle vibrates, and the shorter the wavelength produced.

Figure 22.2 Electromagnetic Waves Are Produced by Vibrating Particles.

around particles can travel through space as they move outward, or radiate, through the field. In this way, energy is transferred from the Sun to Earth without the existence of a physical medium between the two.

The Electromagnetic Spectrum

Electromagnetic waves are classified by their lengths and range from short waves, such as X rays, to long waves, such as radio waves. The **electromagnetic spectrum** (see Figure 22.3) is a continuum in which electromagnetic waves are arranged in order, from longest to shortest wavelength. Infrared (heat) waves and visible-light waves fall roughly in the middle range of wavelengths. The wavelength, or distance between "ripples" of visible light waves, falls between 10^{-6} and 10^{-7} meter.

The Atmosphere

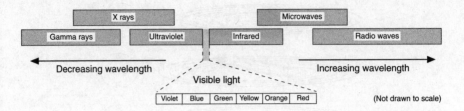

Figure 22.3 Electromagnetic Spectrum. This shows the categories into which electromagnetic waves are classified according to their lengths. Source: *Earth Science Reference Tables*—2011 Edition.

Many different waves emanate from the Sun (see Figure 22.4) because it contains particles moving at many different speeds. However, most waves coming from the Sun are in the visible-light and infrared ranges. To move at the speed needed to emit waves of this length, particles at the Sun's surface must have temperatures between 5,000K and 7,000K. Earth's magnetic field and the Van Allen belts of charged particles in the upper atmosphere deflect many of the waves with short lengths. This circumstance is fortunate, since most short-wavelength radiation is harmful to living things.

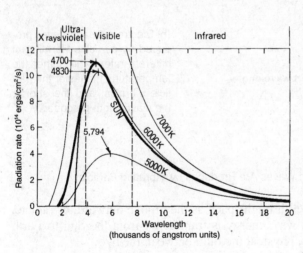

Figure 22.4 Makeup of Solar Radiation. The Sun emits electromagnetic radiation ranging in wavelength from infrared to X rays. The majority of the waves are in the visible-light range with a peak at 4,700 angstrom units (1 angstrom unit = 1×10^{-9} meter—one billionth of a meter). This is very close to the theoretical output of a body at a temperature at 6,000K, which is why the Sun's surface temperature is thought to be between 5,000 and 7,000K.

The Global Radiation Budget

When Earth intercepts electromagnetic waves from the Sun, the moving waves exert a force on the fields surrounding particles of matter in the atmosphere, hydrosphere, and lithosphere, causing the particles to move. This increased motion shows up as an increase in the level of Earth's heat energy. At the same time, Earth's matter radiates energy in the form of electromagnetic waves back out into space, causing a decrease in the level of Earth's heat energy. Thus, both incoming solar radiation and Earth's outgoing, or terrestrial, radiation pass through the atmosphere. Table 22.1 shows the global radiation budget.

TABLE 22.1 GLOBAL RADIATION BUDGET*

Incoming Solar Radiation	Percent Gain	Percent Loss
Reflection from clouds to space		21
Diffuse reflection (scattering) to space		5
Direct reflection from Earth's surface		6
Net energy loss back to space		32
Absorbed by clouds	3	
Absorbed by molecules, dust, water vapor, and CO_2	15	
Absorbed by Earth's surface	50	
Net incoming energy gained by Earth-atmosphere system	**68%**	

Outgoing Terrestrial Radiation	Percent Gain	Percent Loss
Total infrared radiation emitted by Earth's surface	98	
Absorbed by atmosphere	90	
Lost to space		8
Total infrared radiation emitted by the atmosphere	137	
Absorbed by Earth's surface	77	
Lost to space		60
Net outgoing energy lost from entire Earth-atmosphere system		**68%**
Net energy leaving *Earth's surface* (98% emitted − 77% reabsorbed)		21
Net energy leaving the *atmosphere* (137% emitted − 90% reabsorbed)		47

*Source: *The Earth Sciences*, Arthur N. Strahler, Harper & Row, 1971.

Radiative Balance

Earth's average global temperature depends upon the balance between the energy gained by absorbing sunlight and the energy lost by radiating heat. As you can see in Table 22.1, the Earth-atmosphere system's net energy gain from incoming solar radiation is the same as its net energy loss by terrestrial radiation—68 percent. Over long periods of time, then, the average level of Earth's heat energy remains fairly constant, and the average temperature of Earth's surface hardly changes. Therefore, we say that Earth is in **radiative balance**; it is reradiating as much energy as it absorbs.

This fact, however, *does not mean that all points on Earth's surface are in radiative balance at all times*. Most places on Earth go through temperature changes on both a daily and a yearly basis. These temperature changes indicate that there are times when a place is gaining more energy than it is losing and times when it is losing more energy than it is gaining.

There is also much evidence that over long periods of time Earth experiences warming and cooling trends. These may be triggered by changes in Earth's tilt due to precession as it spins on its axis and to increases or decreases in certain gases in the atmosphere.

Let us now consider some of the factors that influence the way in which Earth absorbs incoming solar radiation.

FACTORS AFFECTING INSOLATION

The Sun provides nearly all of the energy received at Earth's surface. Solar radiation reaches the upper atmosphere at a fairly constant rate of about 200 kilocalories per minute per square meter. About one-third of this radiation is reflected back into space, mostly by clouds. As the remaining radiation passes through the atmosphere, some is absorbed by gas molecules. Some is refracted as it crosses boundaries between layers of differing densities. Some is scattered by particles of dust or aerosols in the air. The radiation that reaches Earth's surface, called **insolation**, short for *in*coming *sol*ar radi*ation*, is either reflected or absorbed.

A number of factors control the amount of solar energy that an area absorbs or reflects, including the angle at which insolation strikes the surface, the length of time each day that insolation is received (the duration of insolation), and the nature of the surface.

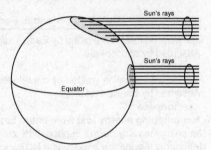

Angle of Insolation

Latitude

Figure 22.5 Insolation near the Equator and the North Pole. Note that the two beams of sunlight approaching Earth are identical, but the angle at which they strike Earth's surface causes insolation received near the Equator to be more concentrated than that received near the poles. The result is greater heating at the Equator.

Since Earth is a sphere, insolation does not strike all points on Earth's surface at the same angle. Near the Equator insolation strikes the surface almost vertically, but near the poles it strikes at a more glancing angle. Therefore, the insolation reaching the tropics is more concentrated than that reaching polar regions (see Figure 22.5).

Daily and Annual Cycles

The angle at which insolation strikes Earth's surface at any location also varies in both daily and annual cycles. The daily cycle starts at dawn with insolation striking the surface at a very low angle; it then becomes increasingly direct throughout the morning, reaches its greatest directness at noon, and becomes increasingly indirect again throughout the afternoon until sunset.

The annual cycles vary with latitude and Earth's position in its orbit around the Sun. In the United States, insolation is most direct on June 21, the summer solstice, and least direct on December 21, the winter solstice.

Figure 22.6a shows the daily cycle of intensity of solar radiation; Figure 22.6b, the yearly cycle of intensity of solar radiation at noon.

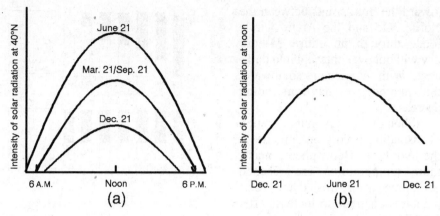

Figure 22.6 Daily and Yearly Cycles of Solar Radiation Intensity. (a) Daily cycle of intensity of solar radiation from 6 A.M. to 6 P.M. at 40°N on the equinox and solstice days. Note that for each day there is a cyclic change from low intensity in early morning to higher intensity at noon and then back to low intensity in late afternoon. (b) Yearly cycle of intensity of solar radiation at noon for an entire year. Note the cyclic change from low intensity at noon on the winter solstice to high intensity at noon on the summer solstice and back to low intensity at the next winter solstice.

Duration of Insolation

The length of time that the surface receives insolation each day, or the **duration of insolation**, depends on the season and the latitude of the location.

Season

The length of daylight varies in an annual cycle with the season (see Figure 22.7). In the United States, the greatest number of hours of insolation is received on June 21, the summer solstice. Each day thereafter, the number of daylight hours decreases, reaching a minimum on December 21, the winter solstice. Then the number of daylight hours increases daily until it peaks again the following June 21. At the fall and spring equinoxes, the number of daylight hours equals the number of hours of darkness. New York State varies from roughly 15 hours of daylight at the summer solstice to only 9 hours at the winter solstice.

Latitude

The length of daylight on any given day also varies with location north or south of the Equator (see Figure 22.8). Since Earth's axis of rotation is tilted, the circles that form the parallels of latitude are also tilted. At different latitudes, different fractions of the parallels are in daylight and darkness. Thus, observers at different latitudes spend different fractions of each 24-hour rotation in daylight and in darkness. When the Northern Hemisphere is tilted

The Atmosphere

toward the Sun, points between the North Pole and the Arctic Circle rotate through an entire 24-hour day without ever entering into darkness. With movement southward, the number of daylight hours decreases.

Although the maximum duration of insolation occurs on June 21 in the Northern Hemisphere, maximum temperatures are reached sometime after this date. Temperatures continue to increase after June 21 because the number of daylight hours still exceeds the number of nighttime hours for many weeks after this date. As long as more energy is received each day than is lost overnight, the temperature will continue to increase.

Nature of Earth's Surface

Earth's surface consists of a wide variety of substances, including ice, water, soil of many types, and vegetation. All interact differently with the insolation that strikes them.

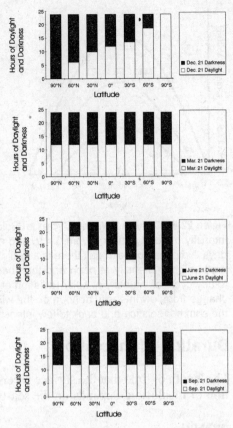

Figure 22.7 Duration of Day and Night at Different Locations on Key Dates Throughout the Year.

Color and Texture

Just as a ball is more likely to bounce off a concrete wall than off a pillow, insolation striking different substances will have a greater or lesser tendency to be absorbed or reflected. Which process will predominate depends on the molecular structure of the substance. The color of a substance is a fairly good indicator of whether the substance absorbs more insolation than it reflects or vice-versa.

The lighter the color of a substance, the more light is being reflected by it; the darker the color, the more light is being absorbed. When insolation strikes substances such as sand and snow, their light color indicates that much of the insolation is being reflected. On the other hand, when insolation strikes rich black soil or deep green leaves, their dark color indicates that much of the insolation is being absorbed. In general, dark-colored substances absorb more insolation than light-colored ones, and the more insolation a substance absorbs the more it is heated.

The word **texture** refers to the smoothness or roughness of a surface. On a smooth surface, insolation is more likely to be reflected away from the surface than on a rough surface. The irregularities and indentations on a rough surface cause some of the insolation to be reflected in such a direction that it hits the surface a second, or even a third, time before leaving (see Figure 22.9). Each time the insolation strikes the surface, a little more of its energy is absorbed. Therefore, if all other characteristics of two surfaces are the same, a rough surface will absorb more insolation than a smooth one.

Figure 22.8 Variations in the Lengths of Day and Night Because of Earth's Tilted Axis of Rotation.

Specific Heat

When different substances absorb equal amounts of heat energy, they do not all change temperature by the same number of degrees. **Specific heat** is the number of joules of heat needed to cause a 1°C temperature change in 1 g of a substance. The difference can be thought of as the result of a kind of molecular inertia. Heavy molecules and tightly held molecules require more energy to get them moving than light and loosely held molecules. The temperature of a substance is strictly a result of how fast the molecules are moving. If you put the same amount of energy into two substances, one with tightly held molecules and one with loosely held molecules, the loosely held molecules would move faster and that substance would register a higher temperature.

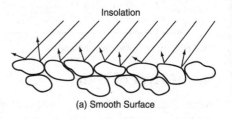
(a) Smooth Surface

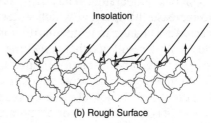
(b) Rough Surface

Figure 22.9 Insolation Striking (a) a Smooth Surface and (b) a Rough Surface.

The Atmosphere

Table 22.2 shows the specific heat of different Earth materials and also the number of degrees each substance will increase in temperature if 1 joule of heat energy is added to it. Notice that water increases in temperature much less than rock material such as granite or basalt with the addition of the same amount of heat. Therefore, if the same amount of insolation is absorbed by an ocean and by the sand-sized particles of broken rock that make up the adjoining beach, the sand will become much hotter than the water. (This is the reason why we seek out bodies of water to swim in when temperatures on land are high.) In general, land increases in temperature more than water when exposed to the same insolation.

TABLE 22.2 SPECIFIC HEAT OF COMMON EARTH MATERIALS

Substance	Specific Heat (Joules/g · °C)	Temperature Increase if 1 Joule of Heat Is Added to 1 Gram of the substance (°C)
Water		
Solid	2.11	0.47
Liquid	4.18	0.24
Gas	2.0	0.5
Dry air	1.01	1.0
Basalt	0.84	1.2
Granite	0.79	1.3
Marble	0.88	1.1
Iron	0.45	2.2
Copper	0.38	2.6
Lead	0.13	7.7

Specific heat of common substances

Latent Heat of Water

When water is melting or evaporating, it absorbs energy with no resulting change in temperature. The energy is used to break internal bonds rather than to increase the speed at which water molecules are moving. Since this heat causes no change in temperature, it is called **latent heat** (*latent* means "hidden"). As a result, ice that is melting or water that is evaporating from ocean surfaces absorbs insolation without increasing in temperature. Therefore, ice-covered land and oceans remain cooler than adjacent land areas.

Latent heat is not an issue on land because the substances of which land is composed do not melt or evaporate in the normal range of temperatures at Earth's surface. However, since water is so abundant and exists in all three phases on Earth, latent heat has an important effect on the temperature changes that occur when insolation strikes water.

HOW ENERGY ENTERS THE ATMOSPHERE

As you have learned, the atmosphere does not absorb much solar energy directly. How, then, is the atmosphere heated? Most insolation passes right through the atmosphere to Earth's surface, where it is absorbed and changed into forms of energy that the atmosphere can absorb. Earth's surface, then, heats the atmosphere. Let us now consider how energy moves from Earth's surface into the atmosphere and, once there, how it moves within the atmosphere.

Conduction

When Earth's surface absorbs solar energy, the absorbed energy causes the molecules at the surface to move more rapidly. We perceive this absorbed energy as heat and measure it in terms of the temperature of the substance as shown by a thermometer. Since Earth's surface is in continuous contact with the base of the atmosphere, its surface molecules strike molecules of gases in air. During these collisions, some heat energy from Earth's surface molecules is transferred to the gas molecules of the atmosphere. This kind of energy transfer, in which heat energy moves from molecule to molecule by collisions, is known as **conduction**. Conduction is important because it is the chief method by which energy moves from Earth's surface to the atmosphere.

Convection

Within the atmosphere, heat spreads very slowly by conduction. Therefore, the air touching the ground becomes warmer and warmer. As this air warms, it expands and becomes less dense than the surrounding air. Finally, a bubble of warm air rises, and cooler air settles to the ground in its place, a type of movement known as **convection**. Since the rising bubble of warm air carries the heat it contains upward with it, convection transfers heat energy *within* the atmosphere. Convection is important because it is the chief method by which energy conducted into the atmosphere from Earth's surface is then carried throughout the atmosphere.

Radiation

Energy from the Sun reaches Earth by **radiation**, the release and transfer of energy in the form of electromagnetic waves. As described earlier, all moving atoms and molecules emit electromagnetic radiation, and the wavelength emitted depends upon how fast these particles are moving. However, since both Earth and its atmosphere are much cooler than the Sun, the energy they radiate has much longer wavelengths. Most of the energy radiated by Earth's surface is infrared radiation.

This fact is important because, although the atmosphere is transparent to most of the Sun's short-wavelength radiation, it is not as transparent to infrared radiation. Ozone, carbon dioxide, and water vapor in the atmosphere absorb or reflect most of Earth's infrared radiation; the rest goes through the atmosphere and out into space. Thus, short wavelengths can readily enter the atmosphere, but long wavelengths cannot readily escape, a phenomenon known as the **greenhouse effect** (see Figure 22.10).

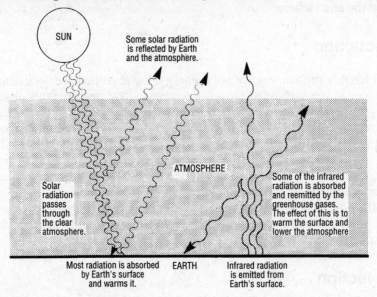

Figure 22.10 The Greenhouse Effect. This sequence of events is called the *greenhouse effect* because the glass in a greenhouse allows light to enter but blocks heat from escaping. Source: *Climate Change*, The Intergovernmental Panel on Climate Change, World Meteorological Organization/UN Environmental Programme, Cambridge University Press.

This term is appropriate because, just as glass surrounds a greenhouse, the atmosphere surrounds Earth. Glass, like the atmosphere, allows visible light rays to pass through it but blocks infrared rays. Similarly, a greenhouse lets sunlight in but does not allow heat to escape. The greenhouse effect keeps the greenhouse—and Earth—warmer than they would otherwise be.

Latent Heat

Heat also enters the atmosphere as energy stored in molecules of water that have evaporated. Try wetting your finger and waving it in the air. Notice that it feels cooler? The reason is that water is evaporating from it. Each molecule of liquid water has to absorb a certain amount of heat from your finger before it can evaporate. Then, when it evaporates and becomes water vapor, each water vapor molecule carries with it into the air the heat energy it absorbed from your finger. The energy stored in water vapor molecules in the air moves along with the air as it circulates in the atmosphere.

Usually, when matter gains heat energy, its temperature increases; and when matter loses heat energy, its temperature decreases. However, when matter changes phase, heat is absorbed or given off in the process of forming or breaking bonds between molecules, rather than in causing molecules to move faster or slower. Thus, as mentioned earlier, the heat energy gained or lost during a phase change does not cause a change in temperature and is therefore latent heat.

The energy *gained* when ice melts into liquid water amounts to about 334 joules per gram of water. The same amount of energy, 334 joules per gram of water, is *released* when liquid water freezes into ice. The energy *gained* when liquid water evaporates into water vapor amounts to about 2,260 joules per gram of water. Here, too, the same amount of energy, 2,260 joules per gram of water, is *released* when the water vapor condenses back into liquid water. The changes of phase that water undergoes and the heat given off or absorbed in each process are summarized in Figure 22.11a.

Water covers more than 70 percent of Earth's surface, so most of the lower atmosphere is in contact with water. Each day solar energy absorbed by this water causes evaporation, sending vast amounts of water vapor into the atmosphere. Every gram of water vapor in the atmosphere contains the 2,260 joules of latent heat energy that was gained during the change of water from a liquid to a gas.

Remember, though, that latent heat stored during evaporation does not show up as a temperature change. The water vapor starts out no warmer than the water from which it formed. How, then, is the atmosphere warmed by latent heat? The heat stored in water molecules when they evaporated is released when the water changes back to water or ice. At the moment the water vapor condenses, it releases its latent heat into the air. As a result, condensation is a warming process that causes the temperature of air to rise as droplets of water in clouds grow in size. This increase in temperature causes the air to expand, become less dense, and rise. Thus, latent heat is an important energy source for violent storms. The release of latent heat during large-scale condensation produces localized heating, which causes updrafts. At the same time, precipitation associated with the heavy condensation causes downdrafts. The result is the violent circulations of air that occur in storms such as hurricanes and tornadoes.

Circulation of the Atmosphere

The transfer of heat energy from solar radiation, and water vapor, into and out of the atmosphere causes regions of different densities to form. The equatorial regions of the Earth are predominantly water and receive more solar energy per year than the polar regions. Air over the Equator becomes warm and moist through contact with the Earth's surface. Air that is warm and moist is less dense than cooler, drier air. Thus, the air at the Equator is surrounded by air of higher density that pushes inward toward the Equator and displaces the warm, moist equatorial air upward. As this air is forced upward it expands outward and cools, resulting in condensation and precipitation. The end result is cooler, dryer air that is denser than the air beneath it and begins to sink back toward the surface. Together, the rising warm, moist air

The Atmosphere

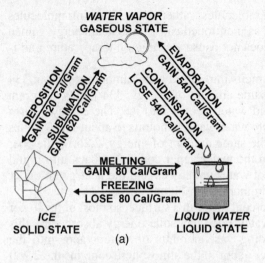

(a) Water gains or loses energy when it undergoes a change in state of matter. Since this gain or loss of heat energy does not result in a change in temperature, it is hidden or *latent heat*.

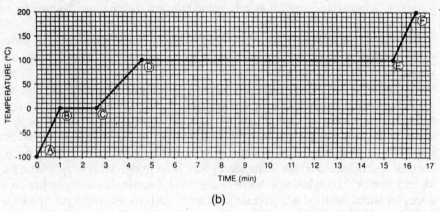

(b) The heating curve for a sample of water heated from a starting temperature of −100°C to a final temperature of +200°C. The same amount of heat is added to the sample every minute. From A to B the ice increases in temperature until it reaches its melting point. From B to C there is a change in state from solid ice to liquid water with no increase in temperature (the curve is flat, like a plateau). From C to D liquid water increases in temperature until it reaches its boiling point. From D to E there is a change in state from liquid water to water vapor, again with no increase in temperature. From E to F water vapor is increasing in temperature. Note that the flat area from B to C is shorter than the flat area from D to E since less energy is required to change water from a solid to a liquid than to change water from a liquid to a gas.

Figure 22.11 Changes in State of Water (a) and Heating Curve for Water (b).

and sinking cool, dry air form a circular pattern of motion called a **convection cell**. See Figure 22.12. A similar convection cell forms over the poles, although it begins with sinking air, which spreads southward when it reaches the Earth's surface, warms, and rises as shown in Figure 22.13. In between the polar and equatorial convection cells lies a third cell, which is set in motion by westerly winds in the middle latitudes. Thus, three convection cells girdle both Earth's Northern and Southern Hemispheres. This *three-cell theory* of global atmospheric circulation explains how the atmosphere distributes solar energy over the whole Earth. See Figure 22.14. The interaction of these processes results in the complex atmospheric occurrence known as weather.

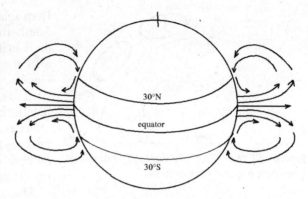

Figure 22.12 Convection Cells Near the Equator. Air near the Equator is heated by more intense insolation, becomes less dense, and floats upward in the surrounding denser air. As the air rises, it expands and cools, causing it to sink back toward the surface. The result is a circular movement of air, or convection cell.

The Coriolis Effect and Global Atmospheric Circulation

The rise or fall of regions of different densities as a result of the action of gravitational force produces atmospheric circulation, which is affected by Earth's rotation. One of the consequences of Earth's rotation is a tendency of all matter that is in motion on Earth's surface to be deflected to the right from its point of origin in the Northern

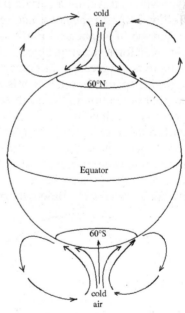

Figure 22.13 Convection Cells Near the Poles. Air near the poles cools because of less intense insolation, becomes more dense, and sinks downward. The sinking air spreads outward at the surface, is warmed as it moves away from the poles, and rises. The result is a circular movement of air, or convection cell.

The Atmosphere

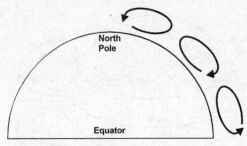

Figure 22.14 The Three-Cell Theory of Convection in the Atmosphere.

Hemisphere and to the left in the Southern Hemisphere, called the **Coriolis effect**. It is named after the French mathematician Gustave Gaspard Coriolis, who first analyzed it in the nineteenth century. The Coriolis effect is not an actual force but rather is the *apparent* effect of a number of different forces acting on any particles in motion.

The main elements of the Coriolis effect are the curvature of Earth's surface, the rotation of Earth, and the tendency of objects in motion to remain in motion in a straight line (Newton's first law of motion). The east-west path between any two points on Earth's surface is not a straight line, but a curve—a segment of a parallel (line of latitude). As a result, any object following a straight path will appear, to an observer on Earth, to curve from its path. We say "appear" because actually it is the observer who is traveling a curved path while the object travels in a straight line. In Figure 22.15 the solid line represents the actual path of an observer on the surface of a spherical Earth as it rotates. The dotted line represents a straight path for a moving object. The Coriolis effect is most pronounced where there is the greatest difference between the curvature of a parallel and a straight-line path—in other words, near the poles.

A similar deflection in the north-south direction is due to the difference in the speeds at which points on Earth's surface are moving because of rotation. As Earth rotates, every object on its surface rotates with it. In one 24-hour rotation, objects near the poles travel less distance than objects near the Equator and thus are moving slower (see Figure 22.16).

Now, let's suppose a rocket at the Arctic Circle (60°N) is aimed due south at New York City (40°N). Before the rocket is even launched, it is moving west

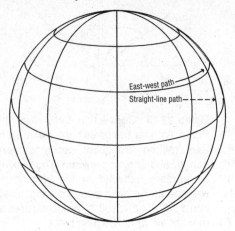

Figure 22.15 Deflection Due to Curvature of Earth's Surface. An object moving in a straight line travels along the dashed line. An observer on Earth's surface moves due east-west as Earth rotates, following a path that curves. To an observer on Earth's surface, the straight-line path seems to veer off, or be deflected, to the right in the Northern Hemisphere. In the Southern Hemisphere, the deflection is to the left.

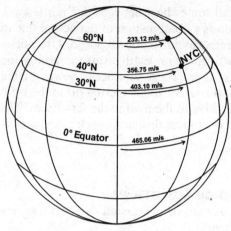

Figure 22.16 Deflection Due to Different Velocities at Different Latitudes. As Earth rotates, objects on its surface move at different speeds, depending on their latitudes. Objects at the Equator move faster because they cover more distance in one 24-hour rotation than objects near the poles.

to east at the speed of Earth's surface at the Arctic Circle—233 meters per second. The target, New York City, is also moving west to east at the speed of Earth's surface at New York City—356 meters per second. Note that the target is moving west to east faster than the rocket! When the rocket is fired, it moves due south and west to east; but since the target is moving west to east faster than the rocket, the rocket lags behind the target and actually hits the ground behind the target. To an observer, the rocket appears to follow a path that curves to the *right* of the target. Similarly, a rocket fired from New York City toward the Arctic Circle would land ahead of the target because New York City is moving west to east faster than the target. Nevertheless, the deflection would still be to the *right*. As seen in Figure 22.17, because of the Coriolis effect no matter what direction the motion, deflection is always to the right in the Northern Hemisphere.

If we repeat this experimentation in the Southern Hemisphere, we find that the direction of deflection is reversed—the rocket appears to curve to the *left* of the target. Thus, deflection is to the right of the wind's point of origin in the Northern Hemisphere and to the left of its point of origin in the Southern Hemisphere, as shown in Figure 22.17. (A good way to recall the direction of deflection is to remember that someone who is *left*-handed is called a *south*paw.)

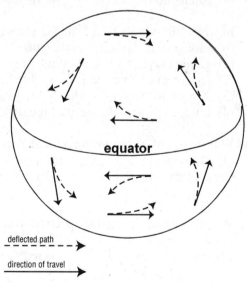

Figure 22.17 The Coriolis Effect. Objects are deflected to the right from their points of origin in the Northern Hemisphere and to the left from their points of origin in the Southern Hemisphere.

The Atmosphere

For objects following paths that fall somewhere between due north-south and due east-west, the total deflection is a combination of deflection due to curvature and deflection due to differences in rotational speed. Of course, frictional forces within the atmosphere and with Earth's surface modify the Coriolis effect, but it still persists. Over short distances, however, the Coriolis effect is imperceptible, so don't worry about veering off to the right when you walk home from school or drive to the mall in the family car. The Coriolis effect becomes a factor only over great distances, such as those covered by planetary winds.

MULTIPLE-CHOICE QUESTIONS

In each case, write the number of the word or expression that best answers the question or completes the statement.

1. Energy is transferred from the Sun to Earth mainly by
   (1) molecular collisions
   (2) density currents
   (3) electromagnetic waves
   (4) red shifts

2. By which process does most of the Sun's energy travel through space?
   (1) absorption   (3) convection
   (2) conduction   (4) radiation

3. In which list are the forms of electromagnetic energy arranged in order from longest to shortest wavelengths?
   (1) gamma rays, X rays, ultraviolet rays, visible light
   (2) radio waves, infrared rays, visible light, ultraviolet rays
   (3) X rays, infrared rays, blue light, gamma rays
   (4) infrared rays, radio waves, blue light, red light

4. Which type of electromagnetic radiation has the longest wavelength?
   (1) ultraviolet   (3) visible light
   (2) gamma rays   (4) radio waves

5. Which color of the visible spectrum has the *shortest* wavelength?
   (1) violet   (3) yellow
   (2) blue   (4) red

6. When is an object in radiative balance?
   (1) when the radiation emitted by the object is equal to that absorbed by the object
   (2) when the radiation emitted by the surroundings is equal to that absorbed by the object
   (3) when the radiation emitted by the object is equal to that absorbed by the surroundings
   (4) when the wavelength of the radiation emitted by the object is equal to that absorbed by the object

7. Most of the solar radiation absorbed by Earth's surface is later radiated back into space as which type of electromagnetic radiation?
   (1) X ray
   (2) ultraviolet
   (3) infrared
   (4) radio wave

8. When Earth cools, most of the energy transferred from Earth's surface to space is transferred by the process of
   (1) conduction
   (2) reflection
   (3) refraction
   (4) radiation

9. What is the usual cause of the drop in temperature that occurs between sunset and sunrise at most New York State locations?
   (1) strong winds
   (2) ground radiation
   (3) cloud formation
   (4) heavy precipitation

10. In New York State, the risk of sunburn is greatest between 11 A.M. and 3 P.M. on summer days because
    (1) the air temperature is hot
    (2) the angle of insolation is high
    (3) Earth's surface reflects most of the sunlight
    (4) the Sun is closest to Earth

11. The average temperature at Earth's equator is higher than the average temperature at Earth's South Pole because the South Pole
    (1) receives less intense insolation
    (2) receives more infrared radiation
    (3) has less land area
    (4) has more cloud cover

The Atmosphere

Base your answers to questions 12 through 16 on your knowledge of Earth science and the graph below, which shows measurements of the insolation above Earth's atmosphere and at Earth's surface on a clear day. [Note that the graph does not show the entire solar spectrum at the longer wavelengths.]

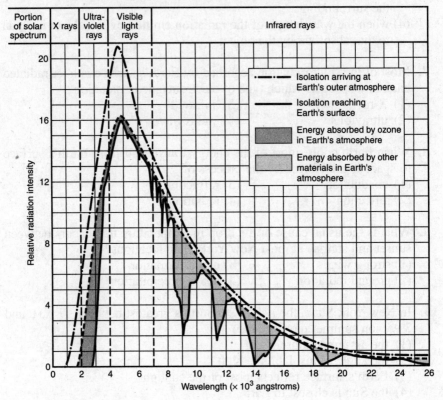

12. Which of the following types of solar radiation has the longest wavelength?
    (1) X rays
    (2) ultraviolet rays
    (3) visible light rays
    (4) infrared rays

13. The greatest intensity of energy reaching the outer atmosphere of Earth from the Sun has a wavelength of approximately
    (1) 4.5×10^0 angstroms
    (2) 4.5×10^3 angstroms
    (3) 3.0×10^3 angstroms
    (4) 9.1×10^2 angstroms

14. In which portion of the solar spectrum does ozone absorb the greatest amount of energy?
    (1) X rays
    (2) ultraviolet rays
    (3) visible light rays
    (4) infrared rays

15. What quantity is most likely represented by the area between the "dot-dash" curve (- · -) and the "dash" curve (- - -)?
    (1) the amount of radiation given off by the Sun
    (2) the amount of radiation absorbed in outer space
    (3) the amount of insolation reflected by the atmosphere back into space
    (4) the amount of insolation absorbed by Earth's surface

16. According to the graph, in which portion of the solar spectrum is the greatest total amount of energy absorbed by ozone and other materials in Earth's atmosphere?
    (1) ultraviolet rays and infrared
    (2) X rays and visible light
    (3) visible light and infrared
    (4) X rays and infrared

Base your answers to questions 17 and 18 on the graph below, which shows the duration of daylight hours throughout the year for five cities located in the Northern Hemisphere.

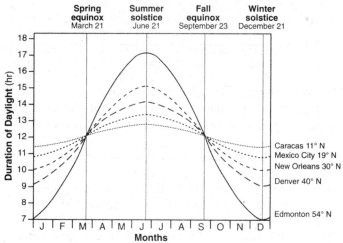

The Atmosphere

17. Which city experiences the greatest variation in daylight hours during one year?
(1) Caracas
(2) Mexico City
(3) New Orleans
(4) Edmonton

18. What is the primary reason each city's duration of daylight hours changes throughout the year?
(1) Earth's axis is tilted 23.5° to the plane of its orbit.
(2) Earth's rotation rate is 15° per day.
(3) The cities are located at different longitudes.
(4) The cities are located at different elevations.

Base your answers to questions 19 through 21 on the graph below, which shows the amount of insolation during one year at four different latitudes on Earth's surface.

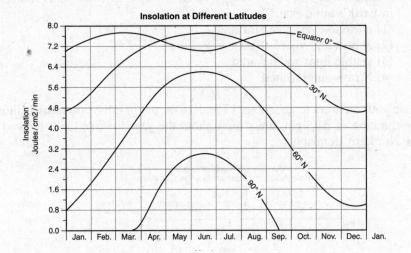

19. This graph shows that insolation varies with
(1) latitude and time of day
(2) latitude and time of year
(3) longitude and time of day
(4) longitude and time of year

20. Why is less insolation received at the equator in June than in March or September?
(1) The daylight period is longest at the equator in June.
(2) Winds blow insolation away from the equator in June.
(3) The Sun's vertical rays are north of the equator in June.
(4) Thick clouds block the Sun's vertical rays at the equator in June.

21. Why is insolation 0 joule/cm²/min from October through February at 90° N?
    (1) Snowfields reflect sunlight during that time.
    (2) Dust in the atmosphere blocks sunlight during that time.
    (3) The Sun is continually below the horizon during that time.
    (4) Intense cold prevents insolation from being absorbed during that time.

22. The diagram below shows four surfaces of equal area that absorb insolation.

    Which letter represents the surface that most likely absorbs the greatest amount of insolation?
    (1) A   (3) C
    (2) B   (4) D

23. Equal volumes of the four samples shown below were placed outside and heated by energy from the Sun's rays for 30 minutes.

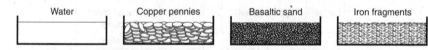

    The surface temperature of which sample increased at the slowest rate?
    (1) water   (3) basaltic sand
    (2) copper pennies   (4) iron fragments

24. Which change would cause a *decrease* in the amount of insolation absorbed at Earth's surface?
    (1) a decrease in cloud cover
    (2) a decrease in atmospheric transparency
    (3) an increase in the duration of daylight
    (4) an increase in nitrogen gas

25. The diagram below shows a student heating a pot of water over a fire. The arrows represent the transfer of heat. Letter A represents heat transfer through the metal pot, B represents heat transfer by currents in the water, and C represents heat that is felt in the air surrounding the pot.

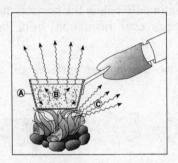

Which table correctly identifies the types of heat transfer at A, B, and C?

Letter	Type of Heat Transfer
A	conduction
B	radiation
C	convection

(1)

Letter	Type of Heat Transfer
A	radiation
B	conduction
C	convection

(3)

Letter	Type of Heat Transfer
A	conduction
B	convection
C	radiation

(2)

Letter	Type of Heat Transfer
A	radiation
B	convection
C	conduction

(4)

26. Which model best represents how a greenhouse remains warm as a result of insolation from the Sun?

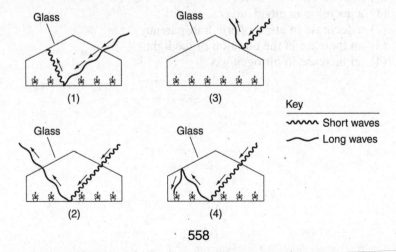

27. Which component of Earth's atmosphere is classified as a greenhouse gas?
    (1) oxygen
    (2) carbon dioxide
    (3) helium
    (4) hydrogen

28. Deforestation increases the greenhouse effect on Earth because deforestation causes the atmosphere to contain
    (1) more carbon dioxide, which absorbs infrared radiation
    (2) less carbon dioxide, which absorbs shortwave radiation
    (3) more oxygen, which absorbs infrared radiation
    (4) less oxygen, which absorbs shortwave radiation

Base your answers to questions 29 through 31 on the map below, which shows Earth's planetary wind belts.

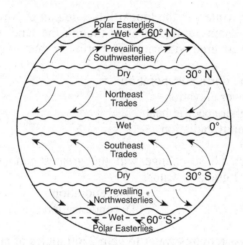

29. The curving of these planetary winds is the result of
    (1) Earth's rotation on its axis
    (2) the unequal heating of Earth's atmosphere
    (3) the unequal heating of Earth's surface
    (4) Earth's gravitational pull on the Moon

30. Which wind belt has the greatest effect on the climate of New York State?
    (1) prevailing northwesterlies
    (2) prevailing southwesterlies
    (3) northeast trades
    (4) southeast trades

31. Which climatic conditions exist where the trade winds converge?
    (1) cool and wet
    (2) cool and dry
    (3) warm and wet
    (4) warm and dry

The Atmosphere

32. The planetary wind belts in the troposphere are primarily caused by the
    (1) Earth's rotation and unequal heating of Earth's surface
    (2) Earth's revolution and unequal heating of Earth's surface
    (3) Earth's rotation and Sun's gravitational attraction on Earth's atmosphere
    (4) Earth's revolution and Sun's gravitational attraction on Earth's atmosphere

33. The Coriolis effect would be influenced most by a change in Earth's
    (1) rate of rotation
    (2) period of revolution
    (3) angle of tilt
    (4) average surface temperature

34. During some winters in the Finger Lakes region of New York State, the lake water remains unfrozen even though the land around the lakes is frozen and covered with snow. The primary cause of this difference is that water
    (1) gains heat during evaporation
    (2) is at a lower elevation
    (3) has a higher specific heat
    (4) reflects more radiation

35. During which phase change will the greatest amount of energy be absorbed by 1 gram of water?
    (1) melting
    (2) freezing
    (3) evaporation
    (4) condensation

36. Which process requires water to gain 2260 joules of energy per gram?
    (1) vaporization
    (2) condensation
    (3) melting
    (4) freezing

37. Why is the condensation of water vapor considered to be a process that heats the air?
    (1) Liquid water has a lower specific heat than water vapor.
    (2) Energy is released by water vapor as it condenses.
    (3) Water vapor must absorb energy in order to condense.
    (4) Air can hold more water in the liquid phase than in the vapor phase.

CONSTRUCTED RESPONSE QUESTIONS

Base your answers to questions 38 through 41 on the data table below. The data table shows the latitude of several cities in the Northern Hemisphere and the duration of daylight on a particular day.

Data Table

City	Latitude (°N)	Duration of Daylight (hr)
Panama City, Panama	9	11.6
Mexico City, Mexico	19	11.0
Tampa, Florida	28	10.4
Memphis, Tennessee	35	9.8
Winnipeg, Canada	50	8.1
Churchill, Canada	59	6.3
Fairbanks, Alaska	65	3.7

38. On the grid below, plot with an **X** the duration of daylight for each city shown in the data table. Connect your **X**s with a smooth, curved line. [1]

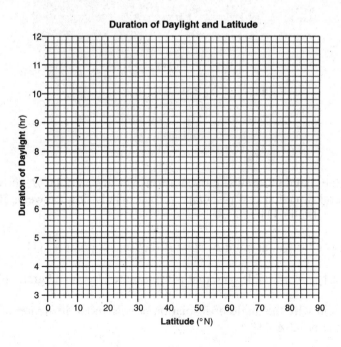

39. Based on the data table, state the relationship between latitude and the duration of daylight. [1]

40. Use your graph to determine the latitude at which the Sun sets 7 hours after it rises. [1]

41. The data were recorded for the first day of a certain season in the Northern Hemisphere. State the name of this season. [1]

Base your answers to questions 42 through 45 on the diagram below, which shows Earth as seen from above the North Pole. The curved arrows show the direction of Earth's motion. The shaded portion represents the nighttime side of Earth. Some of the latitude and longitude lines have been labeled. Points *A* and *B* represent locations on Earth's surface.

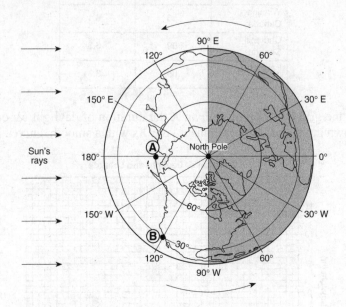

42. On the diagram, draw a curved arrow, starting at point *B*, showing the general direction that planetary surface winds flow between 30°N and 60°N latitude. [1]

43. If it is 4:00 P.M. at point *B*, what is the time at point *A*? [1]

44. Identify *one* possible date that is represented by the diagram. [1]

45. Explain why the angle of insolation at solar noon is greater at point *B* than at point *A*. [1]

Base your answers to questions 46 and 47 on the passage below.

    Average temperatures on Earth are primarily the result of the total amount of insolation absorbed by Earth's surface and atmosphere compared to the amount of long-wave energy radiated back into space. Scientists believe that the addition of greenhouse gases into Earth's atmosphere gradually increases global temperatures.

46. Identify *one* major greenhouse gas that contributes to global warming. [1]

47. Explain how increasing the amount of greenhouse gases in Earth's atmosphere increases global temperatures. [1]

Base your answers to questions 48 and 49 on the field map below, which shows temperatures, in degrees Fahrenheit, taken at several locations on a blacktop parking lot in New York State. The temperatures were recorded at 11:00 A.M. in early June.

77	75	73	71	68	65	63
78	78	77	73	69	65	62
82	80	78	74	70	68	65
81	80	77	75	70	67	64
79	77	76	73	68	66	62
75	75	72	69	65	63	61

48. On the field map, draw the 70°F and 80°F isotherms. The isotherms should be extended to the edges of the map. [1]

49. Explain why the surface of this parking lot usually becomes warmer from 11:00 A.M. to 12 noon each day. [1]

The Atmosphere

Base your answers to questions 50 through 53 on the graph below, which shows the temperatures recorded when a sample of water was heated from −100°C to +200°C. The water received the same amount of heat every minute.

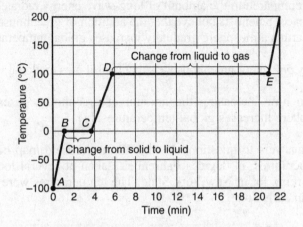

50. In one or more complete sentences, explain why the water did not change temperature between points *B* and *C* or between points *D* and *E* even though the water received the same amount of heat energy each minute. [1]

51. State the lettered points between which the water gained the greatest amount of heat energy. [1]

52. State the lettered points between which the water was changing from a solid to a liquid. [1]

53. The water received the same amount of heat every minute. In one or more complete sentences, explain why the water temperature remained unchanged between points *B* and *C* for less than 2 minutes but remained unchanged between points *D* and *E* for more than 10 minutes. [1]

EXTENDED CONSTRUCTED RESPONSE QUESTIONS

Base your answers to questions 54 through 56 on the passage below and on your knowledge of Earth science.

Ozone in Earth's Atmosphere

Ozone is a special form of oxygen. Unlike the oxygen we breathe, which is composed of two atoms of oxygen, ozone is composed of three atoms of oxygen. A concentrated ozone layer between 10 and 30 miles above Earth's surface absorbs some of the harmful ultraviolet radiation coming from the Sun. The amount of ultraviolet light reaching Earth's surface is directly related to the angle of incoming solar radiation. The greater the Sun's angle of insolation, the greater the amount of ultraviolet light that reaches Earth's surface. If the ozone layer were completely destroyed, the ultraviolet light reaching Earth's surface would most likely increase human health problems, such as skin cancer and eye damage.

54. State the name of the temperature zone of Earth's atmosphere where the concentrated layer of ozone gas exists. [1]

55. Explain how the concentrated ozone layer above Earth's surface is beneficial to humans. [1]

56. Assuming clear atmospheric conditions, on what day of the year do people in New York State most likely receive the most ultraviolet radiation from the Sun? [1]

Base your answers to questions 57 and 58 on the diagram below, which shows incoming solar radiation passing through the glass of a greenhouse and then striking the floor.

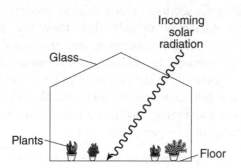

57. Some of the incoming solar radiation is absorbed by the floor. Identify the type of electromagnetic energy reradiated by the floor. [1]

58. Describe *one* way the glass in the greenhouse acts like the greenhouse gases in Earth's atmosphere. [1]

CHAPTER 23
WEATHER

> **KEY IDEAS** Weather, the present condition of the atmosphere at any given location, is described by the physical characteristic of the atmosphere. Weather patterns become evident when atmospheric variables are observed, measured, and recorded. Atmospheric variables include air temperature, air pressure, moisture (relative humidity and dew point), precipitation (such as rain, snow, hail, and sleet), wind speed and direction, and cloud cover. Atmospheric variables are measured using instruments such as thermometers, barometers, psychrometers, precipitation gauges, anemometers, and wind vanes.
>
> Atmospheric variables can be represented in a variety of formats including radar and satellite images, synoptic weather maps, atmospheric cross-sections, and computer models. Analysis of atmospheric variables reveals that they are interrelated. Atmospheric moisture, temperature and pressure distributions, winds and jet streams, air masses, frontal boundaries, and the movement of cyclonic systems and associated tornadoes, thunderstorms, and hurricanes occur in observable patterns. Loss of property, personal injury, and loss of life can be reduced by effective emergency preparedness.

KEY OBJECTIVES
Upon completion of this unit, you will be able to:

- Define weather, and identify the atmospheric characteristics used to describe it.
- Describe the relationships between important atmospheric characteristics.
- Explain how clouds and precipitation form.
- Construct weather maps, given weather information at numerous locations, and identify the positions of air masses and fronts.
- Predict weather and the movement of weather systems based on current weather maps.
- Describe the weather conditions associated with the various types of frontal boundaries.

- Explain how weather information can be used to make forecasts.
- Identify the conditions that typically lead to hazardous weather such as hurricanes and tornadoes.

HOW CAN WEATHER BE DESCRIBED?

Weather is the present condition of the atmosphere at any location. Weather changes are mainly the result of unequal heating, by solar radiation, of Earth's landmasses, oceans, and atmosphere. As heat energy is transferred at the boundaries between the atmosphere, oceans, and landmasses, layers of different temperatures and densities form in the atmosphere. When gravity acts on these layers of different densities, they rise and fall, producing circulation of air in the atmosphere. This circulation causes the air at any one location to constantly be moved away and replaced by air from a different location—with a resulting change in weather.

The constantly changing weather is described in terms of the characteristics of the atmosphere that change, or atmospheric variables. The **atmospheric variables** used to describe weather are measured with weather instruments and include the temperature, pressure, humidity, movements, and transparency of the air. Atmospheric variables are interrelated, so that a change in one is likely to cause a change in another. This interrelatedness makes weather prediction a complex task. However, measurement of atmospheric variables and the creation of field maps based on these data reveal large-scale patterns that can be used to make predictions.

Air Temperature

The air temperature at any given location is related to the heat energy present in the atmosphere at that location. Solar radiation is the main source of heat energy in the atmosphere; therefore anything that influences solar radiation will ultimately influence air temperature. Although the radiation emitted by the Sun is fairly constant, the amount that reaches Earth's surface and is converted to heat energy in the atmosphere is influenced by many factors. These include the angle at which solar radiation strikes Earth's surface; the number of hours of solar radiation per day; the reflection, refraction, or absorption of solar radiation by the atmosphere (e.g., cloud cover); and the nature of the surface materials absorbing solar radiation. In general, the more direct the solar radiation and the longer it occurs, the higher the temperature.

The angle at which solar radiation strikes Earth's surface varies during the course of a day. Because of Earth's rotation, radiation is received during the day but not at night. Therefore, air temperature often varies in a daily cycle. Similar changes occur as Earth moves through its orbit around the Sun, resulting in seasonal temperature variations.

Temperature is measured with thermometers. A **thermometer** works on a simple principle—matter expands when heated and contracts when cooled, and the amount of expansion or contraction depends on the amount of heat gained or lost. A typical thermometer measures the height of a column of alcohol or mercury in a glass tube. Some modern devices, however, use the change in electrical conductivity of certain metals when heated or cooled to measure temperature. Several scales, named for their creators, have been developed over the years to measure temperature: Fahrenheit, Celsius, and Kelvin. Meteorologists in the United States still use the Fahrenheit scale but are changing over to the Celsius.

Meteorologists use several types of specialized thermometers when measuring air temperature. Simple thermometers are mounted in shaded, vented boxes so that they will measure the temperature of the air and not be heated additionally by absorption of sunlight. Maximum-minimum thermometers have pinched tubes and markers that are pushed or dragged back as the fluid in them rises and falls. This feature enables them to record the highest and lowest temperatures during a time period by the positions of the markers. Continuous temperature readings are made with a thermograph, a thermometer in which a coil of heat-sensitive metal moves a pen against a rotating drum.

Air Pressure

Air is a mixture of gases. A gas consists of many tiny individual molecules that are far apart and moving rapidly. As they whiz to and fro, they are kept from escaping into space by the walls of a container or, in the case of the atmosphere, by Earth's gravity.

Gases exert pressure on surfaces with which they come into contact. **Pressure** is a force that acts on a surface area. The pressure that a gas exerts on a surface is the result of gas particles colliding with the surface. Since gas particles move randomly in all directions, they exert pressure equally in all directions. Figure 23.1 shows gas molecules speeding around in a container and striking surfaces held at different angles.

Air pressure is not an atmospheric variable that you can sense, in the way that you feel temperature or see cloud cover. By itself, knowledge of the air pressure in one location is of little use in weather forecasting. However, *field maps of air pressure in many locations* reveal distinct regions within the atmosphere, and *changes* in air pressure signal movements of these large masses of air and their associated weather conditions. Rising pressure

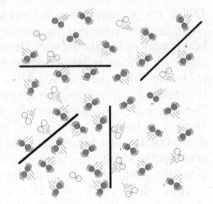

Figure 23.1 Gas Molecules in a Container. Regardless of the angle at which a surface is held, air molecules collide with it, exerting the same pressure in all directions.

usually signals the approach or continuation of fair weather; falling pressure, the approach of a storm.

Air pressure is measured with a **barometer**. Two commonly used barometers are the mercury barometer and the aneroid barometer.

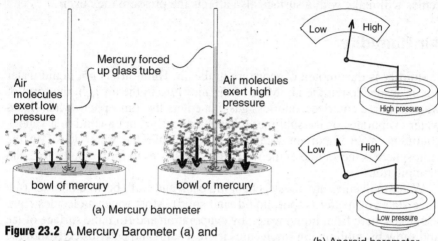

Figure 23.2 A Mercury Barometer (a) and an Aneroid Barometer (b).

In a mercury barometer (see Figure 23.2a), air pressure forces liquid mercury up a tube. The mercury rises in the tube because the tube contains a vacuum and no pressure is exerted on the mercury inside the tube. The higher the air pressure outside the tube, the higher the mercury is pushed up in the tube. Since pressure is measured by the height to which the mercury is pushed, pressure is often described as "inches or millimeters of mercury." The word *aneroid* means no liquid, and an aneroid barometer (see Figure 23.2b) is a can with no liquid inside it. The can contains a spring scale. As air pressure on the sides of the can increases or decreases, the spring compresses and expands and the scale records the magnitude of the force.

Pressure is measured in units of force per unit area. A variety of units are used to describe atmospheric pressure. The **atmosphere** (atm) is the average pressure exerted by Earth's atmosphere at sea level and is equal to 1,013.2 millibars or 14.7 pounds per square inch. The bar is equal to 1×10^5 newtons per square meter, and the *millibar* is 1/1,000 of a bar or 100 newtons per square meter. (1 newton = 0.225 lb.) In meteorology, atmospheric pressure is measured in millibars. **Normal atmospheric pressure** is 1,013.2 millibars.

Air pressure is often misunderstood because much of our personal experience with pressure involves solids pressing against us, or our bodies pressing against something because of gravity acting on us. Since the molecules of solids are bonded together in a single unit, they act as a single unit. Under the influence of gravity the entire object is pulled downward and exerts a downward pressure. In gases, on the other hand, each tiny molecule acts independently. Gravity pulls the molecule down; but since it is hurtling through space, the effect is like that of gravity acting on a baseball speeding toward home. The ball sinks in a downward trajectory but still exerts a sizable horizontal force, as any professional catcher will attest. Think of gases as millions of tiny baseballs flying in all directions at the same time. Anything that affects the strength and the number of collisions that gas molecules will make with a surface also affects the pressure they exert.

Air Humidity

Humidity is the amount of moisture in the air. Humidity is not liquid water that is suspended in the air. Most of the moisture in the air is in the form of water vapor, a colorless, odorless gas that enters the atmosphere when liquid water evaporates or ice sublimes (changes directly from a solid to a vapor). Humidity is an important weather factor because the water vapor in the atmosphere is the source of the water that condenses to form clouds, fog, and precipitation.

Water molecules are always changing back and forth between phases (the three states of matter: vapor, liquid, and solid). Most water molecules enter the atmosphere from liquid water by evaporation and from the surface of ice and snow by sublimation (molecules leave a solid and enter directly into the vapor phase). Water leaves the atmosphere by condensation or deposition (molecules of water vapor form solid crystals without first becoming a liquid). This constant flow of water molecules between phases results in constantly changing levels of humidity in the atmosphere. By far, the greatest amount of activity occurs at the boundary between the atmosphere and the hydrosphere. Molecules of liquid water in the hydrosphere that gain energy from sunlight and their surroundings may become energetic enough to escape the liquid phase and enter the atmosphere as water vapor. When water vapor molecules in the atmosphere come into contact with a liquid water surface, they may be trapped there, so there is a constant two-way traffic of molecules to and from the liquid. The flow of molecules leaving the surface of the liquid and entering the atmosphere as water vapor (**evaporation**) is strongly influenced by temperature. At higher temperatures, molecules have more energy, so a greater number are likely to escape and enter the vapor phase. The flow of molecules arriving at the surface to exit the atmosphere as a liquid (**condensation**) depends on how densely crowded with molecules the water vapor above the liquid becomes. The more densely crowded the molecules, the more likely they are to come into contact with the surface and be trapped.

If more water molecules are leaving the surface than are arriving, there is net evaporation. If more water molecules are arriving at the surface than leaving, there is net condensation. If there is net evaporation, the amount of water vapor in the atmosphere increases. However, as you just learned, if the amount of water vapor in the atmosphere increases, molecules are more likely to come into contact with the surface and be trapped, so condensation also increases. Eventually, a point is reached where condensation and evaporation are in equilibrium, and the amount of water vapor above the surface remains unchanged. See Figure 23.3. The higher the temperature, the greater the amount of water vapor present in the atmosphere before equilibrium is reached. At 0°C, equilibrium is reached when air contains 5 grams of water vapor per cubic meter; at 20°C the amount is 17 grams per cubic meter, and at 100°C it is 598 grams per cubic meter! If for some reason (such as a sudden drop in temperature) the amount of water vapor exceeds the equilibrium amount, condensation will occur at a faster rate than evaporation until equilibrium is reestablished.

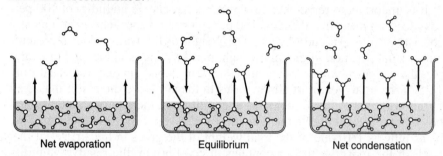

Figure 23.3 Net Evaporation and Net Condensation in Equilibrium. Equilibrium occurs when there are equal flows of molecules leaving and arriving at the surface of the liquid. Net evaporation occurs when the water-vapor content of the atmosphere is less than the equilibrium amount. Net condensation occurs when the water-vapor content of the atmosphere is greater than the equilibrium amount.

The amount of water vapor in the atmosphere is often expressed as absolute humidity or relative humidity. **Absolute humidity** is the number of grams of the water vapor molecules in 1 cubic meter of air. Since it is difficult to isolate the water vapor in air and to measure its mass, absolute humidity is seldom directly measured.

Relative humidity is the ratio of the water vapor now present in the atmosphere to the water vapor present when evaporation and condensation are in equilibrium at that temperature, times 100 percent.

$$\text{Relative humidity} = \frac{\text{Water vapor in atmosphere}}{\text{Water vapor in atmosphere at equilibrium}} \times 100\%$$

Relative humidity is a way of comparing the flows of water molecules leaving and arriving at a surface, and is useful in predicting how the air will feel to a person and whether condensation or evaporation is more likely to occur in the atmosphere.

A relative humidity of 20 percent means that the atmosphere currently contains only 20 percent of the water vapor it contains at equilibrium. As you have learned, at a temperature of 0°C, condensation and evaporation reach equilibrium when the amount of water vapor in the air reaches 5 grams per cubic meter. A relative humidity of 20 percent means that the air currently contains only 1 gram per cubic meter (20% of 5 g/m^3 = 1 g/m^3.) Therefore, there will be net evaporation until the air gains an additional 4 grams per cubic meter and reaches equilibrium (at which point there is no net evaporation).

A relative humidity of 100 percent means that the atmosphere contains 100 percent of the water vapor it contains at equilibrium. Therefore, there will be no net evaporation *or* net condensation. The higher the relative humidity, the more uncomfortable people feel because less moisture evaporates from their bodies before equilibrium is reached. Since less water evaporates from the skin, a wet and "sticky" feeling results.

It is important to remember, however, that a relative humidity of 100 percent does not mean that the air is 100 percent water vapor. The air still mostly consists of nitrogen and oxygen. The weight of the water vapor represents only a small fraction of the total weight of all the gases in the air. For example, the weight of water vapor in very warm, moist air may represent only 1/25 of the total weight of all the gases in the air. Nevertheless, if that is all of the water vapor the air contains at equilibrium, the relative humidity will be 100 percent. Similarly, the weight of water vapor in frigid arctic air may be as little as 1/10,000 of the total weight of the gases in the air; yet, if that is all of the moisture the air contains at equilibrium, the relative humidity will also be 100 percent.

High relative humidity also means that the water vapor in the atmosphere is close to its equilibrium amount. A decrease in temperature, which, in turn, decreases the equilibrium amount, could trigger condensation, which forms clouds, fog, or precipitation. A commonly used value in this connection is the **dew point**, that is, the temperature at which the water vapor in the air will begin to condense into liquid water. This is the temperature at which the amount of water vapor currently in the air equals the equilibrium amount of water vapor.

Humidity is measured with a hygrometer or a psychrometer. A **hygrometer** consists of strands of human hair attached to a pointer. Human hair lengthens slightly as humidity increases. As the hair lengthens and shortens because of changing humidity, the pointer changes position.

A **psychrometer** consists of two thermometers, one whose bulb is kept dry and another whose bulb is kept wet by covering it with a cloth wick soaked with water. Evaporation from the wet wick causes cooling because the fastest molecules in the liquid water are the ones with enough energy to escape. The slower molecules left behind have a lower average energy; hence the liquid's temperature decreases as measured on the wet-bulb thermometer. See Figure 23.4. This cooling effect is the key to measuring relative humidity. The lower the relative humidity, the more water can evaporate from the wet-bulb thermometer and the more the thermometer is cooled. The dry-bulb thermometer, however, remains unchanged and registers the temperature of the air. Therefore, the difference in temperature between the wet-bulb thermometer and the dry-bulb thermometer is directly related to the relative humidity of the air.

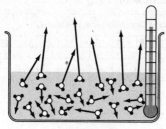
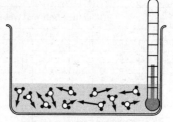

The fastest molecules escape.

The remaining molecules have lower average energies, hence the liquid temperature is lower.

Figure 23.4 After evaporation, the remaining liquid is cooler than before.

The relationship among dry-bulb temperature, wet-bulb temperature, and relative humidity is shown in Table 23.1. To find the relative humidity, first calculate the difference between the wet-bulb and dry-bulb temperatures. Then find on the table the point where the dry-bulb temperature and the difference between the two temperatures intersect. The number at the intersection is the percent relative humidity.

Example:

Find the relative humidity if the dry-bulb temperature is 10°C and the wet bulb temperature is 5°C.

To find the relative humidity you need the *dry-bulb temperature* and the *difference between the dry-bulb and wet bulb temperatures*.

The dry-bulb temperature is given as 10°C.

The difference between the dry-bulb and wet-bulb temperatures must be calculated. It is 10°C – 5°C = 5°C.

The number on the table where these two temperatures intersect is 43. The relative humidity is 43%.

Weather

TABLE 23.1 RELATIVE HUMIDITY (%)

Dry-Bulb Temperature (°C)	Difference Between Wet-Bulb and Dry-Bulb Temperatures (C°)														
	0	1	2	3	4	5	6	7	8	9	10	11	12	13	14	15
−20	100	28														
−18	100	40														
−16	100	48														
−14	100	55	11													
−12	100	61	23													
−10	100	66	33													
−8	100	71	41	13												
−6	100	73	48	20												
−4	100	77	54	32	11											
−2	100	79	58	37	20	1										
0	100	81	63	45	28	11										
2	100	83	67	51	36	20	6									
4	100	85	70	56	42	27	14									
6	100	86	72	59	46	35	22	10								
8	100	87	74	62	51	39	28	17	6							
10	100	88	76	65	54	43	33	24	13	4						
12	100	88	78	67	57	48	38	28	19	10	2					
14	100	89	79	69	60	50	41	33	25	16	8	1				
16	100	90	80	71	62	54	45	37	29	21	14	7	1			
18	100	91	81	72	64	56	48	40	33	26	19	12	6			
20	100	91	82	74	66	58	51	44	36	30	23	17	11	5		
22	100	92	83	75	68	60	53	46	40	33	27	21	15	10	4	
24	100	92	84	76	69	62	55	49	42	36	30	25	20	14	9	4
26	100	92	85	77	70	64	57	51	45	39	34	28	23	18	13	9
28	100	93	86	78	71	65	59	53	47	42	36	31	26	21	17	12
30	100	93	86	79	72	66	61	55	49	44	39	34	29	25	20	16

Source: The State Education Department, *Earth Science Reference Tables.* 2011 ed. (Albany, New York; The University of the State of New York).

The difference between the wet-bulb temperature and the dry-bulb temperature is also directly related to the dew point of the air. The higher the relative humidity, the closer the air is to being filled to its capacity. Cooling air decreases its capacity to hold moisture. The closer the air is to being filled to capacity, the less it has to be cooled to reach the point of maximum moisture. Table 23.2, which shows the relationship among wet-bulb temperature, dry-bulb temperature, and dew point, can be used to find the dew point. The number on the table at the intersection of the dry-bulb temperature and the difference between the wet-bulb and dry-bulb temperatures is the dew point in degrees Celsius.

Example:

Find the dew point if the dry-bulb temperature is 10°C and the wet-bulb temperature is 5°C.

The dry-bulb temperature is given as 10°C.

The difference between the dry-bulb and wet-bulb temperatures must be calculated. It is 10°C − 5°C = 5°C.

The number on the table where these two temperatures intersect is −2. The dew point is −2°C.

TABLE 23.2 DEW POINT TEMPERATURES (°C)

Dry-Bulb Temperature (°C)	Difference Between Wet-Bulb and Dry-Bulb Temperatures (C°)														
	0	1	2	3	4	5	6	7	8	9	10	11	12	13	14	15
−20	−20	−33														
−18	−18	−28														
−16	−16	−24														
−14	−14	−21	−36													
−12	−12	−18	−28													
−10	−10	−14	−22													
−8	−8	−12	−18	−29												
−6	−6	−10	−14	−22												
−4	−4	−7	−12	−17	−29											
−2	−2	−5	−8	−13	−20											
0	0	−3	−6	−9	−15	−24										
2	2	−1	−3	−6	−11	−17										
4	4	1	−1	−4	−7	−11	−19									
6	6	4	1	−1	−4	−7	−13	−21								
8	8	6	3	1	−2	−5	−9	−14								
10	10	8	6	4	1	−2	−5	−9	−14	−28						
12	12	10	8	6	4	1	−2	−5	−9	−16						
14	14	12	11	9	6	4	1	−2	−5	−10	−17					
16	16	14	13	11	9	7	4	1	−1	−6	−10	−17				
18	18	16	15	13	11	9	7	4	2	−2	−5	−10	−19			
20	20	19	17	15	14	12	10	7	4	2	−2	−5	−10	−19		
22	22	21	19	17	16	14	12	10	8	5	3	−1	−5	−10	−19	
24	24	23	21	20	18	16	14	12	10	8	6	2	−1	−5	−10	−18
26	26	25	23	22	20	18	17	15	13	11	9	6	3	0	−4	−9
28	28	27	25	24	22	21	19	17	16	14	11	9	7	4	1	−3
30	30	29	27	26	24	23	21	19	18	16	14	12	10	8	5	1

Source: The State Education Department, *Earth Science Reference Tables*. 2011 ed. (Albany, New York; The University of the State of New York).

Air Movements

In the atmosphere, air circulates because of density differences. The vertical movements of rising and sinking air are called **air currents**. When sinking air reaches Earth's surface, it spreads out horizontally. When rising air expands, it also spreads out horizontally. Horizontal movements of air parallel to Earth's surface are called **winds**. Wind is described by both its speed and its direction.

Wind speed is measured with an **anemometer**. An anemometer consists of three or four cups mounted on a vertical axis and driven by the wind, causing the axis to spin. The speed of rotation of the axis varies with wind speed. Wind speed is measured in knots, a nautical measure of speed; 1 knot = 1.85 kilometers per hour, or 1.15 miles per hour.

Wind direction is determined with a **wind vane**. A wind vane is a pointer mounted on an axis that is attached to a compass rose. The tail of the pointer is larger in surface area than the tip. Thus, the wind exerts more pressure on the tail, causing it to swing around so that the tip points in the direction from which the wind is blowing. A wind is named according to the direction from which it blows. Just as a person who comes from the South is called a Southerner, a wind that blows from the south is called a south wind.

Atmospheric Transparency

All of the gases of which air is composed are transparent. However, there are many substances that become suspended in the atmosphere and decrease its transparency. These substances include dust, volcanic ash, smoke, salt particles, aerosols (droplets of liquids), ice particles, and water droplets. By far, water droplets have the greatest effect on atmospheric transparency. Clouds and fog consist of tiny, but visible, droplets of water. Atmospheric transparency is expressed in three ways: visibility, cloud ceiling, and cloud cover.

Visibility is the horizontal distance through which the eye can distinguish objects. It is usually expressed in miles.

Cloud ceiling is the base height of cloud layers. It is measured with a ceilometer, which uses pulses of light and a photoelectric telescope.

Cloud cover is the fraction of the sky that is obscured by clouds.

WHAT ARE THE RELATIONSHIPS AMONG THE SEVERAL ATMOSPHERIC VARIABLES?

Atmospheric variables are interrelated, so that a change in one is likely to cause a change in another. Since the atmosphere is gaseous, this interrelatedness is best understood in light of the kinetic theory of gases. According to the kinetic theory, gases consist of many tiny, individual molecules that do not interact with each other except when they collide. The molecules are far apart and are in constant random motion, and the temperature of the gas is proportional to the speed at which they are moving.

Air Pressure and Air Temperature

As explained earlier, air pressure is the result of the forces exerted by gas molecules colliding with a surface or each other. Air temperature depends on the speed of the gas molecules. Air pressure and temperature are related because both involve the motion of molecules. The higher the temperature, the faster the gas molecules are moving. The faster the gas molecules are moving, the more force they exert in a collision. Thus, you would expect that an increase in temperature would result in an increase in pressure, but in the atmosphere this is not the case. The reason is that the atmosphere is not contained by rigid walls but is an open system free to expand.

When the temperature of the atmosphere increases, its molecules move faster and collide more vigorously. As a result, the molecules bounce farther apart and the atmosphere expands. Even though the molecules are moving faster, and each individual impact exerts greater force, the total number of collisions decreases because the molecules are spread farther apart. The decrease in collisions far exceeds any increase in force due to faster moving

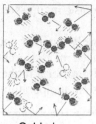
Cold air

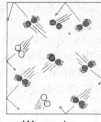
Warm air

Figure 23.5 Relationship Between Air Temperature and Air Pressure. Even though the molecules are moving slower and collide with less force, the cold air on the left exerts more pressure because its closely spaced molecules collide much more frequently.

molecules, so the net effect is a *decrease* in air pressure when air temperature increases. See Figure 23.5.

| Air temperature ↑, Air pressure ↓ | Air temperature ↓, Air pressure ↑ |

Now let us consider what happens to temperature if pressure is changed. See Figure 23.6. The main cause of pressure changes in the atmosphere is the rising or sinking of air in convection currents. When a parcel of air rises, the surrounding air exerts less confining pressure on the parcel and it expands. Conversely, when a parcel of air sinks, the surrounding air exerts more pressure on the parcel and it is compressed.

Any gas that is allowed to expand because the surrounding pressure is reduced will become cooler, and any gas compressed into a smaller volume will become warmer. Such temperature changes, which occur without heat being added from outside sources or heat being lost to the surroundings, are known as **adiabatic** changes.

In adiabatic temperature changes, the change in the temperature of the air reflects a change in the degree of crowding of its molecules. If a gas is allowed to expand into a larger volume, its molecules spread farther apart, and collide with one another less frequently. If a thermometer is placed in such air, fewer molecules will collide with it. Fewer collisions means less energy transferred to the thermometer, causing less expansion of the fluid in the thermometer, and the

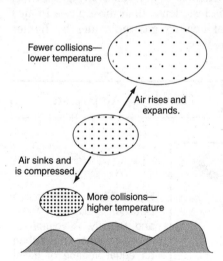

Figure 23.6 Relationship Between Air Pressure and Air Temperature. As air rises, pressure decreases, air molecules spread out and collide less frequently, and temperature decreases. As air sinks, pressure increases, air molecules crowd closer together and collide more frequently, and temperature increases.

Weather

thermometer registers a lower temperature. Thus, a decrease in pressure causes a decrease in temperature.

Compression forces the air molecules to crowd more closely together and causes more frequent collisions. If a thermometer is placed in the compressed air, more molecules will collide with it, and it will register a higher temperature. Thus an increase in pressure causes an increase in temperature.

| Air pressure ↑, Air temperature ↑ | Air pressure ↓, Air temperature ↓ |

Air Pressure and Humidity

The greater the mass of the molecules, the more force they exert during a collision. The mass of a molecule in air depends on the element(s) of which it is composed. Nitrogen and oxygen molecules have more mass than water molecules (molecular mass of $N_2 = 28$, of $O_2 = 32$, of $H_2O = 18$). As water vapor builds up in the atmosphere, lighter water molecules displace heavier nitrogen or oxygen molecules. Since the lighter water molecules exert less force during a collision than the heavier nitrogen and oxygen molecules, air pressure decreases as humidity increases. (Don't be confused by thinking of moist air as containing *liquid* water, which is denser than any gas in air; moist air contains water *vapor*, which is less dense than most gases in air.) Figure 23.7 shows the displacement of heavier gas molecules by lighter water vapor molecules as humidity increases.

| Humidity ↑, Air pressure ↓ | Humidity ↓, Air pressure ↑ |

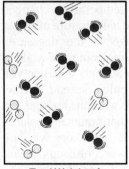

Total Weight of All Molecules = 352
Dry Air

Total Weight of All Molecules = 308
Moist Air

Oxygen M.W. = 32   Nitrogen M.W. = 28   Water M.W. = 18

Figure 23.7 Relationship Between Molecular Mass and Density. A volume of dry air has more mass than an equal volume of moist air because water molecules have less mass than the nitrogen and oxygen molecules they displace. Therefore, moist air is less dense than dry air.

Air Temperature and Humidity

The higher the temperature, the greater the amount of water vapor present in the atmosphere before equilibrium is reached. Thus, warm air contains more water vapor at equilibrium than cool air. Why is this so? Imagine a sealed jar of dry air half filled with water and kept at a certain temperature. All along the water surface, water molecules evaporate and enter the air as water vapor. At the same time, some of the molecules of water vapor in the air lose energy, drop back to the water surface, and return to the liquid. At first, more water molecules leave by evaporation than return to the liquid, and the amount of water vapor in the jar increases. However, over time, the number of water vapor molecules returning to the liquid balances the number of water molecules leaving the liquid. Then, the amount of water vapor in the jar no longer increases. At any given temperature, a fixed amount of water vapor will be present in the air when equilibrium is reached. See Figure 23.8a.

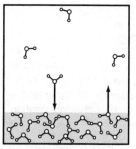
(a) Equilibrium at low temperature

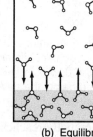
(b) Equilibrium at high temperature

Figure 23.8 At lower temperatures (a) less water vapor is present in the atmosphere when equilibrium is reached than is present at higher temperatures (b).

Now let's consider what happens if we heat the sealed jar. In order for molecules of water in the liquid to break free and become water vapor, they require energy. At a higher temperature, more of the molecules in the liquid water will have enough energy to break free and become a gas. As the water vapor becomes more crowded with molecules, it becomes more likely that some will be trapped at the liquid surface and condense. Thus, the amount of water vapor in the warmed jar increases until a new balance between water molecules leaving and water molecules returning to the liquid is reached. The amount of water vapor in the jar at the higher temperature is greater than it was at the lower temperature. The same volume of space contains more water vapor at equilibrium! See Figure 23.8b. Roughly speaking, for every 10°C increase in temperature, the amount of water vapor in a given volume of air at equilibrium is doubled.

Since the amount of water vapor in the air at equilibrium changes with temperature, the relative humidity is also influenced by temperature. If the temperature increases, causing the equilibrium amount to increase, but the moisture in the air remains the same, the relative humidity will decrease. Conversely, if the temperature decreases, causing the equilibrium amount to decrease, but the moisture in the air remains the same, the relative humidity will increase.

Weather

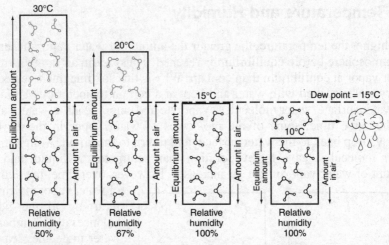

Figure 23.9 Cooling Decreases the Equilibrium Amount of Water Vapor, Leading to Condensation. As air is cooled, the amount of water vapor the air contains approaches the equilibrium amount. At the dew point, evaporation and condensation are in equilibrium. If the air is cooled below the dew point, net condensation results in formation of clouds and precipitation.

As air temperature decreases, it eventually reaches the **dew point**, the temperature at which the moisture the air contains equals the equilibrium amount. At that point, the relative humidity is 100 percent. If the temperature decreases further, moisture in the air will exceed the equilibrium amount, and condensation will occur at a faster rate than evaporation until equilibrium is reestablished. The condensed moisture forms droplets of water or directly forms ice crystals, which, in turn, may form clouds or precipitation. Therefore, as air temperature approaches the dew point, precipitation becomes more likely. See Figure 23.9.

Air Pressure and Winds

Winds blow from regions of high air pressure to regions of low air pressure. This is easy to understand if you think of two people pushing against each other. The person pushing harder will advance against the person pushing with less force. In much the same way, air exerting high pressure will advance against air exerting low pressure. The net movement of the air is from high pressure toward low pressure. See Figure 23.10.

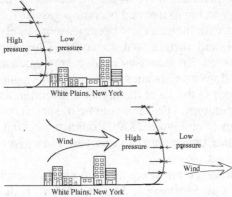

Figure 23.10 Air Pressure and Winds. Winds Blow from High to Low. High-pressure air exerts more force than low-pressure air and pushes the low-pressure air ahead of it, resulting in a horizontal movement of air—a wind.

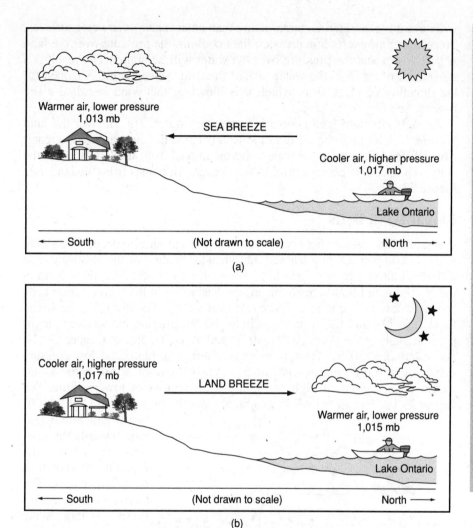

Figure 23.11 Land and Sea Breezes. (a) During the day, land surfaces heat up faster than water. The higher temperatures over the land result in lower air pressures, while the lower temperatures over the water result in higher air pressures. A wind develops that blows from the high-pressure area over the water toward the low-pressure area over the land. (b) At night, land surfaces cool off faster than water and the situation reverses, causing a wind to blow from the land toward the water.

Land and Sea Breezes

Land and sea breezes illustrate this air movement from high to low pressure regions very clearly. Differences in air pressure between adjacent regions occur wherever land and water meet. During the day, when exposed to sunlight, land surfaces increase in temperature more than water surfaces because water has a higher specific heat than soil. As a result, air over land surfaces will have a higher daytime temperature than adjacent air over water surfaces. Since warm air exerts less pressure than cool air, the pressure over the land will be lower and the pressure over the water will be higher. The result is a movement of air from the water toward the land. Since a wind is named for the direction or place from which it is blowing, this wind is called a **sea breeze**. See Figure 23.11a.

At night, though, land cools off faster than water. The air over the land becomes cooler than the air over the water. Thus, the air pressure is greater over the land than over the water, and air moves from the land toward the water. This wind is called a **land breeze** because it comes from the land. See Figure 23.11b.

Global Wind Belts

Differences in pressure between adjacent regions of air also occur on a global scale. In Chapter 22, you learned that unequal heating of the atmosphere at different latitudes causes three large atmospheric convection cells to form in the Northern and Southern Hemispheres. Rising air in these convection cells produces low-pressure belts that circle Earth at the Equator (0°), the Arctic Circle (60°N), and the Antarctic Circle (60°S). Sinking air produces high-pressure belts at the Poles (90°N, 90°S) and at the Tropics of Cancer (30°N) and Capricorn (30°S). Thus, a series of alternating high- and low-pressure belts circle Earth, giving rise to a series of **global wind belts** as air moves from regions of high pressure toward regions of low pressure. See Figure 23.12. The global winds that blow from the high pressure belts at 30° North and South latitude towards the low-pressure belt at the Equator are known as the **trade winds**. The Coriolis effect causes these large-scale winds to curve to the right in the Northern Hemisphere (so they blow from the northeast) and to the left in the Southern Hemisphere (so they blow from the southeast).

Global winds also blow from the high-pressure belts at 30° North and South latitudes toward the low-pressure belts at 60° North and South lati-

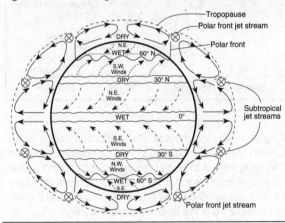

Figure 23.12 Planetary Wind and Moisture Belts in the Troposphere. The drawing shows the locations of the belts near the time of an equinox. The locations shift somewhat with the changing latitude of the Sun's vertical ray. In the Northern Hemisphere, the belts shift northward in summer and southward in winter. Source: The State Education Department, *Earth Science Reference Tables*, 2011 ed. (Albany, New York; The University of the State of New York).

tudes and are known as the **prevailing winds**. The Coriolis effect causes these large-scale winds to blow from the southwest in the Northern Hemisphere and from the southeast in the Southern Hemisphere.

The global winds that blow form the high-pressure regions over the poles toward the low pressure regions at 60°N and 60°S are known as the **polar easterlies**. The Coriolis effect causes these winds to blow from the northeast in the Northern Hemisphere and from the southeast in the Southern Hemisphere.

HOW DO CLOUDS AND PRECIPITATION FORM?

Clouds and precipitation form when air is cooled below its dew point and water vapor condenses into tiny water droplets or ice crystals. The conditions under which this cooling takes place determine what type of cloud or precipitation will form.

Condensation is the change of phase from gas to liquid. To condense into a liquid, a gas must have a surface to condense on. In the atmosphere, this surface is provided by particles suspended in the air, such as salt or ice crystals, dust, or smoke, called **condensation nuclei**. Water vapor in the air condenses when the air is cooled to the dew point and there are condensation nuclei on which condensation can form. **Deposition** is the change of phase from a gas directly to a solid. Water vapor may form ice crystals if moisture is released from the air when its temperature is below the freezing point of water.

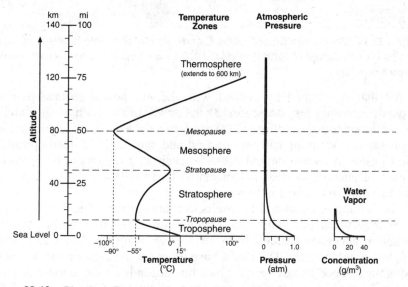

Figure 23.13 Physical Properties of the Atmosphere. Source: The State Education Department, *Earth Science Reference Tables*, 2011 ed. (Albany, New York; The University of the State of New York).

Weather

As you can see from Figure 23.13, temperature drops steadily with altitude in the troposphere. In the upper portion of the troposphere, temperatures below –40°C are the norm year-round. For this reason, even in the summer, cirrus clouds are composed of tiny ice crystals, and thunderstorms whose cloud tops extend into the upper troposphere may produce hail. Clouds are important components of the atmosphere for three reasons. They are the producers of precipitation, such as rain, snow, sleet, and hail. They are also major reflectors of solar radiation, and they serve as key indicators of overall weather conditions.

As shown in Figure 23.14, clouds are categorized by the altitudes at which they exist and their degrees of vertical development.

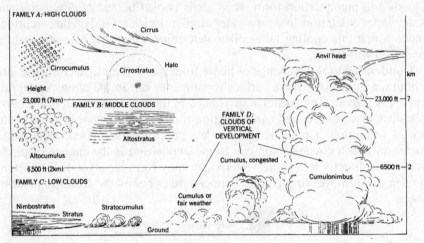

Figure 23.14 Cloud-Form Classification. Clouds are classified into families according to altitude and degree of vertical development. Source: Arthur N. Strahler, *The Earth Sciences*, Harper & Row, 1971.

Air that is warmed by contact with the Sun-heated surface of Earth expands, becomes less dense, and floats upward in the denser surrounding air. As the air rises, it continues to expand, causing a decrease in pressure and temperature. Warm air will also expand and cool if it is forced upward as wind pushes it over a natural obstacle, such as a mountain, or by cooler, denser air pushing its way underneath it. These three conditions that cause air to rise are diagramed in Figure 23.15.

The cooling of air as it rises results in condensation when the air temperature drops to the dew point and condensation nuclei are present. This condensation generates a **cloud** made of a great many tiny but visible water droplets or ice crystals. The **cloud base** indicates the elevation at which the rising air reached its dew point. Since this depends on the amount of moisture in the air and the starting temperature, clouds may form at many different elevations.

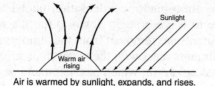
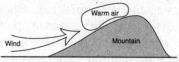

Air is warmed by sunlight, expands, and rises.   Warm air is forced upward by wind.

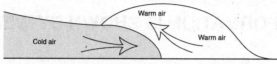

Dense, cold air mass pushes under warm air, forcing it to rise.

Figure 23.15 Three Conditions That Cause Air To Rise. Air rises when it is heated and becomes less dense, when it is forced upward by winds over rising land surfaces, and when it is forced upward by colder, denser air pushing under it.

Fog is a cloud whose base is at ground level. It forms when moist air at ground level is cooled below its dew point. When moist air comes into contact with a surface such as grass, or with a car or other object that has cooled below the dew point because heat has radiated out of it overnight, droplets of water called **dew** condense on the surface. If the temperature is below freezing, tiny ice crystals called **frost** will form instead of dew.

The water droplets or ice crystals in clouds are very tiny (1/50 millimeter) and often remain suspended in the air. Thus, all clouds do not form precipitation. **Precipitation** occurs only when cloud droplets or ice crystals join together and become heavy enough to fall. Table 23.3 summarizes the different types of precipitation and their origins.

TABLE 23.3 TYPES OF PRECIPITATION AND ORIGINS

Name	Description	Origin
Rain	Droplets of water up to 4 mm in diameter	Cloud droplets coalesce when they collide.
Drizzle	Very fine droplets falling slowly and close together	Cloud droplets coalesce when they collide.
Sleet	Clear pellets of ice	Raindrops freeze as they fall through layers of air at below-freezing temperatures.
Glaze	Rain that forms a layer of ice on surfaces it touches	Supercooled raindrops freeze as soon as they come in contact with below-freezing surfaces.
Snow	Hexagonal crystals of ice, or needlelike crystals at very low temperatures	Water vapor sublimes, forming ice crystals on condensation nuclei at temperatures below freezing.
Hail	Balls of ice ranging in size from small pellets to as large as a softball, with an internal structure of concentric layers of ice and snow	Again and again hailstones are hurled up by updrafts in thunderstorms and then fall through layers of air that alternate above and below freezing. Each cycle adds a layer to the hailstone. The more violent the updrafts, the larger and heavier the hailstone can become before falling.

Weather

As precipitation falls through the atmosphere, condensation nuclei and other suspended material that adhere to it are brought down to Earth's surface. In this way dust and many pollutants are removed from the atmosphere by precipitation. However, some pollutants react with the water in precipitation to form harmful solutions such as acid rain, which has had many negative effects on the environment.

WHAT INFORMATION IS SHOWN BY WEATHER MAPS?

Synoptic Weather Maps

One of the most widely used methods of forecasting weather is known as **synoptic forecasting**. Synoptic forecasting is based on examination of a synopsis, or summary of the total weather picture at a particular time.

A **synoptic weather map** is made by measuring atmospheric variables at thousands of weather stations around the world four times a day. These data are then used to create field maps, which reveal large-scale weather patterns. By looking at a sequence of synoptic weather maps, meteorologists can perceive the developments and movements of weather systems and make predictions.

Synoptic weather maps use a symbol called a **station model** to show a summary of the weather conditions at a particular weather station. Figure 23.16 shows a typical station model and explains what each of its elements represents.

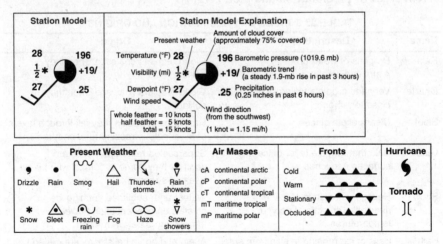

Figure 23.16 A Station Model. The symbol uses shorthand notation to summarize atmospheric variable for a particular location on a weather map. Source: The State Education Department, *Earth Science Reference Tables*, 2011 ed. (Albany, New York; The University of the State of New York).

Field Maps

Station models plotted on a map summarize a wide range of weather data and can be used to create many different field maps. A **field** is defined as a region of space that has a measurable quantity at every point. Field maps can be used to represent any quantity that varies in a region of space. One way to represent field quantities on a two-dimensional field map is to use **isolines**, which connect points of equal field value. For example, a temperature field map contains lines connecting points of equal temperature, or **isotherms**. A pressure field map contains lines connecting points of equal pressure, or **isobars**. Field maps clearly show the weather patterns in the atmosphere.

Figure 23.17 shows a synoptic weather map, followed by the same map with isotherms and isobars drawn in. (Figures 23.17b and 23.17c).

As you can see in Figure 23.17a, there are distinct regions in the atmosphere that have similar conditions. For example, a region of high pressure, cool temperature, and clear skies is centered near Salt Lake City, Utah. You can also see a region of low pressure, slightly warmer temperature, and rain centered near Cincinnati, Ohio.

Within a field, a field value changes as you move from place to place. The rate at which the field value changes is called its **gradient**. Rapid changes in field values, or steep gradients, show up as closely spaced isolines on a field map. In Figure 23.17c, you can see that the isobars are widely spaced within each of the two air masses, but are closely spaced between them. This means that there is little change in pressure within each of the regions, but there is a boundary between them where the pressure changes rapidly. You can see the same pattern repeated for temperature: there are regions of the atmosphere with relatively uniform characteristics, which have boundaries separating them from adjacent regions.

HOW CAN WEATHER INFORMATION BE USED TO MAKE FORECASTS?

Most weather forecasts are based on the movements of large regions of air with fairly uniform characteristics, called **air masses**. By examining a series of synoptic weather maps, meteorologists can track the current movements of air masses and predict future movements.

Air Masses

As stated above, a region of the atmosphere in which the characteristics of the air at any given level are fairly uniform is called an *air mass*. An air mass is identified by its average air pressure, air temperature, moisture content, and winds. The boundaries between air masses, where rapid changes occur, are called **frontal boundaries**, or **fronts**.

Weather

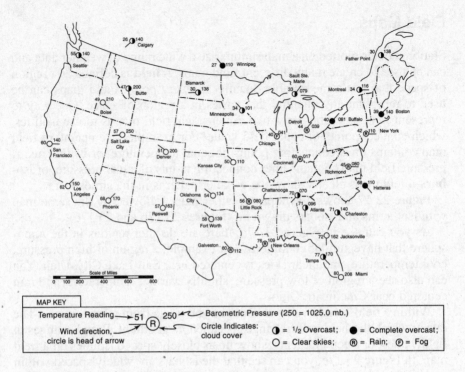

Figure 23.17a Synoptic Weather Map of the United States.

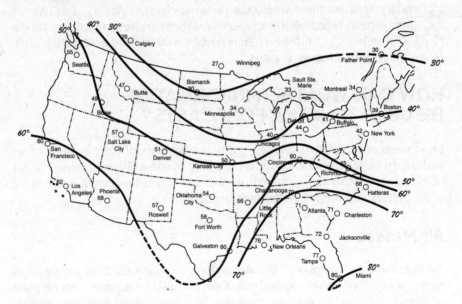

Figure 23.17b Isotherms.

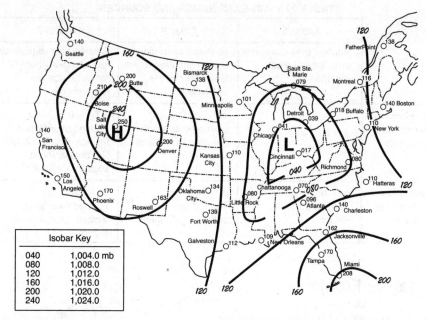

Figure 23.17c Isobars.

The characteristics of an air mass are the result of the geographical region over which it formed, or its **source region**. Air resting on, or moving very slowly over, a region tends to take on the characteristics of that region. For example, air that sits over the Gulf of Mexico in the summer is resting on very warm water. The air is warmed by contact with the water, and evaporation causes much humidity to enter the air. As a result the air becomes warm and moist, just like the Gulf of Mexico. In general, air masses that form near the poles are cold, and those that form near the Equator are warm. Air masses that form over water are moist; those that form over land, dry. The longer the air remains stationary over a region, the larger it becomes and the closer it matches the characteristics of the region. On weather maps, air masses are named for their source regions and are designated by a letter or letters as shown in Table 23.4.

Air masses are moved by planetary winds and jet streams (study Figure 23.12 carefully). In general, arctic and polar air masses are moved south and west over the United States by the northeast wind belt near the North Pole, and tropical air masses are moved to the north and east by the southwest wind belt between 30° and 60°N. As the air masses are carried over Earth's surface, they slowly change and take on the characteristics of the regions over which they are moving. For example, a cP air mass moving south out of Canada will gradually become warmer as it moves farther and farther south.

Air masses that affect U.S. weather and the regions in which they form are shown in Figure 23.18.

Weather

TABLE 23.4 AIR-MASS NAMES AND SOURCES

	Arctic A	Polar P	Tropical T
	Formed over extremely cold, ice-covered regions	Formed over regions at high latitudes where temperatures are relatively low	Formed over regions at low latitudes where temperatures are relatively high
maritime m Formed over water, moist		mP—cold, moist Formed over North Atlantic, North Pacific	mT—warm, moist Formed over Gulf of Mexico, middle Atlantic, Caribbean, Pacific south of California
continental c Formed over land, dry	cA—dry, frigid Formed north of Canada	cP—cold, dry Formed over northern and central Canada	cT—warm, dry Formed over southwestern United States in summer

Weather Fronts

At any given point in time, several air masses will be moving across the United States. They generally move from west to east, driven by the prevailing southwesterlies. When different air masses meet, very little mixing of the air takes place and a sharp transition zone forms between them. This zone is termed a **weather front** because it marks the leading edge, or front, of an air mass that is pushing against another air mass. The whole surface along which the air masses meet is called the **frontal surface**, and the line on the ground marking the transition from one air mass to the other is the **front**. See Figure 23.19. Fronts are areas of rapid changes in weather conditions and are often sites of unsettled and rainy weather. Different types of frontal surfaces, such as cold, warm, stationary, and occluded fronts, will form, depending on which type of air mass is advancing against another.

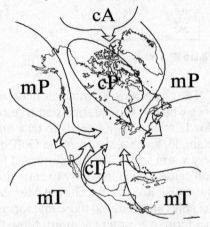

Figure 23.18 Air Masses That Influence Weather in the United States and Their Source Regions.

Cold Fronts

A **cold front** forms along the leading edge of a cold air mass that is advancing against a warmer air mass. The cold air mass is denser than the warmer air ahead of it, so it pushes against and under it like a wedge. This *forces* the warmer air upward rapidly and results in turbulence, rapid condensation forming heavy, vertically developed clouds, and heavy precipitation or thunderstorms. Cold fronts are often preceded by a zone of thunderstorms in the summer and of snow flurries in the winter. After the front passes, the temperature drops sharply and the pressure rises rapidly. Figure 23.20 shows a typical cold front.

Warm Fronts

A **warm front** forms where a warm air mass overrides the trailing edge of a cold air mass

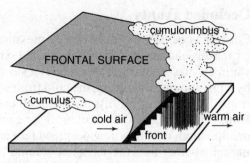

Figure 23.19 Perspective View of a Frontal Surface and a Front Line on the Ground. A *frontal surface* forms along the boundary between two different air masses. The line along which the boundary touches the surface is the *front*.

ahead of it. The less dense warm air mass rides up and over the denser cold air. As the warm air mass rises above the denser cold air mass, it expands and cools, causing condensation to occur over the wide, gently sloping boundary. The results are thickening, lowering clouds and widespread precipitation. Figure 23.21 shows a typical warm front.

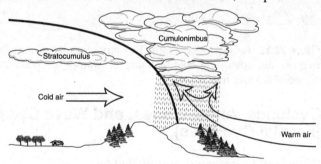

Figure 23.20 A Typical Cold Front.

Stationary Fronts

A **stationary front** forms along the boundary between a warm air mass and a cold air mass when neither moves appreciably in any

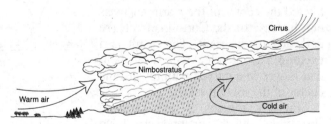

Figure 23.21 A Typical Warm Front.

direction. A stationary front slowly takes on the shape and characteristics of a warm front as the denser cold air slowly slides beneath the less dense, warmer air. Although stationary fronts develop the same widespread rain and cloudiness as do warm fronts, the rain and clouds may persist for a longer period, perhaps many days, until another air mass comes along with enough impetus to get the stalled air masses moving.

Occluded Fronts

Cold air is denser and exerts more pressure than warm air. Therefore, cold air masses and the cold fronts associated with them tend to move faster than warm air masses and warm fronts. If a slow-moving warm front is followed by a fast-moving cold front, the cold front will sometimes overtake the warm front. Then the cold air lifts the warm air entirely aloft, forming an **occluded front**. The lifting of the warm air mass causes large-scale condensation and precipitation, resulting in widespread rain and thunderstorms. Then, depending on whether the cold air mass coming from behind is colder or warmer than the cold air mass now ahead, a cold-front occlusion or a warm-front occlusion may form, as shown in Figure 23.22.

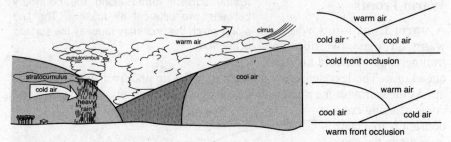

Figure 23.22 An Occluded Front. Two types of occlusions can form, depending on whether the advancing cold air mass is warmer or cooler than the cold air mass ahead of the warm air mass.

Cyclones, Anticyclones, and Wave Cyclones (Frontal Cyclones)

Regions of low pressure surrounded by higher pressure form winds that blow toward the center of the warm air mass but, because of the Coriolis effect, are deflected to the right. The result is a **cyclone**, a counterclockwise flow of air, that moves in a curved path toward the center of the low pressure.

Similarly, regions of high pressure surrounded by lower pressure form winds that blow outward from the center. The winds are deflected by the Coriolis effect into a flow of air, an **anticyclone**, that is clockwise and outward from the center.

A cyclone and an anticyclone are diagramed in Figure 23.23.

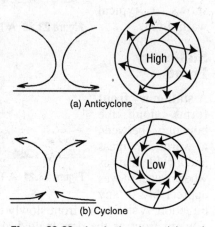

Figure 23.23 An Anticyclone (a) and a Cyclone (b).

Most of the weather systems that move across the United States are *wave cyclones*, or *frontal cyclones*. **Wave cyclones** are warm and cold fronts that move in a counterclockwise direction around a low-pressure center. They typically form in the westerly wind belts along the boundary between polar and tropical air, called the **polar front** (see Figure 23.12). In the United States, this boundary falls just about along our northern border with Canada.

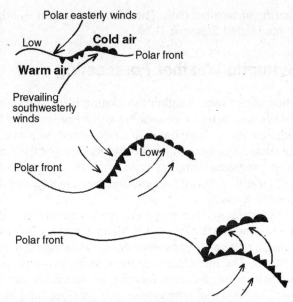

Figure 23.24 Formation of a Wave Cyclone.

A wave cyclone forms when southwesterly winds push warm, low-pressure air against cold air along the polar front (see Figure 23.24). The result is to lower the pressure along the polar front, and the boundary bends to form a wave of cold air advancing from west to east against the warm air—a cold front. The cold air pushes the warm air ahead of it into the cold air on the other side of it, forming a warm front. The end result is a wavelike structure consisting of a cold front overtaking a warm front. The counterclockwise flow in the low-pressure center causes the cold front to swing around the equatorial side of the low-pressure center as the wave moves from west to east. The wave cyclone may even break loose from the polar front and be carried, swirling along, by the jet stream. In the United States, wave cyclones generally move from west to northeast and get carried out over the north Atlantic. The stormy weather associated with a wave cyclone may affect millions of square miles and last for several days.

HOW CAN WE FORECAST THE WEATHER?

Until relatively recently, weather forecasts were based on local observations made directly by human senses. Through first-hand experience and passed-down weather lore, people learned to recognize certain patterns of weather changes associated with cloud types or shifts in wind direction. It was not until the invention of the barometer and the thermometer in the 1600s, however, that accurate measurements of pressure and temperature changes could be made. Even then, forecasting was not practicable until communication among many points by telegraph made possible the rapid collection and sharing of weather data. The first government weather forecasts were issued in the United States in 1870.

Synoptic Weather Forecasting

The basis of most weather forecasting is the systematic collection of weather data from a network of weather-reporting sites that blanket the globe. At the simplest level, these data are summarized on a synoptic weather map using shorthand symbols called *station models*. At their most sophisticated, computer programs compile, store, analyze, and plot weather data, plot isolines for various weather factors, and can display weather data in a variety of graphic formats.

Synoptic weather maps are made four times a day—at midnight, noon, 6 A.M., and 6 P.M. Greenwich Mean Time. By looking at a sequence of these maps, a meteorologist can detect the development and movement of weather systems. The movements of these systems are projected into the future, and predictions of weather for various locations are made. Computer-drawn maps show wind, temperature, air pressure, and humidity patterns at many different atmospheric levels. Statistical analysis is then used to map projections of weather factors.

Synoptic forecasting is fairly accurate for large scale, short-term forecasts. However, there are always locally unique conditions that modify the general behavior of weather systems. For example, cities tend to absorb and reradiate more heat energy than rural areas, creating urban "heat islands." Local meteorologists must take these specific conditions into account when making forecasts, for they have a significant effect on such factors as the exact path a storm will take locally, the times when rain will begin and end, and the extremes that high and low temperatures will reach. For this reason, most local meteorologists have a network of weather hobbyists who feed them weather data in addition to those collected by official Weather Service stations.

Statistical Weather Forecasting

Statistical weather forecasting is based on the past behavior of the atmosphere. Mathematical equations are used to predict the probability that the atmosphere will behave in a certain way. Basically the system works something like this: Suppose you have math first period on Mondays, Tuesdays, and Wednesdays, and science first period on Thursdays. A person observing you would soon see a pattern in which three days of first-period math are followed by a first period of science. However, there would be exceptions. There might be a holiday on a Monday, so the sequence would be science after *two* maths. Or Thursday might be a holiday, so the sequence would be six maths before the next science. However, such deviations in the schedule would probably be the exception rather than the rule. Let's assume that in 10 weeks there was one deviation. Statistical forecasting would then say that for the next week there was a 9 in 10, or 90 percent, chance that three maths would be followed by a science.

Now let's apply this reasoning to weather. Suppose we observed that, nine out of the past ten times the wind shifted to blow from the northeast, we had rain within 6 hours. Then, if the wind shifted to the northeast, statistical forecasting would say that there was a 90 percent chance of rain in the next 6 hours.

The process of statistical forecasting is threefold: maintaining detailed historical records, identifying all previous occurrences of a pattern, and then calculating the probabilities. However, there is no way to anticipate unusual behavior, nor is past behavior a guarantee of future behavior.

Numerical Weather Forecasting

Numerical forecasting is based on fluid dynamics. Theoretically, if the initial states of the atmosphere, land surfaces, and oceans are known, we should be able to predict the behavior of the atmosphere based on the laws of physics governing heat and moisture transfer and fluid behavior. However, such complete information is not currently available.

The sheer quantity of data, coupled with the number and complexity of the required calculations, made numerical forecasting impracticable until computers capable of high-speed calculations were developed in the 1940s. Now, computers solve six equations simultaneously—one for each of the three dimensions of motion, and one each for the conservation of moisture, heat, and mass. These equations are solved for thousands of points on a map grid, and changes at each point are projected. These changes are then fed back into the loop for another round of calculations, producing a projection further into the future. As new actual data are entered into the equations throughout the day, the revised calculations serve to correct the predictions. Calculations for several levels in the atmosphere, projecting changes over the next 2 days, are made every 12 hours. A 5-day forecast is calculated once a day.

Long-Range Weather Forecasting

No matter what method is used, however, forecasting reliability decreases as the time frame increases. Short-term forecasts (1–3 days) of weather are usually far more accurate than long-term forecasts (a week or more). The decrease in forecast reliability over time is due to a number of factors: the wide spacing of initial data points, the unreliability in measurements taken over many areas, and a lack of understanding of the behavior of the atmosphere. A small error in an initial measurement can cause errors in computer-calculated forecasts.

For example, let's make a simple mathematical model that assumes that temperature increases or decreases in direct proportion to the ratio of day to night:

$$\text{Temperature today} \times \frac{\text{Length of night}}{\text{Length of day}} = \text{Temperature tomorrow}$$

Let's compare the temperatures predicted by this model, using a correct temperature reading and one in which an error was made:

Starting Temperature	Prediction When Daylight = 13 hours, Night = 11 hours	Prediction When Daylight = 14 hours, Night = 10 hours
21°C (correct)	21 × 13/11 = 24.8°C	24.8 × 14/10 = 34.7°C
22°C (incorrect)	22 × 13/11 = 26°C	26 × 14/10 = 36.4°C

As you can see, an initial error of 1°C has almost *doubled* as it was compounded in the calculations. Errors like these multiply as forecasts are extended farther into the future, and at some point the forecast becomes useless.

Chaos theory shows that small, unpredictable changes can result in large-scale changes in a system over time. For example, a forest fire set by lightning may cause a localized increase in temperature that may set in motion large-scale changes in air-flow patterns. For this reason, chaos theory seems to indicate that long-term forecasts may never be reliable.

WHERE AND HOW ARE THE MOST HAZARDOUS WEATHER SITUATIONS LIKELY TO OCCUR?

Thunderstorms

One of the most common weather hazards is a thunderstorm (see Figure 23.25). A thunderstorm begins as a convection cell in the atmosphere. Warm air rises, expands, cools, and then sinks back to Earth's surface. If the convection in the cell is strong, because of intense heating, the air may rise very rapidly, reaching heights of many kilometers in just a few minutes. A cloud forms in the updraft of the convection cell and grows as more and more warm, moist air is carried upward. The strong updrafts in the rising air support water droplets and ice crystals in the cloud so that they grow in size. When the updrafts can no longer support the moisture, it falls as rain or even hail. The falling rain sets up downdrafts that cause internal friction with updrafts, and the internal friction builds up static electric charges that may discharge as lightning. Thunderstorms are likely to occur wherever and whenever there is strong heating of Earth's surface.

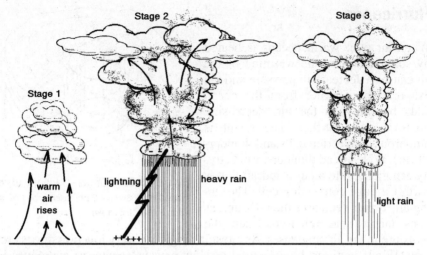

Figure 23.25 Formation of a Thunderstorm. Stage 1: Intense heating causes warm air to rise rapidly, forming tall, vertically developed clouds. Stage 2: Strong updrafts keep precipitation aloft until it is heavy. Falling rain creates downdrafts, and internal friction with updrafts causes buildup of static charge, leading to lightning and thunder. Stage 3: Downdrafts cause air to cool and updrafts subside, so rain becomes lighter.

Tornadoes

Late in the day, when Earth's surface is warmest and convection is strongest, a tornado may form (see Figure 23.26). Tornadoes are small (most are less than 100 meters in diameter), brief (most last only a few minutes) disturbances that usually develop over land from intense thunderstorms. When heating is very intense, warm air rises in strong convection currents. The upward movement of the air causes a sharp decrease in pressure. Air rushes into the low-pressure region from the sides and, deflected by the Coriolis effect, is given a spin. As air moves toward the center of the updraft, it reaches high speeds because of the big difference in pressure. The rapidly moving air decreases the pressure even more, further feeding the updraft. The whole process spirals upward in intensity, and a funnel forms that eventually touches the ground. Wind speeds near the center of a tornado may reach 500 kilometers per hour or more.

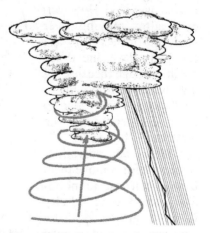

Figure 23.26 Formation of a Tornado.

Hurricanes

When air over warm oceans is heated by solar radiation, the warming of the air causes a decrease in pressure and the evaporation of water from the ocean adds humidity to the air, decreasing pressure even further. The result is numerous convection cells and thunderstorms. Widespread thunderstorm activity can merge into a large updraft that is part of a huge convection cell. The rising air further decreases the pressure, as does latent heat released when the moisture in the air condenses. As a result a region of very low pressure is created. Winds begin to blow toward the low-pressure center as air moves in from surrounding areas of higher pressure. The Coriolis effect deflects the winds, and a cyclone forms. Heat, released when fresh moisture brought in by winds condenses, feeds the convection cell with new energy, and it grows larger and stronger. When the winds in the cyclone exceed 119 kilometers per hour, we have a hurricane (see Figure 23.27). Fully formed hurricanes are huge cyclones, often exceeding 500 kilometers in diameter.

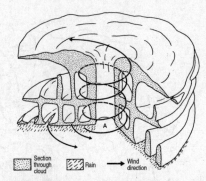

Figure 23.27 Cross Section of a Hurricane. Source: Earth Science on File. © Facts on File, Inc., 1988.

Hurricanes tend to form over oceans near the Equator during summer months, when the ocean surface is the warmest. Once the hurricane has formed, heat and moisture from the ocean provide the energy that maintains it. Anything that cuts off the supply of heat or moisture will diminish the strength of the hurricane, so hurricanes weaken as they move over land or cool water. Therefore a hurricane is usually most destructive when it first moves over land. However, even after moving over land, hurricanes can last for many days and cause much damage because of high winds and flooding.

Planetary winds tend to move hurricanes to the west until they reach 30° N; then the prevailing southwesterlies move them from southwest to northeast (see Figure 23.28). However, air masses that a hurricane encounters may deflect its path.

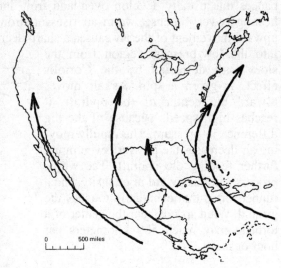

Figure 23.28 Typical Tracks of Tropical Cyclones—Hurricanes—near the United States.

Emergency Preparedness for Hazardous Weather

Hazardous weather, such as tornadoes, severe thunderstorms, and hurricanes, claims many lives and causes much property damage each year in the United States. Hazardous weather happens; there's not much humans can do about that. Measures can be taken, however, to reduce the risk of damage and loss of life. Reducing the risk is called *mitigation*. Some forms of mitigation are expensive and complicated, such as moving to a safer location or making structural changes to a house. Some measures are easier, such as developing an emergency plan, having disaster supplies on hand, protecting windows with shutters, and securing loose outdoor items that could cause damage when hurled by strong winds.

Emergency preparedness is an important responsibility of local government. Communities can help reduce risks by taking such actions as developing disaster plans, establishing evacuation routes and public shelters, enacting zoning regulations that limit development of high-risk areas, and enforcing building codes that require structures to withstand hazardous weather forces.

Watches and Warnings

Advance planning and quick response are the keys to surviving hazardous weather. Toward this end, the National Weather Service issues two levels of alerts about life-threatening weather such as tornadoes, severe thunderstorms, and hurricanes: watches and warnings. *Watches* are issued for tornadoes and severe thunderstorms when weather conditions are such that these hazards are likely to develop in your area. Hurricane watches are issued when there is a threat of hurricane conditions within 24–36 hours. *Warnings* are issued when a tornado or severe thunderstorm has been sighted or indicated on radar in your area, or when hurricane conditions are expected in 24 hours or less.

When a severe thunderstorm or tornado is coming, there is very little time to make life-or-death decisions, so the National Weather Service issues a warning as soon as they are spotted. On the other hand, hurricanes are easier to track because they are so large and last so long. A 1-mile-wide tornado is huge and may last for minutes or hours; a 100-mile-wide hurricane is small, and hurricanes usually last for a week or more. Although hurricane warnings are easier to issue, the decisions involved are more complicated. An average hurricane warning along several hundred miles of United States coastline costs tens of millions of dollars spent for boarding up homes, closing down schools, businesses, and manufacturing plants, and evacuating people. Then, just a small deviation in a hurricane's path may mean the difference between a disaster and a very rainy day for a city under a hurricane warning. Forecasting is improving, however, and in the next few years meteorologists hope to make 30-hour forecasts as accurate as current 24-hour forecasts.

Weather

What to Do During Watches and Warnings

If a watch is issued for your area, you should stay alert to the weather by listening to the radio or television and be prepared to take shelter. This is the time to check emergency supplies and to remind family members of disaster plans and ways to respond during an emergency. Check for hazards outdoors and secure your home.

If a warning is issued, the danger is immediate and very serious, and everyone should go to a safe place. Once there, listen to a battery-operated radio or television for further instructions and for an official "all clear" before resuming normal activities. The Federal Emergency Management Agency has developed severe thunderstorm, tornado, and hurricane fact sheets that describe specific steps to be taken when watches and warnings are issued for these weather hazards. These fact sheets are available online at *www.fema.gov*.

MULTIPLE-CHOICE QUESTIONS

In each case write the number of the word or expression that best answers the question or completes the statement.

1. Which list correctly matches each instrument with the weather variable it measures?
   (1) wind vane; wind speed thermometer; temperature precipitation gauge; relative humidity
   (2) wind vane; wind direction thermometer; dewpoint psychrometer; air pressure
   (3) barometer; relative humidity anemometer; cloud cover precipitation gauge; probability of precipitation
   (4) barometer; air pressure anemometer; wind speed psychrometer; relative humidity

2. Which weather variable can be determined by using a psychrometer?
   (1) barometric pressure   (3) relative humidity
   (2) cloud cover   (4) wind speed

3. A barometric pressure of 1021.0 millibars is equal to how many inches of mercury?
   (1) 29.88   (3) 30.25
   (2) 30.15   (4) 30.50

4. A temperature of 80° Fahrenheit would be approximately equal to how many degrees on the Celsius scale?
   (1) 34   (3) 27
   (2) 299   (4) 178

5. Most water vapor enters Earth's atmosphere by the processes of
   (1) radiation and cementation
   (2) conduction and convection
   (3) condensation and precipitation
   (4) evaporation and transpiration

6. Under which atmospheric conditions will water most likely evaporate at the fastest rate?
   (1) hot, humid, and calm
   (2) hot, dry, and windy
   (3) cold, humid, and windy
   (4) cold, dry, and calm

7. The diagram below shows the temperature readings on a weather instrument.

   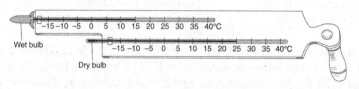

   Based on these readings, the relative humidity of the air is closest to
   (1) 8%
   (2) 11%
   (3) 32%
   (4) 60%

8. A parcel of air has a dry-bulb temperature of 24°C and a relative humidity of 55%. What is the dewpoint of this parcel of air?
   (1) 6°C
   (2) 14°C
   (3) 24°C
   (4) 29°C

9. What is the difference between the dry-bulb temperature and the wet-bulb temperature when the relative humidity is 28% and the dry-bulb temperature is 0°C?
   (1) 11°C
   (2) 2°C
   (3) 28°C
   (4) 4°C

10. Earth's surface winds generally blow from regions of higher
    (1) air temperature toward regions of lower air temperature
    (2) air pressure toward regions of lower air pressure
    (3) latitudes toward regions of lower latitudes
    (4) elevations toward regions of lower elevations

Weather

11. Which graph best represents the change in air pressure as air temperature increases at Earth's surface?

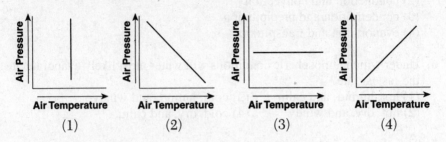

12. Air pressure is usually highest when the air is
(1) cool and humid
(2) cool and dry
(3) warm and humid
(4) warm and dry

13. An observer measured the air temperature and the dewpoint and found the difference between them to be 12°C. One hour later, the difference between the air temperature and the dewpoint was found to be 4°C. Which statement best describes the changes that were occurring?
(1) The relative humidity was decreasing, and the chance of precipitation was decreasing.
(2) The relative humidity was decreasing, and the chance of precipitation was increasing.
(3) The relative humidity was increasing, and the chance of precipitation was decreasing.
(4) The relative humidity was increasing, and the chance of precipitation was increasing.

14. Which statement best explains why an increase in the relative humidity of a parcel of air generally increases the chance of precipitation?
(1) The dewpoint is farther from the condensation point, causing rain.
(2) The air temperature is closer to the dewpoint, making cloud formation more likely.
(3) The amount of moisture in the air is greater, making the air heavier.
(4) The specific heat of the moist air is greater than the drier air, releasing energy.

15. The upward movement of air in the atmosphere generally causes the temperature of that air to
(1) decrease and become closer to the dewpoint
(2) decrease and become farther from the dewpoint
(3) increase and become closer to the dewpoint
(4) increase and become farther from the dewpoint

16. Which weather condition most directly determines wind speeds at Earth's surface?
    (1) visibility changes
    (2) amount of cloud cover
    (3) air-pressure gradient
    (4) dewpoint differences

17. The table below shows air-pressure readings taken at two cities, in the same region of the United States, at noon on four different days.

Air-Pressure Readings

Day	City A Air Pressure (mb)	City B Air Pressure (mb)
1	1004.0	1004.0
2	1000.1	1002.9
3	1000.2	1011.1
4	1010.4	1012.3

The wind speed in the region between cities A and B was probably the greatest at noon on day
(1) 1
(2) 2
(3) 3
(4) 4

Base your answers to questions 18 and 19 on the weather map below, which shows a low-pressure system centered near Poughkeepsie, New York. Isobars shown are measured in millibars.

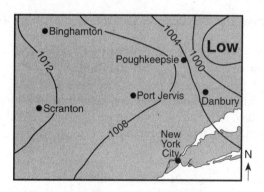

18. Which city is most likely experiencing winds of the greatest velocity?
    (1) New York City
    (2) Binghamton
    (3) Poughkeepsie
    (4) Scranton

19. Surface winds are most likely blowing from
    (1) Danbury toward New York City
    (2) Poughkeepsie toward Scranton
    (3) Binghamton toward Danbury
    (4) Port Jervis toward Binghamton

20. On sunny summer days, a breeze often develops that blows from large bodies of water toward nearby landmasses because the
    (1) temperature of the air above the landmasses is greater
    (2) specific heat of the landmasses is greater
    (3) temperatures of the bodies of water are greater
    (4) air over the bodies of water becomes heavier with additional water vapor

21. The cross section below shows a sea breeze blowing from the ocean toward the land. The air pressure at the land surface is 1013 millibars.

    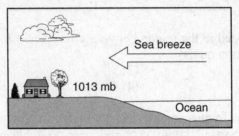

    (Not drawn to scale)

    The air pressure at the ocean surface a few miles from the shore is most likely
    (1) 994 mb   (3) 1013 mb
    (2) 1005 mb   (4) 1017 mb

22. Adjacent water and landmasses are heated by the morning Sun on a clear, calm day. After a few hours, a surface wind develops. Which map best represents this wind's direction?

    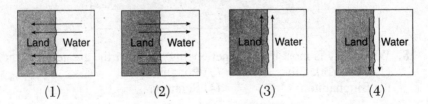

23. Which cross section below best represents the conditions that cause early winter lake-effect snowstorms in New York State?

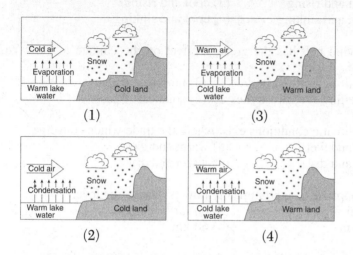

24. The planetary wind belts in the troposphere are primarily caused by the
(1) Earth's rotation and unequal heating of Earth's surface
(2) Earth's revolution and unequal heating of Earth's surface
(3) Earth's rotation and Sun's gravitational attraction on Earth's atmosphere
(4) Earth's revolution and Sun's gravitational attraction on Earth's atmosphere

Base your answers to questions 25 through 28 on the diagram below, which represents the planetary wind and moisture belts in Earth's Northern Hemisphere.

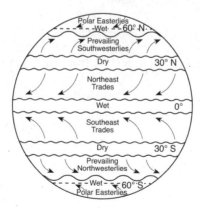

Weather

25. The climate at 90° north latitude is dry because the air at that location is usually
    - (1) warm and rising
    - (2) warm and sinking
    - (3) cool and rising
    - (4) cool and sinking

26. Which wind belt has the greatest effect on the climate of New York State?
    - (1) prevailing northwesterlies
    - (2) prevailing southwesterlies
    - (3) northeast trades
    - (4) southeast trades

27. Which climatic conditions exist where the trade winds converge?
    - (1) cool and wet
    - (2) cool and dry
    - (3) warm and wet
    - (4) warm and dry

28. The tropopause is approximately how far above sea level?
    - (1) 12 mi
    - (2) 12 km
    - (3) 60 mi
    - (4) 60 km

29. Cloud formation is likely to occur in rising air because rising air
    - (1) expands and cools
    - (2) expands and warms
    - (3) contracts and cools
    - (4) contracts and warms

30. Condensation of water vapor in the atmosphere is most likely to occur when a condensation surface is available and
    - (1) a strong wind is blowing
    - (2) the temperature of the air is below 0°C
    - (3) the air temperature reaches the dewpoint
    - (4) the air pressure is rising

31. Which is a form of precipitation?
    - (1) frost
    - (2) snow
    - (3) dew
    - (4) fog

32. What form of precipitation results from rain that falls through a freezing air mass?
    - (1) fog
    - (2) snow
    - (3) sleet
    - (4) frost

Base your answers to questions 33 through 35 on the station models below, which show various weather conditions recorded at the same time on the same day at four different cities.

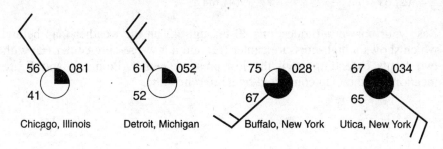

33. Which wind speed was recorded at Detroit?
    (1) 15 knots   (3) 35 knots
    (2) 25 knots   (4) 45 knots

34. Which city had the lowest relative humidity?
    (1) Chicago   (3) Buffalo
    (2) Detroit   (4) Utica

35. Which weather symbol best represents the type of precipitation that was most likely occurring in Utica?

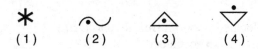

36. Air masses are identified on the basis of temperature and
    (1) type of precipitation   (3) moisture content
    (2) wind velocity   (4) atmospheric transparency

37. The properties of an air mass are mostly determined by the
    (1) rate of Earth's rotation
    (2) direction of Earth's surface winds
    (3) source region where the air mass formed
    (4) path the air mass follows along a land surface

38. Which geographic area is a common source region for cP air masses that move into New York State?
    (1) southwestern United States   (3) the north Pacific Ocean
    (2) central Canada   (4) the Gulf of Mexico

39. Which type of air mass is associated with warm, dry atmospheric conditions?
    (1) cP
    (2) cT
    (3) mP
    (4) mT

Base your answers to questions 40 through 42 on the weather map below, which shows a high-pressure center (**H**) and a low-pressure center (**L**) with two fronts extending from the low-pressure center. Points *X* and *Y* are locations on the map connected by a reference line.

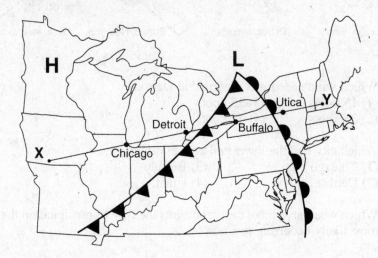

40. Which type of front is located between Buffalo and Detroit?
    (1) stationary
    (2) warm
    (3) occluded
    (4) cold

41. Which cross section best represents the fronts and air movements in the lower atmosphere along line *XY*?

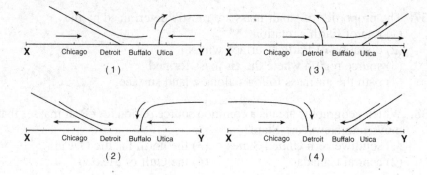

42. Which map best shows the most probable areas of precipitation associated with these weather systems?

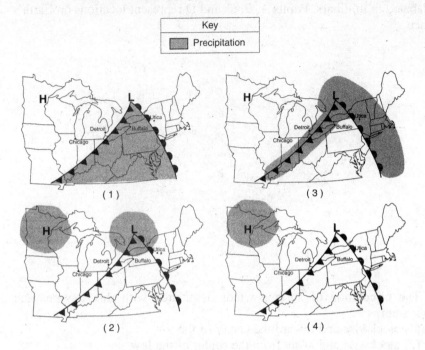

43. In which direction do the surface winds blow around a high-pressure system in the Northern Hemisphere?
 (1) clockwise and inward
 (2) clockwise and outward
 (3) counterclockwise and inward
 (4) counterclockwise and outward

Weather

Base your answers to questions 44 through 47 on the weather map below, which shows a low-pressure system over the central United States. Isobars are labeled in millibars. Points A, B, C, and D represent locations on Earth's surface.

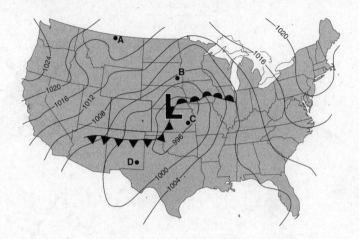

44. The circulation of surface winds associated with this low-pressure system is
(1) clockwise and toward the center of the low
(2) clockwise and away from the center of the low
(3) counterclockwise and toward the center of the low
(4) counterclockwise and away from the center of the low

45. The air pressure at the center of this low is
(1) 991 mb   (3) 997 mb
(2) 994 mb   (4) 1001 mb

46. Which location is most likely experiencing the fastest wind speed?
(1) A   (3) C
(2) B   (4) D

47. Which map shows the most likely path this low-pressure center will follow during the next 12 hours?

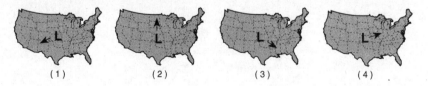

48. Which weather variable would most likely *decrease* ahead of an approaching storm system?
 (1) wind speed
 (2) air pressure
 (3) cloud cover
 (4) relative humidity

49. The weather satellite image below shows two large, swirl-shaped cloud formations, labeled *A* and *B*, over the Pacific Ocean.

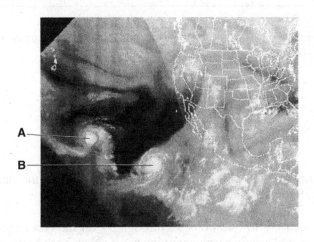

These large, swirl-shaped formations most likely represent
 (1) polar air masses
 (2) warm fronts
 (3) tornadoes
 (4) hurricanes

50. Present-day weather predictions are based primarily upon
 (1) land and sea breezes
 (2) cloud heights
 (3) ocean currents
 (4) air mass movements

51. In the United States, most tornadoes are classified as intense
 (1) low-pressure funnel clouds that spin clockwise
 (2) low-pressure funnel clouds that spin counterclockwise
 (3) high-pressure funnel clouds that spin clockwise
 (4) high-pressure funnel clouds that spin counterclockwise

52. On a certain day, the isobars on a weather map are very close together over eastern New York State. To make the people of this area aware of possible risk to life and property in this situation, the National Weather Service should issue
 (1) a dense-fog warning
 (2) a high-wind advisory
 (3) a heat-index warning
 (4) an air-pollution advisory

Weather

CONSTRUCTED RESPONSE QUESTIONS

Base your answers to questions 53 and 54 on the table below, which shows weather data recorded at Albany, New York.

Data Table

Location	Temperature (°F)	Dewpoint (°F)	Cloud Cover (%)	Pressure (mb)	Wind Direction	Wind Speed (knots)
Albany	58	36	25	1017.0	from the west	20

53. Complete the station model below, using the proper format to represent these *six* weather conditions accurately. [1]

54. State one reason why rain was unlikely at the time the data was collected. Support your answer by using the data. [1]

Base your answers to questions 55 through 57 on the cross section below, which shows two weather fronts moving across New York State. Lines *X* and *Y* represent frontal boundaries. The large arrows show the general direction the air masses are moving. The smaller arrows show the general direction warm, moist air is moving over the frontal boundaries.

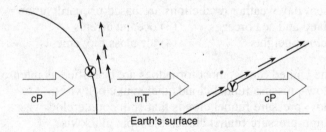

55. Which type of front is represented by letter *X*? [1]

56. Explain why the warm, moist air rises over the frontal boundaries. [1]

57. Which type of front forms when front *X* catches and overtakes front *Y*? [1]

Base your answers to questions 58 through 62 on the following satellite image. The satellite image shows a low-pressure system over a portion of the United States. Air mass symbols and frontal boundaries have been added. Line *XY* is one frontal boundary. Points *A*, *B*, *C*, and *D* represent surface locations. White areas represent clouds.

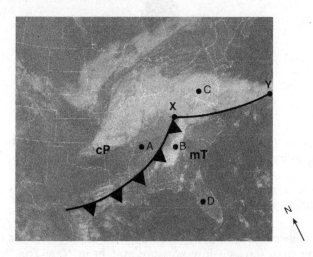

58. Draw the proper symbol to represent the most probable front on line *XY*. [1]

59. State *one* process that causes clouds to form in the moist air along the cold front. [1]

60. Describe *one* piece of evidence shown on the map that suggests location *A* has a lower relative humidity than location *B*. [1]

61. Explain why location *C* most likely has a cooler temperature than location *D*. [1]

62. State the compass direction that the center of this low-pressure system will move over the next few days if it follows a normal storm track. [1]

Weather

Base your answers to questions 63 and 64 on the station model below, which shows the weather conditions at Rochester, New York, at 4 P.M. on a particular day in June.

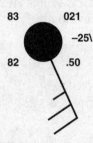

63. What was the actual barometric pressure, according to the station model, to the nearest tenth of a millibar? [1]

64. The winds shown by this station model were blowing from which compass direction and at what wind speed? [1]

Base your answers to questions 65 through 67 on the weather map below. The weather map shows a low-pressure system in New York State during July. The **L** represents the center of the low-pressure system. Two fronts extend from the center of the low. Line *XY* on the map is a reference line.

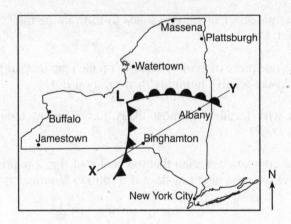

65. The cross section shows a side view of the area along line *XY* on the map. On lines 1 and 2 in the cross section, place the appropriate two-letter air mass symbols to identify the most likely type of air mass at *each* of these locations. [1]

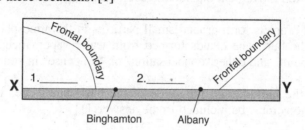

66. The forecast for one city located on the map is given below:

"In the next hour, skies will become cloud covered. Heavy rains are expected with possible lightning and thunder. Temperatures will become much cooler."

State the name of the city for which this forecast was given. [1]

67. Identify *one* action that people should take to protect themselves from lightning. [1]

EXTENDED CONSTRUCTED RESPONSE QUESTIONS

Base your answers to questions 68 through 72 on the weather satellite photograph of a portion of the United States and Mexico provided below. The photograph shows the clouds of a major hurricane approaching the eastern coastline of Texas and Mexico. The calm center of the hurricane, the eye, is labeled.

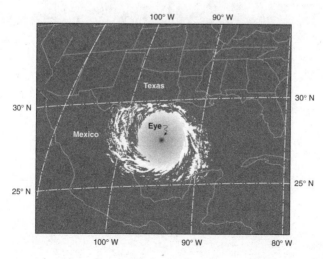

Weather

68. This hurricane has a pattern of surface winds typical of all low-pressure systems in the Northern Hemisphere. On the satellite photograph provided, draw *three* arrows on the clouds to show the direction of the surface wind movement outside the eye of the hurricane. [1]

69. Cloud droplets form around small particles in the atmosphere. Describe how the hurricane clouds formed from water vapor. Include the terms "dewpoint" and either "condensation" or "condense" in your answer. [1]

70. State the latitude and longitude of the hurricane's eye. The compass directions must be included in the answer. [1]

71. At the location shown in the photograph, the hurricane had maximum winds recorded at 110 miles per hour. Within a 24-hour period, the hurricane moved 150 miles inland and had maximum winds of only 65 miles per hour. State why the wind velocity of a hurricane usually decreases when the hurricane moves over a land surface. [1]

72. (a) State *two* dangerous conditions, other than hurricane winds, that could cause human fatalities as the hurricane strikes the coast. [2]
    (b) Describe *one* emergency preparation humans could take to avoid a problem caused by one of these dangerous conditions. [1]

73. The map provided below shows six source regions for different air masses that affect the weather of North America. The directions of movement of the air masses are shown. Using the standard two-letter air-mass symbols from the *Earth Science Reference Tables*, label the air masses by writing the correct symbol in each circle on the map. [2]

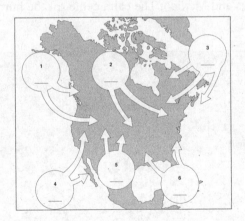

Base your answers to questions 74 and 75 on the cross section provided below, which represents a house at an ocean shoreline at night. Smoke from the chimney is blowing out to sea.

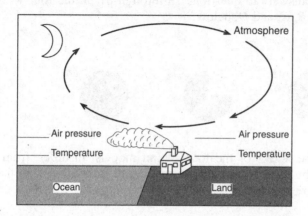

74. Label the *two* lines provided on the cross section to show where air pressure is relatively "high" and where it is relatively "low." [1]

75. Assume that the wind blowing out to sea on this night is caused by local air-temperature conditions. Label the *two* lines provided on the cross section to show where Earth's surface air temperature is relatively "warm" and where it is relatively "cool." [1]

Base your answers to questions 76 and 77 on the diagram below, which represents water molecules attached to salt and dust particles within a cloud in the atmosphere.

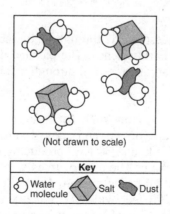

(Not drawn to scale)

76. Explain why salt and dust particles are important in cloud formation. [1]

Weather

77. State *one* natural process that causes large amounts of dust to enter Earth's atmosphere. [1]

Base your answers to questions 78 through 81 on the four weather station models, *A*, *B*, *C*, and *D*, below.

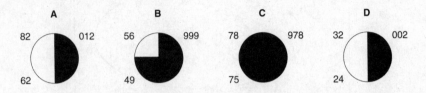

78. List the letters of the four station models, in order, from the station with the highest air pressure reading to the station with the lowest air pressure reading. [1]

79. Convert the air temperature at station *A* into degrees Celsius. [1]

80. What evidence indicates that station *C* has the highest relative humidity? [1]

81. On station model *D* below, draw the proper symbol to indicate a 25-knot wind coming from the west. [1]

Base your answers to questions 82 through 86 on the following passage and on your knowledge of Earth science. The passage describes a tornado produced from a thunderstorm that moved through a portion of New York State on May 31, 1998.

New York Tornado

A small tornado formed and moved through the town of Apalachin, New York, at 5:30 P.M., producing winds between 40 and 72 miles per hour. The tops of trees were snapped off, and many large limbs fell to the ground. The path of the destruction measured up to 200 feet wide. At 5:45 P.M., the tornado next moved through the town of Vestal, where winds ranged between 73 and 112 miles per hour. Many people experienced personal property damage as many homes were hit with flying material.

At 6:10 P.M., the tornado moved close to Binghamton, producing winds between 113 and 157 miles per hour. A 1000-foot television tower was pushed over, and many heavy objects were tossed about by the strong winds. Then the tornado lifted off the ground for short periods of time and bounced along toward the town of Windsor. At 6:15 P.M., light damage was done to trees as limbs fell and small, shallow-rooted trees were pushed over in Windsor. The tornado increased in strength again at 6:20 P.M. as it moved into Sanford. Some homes were damaged as their roof shingles and siding were ripped off. One mobile home was turned over on its side.

The tornado moved through the town of Deposit at 6:30 P.M., creating a path of destruction 200 yards wide. The tornado skipped along hilltops, touching down occasionally on the valley floors. However, much damage was done to homes as the tornado's winds reached their maximum speeds of 158 to 206 miles per hour. The tornado weakened and sporadically touched down after leaving Deposit. By 7:00 P.M., the tornado had finally ended its 1-hour rampage.

82. On the map below, draw the path of the tornado and the direction the tornado moved, by following the directions below. [2]
    • Place an **X** through the point for *each of the six* towns mentioned in the passage.
    • Connect the **X**s with a line in the order that each town was mentioned in the passage.
    • Place an arrow at one end of your line to show the direction of the tornado's movement.

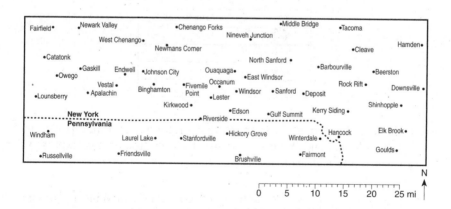

Weather

83. The tornado mentioned in this passage was produced by cold, dry air from Canada quickly advancing into warm, moist air already in place over the northeastern United States. List the two-letter air mass symbols that would identify *each* of the two air masses responsible for producing this tornado. [1]

84. Which type of front was located at the boundary between the advancing cold, dry air mass and the warm, moist air mass? [1]

85. Using the Fujita Scale shown below and the information in the passage, complete the table by assigning an F-Scale number for the tornado as it passed through each town given in the table. [1]

Fujita Scale

F-Scale Number	Wind Speed (mph)	Type of Damage Done
F–0	40–72	some damage to chimneys; breaks branches off trees; pushes over shallow-rooted trees; damages sign boards
F–1	73–112	peels surface off roofs; mobile homes pushed off foundations or overturned; moving autos pushed off the roads; attached garages may be destroyed
F–2	113–157	considerable damage; roofs torn off frame houses; mobile homes demolished; boxcars pushed over; large trees snapped or uprooted; light-object missiles generated
F–3	158–206	roof and some walls torn off well-constructed homes; trains overturned; most trees in forest uprooted
F–4	207–260	well-constructed houses leveled; structures with weak foundations blown off some distance; cars thrown and large missiles generated
F–5	261–318	strong frame houses lifted off foundations and carried considerable distances to disintegrate; automobile-sized missiles fly through the air in excess of 100 meters; trees debarked; steel-reinforced concrete structures badly damaged

Town	F-Scale Number
Vestal	
Windsor	
Sanford	
Deposit	

86. Calculate the tornado's average rate of travel, in miles per minute, between Vestal and Windsor by using the equation below. Express your answer to the *nearest tenth*. [1]

$$\text{tornado's rate of travel} = \frac{\text{distance between Vestal and Windsor (miles)}}{\text{time (minutes)}}$$

CHAPTER 24
CLIMATE

> **KEY IDEAS** *Climate* is the term used to describe the general weather conditions in a given area over a long period of time. Global climate is determined by the interaction of solar energy with Earth's surface and atmosphere. Dynamic processes such as cloud formation and Earth's rotation, as well as the positions of mountain ranges and oceans, influence this energy transfer. Climate is important in determining the habitability of a location and the most reasonable land uses within the region. One of the most important effects of climate is the availability of water at the surface and within the ground.

KEY OBJECTIVES
Upon completion of this unit, you will be able to:

- Differentiate between weather and climate.
- Identify the various factors that control climate.
- Describe the role that the cycling of water and energy in and out of the atmosphere plays in determining climatic patterns.
- Describe several patterns of global climatic change.

WHAT IS CLIMATE?

Definition of Climate

Climate is the characteristic weather of a region, particularly **temperature** and **precipitation**, averaged over an extended period of time. We recognize that weather phenomena, when viewed against the long-term, stable backdrop of global climate, are temporary, local changes.

The climate of a region cannot be determined by looking at the weather over a single year. In 1988, the Midwest suffered a nearly rainless summer. Fields of grain in the nation's heartland turned brown and shriveled, water levels dropped in the Mississippi River, and wildfires blazed through millions of acres of forested land. Yet, in the summer of 1993, many of these same places were drenched by torrential rains, farmers watched floodwaters wipe out their fields, and the Mississippi River flooded communities all along its course. Many people wondered, are we seeing Earth's global climate change? One or two extreme summers can't answer that question. Our picture of climate develops slowly as we watch scores of seasons pass. Some winters are warmer than others, some summers dryer, some falls colder. We can gain a sense of the shifting patterns of climate only by comparing measurements taken over many years and even decades.

At any given point in time, a wide variety of climates can exist on Earth simultaneously. Earth's climates range from steamy rain forests to icy glaciers, hot deserts, and cool forests of conifers. However, as shown by the summers of 1988 and 1993 in the Midwest described above, any given region may also be subject to wide seasonal changes in temperature and rainfall. One of the challenges in studying climate is deciding just how to measure it.

The Main Elements of Climate: Temperature and Precipitation

Climate is the result of the interplay of a number of factors. One of the most important elements in determining the climate of a region is energy. As you know, Earth's main source of energy is sunlight, which warms the land, which, in turns, heats the atmosphere. Globally, the amount of heating that occurs depends on the amount of sunlight reaching Earth's surface. A key way of tracking energy flow in a region is by monitoring temperatures. Therefore, scientists who study climate keep track of air temperatures over land and sea, air temperatures at various altitudes, and ocean water temperatures around the globe.

Another major climate element is water. Water is important to all living things, and controls to a large extent the types of plant and animal life that can inhabit a region. It also plays an important role in weathering and is the main agent of erosion and deposition on Earth. Thus, the distribution of atmospheric moisture and the amount of precipitation are important factors in characterizing climates. Some other measurements used to monitor climates and climate changes are the amount of snow and ice cover on land, the extent of sea ice, and the concentrations of various gases in the atmosphere.

CLIMATE FACTORS

You have learned that the main elements of a region's climate are its temperature and precipitation patterns. However, various factors influence the cycling of water and energy in Earth's atmosphere and thereby produce different climates. Temperature and precipitation patterns are controlled by such factors as latitude, the distribution of land and water, high- and low-pressure belts and prevailing winds, monsoons, ocean currents, vegetative cover, elevation, mountain ranges, clouds, and cyclonic storm activity.

Latitude

Latitude is the primary factor of temperature control in climate because it determines both the angle and duration of insolation. Much more insolation is received in low latitudes near the Equator than in high latitudes near the poles.

Only the areas that lie between 23½° N (the Tropic of Cancer) and 23½° S (the Tropic of Capricorn) ever receive the vertical rays of the Sun. The average angle of insolation decreases toward the poles, causing the insolation to be spread over a larger area and therefore to be less intense. As the angle of insolation decreases, the solar energy passes through more of the atmosphere, so more is reflected or absorbed and insolation is further decreased.

While the average length of daylight, or duration of insolation, is 12 hours everywhere on Earth at the time of the equinoxes, at other times it varies. It is always 12 hours at the Equator; but as one moves north or south of the Equator, the duration of insolation varies in a cyclic manner. It increases toward the pole tilted toward the Sun, ranging from 12 hours at the Equator to 24 hours at the pole. It decreases toward the pole tilted away from the Sun, ranging from 12 hours at the Equator to 0 hours at the pole. The result is that the higher the latitude, the greater the cyclic change in duration of insolation over the course of a year. The poles range from 24 hours of insolation in the summer to no insolation in the winter. At the latitudes of New York State, insolation ranges from about 15 hours in the summer to about 9 hours in the winter.

Differences in the angle and duration of insolation with latitude result in three main temperature zones: the always-hot **torrid zone** near the Equator, the always-cold **frigid zone** near the poles, and the seasonally changing, intermediate **temperate zone** in between.

Distribution of Land and Water

The irregular distribution of land and water surfaces on Earth is another major factor controlling climate. As discussed earlier, land surfaces increase in temperature more than water surfaces when insolation strikes them, and they also cool off more rapidly than water surfaces. As a result, air temperatures are warmer in the summer and cooler in the winter over landmasses than they are over oceans at the same latitude. Large bodies of water tend to moderate the temperatures of nearby landmasses by warming them in winter and cooling them in summer. For this reason cities in the interior United States have a greater annual temperature range than coastal cities (see Figure 24.1).

Large areas permanently covered by ice also affect climate. The light color, high specific heat capacity, and latent heat of fusion of ice combine to keep it from melting completely in certain areas. The air over ice surfaces is chilled by contact with the ice, and this cold air, in turn, chills the land and water surrounding these surfaces.

Climate

	J	F	M	A	M	J	J	A	S	O	N	D
San Francisco, CA	49	51	53	56	58	61	63	63	64	61	55	50
St. Louis, MO	32	35	43	55	64	74	78	77	70	58	44	35

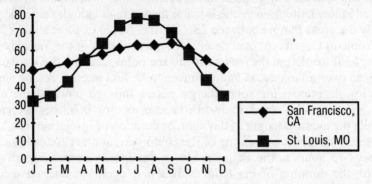

Figure 24.1 Average Monthly Temperatures at St. Louis, Missouri, an Inland City, and San Francisco, California, a Coastal City, Both at Approximately the Same Latitude.

Pressure and Wind Belts

The huge convection cells that form in the atmosphere because of unequal heating give rise to global pressure and wind belts. A belt of low pressure caused by rising warm air lies over the Equator. Low pressure favors evaporation, so the air there tends to be both warm and moist. The rising air spreads outward from the Equator, cools, loses moisture because of condensation, and then sinks back to Earth. This descending air creates a belt of high pressure on either side of the equatorial low between 25° and 30° north or south latitude. As the air sinks toward the ground, it is compressed and warms slightly. This warm, dry air is able to absorb moisture, and land surfaces in these regions tend to be dry. Most of Earth's deserts are located in these subtropical, high-pressure belts.

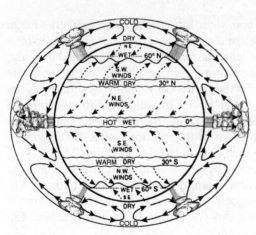

Figure 24.2 The Climatic Effects of Convection Cells in the Atmosphere.

The adjacent high- and low-pressure belts give rise to winds that carry warm, moist air outward from the Equator. These are known as the tropical easterlies or **trade winds**.

The poles are regions of high pressure because of their frigid, dry air. As the frigid air moves outward from the poles, it warms up and begins to rise, creating the subpolar low-pressure belts at 60° north and south latitude. The 30° high-pressure belts and the 60° low-pressure belts give rise to the moist but cooler prevailing westerlies.

In general, the high-pressure belts create regions of dry air, and the low-pressure belts create regions of moist air. The planetary wind belts carry moist or dry air over landmasses, greatly influencing their climates. These winds also transfer heat energy from the Equator toward the poles, modifying temperature patterns.

Figure 24.2 shows the planetary wind and pressure belts and their climatic effects.

Monsoons

Some landmasses are so large that they create their own seasonal convection cells. A **monsoon** is a seasonally reversing system of surface winds caused by temperature differences between land and ocean. A monsoon is a large-scale version of the type of circulation seen in land and sea breezes. The largest monsoon circulation occurs in southern Asia.

In land and sea breezes, we see *daily* changes in wind direction along a coastline. During the day, land heats up more quickly than water. Air above the land warms up and rises, forming a convection cell that draws cooler air from over the sea inland—a sea breeze. During the night, the land cools off more quickly than the water. The air above the warmer water rises, forming a convection cell that draws cooler air from over the land out to sea—a land breeze. (Refer to Figure 23.11: Land and Sea Breezes.)

In a monsoon, we see *seasonal* changes in wind direction between a large landmass and surrounding oceans. During the summer, the large Asiatic landmass warms up and a low-pressure center develops in the southern part of the continent. This low-pressure center results in winds that blow inland from the surrounding oceans. These moist summer monsoon winds bring clouds and heavy rains. During the winter, intense cooling produces a mass of very cold, very dry air over Siberia. The result is a high-pressure center in the northern part of the landmass that causes the winds to reverse direction and blow from the continental interior outward to the coast. Although these winter monsoon winds may warm up as they descend toward the coast, they pick up little moisture over the land and they bring clear, dry weather. See Figure 24.3.

A monsoon wind system also affects northern Australia. Since the seasons in Australia are opposite to those in Asia, however, winds blow in from the sea in January and outward from the continental interior in July. There is also some indication of a seasonal monsoon in the central part of the United States. During the summer, moist air generally flows from the Gulf of Mexico northward into the central plains. In the winter, cool, dry winds from Canada reverse this flow most of the time.

Climate

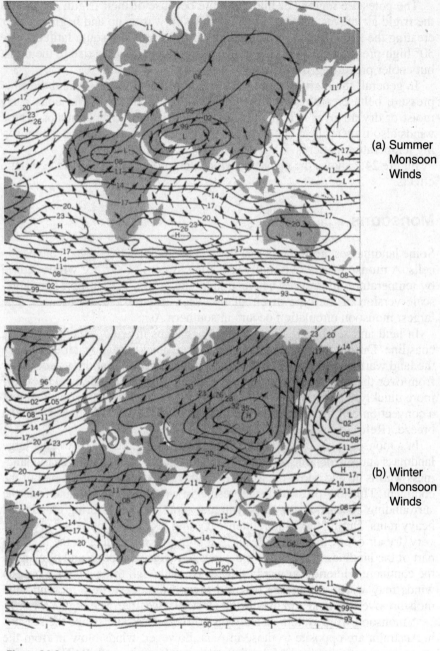

Figure 24.3 A Monsoon. A monsoon acts like a large-scale sea breeze in the summer (a) and a land breeze in the winter (b). Source: *The Earth Sciences*, Arthur N. Strahler, Harper & Row, 1971.

Ocean Currents

Ocean currents control climates by transferring heat from the Equator toward the poles, thereby cooling the equatorial regions, warming the polar regions, and influencing the temperature of adjacent landmasses these currents pass. The major surface ocean currents (see Figure 24.4) follow the prevailing winds that blow out of the subtropical high-pressure belts, giving rise to a clockwise flow in the Northern Hemisphere and a counterclockwise flow in the Southern Hemisphere.

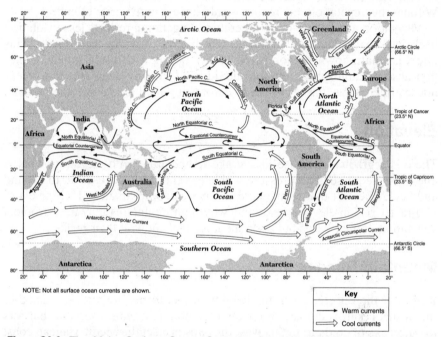

Figure 24.4 The Major Surface Ocean Currents of the World. Source: The State Education Department, *Earth Science Reference Tables*, 2011 ed. (Albany, New York: The University of the State of New York).

Wherever the ocean currents flow toward the poles, they carry warm water from the Equator. Wherever they flow toward the Equator, they carry cold water from the polar regions. Ocean currents such as the Gulf Stream and the North Atlantic Current carry warm water northward and eastward. The warm air over these ocean currents moderates the climate of the eastern coast of the United States, the Azores, and western Europe. The cold air over the icy waters of the Peru Current modifies the climate all along the western coast of South America, causing it to be much cooler and dryer than places with the same latitudes on the eastern coast of South America, where the Brazil Current and South Equatorial Current bring warm, moist air with their waters.

Vegetative Cover

Climate influences a region's natural vegetation; conversely, natural vegetation influences climate by affecting rainfall and temperature. Vegetation affects the water cycle by influencing the processes of evapotranspiration and surface runoff, which, in turn, influence rainfall. **Evapotranspiration** is a process that combines evaporation and transpiration, the two mechanisms by which liquid water is returned to the atmosphere as water vapor. **Transpiration** is the process by which plants release water vapor into the atmosphere through their leaves. It has been estimated that approximately half the rainfall in the Amazon Basin is derived from local evapotranspiration. Without the vegetation, the region would have a much drier climate.

Vegetation also influences temperature. Vegetation reflects less insolation back into space than bare surfaces, which tend to warm the climate. This effect is small, however, compared to the cooling that results when vegetation absorbs atmospheric carbon dioxide and thereby decreases the greenhouse effect. Thus, the net effect of deforesting land is to warm the climate.

Elevation

There is a gradual decrease in average temperature with elevation. This is due to the decrease in air pressure with elevation, which causes air to expand and cool. The fact that temperature decreases approximately 1°C for every 100-meter rise in elevation explains why high mountains may have tropical vegetation at their bases but permanent ice and snow at their peaks.

Mountain Ranges

Relief, or the differences in the heights of landforms in an area, is another factor that controls climate. For example, mountain ranges serve as barriers to outbreaks of cold air. In this way, the Alps protect the Mediterranean coast and the Himalayas protect India's lowlands.

In a process called the **orographic effect**, the windward side of a mountain facing winds carrying moisture-laden air usually receives much more precipitation than the leeward side. Mountain ranges force the air that is blown over them by winds to rise, and cool by expansion. This decreases the air's capacity to hold moisture and causes condensation, which forms clouds and precipitation. By the time the air reaches the top of the mountain, it has lost much of its moisture. When the air descends on the leeward side of the mountain, it is warmed by compression, its capacity to hold moisture is decreased, and precipitation is less likely. This orographic effect explains why cities along the Oregon coast west of the Cascades (e.g., Tillamook) have much precipitation while cities in the state's interior (e.g., Bend) have a dry climate (see Figure 24.5).

Clouds and Cyclonic Storm Tracks

Clouds control climate by reducing the amount of insolation gained or lost by a region. During daylight, clouds block sunlight and keep the temperature from rising as high as it would if the skies were clear. At night, the water droplets and water vapor in clouds absorb long-wavelength heat being radiated into space and thus keep the region from getting as cold as it would without them.

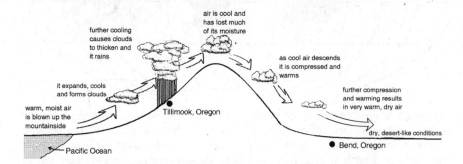

Figure 24.5 The **Orographic Effect**. On the windward side of the mountains, moist air is forced to rise and then cools by expansion, resulting in clouds and precipitation. On the leeward side, the cool air, which has lost much of its moisture, descends and warms by compression, resulting in warm, dry air that creates desertlike conditions.

The warm and cold fronts of wave cyclones that form along the polar front produce the changing weather of middle latitudes. These fronts produce precipitation and carry it over a wide area, causing changes in temperature as both warm and cold fronts pass through.

TYPES OF CLIMATES: THREE PATTERNS OF CLASSIFICATION

Climates are classified by moisture, temperature, and vegetation patterns. There are several systems in use today. The accompanying table summarizes terms often used to describe different patterns of moisture, temperature, and vegetation.

TABLE 24.1 DESCRIPTIVE TERMS FOR CLIMATIC PATTERNS

Moisture	Temperature	Vegetation
Arid	Polar	Desert
Semiarid	Subpolar	Grassland, steppe, taiga
Subhumid	Subtropical	Deciduous forest
		Coniferous forest
Humid	Tropical	Rain forest

Climate

Figure 24.6 is a map of the major climates of the world.

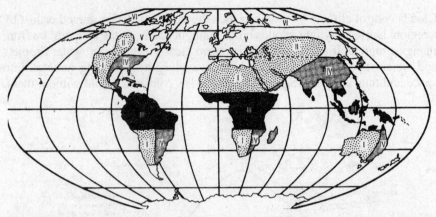

Figure 24.6 Major Climates of the World. I. Dry—hot summers, warm or cool winters. II. Dry—hot summers, cold winters. III. Humid—hot all year. IV. Humid—hot summers, cool winters. V. Humid—warm summers, cold winters. VI. Humid—cold summers, cold winters. Source: Part I: *Teacher's Guide—Investigating the Earth*, Earth Science Curriculum Project, Sponsored by the American Geological Institute and the National Science Foundation, Copyright © 1967, American Geological Institute.

EFFECTS OF CLIMATE

Climate has its greatest effect on vegetation. Individual plant species can generally survive only in a certain range of sunlight, temperature, precipitation, humidity, soil type, and wind. There is such a strong relationship between climate types and vegetation that climate regions are often named for their dominant vegetations.

Climate also affects the type of soil that will develop in an area. Warm, wet climates break down rock into soil faster than cool, dry climates. Soils in wet climates tend to be less fertile, though, because heavy precipitation and infiltration dissolve out nutrients needed by plants. Climate also affects soil by determining the types of plants that will become part of the soil when they eventually decay.

Another effect of climate is the shape into which landforms are eroded. In humid climates, where abundant rainfall makes running water the primary agent of erosion, landforms tend to have rounded contours. In arid climates, where wind is dominant and water is likely to erode Earth's surface in violent bursts, landforms are likely to be more jagged and angular. See Figure 24.7.

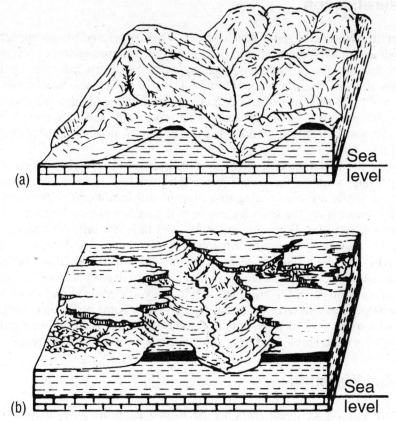

Figure 24.7 The Effects of Climate on Landforms. (a) In humid climates, abundant rainfall promotes a vegetative cover that protects the soil from rapid erosion by runoff and produces gently rounded slopes. (b) In arid climates, with little protective vegetation, erosion by rapid runoff and wind leads to steep slopes and exposed bedrock.

GLOBAL CLIMATIC TRENDS

Earth is an active planet. Plate motions move landmasses to different latitudes and change topography. Global temperatures change because of a variety of factors, including changes in the output of energy by the Sun, variation in the tilt of Earth's axis of rotation, and the blocking of incoming solar radiation by some pressure in the atmosphere. As a result, climatic zones shift over time. Some of the climatic trends noted in recent years include desertification, global warming, and the cyclic climate changes associated with El Niño.

Desertification

Desertification is the rapid development of deserts caused by the impact of human activities. Desertification does not occur because of forces originating within the desert. Rather, this patchy changing of dry, habitable land to unhabitable desert is the result of unwise land use by humans, accelerated by natural factors such as drought. Vegetation in dry lands is naturally scarce. Nevertheless, it is a valuable resource that protects the soil from erosion and provides food for people and for livestock. It may also be used for shelter and fuel.

When the sparse natural vegetation of dry land is severely disturbed by overgrazing, clearing for crops, or drought, the land deteriorates. Erosion removes fertile topsoil, leading to a drop in soil fertility. With fewer plant roots to break it up, the soil crusts over in the baking Sun and becomes less permeable to water. When precipitation does fall, less infiltrates and more runs off, leading to increased erosion and a decrease in soil moisture and groundwater. The result is a permanent conversion of marginal dry lands to desert. Natural cycles of drought do play a role in desertification; but without human influence, the arid land is less severely damaged and its natural systems generally recover when the drought ends. Some projections suggest that within the next 20 years human activities will have caused one-third of the world's once-arable land to become useless for growing food crops.

Global Warming

Modern burning of fossil fuels has been increasing the amount of carbon dioxide in the air. In the past century, carbon dioxide levels in the atmosphere have increased by an estimated 25 percent, and these levels continue to climb. It is feared that, if the heat trapped by carbon dioxide increases with its concentration, rising levels of this gas will cause increased greenhouse-effect heating of the atmosphere. Global temperatures have shown an upward trend since 1850; it is difficult to tell, however, whether this increased warming is due to human impact or is part of the natural fluctuations in global temperatures.

One of the best known changes in global temperatures in modern times was the "little Ice Age," which lasted from 1450 to 1850. Paintings from this time show alpine glaciers in the Alps advancing dramatically. The high mountains of Ethiopia were covered in snow. There were times when a person could walk across New York Harbor from Manhattan to Staten Island. Then, about 1850, steady warming began.

Only in the mid-1990s did climatologists believe that they had enough evidence to state that greenhouse-effect heating has occurred and that the warming trend has moved beyond the probable range of normal changes in global temperatures. Global warming has probably accelerated desertification. It also has the potential to severely affect agriculture worldwide, increase sea level as glacial ice caps melt, and change global weather patterns. Recent projections suggest that soil-moisture levels in U.S. farmlands could drop by 40 percent if carbon dioxide levels double—an alarming prospect for farmers in areas that get barely enough rainfall now.

The El Niño-Southern Oscillation

The water in the ocean can be thought of as a series of thin layers, almost like pages in a book. When a steady wind blows over the surface of the ocean, the wind pushes the top layer, which begins to move. Due to the Coriolis effect, the top layer of water moves 45° to the right of the wind direction in the Northern Hemisphere (or to the left in the Southern Hemisphere). The top layer then pushes on the layer below it; and again the Coriolis effect comes into play, causing the second layer to move slightly, and a bit slower, to the right. As each layer pushes the one below it the direction of motion of the underlying layer continues to be successively further to the right. With increasing depth, each layer also moves more slowly until, at a depth of no more than a few hundred meters, the wind's effects are not felt at all. Though each layer moves in a slightly different direction, the net effect is that the upper part of the ocean is transported 90° to the wind direction. See Figure 24.8. When global winds blow steadily parallel to a coastline, this process carries warm surface waters offshore. As this warmer water is carried off, cold, nutrient-rich water rises to the surface to replace it—a process called upwelling. See Figure 24.9.

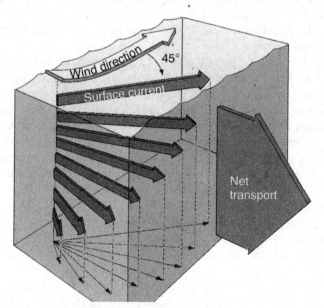

Figure 24.8 The Effects of Winds and Surface Currents. The net effect of winds blowing over surface waters is a current flowing at 90° to the wind direction.

Climate

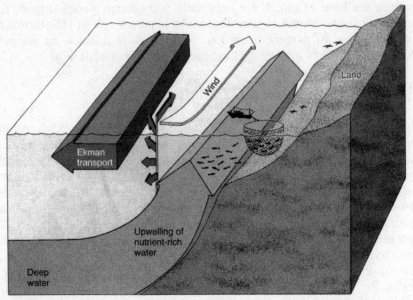

Figure 24.9 When prevailing winds blow parallel to a coast, warm surface water is carried offshore. Deep water that is rich in nutrient rises to the surface. This is called coastal upwelling. Source: *Marine Biology*, 3rd Ed., Peter Castro and Michael Huber, McGraw-Hill, 2000.

Along the eastern sides of many ocean basins, the global trade winds blow parallel to the coast and warm surface waters are carried offshore, producing strong coastal upwelling. As a result of the nutrient-rich water brought to the surface by this upwelling, these coastal areas rank among the ocean's richest fishing grounds.

The term **El Niño** originally referred to the changes in surface currents along the coasts of Peru and Chile. Trade winds blowing north along these coasts produce upwelling, making these waters rich fisheries. Every year, usually in December, the trade winds subside, upwelling slacks off, and the water becomes warmer. Local fisherman have known for centuries about this change in currents, which marks the end of the peak fishing season. Since the change comes around Christmas, they call it El Niño—"The Child." Every few years, however, the change is greater than usual. The surface water gets much warmer, the upwelling of nutrients ceases, and fish that depend on these food sources disappear.

The warming of the surface waters during El Niño causes a large region of low pressure to form over the southeastern Pacific Ocean. This, in turn, causes changes in the jet stream winds, which profoundly affect global weather patterns.

People in India depend on the summer monsoon to water their crops. When the monsoon fails, famine is widespread. In the early 1900s, analysis of weather records in an attempt to predict when India's summer monsoon would fail led to an important discovery: the air pressure over the Indian

Ocean fluctuates. When it is extremely low, the normal monsoon flow carrying moist air from the ocean over the land is disrupted. Also, when the air pressure is unusually high over the Pacific Ocean, it tends to be unusually low over the Indian Ocean, and vice-versa. Over time, the air pressure seesaws back and forth over these two regions, a phenomenon called the **Southern Oscillation**. The changes in air pressure bring dramatic changes in wind and rainfall, and failure of the summer monsoon in India. Exactly what triggers the process is not yet clear; but when the air pressure starts to seesaw, failure of the monsoon can be expected.

By the 1950s it was clear that El Niño and the Southern Oscillation were not isolated events, but part of a complex interaction between ocean and atmosphere that spans half the globe. See Figure 24.10. The El Niño-Southern Oscillation, or **ENSO**, is not really an abnormal event, but simply one extreme of a regular climate cycle. Nevertheless, the weakening or reversal of wind and ocean current patterns drastically changes normal regional weather systems.

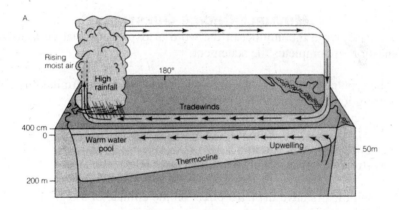

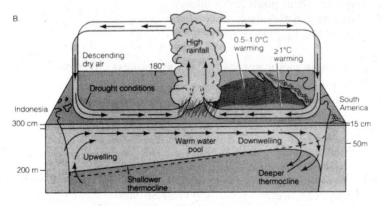

Figure 24.10 The El Niño-Southern Oscillation (ENSO). Source: *Environmental Geology*, Barbara W. Murck, Brian J. Skinner, and Stephen C. Porter, John Wiley, 1996.

Climate

The ENSO has had profound effects on weather, wildlife, and society. El Niño years bring drought, forest fires, and heat waves to some regions and storms and floods to others. One of the major El Niño events in the last century occurred in 1982–83; milder events followed in 1986–87 and 1991–95. By analyzing oceanographic and atmospheric data, scientists detected signs of an El Niño early in 1997 and predicted that another event was on its way. The 1997–98 event turned out to be even stronger than the 1982–83 El Niño. In Peru, Chile, and Ecuador, it brought torrential rains, flooding, and failure of the fishing industry. There were also floods in China, East Africa, and southern California. At the other extreme, Indonesia, Australia, southern Africa, and Central America suffered droughts. The weather changes were so severe that Florida, which rarely experiences tornadoes, was hit by a series of killer tornadoes. Estimated losses from the 1997–98 El Niño totaled more than $20 billion, but could have been much higher. The scientists' prediction gave the world advance warning and allowed officials in many countries to build up food reserves and prepare for impending disasters.

MULTIPLE-CHOICE QUESTIONS

In each case, write the number of the word or expression that best answers the question or completes the statement.

1. The type of climate in a particular location can be determined by comparing long-term monthly averages for
   (1) precipitation and temperature
   (2) wind speed and wind direction
   (3) infiltration and runoff
   (4) stream discharge and stream velocity

2. Which factors have the *least* effect on the climate of a region?
   (1) mountain barriers and nearness to large bodies of water
   (2) longitude and population density
   (3) latitude and elevation
   (4) wind belts and storm tracks

3. Which single factor generally has the *greatest* effect on the climate of an area on the Earth's surface?
   (1) the distance from the equator
   (2) the extent of vegetative cover
   (3) the month of the year
   (4) the degrees of longitude

4. Which surface ocean current transports warm water to higher latitudes?
   (1) Labrador Current   (3) Gulf Stream
   (2) Falkland Current   (4) West Wind Drift

5. Which graph best represents the general relationship between latitude and average surface temperature?

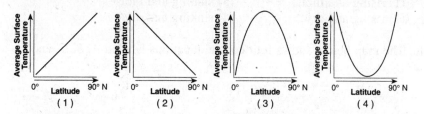

6. Large oceans moderate the climatic temperatures of surrounding coastal land areas because the temperature of ocean water changes
   (1) rapidly due to water's low specific heat
   (2) rapidly due to water's high specific heat
   (3) slowly due to water's low specific heat
   (4) slowly due to water's high specific heat

Base your answers to questions 7 and 8 on the map below, which represents an imaginary continent. Locations *A* and *B* are on opposite sides of a mountain range on a planet similar to Earth. Location *C* is on the planet's equator.

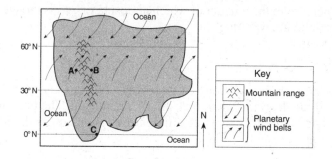

7. Compared with the climate at location *A*, the climate at location *B* would most likely be
   (1) warmer and more humid
   (2) warmer and less humid
   (3) cooler and more humid
   (4) cooler and less humid

8. Location *C* most likely experiences
   (1) low air pressure and low precipitation
   (2) low air pressure and high precipitation
   (3) high air pressure and low precipitation
   (4) high air pressure and high precipitation

Climate

9. Snowfall is rare at the South Pole because the air over the South Pole is usually
 (1) rising and moist
 (2) rising and dry
 (3) sinking and moist
 (4) sinking and dry

10. The map below shows four coastal locations labeled A, B, C, and D.

The climate of which location is warmed by a nearby major ocean current?
 (1) A
 (2) B
 (3) C
 (4) D

11. An increase in the amount of which atmospheric gas is thought to cause global climate warming?
 (1) oxygen
 (2) hydrogen
 (3) nitrogen
 (4) carbon dioxide

Base your answers to questions 12 through 16 on your knowledge of Earth science and the diagram below that shows a mountain. The prevailing wind direction and air temperatures at different elevations on both sides of the mountain are indicated.

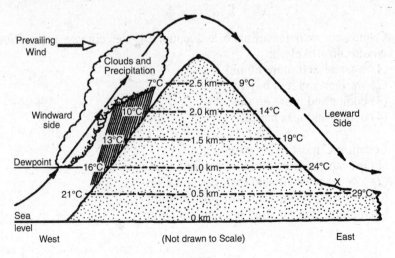

12. What would be the approximate air temperature at the top of the mountain?
    (1) 12°C　　　　　　　　(3) 0°C
    (2) 10°C　　　　　　　　(4) 4°C

13. On which side of the mountain and at what elevation is the relative humidity probably 100%?
    (1) on the windward side at 0.5 km
    (2) on the windward side at 1.5 km
    (3) on the leeward side at 1.0 km
    (4) on the leeward side at 2.5 km

14. How does the temperature of the air change as the air rises on the windward side of the mountain between sea level and 0.5 kilometer?
    (1) The air is warming because of compression of the air.
    (2) The air is warming because of expansion of the air.
    (3) The air is cooling because of compression of the air.
    (4) The air is cooling because of expansion of the air.

15. Which feature is probably located at the base of the mountain on the leeward side (location X)?
    (1) an arid region　　　　(3) a glacier
    (2) a jungle　　　　　　　(4) a large lake

16. The air temperature on the leeward side of the mountain at the 1.5-kilometer level is higher than the temperature at the same elevation on the windward side. What is the probable cause of this difference?
    (1) Heat stored in the ocean keeps the windward side of the mountain warmer.
    (2) The insolation received at sea level is greater on the leeward side of the mountain.
    (3) The air on the windward side of the mountain has a lower adiabatic lapse rate than the air on the leeward side of the mountain.
    (4) Potential energy is lost as rain runs off the windward side of the mountain.

Climate

Base your answers to questions 17 through 20 on the climate graphs below, which show average monthly precipitation and temperatures at four cities, *A*, *B*, *C*, and *D*.

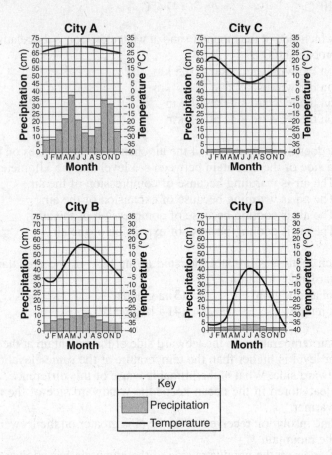

17. City *A* has very little variation in temperature during the year because city *A* is located
    (1) on the dry side of a mountain
    (2) on the wet side of a mountain
    (3) near the center of a large landmass
    (4) near the equator

18. During which season does city *B* usually experience the month with the highest average precipitation?
    (1) spring   (3) fall
    (2) summer   (4) winter

19. It can be concluded that city *C* is located in the Southern Hemisphere because city *C* has
    (1) small amounts of precipitation throughout the year
    (2) large amounts of precipitation throughout the year
    (3) its warmest temperatures in January and February
    (4) its warmest temperatures in July and August

20. Very little water will infiltrate the soil around city *D* because the region usually has
    (1) a frozen surface
    (2) nearly flat surfaces
    (3) a small amount of runoff
    (4) permeable soil

Base your answers to questions 21 through 24 on the maps and the passage below. The maps show differences in trade wind strength, ocean current direction, and water temperature associated with air pressure changes from normal climate conditions to El Niño conditions.

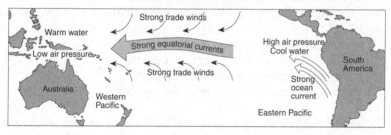

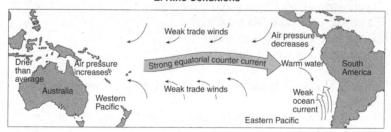

El Niño Conditions

El Niño conditions occur with a buildup of warm water in the equatorial Pacific Ocean off the coast of South America. The immediate cause of this buildup is a change in air pressure that weakens the southern trade winds. These are the planetary winds that move air from 30° S to the equator. Normally, these strong, steady winds, with the help of their counterparts in the Northern Hemisphere, push equatorial water westward away from South America. However, at intervals of two to seven years, these winds weaken, causing the westward water flow to reverse. This results in an accumulation of unusually warm water on the east side of the equatorial Pacific Ocean. This warm water not only changes the characteristics of the air above it but is also thought to be the cause of weather changes around the world. El Niño conditions may last only a few months but often last a year or two.

21. The trade winds between 30° S and the equator usually blow from the
    (1) northeast
    (2) southeast
    (3) northwest
    (4) southwest

22. Under normal climate conditions, what are the characteristics of the surface ocean current that flows along most of the west coast of South America?
    (1) cool water moving toward the equator
    (2) cool water moving away from the equator
    (3) warm water moving toward the equator
    (4) warm water moving away from the equator

23. During El Niño conditions, air above the Pacific Ocean moving over the land on the equatorial west coast of South America is likely to be
    (1) cooler and drier than usual
    (2) cooler and wetter than usual
    (3) warmer and drier than usual
    (4) warmer and wetter than usual

24. Equatorial Pacific trade winds weaken during El Niño conditions when air pressure
    (1) falls in the western Pacific and rises in the eastern Pacific
    (2) falls in both the western and eastern Pacific
    (3) rises in the western Pacific and falls in the eastern Pacific
    (4) rises in both the western and eastern Pacific

CONSTRUCTED RESPONSE QUESTIONS

The diagram below shows warm, moist air moving off the ocean and over a mountain, causing precipitation between points 1 and 2.

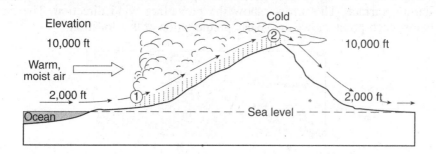

25. Describe *two* changes that occur to the warm, moist air between points 1 and 2 that would cause cloud formation. [2]

26. Identify by name the surface ocean current that cools the climate of locations on the western coastline of North America. [1]

Base your answers to questions 27 through 29 on the map below, which shows an imaginary continent on a planet that has climate conditions similar to Earth. The continent is surrounded by oceans. Two mountain ranges are shown. Points *A* through *D* represent locations on the continent.

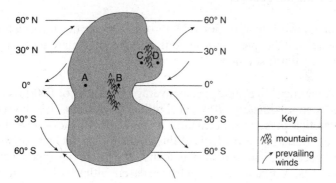

27. Identify one labeled latitude on this continent where a high-pressure zone exists and dry air is sinking to the surface. Include both the unit and compass direction in your answer. [1]

28. Identify one factor that causes a colder climate at location *B* than at location *A*. [1]

29. Explain why location *C* has a warmer and drier climate than location *D*. [1]

Climate

EXTENDED CONSTRUCTED RESPONSE QUESTIONS

Base your answers to questions 30 through 32 on the cross section and bar graph below. The cross section shows a portion of Earth's crust along the western coast of the United States. The points show different locations on Earth's surface. The arrows show the prevailing wind direction. The bar below each point shows the yearly precipitation at that location.

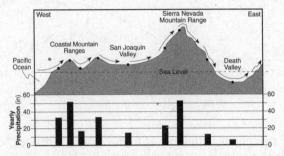

30. Explain why the valleys have *lower* amounts of precipitation than points on the western slopes of the mountain ranges. [1]

31. What is the yearly precipitation total for the four points located in the Coastal Mountain Ranges? [1]

32. State *one* reason why colder temperatures would be recorded at the top of the Sierra Nevada Mountain Range than at the top of the Coastal Mountain Ranges. [1]

Base your answers to questions 33 through 35 on the information below, which describes the past and present climate of Antarctica, and on your knowledge of Earth science.

> Antarctica's ice sheet has an average thickness of 6600 feet and holds approximately 70% of Earth's freshwater. Ice layers in Antarctica preserve information about Earth's history. Fossil evidence found in the bedrock of this continent shows that Antarctica was once tropical and is a potential source of untapped natural resources. Antarctica is now a frozen desert with very little snowfall.

33. Explain why Antarctica's cold climate is responsible for its very low amount of yearly precipitation. [1]

34. What evidence is preserved in Antarctica that provides information about Earth's past climates? [1]

35. Scientists are concerned that the Antarctic ice may melt as the result of global warming. State one effect that this melting would most likely have on Long Island, New York. [1]

Base your answers to questions 36 through 40 on the map and data tables below. The map shows the location of Birdsville and Bundaberg in Australia. Data table 1 shows the average monthly high temperatures for Birdsville. Data table 2 includes the latitude and longitude, elevation above sea level, and the average rainfall in January for Birdsville and Bundaberg.

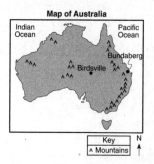

Map of Australia

Data Table 1
Average Monthly High Temperatures for Birdsville, Australia

Month	Temperature (°C)
January	39
February	38
March	35
April	30.5
May	25
June	22
July	21
August	23.5
September	28
October	32.5
November	36
December	38

Data Table 2
Information about Two Australian Cities

City	Latitude (° S)	Longitude (° E)	Elevation (m)	Average January Rainfall (mm)
Birdsville	25.9	139.4	47	25
Bundaberg	24.9	152.4	14	105

36. On the grid below, plot with an **X** the average monthly high temperatures for Birdsville, Australia. Connect the **X**s with a line. The average monthly high temperatures for Bundaberg have already been plotted on the graph for you. [1]

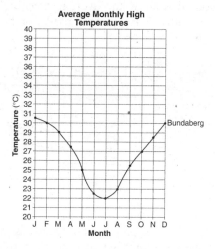

37. State *one* factor that could account for the difference between the average high temperatures recorded in December for Birdsville and for Bundaberg. [1]

38. State *one* reason for the difference in the average January rainfall for Birdsville and for Bundaberg. [1]

39. Explain why Bundaberg will experience solar noon before Birdsville each day. [1]

40. On the map below, draw the 30° S latitude line. [1]

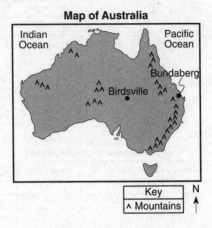

Map of Australia

Base your answers to questions 41 and 42 on the map below, which shows one method of classifying Earth's surface into latitudinal climate belts. In the tropical climate belt, the average monthly temperatures never drop below 18°C. In the polar climate belts, the average monthly temperatures never rise above 10°C. The isotherms show the average monthly temperature of the coolest and warmest months. The effects of elevation have been omitted.

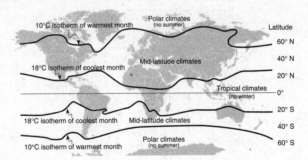

41. According to the isotherms on the map, locations in the midlatitude climate belts have average monthly temperatures between what values? [1]

42. Describe a specific characteristic of insolation received in the tropical climate belt region that causes the average monthly temperature to remain warm all year. [1]

Appendix A ANSWERS TO REVIEW QUESTIONS

CHAPTER 1

Multiple-Choice Questions

1.	3	5.	3	9.	4	13.	2	17.	4
2.	2	6.	1	10.	1	14.	4	18.	4
3.	4	7.	4	11.	1	15.	2	19.	2
4.	1	8.	1	12.	3	16.	2		

Constructed Response Questions

20. 3:00 P.M.
21.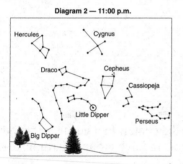
22. North
23. Hercules: down and to the left (west) *and* Perseus: up and to the right (east)
24. (a) zenith   (b) celestial meridian   (c) celestial horizon

Extended Constructed Response Questions

25. Stars appear to move from east to west at 15° per hour. Therefore, to create an arc of 60° would require 60° ÷ 15°/hr = 4 hr.
26. Real motion is an actual change in position, apparent motion is an "observed" change in position.
27. The stars appear to move because Earth's rotation causes an observer to move in relation to the stars.
28. Stars do not change position in relation to other stars on the celestial sphere; planets do.
    Stars do not appear as disks when viewed with a telescope; planets do.
    Stars twinkle; planets do not.
    Stars are scattered throughout the celestial sphere; all the planets appear near the ecliptic.
29. The North Star is located almost directly over Earth's axis of rotation.

CHAPTER 2

Multiple-Choice Questions

1.	1	5.	4	9.	1	13.	1
2.	1	6.	2	10.	2	14.	2
3.	4	7.	1	11.	2	15.	1
4.	1	8.	3	12.	1	16.	3

Constructed Response Questions

17. (a) 3 (b) 4 (c) 7 (d) 2 (e) 5
18. Jupiter
19. Earth's rotation causes day and night.
20. The line graph shows a direct relationship.

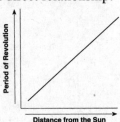

21. The geocentric model has Earth in the center. In a geocentric model, Earth does not rotate. Instead, planets revolve around Earth instead of the Sun.
22. The distance from Earth to the Sun varies. There are two foci instead of one center. Earth's orbit is an oval shape; Earth's eccentricity of orbit is 0.017.
23. The force of gravity decreases and then increases.
24. Earth's orbit would become more eccentric, i.e., more elliptical.
25. 0.333
26. The given ellipse has a higher eccentricity than the orbit of Mars.

Extended Constructed Response Questions

27.

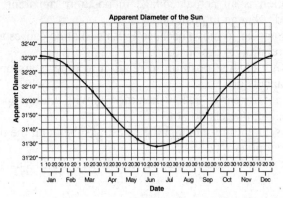

28. Earth has an elliptical orbit; therefore, the distance between the Sun and Earth varies in a cyclic manner.
29. The comet orbits the Sun; the comet doesn't orbit Earth.
30. The comet's average distance from the Sun is greater. Therefore, the comet has a larger orbit. During most of its orbit, the comet is moving slower than Earth.
31. 32.

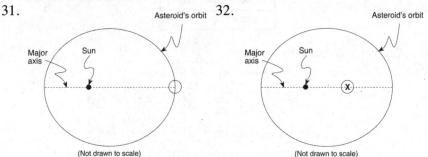

CHAPTER 3

Multiple-Choice Questions

1.	4	6.	1	11.	4	16.	4	21.	1
2.	4	7.	1	12.	3	17.	3		
3.	1	8.	1	13.	1	18.	3		
4.	4	9.	3	14.	4	19.	1		
5.	1	10.	4	15.	3	20.	1		

Constructed Response Questions

22.

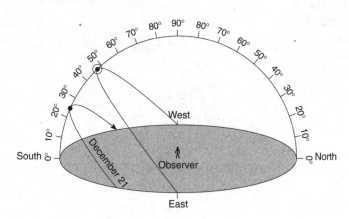

23. Any date from June 19 to June 23

Appendix A: Answers to Review Questions

24.

Season	Earth's Position
spring	A
summer	C
fall	E
winter	G

25. Any value from 88 to 94 days

26.

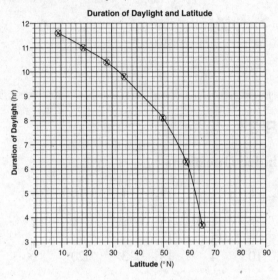

27. As latitude increases, the duration of daylight decreases.
28. 56° ± 1°N
29. winter
30., 31.

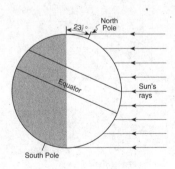

32. June

Extended Constructed Response Questions

33.–35.

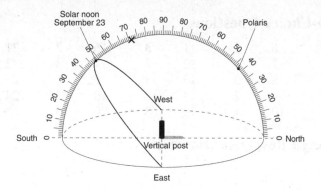

36. 30°
37. 42° N
38. An **X** should be placed within the shaded area on Earth's surface as shown in the diagram below.

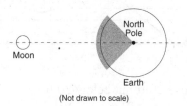

39. The distance to the Moon is increasing. As distance between objects increases, the gravitational attraction decreases.
40. 2 seconds
41. The Moon is closer to Earth than the Sun. Therefore the Moon's gravitational attraction to Earth is greater than the Sun's.
42.

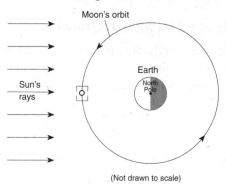

43. Tides occur in a regularly repeating pattern.
44. 27.3 days
45. Three answers are acceptable. During a lunar eclipse, Earth blocks the sunlight from reaching the Moon. The Moon must move into Earth's shadow. Earth's shadow must fall on the Moon.

CHAPTER 4

Multiple-Choice Questions

1.	1	6.	1	11.	3	16.	2	21.	2
2.	1	7.	1	12.	2	17.	3	22.	3
3.	4	8.	3	13.	1	18.	1	23.	3
4.	2	9.	2	14.	2	19.	3	24.	4
5.	2	10.	1	15.	3	20.	3		

Constructed Response Questions

25. $15°$
26. $90°$
27. 12 hours
28.

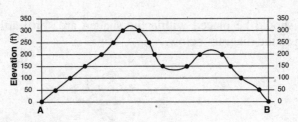

29. 25 ft/mi (±1).
30. Any value above 20 ft but below 30 ft
31.

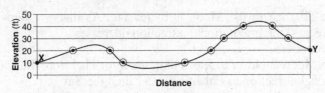

Extended Constructed Response Questions

32., 33. An example of correctly drawn arrows and the apparent September 23 path are shown below.

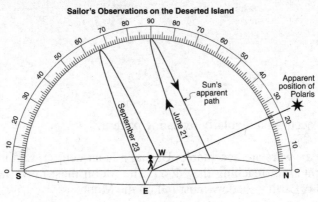

34. 23°30'N (±1°) or 23.5°N (±1°) or 23½°N (±1°)
35. 15 or 15°E or 15°W
36., 37.

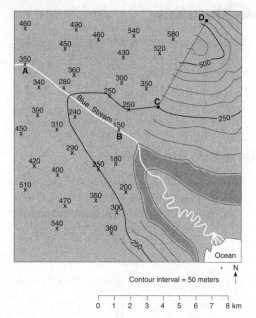

The center of the **X** must be within the shaded sections between the 50- and 100-meter contour lines shown in the diagram.
38. Any value from 28.0 to 29.0 meters/kilometer
39.

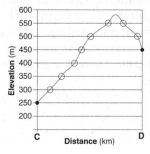

40. An example of a correctly drawn island is shown below.

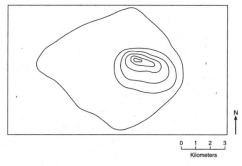

653

CHAPTER 5

Multiple-Choice Questions

1.	1	7.	4	13.	3	19.	4	25.	3
2.	4	8.	1	14.	2	20.	3	26.	1
3.	3	9.	1	15.	2	21.	3	27.	4
4.	3	10.	3	16.	1	22.	3		
5.	2	11.	1	17.	1	23.	2		
6.	1	12.	2	18.	3	24.	4		

Constructed Response Questions

28. Three answers are acceptable. Earth is revolving around the Sun. Tilt of Earth's axis. Parallelism of Earth's axis.
29. The center of the **X** must be between the brackets indicated on Earth's orbit, as shown in the diagram below.

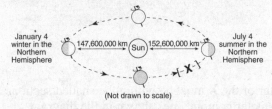

30. Acceptable responses include but are not limited to these examples. The North Pole is tilted toward the Sun in the summer. In summer, the Sun is higher in the sky due to the tilt of Earth's axis. New York State receives higher angles of insolation in summer when Earth is farthest from the Sun. New York State receives lower angles of insolation in winter when Earth is closest to the Sun. There is a greater duration of insolation.
31.

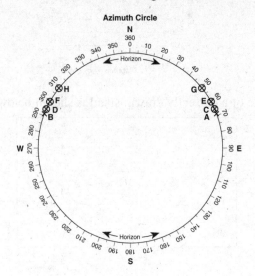

32. Two answers are acceptable. As the latitude of the observer increases, the azimuth decreases. As the latitude increases, the sunrise is farther north of east.
33.

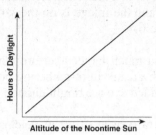

34., 35.

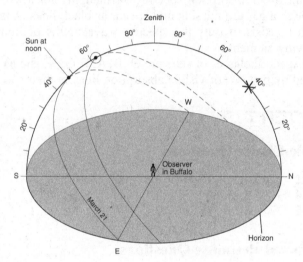

CHAPTER 6

Multiple-Choice Questions

1.	2	4.	1	7.	3	10.	4	13.	4
2.	4	5.	2	8.	1	11.	3	14.	1
3.	1	6.	2	9.	2	12.	1		

Constructed Response Questions

15. Light-gathering power is proportional to the square of the aperture of the telescope. Thus, the light-gathering power of a 200-inch telescope is 40,000 (200^2), and the light-gathering power of a 20-inch telescope is 400 (20^2). Therefore, the light-gathering power of a 200-inch telescope is 1,000 times greater than that of a 20-inch telescope (400,000 ÷ 400 = 1,000).
16. Light-gathering power; resolution; magnification

17. In a refracting telescope, a lens is used to create the image that is examined with the objective. The image is formed on the opposite side of the lens from the object. In a reflecting telescope, a curved mirror is used to create the image and the image is on the same side of the mirror as the object. See Figure 6.4.
18. See Figure 6.2d.
19. Telescopes have a much larger aperture than the eye. Therefore, telescopes have much greater light-gathering power. Objects too dim to be seen with the unaided eye are bright enough to be visible through a telescope.
20. Continuous spectra consist of all wavelengths of light dispersed by a prism. In an absorption spectra, wavelengths absorbed as light passes through a gas are missing and appear as black lines. A bright-line spectrum consists of only the specific wavelengths of light emitted by a glowing element.
21. The spectral classes of stars are O, B, A, F, G, K, and M. Stars are classified by patterns of visible absorption lines in their spectra.

CHAPTER 7

Multiple-Choice Questions

1. 2
2. 4
3. 2
4. 1
5. 4
6. 1
7. 2
8. 1
9. 3
10. 3
11. 1
12. 1
13. 3
14. 2
15. 3

Constructed Response Questions

16. As temperature increases, luminosity increases.
17. red giants or giants
18.

Star	Color	Classification
Sun	yellow	main sequence
Procyon B	white	white dwarf

19.

universe	galaxy	star

Largest ————————————————→ Smallest

Appendix A: Answers to Review Questions

20.

Stars	Temperature		Luminosity	
	Hotter	Cooler	Brighter	Dimmer
Procyon B	X			X
Barnard's Star		X		X
Rigel	X		X	

21. Betelgeuse is larger. Betelgeuse is more massive than Aldebaran.
22. Stars with larger masses reach the main sequence faster.
23. The luminosity will decrease.
24. gravity or gravitational

Extended Constructed Response Question

25. The correctly plotted graph is shown below:

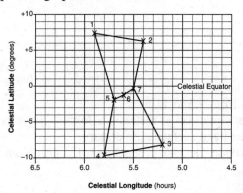

26. Betelgeuse is cooler and about the same or slightly less luminous than Rigel.
27. Milky Way or Milky Way Galaxy
28. Acceptable responses include but are not limited to these examples. Earth revolves in its orbit. Orion is visible only during the daytime in July. The Sun is between Earth and Orion in July.

CHAPTER 8

Multiple-Choice Questions

1. 2
2. 3
3. 4
4. 4
5. 1
6. 4
7. 3
8. 3
9. 3
10. 1
11. 1
12. 3
13. 3
14. 2
15. 3
16. 2
17. 4
18. 3
19. 3
20. 4
21. 3
22. 4

Constructed Response Questions

23., 24.

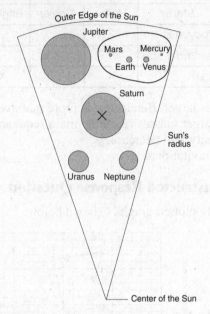

25. Any value from 9.5 to 11.5
26. The greater the average distance a Jovian planet is from the Sun, the colder its average surface temperature.
27. Carbon dioxide causes a greenhouse effect on Venus.
28.

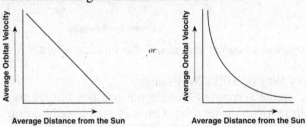

29. Mercury
30. The geocentric model has all celestial objects revolving around Earth. These moons orbit Jupiter, not Earth.

Extended Constructed Response Questions

31. Three answers are acceptable: Saturn or Uranus or Neptune
32. Earth's diameter is smaller, and its density is greater.
33. Color: yellow; Luminosity: 1
34. frozen gases
35. any value from 150 K to 200 K
36. As distance from the Sun increases, temperature decreases.
37. 5.2 AU

CHAPTER 9

Multiple-Choice Questions

1. 2
2. 2
3. 4
4. 2
5. 1
6. 1
7. 1
8. 3
9. 1
10. 1
11. 4
12. 4
13. 2

Constructed Response Questions

14. Acceptable responses include but are not limited to cosmic background radiation remains, a red shift in the light from stars in distant galaxies, the apparent expansion of the universe, and more-distant stars are moving away from Earth at a greater rate than nearby stars.
15. 1300 (±200) million years
16. Background radiation; all galaxies receding
17. (a) Universe contracts, leading to another big bang
    (b) Universe continues to expand and cool indefinitely

Extended Constructed Response Questions

18.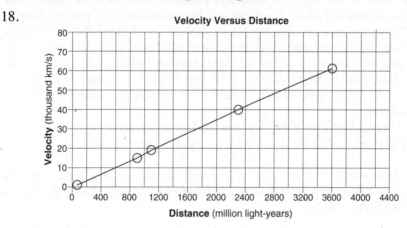

19. Acceptable responses include but are not limited to as the distance increases, the velocity increases; the farther from Earth, the faster it moves; the farther a galaxy is from Earth, the greater the velocity; and direct relationship.
20. Any value from 3,900 to 4,300 million light-years
21. Acceptable responses include but are not limited to these examples. A shift of light from distant galaxies toward the red end of the spectrum shows that galaxies are moving away from Earth. The red shift shows that the universe is expanding.
22. Acceptable responses include but are not limited to these examples. Earth and our solar system are younger than the Milky Way galaxy. The estimated age of Earth and our solar system is 4.6 billion years, and these

distant galaxies are 12 billion years old. Our solar system is about 5 billion years old, much younger than these 12-billion-year-old galaxies.

CHAPTER 10

Multiple-Choice Questions

1. 2
2. 1
3. 1
4. 3
5. 3
6. 2
7. 1
8. 1
9. 3
10. 4
11. 3
12. 3
13. 1
14. 4
15. 1
16. 4

Constructed Response Questions

17. Differentiation occurred while Earth's interior was a molten fluid. In a fluid, two of the forces acting on an object are weight and buoyancy. Weight is a downward force because gravity is attracting the object's mass toward Earth's center. The denser the object, the greater its weight. Buoyancy is an upward force and is equal to the weight of the fluid displaced by the object. If an object is denser than the surrounding fluid, the downward force of weight will be greater than the upward force of buoyancy, and the object will sink downward. If the object is less dense than the surrounding fluid, the downward force of weight will be less than the upward force of buoyancy, and the object will float upward. Thus, denser materials (e.g., iron) sank toward the center of the molten Earth, while less dense materials (e.g., water) floated toward the surface.
18. It is highly unlikely that Earth would have captured such a large object and even less likely that it would go into a circular orbit.
    Or
    Moon rocks are more similar to Earth's mantle than to meteorites like those that would have accreted to form an object the size of the Moon.
19. See Figure 10.2.

Extended Constructed Response Questions

20. (a) 1. Crust   2. Mantle   3. Core
    (b) 1. Core—3,470 km   2. Mantle—2,860–2,890 km
        3. Crust—10–40 km
    (c) When Earth was mostly molten, denser materials sank toward the center while less dense materials floated toward the surface—a process called differentiation.
21. 1 credit for *each* of two correct responses. Acceptable responses include but are not limited to these examples:
    Friction from Earth's atmosphere causes many meteors to burn up.

Many meteors fall into the ocean, so craters aren't visible.
Old craters are eroded away.
Vegetative growth may hide evidence of a crater.
Plate tectonics has destroyed some craters.
Deposition has buried some craters.

CHAPTER 11

Multiple-Choice Questions

1. 1
2. 4
3. 1
4. 3
5. 3
6. 1
7. 4
8. 3
9. 3
10. 1
11. 2
12. 1
13. 3
14. 1
15. 3
16. 2
17. 1
18. 4
19. 2
20. 2

Constructed Response Questions

21. Acceptable responses include but are not limited to cooling and solidification are processes that form igneous rocks; as early Earth cooled and solidified, igneous rocks were formed; the once-molten Earth formed igneous rocks as it solidified.
22. hematite
23. stratosphere
24.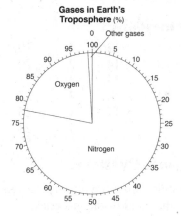

Extended Constructed Response Questions

25. stratosphere
26. Acceptable responses include but are not limited to the following. The ozone layer absorbs some of the harmful ultraviolet radiation from the Sun. The layer decreases the amount of ultraviolet radiation reaching Earth. The ozone protects humans from skin cancer and eye damage.
27. Acceptable responses include but are not limited to June 20, 21, or 22; the first day of summer; and the summer solstice.

28.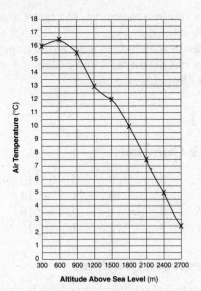

29. Acceptable responses include but are not limited to these examples. Inverse relationship, indirect relationship, or as elevation increases, air pressure decreases.

CHAPTER 12

Multiple-Choice Questions

1.	1	7.	3	13.	4	19.	1	25.	3
2.	4	8.	3	14.	2	20.	1	26.	2
3.	3	9.	3	15.	3	21.	4		
4.	4	10.	4	16.	4	22.	1		
5.	1	11.	3	17.	1	23.	4		
6.	3	12.	3	18.	4	24.	4		

Constructed Response Questions

27. Liquid. The molecules are more closely packed in the liquid phase than they are in either the solid or the gas phase.
28. The molecules would be spaced even farther apart.
29. As the water changes from water to ice, the volume of the water sample increases.
30. Any value from 1,408 to 1,409 millions of cubic kilometers.
31. Acceptable responses include but are not limited to the following. The oceans cover a larger portion of Earth's surface than the continents. Air over oceans has more moisture than air over land. More evaporation occurs over the oceans.
32. Acceptable responses include but are not limited to any two of the following: slope of the land surface, soil type or composition, vegetation

or lack of vegetation, land use/a paved surface, degree of soil saturation, porosity of the soil, and permeability or impermeability of the surface.

Extended Constructed Response Questions

33. evaporation
34. Acceptable responses include but are not limited to condensation is the phase change from water vapor (gas) to water (liquid).
35. Acceptable responses include but are not limited to the soil is saturated, the rate of rainfall exceeds the rate of infiltration, the ground is frozen, and the land has a steep slope.
36. Acceptable responses include but are not limited to plants release water into the air by transpiration, and runoff is slowed by plants so more infiltration can occur.
37. 334 Joules/gram
38.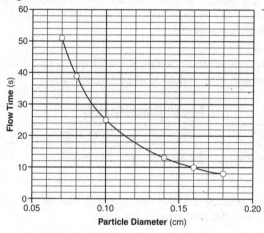
39. Any value from 14 s to 16 s
40. Acceptable responses include but are not limited to the following. Larger particles have larger pore spaces between them. Larger particles have less total surface area than smaller particles and, therefore, less friction with the moving water.

CHAPTER 13

Multiple-Choice Questions

1.	1	9.	3	17.	4	25.	3	33.	2
2.	4	10.	4	18.	3	26.	2	34.	3
3.	1	11.	4	19.	1	27.	3	35.	3
4.	3	12.	1	20.	3	28.	1	36.	4
5.	2	13.	3	21.	2	29.	3	37.	3
6.	3	14.	4	22.	2	30.	2	38.	4
7.	4	15.	1	23.	1	31.	2	39.	4
8.	3	16.	2	24.	3	32.	3	40.	2

Appendix A: Answers to Review Questions

Constructed Response Questions

41. Ordovician Period
42. Acceptable responses include but are not limited to the following. The basalt intrusion cuts across the fault. The intrusion is not displaced by the fault. The fault does not cut across the basalt intrusion.
43. Acceptable responses include but are not limited to the fossil is too old, Carbon-14 dating is inaccurate because very little Carbon-14 is present, and Carbon-14 has a short half-life.
44. Cambrian Period
45. Acceptable responses include but are not limited to any *two* of the following: uplift, erosion, weathering, subsidence, deposition, and burial.
46. Acceptable responses include but are not limited to widespread geographic distribution and short existence in geologic time.
47. Acceptable responses include but are not limited to the following. Carbon-14's half-life is too short. Not enough carbon-14 is left to measure. The fossils are too old.
48. Silurian Period
49. Acceptable responses include but are not limited to it existed for a short geologic time and it is widespread geographically.
50.

Fossil Classification

Index Fossil	Eospirifer	Manticoceras	Phacops
General Fossil Group	Brachiopod	Ammonoid	Trilobite

51. Acceptable responses include but are not limited to intrusion of the Palisades sill and the breakup of Pangaea.
52.

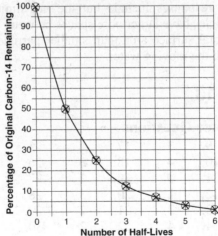

53. 22,800 years
54. Tree trunk and an acceptable explanation. Acceptable explanations include but are not limited to the tree trunk is a recent organic remain and carbon-14 is used to date recent remains.
55. Acceptable responses include but are not limited to wide geographic distribution and existed for a short period of geologic time
56.

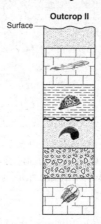

57.

				unconformity		
D	'C	fault GH		XY	B	A

Oldest ⟶ Youngest

58. One credit is allowed for circling the letter *I*.
59. One credit is allowed if the center of the **X** is located anywhere in the contact metamorphic zone in the limestone layer.
60. Acceptable responses include but are not limited to the following. No contact metamorphism is shown in rock unit *D*. Rock unit *F* was eroded and then rock unit *D* was formed. A buried erosional surface is between *F* and *D*. Rock unit *D* is on top of rock unit *F*.
61. Acceptable responses include but are not limited to rock unit *H* was displaced by movement along a fault; rock unit *H* was broken when an earthquake occurred.

Appendix A: Answers to Review Questions

Extended Constructed Response Questions

62. One credit is allowed if *two* **X**s are located on *two* of the three boundaries shown below. Credit is *not* allowed if both **X**s are along the same unconformity.

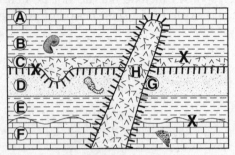

63. Acceptable responses include but are not limited to Devonian Period or Mississippian Period, and Carboniferous Period or Devonian Period.
64. Acceptable responses include but are not limited to *C* is on top of *D* and *C* metamorphosed *D*.
65. *G*
66. Acceptable responses include but are not limited to Cretaceous Period, Paleogene Period, Neogene Period, and Quaternary Period.
67. Acceptable responses include but are not limited to the fossils are too old for carbon-14 dating and carbon-14 has a very short half-life.
68. One credit is allowed for correctly drawing the line of unconformity as shown below.

69. Silurian Period
70. Acceptable responses include but are not limited to the following. The graywacke layers are tilted. The layers are now vertical. The unconformity indicates that the graywacke layers were uplifted and eroded.
71. One credit is allowed for *two* acceptable responses. Acceptable responses include but are not limited to uplift, weathering, erosion, tilting, submergence, burial, and deposition.
72. Any latitude from 40°S to 44°S and any longitude from 65°W to 69°W are acceptable. The correct units and compass directions must be included.
73. clay

Appendix A: Answers to Review Questions

74. Acceptable responses include but are not limited to earliest birds and birds.
75. Acceptable responses include but are not limited to radioactive dating, identifying an index fossil in the layer containing this fossil, and correlating rock layers or fossils.
76. Acceptable responses include but are not limited to the following. Mountain barriers changed the flow of winds. The air sinks on the Patagonia side of the Andes. Patagonia is located on the leeward side of the mountains. Patagonia is located in the rain shadow.
77. 5,700 years or 5.7×10^3 years
78. Acceptable responses include but are not limited to ^{238}U has a longer half-life, ^{238}U can be used to date older geologic events, and ^{14}C is used to date organic remains while ^{238}U is not.

CHAPTER 14

Multiple-Choice Questions

1.	3	5.	4	9.	3	13.	2	17.	1
2.	4	6.	2	10.	1	14.	4	18.	3
3.	3	7.	2	11.	2	15.	1		
4.	1	8.	4	12.	1	16.	2		

Constructed Response Questions

19.

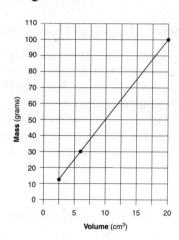

Appendix A: Answers to Review Questions

20. 50 grams
21. Many minerals have the same color; few minerals have exactly the same cleavage.
22. Minerals form by natural processes over long periods of time. Once used, they may not form again during your lifetime or even during a thousand lifetimes.
23.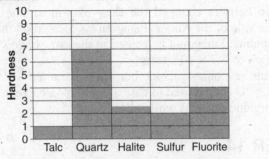
24. One credit is allowed for quartz and an acceptable reason. Acceptable reasons include but are not limited to the following. Quartz is the hardest mineral shown. Quartz has a hardness of 7. Quartz has the same hardness as garnet, which is used as an abrasive.
25. luster
26. The color of the dust or powdered form of the mineral; the color of the mark left when a mineral is rubbed on an unglazed porcelain tile
27. Acceptable responses include but are not limited to quartz, garnet, diamond, and pyrite.

Extended Constructed Response Questions

28. Acceptable responses include but are not limited to the following. The dangers of asbestos fibers were realized. The concern over the health risk of asbestos resulted in less use. Exposure to high concentrations of asbestos leads to health problems.
29. Adirondacks or Adirondack Mountains or Grenville Province
30. Acceptable responses include but are not limited to the internal arrangement of atoms, chemical composition, the environment in which they form, and chains of silicate tetrahedra.
31. talc
32. zone D
33. Acceptable responses include but are not limited to sulfur and hematite.
34. chromium or Cr
35. Acceptable responses include but are not limited to hardness, luster, and crystal shape.
36. marble
37. Eocene Epoch
38. Acceptable responses include but are not limited to convergent plate boundary, subduction zone, and collision boundary.

CHAPTER 15

Multiple-Choice Questions

1.	1	9.	4	17.	4	25.	2	33.	1
2.	3	10.	4	18.	3	26.	2	34.	4
3.	3	11.	1	19.	4	27.	3	35.	3
4.	1	12.	4	20.	4	28.	1	36.	2
5.	4	13.	2	21.	3	29.	3	37.	1
6.	4	14.	1	22.	3	30.	2	
7.	2	15.	3	23.	1	31.	2	
8.	4	16.	3	24.	4	32.	2	

Constructed Response Questions

38. Acceptable responses include but are not limited to *A* is cooling slower than *B*, *B* is cooling faster than *A*, intrusive rock forms from molten rock that cools slowly, and extrusive rock forms from molten rock that cools rapidly.
39. 1 mm to 10 mm or 1 to 10 mm. Correct units must be included in the answer.
40. Acceptable responses include but are not limited to obsidian, basaltic glass, pumice, and vesicular basalt glass.
41. Acceptable responses include but are limited to it shows banding, the rock is foliated, the minerals are segregated into layers, and distortion.
42. Any *two* of the following three responses must be included: pyroxene (augite), mica (biotite), and amphibole (hornblende).
43. Garnet and one acceptable use. Acceptable responses include but are not limited to jewelry and abrasives.
44. sandstone
45. gneiss
46. Credit is allowed for a correct description of *two* or more characteristics. Acceptable responses include but are not limited to:

Characteristics of Granite	Description
Texture	coarse nonvesicular 1 mm to 10 mm
Color	light colored white pink gray
Density	low 2.7 g/cm³

47. Acceptable responses include but are not limited to calcite and $CaCO_3$.

48. Acceptable responses include but are not limited to the following. Limestone reacts with acids in groundwater. Acids in water cause limestone to dissolve. Chemicals weather limestone. Water flowing through cracks removes limestone.
49.

50. less than 0.0004 cm or any number given that is less than 0.0004 cm
51. slate
52. Acceptable responses include but are not limited to nonvesicular, coarse, and large crystal.
53. One credit is allowed for placing an **X** whose center falls within the shaded zone of contact metamorphism shown in the diagram below.

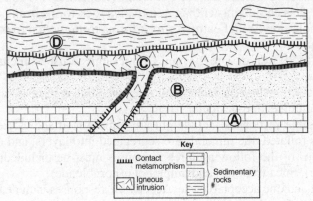

54. calcite
55. Acceptable responses include but are not limited to the following. Intrusions are younger than any rock they metamorphose. Contact metamorphism can be seen between rock layer *D* and the igneous intrusion.
56. Sedimentary or any clastic sedimentary rock or any specific clastic sedimentary rock name
57. Form of coal: anthracite or hard coal or metamorphic coal. Explanation: It forms under great pressures, which increases density. Anthracite is the metamorphic form of coal.
58. Acceptable responses include but are not limited to the earliest land plants did not occur until the Silurian and extensive coal-forming forests didn't exist until the Carboniferous Period.
59. Acceptable responses include but are not limited to the following. A "clean" sandstone contains mostly quartz, while a "dirty" sandstone can contain other minerals, such as plagioclase feldspar and calcite. A "dirty" sandstone contains many different minerals, while a "clean" sandstone contains mostly one kind of mineral.

60. Acceptable responses include but are not limited to burial, compaction, and cementation.
61. Acceptable responses include but are not limited to noise or dust from drilling, blasting, grinding, and/or truck traffic; pollution of streams and groundwater; increased erosion; and habitat destruction/deforestation.
62. calcite or dolomite
63. Acceptable responses include but are not limited to heat and pressure increase from B to C, regional metamorphism is greatest at C, and different grades of metamorphism.
64. fine
65. The oceanic crust is more dense than the continental crust.

CHAPTER 16

Multiple-Choice Questions

1.	1	8.	2	15.	2	22.	1	29.	4
2.	3	9.	3	16.	3	23.	1	30.	4
3.	1	10.	1	17.	3	24.	2	31.	3
4.	3	11.	2	18.	3	25.	2	32.	1
5.	3	12.	3	19.	3	26.	1	33.	4
6.	3	13.	2	20.	1	27.	1		
7.	2	14.	2	21.	4	28.	1		

Constructed Response Questions

34. Acceptable responses include but are not limited to tectonic plate movement, movement along a fault, and volcanic eruption.
35. Acceptable responses include but are not limited to P-wave, primary wave, and compressional wave.
36. Any response from 12 min 30 sec to 12 min 50 sec
37. Acceptable responses include but are not limited to the following. The arrival time of the P-wave at station A is later than the arrival time of the P-wave at station B. The arrival time difference between the P-wave and S-wave is greater at station A. The amplitudes of the P-wave and S-wave tracings are greater on the seismogram at station B.
38. 15 minutes 50 seconds (±10 seconds)
39. 44.5°N (latitude) and 73.7°W (longitude). The correct compass directions and degrees must be included for both latitude and longitude to receive credit.
40. 3 or three
41. Peru is closer to the epicenter.
42. 3 min 0 sec (±20 sec)

Appendix A: Answers to Review Questions

Extended Constructed Response Questions

43.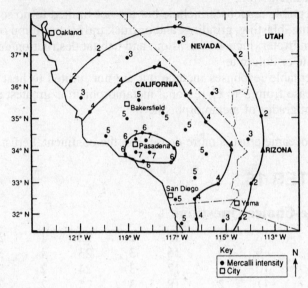

44. Pasadena
45. 35°30'N, 119°W
46. Movement along the San Andreas fault or movement along the boundaries of the North American and Pacific Plates
47. Acceptable responses include but are not limited to chimneys fell, heavy furniture overturned, and Anchorage suffered much damage to substantial structures.
48. Both the North American Plate and the Pacific Plate
49. Acceptable responses include but are not limited to S-waves were absorbed through the liquid outer core and S-waves cannot travel through the liquid outer core.
50. Acceptable responses include but are not limited to tsunami and coastal flooding.
51. Credit is allowed if both responses are correct. Latitude: any value from 61°N to 62°N; Longitude: any value from 147°W to 148°W
52. Credit allowed if all *three* isolines are drawn correctly. See example below.

53. Credit allowed if the center of the **X** is located within the crosshatched area below.
Example of a 2-credit response for questions 52 and 53:

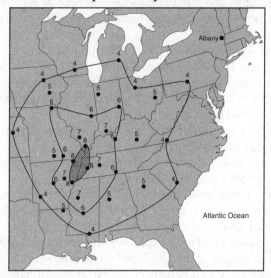

54. 3 min 0 sec (±10 sec)
55. Acceptable responses include but are not limited to the following. The western coast of the United States is near plate boundaries. More major faults are located on the western coast of the United States. Fewer active faults are located in the central portion of the United States compared with the western coast of the United States. The central portion of the United States is in the middle of a tectonic plate.
56. Credit is allowed for two correct responses. Acceptable responses include but are not limited to the following. Plan evacuation routes. Identify earthquake hazard zones or areas that are subject to damage during an earthquake. Plan emergency communication procedures. Develop emergency information brochures. Store food, supplies, and fresh water. Build earthquake-resistant structures. Identify shelter locations.

CHAPTER 17

Multiple-Choice Questions

1.	3	6.	1	11.	1	16.	1	21.	4
2.	3	7.	1	12.	1	17.	3	22.	4
3.	4	8.	1	13.	2	18.	4	23.	2
4.	2	9.	1	14.	3	19.	1	24.	3
5.	4	10.	1	15.	2	20.	2		

Appendix A: Answers to Review Questions

Constructed Response Questions

25. Volcanic eruptions, geothermal activity (geysers, hot springs, etc.), temperature readings taken in deep mines and bore holes
26. Magma is molten rock beneath Earth's surface. Lava is molten rock at Earth's surface.
27. Cinder cone—solid particles ejected during a volcanic eruption build up around the vent in a steep, narrow cone.
    Composite cone—large, symmetrical cones consisting of alternate layers of solidified lava and volcanic rock particles are formed by alternating explosive and quiet eruptions.
    Shield cone—a broad, flat cone is formed when lava flows quietly from the vent, spreading out in wide, flat sheets.
    Lava plateau—lava erupts from a fissure and spreads out in wide, thin sheets that blanket the surrounding land.
28. The longer molten rock takes to cool, the more time mineral crystals have to grow in size. Igneous intrusions are insulated by surrounding rock and cool slowly, allowing large mineral crystals to form.
29. The magma associated with explosive eruptions has a composition similar to that of granite and is thick and pasty. The magma associated with quiet eruptions has a composition similar to that of basalt and is thinner and more fluid.

    The material ejected during explosive eruptions typically consists of particles of molten rock that solidify as they "rain" down on the surrounding land. The material ejected during quiet eruptions is fluid rock that spreads out over the surrounding land in sheets and then solidifies.

Extended Constructed Response Questions

30. Layer *I*
31. Layer *B*
32.

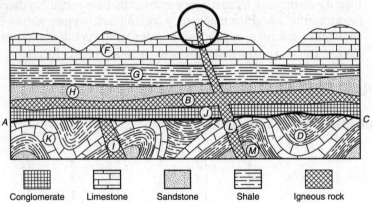

674

33. The high temperature of the molten rock that intruded to form layer *B* most likely caused the rock in layer *H* to undergo contact metamorphism.
34. Radioactive decay; heat transferred outward from Earth's core
35. (1) South American Plate; (2) Cocos Plate; (3) Caribbean Plate; (4) Nazca Plate
36. Acceptable responses include but are not limited to mass movement of mud down the mountain; a mud avalanche; it melted snow, causing mudslides; and hot ash and pumice melted snow, creating landslides.
37. Acceptable responses include but are not limited to a drop in pressure on the magma; steam and gases that were dissolved in the magma violently expanded; cracks in Earth's crust lowered pressure on the magma; and magma pressure cracked the overlying rocks, releasing the gases.
38. Acceptable responses include but are not limited to these examples. Escaping gas bubbles are trapped in the rapidly cooling magma. Gas/air pockets form in the rock as it cools.
39. Acceptable responses include but are not limited to these examples. Hawaiian magma is mafic, and the magma of Nevado del Ruiz volcano is andesitic. Hawaiian magma is runny, and the magma of Nevado del Ruiz is thick and slow moving. Hawaii is located at a hot spot in the center of the Pacific Plate. Nevado del Ruiz is near a subduction plate boundary.
40. Acceptable responses include but are not limited to these examples. Geologists should monitor conditions and provide early warning. People should leave their houses when early warning of an eruption is given. Avoid building homes in valleys. People should be discouraged from building near the volcano. Evacuation routes should be publicized. Predicted mudslide routes should be identified.

CHAPTER 18

Multiple-Choice Questions

1.	2	10.	1	19.	2	28.	4	37.	3
2.	2	11.	3	20.	3	29.	2	38.	1
3.	3	12.	2	21.	1	30.	1	39.	3
4.	4	13.	4	22.	1	31.	1	40.	2
5.	1	14.	2	23.	2	32.	3	41.	1
6.	1	15.	1	24.	2	33.	3	42.	1
7.	4	16.	1	25.	3	34.	2	43.	1
8.	2	17.	1	26.	3	35.	1	44.	4
9.	1	18.	3	27.	2	36.	3		

Appendix A: Answers to Review Questions

Constructed Response Questions

45. The **X** should be drawn somewhere on the Nazca Plate shaded below.

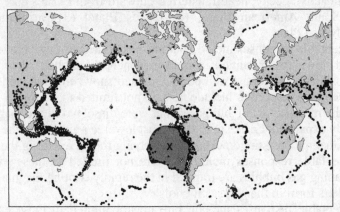

46. Acceptable responses include but are not limited to the following. Most major earthquakes occur at tectonic plate boundaries. Most earthquakes occur at the location of major fault zones. Crustal movement at plate boundaries causes frequent earthquake activity.
47. Acceptable responses include but are not limited to a hot spot, a magma plume, and the mantle.
48. Acceptable responses include but are not limited to divergent, diverging lithospheric plates, seafloor spreading, and rifting.
49. Acceptable responses include but are not limited to subduction and convergence.
50. Acceptable responses include but are not limited to transform movement, faulting, and the plates slide past each other.
51. Acceptable responses include but are not limited to the following. From point A to point B, the age of the surface bedrock decreases; from B to C, the age of the surface bedrock increases. The surface bedrock at point B is younger than the surface bedrock at point A and at point C. It gets younger, and then gets older.
52. Cenozoic Era
53. basalt or vesicular basalt or diabase
54. North American Plate and Eurasian Plate
55. Acceptable responses include but are not limited to hot spot, rising convection currents, magma chamber, and rising magma.

Extended Constructed Response Questions

56. Acceptable responses include but are not limited to transform boundary and transform fault.
57. Acceptable responses include but are not limited to subduction of Arabian Plate and convergence.

58.

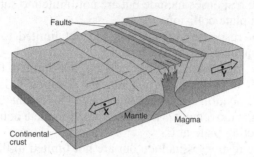

59. One credit allowed if both texture and color are correct. Acceptable responses include but are not limited to the following. Texture—fine grained; nonvesicular or vesicular; glassy; noncrystalline; grain size less than 1 mm. Color—dark colored; black; green.
60. One credit allowed if both minerals are correct. Acceptable responses include but are not limited to plagioclase feldspar, potassium feldspar (orthoclase), quartz, amphibole (hornblende), and biotite (mica).
61. Acceptable responses include but are not limited to WSW, SW, and southwest.
62. Any value from 23.5 to 26.5 miles per million years
63. One credit allowed if the center of the **X** is within the stippled area shown in the example below.
64. One credit allowed for an arrow that shows Laurentia moving to the southeast, south, or east as shown in the example below.

65. Ordovician Period
66. Any value from 500°C to 1,200°C

Appendix A: Answers to Review Questions

67. Indian-Australian Plate
68. Acceptable responses include but are not limited to subduction, convergence, and plate collision.
69. Acceptable responses include but are not limited to move to higher ground, evacuate, and move inland.
70. Acceptable responses include but are not limited to the following. Location *B* is at a transform fault, and location *C* is at a subduction boundary. Location *B* has horizontal plate movement, but location *C* has vertical plate movement. There is a transform plate boundary at *B*. There is a subducting plate at *C*.
71. Acceptable responses include but are not limited to the following. *A* is located at a plate boundary, and *D* is not located at a plate boundary. Crustal plates are colliding at *A*, and no plate collision is occurring at *D*. Location *A* is on the Pacific Ring of Fire.
72. Acceptable responses include but are not limited to the following. *E* is located above a mantle hot spot. *E* is the Canary Islands Hot Spot. *F* is near the center of a tectonic plate.
73. Acceptable responses include but are not limited to the following: Location *G* is at the ridge and is presently forming while *H* was at the ridge in the past. Location *H* is moving away from the new crust forming in the region at *G*. The youngest ocean floor bedrock is at the mid-ocean ridge. The plates are diverging at the Southeast Indian Ridge.
74. One credit allowed if the center of the **X** falls within the circle shown.

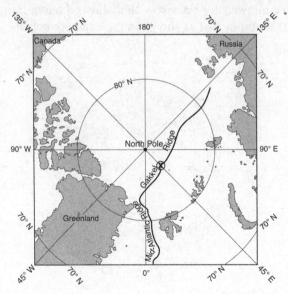

75. Acceptable responses include but are not limited to the following. The plates are moving apart or spreading. The tectonic plates are moving away from each other. The ridge is a diverging plate boundary. Rifting is occurring.

76. North American Plate and Eurasian Plate
77. Acceptable responses include but are not limited to magma/lava, volcanoes, and smoker vents.
78. pyroxene (augite) and olivine

CHAPTER 19

Multiple-Choice Questions

1.	1	6.	4	11.	4	16.	4	21.	3
2.	1	7.	4	12.	4	17.	3		
3.	2	8.	4	13.	4	18.	4		
4.	2	9.	4	14.	4	19.	1		
5.	4	10.	2	15.	1	20.	3		

Constructed Response Questions

22.

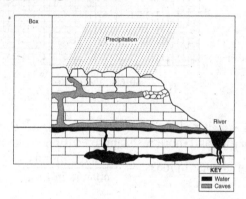

23. Acceptable responses include but are not limited to the following. The acid rain dissolves the limestone. The calcite in limestone chemically reacts with the acid.
24. Acceptable responses include but are not limited to burning fossil fuels, exhaust emission from automobiles, and smoke from factories.
25. Soil horizon A
26. (a) leaching
    (b) The soil in horizon B is less fertile than the soil in horizon A because it contains less organic matter and its high clay content inhibits the penetration of water and air.
27. Climate; type of bedrock

Extended Constructed Response Questions

28. (a–c)

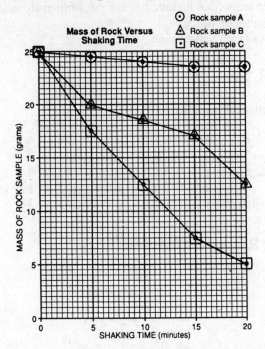

29. Sample *A* weathered more slowly than samples *B* and *C* because sample *A* is harder and more resistant to erosion.

30. (a) Rate of change in mass = $\dfrac{\text{change in mass}}{\text{change in time}}$

   (b) Rate of change in mass = $\dfrac{25.0 \text{ g} - 5.0 \text{ g}}{20 \text{ min}}$

   (c) Rate of change in mass = 1.0 g/min

31. During the process of abrasion, any part of a rock surface that juts out, such as an edge or corner, is more likely to be struck by another rock particle and be broken off. Thus, after 20 minutes of shaking, the corners and edges of rock material *C* will be smoother and more rounded than they were before shaking.

32. Acceptable responses include but are not limited to these examples. The physically weathered sediments are larger in particle size than the chemically weathered particles. The sand fragments are larger than the clay fragments. The sand fragments range from 0.006 cm to 0.2 cm in diameter, and the clay fragments are less than 0.0004 cm in diameter.

33. Moisture and temperature should both increase.

34. Answers must have *three* of the following mineral grains: plagioclase feldspar, biotite, amphibole, quartz, or pyroxene.

35. The answer must state granite and describe the greater hardness of the minerals found in granite. Acceptable responses include but are not limited to this example. Granite is composed mainly of quartz and feldspar, which are resistant to abrasion because of their hardness (7 and 6, respectively), while marble is made of calcite, which is softer (hardness of 3).

CHAPTER 20

Multiple-Choice Questions

1.	3	7.	1	13.	3	19.	3	25.	1
2.	4	8.	2	14.	3	20.	4	26.	2
3.	3	9.	1	15.	2	21.	4	27.	4
4.	3	10.	4	16.	2	22.	2	28.	1
5.	3	11.	2	17.	4	23.	4	29.	3
6.	1	12.	2	18.	2	24.	2		

Constructed Response Questions

30. pebble
31. Acceptable responses include but are not limited to abrasion, weathering, erosion, and particles were worn down as they were scraped along the bedrock.
32. shale
33. Acceptable responses include but are not limited to the following. A glacier forms a U-shaped valley. Glaciers form U-shaped valleys, and streams form V-shaped valleys.
34. Acceptable responses include but are not limited to the following. The type 3 stream meanders more. The type 3 stream occupies a wider floodplain. The type 1 stream has a straighter course.
35. Acceptable responses include but are not limited to the following. Stream velocity is greater on the outside of the meandering channel. Stream flow is slower on the inside of the meandering channel. Water is moving faster on the outside of a meandering curve.
36. Acceptable responses include but are not limited to the following. These tumbling cobbles and pebbles were abraded against other transported rocks and the stream channel. Abrasion occurred as the rocks bounced and rolled along the bottom of the streambed. Sharp corners and edges were knocked off, scraped, and/or worn down. They ground against other sediment and rocks. **Note:** Credit is *not* allowed for statements that describe the water alone as the primary cause of rounding.
37. Pleistocene Epoch
38. Acceptable responses include but are not limited to glacial sediment is unsorted and piles of mixed sediment sizes.
39. Acceptable responses include but are not limited to southwest (SW) and south southwest (SSW).

Appendix A: Answers to Review Questions

40. Acceptable responses include but are not limited to parallel scratches, grooves, or striations; and orientation of glacial features, such as drumlins and lateral moraines.

Extended Constructed Response Questions

41. Credit is allowed if the center of 10, 11, or 12 **X**s are plotted within the circles shown on the graph below and are correctly connected with a line that passes within the circles.

42. Acceptable responses include but are not limited to the following. As stream discharge increases, suspended sediment increases. There is a direct relationship between stream discharge and suspended sediment.
43. Acceptable responses include but are not limited to snowmelt in April results in a greater discharge, greater rainfall in April, and saturated ground would cause more runoff in April.
44. Two credits are earned for correctly listing three agents of erosion and identifying a characteristic surface feature formed by each of the three agents of erosion. Only one credit is earned for correctly listing only two agents of erosion and identifying a characteristic surface feature formed by each of the two agents of erosion. Only one credit is earned for correctly listing three agents of erosion even if the surface feature is incorrectly identified. Acceptable responses include but are not limited to:

Agent of Erosion	Surface Feature Formed
Waves	beach, sandbars, barrier islands
Wind	loss of topsoil, dunes
Glacier	U-shaped valley, moraines, drumlins
Running water (streams)	V-shaped valley, deltas, meanders
Mass movement	landslides, slumps

Appendix A: Answers to Review Questions

45. 450 ft (±50 feet)
46. Acceptable responses include but are not limited to the following. The continental ice sheet generally moved from north to south. Glacial erosion occurred. The original stream valleys had a north-south orientation.
47. Acceptable responses include but are not limited to the following. Water has a higher specific heat than land. Water heats and cools slowly because of its higher specific heat. Bodies of water change temperature more slowly than surrounding land.
48. Acceptable responses include but are not limited to sediments were deposited in water and compressed, compaction and deposition, burial and cementation, and chemical precipitation and evaporation.

CHAPTER 21

Multiple-Choice Questions

1.	3	10.	2	19.	2	28.	1	37.	2
2.	2	11.	1	20.	4	29.	1	38.	1
3.	2	12.	4	21.	3	30.	3	39.	2
4.	1	13.	3	22.	1	31.	3	40.	2
5.	4	14.	2	23.	1	32.	1	41.	2
6.	1	15.	3	24.	4	33.	1	42.	3
7.	3	16.	4	25.	2	34.	4	43.	2
8.	3	17.	3	26.	4	35.	4		
9.	2	18.	3	27.	3	36.	2		

Constructed Response Questions

44. Acceptable responses include but are not limited to these examples. The stream velocity decreases. The still water of the lake slows the stream current.
45. Acceptable responses include but are not limited to silt and clay.
46. The answer must include a correct bedrock characteristic. Acceptable responses include but are not limited to these examples. The Adirondacks have faulted, folded, and deformed bedrock. The Adirondacks have intensely memorphosed bedrock. The oldest bedrock is near the center of the Adirondacks.

    The answer must also include a correct land surface characteristic. Acceptable responses include but are not limited to these examples. The Adirondacks have high elevations. The Adirondacks have steep slopes. The Adirondacks are a partially eroded dome.
47. Acceptable responses include but are not limited to southeastward, from NW to SE, south-southeastward, and from N to S.

48.

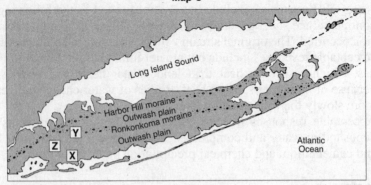

49. Hudson Highlands or Adirondack Mountains
50. Acceptable responses include but are not limited to the following. Global warming will cause glaciers to melt, which will raise the sea level. New York City and Long Island could be flooded when the sea level rises.
51. One credit is allowed if the center of the **X** is located within the shaded area shown. Credit is also allowed if a symbol other than **X** is used.

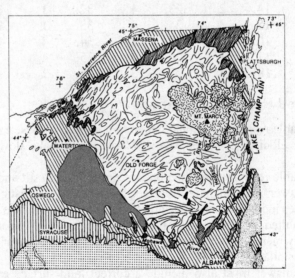

52. 74°W Both the correct unit and compass direction must be included in the answer.
53. One credit is allowed if both the name of the geologic age and the name of the bedrock are correct. Acceptable responses include but are not limited to:
Geologic age—Proterozoic, Middle Proterozoic, Precambrian, about 1,000 million years.
Name of bedrock—anorthosite, anorthositic.

Extended Constructed Response Questions

54. One credit is allowed if the cross section shows that the water is deeper near point A.

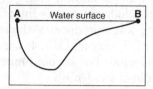

55. Acceptable responses include but are not limited to a direct relationship, and as the stream velocity increases, the stream can carry bigger sediment.
56. One credit is allowed if both the size and shape changes are correctly described. Acceptable responses include but are not limited to:
Size—The pebbles become smaller. The size of the pebbles decreases.
Shape—The pebbles become rounder. The pebbles become more spherical.
57. pebbles or sand
58. One credit is allowed for a straight or curved line that shows a direct relationship.

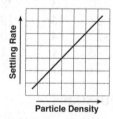

59.

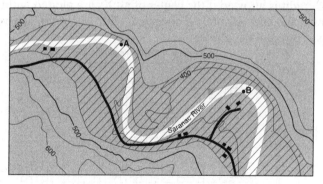

60. Acceptable responses include but are not limited to the following. The contour line bends upstream when crossing the river. The elevation of the river near the western edge of the map is 450 ft but is only 400 ft farther east.
61. Acceptable responses include but are not limited to the following. Erosion is greater on the outside of the meander curve. The velocity of the stream is greater at point A.

685

Appendix A: Answers to Review Questions

62. Acceptable responses include but are not limited to move to higher ground, build levees, build flood-control dams upriver, and plan evacuation routes.
63. any response from 490 to 443 million years.
64. Acceptable responses include but are not limited to the channel at *A* has a V shape, and running water produces V-shaped channels.
65. 1. Till, 2. stratified drift, 3. clay and silt, 4. wind blown sand
66. Acceptable responses include but are not limited to glacial deposits are unsorted and till is a direct ice deposit.
67. Acceptable responses include but are not limited to the following. The gentle slope of the dune is on the southwest side. The windward side has a less steep slope. The steeper side is leeward.

CHAPTER 22
Multiple-Choice Questions

1.	3	9.	2	17.	4	25.	2	33.	1
2.	4	10.	2	18.	1	26.	4	34.	3
3.	2	11.	1	19.	2	27.	2	35.	3
4.	4	12.	4	20.	3	28.	1	36.	1
5.	1	13.	2	21.	3	29.	1	37.	2
6.	1	14.	2	22.	4	30.	2		
7.	3	15.	3	23.	1	31.	3		
8.	4	16.	1	24.	2	32.	1		

Constructed Response Questions

38. One credit is allowed if the center of all **X**s are plotted within the circles shown and are correctly connected with a smooth, curved line that passes through the circles.

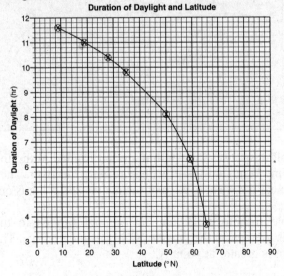

39. Acceptable responses include but are not limited to the following. As latitude increases, the duration of daylight decreases. Higher latitudes have shorter daylight periods. Lower latitudes have longer daylight periods. It is an inverse relationship.
40. One credit is allowed for a correct answer ±1° based on the student-drawn graph. For example, on the graph shown, the answer should be 56° ±1°N.
41. winter
42. One credit is allowed for an arrow beginning at point *B* and curving to the northeast or curving to the right. Credit is allowed even if the arrow is not curved or if it does not start at point *B* but is drawn from southwest to northeast.

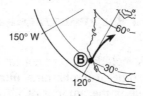

43. 12 noon or noon or 12 P.M.
44. Acceptable responses include but are not limited to March 20, 21, or 22; September 21, 22, 23, or 24; and equinox.
45. Acceptable responses include but are not limited to the following. *B* is located at a lower latitude. *B* is located closer to the equator. *A* is farther north. *A* is located at a greater distance from the latitude receiving direct Sun rays on this day.
46. Acceptable responses include but are not limited to methane, water vapor, carbon dioxide (CO_2), and ozone.
47. Acceptable responses include but are not limited to the following. Greenhouse gases absorb the longer-wave radiation coming from Earth's surface. Greenhouse gases trap the heat given off by Earth. Greenhouse gases absorb infrared energy.
48. One credit is allowed if both isotherms are correctly drawn to the edges of the map. If additional isotherms are drawn, all isotherms must be correct to receive credit.

Appendix A: Answers to Review Questions

49. Acceptable responses include but are not limited to the following. The intensity of insolation increases from 11:00 A.M. to 12 noon. The Sun's energy becomes more concentrated. Sunlight becomes more direct. The Sun rises higher in the sky.
50. The energy absorbed during a phase change, as between points B and C or between points D and E, is used to break molecular bonds rather than to make molecules move faster. Since molecular speed doesn't increase, temperature doesn't increase.
51. D and E
52. B and C
53. More energy (2,260 J/g) is required to evaporate water between points D and E than to melt water (334 J/g) between points B and C.

Extended Constructed Response Questions

54. stratosphere
55. Acceptable responses include but are not limited to the following. The ozone layer absorbs some of the harmful ultraviolet radiation from the Sun. The layer decreases the amount of ultraviolet radiation reaching Earth. The ozone protects humans from skin cancer and eye damage.
56. Acceptable responses include but are not limited to June 20, 21, or 22; the first day of summer; and summer solstice
57. Acceptable responses include but are not limited to infrared, microwaves, and radio waves.
58. Acceptable responses include but are not limited to allows most solar radiation to pass through, prevents some terrestrial radiation from escaping, and traps heat.

CHAPTER 23
Multiple-Choice Questions

1.	4	12.	2	23.	1	34.	1	45.	2
2.	3	13.	4	24.	1	35.	4	46.	2
3.	2	14.	2	25.	4	36.	3	47.	4
4.	3	15.	1	26.	2	37.	3	48.	2
5.	4	16.	3	27.	3	38.	2	49.	4
6.	2	17.	3	28.	2	39.	2	50.	4
7.	3	18.	3	29.	1	40.	4	51.	2
8.	2	19.	3	30.	3	41.	3	52.	2
9.	4	20.	1	31.	2	42.	3		
10.	2	21.	4	32.	3	43.	2		
11.	2	22.	1	33.	2	44.	3		

Appendix A: Answers to Review Questions

Constructed Response Questions

53. One credit is earned if all six weather conditions are correctly located and in the proper format.

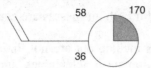

54. Acceptable responses include but are not limited to the following. Temperature and dewpoint values are far apart. Relative humidity is very low. Cloud cover is only 25%. Air pressure of 1017.0 mb most likely indicates the presence of a high-pressure system.
55. cold front
56. Acceptable responses include but are not limited to the following. The warm, moist air is less dense. The warm, moist air is lighter. Warm air is overriding the more dense cold air.
57. occluded front
58.

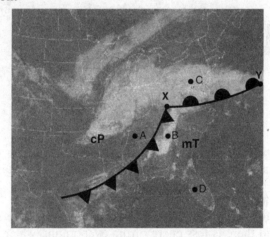

59. Acceptable responses include but are not limited to condensation, expanding air, cooling to the dewpoint, rising air, and deposition (sublimation).
60. Acceptable responses include but are not limited to the following. Location A is influenced by a cold, dry air mass. Location A has clear skies. Location B is in a warm, moist air mass. Location B has cloud cover.
61. Acceptable responses include but are not limited to the following. Location C is cooler because it is farther north. C is in a continental polar air mass, which is cold, dry air. Location C has clouds that block some of the sunlight.
62. Acceptable responses include but are not limited to east, northeast, and ENE.

63. 1002.1 mb
64. One credit is allowed if both the compass direction and wind speed are correct: from the south southeast (SSE) or southeast (SE) at 25 knots (±2).
65. One credit is allowed if both air masses are correct: (1) cP (2) mT.
66. Binghamton
67. Acceptable responses include but are not limited to move indoors, do not use electrical equipment or telephones, and do not stand under tall objects.

Extended Constructed Response Questions

68.

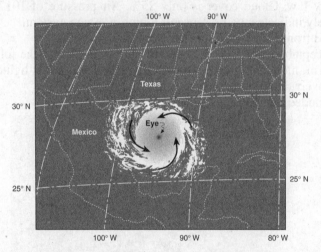

69. Acceptable responses include but are not limited to these examples. Rising air cools to the dewpoint, and water vapor condenses. Condensation occurs when the dewpoint is reached.
70. 27°30'N or 27.5°N (±1°) and 95°W (±1°)
71. Acceptable responses include but are not limited to these examples. Over land there is less energy from evaporating water. Winds decrease in strength due to friction with the land.
72. (a) Response should include two dangerous conditions. Acceptable responses include but are not limited to these examples: flooding and tornadoes, storm surge and collapsing structures, hail and lightning, and downed electrical wires and flying debris.
    (b) The response must be an emergency preparation that can be taken before the approaching hurricane hits the area. Acceptable responses include but are not limited to these examples. Evacuate to a higher elevation. Take shelter. Board up windows. Build a seawall.

73.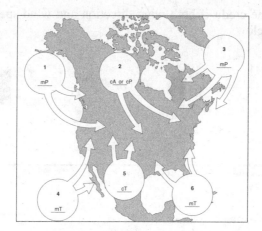
74. One credit is allowed for showing "high" air pressure over the land and "low" air pressure over the ocean, as shown below.
75. One credit is allowed for showing "warm" air temperature over the ocean and "cool" air temperature over the land, as shown below.

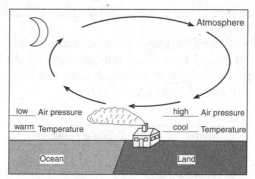

76. Acceptable responses include but are not limited to the following. Water droplets form on the surfaces provided by the salt and dust particles. Salt and dust particles are condensation nuclei, allowing the water vapor to change into liquid drops and to form clouds.
77. Acceptable responses include but are not limited to dust particles can be blown into the atmosphere by winds, a volcanic eruption, and a forest fire.
78. Highest air-pressure station: A
                                    D
                                    B
    Lowest air-pressure station: C

79. Any value from 27°C to 28°C
80. Acceptable responses include but are not limited to the following. The difference between air temperature and dewpoint is smallest at station C. Station C has the lowest air pressure. Station C has 100% overcast skies.

Appendix A: Answers to Review Questions

81. Allow one credit if both wind direction and wind speed are correct as shown below.

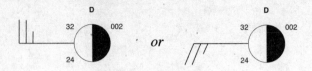

82.

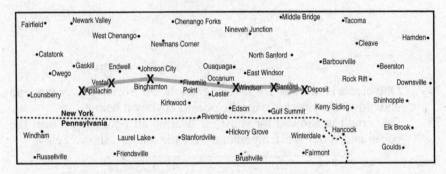

83. One credit is allowed for two correct responses. Allow credit for either upper- or lowercase letters. For example, allow credit for MT, Mt, mt, or mT. Acceptable responses include but are not limited to cP and mT, mT and cP, mT and cA, and cA and mT.
84. cold front
85.

Town	F-Scale Number
Vestal	1
Windsor	0
Sanford	1
Deposit	3

86. 0.7 mi/m

CHAPTER 24

Multiple-Choice Questions

1.	**1**	6.	**4**	11.	**4**	16.	**3**	21.	**2**
2.	**2**	7.	**2**	12.	**4**	17.	**4**	22.	**1**
3.	**1**	8.	**2**	13.	**2**	18.	**2**	23.	**4**
4.	**3**	9.	**4**	14.	**4**	19.	**3**	24.	**3**
5.	**2**	10.	**3**	15.	**1**	20.	**1**		

Constructed Response Questions

25. Two credits are allowed, one credit for each of two correct responses. Acceptable responses include but are not limited to these examples: air rises, air expands, air cools, the temperature reaches the dewpoint, and water vapor condenses.
26. California Current
27. either 30°N or 30°S
28. Acceptable responses include but are not limited to location B is located high in the mountains and location A is located at a lower elevation.
29. Acceptable responses include but are not limited to the following. Location C is located in air that is sinking, compressing, and warming. Location C is on the leeward side of a mountain. Location D is near a large body of water. Air traveling over the mountains loses its moisture at D.

Extended Constructed Response Questions

30. Acceptable responses include but are not limited to the following. The air on the western slopes of the mountains is rising. The valleys are located on the eastern side of mountain ranges where air is sinking. Air is warmed by compression as it descends the mountain slopes, so relative humidity decreases.
31. 134 in (±4 in)
32. Acceptable responses include but are not limited to the following. The Sierra Nevada Mountain Range is higher in elevation. Higher elevations have lower temperatures. Expansional cooling increases with higher mountains.
33. Acceptable responses include but are not limited to the following. Cold air holds very little water vapor. Very little evaporation takes place in Antarctica. Antarctica is in a region where air is sinking, therefore clouds seldom form. Very little precipitation occurs in a high-pressure area.
34. Acceptable responses include but are not limited to fossils, volcanic dust, pollen, trapped gases, and microbes.
35. Acceptable responses include but are not limited to the following. The sea level would most likely rise. The shape of Long Island would change. Submergence would occur. Long Island would become smaller. Buildings would be flooded.

36.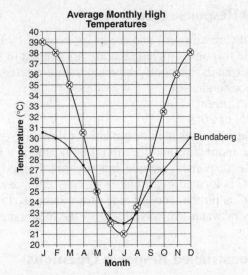

37. Acceptable responses include but are not limited to the following. Birdsville is located inland near the center of the continent. Bundaberg is located near a large body of water (the ocean) that moderates climate temperatures.
38. Acceptable responses include but are not limited to the following. Bundaberg is located near the ocean. Birdsville is located inland. The warm ocean current affects the climate of Bundaberg. Bundaberg is located on the windward side of the mountain.
39. Acceptable responses include but are not limited to the following. Bundaberg is located east of Birdsville. Birdsville is west of Bundaberg. Earth rotates west to east.
40. One credit is allowed for drawing a horizontal line at 30°S latitude within the shaded area shown below.

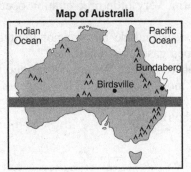

41. 10°C and 18°C or 18°C and 10°C
42. Acceptable responses include but are not limited to the following. This region receives a high angle of insolation each day. High-intensity insolation is received all year. The Sun is higher in the sky all year. The tropical region receives more intense sunlight.

Appendix B

REFERENCE TABLES FOR PHYSICAL SETTING/ EARTH SCIENCE

The University of the State of New York • THE STATE EDUCATION DEPARTMENT • Albany, New York 12234 • www.nysed.gov

Reference Tables for Physical Setting/EARTH SCIENCE

Radioactive Decay Data

RADIOACTIVE ISOTOPE	DISINTEGRATION	HALF-LIFE (years)
Carbon-14	$^{14}C \rightarrow {}^{14}N$	5.7×10^3
Potassium-40	$^{40}K \rightarrow {}^{40}Ar$ $^{40}K \rightarrow {}^{40}Ca$	1.3×10^9
Uranium-238	$^{238}U \rightarrow {}^{206}Pb$	4.5×10^9
Rubidium-87	$^{87}Rb \rightarrow {}^{87}Sr$	4.9×10^{10}

Specific Heats of Common Materials

MATERIAL	SPECIFIC HEAT (Joules/gram • °C)
Liquid water	4.18
Solid water (ice)	2.11
Water vapor	2.00
Dry air	1.01
Basalt	0.84
Granite	0.79
Iron	0.45
Copper	0.38
Lead	0.13

Equations

$$\text{Eccentricity} = \frac{\text{distance between foci}}{\text{length of major axis}}$$

$$\text{Gradient} = \frac{\text{change in field value}}{\text{distance}}$$

$$\text{Rate of change} = \frac{\text{change in value}}{\text{time}}$$

$$\text{Density} = \frac{\text{mass}}{\text{volume}}$$

Properties of Water

Heat energy gained during melting	334 J/g
Heat energy released during freezing	334 J/g
Heat energy gained during vaporization	2260 J/g
Heat energy released during condensation	2260 J/g
Density at 3.98°C	1.0 g/mL

Average Chemical Composition of Earth's Crust, Hydrosphere, and Troposphere

ELEMENT (symbol)	CRUST Percent by mass	CRUST Percent by volume	HYDROSPHERE Percent by volume	TROPOSPHERE Percent by volume
Oxygen (O)	46.10	94.04	33.0	21.0
Silicon (Si)	28.20	0.88		
Aluminum (Al)	8.23	0.48		
Iron (Fe)	5.63	0.49		
Calcium (Ca)	4.15	1.18		
Sodium (Na)	2.36	1.11		
Magnesium (Mg)	2.33	0.33		
Potassium (K)	2.09	1.42		
Nitrogen (N)				78.0
Hydrogen (H)			66.0	
Other	0.91	0.07	1.0	1.0

2011 EDITION°
This edition of the Earth Science Reference Tables should be used in the classroom beginning in the 2011–12 school year. The first examination for which these tables will be used is the January 2012 Regents Examination in Physical Setting/Earth Science.

Eurypterus remipes
New York State Fossil

°Note that the 2011 edition of the *Earth Science Reference Tables* has the same content as the 2010 edition but it does not have the ruler.

Appendix B: Earth Science Reference Tables

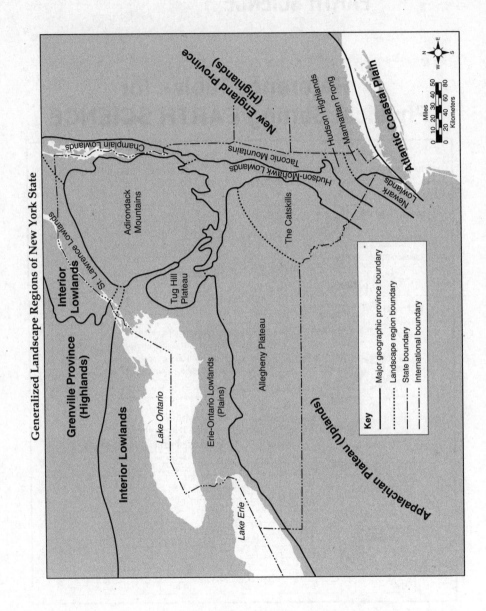

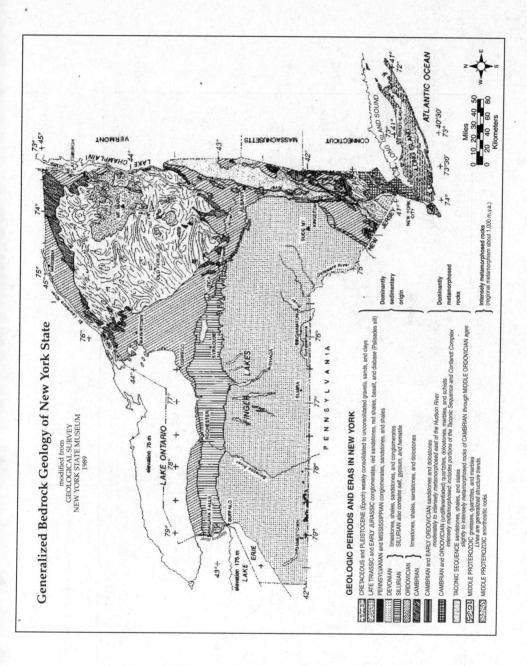

Appendix B: Earth Science Reference Tables

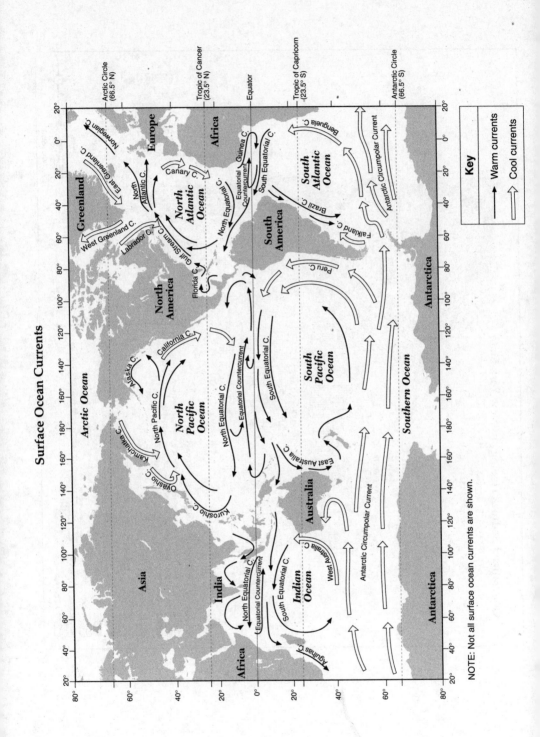

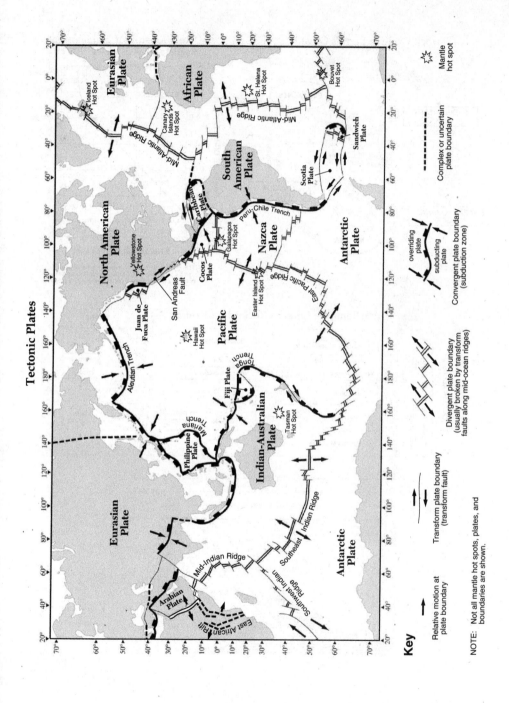

Appendix B: Earth Science Reference Tables

Rock Cycle in Earth's Crust

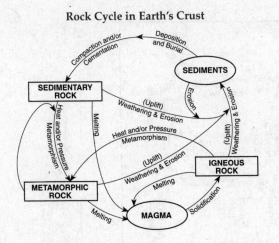

Relationship of Transported Particle Size to Water Velocity

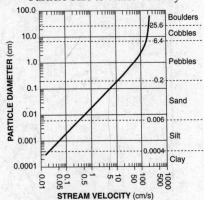

This generalized graph shows the water velocity needed to maintain, but not start, movement. Variations occur due to differences in particle density and shape.

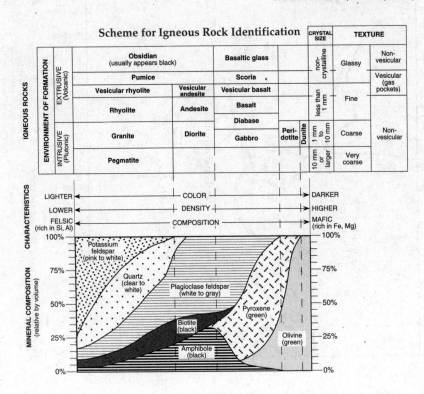

700

Appendix B: Earth Science Reference Tables

Scheme for Sedimentary Rock Identification

INORGANIC LAND-DERIVED SEDIMENTARY ROCKS

TEXTURE	GRAIN SIZE	COMPOSITION	COMMENTS	ROCK NAME	MAP SYMBOL
Clastic (fragmental)	Pebbles, cobbles, and/or boulders embedded in sand, silt, and/or clay	Mostly quartz, feldspar, and clay minerals; may contain fragments of other rocks and minerals	Rounded fragments	Conglomerate	
			Angular fragments	Breccia	
	Sand (0.006 to 0.2 cm)		Fine to coarse	Sandstone	
	Silt (0.0004 to 0.006 cm)		Very fine grain	Siltstone	
	Clay (less than 0.0004 cm)		Compact; may split easily	Shale	

CHEMICALLY AND/OR ORGANICALLY FORMED SEDIMENTARY ROCKS

TEXTURE	GRAIN SIZE	COMPOSITION	COMMENTS	ROCK NAME	MAP SYMBOL
Crystalline	Fine to coarse crystals	Halite	Crystals from chemical precipitates and evaporites	Rock salt	
		Gypsum		Rock gypsum	
		Dolomite		Dolostone	
Crystalline or bioclastic	Microscopic to very coarse	Calcite	Precipitates of biologic origin or cemented shell fragments	Limestone	
Bioclastic		Carbon	Compacted plant remains	Bituminous coal	

Scheme for Metamorphic Rock Identification

TEXTURE	GRAIN SIZE	COMPOSITION	TYPE OF METAMORPHISM	COMMENTS	ROCK NAME	MAP SYMBOL
FOLIATED — MINERAL ALIGNMENT	Fine	MICA, QUARTZ, FELDSPAR, AMPHIBOLE, GARNET, PYROXENE	Regional (Heat and pressure increases)	Low-grade metamorphism of shale	Slate	
	Fine to medium			Foliation surfaces shiny from microscopic mica crystals	Phyllite	
				Platy mica crystals visible from metamorphism of clay or feldspars	Schist	
BANDING	Medium to coarse			High-grade metamorphism; mineral types segregated into bands	Gneiss	
NONFOLIATED	Fine	Carbon	Regional	Metamorphism of bituminous coal	Anthracite coal	
	Fine	Various minerals	Contact (heat)	Various rocks changed by heat from nearby magma/lava	Hornfels	
	Fine to coarse	Quartz	Regional or contact	Metamorphism of quartz sandstone	Quartzite	
		Calcite and/or dolomite		Metamorphism of limestone or dolostone	Marble	
	Coarse	Various minerals		Pebbles may be distorted or stretched	Metaconglomerate	

Appendix B: Earth Science Reference Tables

GEOLOGIC HISTORY

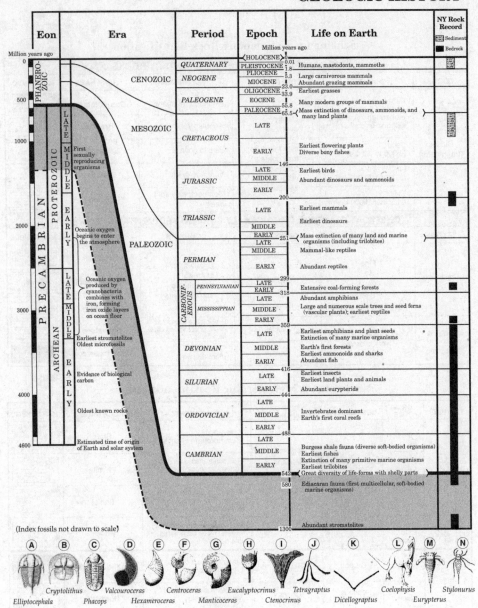

Appendix B: Earth Science Reference Tables

OF NEW YORK STATE

Time Distribution of Fossils (including important fossils of New York)	Important Geologic Events in New York	Inferred Positions of Earth's Landmasses
The center of each lettered circle indicates the approximate time of existence of a specific index fossil (e.g. Fossil (A) lived at the end of the Early Cambrian).	Advance and retreat of last continental ice	59 million years ago
NAUTILOIDS, DINOSAURS, MAMMALS, BIRDS (O, S)	Sands and clays underlying Long Island and Staten Island deposited on margin of Atlantic Ocean	
	Dome-like uplift of Adirondack region begins	119 million years ago
AMMONOIDS, CRINOIDS, VASCULAR PLANTS (L), GASTROPODS, CORALS, BRACHIOPODS	Initial opening of Atlantic Ocean North America and Africa separate ⟨Intrusion of Palisades sill⟩ Pangaea begins to break up	
		232 million years ago
TRILOBITES	Alleghenian orogeny caused by collision of North America and Africa along transform margin, forming Pangaea	
EURYPTERIDS, GRAPTOLITES, PLACODERM FISH (C, F, G, N, Q, R, X, Z; I, H, M, P, V, U, Y; E, K, T, W; B, D, J)	Catskill delta forms Erosion of Acadian Mountains Acadian orogeny caused by collision of North America and Avalon and closing of remaining part of Iapetus Ocean	359 million years ago
	Salt and gypsum deposited in evaporite basins	
	Erosion of Taconic Mountains; Queenston delta forms Taconian orogeny caused by closing of western part of Iapetus Ocean and collision between North America and volcanic island arc	458 million years ago
(A)	Widespread deposition over most of New York along edge of Iapetus Ocean	
	Rifting and initial opening of Iapetus Ocean Erosion of Grenville Mountains Grenville orogeny: metamorphism of bedrock now exposed in the Adirondacks and Hudson Highlands	

(O) Mastodont — Beluga Whale
(P) Cooksonia — Aneurophyton
(Q) Naples Tree
(R) Bothriolepis
(S) Condor
(T) Lichenaria
(U) Cystiphyllum — Pleurodictyum
(V) Maclurites
(W) Platyceras
(X) Eospirifer
(Y) Mucrospirifer
(Z)

ESC/BW/TN (2009)

Appendix B: Earth Science Reference Tables

Inferred Properties of Earth's Interior

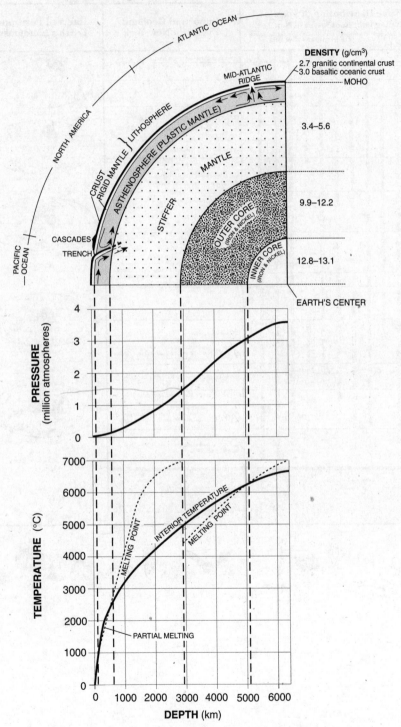

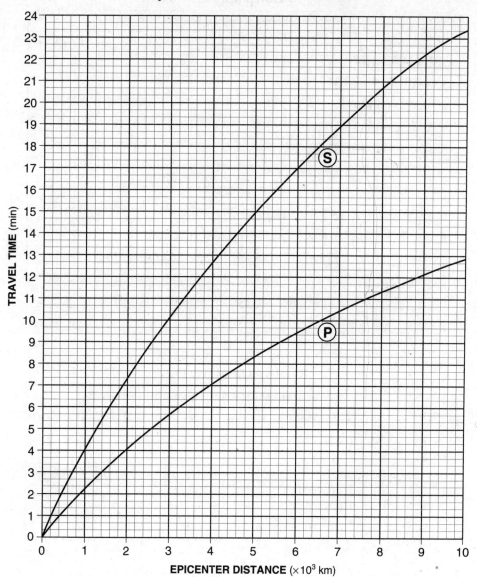

Appendix B: Earth Science Reference Tables

Dewpoint (°C)

Dry-Bulb Temperature (°C)	Difference Between Wet-Bulb and Dry-Bulb Temperatures (C°)															
	0	1	2	3	4	5	6	7	8	9	10	11	12	13	14	15
−20	−20	−33														
−18	−18	−28														
−16	−16	−24														
−14	−14	−21	−36													
−12	−12	−18	−28													
−10	−10	−14	−22													
−8	−8	−12	−18	−29												
−6	−6	−10	−14	−22												
−4	−4	−7	−12	−17	−29											
−2	−2	−5	−8	−13	−20											
0	0	−3	−6	−9	−15	−24										
2	2	−1	−3	−6	−11	−17										
4	4	1	−1	−4	−7	−11	−19									
6	6	4	1	−1	−4	−7	−13	−21								
8	8	6	3	1	−2	−5	−9	−14								
10	10	8	6	4	1	−2	−5	−9	−14	−28						
12	12	10	8	6	4	1	−2	−5	−9	−16						
14	14	12	11	9	6	4	1	−2	−5	−10	−17					
16	16	14	13	11	9	7	4	1	−1	−6	−10	−17				
18	18	16	15	13	11	9	7	4	2	−2	−5	−10	−19			
20	20	19	17	15	14	12	10	7	4	2	−2	−5	−10	−19		
22	22	21	19	17	16	14	12	10	8	5	3	−1	−5	−10	−19	
24	24	23	21	20	18	16	14	12	10	8	6	2	−1	−5	−10	−18
26	26	25	23	22	20	18	17	15	13	11	9	6	3	0	−4	−9
28	28	27	25	24	22	21	19	17	16	14	11	9	7	4	1	−3
30	30	29	27	26	24	23	21	19	18	16	14	12	10	8	5	1

Relative Humidity (%)

Dry-Bulb Temperature (°C)	Difference Between Wet-Bulb and Dry-Bulb Temperatures (C°)															
	0	1	2	3	4	5	6	7	8	9	10	11	12	13	14	15
−20	100	28														
−18	100	40														
−16	100	48														
−14	100	55	11													
−12	100	61	23													
−10	100	66	33													
−8	100	71	41	13												
−6	100	73	48	20												
−4	100	77	54	32	11											
−2	100	79	58	37	20	1										
0	100	81	63	45	28	11										
2	100	83	67	51	36	20	6									
4	100	85	70	56	42	27	14									
6	100	86	72	59	46	35	22	10								
8	100	87	74	62	51	39	28	17	6							
10	100	88	76	65	54	43	33	24	13	4						
12	100	88	78	67	57	48	38	28	19	10	2					
14	100	89	79	69	60	50	41	33	25	16	8	1				
16	100	90	80	71	62	54	45	37	29	21	14	7	1			
18	100	91	81	72	64	56	48	40	33	26	19	12	6			
20	100	91	82	74	66	58	51	44	36	30	23	17	11	5		
22	100	92	83	75	68	60	53	46	40	33	27	21	15	10	4	
24	100	92	84	76	69	62	55	49	42	36	30	25	20	14	9	4
26	100	92	85	77	70	64	57	51	45	39	34	28	23	18	13	9
28	100	93	86	78	71	65	59	53	47	42	36	31	26	21	17	12
30	100	93	86	79	72	66	61	55	49	44	39	34	29	25	20	16

Appendix B: Earth Science Reference Tables

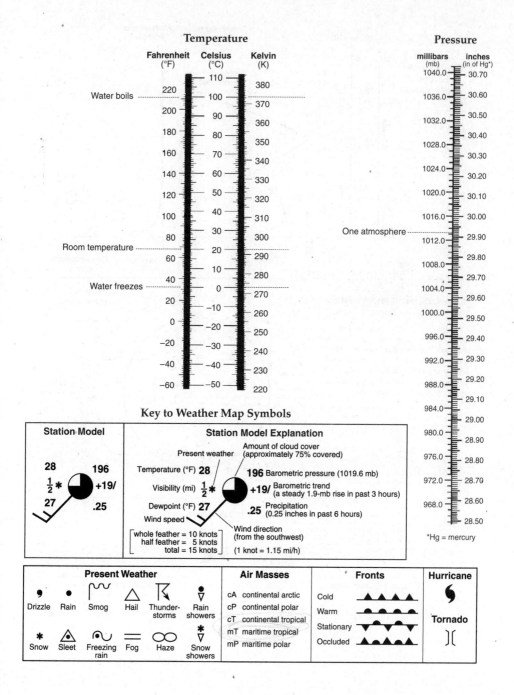

Appendix B: Earth Science Reference Tables

Selected Properties of Earth's Atmosphere

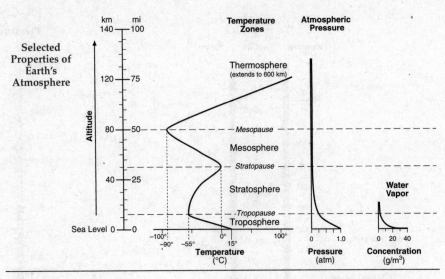

Planetary Wind and Moisture Belts in the Troposphere

The drawing on the right shows the locations of the belts near the time of an equinox. The locations shift somewhat with the changing latitude of the Sun's vertical ray. In the Northern Hemisphere, the belts shift northward in the summer and southward in the winter.

(Not drawn to scale)

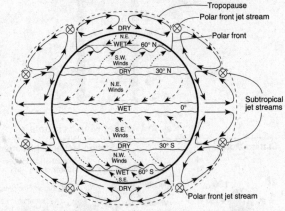

Electromagnetic Spectrum

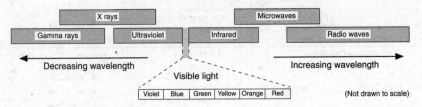

(Not drawn to scale)

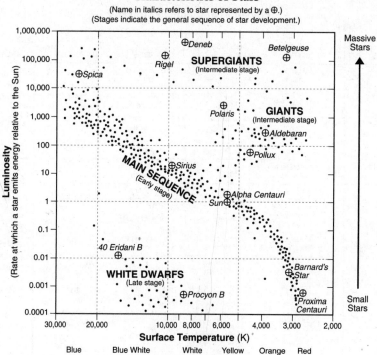

Characteristics of Stars

(Name in italics refers to star represented by a ⊕.)
(Stages indicate the general sequence of star development.)

Solar System Data

Celestial Object	Mean Distance from Sun (million km)	Period of Revolution (d=days) (y=years)	Period of Rotation at Equator	Eccentricity of Orbit	Equatorial Diameter (km)	Mass (Earth = 1)	Density (g/cm³)
SUN	—	—	27 d	—	1,392,000	333,000.00	1.4
MERCURY	57.9	88 d	59 d	0.206	4,879	0.06	5.4
VENUS	108.2	224.7 d	243 d	0.007	12,104	0.82	5.2
EARTH	149.6	365.26 d	23 h 56 min 4 s	0.017	12,756	1.00	5.5
MARS	227.9	687 d	24 h 37 min 23 s	0.093	6,794	0.11	3.9
JUPITER	778.4	11.9 y	9 h 50 min 30 s	0.048	142,984	317.83	1.3
SATURN	1,426.7	29.5 y	10 h 14 min	0.054	120,536	95.16	0.7
URANUS	2,871.0	84.0 y	17 h 14 min	0.047	51,118	14.54	1.3
NEPTUNE	4,498.3	164.8 y	16 h	0.009	49,528	17.15	1.8
EARTH'S MOON	149.6 (0.386 from Earth)	27.3 d	27.3 d	0.055	3,476	0.01	3.3

Appendix B: Earth Science Reference Tables

Properties of Common Minerals

LUSTER	HARD-NESS	CLEAVAGE	FRACTURE	COMMON COLORS	DISTINGUISHING CHARACTERISTICS	USE(S)	COMPOSITION*	MINERAL NAME
Metallic luster	1–2	✓		silver to gray	black streak, greasy feel	pencil lead, lubricants	C	Graphite
Metallic luster	2.5	✓		metallic silver	gray-black streak, cubic cleavage, density = 7.6 g/cm³	ore of lead, batteries	PbS	Galena
Metallic luster	5.5–6.5		✓	black to silver	black streak, magnetic	ore of iron, steel	Fe_3O_4	Magnetite
Metallic luster	6.5		✓	brassy yellow	green-black streak, (fool's gold)	ore of sulfur	FeS_2	Pyrite
Either	5.5–6.5 or 1		✓	metallic silver or earthy red	red-brown streak	ore of iron, jewelry	Fe_2O_3	Hematite
Nonmetallic luster	1	✓		white to green	greasy feel	ceramics, paper	$Mg_3Si_4O_{10}(OH)_2$	Talc
Nonmetallic luster	2		✓	yellow to amber	white-yellow streak	sulfuric acid	S	Sulfur
Nonmetallic luster	2	✓		white to pink or gray	easily scratched by fingernail	plaster of paris, drywall	$CaSO_4 \cdot 2H_2O$	Selenite gypsum
Nonmetallic luster	2–2.5	✓		colorless to yellow	flexible in thin sheets	paint, roofing	$KAl_3Si_3O_{10}(OH)_2$	Muscovite mica
Nonmetallic luster	2.5	✓		colorless to white	cubic cleavage, salty taste	food additive, melts ice	NaCl	Halite
Nonmetallic luster	2.5–3	✓		black to dark brown	flexible in thin sheets	construction materials	$K(Mg,Fe)_3$ $AlSi_3O_{10}(OH)_2$	Biotite mica
Nonmetallic luster	3	✓		colorless or variable	bubbles with acid, rhombohedral cleavage	cement, lime	$CaCO_3$	Calcite
Nonmetallic luster	3.5	✓		colorless or variable	bubbles with acid when powdered	building stones	$CaMg(CO_3)_2$	Dolomite
Nonmetallic luster	4	✓		colorless or variable	cleaves in 4 directions	hydrofluoric acid	CaF_2	Fluorite
Nonmetallic luster	5–6	✓		black to dark green	cleaves in 2 directions at 90°	mineral collections, jewelry	(Ca,Na)(Mg,Fe,Al) $(Si,Al)_2O_6$	Pyroxene (commonly augite)
Nonmetallic luster	5.5	✓		black to dark green	cleaves at 56° and 124°	mineral collections, jewelry	$CaNa(Mg,Fe)_4(Al,Fe,Ti)_3$ $Si_6O_{22}(O,OH)_2$	Amphibole (commonly hornblende)
Nonmetallic luster	6	✓		white to pink	cleaves in 2 directions at 90°	ceramics, glass	$KAlSi_3O_8$	Potassium feldspar (commonly orthoclase)
Nonmetallic luster	6	✓		white to gray	cleaves in 2 directions, striations visible	ceramics, glass	$(Na,Ca)AlSi_3O_8$	Plagioclase feldspar
Nonmetallic luster	6.5		✓	green to gray or brown	commonly light green and granular	furnace bricks, jewelry	$(Fe,Mg)_2SiO_4$	Olivine
Nonmetallic luster	7		✓	colorless or variable	glassy luster, may form hexagonal crystals	glass, jewelry, electronics	SiO_2	Quartz
Nonmetallic luster	6.5–7.5		✓	dark red to green	often seen as red glassy grains in NYS metamorphic rocks	jewelry (NYS gem), abrasives	$Fe_3Al_2Si_3O_{12}$	Garnet

*Chemical symbols:
Al = aluminum Cl = chlorine H = hydrogen Na = sodium S = sulfur
C = carbon F = fluorine K = potassium O = oxygen Si = silicon
Ca = calcium Fe = iron Mg = magnesium Pb = lead Ti = titanium

✓ = dominant form of breakage

Glossary of Earth Science Terms

ablation The melting and evaporation of glacial ice.

ablation debris Material on the surface of a glacier that was suspended in the ice but is now exposed due to ablation.

abrasion The breaking down of rocks by rubbing together.

absolute brightness Any measure of the intrinsic (not dependent on the position or distance of the observer) brightness or luminosity of a celestial object.

absolute humidity The number of grams of water vapor in a cubic meter of air.

absolute magnitude The intrinsic brightness or luminosity of a celestial object expressed as the magnitude that a star would appear to have if it were 10 parsecs away.

absorption spectrum A spectrum in which certain regions are dark or dim because those wavelengths of light were absorbed by atoms between the source and the observer.

abyssal plain Wide, flat areas adjacent to the mid-ocean ridges.

accretion The sticking together of particles to form larger particles.

acid rain Rain that is acidic due to substances that have become dissolved in it.

adiabatic A change that does not involve the addition or removal of heat.

adiabatic lapse rate The rate at which the temperature of air changes due to a decrease in air pressure rather than the addition or removal of heat.

aerobic atmosphere An atmosphere with sufficient oxygen to support aerobic respiration by living things.

agent of erosion A medium set in motion by gravity that can exert a force on sediments causing them to be moved.

air current Air that is moving in a vertical direction.

air mass A large region of air with uniform characteristics at any given level.

air pressure The force exerted on a unit of area by the air.

alluvial fan A flat, fan-shaped mound of sediment deposited where a stream slows down because its gradient suddenly becomes less steep (i.e., where a mountain valley empties onto a plain).

altitude 1. The angle between the horizon and an object seen in the sky with the observer at the vertex. 2. Distance above the earth's surface.

anemometer An instrument that measures wind speed.

angle of repose The greatest angle at which a pile of sediments is stable.

angstrom (Å) A unit of length. There are 10^8 angstroms in a centimeter.

angular unconformity A place where sedimentary layers at one angle are in contact with sedimentary layers at a different angle.

annular eclipse A solar eclipse in which the moon's shadow does not actually reach the surface of the Earth; to an observer on Earth, the sun's disk is obscured by the Moon except for a narrow ring around the perimeter.

anticline An upward, arched fold.

anticyclone A flow of air that is clockwise and outward from a high pressure center.

apparent brightness The brightness of a celestial object as seen by human eyes on Earth.

apparent daily motion The apparent motion of all celestial objects along a circular path at a constant rate of 15° per hour, or one complete circle every day.

apparent magnitude The apparent brightness of a celestial object expressed in the astronomical magnitude scale; the larger the number, the fainter the star.

aquiclude A rock or sediment layer that is impermeable to water.

Glossary

aquifer A rock or sediment layer in which the pore spaces are saturated with water.

arrival time The precise time at which a seismic wave reaches a seismograph and is recorded.

artesian well A well in which water rises to the surface unassisted.

asteroid Solid body having no atmosphere that orbits the sun.

asthenosphere A region of the upper mantle between 100 and 350 kilometers in depth that behaves like a fluid.

astronomical unit The average distance from the Earth to the Sun; 149,598,000 kilometers.

atmosphere The thin shell of gases bound to the Earth by gravity; it consists of gases, water, ice, dust, and other particles.

atmospheric variables The characteristics of the atmosphere that change.

aurora Glowing, often shimmering light forms seen near the poles that are caused by radiation from high altitude air molecules excited by collisions with particles from the Sun.

axis of rotation The central line around which all parts of an object move during rotation.

banding A structure in a metamorphic rock of nearly parallel bands of different textures or minerals; i.e., the nearly parallel bands of light and dark minerals seen in gneiss.

barrier bar A sandbar running parallel to a shoreline that shelters the water behind it from waves.

batholith A large, irregularly shaped pluton covering an area greater than 75 square kilometers.

beach A strip of loose materials along the shore of a body of water formed and washed by waves.

bed A layer of sediment with a particular physical or structural character.

bedrock A general term for the solid rock underlying the soil or loose material covering the Earth's surface.

bench mark A permanent metal marker set into the ground that gives the exact location and precisely measured elevation of that point.

big bang theory The idea that the universe started out with all of its matter in a small volume and then expanded outward in all directions.

body waves Seismic waves that travel through the body of the earth; P-waves and S-waves are body waves.

bright-line spectrum A spectrum consisting of bright, narrow lines of color. Each color corresponds to a specific wavelength of light emitted by an atom when an electron jumps from a specific higher energy level to a specific lower energy level. Since the atoms of each element have a unique structure, each produces a unique sequence of bright lines.

capillary fringe A narrow region just above the water table in which narrow pore spaces are filled with water that is drawn up by capillary action.

capillarity The action by which a fluid, such as water, is drawn upward in narrow spaces as a result of surface tension.

carbonaceous chondrite A stony meteorite containing hydrated claylike minerals and a great variety of organic compounds.

carbonation A chemical reaction in which carbon dioxide combines with a substance.

celestial meridian A meridian on the celestial sphere that passes through the zenith of a given place and terminates at the celestial poles.

celestial objects Objects that can be seen in the sky, but that are beyond the Earth's atmosphere.

celestial sphere An imaginary sphere of infinite radius around Earth's center upon whose "inner surface" all celestial objects appear to be projected. The apparent dome of the visible sky forms half of the celestial sphere.

cementation The binding together of sediment particles by a cementing material.

chaos theory The idea that small, unpredictable changes can cause large-scale changes in a system over time.

chemical sediments Particles (usually crystals) that crystallized out of solutions at or near the Earth's surface.

chemical weathering Processes that break down rock by changing its chemical composition.

cinder cone volcano A narrow, steep-sided volcano formed from the accumulation of tephra around the vent.

clastic sediments Rock or mineral fragments formed by the breakdown of rock due to weathering.

cleavage The tendency of a mineral to break along flat surfaces parallel to atomic planes in its crystalline structure.

climate The weather in a region averaged over a long period of time.

closed universe A model of the universe of fixed volume with curved boundaries. If closed, the universe will eventually contract and give rise to another big bang.

cloud base The elevation of the base of a cloud; the elevation at which condensation begins.

cloud ceiling The elevation of the base of a layer of cloud.

cloud cover The fraction of the sky obscured by clouds.

coastal zone The part of a continent that is adjacent to the ocean; it includes features such as the coast, shoreline, beaches, marshes, estuaries, lagoons, and deltas.

comet Bodies composed of meteoroids embedded in ice that revolve around the Sun in highly elliptical orbits.

comet coma The cloud of gas surrounding a comet's nucleus when ice in the nucleus sublimes as the comet approaches the Sun.

comet nucleus The solid center of a comet thought to consist of meteoroids embedded in ice.

comet tail The elongation of the cloud of gas surrounding a comet caused by collisions with particles emitted by the Sun.

compaction The reduction in volume or thickness of a sediment layer due to overlying weight or pressure.

composite cone volcano A large symmetrical volcano consisting of alternating layers of lava flows and tephra.

compound A substance that consists of two or more elements joined together in a definite proportion and have properties different from the elements that compose it.

compression Forces acting toward each other along the same line of action causing rock to be squeezed.

conchoidal fracture A tendency to break along a curved, concave surface that looks somewhat like a mussel shell.

condensation The process by which a gas changes into a liquid.

condensation nuclei Particles suspended in the air that provide a surface upon which condensation can occur.

conic section A family of figures obtained by cutting a circular cone with a plane.

conservation The act of decreasing the consumption of a natural resource, usually by more efficient use.

constellation A group of stars in an area of the sky that form imaginary patterns, supposedly in the form of an animal, person, or thing.

contact metamorphism Changes in rock that result from the extreme heat produced by contact with magma or lava.

continental drift The theory that the continents drift slowly across the Earth's surface, sometimes colliding and breaking into pieces.

continental glacier Immense masses of ice that form at high latitudes where temperatures are always cold and that spread out over the entire land surface rather than being confined to valleys.

Glossary

continental rise A gently sloping feature between the continental slope and the abyssal plain.

continental shelf The shallow part of the ocean floor adjacent to the coastal zone.

continental slope The portion of the ocean floor that slopes steeply from the continental shelf toward the abyssal plain.

continuous spectrum A spectrum that contains light of all colors.

contour interval The difference in elevation between two adjacent contour lines.

convection cell A circular pattern of motion in which warm, moist air rises, expands outward, cools, and sinks back to the surface.

convection current A fluid current set in motion by differences in density within the fluid.

convergent boundary Places where edges of adjacent plates are colliding.

coordinates Two numbers that describe the position of any point on a map in terms of the intersection of two lines.

coordinate system A system for assigning two numbers to every point on a surface.

core The innermost zone of the Earth's interior.

Coriolis effect A deflection of motion from a straight-line path due to the rotation of the Earth.

correlation The process of determining that rock layers in different areas are the same age.

cosmic background radiation A remnant of radiation left over from the original big bang that fills the universe.

cosmology The study of the universe.

crevasse A deep, nearly vertical fissure in glacial ice.

cross-bedding Layers of sediment lying at an angle to the plane in which sediment layers are normally deposited.

crystal A solid whose internal structural pattern is expressed as plane faces that can be seen with the unaided eye.

crystalline Having a definite internal structural pattern.

crystallization The process of forming crystals.

cutoff A new channel formed when a stream cuts through a narrow strip of land between adjacent meanders.

cyclone A flow of air in which motion is counterclockwise and toward a low pressure center.

decarbonation A chemical reaction in which carbon dioxide is given off.

decay product The new element formed when a radioisotope decays.

deferent In Ptolemy's geocentric model of the universe, a circle with the Earth near its center. The center of a planet's epicycle moves around the circumference of a deferent.

deflation The pick up and removal of fine sediments from the Earth's surface by wind.

deflation hollow A shallow depression formed by the removal of sediments by wind.

degassing The process by which water is released from the molecules of minerals when heated to high temperatures.

dehydration A chemical reaction in which water is given off.

delta A flat, fan-shaped mound of sediment deposited where a stream enters a body of standing water and slows down.

density The amount of matter in a given amount of space; usually expressed as the mass per unit of volume of a substance.

desertification The rapid development of deserts caused by the impact of human activities.

desert pavement A continuous layer of sediments too heavy to be moved by wind that remains after finer sediments have been removed by wind, and shields underlying materials from further deflation.

dew Droplets of water that condense on ground surfaces due to radiative cooling of the ground overnight.

dew point The temperature at which the water vapor in air will begin to condense.

differentiation The process by which planets develop concentric layers or zones due to density differences.

dike A pluton that cuts across existing rock layers.

disconformity An irregular erosional surface between parallel layers of rock.

divergent boundary Places where edges of adjacent plates are spreading apart.

Doppler effect The shift in wavelength of a spectrum line from its normal position due to relative motion between the source and the observer.

drainage basin The area from which precipitation drains into a stream or system of streams.

drift A general term for all sediments transported and deposited by a glacier.

drizzle Very fine droplets of water falling slowly and close together.

drumlin Low mounds of till having an elongated teardrop shape formed when a glacier rides over a previously deposited pile of sediment.

dry bulb temperature The temperature recorded by a thermometer whose bulb is kept dry.

ductile deformation A permanent change in rock that occurs when stress exceeds the strength of a rock's internal bonds.

dune A mound of sand deposited by wind.

earthflow The slow, downhill movement of a water-saturated layer of soil and vegetation.

earthquake A sudden trembling of the ground.

eccentricity The flatness of an ellipse, expressed as the ratio of the distance between the foci of the ellipse to the length of its major axis.

ecliptic The apparent annual path of the Sun around the sky; the intersection of the plane of Earth's orbit and the celestial sphere.

elastic deformation A temporary change in which rock strains due to stress but returns to its original condition after the stress is removed.

electromagnetic spectrum A continuum in which electromagnetic waves are arranged in order of wavelength.

electromagnetic wave An oscillation in the electromagnetic field surrounding a particle.

element A substance that cannot be broken down into simpler substances by ordinary chemical means.

ellipse A closed figure obtained by cutting a circular cone with a plane; its appearance is that of a flattened circle.

El Niño A current of warm water flowing southward along the coast of Ecuador, which about every seven to ten years, extends to the coast of Peru, where it interrupts the upwelling of nutrient-rich water, causing plankton and fish to die.

end moraine A pile of till that forms at the melting edge of a glacier.

eon The largest unit of geologic time.

epicenter The point on the Earth's surface directly above the focus of an earthquake.

epicycle In Ptolemy's geocentric model of the universe, the circular orbit of a planet. The center of an epicycle travels around the circumference of another circle called the deferent.

epoch The unit of geologic time into which periods are divided.

equator A circle on which all points are midway between the Earth's poles.

equatorial diameter Diameter of the Earth measured from a point on the Equator through the center of the Earth to the Equator (12,756 kilometers).

equinox Either of the two times during a year when the Sun crosses the celestial Equator and when the length of day and night are approximately equal.

Glossary

era The unit of geologic time into which eons are divided.

erosion Any process that transports sediments.

erratic An isolated boulder deposited by a glacier.

escape velocity The smallest velocity a body must attain in order to escape from the gravitational attraction of another body.

evaporite Sediments, rocks, or minerals formed when water containing dissolved minerals evaporates.

evapotranspiration The combined processes of evaporation and transpiration, the two mechanisms by which water is returned to the atmosphere.

evolution The idea that current life-forms have developed from earlier, different life-forms.

exfoliation The scaling off, or peeling, of successive layers from the surface of rocks.

extrusive igneous rock A rock formed by the rapid cooling of lava at the surface of the Earth.

faulting A sudden movement of rock along planes of weakness in the Earth's crust called faults.

felsic Rocks rich in the minerals feldspar and silica (quartz).

field A region of space in which a measurable quantity is present at every point.

field map Represents any quantity that varies in a region of space.

floodplain A broad, flat strip of land adjacent to a stream consisting of sediments deposited when the stream is in flood.

focus The point where rock first breaks or moves in an earthquake.

focus of an ellipse One of two points along the major axis of an ellipse with the characteristic that the total distance from one focus to any point on the ellipse and back to the other focus is always the same.

fog A cloud whose base is at ground level.

fold A bend or warp in layered rock.

fossil Any remains, trace, or imprint of a living thing that has been preserved in the Earth's crust.

fossil fuel A fuel derived from the remains of once-living things; e.g., coal, petroleum, and natural gas.

Foucault pendulum A freely swinging pendulum whose path appears to change direction relative to the Earth's surface in a predictable manner due to the Earth's rotation.

fracture The tendency not to break along any particular direction.

fragmental sedimentary rock A rock composed of rock fragments cemented or compacted together in a solid mass.

frame of reference A specific point in time or space that serves as the basis of comparison for describing change; in a coordinate system, the coordinate axes, in terms of which position is specified.

frequency The number of waves that pass a given point in a unit of time.

front The boundary between two adjacent air masses.

frost Ice crystals that form on ground surfaces by sublimation due to radiative cooling of the ground overnight.

frost action The breaking down of rock by the force exerted by water expanding as it freezes into ice.

galaxy A system consisting of hundreds of billions of stars.

gas giant Planets that have thick atmospheres surrounding a small, rocky core.

geocentric model A model of the universe in which the Earth is at the center and all other objects revolve around the Earth.

geologic column The correlation of rocks worldwide into a single sequence showing their relative ages.

geologic time scale The division of the Earth's past into a sequence of time units.

geothermal gradient The rate at which temperatures within Earth rise with depth.

glacial advance The forward and downslope movement of a glacier that results when accumulation exceeds ablation.

glacial basin Depressions formed when fast-moving glaciers remove large quantities of rock material plucked from the bedrock surface.

glacial polish A bedrock surface smoothed by abrasion as a glacier flows over it.

glacial retreat A decrease in the length of a glacier that results from ablation exceeding accumulation.

glacial striations Long, narrow grooves inscribed in bedrock by rock fragments embedded in a glacier as the glacier moves over the bedrock.

glacial till Unsorted sediment deposited by a glacier.

glacier A solid mass of ice formed by the compaction of accumulated snow that flows downslope and outward under the influence of gravity.

glaze Rain that forms a layer of ice as soon as it comes into contact with below-freezing surfaces.

global wind belt A region of prevailing winds caused by a difference in air pressure between adjacent regions on a global scale.

graded bedding A layer of sediments in which the particle size changes gradually from large on the bottom to small on the top.

gradient The rate at which a field changes; usually measured in terms of change in field value per unit of distance.

gravitational differentiation The process by which the material from which the solar system formed became separated into layers according to density, because atoms of lighter elements do not exert as strong a gravitational attraction as atoms of heavier elements.

gravity The force of attraction that exists between all matter.

great circle A circle that bisects the Earth.

greenhouse effect The process by which solar radiation that is re-radiated at a longer wavelength is absorbed by carbon dioxide, water vapor, and other gases in the atmosphere rather than escaping into space.

ground moraine A thin, widespread layer of till deposited beneath a glacier.

groundmass A mass of fine mineral crystals surrounding the large, isolated phenocrysts in a porphyry.

gully A long, narrow trough eroded in soil by running water.

hail Balls of ice with an internal structure of concentric layers of ice and snow formed when ice pellets fall through turbulent air and are alternately carried through layers of air that are above and below freezing.

half-life The time required for half of a mass of radioactive atoms to decay.

hanging valley A valley of a tributary glacier whose floor is above, or hanging over, the floor of the main glacier that it feeds.

hardness A mineral's ability to resist being scratched.

heliocentric model A model of the universe in which the Sun is at the center of our solar system and the planets revolve around the Sun.

Hertzsprung-Russell Diagram A plot of the luminosity versus spectral type for a group of stars; also called the H-R Diagram.

horizon The visible line of demarcation between land or sea and sky.

horizontal sorting The gradual change in the size of sediments deposited horizontally along the bed of a stream as the water slows down.

humidity The amount of water vapor in the air.

Glossary

humus The dark, highly decomposed organic matter in soil.

hurricane Large, cyclonic storms in which the winds exceed 119 kilometers per hour.

hydration A chemical reaction in which water combines with a substance.

hydrogen bonding A weak bond between adjacent water molecules due to the attraction between the slightly positive end of one water molecule and the slightly negative end of another water molecule.

hydrolysis A reaction in which water separates into hydrogen ions and hydroxide ions that replace ions in a mineral.

hydrosphere All of the water that rests on the lithosphere; it includes all oceans, lakes, streams, groundwater, and ice.

ice age A period of time during which conditions exist that cause ice sheets more than 1 million square kilometers in area and thousands of meters thick to develop on nonpolar continents.

iceberg A mass of floating ice that detached from a glacier into a body of water.

igneous intrusion A mass of magma that penetrates an opening in an existing rock and solidifies.

igneous rock Rock formed by the cooling and crystallization of molten minerals (magma or lava).

immature soil A soil in which boundaries between forming horizons are indistinct.

impact crater A circular depression formed on a surface by the impact of a projectile, such as a meteorite or a comet.

in solution Sediments that are transported as a consequence of being dissolved in the water flowing in a stream.

index fossil A fossil of an organism that had distinctive body features and was abundant over a wide geographical area, but only existed for a short period of time.

infiltrate The process by which water seeps into the ground and through interconnected pore spaces.

inorganic Non-living and not derived from a living thing.

insolation Solar radiation that reaches the surface of the Earth; short for incoming solar radiation.

interglacial Time periods during an ice age when glaciers retreat or even disappear.

international date line The 180° meridian; it marks the transition from one date to another.

intrusive igneous rock A rock formed by the slow cooling of magma within the earth.

ion An atom or molecule in which the electrical charges are unbalanced resulting in a net positive or negative electrical charge.

isobar A line connecting points of equal air pressure.

isoline Lines connecting points of equal field value.

isostasy The idea that the Earth's crust floats on hot fluid rock in the mantle.

isosurface Surfaces on which every point has the same field value.

isotherm A line connecting points of equal temperature on a field map.

isotope Varieties of the same element that differ slightly in mass.

joint A fracture in a rock along which movement has not occurred.

kettle A depression formed when a block of ice buried in till melts.

key bed Well-defined, easily identifiable rock layers that have distinctive characteristics or fossil content that allow(s) them to be used in correlation.

Kuiper Belt A thick belt of debris left over from the formation of the outer planets, extending from Neptune's orbit (29 AU) to the outer edge of

Pluto's orbit (50 AU). Considered the probable source of short-period comets, over 60 objects with diameters over 100 kilometers have been detected orbiting the Sun within this region. Pluto and its moon Charon are considered Kuiper Belt objects.

laccolith A pluton formed when magma intrudes between rock layers and pushes upward and solidifies creating a rock structure that has a flat bottom and an arched top.

lagoon The sheltered body of water between a barrier bar and the mainland.

land breeze A local wind blowing from land toward an adjacent body of water because nighttime cooling increases the air pressure over the land.

landform Physical features of Earth's surface that have a characteristic shape and are produced by natural processes.

landscape A region in which the landforms are related by their structure, the processes that formed them, and the way in which they developed.

landscape region A grouping of landscapes with similar relief, stream patterns and soil associations.

latent heat of fusion The amount of heat energy given off or absorbed in causing a change between the liquid and solid states that does not result in a change in temperature.

latent heat of vaporization The amount of heat energy given off or absorbed in causing a change between the liquid and gaseous states that does not result in a change in temperature.

lateral moraine A mound of till deposited along the sides of a glacier.

latitude The angular distance north or south of the equator.

Laurentide ice sheet The glacier that advanced several times across the northern United States during the Pleistocene epoch.

lava Magma (molten rock) that has emerged onto the Earth's surface.

lava plateau A flat sheet of igneous rock formed when a very fluid lava erupts from a long fissure instead of from a central vent.

leeward The side of a feature facing away from the wind.

lens A transparent object of regular form that changes the direction of travel of light waves that pass through it.

levee A low, thick, ridgelike deposit along the banks of a stream consisting of coarse sediments deposited when a stream overflows its banks.

light year The distance light travels in one year; roughly 9.5 trillion kilometers, or 6 trillion miles.

lithification The process of converting sediments into coherent, solid rock.

lithosphere Earth's solid outer layer of soil and solid brittle rock extending from the surface to a depth of about 100 kilometers; it includes the crust and the upper mantle.

loess A fine, buff-colored sediment deposited by wind.

longitude The angular distance east or west of the prime meridian.

longitudinal waves Waves in which the medium oscillates parallel to the direction of wave motion.

longshore current An ocean current flowing parallel to a shore caused by the approach of waves at an angle to the coast.

luminosity The total amount of energy radiated by a star in 1 second.

lunar eclipse An event in which the Moon passes into the Earth's shadow during the full moon phase.

luster The way light reflects from the surface of a mineral.

mafic Rocks rich in minerals containing magnesium and Fe (iron).

magma Molten rock beneath the Earth's surface.

magma chamber A pocket of molten magma in the crust or upper mantle.

Glossary

magnetosphere The region surrounding Earth that is influenced by its magnetic field

magnitude scale The astronomical system for measuring the brightness of a star; the higher the magnitude, the fainter the star.

main sequence A diagonal line across the H-R diagram on which the majority of stars lie; the main phase of the life of a star.

mantle The zone of the Earth's interior between the crust and the core.

map A model of the Earth's surface.

map scale The ratio between distance on a map and actual distance on the ground; usually expressed as a ratio.

mass The amount of matter in an object.

mass wasting The downhill movement of sediments under the direct influence of gravity.

mature soil A soil in which distinct horizons have formed.

meander A curving loop in a stream channel.

meridian Semicircles on which all points have the same longitude.

mesopause The boundary between the mesosphere and the thermosphere.

mesosphere The layer of the atmosphere ranging from 30 to 50 kilometers; temperatures in this layer decrease with altitude.

metamorphic rock Rocks that form as a result of physical and chemical changes in existing rocks.

metamorphism The processes that change the physical and chemical properties of existing rocks.

meteor The streak of light seen in the sky as a meteoroid is vaporized by frictional heating as it falls through the atmosphere.

meteorite The portion of a meteoroid that survives to strike the ground.

meteoroid A chunk of matter moving through space.

mid-ocean ridge A chain of undersea mountains running through the center of an ocean.

Milky Way galaxy The galaxy in which our solar system is located.

mineral A naturally occurring, inorganic, crystalline solid with a composition that can be expressed by a chemical symbol or a formula.

mineral composition The minerals present in a rock.

mixture Consisting of one or more substances that retain their characteristic properties and can be separated by relatively simple means.

modified Mercalli scale A system that measures the strength of an earthquake based upon perception of motion by human observers and damage to structures built by humans.

Moho The boundary between the dense rock of the mantle and the less dense rock of the crust. It was named for the seismologist Andrija Mohorovicic, who recognized its existence after analyzing seismic wave behavior inside the Earth.

Mohorovicic discontinuity The boundary between the Earth's crust and mantle. It marks the interface between rocks within the Earth of two different average densities.

monocline A fold in which only one side of the fold is inclined.

moraine A deposit of unsorted sediments deposited directly from glacial ice.

mountain Any part of Earth's crust that projects at least 300 m (1,000 ft) above the surrounding land, has a limited summit area (as opposed to a plateau), steep sides, and considerable bare-rock surface.

mudflow The rapid downhill flow of a mixture of rock fragments, soil, and water.

natural selection The process by which organisms that are better suited to their environment survive and reproduce while poorly suited organisms do not.

Glossary

neap tide A tidal cycle in which there is very little difference between the water level at high and low tides; caused by the canceling out effect of the Moon's gravity acting at right angles to the Sun's.

nebula A cloud of interstellar matter.

neutrino A fundamental particle produced in certain nuclear reactions that has no charge and no mass, but which does have momentum.

neutron star An incredibly dense (10^{16}-10^{18} kg/m^3) star whose interior is composed mostly of neutrons.

nonconformity A place where sedimentary rock lies directly on another type of rock, such as igneous or metamorphic rock.

nuclear fusion A nuclear reaction in which the nuclei of atoms combine; during fusion, some of the mass of the combining nuclei is converted into energy.

numerical forecasting Weather forecasting based upon mathematical models of the physical behavior of the atmosphere.

oasis A place where deflation has lowered the surface in a desert to the water table allowing plants to take root and grow.

oblate spheroid A slightly flattened sphere; the Earth's shape.

ocean-floor spreading The idea that the ocean floors were spreading sideways away from the mid-ocean ridges.

Oort Cloud A vast shell of icy objects (i.e., comets) surrounding the solar system between 50,000 and 100,000 AU; thought to be the source of long-period comets.

open universe A model of the universe with infinite volume and no boundaries. If open, the universe will continue expanding forever.

orbit The path along which a satellite revolves around a primary.

orbital velocity The smallest velocity a body must attain in order to enter a stable orbit around another body.

ordinary well A hole dug in the ground so that it penetrates the water table in which water accumulates and can be pumped to the surface.

organic sediments Particles produced by the life activities of plants or animals.

origin time The time at which an earthquake occurred.

original horizontality, law of The observation that sediments deposited in water form flat level layers.

orographic effect The net increase in temperature of a parcel of air that is forced over a mountain range due to adiabatic cooling on the windward side followed by adiabatic heating on the leeward side.

outgassing The release of gases and water vapor from molten rocks, leading to the formation of Earth's atmosphere and oceans.

outwash plain A flat plain next to the leading edge of a glacier formed by the deposition of sediments from meltwater streams.

oxbow lake A crescent-shaped body of water formed when a cutoff isolates the water in a meander from the main channel of the stream.

oxidation A chemical reaction in which a substance combines with oxygen.

oxidizing atmosphere An atmosphere that contains enough oxygen to cause elements and compounds to oxidize, but not enough to support aerobic life.

ozone A gas whose molecules consist of three atoms of oxygen joined together.

P-waves Longitudinal seismic waves produced by an earthquake that travel the fastest and are therefore the first to be recorded by a seismograph.

Pacific ring of fire A narrow zone of active volcanoes surrounding the Pacific Ocean.

paleomagnetism A record of the orientation of the Earth's magnetic field preserved in the crystalline structure of igneous rock containing magnetic minerals.

parallel Circles on which all points have the same latitude north or south of the Equator.

parsec The distance at which a star's parallax would be 1 second of arc. 1 parsec (pc) = 206,265 AU = 3.26 ly.

parting The tendency to break along flat surfaces that follow structural weaknesses.

path of totality The path of the Moon's umbra over the surface of the Earth during a solar eclipse.

pedalfer Soils rich in aluminum and iron compounds.

pedocal Soils rich in calcium compounds.

penumbra The portion of a shadow in which only part of the light has been blocked, so the light is dimmed but not totally absent.

period The unit of geologic time into which eras are divided.

period of revolution The length of time it takes for a satellite to complete one revolution around its primary.

permeability The rate at which water can infiltrate a material.

phases of matter The form in which matter may exist; solid, liquid, and gas.

phases of the Moon The changing shape of the illuminated portion of the Moon that can be seen by an observer on Earth.

phenocryst Isolated large crystals in a rock.

photosynthesis The process by which cells convert water and carbon dioxide into carbohydrates and oxygen using light as an energy source.

physical weathering Processes that break down rock without changing their chemical composition.

physiographic province A large-scale region in which all parts are similar in geologic structure and climate, and thus has had a unified history of landform development.

phytoplankton Plankton that carry out photosynthesis.

plain A flat area at low elevation.

planet A body that is at least partly solid, orbits the Sun, and emits mainly reflected sunlight.

planetesimal One of the small bodies from which planets formed.

plant action The breaking down of rock by the force exerted by the growth of plant roots and stems or by chemicals produced as a by-product of plant growth.

plateau A large, flat region elevated more than 150–300 m (500–1000 ft) above the surrounding land or above sea level.

plate tectonics The theory that the Earth's crust consists of large, rigid slabs of rock, or plates, resting on a fluidlike layer of denser rock and that these plates slide around causing their edges to collide, separate, and slide past each other.

Pleistocene The time period, which began 1.6 million years ago, during which climates grew colder worldwide.

plucking The tearing loose of rock frozen to the base of a glacier as the glacier moves along.

pluton A rock structure formed by the solidification of an igneous intrusion.

polar diameter Diameter of the Earth measured from a pole through the center of the Earth to the opposite pole (12,714 kilometers).

polar front A boundary between polar and tropical air that forms in the prevailing westerly global wind belts.

Polaris A star whose position is almost directly over the Earth's north geographic pole.

pollutant The particular materials or forms of energy that are harmful to humans.

pollution The concentration of any material or energy form that is harmful to humans.

Glossary

porosity The percentage of empty space in a material.

porphyry A rock consisting of large crystals called phenocrysts embedded in a mass of fine mineral grains.

potential evapotranspiration The maximum amount of water that could be returned to the atmosphere by evapotranspiration under a given set of conditions.

precession The very slow change in the orientation of a rotating body's axis of rotation.

precipitate A solid that crystallizes out of a solution and settles to the bottom of the solution.

precipitation Condensed moisture that falls to the ground.

pressure unloading The release of stress on rock as the stress of the weight of overlying rock is removed as the latter is worn away, or the removal of the stress as the weight of the ice in a glacier is removed as the glacier melts away.

primary The body a satellite orbits.

prime meridian The zero meridian, or starting point, for measuring the angular distance east or west of any other meridian.

profile What a cross section of the land between two points would look like viewed from the side.

protostar A collapsing cloud of dust and gas destined to become a star.

radiation curve A plot of intensity versus wavelength of the radiation emitted by a source, e.g. a star. The resulting graph illustrates the rate at which a source is emitting energy at different wavelengths.

radiative balance The condition in which the Earth's average level of heat energy remains constant over a long period of time.

radioactive decay The process by which unstable isotopes break apart spontaneously giving off energy, subatomic particles, or both.

radioisotope Isotopes that are unstable and undergo radioactive decay.

recharge Water that seeps into the ground and replaces soil moisture that was lost during a period of drought.

red giant A large red star that has consumed most of its hydrogen and developed a helium core; heat from contraction of the helium core triggered fusion in the outer shell of hydrogen, causing the star to expand to hundreds of times its initial size, and its outer surface to cool.

reducing atmosphere An atmosphere with so little oxygen that elements and compounds do not oxidize.

refraction The bending of light as it passes from one medium to another.

regional metamorphism Changes in rock over an extensive area that occur due to the pressure and high temperatures associated with either deep burial or movements of the Earth's crust.

relative age The age of an object or event relative to another object or event.

relative humidity The ratio of water vapor in the air to the maximum amount of water vapor the air can hold at that temperature.

relief The differences in the height of landforms in an area.

renewable resource A material that can be renewed, or restored to its original form after being used, by means of a natural process; i.e., pure water that is polluted by human use is purified by the natural processes of the water cycle.

replacement reaction A chemical reaction in which one element replaces another element in a molecule.

residual soil A soil formed by the weathering of bedrock in place.

respiration The oxidation of food within living cells by which the chemical energy of food molecules is released in a series of metabolic steps involving the consumption of oxygen and the liberation of carbon dioxide and water.

Glossary

retrograde motion An apparent motion of a planet in a direction opposite to its normal motion.

revolution The motion of one body around another body.

Richter scale A system that measures the strength of an earthquake by the amount of motion of the Earth's crust that occurs during the earthquake.

rill Tiny grooves eroded in soil by a trickling stream of water flowing downhill.

rock The naturally formed, solid material that makes up the Earth's crust.

rock cycle A sequence of events that shows how rocks are formed, changed, destroyed, and reformed from the same parent material.

rock-forming minerals The one hundred or so common minerals that make up more than 95 percent of the rock in the Earth's crust.

rockfall Mass wasting process in which rock fragments broken loose by weathering fall from cliffs or bounce by leaps down steep slopes.

rockslide Mass wasting process in which rock fragments slide downhill.

rotation Motion in which every part of an object is moving in a circular path around a central line called the axis of rotation.

runoff Runoff is precipitation that does not evaporate or sink into the ground, but runs downhill along the Earth's surface.

S-waves Transverse seismic waves produced by an earthquake that travel slower than P-waves and are therefore the second to be recorded by a seismograph.

salinity The total amount of dissolved matter in seawater expressed in parts per thousand.

saltation A type of motion in which sediments transported in a stream move along the stream bed in a series of bounces, hops, or leaps.

sandbar A long, narrow pile of sand deposited in open water.

satellite A solid body that orbits another body; i.e., the Moon is a satellite of the Earth.

sea breeze A local wind blowing from a body of water toward land that develops because daytime heating lowers the air pressure over the land.

sea cliff A steep cliff formed by the undercutting of steep slopes by waves and subsequent collapse of the overhanging material.

sea stack An isolated pillar of resistant rock that was detached from a headland by wave erosion.

sea terrace A flat platform of sediments deposited by wave action on the ocean floor adjacent to a steep coastline.

sea waves Wind-generated sea waves that are directly affected by the wind.

sediment Solid particles that have been transported and then deposited by agents of erosion.

sedimentary rock Rock formed by the lithification of sediments.

seismic wave A vibration of the Earth's crust produced by faulting.

seismogram A line recorded on paper by a seismograph that represents the motion during an earthquake.

seismograph A device that detects the motion of the Earth's crust during an earthquake.

seismology The study of earthquakes and their effects.

shadow zone A region in which no seismic waves from an earthquake are received by seismograph stations because the waves have been refracted or blocked as they traveled through the Earth.

shear Forces acting along different lines of action causing rock to twist or tear.

shield volcano A wide, gently sloping volcano formed from successive layers of solidified lava flows.

sidereal day The time from when a star crosses an observer's meridian until the same star next crosses the observer's meridian.

sidereal month The time it takes the Moon to complete one 360° orbit around Earth, measured relative to a distant star; 27.32166 mean solar days, or 27 days, 7 hours, forty-three minutes and 11 seconds.

silicates Minerals composed of compounds of silicon and oxygen together with other elements; the most abundant mineral group on Earth.

silicon tetrahedron An ion of silicon joined with four ions of oxygen in the shape of a tiny tetrahedron (pyramid-like shape).

sill A pluton that forms between, and parallel to, existing rock layers.

sleet Clear pellets of ice that form when raindrops freeze as they fall through layers of air at below-freezing temperatures.

slump Mass wasting process in which a mass of bedrock or soil slides downward along a curved plane of weakness.

snow Hexagonal crystals of ice or needlelike ice crystals formed by the sublimation of water vapor in the atmosphere.

soil The accumulation of loose, weathered material that covers much of the land surface of the Earth.

soil association A group of two or more soils occurring together in a characteristic pattern in a given geographical area.

soil creep The invisibly slow, downhill movement of soil.

soil horizon Recognizable layers in soil with characteristic properties.

soil moisture storage Water that is retained in the soil after water seeps into the ground.

soil moisture zone A moist layer of soil in which water clinging to soil particles supplies plant roots with water.

soil profile A cross section of soil from surface to bedrock.

solar day The time it takes Earth to rotate from one solar noon until the next solar noon.

solar eclipse An event in which the Moon passes directly between the Earth and the Sun, casting a shadow on the Earth and partly or completely blocking an observer's view of the Sun.

solar noon The moment when the Sun is at its highest point in the sky for the day.

solar wind The flow of gas from the Sun out through the solar system; near Earth the solar wind travels at speeds of 600–1000 km/s.

solstice The two times of the year when the Sun's altitude at noon reaches a maximum or a minimum. The summer solstice in the Northern Hemisphere occurs about June 21, when the Sun is in the zenith at the Tropic of Cancer; the winter solstice occurs about December 21, when the Sun is over the Tropic of Capricorn. The summer solstice is the longest day of the year and the winter solstice is the shortest.

source region The geographical region in which an air mass originated.

specific gravity The density of a mineral compared to the density of water.

specific heat capacity The number of degrees the temperature of one gram of a material will increase if one calorie of heat is added to it.

spectrograph An instrument for making a photographic record of a spectrum.

spectrum The band of colors formed when a beam of white light is passed through a prism; each wavelength of light in the beam is refracted at a slightly different angle, causing the light to be arrayed in order of its constituent wavelengths.

spit A sandbar attached at one end to the mainland.

spring A place where groundwater seeps out of the ground.

spring tide A tidal cycle in which there are very high and very low tides; caused by the additive effect of the Sun and Moon's gravity acting together along the same line of action.

Glossary

station model A shorthand symbolic representation of the weather conditions at a particular location.

statistical forecasting Weather forecasting based upon the statistical analysis of historical weather data.

stellar parallax The change in the position of a near object relative to distant objects as an observer's position changes.

stock An irregularly shaped pluton covering an area less than 75 square kilometers.

strain The change in size or shape of a rock due to stress.

stratopause The boundary between the stratosphere and the mesosphere.

stratosphere The layer of the atmosphere ranging from about 6 to 30 kilometers above the Earth's surface; it contains ozone that absorbs ultraviolet light; temperatures in this layer rise with altitude.

streak The color of a powdered mineral; created by rubbing a mineral against an unglazed piece of porcelain.

stream A body of water that moves under the influence of gravity to progressively lower levels in a narrow, but clearly defined channel.

stream banks The sides of the channel containing the water of a stream.

stream bed The bottom of the channel containing the water of a stream.

stream channel Clearly defined path along which the water in a stream flows.

stream gradient The angle between the stream bed and the horizontal; the slope of the stream.

stream load The sediments transported by the water flowing in a stream.

stream mouth The site at which a stream flows into a standing body of water.

stream pattern The pattern formed by the system of streams in an area as they flow down slopes and join with other streams, e.g. dendritic, trellis, radial, rectangular, annular.

stream source The site at which the water in a stream originates.

stress Forces acting on a rock.

subduction The sliding of a denser ocean plate beneath a less dense continental plate resulting in the melting of the ocean plate as it plunges into the hot mantle.

sublimation The process by which a solid changes directly into a gas without passing through a liquid stage.

supercluster A cluster of galaxy clusters.

supergiant star An exceptionally luminous star that is 10 to 1000 times the Sun's diameter.

supernova A rare stellar explosion in which the entire outer envelope of a star is blown away in a violent outburst, leaving behind a dense core.

superposition The idea that the bottom layer of a sedimentary series is the oldest, unless it was overturned or had older rock thrust over it, because the bottom layer was deposited first.

surf Waves that are breaking along a shoreline.

suspended load The sediments carried in suspension in the water flowing in a stream.

swells Long wavelength sea waves that settle into a uniform pattern once they leave the area of turbulent winds that formed them.

syncline A downward, valleylike fold.

synodic month The time between successive new moon phases; 29.530588 mean solar days, or 29 days, 12 hours, 44 minutes and three seconds.

synoptic forecasting Weather forecasting based upon analysis of a sequence of synoptic weather maps and charts.

synoptic weather map A map that summarizes weather data from many locations using station models.

system A group of things or processes that interact to perform some function.

talus A mound of rock fragments that accumulates at the base of a rock cliff that is weathering.

Glossary

technology The application of knowledge to do things.

telescope An instrument for collecting electromagnetic radiation and producing magnified images of distant objects.

temperature A specific degree of hotness or coldness as indicated on or referred to a standard scale.

tension Forces acting away from each other along the same line of action causing rock to stretch.

tephra A collective name for particles formed when globs of lava ejected during an eruption cool and solidify as they fall to the ground.

terminal moraine A mound of till deposited along the leading edge of a glacier.

terrigenous Sediments derived from the land or a continent.

texture The size, shape, and arrangement of mineral crystals or grains in a rock.

thermocline A layer of seawater in which temperature drops sharply with depth.

thermosphere The uppermost layer of the atmosphere ranging from about 50 to 75 kilometers in altitude where it grades into space; this layer contains the ionosphere, which consists of charged particles or ions that deflect harmful radiation and reflect radio waves.

thunderstorm A vigorous, turbulent convection cell in the atmosphere resulting in a transient, violent storm of thunder and lightning, often accompanied by rain and sometimes hail.

time zone A region in which the same time is used throughout, instead of the local time at each place within it.

topographic map A field map in which the field value is elevation above sea level; it shows the three-dimensional shape of the Earth's surface in two dimensions.

tornado A rotating column of air whirling at speeds of up to 800 kilometers per hour, usually accompanied by a funnel-shaped downward extension of a cumulonimbus cloud.

transform boundary Places where edges of plates are sliding laterally past each other.

transpiration The process by which plants release water into the atmosphere through their leaves.

transported soil A soil that formed from material that was deposited in a region after being transported from another place.

transverse waves Waves in which the medium oscillates perpendicular to the direction of wave motion.

travel time The time it takes for a seismic wave to travel from the epicenter to a seismograph.

trench A deep crevice in the ocean floor produced by downward warping of the ocean floor where plates collide.

tropopause The boundary between the troposphere and the stratosphere.

troposphere The layer of the atmosphere ranging from the Earth's surface to about 6 kilometers in altitude; it contains most of the air in the atmosphere; temperatures in this layer decrease with altitude.

tsunami Waves caused by undersea movements of the Earth's crust such as slumping, earthquakes, or volcanic eruptions.

umbra The portion of a shadow in which all light has been blocked.

unconformity A break or gap in the sequence of a series of rock layers.

uniformitarianism The idea that the Earth's features have formed gradually by natural processes that are still occurring, not by instantaneous creation or by catastrophic events.

usage Moisture lost from the soil during a period of drought.

valley glacier A glacier that forms between mountain slopes at high elevations.

727

Glossary

velocity The speed and direction in which something is moving.

vent The central opening of a volcano out of which lava erupts.

ventifact A rock particle that has developed a flat side facing the prevailing winds due to abrasion of its exposed surface.

vertical sorting The separation and deposition of sediments in successive layers caused by differences in the settling rate of the sediment particles; particles that settle fastest form the bottom-most layer, whereas those that settle most slowly form the top layer.

vesicular A rock texture characterized by cavities formed by gas bubbles escaping from a lava as it cools and solidifies.

visibility The horizontal distance through which the eye can distinguish objects.

volcanic glass A rock formed by cooling that was so rapid that no crystalline structure was able to form; such rocks have a glassy appearance.

volcanism The processes by which magma rises into the crust and is extruded onto the Earth's surface.

volcano The opening in the Earth's crust through which lava emerges and the moundlike structure formed by the accumulation of lava.

walking the outcrop Physically following layers of rock from one place to another in order to correlate them.

water cycle The process by which the Earth's water cycles in and out of the atmosphere.

watershed The area drained by a stream and its tributaries.

water table The top of the subsurface zone in which the pore spaces in a soil are saturated with water.

wave cyclone Warm and cold fronts that move in a counterclockwise direction around a low pressure center.

wavelength The distance between two successive wave crests or troughs.

weather The present condition of the atmosphere at any location.

weathering The breakdown of rocks into smaller particles by natural processes.

weight The gravitational force exerted on an object by the Earth.

Wentworth Scale A scheme for classifying and naming sediment particles according to their size.

wet-bulb temperature The temperature recorded by a thermometer whose bulb is kept wet; it is usually lower than dry-bulb temperature due to evaporative cooling.

white dwarf A star, below the main sequence in the H-R diagram, which is small and dense; a dying star that has collapsed to about Earth's size and is slowly cooling off.

wind Air that is moving horizontally.

windward The side of a feature facing the wind.

zenith A point directly overhead to an observer.

zone of aeration The subsurface zone in which the pore spaces in a soil are filled mainly with air.

zone of saturation The subsurface zone in which the pore spaces in a soil are saturated with water.

Examination June 2016
Physical Setting/Earth Science

PART A

Answer all questions in this part.

Directions (1–35): For *each* statement or question, choose the word or expression that, of those given, best completes the statement or answers the question. Some questions may require the use of the *2011 Edition Reference Tables for Physical Setting/Earth Science*. Record your answers in the space provided.

1 Earth's approximate rate of revolution is
   (1) 1° per day
   (2) 15° per day
   (3) 180° per day
   (4) 360° per day   1 ____

2 Planetary winds in the Northern Hemisphere are deflected to the right due to the
   (1) Doppler effect
   (2) Coriolis effect
   (3) tilt of Earth's axis
   (4) polar front jet stream   2 ____

3 Which star is hotter, but less luminous, than *Polaris*?
   (1) *Deneb*
   (2) *Aldebaran*
   (3) *Sirius*
   (4) *Pollux*   3 ____

4 Which statement best explains why Earth and the other planets of our solar system became layered as they were being formed?

(1) Gravity caused less-dense material to move toward the center of each planet.
(2) Gravity caused more-dense material to move toward the center of each planet.
(3) Materials that cooled quickly stayed at the surface of each planet.
(4) Materials that cooled slowly stayed at the surface of each planet.

4 _____

5 Which conditions on Earth's surface will allow for the greatest amount of water to seep into the ground?

(1) gentle slope and permeable
(2) gentle slope and impermeable
(3) steep slope and permeable
(4) steep slope and impermeable

5 _____

6 The photograph below shows a Foucault pendulum at a museum. The pendulum knocks over pins in a regular pattern as it swings back and forth.

This pendulum movement, and the pattern of knocked-over pins, is evidence of Earth's

(1) nearly spherical shape
(2) gravitational attraction to the Sun
(3) rotation on its axis
(4) nearly circular orbit around the Sun   6 ____

7 Earth's early atmosphere contained carbon dioxide, sulfur dioxide, hydrogen, nitrogen, water vapor, methane, and ammonia. These gases were present in the atmosphere primarily because

(1) radioactive decay products produced in Earth's core were released from Earth's surface
(2) evolving Earth life-forms produced these gases through their activity
(3) Earth's growing gravitational field attracted these gases from space
(4) volcanic eruptions on Earth's surface released these gases from the interior   7 ____

8 The diagram below represents the apparent positions of the Big Dipper, with respect to *Polaris*, as seen by an observer in New York State at midnight on the first day of summer and on the first day of winter.

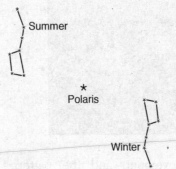

The change in the apparent position of the Big Dipper between the first day of summer and the first day of winter is best explained by Earth

(1) rotating for 12 hours
(2) rotating for 1 day
(3) revolving for 6 months
(4) revolving for 1 year

8 _____

9 The weather station model shown below indicates that winds are coming from the

(1) southeast at 10 knots
(2) northwest at 10 knots
(3) southeast at 20 knots
(4) northwest at 20 knots

9 _____

10 Which type of air mass most likely has high humidity and high temperature?
   (1) cP   (3) mT
   (2) cT   (4) mP   10 _____

11 What is the relative humidity if the dry-bulb temperature is 16°C and the wet-bulb temperature is 10°C?
   (1) 45%   (3) 14%
   (2) 33%   (4) 4%   11 _____

12 The table below shows the air temperature and dewpoint at each of four locations, A, B, C, and D.

Location	A	B	C	D
Air temperature (°F)	80	60	45	35
Dewpoint (°F)	60	43	35	33

Based on these measurements, which location has the greatest chance of precipitation?
   (1) A   (3) C
   (2) B   (4) D   12 _____

13 Which type of electromagnetic radiation has the shortest wavelength?
   (1) ultraviolet   (3) radio waves
   (2) gamma rays   (4) visible light   13 _____

14 Which gas is considered a major greenhouse gas?
   (1) methane   (3) oxygen
   (2) hydrogen   (4) nitrogen   14 _____

15 The diagram below represents Earth and the Sun's incoming rays. Letters A, B, C, and D represent locations on Earth's surface.

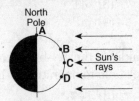

Which two locations are receiving the same intensity of insolation?

(1) A and B
(2) B and C
(3) C and D
(4) D and B

15 _____

16 Most of the sand that makes up the sandstone found in New York State was originally deposited in which type of layers?

(1) tilted
(2) horizontal
(3) faulted
(4) folded

16 _____

17 The map below shows the current location of New York State in North America.

Approximately how many million years ago (mya) was this New York State region located at the equator?

(1) 59 mya
(2) 119 mya
(3) 359 mya
(4) 458 mya

17 _____

18 Many scientists infer that one cause of the mass extinction of dinosaurs and ammonoids that occurred approximately 65.5 million years ago was

(1) tectonic plate subduction of most of the continents
(2) an asteroid impact that resulted in climate change
(3) a disease spreading among many groups of organisms
(4) severe damage produced by worldwide earthquakes   18 _____

19 During which geologic epoch do scientists infer that the earliest grasses first appeared on Earth?

(1) Holocene   (3) Oligocene
(2) Pleistocene   (4) Eocene   19 _____

20 What are the inferred pressure and temperature at the boundary of Earth's stiffer mantle and outer core?

(1) 1.5 million atmospheres pressure and an interior temperature of 4950°C
(2) 1.5 million atmospheres pressure and an interior temperature of 6200°C
(3) 3.1 million atmospheres pressure and an interior temperature of 4950°C
(4) 3.1 million atmospheres pressure and an interior temperature of 6200°C   20 _____

21 A seismic P-wave is recorded at 2:25 p.m. at a seismic station located 7600 kilometers from the epicenter of an earthquake. At what time did the earthquake occur?

(1) 2:05 p.m.   (3) 2:14 p.m.
(2) 2:11 p.m.   (4) 2:36 p.m.   21 _____

22 A seismic station recorded the *P*-waves, but no *S*-waves, from an earthquake because *S*-waves were

(1) absorbed by Earth's outer core
(2) transmitted only through liquids
(3) weak and detected only at nearby locations
(4) not produced by this earthquake   22 ____

23 The Catskills of New York State are best described as a plateau, while the Adirondacks are best described as mountains. Which factor is most responsible for the difference in landscape classification of these two regions?

(1) climate variations   (3) vegetation type
(2) bedrock structure   (4) bedrock age   23 ____

24 An elongated hill that is composed of unsorted sediments deposited by a glacier is called

(1) a delta   (3) a sand dune
(2) a drumlin   (4) an outwash plain   24 ____

25 Which rock was subjected to intense heat and pressure but did *not* solidify from magma?

(1) sandstone   (3) gabbro
(2) schist   (4) rhyolite   25 ____

26 The map below shows a stream drainage pattern where the streams radiate outward from the center.

Which landscape feature would produce this stream drainage pattern?

(1) steep cliff
(2) glacial kettle lake
(3) volcanic mountain
(4) flat plain

26 ____

27 The map below shows the area that, at one time, was covered by ancient Lake Bonneville. Evidence of ancient shorelines indicates that, near the end of the last ice age, Lake Bonneville existed in western Utah and eastern Nevada. The Great Salt Lake in Utah is a remnant of the former Lake Bonneville.

Which material that was formerly on the bottom of Lake Bonneville is most likely exposed on the land surface today?

(1) folded metamorphic bedrock
(2) flat-lying evaporite deposits
(3) coarse-grained coal beds
(4) fine-grained layers of volcanic lava

27 ____

28 The cross section below represents a portion of a meandering stream. Points X and Y represent two positions on opposite sides of the stream.

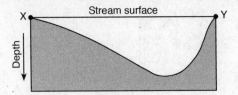

Based on the cross section, which map of a meandering stream best shows the positions of points X and Y?

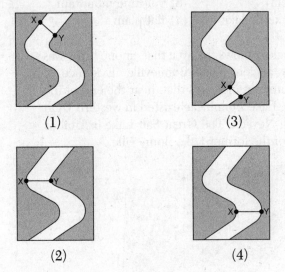

28 _____

29 When wind and running water gradually decrease in velocity, the transported sediments are deposited
  (1) all at once, and are unsorted
  (2) all at once, and are sorted by size and density
  (3) over a period of time, and are unsorted
  (4) over a period of time, and are sorted by size and density   29 _____

30 The graph below shows ocean water levels for a shoreline location on Long Island, New York. The graph also indicates the dates and times of high and low tides.

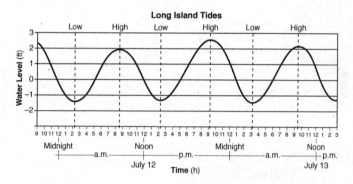

Based on the data, the next high tide occurred at approximately
  (1) 4 p.m. on July 13   (3) 4 p.m. on July 14
  (2) 10 p.m. on July 13   (4) 10 p.m. on July 14   30 _____

31 Which diagram best represents heat transfer mainly by the process of conduction?

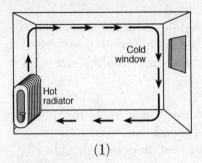
(1)

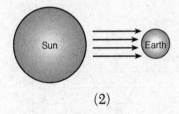
(2)

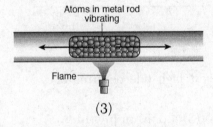
(3)

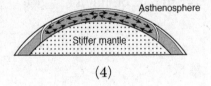
(4)

31 _____

32 The diagram below represents the position of Earth in its orbit and the position of a comet in its orbit around the Sun.

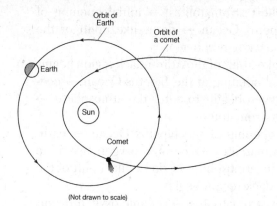

(Not drawn to scale)

Which inference can be made about the comet's orbit, when it is compared to Earth's orbit?

(1) Earth's orbit and the comet's orbit have the same distance between foci.
(2) Earth's orbit has a greater distance between foci than the comet's orbit.
(3) The comet's orbit has one focus, while Earth's orbit has two foci.
(4) The comet's orbit has a greater distance between foci than Earth's orbit.

32 _____

33 Which sequence of geologic events is in the correct order, from oldest to most recent?

(1) oceanic oxygen begins to enter the atmosphere → earliest stromatolites → initial opening of the Iapetus Ocean → dome-like uplift of the Adirondack region begins
(2) dome-like uplift of the Adirondack region begins → initial opening of the Iapetus Ocean → oceanic oxygen begins to enter the atmosphere → earliest stromatolites
(3) initial opening of the Iapetus Ocean → earliest stromatolites → oceanic oxygen begins to enter the atmosphere → dome-like uplift of the Adirondack region begins
(4) earliest stromatolites → oceanic oxygen begins to enter the atmosphere → initial opening of the Iapetus Ocean → dome-like uplift of the Adirondack region begins

33 _____

34 The cross section of the atmosphere below represents the air motion near two frontal boundaries along reference line *XY* on Earth's surface.

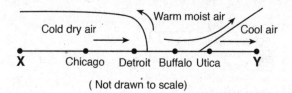

(Not drawn to scale)

Which weather map correctly identifies these fronts and indicates the direction that these fronts are moving?

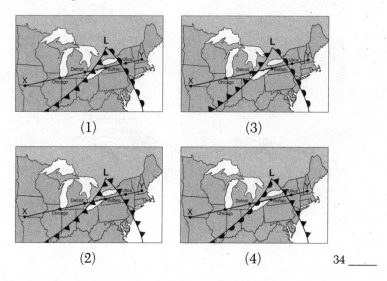

34 _____

35 Which block diagram represents the plate motion that causes the earthquakes that occur along the San Andreas Fault in California?

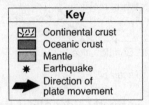

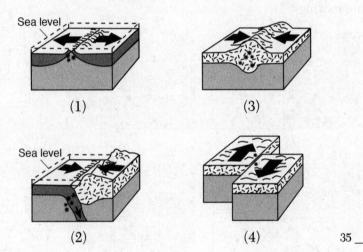

35 _____

PART B–1
Answer all questions in this part.

Directions (36–50): For *each* statement or question, choose the word or expression that, of those given, best completes the statement or answers the question. Some questions may require the use of the *2011 Edition Reference Tables for Physical Setting/Earth Science*. Record your answers in the space provided.

Base your answers to questions 36 through 39 on the map and the passage below and on your knowledge of Earth science. The map shows four different locations in India, labeled *A*, *B*, *C*, and *D*, where vertical sticks were placed in the ground on the same clear day. The locations of two cities in India are also shown.

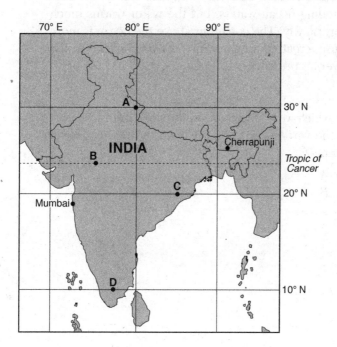

Examination June 2016

Monsoons in India

A monsoon season is caused by a seasonal shift in the wind direction, which produces excessive rainfall in many parts of the world, most notably India. Cherrapunji, in northeast India, received a record 30.5 feet of rain during July 1861. During the monsoon season from early June into September, Mumbai, India averages 6.8 feet of rain. Mumbai's total average rainfall for the other eight months of the year is only 3.9 inches.

Monsoons are caused by unequal heating rates of land and water. As the land heats throughout the summer, a large low-pressure system forms over India. The heat from the Sun also warms the surrounding ocean waters, but the water warms much more slowly. The cooler air above the ocean is more dense, creating a higher air pressure relative to the lower air pressure over India.

36 At which map location would no shadow be cast by the vertical stick at solar noon on the first day of summer?

(1) A
(2) B
(3) C
(4) D

36 _____

37 Which map shows both the dominant air pressure system that forms over India in the summer and the direction of surface winds around this air pressure system? [High pressure = **H**, Low pressure = **L**]

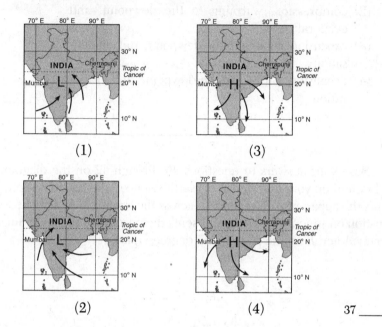

37 _____

38 The unequal heating rates of India's land and water are caused by

(1) land having a higher density than water
(2) water having a higher density than land
(3) land having a higher specific heat than water
(4) water having a higher specific heat than land

38 _____

39 Which processes lead to cloud formation when humid air rises over India?

(1) compression, warming to the dewpoint, and condensation
(2) compression, warming to the dewpoint, and evaporation
(3) expansion, cooling to the dewpoint, and condensation
(4) expansion, cooling to the dewpoint, and evaporation

39 _____

Base your answers to questions 40 through 42 on the diagram below and on your knowledge of Earth science. The diagram represents the apparent path of the Sun across the sky at a New York State location on June 21. Point A represents the position of the noon Sun. Points A and B on the path are 45 degrees apart.

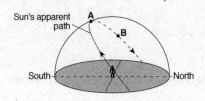

40 How many hours (h) will it take for the apparent position of the Sun to change from point A to point B?

(1) 1 h   (3) 3 h
(2) 2 h   (4) 4 h

40 _____

41 Compared to the Sun's apparent path on June 21, the Sun's apparent path on December 21 at this location will

(1) be shorter, and the noon Sun will be lower in the sky
(2) be longer, and the noon Sun will be higher in the sky
(3) remain the same length, and the noon Sun will be lower in the sky
(4) remain the same length, and the noon Sun will be higher in the sky   41 _____

42 Which diagram represents the correct position of *Polaris* as viewed from this New York State location on a clear night?

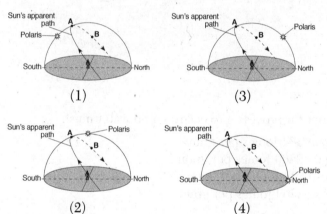

42 _____

Base your answers to questions 43 and 44 on the diagram below and on your knowledge of Earth science. The diagram represents the water cycle. Letters A through C represent different processes in the water cycle.

(Not drawn to scale)

43 In order for process A to occur, liquid water must

(1) gain 334 Joules per gram
(2) gain 2260 Joules per gram
(3) lose 334 Joules per gram
(4) lose 2260 Joules per gram

43 _____

44 Which process is represented by letter B?

(1) capillarity
(2) transpiration
(3) infiltration
(4) precipitation

44 _____

Base your answers to questions 45 through 47 on the photograph below and on your knowledge of Earth science. The photograph shows a small waterfall located on the Tug Hill Plateau.

45 During which geologic time period was the surface bedrock at this location formed?

(1) Cretaceous
(2) Triassic
(3) Devonian
(4) Ordovician

45 _____

46 Compared to the bedrock layers above and below the rock ledge shown at the waterfall, the characteristic that is primarily responsible for the existence of the rock ledge is its greater

(1) resistance to weathering
(2) abundance of fossils
(3) thickness
(4) age

46 _____

47 Rock fragments that are tumbled and carried over long distances by this stream are most likely becoming

(1) less dense, harder, and smaller
(2) less rounded, jagged, and larger
(3) more dense, angular, and smaller
(4) more rounded, smoother, and smaller   47 _____

Base your answers to questions 48 through 50 on the rock columns below and on your knowledge of Earth science. The rock columns represent four widely separated locations, W, X, Y, and Z. Numbers 1, 2, 3, and 4 represent fossils. The rock layers have *not* been overturned.

Location W
- Brown siltstone (2)
- Red sandstone
- Black shale (1)
- Gray limestone (4)
- Tan sandstone (2)(4)

Location X
- Tan limestone
- Brown siltstone (2)
- Red sandstone
- Black shale (1)
- Gray limestone

Location Y
- Black shale (1)
- Gray limestone (4)
- Tan sandstone (4)
- Gray siltstone (4)
- Green shale

Location Z
- Gray conglomerate
- Tan limestone (3)
- Brown siltstone
- Red sandstone
- Black shale (1)

48 Which numbered fossil best represents an index fossil?

(1) 1   (3) 3
(2) 2   (4) 4   48 _____

49 Which rock layer is the oldest?

(1) tan sandstone   (3) green shale
(2) gray limestone   (4) black shale   49 _____

50 Which rock layer formed from the deposition of land-derived sediments that had a uniform particle size of about 0.01 cm in diameter?

(1) brown siltstone   (3) gray conglomerate
(2) black shale   (4) red sandstone   50 _____

PART B–2
Answer all questions in this part.

Directions (51–65): Record your answers in the spaces provided. Some questions may require the use of the *2011 Edition Reference Tables for Physical Setting/Earth Science*.

Base your answers to questions 51 through 53 on the data table below and on your knowledge of Earth science. The data table lists four constellations in which star clusters are seen from Earth. A star cluster is a group of stars near each other in space. Stars in the same cluster move at the same velocity. The length of the arrows in the table represents the amount of redshift of two wavelengths of visible light emitted by these star clusters.

Data Table

Constellation in which star cluster is seen from Earth	Redshift of two wavelengths of light absorbed by calcium	Distance from Earth (billion light years)	Velocity of star cluster moving away from Earth (km/s)
Ursa Major	Violet → Red	1.0	15,000
Corona Borealis	Violet → Red	1.4	22,000
Boötes	Violet → Red	2.5	39,000
Hydra	Violet → Red	4.0	61,000

Note: One light year is the distance light travels in one year.

51 Describe the evidence shown by the light from these star clusters that indicates that these clusters are moving away from Earth. [1]

_____

_____

52 Write the chemical symbol for the element, shown in the table, that absorbs the two wavelengths of light. [1]

_____

53 Identify the name of the nuclear process that is primarily responsible for producing energy in stars. [1]

_____

Base your answers to questions 54 through 57 on the diagram below and on your knowledge of Earth science. The diagram represents the Moon in eight positions in its orbit around Earth. One position is labeled *A*.

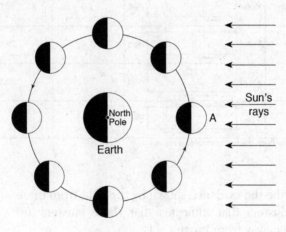

(Not drawn to scale)

54 Circle the type of eclipse that may occur when the Moon is at position A. Explain why this type of eclipse may occur when the Moon is at this position. [1]

Circle one: **lunar eclipse**   **solar eclipse**

Explanation: _____

_____

55 The diagram below represents one phase of the Moon as observed from New York State.

On the diagram *below*, place an **X** on the Moon's orbit to represent the Moon's position when this phase was observed. [1]

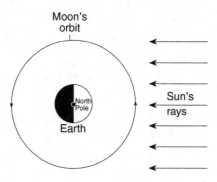

56 State the number of days needed for the Moon to show a complete cycle of phases from one full Moon to the next full Moon when viewed from New York State. [1]

_____ **days**

57 Explain why the Moon's revolution and rotation cause the same side of the Moon to always face Earth. [1]

_____

_____

Base your answers to questions 58 through 61 on the weather map below and on your knowledge of Earth science. The weather map shows atmospheric pressures, recorded in millibars (mb), at locations around a low-pressure center (**L**) in the eastern United States. Isobars indicate air pressures in the western portion of the mapped area. Point *A* represents a location on Earth's surface.

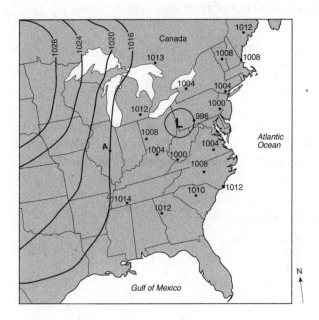

58 On the weather map *above*, draw the 1012 millibar and the 1008 millibar isobars. Extend the isobars to the east coast of the United States. [1]

59 Identify the weather instrument that was used to measure the air pressures recorded on the map. [1]

_____

60 Identify the compass direction toward which the center of the low-pressure system will move if it follows a typical storm track. [1]

_____

61 Convert the air pressure at location A from millibars to inches of mercury. [1]

_____ in of Hg

_____

Base your answers to questions 62 through 65 on the graph below and on your knowledge of Earth science. The graph shows the rate of decay of the radioactive isotope carbon-14 (^{14}C).

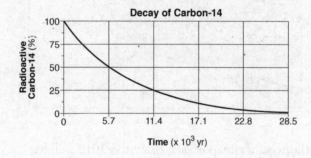

62 Complete the flow chart *below* by filling in the boxes to indicate the percentage of carbon-14 remaining and the time that has passed at the end of each half-life. [1]

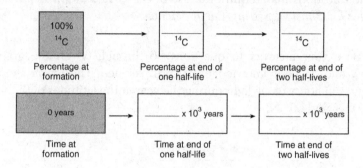

63 Identify the decay product formed by the disintegration of carbon-14. [1]

_____

64 Explain why carbon-14 *cannot* be used to accurately determine the age of organic remains that are 1,000,000 years old. [1]

_____

_____

65 State the name of the radioactive isotope that has a half-life that is approximately the same as the estimated time of the origin of Earth. [1]

_____

Examination June 2016

PART C
Answer all questions in this part.

Directions (66–85): Record your answers in the spaces provided. Some questions may require the use of the *2011 Edition Reference Tables for Physical Setting/Earth Science*.

Base your answers to questions 66 through 69 on the graph below and on your knowledge of Earth science. The graph shows changes in hours of daylight during the year at the latitudes of 0°, 30° N, 50° N and 60° N.

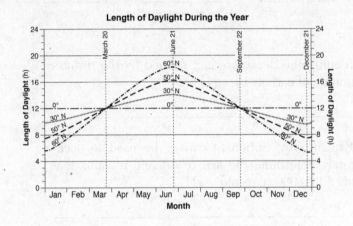

66 Estimate the number of daylight hours that occur on January 1 at 40° N latitude. [1]

_____ h

67 Identify the latitude shown on the graph that has the earliest sunrise on June 21. Include the units and compass direction in your answer. [1]

_____

68 Explain why all four latitudes have the same number of hours of daylight on March 20 and September 22. [1]

_____

_____

69 The graph *below* shows a curve for the changing length of daylight over the course of one year that occurs for an observer at 50° N latitude. On this same graph, draw a line to show the changing length of daylight over the course of one year that occurs for an observer at 50° S latitude. [1]

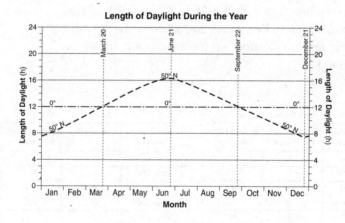

Base your answers to questions 70 through 74 on the passage and data tables below, on the map below, and on your knowledge of Earth science. The data tables show trends (patterns) of two lines of Hawaiian island volcanoes, the Loa trend and the Kea trend. For these trends, ages and distances of the Hawaiian island volcanoes are shown. The map shows the locations of volcanoes, labeled with Xs, that make up each trend line.

Hawaiian Volcano Trends

The Hawaiian volcanic island chain, located on the Pacific Plate, stretches over 600 kilometers. This chain of large volcanoes has grown from the seafloor to heights of over 4000 meters. Geologists have noted that there appear to be two lines, or "trends," of volcanoes—one that includes Mauna Loa and one that includes Mauna Kea. Loihi and Kilauea are the most recent active volcanoes on the two trends shown on the map.

Loa Trend

Loa Trend Volcanoes	Volcano Age (million years)	Distance from Loihi (km)
Kauai	4.6	575
Waianae	3.7	465
Koolau	2.2	375
West Molokai	1.7	350
Lanai	1.2	300
Kahoolawe	1.1	250
Hualalai	0.3	130
Mauna Loa	0.2	70
Loihi	0	0

Kea Trend

Kea Trend Volcanoes	Volcano Age (million years)	Distance from Kilauea (km)
East Molokai	1.7	256
West Maui	1.5	221
Haleakala	0.9	182
Kohala	0.5	100
Mauna Kea	0.4	54
Kilauea	0.1	0

Volcanoes and Islands of Hawaii

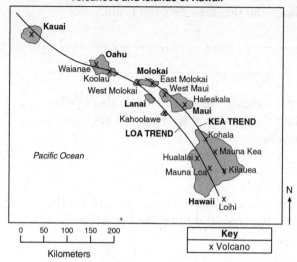

70 The average distance between the volcanoes along the Kea trend is 51.2 kilometers. Place an **X** on the map *above* to identify the location on the seafloor where the next volcano will most likely form as a part of the Kea trend. [1]

71 Identify the *two* volcanoes, one from each trend, that have the same age. [1]

_____ and _____

72 State the general relationship between the age of the volcanoes and the distance from Loihi. [1]

_____

_____

73 Identify the tectonic feature beneath the moving Pacific Plate that caused volcanoes to form in *both* the Loa and Kea trends. [1]

_____

74 Identify the compass direction in which the Pacific Plate has moved during the last 4.6 million years. [1]

_____

Base your answers to questions 75 through 79 on the topographic map below and on your knowledge of Earth science. Lines *AB* and *CD* are reference lines on the map. Letter *E* indicates a location in a stream.

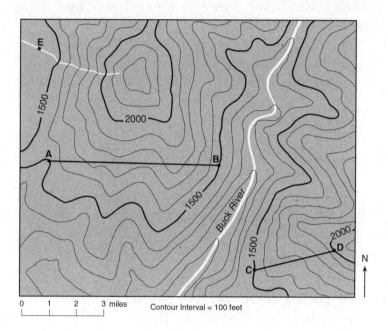

75 On the map *above*, draw an **X** on the location with the highest elevation. [1]

76 Using the grid *below*, construct a topographic profile along line *AB* by plotting the elevation of each contour line that crosses line *AB*. Points *A* and *B* have already been plotted on the grid. Connect all plots with a line from *A* to *B* to complete the profile. [1]

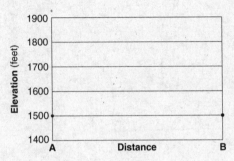

77 Calculate the gradient along line *CD*. [1]

_____ ft/mi

78 Describe how the contour lines indicate the direction in which Buck River flows. [1]

_____

_____

79 Determine the velocity of the stream at location *E* where the largest particle being carried at location *E* has a diameter of 10.0 centimeters. [1]

_____ cm/s

Base your answers to questions 80 through 83 on the passage below and on your knowledge of Earth science.

Dimension Stone: Granite

Dimension stone is any rock mined and cut for specific purposes, such as kitchen countertops, monuments, and the curbing along city streets. Examples of rock mined for use as dimension stone include limestone, marble, sandstone, and slate. The most important dimension stone is granite; however, not all dimension stone sold as granite is actually granite. Two examples of such rock sold as "granite" are syenite and anorthosite. Syenite is a crystalline, light-colored rock composed primarily of potassium feldspar, plagioclase feldspar, biotite, and amphibole, while anorthosite is composed almost entirely of plagioclase feldspar. Like actual granite, both syenite and anorthosite have large, interlocking crystals.

80 Explain why syenite is classified as a plutonic igneous rock. [1]

_____

_____

81 State *one* reason why anorthosite is likely to be white to gray in color. [1]

_____

82 The igneous rock gabbro is sometimes sold as "black granite." Compared to the density and composition of granite, describe how the density and composition of gabbro are different. [1]

Density of gabbro: _____

Composition of gabbro: _____

83 Identify *one* dimension stone mentioned in the passage that is composed primarily of calcite. [1]

_____

Base your answers to questions 84 and 85 on the map of Australia below and on your knowledge of Earth science. Points *A* through *D* on the map represent locations on the continent.

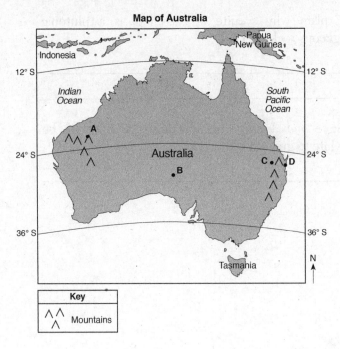

84 Explain why location A has a cooler average yearly air temperature than location B. [1]

_____
_____

85 The cross section below represents a mountain between locations C and D and the direction of prevailing winds.

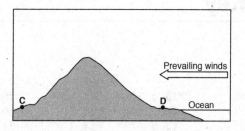

Explain why location D has a wetter climate than location C. [1]

_____
_____

Answers
June 2016
Physical Setting/Earth Science

Answer Key

PART A

1. 1	8. 3	15. 4	22. 1	29. 4
2. 2	9. 3	16. 2	23. 2	30. 2
3. 3	10. 3	17. 3	24. 2	31. 3
4. 2	11. 1	18. 2	25. 2	32. 4
5. 1	12. 4	19. 3	26. 3	33. 4
6. 3	13. 2	20. 1	27. 2	34. 1
7. 4	14. 1	21. 3	28. 4	35. 4

PART B–1

36. 2	39. 3	42. 3	45. 4	48. 1
37. 1	40. 3	43. 2	46. 1	49. 3
38. 4	41. 1	44. 2	47. 4	50. 4

Index

Abrasion, 432, 459, 468–469, 474–475
Absolute age, 255
Absolute humidity, 571
Absolute magnitude, 144
Absorption spectrum, 132
Acid rain, 437
Actual age, 255
Adiabatic changes, 577
Age, 255–257
  determining, 263–269
Air composition, 219
Air humidity, 570–575
Air masses, 587–590
Air movements, 575
Air pressure
  description of, 568–570
  humidity and, 578
  temperature and, 576–578
  winds and, 580–583
Air temperature
  description of, 567–568
  humidity and, 579–580
Alfonsine Tables, 22
Alluvial fans, 498–499
Ancient environments, 271–272
Andesite, 321
Angle of insolation, 48, 540–541
Angle of repose, 453
Angular unconformities, 257
Animal action, 432–433
Annual cycles, 540–541
Annual traverse of the constellations, 45
Annular eclipse, 53
Annular stream pattern, 512
Anticline, 399
Anticyclones, 592–593
Apparent magnitude, 143
Apparent motion, 5–6
Aquicludes, 243
Aquifers, 242
Archean eon, 260–261

Aristotle, 18–20
Artes, 477
Artesians, 243
Asterism, minerals, 298
Asteroids, 174–175
Asthenosphere, 406–407, 409
Astronomy
  early, 1–17
  celestial
    coordinate system, 8–9
    meridian, 7
    motion, 4–6
    objects, 3–4
    sphere, 6–10
  constellations, 5
  Dubhe, 8
  horizon, 7
  Jupiter, 10
  Mars, 10
  Merak, 8
  Mercury, 10
  Moon, 10
  planets, 10
  Polaris, 8
  Saturn, 10
  sky, 1–6
  stars, 8, 10
  Sun, 9–10
  Venus, 10
  zenith, 7
  zodiac, 9
  modern, 18–40
    Alfonsine Tables, 22
    Aristotle, 18–20
    astronomical unit, 28
    Brahe, Tycho, 24–25
    Copernicus, 22–24
    Galilei, Galileo, 29
    gravity, 29–31
    heliocentric model, 18–40
    Kepler, Johannes. *See* Kepler, Johannes
    Moon, 29
    Newton, Isaac, 29–31

  orbital motion, 30–31
  parallax, 19–20, 23
  planetary motion
    Kepler's law, 25–28
    velocity, 26–27
  Ptolemy, 18–19, 21–22
  retrograde motion, 21, 24
  revolution, period of, 28
  spheres, 19–20
  universal gravitation, 29–30
  tools for studying, 125–138
    lenses, 127–128
    Mount Palomar Observatory, 128
    observation instruments, 134–135
    spectrograph, 133
    spectrophotometer, 133
    spectroscopes. *See* Spectroscopes
Atmosphere, 535–565
  aerobic, 218
  circulation, 547–549
  conduction, 545
  convection, 545
  Coriolis effect, 549–552
  development, 217–219
  entering, 544–552
  formation, 216–217
  insolation. *see* Insolation
  latent heat, 546–547
  layers, 220–221
  lost, 215
  origin, 215–219
  oxidizing, 217–218
  radiation, 545–546
  reducing, 217
  solar radiation. *See* Solar radiation
  structure, 219–221
  transparency, 576
  variables, 567

771

Index

Axis, 74
Axis of rotation, 42

Banding, 330
Barite, 303
Barometer, 569
Barrier bars, 501
Basalt, 318, 321
Basaltic glass, 320
Batholiths, 387
Bayer, Johann, 140
Beaches, 499–500
Bench marks, 396
Berms, 499
Big Bang Theory, 194–196
Biogenic sediments, 325
Black hole, 155
Body waves, 354
Borates, 303
Boundaries, 407–408
Brahe, Tycho, 24–25
Breezes, 581–582
Brightline spectrum, 131
Brightness, stellar, 142–144

Calcite, 303
Callisto, 173
Capillarity, 234
Capillary action, 241
Capillary fringe, 241
Carbon-14, 265–269
Carbonate ion, 303
Carbonates, 303
Carbonation, 436
Cartesian system, 74
Catastrophic theory, 165
Celestial coordinate system, 8–9
Celestial meridian, 7
Celestial motion, 4–5
Celestial objects, 3–4
Celestial sphere, 6–10
Celestite, 303
Celsius, 568
Cementation, 322
Cenozoic era, 261–262
Ceres, 170
Chaos theory, 596
Charon, 167
Chatoyancy, minerals, 298

Chemical formulas, 293
Chemical sediments, 325–326
Chemical symbol, 293
Chemical tests, 298
Chemical weathering, 434–437
Chromates, 303
Cinder cone volcano, 385–386
Cinnabar, 303
Circles, great, 75–78
Cirques, 473, 477
Clastic sediments, 324–325
Cleavage, minerals, 296
Climate, 437–438, 621–646
 continental drift theory and, 403–404
 definition, 621
 effects, 630–631
 factors, 622–629
 landscapes and, 513–514
 types of, 629–630
Climatic trends, 631–636
Cloud ceiling, 576
Cloud cover, 576
Clouds, 583–586, 629
Clusters, 191–192
Coastal processes, 465–467
Cold fronts, 590–591
Color
 description of, 542–543
 of minerals, 295
 of stars, 144–148
Cometary impacts, 216–217
Comets, 176–177
Compaction, 322
Composite cone, 386
Compounds, 293–294
Compression stress, 398
Condensation, 238, 570–571, 583
Conduction, 383, 545
Constellations, 5
Contact metamorphism, 329
Continental drift theory, 400–404
Continental glaciers, 474
Continuous spectrum, 130
Contour interval, 82

Contour lines, 81–82
Convection, 383, 545
Convection cell, 549
Convection currents, 461
Convergent boundaries, 407
Coordinate system, 8–9, 74
Copernicus, 22–24
Core, 363–364
Coriolis effect, 43–44, 549–552, 583
Correlation, rock layers, 257–258
Corundum, 303
Cosmic background radiation, 196
Cosmology, 190–203
 Big Bang Theory, 194–196
 clusters, 191–192
 cosmic background radiation, 196
 Doppler effect, 193–194
 light-year, 192
 Olber's Paradox, 191
 superclusters, 192
 universe. See Universe
Crust, 205–206
 definition of, 363
 dynamic, 396–400
 lithosphere, 396
Crustal movements
 causes, 398
 effects, 398–399
 evidence, 396–397
Crystalline form, 294
Crystalline spheres, 19–20
Crystalline structure, 236
Crystalline texture, 329
Currents, 468
 convection, 461
 equatorial, 462
 longshore, 466
 ocean, 461–463
 wind-driven, 461–462
Cyclones, 592–593
Cyclonic storm tracks, 629

Daily cycles, 540–541
Day, 96–106
 sidereal, 101–102

Index

solar, 101–102
Deferents, 21
Deflation, 468
Deltas, 498–499
Dendritic stream pattern, 511
Density, 330
Deposition, 492–534
  glaciers, 503–509
  gravity, 495–497
  moving water, 497–501
  particles, characteristics, 494
  sediment, bedding, 495
  sorting, 493
  speed of medium, 493
  wind, 501–503
Deposits, 505
Desertification, 632
Dessication, 322
Dew, 585
Dew point, 572, 575, 580
Diabase, 318
Dikes, 386
Diorite, 321
Disconformities, 257
Dispersion, 130
Displaced structures, 396
Divergent boundaries, 407
Divides, 510
Domolmite, 303
Doppler effect, 193–194
Double refraction, minerals, 298
Drainage basin, 510
Draper, Henry, 133
Drizzle, 585
Drumlins, 507
Ductile deformation, 398–399
Dunes, 502–503
Dunite, 321
Duration of insolation, 48
Dwarf plants, 169

Earth, 169–171
  aerobic, 218
  air of, 219
  cometary impacts, 216–217
  development, 217–219
  formation, 216–217
  layers, 220–221
  lost, 215
  origin, 215–219
  outgassing, 216
  oxidizing, 217–218
  reducing, 217
  structure, 219–221
  atmosphere of, 215–227
  birth of, 204–205
  coordinate system, 67–97
  core, 205–206
  core of, 363–364
  crustal movements. See Crustal movements
  crust of. See Crust
  differentiation of, 205
  hydrosphere, 228–252
  interior structure, 362–366
  layers of, 205–206
  magnet, 206–207
  mantle, 205–206, 363–364
  motions
    effects, 46–55
    heliocentric. See Heliocentric motions
  origins of, 204–214
  shape of, 68–72
    eclipses, 6
    evidence, 69–72
    gravity, measurements, 71–72
    photographs from space, 70
    Polaris, altitude, 70–71
    roundness, 68–69
    smoothness, 69
  size, determining, 72–73
  surface, 542–544
    mapping, 74–85
Earthflow, 455, 496
Earthquakes, 348–379, 396
  causes, 349
  definition, 348–349
  effects, 359–361
  elastic rebound theory, 349
  epicenter. See Epicenter
  measuring of, 358–359
  Mercalli scales for, 359
  Richter magnitude scale for, 358–359
  P-waves, 354–355
  seismic waves, 349, 352–354
  seismogram, 353–358
  S-waves, 354–355
  waves, 349–354
    frequency, 351
    length, 351
    longitudinal waves, 352
    motion, 349
    periodic waves, 351
    refraction, 351
    transverse waves, 352
    types, 352
    velocity, 351
  zones, 361–362
Eccentricity, 25–26
Eclipse, 51–54
  annular, 53
  evidence of Earth's shape, 69
  lunar, 53–54
  solar, 51–53
Elastic deformation, 398–399
Elastic limit, 398
Elastic rebound theory, 349
Electromagnetic spectrum, 537–538
Electromagnetic waves, 536–537
Elements, 293
Elevation, 628
Elevation field maps, 81
Ellipse, 25–26
El Niño, 631, 633–636
End moraine, 506
Energy, in atmosphere, 544–552
  circulation, 547–549
  conduction, 545
  convection, 545
  Coriolis effect, 549–552
  latent heat, 546–547
  radiation, 545–546
Environments, ancient, 271–272

Index

Eons, 260–262
Epicenter, 349
  distance to, 355–356
  location of, 356–358
Epicycles, 21–22
Epochs, 261–263
Equator, 75
Equatorial currents, 462
Equinoxes, 110–111
Eras, 261–262
Eratoshenes' method, 73
Eris, 170
Erosion, 451–491
  agents of. *see* Erosion agents
  evidence of, 452
Erosion agents, 452–480
  coastal processes, 465–467
  convection currents, 461
  equatorial currents, 462
  glaciers, 470–477, 480
  gravity, 452
    angle of repose, 453
    earthflow, 455
    mass wasting, 453–456, 478
      rapid, 454–455
      slow, 455–456
    mudflow, 455
    rockfalls, 454
    slump, 454–455
    soil creep, 455–456
  longshore currents, 466
  oceans, 461–468
    currents, 461–463
  raindrops, 456
  runoff, 456
  shoreline erosion, 467–468
  streams, 456–461, 478–479
    erosion, 459
    life cycle, 459–461
    transport, 457–459
  swells, 465
  tsunamis, 465
  water in motion, 456–468
  waves, 463–465, 479
    crest, 464
    height, 464
    length, 464

    refraction, 466
    trough, 464
  wind-driven currents, 461–462
  wind-generated, 463–465
  wind, 468–470, 480
Erratics, 507–508
Europa, 173
Evaporation, 235, 238, 570–571
Evaporative cooling, 235–236
Evaporites, 326
Evapotranspiration, 628
Evolution, 270
Exfoliation, 432
Exposure, 438–439
Extrusive rocks, 317–318
Eyepiece, 128

Fahrenheit, 568
Faults, 256, 399–400
Feldspar, 319
Field maps, 79–85, 587
  contour lines, 81–82
  conventions, 85
  direction, 82
  elevation field maps, 81
  gradient, 80
  isolines, 79–80
  isosurfaces, 81
  profiles, 84–85
  scale, 82, 84
  symbols, 82–83
  topographic, 81
Finger lakes, 508
Firn, 470
Flame color, minerals, 298
Floodplains, 497–498
Focal length, 127
Focal point, 127
Focus, 25
Fog, 585
Folded rock layers, 396
Folds, 399
Foliation, 330
Fossils, 397
  continental drift theory, 402

fuels, conservation, 304–305
  index, 259
Foucault Pendulum, 42–43
Fracture, minerals, 296–297
Frame of reference, 74
Freezing, 236
Frequency, 351
Frigid zone, 623
Frontal cyclones, 592–593
Fronts, 587, 590–592
Frost, 585
Frost action, 431–432
Fusion, 148–149

Gabbro, 317–318, 321
Galaxies, 134, 164
Galena, 303
Galilei, Galileo, 29
Ganymede, 173
Geocentric model, 1–17
Geocentric universe, 19–21
Geological record, 254–272
Geological structures, 402
Geologic column, 260
Geologic time scale, 260–263
Geothermal gradient, 380–381
Glacial deposits, 504–505
Glacial drift, 506
Glacial erosion, 474–477
Glacial grooves, 476
Glacial lakes, 508–509
Glacial polish, 475
Glacial striations, 475–476
Glacial transport, 503–504
Glaciers, 470–477, 480
  deposition, 503–509
  landforms, 506–509
  types, 473–474
Global climatic trends, 631–636
Global radiation budget, 538–539
Global warming, 632–633
Global wind belts, 582–583
Gradient, 80, 587
Granite, 317, 321

Index

Gravitational differentiation, 175
Gravity, 29–31, 150–151
  agent of erosion, 452
  measurements, 71–72
Greenhouse effect, 546
Greenwich Mean Time, 105
Groundmass, 318
Groundwater, 238, 240–243
  aquicludes, 243
  artesians, 243
  capillary action, 241
  capillary fringe, 241
  permeability, 241
  porosity, 240
  soil moisture zone, 241
  water table, 241–242
  wells, 242
  zone of saturation, 241
Gypsum, 303

Hachure marks, 82
Hail, 585
Half-life, 266–269
Halides, 303
Halite, 326
Halley's Comet, 168
Hanging valleys, 477
Hardness, minerals, 295–296
Heat, internal
  conduction, 383
  convection, 383
  original heat, 381–382
  radiation, 383
  radioactive decay, 382
  sources of, 381–382
  tidal friction, 382
  transfer of, 382–383
Heat capacity, 235
Heliocentric model, 18–40
Heliocentric motions, 41–66
  constellations, annual traverse, 45
  Coriolis effect, 43–44
  Earth-Moon-Sun System, 50–55
  eclipses, 51–54
  Foucault Pendulum, 42–43
  Moon, phases, 50–51

  parallelism, Earth's axis of rotation, 44
  Polaris, 45
  precession, 45
  revolution, 44–45
  rotation, 42–44
    parallelism, 44
  Sun's path, 46–50
  tides, 54–55
  Vega, 45
Hematite, 303
Hertzsprung-Russell Diagram, 146–148
Hipparchus, 143
Horizon, 7
Horns, 477
Hot spots, 408–409
Hours, 99–101
Human perception, 125–128
Humidity, 570–575
  air, 570–575
  relative, 571–572
Hurricanes, 598
Hutton, James, 254
Huygens, Christian, 173
Hydration, 435–436
Hydrogen bonds, 233
Hydrolysis, 435–436
Hydrosphere
  groundwater, 238, 240–243
  origin, 228–252
    groundwater, 238
    outgassing, 230–231
    runoff, 238
    seawater. *See* Seawater
  water, 229–231
    liquid, 234–235
    solid, 236–237
    states, 234–237
    structure, 233–234
    vapor, 235–236
Hydroxides, 303
Hygrometer, 572

Ice movement, 471–473
Igneous rock, 256, 315
  characteristics, 316
  formation, 315–316
  identification, 319–321

  mineral composition, 319
  porphyrys, 318–319
  texture, 317–319
Immature soil, 441
Impact events, 177–179
Index fossils, 259
Inorganic matter, 292
Insolation, 47–48, 238, 540–544
  angle of, 540–541
  annual cycles, 540–541
  color, 542–543
  daily cycles, 540–541
  duration, 541–542
  Earth's surface, 542–544
  latent heat of water, 544
  latitude, 540–542
  season, 541
  specific heat, 543–544
  texture, 542–543
Internal heat transfer, 382–383
International Date Line, 105–106
Interplanetary bodies, 174–179
Intrusions, 386
Intrusive rocks, 317
Inverse-square law, 143
Io, 173
Isobars, 587
Isolines, 79–80, 587
Isostasy, 401
Isosurfaces, 81
Isotherms, 587
Isotopes, 265–269

Joints, 399
Jovian planets, 166, 169, 172–174
Jupiter, 10, 167, 169–170, 172–173

Kelvin, 568
Kepler, Johannes, 24–28
  law of equal areas, 27
  law of planetary motion, 25–28
Kettle lake, 508
Kettles, 508

Index

Key beds, 260
Kirchhoff, Gustav, 131–132
Kuiper belt, 167

Laccoliths, 387
Lagoons, 501
Land breezes, 581–582
Land distribution, 623–624
Landforms, 509
Landscape development, 509–516
  annular stream pattern, 512
  climate and landscapes, 513–514
  dendritic stream pattern, 511
  drainage basin, 510
  landforms, 509
  landscape regions, 514–516
    New York State, 515–516
  mountains, 510
  plains, 510
  plateaus, 510
  radial stream pattern, 511
  rectangular stream pattern, 512
  relief, 509
  soil associations, 512
  stream drainage system, 510
  stream patterns, 510–512
  topography, 509
  trellis stream pattern, 512
  tributary, 510
Landslide, 454
Latent heat, 546–547
  of fusion, 237
  of vaporization, 235
  of water, 544
Latitude, 540–542, 623
  determining, 78–79
  Sun's path changes, 48–49
Latitude-longitude coordinate system, 74–78
Lava, 315, 385–386
Lava plateau, 386

Laws
  equal areas, 27
  original horizontality, 255
  planetary motion, 25–28
Layers, distortion, 330
Lenses, 127–128
Levees, 497–498
Lichens, 436–437
Life
  evolution, 270
  geological record, 254–272
  natural selection, 270–271
Light, 127–128
  refracted, 127
  visible, 126
Light-year, 141, 192
Limbs, 399
Lithification, 322–323
Lithosphere, 396, 406–407, 409
Load, 458
Local time, 103–106
Loess, 502
Longitude, 74–78
Longitudinal waves, 352
Long-range weather forecasting, 595–596
Longshore currents, 466
Luminescence, minerals, 298
Luminosity, 142, 145, 147
Lunar eclipse, 53–54
Luster, minerals, 295

Magma, 315, 384
Magnetism, 298
Magnetite, 303
Magnetosphere, 206–207
Magnitudes, 143
  absolute, 144
  apparent, 143
  scale, 143
Main sequence, 146–147
Mantle, 363–364
Map
  direction, 82
  profiles, 84–85
  symbols, 82–83

Mapping, 67–97
  axis, 74
  Cartesian system, 74
  coordinate systems, 74
  equator, 75
  field maps. *see* Field maps
  frame of reference, 74
  great circles, 75–76
  latitude, 78–79
  latitude-longitude coordinate system, 74–78
  longitude, 74–78
    determining, 78–79
    meridians, of, 76
  meridians, 75–78
  origin, 74
  parallels, 75
  polar coordinate system, 74
  prime meridian, 77–78
Map scale, 82, 84
Marcasite, 303
Mariner spacecrafts, 171
Mars, 10, 169–172
Mass, 151–155
Mass movements, 495–497
Mass wasting, 453–456, 478
  rapid, 454–455
  slow, 455–456
Mature soil, 441
Meanders, 460
Mediterranean Belt, 387
Merak, 8
Mercalli scales, 359
Mercury, 10, 169–170
Meridians, 75–78
  celestial, 7
  of longitude, 76
  prime, 77–78
Mesozoic era, 261–262
Metamorphic rock, 315, 327–332
  characteristics, 329–331
  classification, 331
  formation, 327–328
    chemical activity, 328
    heat, 327
    pressure, 328

Index

identification, 331–332
texture, 329–330
Metamorphism, 328–329
Meteor, 175
Meteorite, 175
Meteoroids, 175–176
Mid-ocean ridges, 404
Milky Way, 134, 164
Minerals, 291–313. *See also
specific mineral*
  building blocks of rocks,
    299–300
  changes, 330–331
  characteristics of, 292–294
  chemical
    formulas, 293
    symbol, 293
    tests, 298
  cleavage, 296
  color, 295
  composition, 43
  compounds, 293–294
  conservation, 304
  crystalline form, 294
  definition, 292
  double refraction, 298
  elements, 293
  flame color, minerals, 298
  fracture, 296–297
  groups of, 300–304
  hardness, 295–296
  identifying, 294–299
  inorganic matter, 292
  luminescence, 298
  luster, 295
  magnetism, 298
  minor groups of, 303–304
  oxygen, 303
  parting, 296
  piezoelectricity, 298
  play of colors, 298
  polyatomic ions, 303
  pyroelectricity, 298
  resource distribution,
    305–306
  special properties, 298
  specific gravity, 297–298
  streak, 295
Mohorovicic, Andrija, 363
Monoclines, 399

Monsoons, 625–626
Month, 106–108
  sidereal, 107–108
  synodic, 107–108
Moon, 10, 29, 171
  eclipse, 51–54
  origins, 208–210
    Apollo missions, 208
    giant impact, 208–210
  phases, 50–51
Moraines, 506
Motion
  apparent, 5–6
  celestial, 4–5
  Newton's laws, 29–31
  orbital, 30–31
  planetary, Kepler's law,
    25–28
  real, 5–6
  retrograde, 21, 24
  sky, 2
Mountains, 510, 628
Mount Palomar
  Observatory, 128
Mt. Etna, 387
Mt. Vesuvius, 387
Mudflow, 455, 496

Natural Selection, 270–271
Neap tides, 55
Nebula, 135, 150–151
Nebular theory, 166
Neptune, 167, 169–170,
  174
Neutrinos, 153
Neutron star, 154
Newton, Isaac, 29–31
Nitrates, 303
Nonconformities, 257
Nuclear fusion, 148–149
Numerical weather
  forecasting, 595

Oasis, 469
Objective, 128
Observations, instruments,
  134–135
Obsidian, 317, 320
Occluded fronts, 592
Oceans, 461–468

  currents, 461–463,
    626–627
  deposition, 499–501
  floor spreading theory,
    404–406
Olber's Paradox, 191
Oort cloud, 168
Optical axis, 127
Orbital motion, 30–31
Orbits, 25–27
Organic sediments, 325
Original heat, 381–382
Original horizontality, 255
Orion, 169
Orographic effect, 628–629
Outcrop, walking, 258
Outgassing, 216, 230–231
Outwash plains, 508
Overland flow, 456
Overturned, 399
Oxbow lakes, 497
Oxidation, 434–435
Oxides, 303
Oxygen, 303

Paleomagnetism, 404–405
Paleozoic era, 261–262
Parallax, 19–20, 23,
  140–141
Parallelism, Earth's axis of
  rotation, 44
Parallels, 75
Parent material, 440
Parent rock, 328
Parsec, 141–142
Particle, 438, 494
Parting, minerals, 296
Path of totality, 52
Pedalfers, 442
Pegmatite, 317, 321
Peneplain, 461
Penumbra, 52
Perception, human, 125–128
Peridotite, 321
Periodic waves, 351
Periods, 261–262
  of revolution, 28
Permeability, 241
Phanerozoic eon, 260–261
Phenocrysts, 318

Index

Phosphates, 303
Physical weathering, 430–434
  abrasion, 432
  animal action, 432–433
  exfoliation, 432
  frost action, 431–432
  plant action, 432–433, 436–437
  pressure unloading, 434
  temperature, changes, 434
Piezoelectricity, minerals, 298
Plains, 510
Planet(s), 10, 169–170
  dwarf, 169
  Earth, 169–171
  jovian, 169, 172–174
  Jupiter, 167, 169–170, 172–173
  Mars, 169–172
  Mercury, 169–170
  Neptune, 167, 169–170, 174
  Pluto, 167, 170
  Saturn, 167, 169–170, 173
  terrestrial, 170–172
  Uranus, 167, 169–170, 173–174
  Venus, 169–171
Planetary motion
  Kepler's law, 25–28
  velocity, 26–27
Planetesimals, 166
Plant action, 432–433, 436–437
Plateaus, 510
Plate boundaries, 407–408
Plate tectonics, 395–411
  asthenosphere, 406–407, 409
  boundaries, 407–408
  hot spots, 408–409
  key ideas, 409–411
  lithosphere, 406–407, 409
  rigidity, 406
Play of colors, minerals, 298
Plucking, 475
Pluto, 167, 170

Plutonic rock, 317
Plutons, 386
Polar coordinate system, 74
Polar easterlies, 583
Polar front, 593
Polaris, 8, 45
  altitude, 70–71
Polyatomic ions, 303
Porosity, 240
Precambrian eon, 261–262
Precession, 45
Precipitation, 238, 585, 622
Predictions, 21–22
Pressure
  air, 568–570
  belts, 624–625
  unloading, 434
Prevailing winds, 583
Prime meridian, 77–78
Principle of superposition, 255
Prism, 129
Proterozoic eon, 260–261
Protostar, 150–151
Psychrometer, 572
Ptolemy, 18–19, 21–22
Pumice, 318, 320
P-waves, 354–355, 363
Pyrite, 303
Pyroelectricity, minerals, 298

Quartz, 319

Radial stream pattern, 511
Radiation, 383, 545–546
Radiation curve, 144
Radiative balance, 539
Radioactive decay, 266, 382
Radioisotopes, 266–269
Rain, 585
Raindrops, 456
Rapid mass movements, 495–496
Real motion, 5–6
Rectangular stream pattern, 512
Recumbent folds, 399
Red dwarfs, 152
Reflecting telescope, 128

Refracted light, 127
Refracting telescope, 128
Refraction, 351
Regional metamorphism, 329
Relative age, 255–257
Relative humidity, 571, 574
Relief, 509, 628
Residual soil, 440
Resource distribution, 305–306
Retrograde motion, 21, 24
Revolution, 44–45
  period of, 28
Rhyolite, 318, 321
Richter magnitude scale, 358–359
Ring of Fire, 387
Rock drumlins, 476
Rockfalls, 454, 495
Rock(s), 314–347
  age, 263–269
  banding, 330
  crystalline texture, 329
  cycle, 332–333
  definition, 315
  density, 330
  extrusive rocks, 317–318
  folded layers, 396
  foliation, 330
  igneous. *See* Igneous rock
  intrusive rocks, 317
  lava, 315, 317
  layers, 257, 330
  magma, 315, 317
  metamorphic. *see* Metamorphic rock
  metamorphism, 328–329
  mineral
    building blocks, 299–300
    composition, changes, 330–331
  parent rock, 328
  physical characteristics, matching, 258–259
  rigidity, 406
  sedimentary. *See* Sedimentary rock
  tilted layers, 396
  types, 315–332

volcanic glasses, 317
Wentworth scale, 325
young, 404
Rockslide, 454
Rotation, 42–44
  axis of, 42
  parallelism, 44
  Coriolis effect, 43–44
  evidence of, 42–44
  Foucault Pendulum, 42–43
Runoff, 238, 456

Saltation, 458
Sandbar, 500
Satellites, 169
Saturn, 10, 167, 169–170, 173
Scale, map, 82, 84
Sea breezes, 581–582
Sea cave, 467–468
Sea cliff, 467
Season, 541
  Sun's path, 48–49, 108–109
Sea terrace, 468
Seawater, 231–233
  composition, 231–233
  dissolved gases, 233
  inorganic matter, 231–232
  organic matter, 232
  origin, 231
  particulates, 233
Sediment
  bedding, 495
  definition of, 440
  shallow water, 397
Sedimentary rock, 315, 322–327
  cementation, 322
  characteristics, 323–324
  chemical sediments, 325–326
  compaction, 322
  dessication, 322
  identification, 326–327
  layers, high elevation, 396
  lithification, 322–323
  organic sediments, 325
  particles, types, 324–326

Seismic waves, 349, 352–354
Seismogram, 353–358
Seismograph, 352–353
Shear stress, 398
Sheet erosion, 456
Shield volcano, 385–386
Shoreline erosion, 467–468
Short-lived giants, 153–155
Sidereal day, 101–102
Sidereal month, 107–108
Silicates, 300–302
Sills, 386
Sky, 1–6
Sleet, 585
Slow mass movements, 496
Slump, 454–455, 496
Snow, 585
Soil, 440–442
  associations, 512
  creep, 455–456, 496
  formation of, 430–450
  immature, 441
  mature, 441
  parent material, 440
  pedalfers, 442
  residual, 440
  sediments, 440
  transported, 440
Soil moisture zone, 241
Soil profile, 441
Solar day, 101–102
  mean, 103
Solar eclipse, 51–53
Solar radiation, 536–539
  electromagnetic spectrum, 537–538
  electromagnetic waves, 536–537
  global radiation budget, 538–539
  radiative balance, 539
Solar system, 163–189
  asteroids, 174–175
  catastrophic theory of, 165
  comets, 176–177
  evolution of, 165–169
  exploration, 179
  galaxy, 164

  gravitational differentiation, 175
  impact events, 177–179
  interplanetary bodies, 174–179
  Kuiper belt, 167
  meteor, 175
  meteorite, 175
  meteoroids, 175–176
  Milky Way, 164
  moon, 171
  nebular theory of, 166
  origin, 165–169
  planetesimals, 166
  planets. *see* Planet(s)
  structure, 169–179
Solstices, 109–110
Solution, 435–436
Sorting, 493
Southern Oscillation, 633–636
Space, photographs from, 70
Special properties, of minerals, 298
Specific gravity, of minerals, 297–298
Specific heat, 543–544
Speckle interferometry, 148
Spectral line, 131
Spectrograph, 133
Spectrophotometer, 133
Spectroscopes, 129–133
  brightline spectrum, 131
  continuous spectrum, 130
  dispersion, 130
  prism, 130–132
  spectral line, 131
Spectrum, 129
Speed of medium, 493
Sphalerite, 303
Spheres, 19–20
Splash erosion, 456
Spring tides, 55
Stacks, 468
Stars, 10, 134, 139–162
  black hole, 155
  brightness of, 142–144
  characteristics of, 140–148
  color, 144–148

Index

distance to, 140–142
evolution of, 148–155
gravity, 150–151
Hertzsprung-Russell
    Diagram, 146–148
luminosity of, 142, 145,
    147
magnitudes of, 143–144
main sequence, 146–147
mass of, 151–155
names of, 140
nebula, 150–151
neutrinos, 153
nuclear fusion, 148–149
origin of, 148–155
parallax, 140–141
pointer, 8
protostar, 150–151
radiation curve, 144
red dwarfs, 152
short-lived giants,
    153–155
Sun-class, 152–153
supernova, 154
temperature, 144–148
Stationary fronts, 591
Station model, 586
Statistical weather
    forecasting, 594–595
Stellar brightness, 142–144
Stibnite, 303
Strain, 398
Streak, minerals, 295
Streams, 456–461, 478–479
    deposition, 497
    drainage system, 510
    erosion, 459
    life cycle, 459–461
    patterns, 510–512
    transport, 457–459
Stress, 398
Structures, displaced, 396
Subduction, 406
Sublimation, 583
Sulfate ion, 303
Sulfates, 303
Sulfides, 303
Sun
    description of, 10, 169
    eclipses, 51–53

path of, 9–10
    changing, 47–50,
        111–112
    seasons, 108–109
Sun-class stars, 152–153
Sunrise, 100
Sunset, 100
Superclusters, 192
Supernova, 154
Superposition, 255
Surf, 465
Surface tension, 233–234
Surface waves, 354
S-waves, 354–355, 363
Swells, 465
Symbols, map, 82–83
Syncline, 399
Synodic months, 107–108
Synoptic forecasting, 586
Synoptic weather, 586,
    594

Talus, 495
Telescopes, 128–129
Temperate zone, 623
Temperature, 621–622
    air, 567–568
    changes, 434
    star, 144–148
Tension stress, 398
Tephra, 385
Terrestrial planets, 170–172
Texture, 542–543
Thermometer, 567
Thunderstorms, 596–597
Tidal friction, 382
Tides, 54–55
    neap, 55
    spring, 55
Tilted rock layers, 396
Time, 98–124, 440
    day, 96–106
    eons, 260–262
    epochs, 261–263
    equinoxes, 110–111
    eras, 261–262
    geologic scale of, 260–263
    Greenwich Mean Time,
        105
    hours, 99–101

International Date Line,
    105–106
local, 103–106
month, 106–108
periods, 261–262
seasons, Suns path,
    108–109
sidereal day, 101–102
sidereal month, 107–108
solar day, 101–102
    mean, 103
solar noon, 101
solstices, 109–110
sunrise, 100
sunset, 100
Sun's path, 108–109,
    111–112
synodic months, 107–108
year, 108–113
zones, 104–105
Titan, 173
Tombolos, 500
Topographic maps, 81, 85
Topography, 509
Tornadoes, 597
Torrid zone, 623
Totality, path of, 52
Traction, 458
Trade winds, 582, 624
Transform boundaries,
    407–408
Transpiration, 239, 628
Transported soil, 440
Transverse waves, 352
Trellis stream pattern, 512
Trenches, 405–406
Tributary, 457, 510
Tsunamis, 465

Umbra, 52
Unconformities, 257
Uniformitarianism, 254
Universal gravitation,
    29–30
Universe
    closed, 197
    expanding, 193–194
    future, 196
    open, 196
    structure, 191–193

Index

theories of origin, 190–203
Uranus, 167, 169–170, 173–174
U-shaped valleys, 476

Valley glaciers, 473–474
Valleys, 476–477
Van Allen belts, 207
Varve, 506
Vega, 45
Vegetative cover, 628
Velocity, planetary motion, 26–27
Ventifacts, 470
Venus, 10, 169–171
Vesicles, 318
Vesicular rocks, 321
Visibility, 576
Volcanic eruptions, 396
Volcanic glasses, 317
Volcanic rock, 317–318
Volcanoes, 380–394
  batholiths, 387
  cinder cone, 385–386
  composite cone, 386
  dikes, 386
  formation, 385
  geothermal gradient, 380–381
  internal heat of the Earth, 381–383
  intrusions, 386
  intrusive activities, 386–387
  laccoliths, 387
  lava of, 385
  magma of, 384
  Mt. Etna, 387
  Mt. Vesuvius, 387
  plateau, 386
  plutons, 386
  shield volcano, 385–386
  sills, 386
  structures, 385–386
  tephra, 385
von Fraunhofer, Joseph, 130

Warm fronts, 591
Wasting, mass, 453–456, 478

Water
  capillarity of, 234
  condensation of, 238
  crystalline structure of, 236
  distribution, 623–624
  evaporation of, 235, 238
  evaporative cooling, 235–236
  freezing of, 236
  heat capacity of, 235
  hydrogen bonds, 233
  latent heat of
    fusion, 237
    vaporization, 235
  liquid, 234–235
  motion, erosion, 456–468
  origin of, 229–231
  precipitation, 238
  solid, 236–237
  states of, 234–237
  structure of, 233–234
  surface tension of, 233–234
  transpiration, 239
  universal solvent, 237–238
Water cycle, 238–239
Water table, 241–242
Water vapor, 235–236
Wave(s), 463–465, 479
  behavior, 350–352
  crest, 464
  cut notch, 467
  cyclones, 592–593
  height, 464
  motion, 349
  refraction, 466
  trough, 464
  types, 352
  velocity, 351
  wind-generated, 463–465
Wavelength, 351, 464
Weather
  adiabatic changes, 577
  air humidity, 570–575
  air masses, 587–590
  air movements, 575
  air pressure. *See* Air pressure

air temperature, 567–568
  and humidity, 579–580
anticyclones, 592–593
atmospheric
  transparency, 576
  variables, 567
barometer, 569
breezes, 581–582
Celsius, 568
chaos theory, 596
clouds, 576, 583–586
cold fronts, 590–591
condensation, 570–571, 583
Coriolis effect, 583
cyclones, 592–593
dew, 585
dew point, 572, 575
drizzle, 585
evaporation, 570–571
Fahrenheit, 568
field maps, 587
fog, 585
forecasts, 587–596
frontal cyclones, 592–593
fronts, 587, 590–592
frost, 585
global wind belts, 582–583
gradient, 587
hail, 585
hazardous, emergency
  preparedness, 599–600
humidity, 570–575
hurricanes, 598
isobars, 587
isolines, 587
isotherms, 587
kelvin, 568
land breezes, 581–582
long-range weather
  forecasting, 595–596
maps, 586–587
numerical weather
  forecasting, 595
occluded fronts, 592
precipitation, 585
prevailing winds, 583
rain, 585

Index

relative humidity, 571, 574
sea breezes, 581–582
sleet, 585
snow, 585
stationary fronts, 591
statistical weather forecasting, 594–595
sublimation, 583
synoptic forecasting, 594
synoptic maps, 586
thermometer, 567
thunderstorms, 596–597
tornadoes, 597
trade winds, 582
visibility, 576

warm fronts, 591
warnings, 599–600
watches, 599–600
wave cyclones, 592–593
weather, 566–620
Weathering, 430–450
  chemical. *See* Chemical weathering
  factors affecting, 437–440
  physical. *See* Physical weathering
  products of, 440–442
  soils, 440–442
Wegener, Alfred, 258, 400–401
Westerlies, 462

Wind belts, 624–625
Wind deposition, 501–503
Wind-driven currents, 461–462
Wind erosion, 468–470, 480

Year, 108–113
Young rock, 404

Zenith, 7
Zodiac, 9
Zone of aeration, 241
Zone of saturation, 241
Zone of soil moisture, 241

It's finally here—
online Regents exams from the experts!

- Take complete practice tests by date, or choose questions by topic
- All questions answered with detailed explanations
- Instant test results let you know where you need the most practice

Online Regents exams are available in the following subjects:

- Biology–The Living Environment
- Chemistry–The Physical Setting
- Earth Science–The Physical Setting
- English (Common Core)
- Global History and Geography
- Algebra I (Common Core)
- Geometry (Common Core)
- Algebra II (Common Core)
- Physics–The Physical Setting
- U.S. History and Government

Getting started is a point and a click away!

Only $24.99 for a subscription... includes all of the subjects listed above!

Teacher and classroom rates also available.

For more subscription information and to try our demo visit our site at *www.barronsregents.com*.

Prices subject to change without notice.

(#105) R8/16

Barron's Let's Review—
Textbook companions to help you ace your Regents Exams

  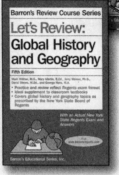 

Let's Review: Algebra I
ISBN 978-1-4380-0604-8, $14.99, Can$16.99

Let's Review: Algebra II
ISBN 978-1-4380-0844-8, $14.99, Can$16.99

Let's Review: Biology—The Living Environment, 6th Ed.
ISBN 978-1-4380-0216-3, $13.99, Can$16.99

Let's Review: Chemistry—The Physical Setting, 5th Ed.
ISBN 978-0-7641-4719-7, $14.99, Can$16.99

Let's Review: Earth Science—The Physical Setting, 4th Ed.
ISBN 978-0-7641-4718-0, $14.99, Can$16.99

Let's Review: English, 5th Ed.
ISBN 978-1-4380-0626-0, $13.99, Can$16.99

Let's Review: Geometry
ISBN 978-1-4380-0702-1, $14.99, Can$17.99

Let's Review: Global History and Geography, 5th Ed.
ISBN 978-1-4380-0016-9, $14.99, Can$16.99

Let's Review: Physics, 5th Ed.
ISBN 978-1-4380-0630-7, $14.99, Can$17.99

Let's Review: U.S. History and Government, 5th Ed.
ISBN 978-1-4380-0018-3, $14.99, Can$16.99

Available at your local book store or visit **www.barronseduc.com**

Barron's Educational Series, Inc.
250 Wireless Blvd.
Hauppauge, N.Y. 11788
Order toll-free: 1-800-645-3476
Order by fax: 1-631-434-3217

In Canada:
Georgetown Book Warehouse
34 Armstrong Ave.
Georgetown, Ontario L7G 4R9
Canadian orders: 1-800-247-7160
Order by fax: 1-800-887-1594

For extra Regents review visit our website: **www.barronsregents.com**

Prices subject to change without notice.

#225 R8/16

Join the half-million students each year who rely on Barron's Redbooks for Regents Exams success

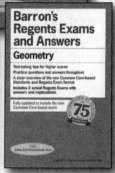

Algebra I
ISBN 978-1-4380-0665-9, $7.99, Can$9.50

Algebra II
ISBN 978-1-4380-0843-1, $7.99, Can$8.99

Biology—The Living Environment
ISBN 978-0-8120-3197-3, $7.99, Can$9.50

Chemistry—The Physical Setting
ISBN 978-0-8120-3163-8, $7.99, Can$9.50

Earth Science—The Physical Setting
ISBN 978-0-8120-3165-2, $7.99, Can$9.50

English
ISBN 978-0-8120-3191-1, $7.99, Can$9.50

Geometry
ISBN 978-1-4380-0763-2, $7.99, Can$9.50

Global History and Geography
ISBN 978-0-8120-4344-0, $7.99, Can$9.50

Physics—The Physical Setting
ISBN 978-0-8120-3349-6, $7.99, Can$9.50

U.S. History and Government
ISBN 978-0-8120-3344-1, $7.99, Can$9.50

Prices subject to change without notice.

Available at your local book store or visit **www.barronseduc.com**

Barron's Educational Series, Inc.
250 Wireless Blvd.
Hauppauge, NY 11788
Order toll-free: 1-800-645-3476
Order by fax: 1-631-434-3217

In Canada:
Georgetown Book Warehouse
34 Armstrong Ave.
Georgetown, Ontario L7G 4R9
Canadian orders: 1-800-247-7160
Order by fax: 1-800-887-1594

For extra Regents review visit our website: **www.barronsregents.com**

(#224) R8/16

ADD REAL POWER
to your preparation for Regents Exams
...with BARRON'S **REGENTS POWER PACKS**

They're concentrated, high-power help for students preparing for the New York State Regents exams. They also offer terrific extra classroom help to students taking final exams in the other 49 states. Combined in one money-saving package is a copy of the latest *Barron's Regents Exams and Answers* and a copy of *Barron's Let's Review* in each of the subjects listed below. *Let's Review Books* feature in-depth reviews of their subjects plus additional Regents exams to help students practice for the real thing. Purchase of *Barron's Regents Power Packs* represents an average savings of $2.69 off the total list price of both books.

Algebra I Power Pack
ISBN 978-1-4380-7574-7, $19.99, *Can$22.99*

Algebra II Power Pack
ISBN 978-1-4380-7659-1, $19.99, *Can$23.99*

Biology–Living Environment Power Pack
ISBN 978-1-4380-7434-4, $18.99, *Can$21.99*

Chemistry–The Physical Setting Power Pack
ISBN 978-1-4380-7242-5, $19.99, *Can$22.99*

Earth Science–Physical Setting Power Pack
ISBN 978-1-4380-7243-2, $19.99, *Can$22.99*

English Power Pack
ISBN 978-1-4380-7599-0, $18.99, *Can$22.99*

Geometry Power Pack
ISBN 978-1-4380-7619-5, $19.99, *Can$23.99*

Global History and Geography Power Pack
ISBN 978-1-4380-7296-8, $19.99, *Can$22.99*

Physics Power Pack
ISBN 978-1-4380-7598-3, $19.99, *Can$23.99*

U.S. History and Gov't. Power Pack
ISBN 978-1-4380-7262-3, $19.99, *Can$22.99*

Prices subject to change without notice.

Available at your local book store or visit **www.barronseduc.com**

Barron's Educational Series, Inc.
250 Wireless Blvd.
Hauppauge, NY 11788
Order toll-free: 1-800-645-3476
Order by fax: 1-631-434-3217

In Canada:
Georgetown Book Warehouse
34 Armstrong Ave.
Georgetown, Ontario L7G 4R9
Canadian orders: 1-800-247-7160
Order by fax: 1-800-887-1594

For extra Regents review visit our website: **www.barronsregents.com**

NOTES